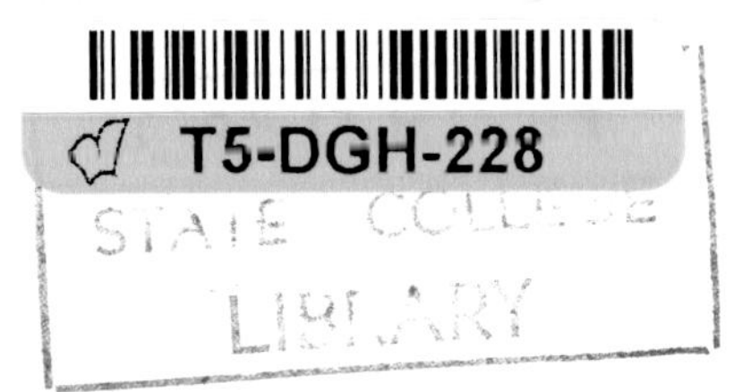

MATHEMATICS IN BIOLOGY:
CALCULUS AND RELATED TOPICS

MATHEMATICS IN BIOLOGY: CALCULUS AND RELATED TOPICS

BY

DUANE J. CLOW
DEPARTMENT OF MATHEMATICS
COLORADO STATE UNIVERSITY
FORT COLLINS, COLORADO

AND

N. SCOTT URQUHART
DEPARTMENT OF EXPERIMENTAL STATISTICS
NEW MEXICO STATE UNIVERSITY
LAS CRUCES, NEW MEXICO

W. W. Norton & Company, Inc. • New York

First Edition

ISBN 0 393 09280 1

Printed in the United States of America

1 2 3 4 5 6 7 8 9 0

PREFACE

The analytical sciences of mathematics, statistics, and computing have become increasingly important tools in biology. This textbook provides material to help students of biology learn how to use some of these analytical tools: functions, the operations of differential and integral calculus on functions, and related topics. The presentation assumes a working knowledge of basic college algebra, but no further mathematics.

The precision of mathematics has provided biologists with a concise means for expressing complex ideas. Then the tools of mathematics open new avenues for examining these ideas and their implications. Some biological problems, even apparently simple ones, require a mathematical approach for their understanding. As more and more biologists use mathematics to state their ideas, it appears likely that comprehension of the literature of biology will demand an increasing understanding of basic mathematics. The trend toward mathematics has occurred for various reasons: One is biologists' increasing use of ideas from chemistry and physics, disciplines that have long used mathematics; another is the growing interface of biology with technology in areas such as environmental problems.

Mathematics does have a varying level of importance across the several biological disciplines. Consequently we have deviated from the familiar order of introducing calculus topics to students from the physical sciences and engineering. Because these students almost always complete an entire sequence in calculus, course content can be ordered for instructional ease and consistency of mathematical perspective. Instead we have organized the material to get to ideas of central importance in biology as soon as feasible; for example, probability and the exponential function appear early and get continuing exposure. Nevertheless, a student who has completed most of this book should be prepared to enter advanced mathematics courses with students who have taken a more traditional calculus course. At some institutions an intervening introduction to linear algebra may be needed.

Biology needs both the analytical perspective of mathematics, and some facts which exist within the body of mathematical knowledge. The first two chapters explore the relation between biological and mathematical problems. Much of biology seeks to characterize groups of organisms; therefore we deal next with sets and their characterization by probability. The exponential and power functions appear throughout biology, so next we introduce functions in general and these specific functions. From there we turn to the tools of calculus to characterize, find, and understand functions. During the middle and later chapters of this book, we introduce those mathematical concepts needed to progress further. Thus we discuss three-dimensional geometry before multiple integration and trigonometry before alternate coordinate systems. This is not a book on statistics, computer science, or matrices, though we do comment on these topics where it seems appropriate.

The material has been organized to leave substantial flexibility for individual situations. Optional material is indicated by *; it may be omitted without any loss of continuity. In particular, much of the material on probability and statistics appears in optional sections. The first eight chapters follow each other in a natural order: functions, limits, derivatives, and integrals. This can be followed by either multidimensional calculus (chapters 10-12), or further topics in univariate calculus (chapters 13, 14, and 16). Chapter 15 presents angular coordinate systems and requires chapters 10-14 as prerequisite. The last chapter sketches other parts of mathematics which a biologist may want to learn about. The entire book can be covered in one year, or the first eight chapters in one semester, assuming five class meetings per week.

We have tried to stay away from two extremes in mathematical instruction: abstraction and drill. Abstraction and rigor have proved essential to the development of mathematics, but this does not imply that they should be emphasized in introductory instruction. Even very eminent mathematicians join biologists in objecting to trends toward undue abstraction at the introductory level. Nevertheless, proofs still have a place. Where the proof of a result may aid understanding, we do not hesitate to give it. Sometimes even a biologist needs to be able to prove a result of interest to himself or herself. Similarly, we approach drill with caution. Drill may develop the formal ability to solve stated problems, but it does not necessarily lead either to real understanding or to the intellectual independence a biologist must have. Some drill is needed so our exercises provide an opportunity for it, but the objective is to learn to solve problems with understanding. To this end we discuss the process of problem solving in Chapter 2. Some of the chapters conclude with review problems. These may require the use of any idea developed up to that point in the book. Similarly some exercises include problems which require the use of tools that have been discussed in earlier sections.

This book has evolved to meet a need expressed by biologists. Over ten years ago the biological sciences at Colorado State University recognized that their graduate students were very deficient in mathematical background. The course developed for these students sought to provide a background in mathematics sufficient for them to read current biological literature. Originally it was hoped that the course would be necessary for only a few years because future graduate students would come better prepared. This hope never materialized so the course was reorganized for undergraduates. During this time the course has grown from a small group of graduate students to a very large, and often unwieldy, undergraduate course. This book has grown out of that experience. Various drafts have been used at Colorado State for about seven years. The students in these courses have been very helpful. They have identified points of difficulty, offered alternative explanations which seemed more suited to biologists, and suggested biological situations which have led to many of the examples.

In addition to suggestions from students, our biological examples and problems have come from a number of sources. Many have come from our experience, either directly or slightly modified to avoid lengthy explanations of minor details. Others have appeared in the open literature, and some were suggested by our colleagues. Several examples receive recurring consideration from different mathematical perspectives. This illustrates how several mathematical tools can provide insight into a specific biological problem. We have omitted many good and interesting biological examples because the text is designed for students who have a serious interest in biology, but perhaps not yet a substantial knowledge. This is in keeping with our intention that freshmen will be able to learn from this book.

Many people have aided the development of this book. The students at Colorado State have been especially helpful in many ways; to them we owe a great deal of thanks. We have received substantial support and encouragement from our respective department heads, Ervin Deal and Morris Finker, and from colleagues, both biologists and other analytical scientists. In particular, we appreciate the perspectives we have gleaned from Ralph Niemann and Douglas Robson. The final draft was read by John Bishir, David Zachmann, and Richard Glaze; their comments proved useful. Professor Bishir's comments were especially detailed, helpful, encouraging, and perceptive. Joseph Janson and Christopher Lang of W. W. Norton & Company have helped in the final stages of preparation. This book has been reproduced from copy skillfully typed by Michal Hakonson. Finally, our respective families have been patient during the several years we have worked on the manuscript. In thanking all of these people, we of course accept full responsibility for all mistakes which remain. As this is a preliminary edition, we especially welcome any comments you may have.

Fort Collins, Colorado
Las Cruces, New Mexico
February, 1974

Duane J. Clow
N. Scott Urquhart

TABLE OF CONTENTS

TABLE OF CONTENTS

TABLE OF CONTENTS

TABLE OF CONTENTS

TABLE OF CONTENTS

Chapter 1

MATHEMATICS IN BIOLOGY

You have every right to ask, "Why should a biologist have to learn anything about mathematics?" If you think back about your previous schooling, you have been learning something about mathematics most of the time you have been in school. Thus, an even more perplexing question might be, "Why should I take more mathematics?"

The next section sketches an answer to this question while the remainder of the chapter develops its background. Even if this introductory chapter fails to thoroughly convince you that more mathematics has a real place in your learning, we hope the rest of this book will demonstrate the utility of mathematics in biology.

This introductory chapter uses a few concepts and terms discussed in later chapters. These chapters will expand your present understanding of terms such as function, variable, and probability. Here we only want to indicate how mathematics is used in biology. This chapter will take on a fuller meaning as you come to understand more of the material which follows. Thus consider rereading it later.

1.1 MATHEMATICS AND MODELS IN BIOLOGY

Most people view mathematics as a cold analytic science removed from reality, not as a tool of action. Ecological and cultural stresses exist throughout the United States and most other countries. These are observable in various efforts associated with energy, pollution, the management of renewable natural resources, working with people to rebuild urban centers, and elsewhere. We have built a complex society which we have to live with and somehow manage. Biologists face a compelling task in the years ahead. They should contribute skills to help resolve present and future problems in ways which can lead to enlightened management.

Mature management requires models which will predict the consequences of various suggested actions. Modeling is a complex process which exists at many levels, but the tools of mathematics provide a powerful means for helping describe and analyze a situation. In a sense, biology involves two perspectives: The population and the individuals which make up the population. For example, anatomy, physiology and related areas deal with how an individual organism functions, particularly how its components operate individually and collectively. Mathematics has found fruitful applications in a number of areas describing physiological and anatomical functions. At the level of a population, concern focuses on the mass behavior of many individuals, whether grasses, trees, insects, humans, or whales. The impact of various environmental stresses on the physiology of individual organisms will influence the behavior of a population, but certain aspects of a population operate as an entity with individual variations tending to cancel out each other so that the population moves ahead in some reasonably discernable fashion.

What does mathematics do? In other words, why does mathematics have a role in understanding and projecting the consequences of various actions on biological entities? At the most primitive level mathematics considers collections of objects, called sets, a term probably familiar to you. Mathematics seeks to do two kinds of things: (i) Characterize a specified set of objects; (ii) relate objects in different sets to each other. So far essentially the same can be said for almost any science, so why is mathematics so special? The power of mathematics, as well as the allied disciplines of statistics and computer science, lies in the precision of its language and operations. Until mathematics gets translated out of its formality and adapted to those problems which exist, its power has not been exploited completely. For biology to utilize this powerful structure some biologists will have to know a respectable amount about mathematics. This applies even when you are fortunate enough to find a mathematician sympathetic with biological applications. Communication simply requires a common ground for discussion.

The development of models occupies a central role in explaining occurrences in biology or any of the natural sciences. Models assume various forms of expression, but three levels of precision can be identified: Verbal, diagramatic, and explicit, as an equation. The explicit statement of a model usually will utilize mathematics due to the natural economy and precision of mathematical notation.

Construction of a mathematical model focuses attention on the central features of the situation at the expense of its peripheral or vague aspects. This identification of essential aspects may prove to be as valuable as the final model. When central features have been chosen wisely, the resulting model may prove valuable in suggesting new avenues for approaching the biological problem. The validity of the resultant model rests upon its ability to correctly predict outcomes under new conditions. An incorrect model may, nevertheless, prove fruitful by suggesting interesting new experiments or pointing to relations which previously were unsuspected.

An artist creating a realistic painting does the same thing. His painting can capture the essence of a colorful maple tree in the fall, of a wheat field ready for harvest or of ducks swimming with a newly hatched brood of ducklings without displaying every leaf, every head of grain or every piece of down on the ducklings. So long as a representation, a model, embodies the essential features, it has done its job of communication. Conversely, it distorts the real situation if it lacks essential features or communicates a misshapen view.

Even though an incorrect model may have great utility, a model can be correct in two senses: Mathematically and biologically. A model may contain no mathematical errors, yet lead to inappropriate predictions because it ignores essential biological features. On the other hand, it may lead to realizable predictions because its central features have been chosen wisely, and accurately translated into mathematics. Clearly you should strive always to have your models contain no mathematical errors. Within a textbook context, however, you cannot isolate when a model fails to be correct in the second sense. In practice when a model is not correct in the second sense, the investigator will seek to improve it by trying to incorporate more features of the actual situation.

As you read on, many things said above will develop a fuller meaning. If you feel that another view of mathematics and models in biology would help you grasp the central thoughts, find the book *Mathematics in Medicine and the Life Sciences* by G. R. Stibitz (Year Book Medical Publisher, 1966). Its first three chapters present a view of mathematics and models in biology which complements the one set out here.

1.2 ORGANISMS: ONE AND MANY

Before we delve further into how mathematics fits into biology, let's talk briefly about biology itself. At this point the discussion must be fairly general: It cannot be precise in every detail. We can, nevertheless, get an overview of biology into which we can work mathematics.

Biology distinguishes itself from the rest of science by its central focus on living organisms, that is, entities having some facility for reproducing themselves. These individuals vary from single fairly primitive cells to large, complex, many-celled organisms like mammals. At one end we get down to very small entities called viruses, whose living quality has been questioned by some biologists. Man appears at the other end of the spectrum of complexity. Biology almost entirely avoids the behavioral aspects of man which it considers in other creatures. Although in a strict sense the behavioral aspects of man are biological, they have been left to the disciplines of psychology and sociology. Some considerations in the sequel reflect on behavioral characteristics of various organisms; many of these ideas also apply to man.

At the risk of oversimplification, biology has these perspectives: (i) Individual organisms, (ii) interactions between individual organisms, (iii) populations of organisms, and (iv) interactions among populations. Although most subjects can be viewed from several perspectives, one perspective may clarify presentation of a topic. The view of biology advanced above proves rather useful in examining the role of mathematics in biology.

An individual organism can be described in absolute or comparative terms. Anatomy does essentially this. Physiology is the study of processes within an organism and how they respond to stimuli. For example, what influences blood pressure, one of the "vital signs" of life in higher animals? Consider your own blood pressure. It varies with the pumping cycle of your heart, with distance from the heart, your sleeping versus waking state, your age, your level of activity, anxiety or fear, etc.; various drugs, both medically prescribed or others, can drastically alter it. A plant physiologist has similar concerns when he studies how photosynthesis, one of the central processes of plants, responds to temperature, light, water, structure of the plant's nutrient transport system, etc. These two brief and somewhat general examples illustrate that physiology seeks to explain how certain responses functionally relate to various stimuli. A particular concern often centers on how these processes accelerate or decelerate in the presence of various stimuli, as well as on the general form of the function which relates the response to the stimuli. The mathematics of rates of change relies on calculus, a topic examined closely in the following chapters. It provides tools for relating functions to their rates of change.

The interaction of one organism with another manifests itself in both physiological and behavioral ways. Pathology includes the study of how one organism affects the physiology of another. Diseases and infections such as polio, chicken pox, the common cold, and dutch elm (tree) disease provide familiar examples. Microorganisms cause these maladies by infecting the host organism and altering its physiological processes. We can think of parasitism, namely, one organism living on and drawing its food from another, such as a tick sucking blood from an elk, as manifesting a less extreme example of infection.

Individual organisms interact in other ways. Competition demonstrates a familiar form. For example, two animals of the same or different species can compete for the same food or living space; animals of the same species can compete for the same mate; plants of the same or different species can compete for the same water and nutrients in the soil. The mathematics involved in studying interactions between organisms varies from the same mathematics used to study a single organism to rather complex mathematics describing structural relations within a system. At the simpler end of the spectrum we find predator-prey relationships. For example, a biologist may need to know how much prey a particular predator consumes for each unit of time expended, an idea often described in terms of searching time. Knowledge of a biological problem may suggest a particular functional form to characterize such a relation. Then as you go from one species of predator to another, the estimation of certain features of that form emerges as a major issue. Such matters can become very important in trying to pick biological control agents for helping protect certain natural resources from pests (like trees from bark beetles).

In considering a single population of organisms, interest usually centers on characterizing its various attributes. Genetics provides clear-cut examples. For example, a particular mating may produce either male or female offspring in higher animals. Nevertheless the sex of an offspring is determined only in probabilistic terms because a random mechanism operates during the development of the sperm and egg. You may know a mother who wants a daughter, a very personal consideration, but the dairyman who only wants heifer (female) calves has a very practical consideration. Thus, a geneticist may want to understand this system so that he can alter the frequency with which particular outcomes occur. This illustrates a kind of mathematics often applied to a single population, namely, frequencies of outcome. Frequencies appear in other contexts too. Various forms such as abnormalities or diseases occur according to probabilities which may be altered by knowing something about circumstances or parentage.

A very complex form of biological endeavor concerns how one population interacts with another. A simple example involves how a disease spreads through a population, the concern of a discipline called epidemiology. For example, when a new strain of Asian flu enters the United States, we need to project its severity and its likely spread through the country in order to prepare enough vaccines. The same problem occurs with diseases in plants, whether commercial crops, trees, or vegetation along a stream.

The most complex form of population to population interaction is called ecology, a word overworked at the present time. Ecology seeks to bring together all other levels of biology to study how an entire community of organisms, consiting of many populations, coexist and interact at one time. In studying this kind of interaction the physical as well as biological interactions must be studied. Two major topics emerge. One centers on the energy flow within the community. Plants transform the incoming energy of sunlight into plant matter. This passes through several levels of animals from the primary consumers up to the carnivorous animals which eat only other animals. The other major flow involves nutrients. Plants take them up from the soil and convert them into compounds as part of photosynthesis. In turn, animals consume them. When the animals give off waste materials or begin to decompose after death, the nutrients recycle to the soil through decomposing microorganisms. These processes are extremely complex, yet they must be studied if we are to understand the impacts of various actions on our environment. The mathematics involved in population ecology takes on varying forms, but basically we need to understand how each process depends on other things in a functional form, that is, in terms of equations. This requires an examination of rates of accumulation and of structures. Each of these ideas has its counterpart in mathematics and as we progress we will try to illuminate them more.

This overview of biology ignores taxonomy, classically a very important area. Taxonomy basically concerns the naming of organisms in a manner which reflects their structure, evolution and role. This perspective can be described as being both very important and not at all important. In studying a particular organism, its identity makes little difference until we want to relate it to similar organisms. Its actual name makes no difference but how it structurally relates to other organisms may be rather important. Such considerations can become important in ecology because it considers the joint effect on the community of each of its components. Thus, we need to identify each component and know what kinds of effects it has on the community before we begin a detailed examination of interactions between components.

1.3 MODELING LARGE BIOLOGICAL SYSTEMS

The physical sciences have had profound success in modeling large and complex processes. A familiar example appears in the space activities which culminated in the landing of men on the moon. These were extremely complex endeavors. How were they developed? A set of ideas called systems analysis played a central role. Stripped to the barest essentials, this concept concerns the division of a large problem into components. For each component appropriate theories are postulated, validated and refined through experimentation and experience. After each part has been resolved, it is interfaced with each other part until finally a model of the entire endeavor emerges. The different parts of the model operate on quantities called parameters which describe its operating constraints or conditions. For example, one part of the space computer model could involve the weight of various fluids in the life support system. Calculation of trajectories require the weight of various components of the system in space. As fluids are consumed or expelled from the life support system in space, the model concerning fluids keeps track of this. Each time the weight of the entire space vehicle is needed, this particular component is interrogated for its contribution to the weight. This merely provides a small example of a small part of a large and complex model.

Within the last decade the research and development technology of the aerospace industry, has moved into other areas of science and business. In biology the most striking examples are associated with the name *systems ecology*. A major project called the International Biological Program (IBP) provides a focus for beginning to work in these directions in ecology. The entire world has been divided into many biomes, that is, areas with reasonably similar ecological characteristics. For example, the biomes in the United States include the eastern hardwood forest, the central grasslands, the desert of the Southwest, and the Pacific coniferous forest. Many other biomes have been identified, but to list them here would detract from our general purpose. The essential thing is that the entire ecosphere, that is, the world, has been divided into components. Selected sites in each biome are being studied intensively to learn how processes take place in that biome. These processes concern interactions of organisms with organisms, with the physical environment, with energy sources, and of populations with populations.

These activities probably will accelerate as the technology moves out of the aerospace industry into other areas. Their application to biology has turned out to be more complex than anticipated. Within the physical sciences fairly straightforward reasoning in a slightly idealized framework produces nearly precise functional relations among components. We frequently describe this by saying that most physical phenomena come reasonably close to obeying the underlying "physical laws". For example, enough is known about gravity to have a reasonable understanding of how a spaceship moving through space will be affected by gravity from the moon and from the earth. By contrast, biology appears much more involved, exhibiting all of the complexities of physical systems compounded with the variation of living organisms. Yet the same kinds of ideas may eventually prevail. Biology has a far greater degree of interdependence among its components than seems to be necessary for functional models in physical science. For example, soil microorganisms occupy an important place in the cycling of nutrients through an ecosystem. Their nature, number and the favorableness of their environment can greatly influence an ecosystem. In fact, each organism depends on each other organism in some manner. These interdependences will have to be understood more fully than they presently are before models can be completed.

A second distinction between the physical and biological sciences involves the theoretical base. Since the physical laws are reasonably well understood, they can be adapted to various environments. Knowledge of biology has not progressed to a stage where we begin to understand the complexity of its processes nearly as well. Part of the reason for this lies in a third important distinction between the physical and biological sciences. While most physical phenomena nearly follow stated laws, biological phenomena exhibit substantial variation from unpredictable and uncontrollable sources. Two important components of the variation arise from the unique genetic constituency of each individual and the impact of the external physical environment on individual organisms. The biological process for controlling inheritance embodies a random mechanism so that each individual receives a unique genetic composition. Thus, a population, whether a population of microbes or people, consists of entirely unique individuals merely sharing common characteristics. For example, consider eye color, stature or susceptibility to various diseases in humans. You certainly know of great variation in these observable characteristics. Equal variation occurs in many less easily observed characteristics, even if the population has a constant environment. A population existing in nature is subject to the weather that impinges on it. From year to year weather varies greatly in ways that presently evade prediction at the local level.

These two facts, variation in the response of biological organisms to specified conditions, and the natural variation of conditions, lead to some real differences between the modeling in the biological sciences from the physical sciences. The distinctions between the biological and physical sciences have at least two implications. We should not expect to be as precise in projecting the consequences of various actions in biology as in the physical sciences. Secondly, it probably will take a much greater period of time to begin to grasp the complexities of a functioning biological system, compared with the recent successes in the physical sciences.

These comments clearly indicate that biologists must take central roles in developing models which concern biological phenomena. They must actively participate in all levels: As technicians, as scientists, and as administrators. Certain events clearly indicate the dangers inherent when the appropriate life scientists are excluded from the processes. About 1970 a severe change in funding shook the aerospace industry, as the amount of money for moonshots and military projects decreased. The aerospace industry began to branch out and to try to work on environmental and urban issues. A number of very glaring failures occurred. Subsequent analysis showed that the failures occurred mainly because the people involved did not understand the processes under investigation. The investigators were basically physical scientists who were somewhat naive about the biological and social sciences. The failures generally could not be attributed to either the management or scientific processes which were used in approaching the problems. This says not only that biologists must become involved, but that they also need the knowledge to become appropriately involved.

The above comments suggest some practical implications for mathematics in biology. For example biologists need to understand what functions are. Functions basically describe the numerical consequences of various actions in specified environments. The biologist needs to understand what kind of functions he has if he is given a particular group; he must be able to reason from what he knows about a system to related functions; he must know how to investigate functions to be sure that they do the things that he wants. Mathematics has a great precision of definition and approach. These features eventually may become more valuable to biologists than the content of mathematics. Perhaps the greatest need involves the biologists' ability to approach problem solving analytically. Thus, in addition to mathematics as content, we will endeavor to give various illustrations of the process of solving problems in this book. In fact, the next chapter is devoted entirely to this idea.

1.4 DESCRIBING BIOLOGICAL RELATIONS

Biological phenomena do not get expressed in mathematical terms by some magical process. A mathematical relationship results from thought, work and imagination. The development of a worthwhile relationship usually involves several iterations through a process that has important stages both before and after formal mathematical considerations. It begins by identifying the biological concept under scrutiny. This may seem mundane, but consider a seemingly clear question such as "How does pollution affect human health?" You must decide what kinds of pollution to consider because you eventually will need a scheme for measuring related variables. Further examination of this question requires a precise definition of health. Is it general well being? Is it the appearance of specific diseases, etc? After this first stage, the next task involves the determination of relevant responses. What response reflects on the biological concept under examination? Until this has been resolved you have no chance of expressing your biological knowledge in terms which can serve as input to the development of a mathematical relation. Once these assumptions have been developed and stated, the problem becomes a mathematical one of what functions relate the variables under consideration. The relation developed should reflect the assumption stated in the preceding step. Once mathematical relations have been established, they should be scrutinized to make sure that they conform to available knowledge of the biological problem. In particular, examine various extreme behaviors of the function, as well as its general behavior range of interest. At this point some recycling through this process probably will be necessary.

You should not hope to develop a useful mathematical relation without thoroughly understanding the biological situation. Relations can be developed from two general perspectives. One perspective develops the model from the exact physiological, chemical and physical characteristics of the process under examination. The other perspective seeks relationships which approximate the biological phenomenon well enough to be useful. The following exemplifies the first perspective:

EXAMPLE 1.1: This situation arises in evaluating radiation damage to the testes of rats. We need to determine the number of different types of cells in the seminiferious tubules of an adult male rat. The following pattern of cellular division has been established experimentally:

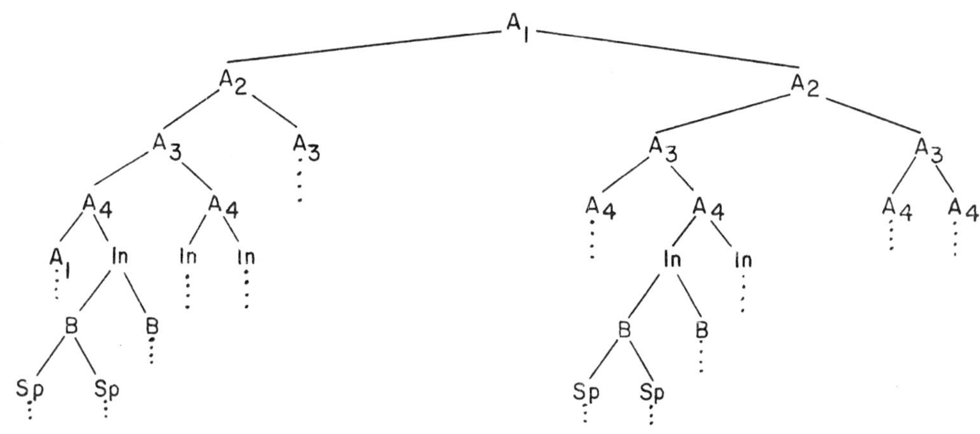

Figure 1.1

Under certain circumstances, it can be described satisfactorily as this: Each type A_1 spermatogonium produces two type A_2 spermatogonia; each of these divides giving rise to two type A_3 spermatogonia; each of these give rise to two type A_4 spermatogonia. All type A_4 spermatogonia, except one, divide and give rise to two intermediate type spermatogonia. Each of these divides to produce two type B spermatogonia which finally divide to yield two resting primary spermatocytes. The remaining type A_4 spermatogonium divides giving rise to one intermediate spermatogonium and one type A_1 spermatogonium to restart the whole process.

This description provides a basis for establishing a relation between the number of cells which exist at each level and the initial number of type A_1 spermatogonia. Since each type A_1 spermatogonium cell will produce two type A_2 spermatogonia, the number of type A_1 and type

A_2 spermatogonia cells, symbolized by $N(A_1)$ and $N(A_2)$, relate through

$$N(A_2) = 2N(A_1) \ . \qquad (1.1)$$

Since each type A_2 spermatogonium will produce two type A_3 spermatogonia cells,

$$N(A_3) = 2N(A_2) \ ,$$

but from Equation 1.1 this can be related to number of type A_1 spermatogonia by

$$N(A_2) = 2(2N(A_1)) = 4N(A_1) \ . \qquad (1.2)$$

Similarly each type A_3 spermatogonium will result in two type A_4 spermatogonia so

$$N(A_4) = 2N(A_3)$$

but from Equation 1.2 this reduces to

$$N(A_4) = 8N(A_1) \ . \qquad (1.3)$$

Examine the cellular diagram again to note that $2N(A_4) - 1$ intermediate spermatogonia cells arise from the $N(A_4)$ type A_4 spermatogonia, or

$$N(In) = 2N(A_4 - 1) \ ,$$

but from Equation 1.3, this reduces to

$$N(In) = 2(8N(A_1)) - 1 = 16N(A_1) - 1 \ . \qquad (1.4)$$

The number of type B spermatogonia is twice the number of intermediate type spermatogonia so

$$N(B) = 2N(In) = 2(16N(A_1) - 1) = 32N(A_1) - 2 \qquad (1.5)$$

after substitution from Equation 1.4 for N(In). Each type B spermatogonium yields two resting primary spermatocytes. Thus

$$N(Sp) = 2N(B) = 2(32N(A_1) - 2) = 64N(A_1) - 4 \qquad (1.6)$$

after substitution of N(B) from Equation 1.5.

Thus by knowing the number of type A_1 spermatogonia originally, we can predict the number of cells in successive steps, if no cells die. Simple models like this become very important in studying the kinetics of cellular response to different types of stresses upon the system such as heat, ionizing radiation and chemicals because deviation from the model reflects damage induced by the stress. ▮

The second perspective for developing a model proceeds this way: The precise expression of a biological idea may not always be possible. Thus, we may seek an approximate relation expressing the general phenomenon, even though it may fail in some detail. Approximation progresses through two different modes. One involves the approximation of an actual relationship by a different function; the other approximates an irregular or discrete relation by a smooth one. Examples in the rest of this section should clarify this point. A greenhouseman concerned with the production of tomatoes in a moderately controlled environment knows that his production depends on many aspects of his environment. One important aspect involves the amount of nitrogen that his plants get. In fact, he knows that as the amount of nitrogen is increased from some minimal level to a very high level, the weight of tomatoes produced will increase up to some point past which production falls because excess nitrogen tends to burn the tomato plants. He may not know, or even need to know, the exact relation between yield and nitrogen level. Any one of several smooth curves may describe the actual relation well enough that he can investigate the phenomenon to get the answer he needs. The following example illustrates this perspective in more detail:

EXAMPLE 1.2: Insects have an important physiological attribute: They cannot control their body temperature; their rate of metabolism and activity depends upon the temperature around them. A few insects, like locusts, increase body temperature by orienting themselves to receive maximum solar radiation, while a few others, like bees, cool themselves by vibrating their wings and evaporating moisture. But for most insects, the body temperature depends directly upon the temperature of their microhabitats. Thus consider testing this hypothesis: Insects have a body temperature equal to that of the surrounding air. A relevant experiment could involve some insects protected from strong direct radiation but kept at different temperatures. Air temperature could be measured with a thermometer and body temperature by use of a thermocouple. We could obtain the following, where each measurement results from the average of six readings on each individual insect-temperature situation.

Air Temperature °C	Insect Temperature °C
25.6	26.2
43.2	41.0
28.7	28.9
31.2	31.0
31.5	31.5
26.2	25.6
30.1	28.4
31.4	31.6
30.4	31.5
28.8	28.9
37.5	36.7
35.6	37.5
23.0	24.5
18.2	18.7

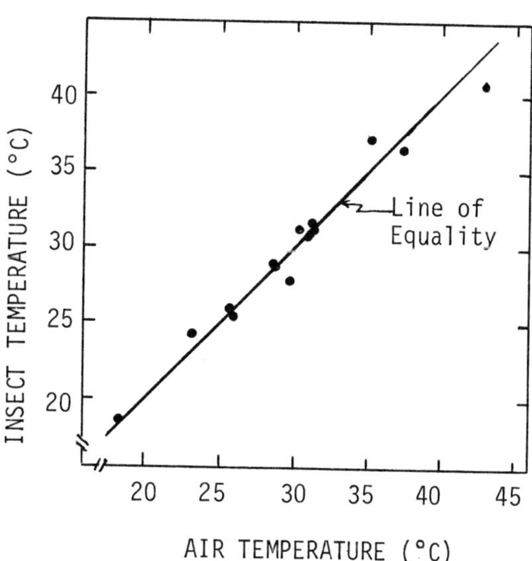

Figure 1.2

A plot of these points gives the graph shown in Figure 1.2.

Notice from the graph that body temperatures generally follow the air temperatures, despite the small scatter of points due to individual variation and some error in measuring true temperatures. From this we could conclude that the body temperature of these insects approximately equals the temperature of the surrounding air provided they are not exposed to strong, direct radiation. (This qualification is required because our experiment was limited to that situation.) ▌

The above illustrates approximation of a relationship by a function which may be slightly inappropriate. Now we explore another kind of approximation.

Biological phenomena sometimes display an element of discreteness. (Discrete responses can assume only isolated values, rather than all values in an interval.) Counting provides a simple illustration. Biologists may find serious interest in counting many different kinds of things. For example, a radiation biologist has to count radioactive decays in order to know something about the intensity of radiation received by a particular organism. A fruit scientist may be very seriously concerned with the number of fruit on a particular tree. Throughout biology interest focuses on the number of offspring from a particular female. This has as much relevance to insects as to fish or to deer. These counts, as most kinds of counts, can vary from a very small number to a very large number. For example, a female white tail deer frequently will have two offspring, but the only reasonable number of offspring would be 0, 1, 2 or occasionally 3. At the other extreme, a fish or an insect may lay thousands of eggs. In studying counts we are examining a population characteristic in order to see how it changes with various external factors. When we approximate a discrete function by a smooth (continuous) function lacking jumps, we may or may not be introducing a fairly substantial error into our analysis. The relative importance of discrete-continuous distinction decreases as either the population size or the number of offspring per individual increases. In fact, discreteness frequently presents more of a distracting than essential feature, particularly in those situations where the exact number is not important. For general changes in numbers, a smooth function often provides a satisfactory approximation.

1.5 MATHEMATICAL TOOLS FOR AN APPRENTICE BIOLOGIST

This brief introduction points out several basic things that a biologist should know about mathematics. First, he should have some grasp of how to express biological problems in mathematical terms. This idea serves as the content of the next chapter. It is an unusual chapter for a mathematics textbook, yet we believe it is a very important topic for biologists to understand. Next the biologist should have some grasp of the mathematical idea of a set, often expressed in biological terms as the population, and things he can do with sets or find out about populations. This material is covered somewhat in Chapter 3 with some emphasis on relating outcomes to probability. Next we call attention to the value of functions as a means of expressing relationships between different biological entities. Chapters 4 and 5 deal with this topic: Chapter 4 presents functions as entities while Chapter 5 introduces the idea of limiting behavior of functions. These two chapters introduce some functions which are very important in biology, as well as approaching functions as a topic. Finally, we hinted at the value of ascertaining things about functions from their rate of change, or conversely of going from a function's rate of change to the function. These ideas involve the mathematical concepts of differentiation and integration, topics to which much of the remainder of the book is devoted. Initially, in Chapters 6, 7 and 8 we consider relationships involving one response of interest and one underlying variable. Subsequent parts of the book concern responses depending on several variables.

This introduction identified other features of the analytic sciences which will not be emphasized. Although we do consider probability, statistics is mentioned only occasionally. Statistics is an important topic, yet at this level, it is not the most essential material. Another possible topic deals with arrays, often referred to as matrices, or associated with linear algebra. This topic has extensive application in intermediate and advanced biological considerations, but was omitted here because it would substantially lengthen the book. Finally, we have alluded to the value of computers in studying biological problems. By largely avoiding computers, we do not mean to imply anything to the contrary. Computers have an essential role in the examination of complex biological problems, but a thorough consideration of this topic also lies outside the scope of this book. The last chapter returns to these topics. For each, it briefly explores the nature of the topic, displays applications in biology and gives references.

Chapter 2

APPROACHING THE PROBLEM

Do you like word problems? If you answered *NO*, stick with us because this chapter should improve your ability to solve them. If you find word problems tantalizing, you may recognize some of your approaches here; if you do not find word problems easy, points brought out here may improve your success rate.

You may wonder why we are concerned about your ability to solve word problems. Successful attacks on either textbook word problems or real biological problems share certain common features. They share a problem solving process for which we identify five steps. A different dissection could produce more or fewer steps. Please realize, however, that we cannot completely explain this complex process in a brief chapter. As you progress, you will gain insights from your solution of problems and from our examples. Thus, you probably will find it worthwhile to reread this chapter later.

The first section in this chapter presents a continuing example while the next five sections examine the five steps in the problem solving process. These five steps are:

1. Diagnosing the Problem -- what is the problem all about
2. Establishing Relationships -- how to establish relationships which will lead to the solution of the problem
3. Picking Useful Mathematical Tools
4. Applying the Selected Tool
5. Interpreting the Consequences

As you read through the example in the next section, try to identify each of these steps.

2.1 AN EXAMPLE

The U. S. Forest Service administers millions of acres of timber lands in national forests. Present policy allows private firms to harvest selected portions of this timber under specified conditions. Firms frequently are selected through competitive bidding. With this context, the following sort of situation does arise.

SITUATION: A chief forester sets out to gather information needed to prepare his firm's bid on timber in a remote area. From a timber cruiser's report he already has an estimate of the number of harvestable trees of each diameter. Due to the remoteness of the area, the chief forester suspects that trees will have to be larger than usual before they will be worth harvesting. Thus, he asks an assistant, "How large must trees be in this area before we can afford to harvest them, assuming we get the bid at two cents per board foot†?"

AN APPROACH: As the assistant turns this problem over in his mind, he might have these thoughts, among others: What does the old grump mean by large? He probably is thinking board feet, but I will have to work with diameter (at chest height, say) in order for my results to be usable in the field. The cost to fell (cut down), delimb, and to transport the trees to the sawmill will increase with the size of the tree. The larger logs are worth slightly more per board foot because we can get larger boards out of them. Now, somehow I have to put this all together in order to get the minimum diameter which will allow the company to realize a profit on its harvest of this stand of timber.

† Board feet is a standard way for describing volume of lumber. One board foot equals one square foot of wood, one inch thick.

From various research reports and accounting records, the assistant establishes the following approximate relations for the particular stand of timber under consideration:

volume in bd. ft. (v) to diameter in inches (d): $v = 10 + .007\ (d-5)^3$, $d > 10$ (2.1)

harvest costs per tree in \$ (c) to volume: $c = 1.20 + 0.1\ v$ (2.2)

sale value per bd. ft. in ¢ (s) to volume: $s = 9 + .03\ v$ (2.3)

These equations do have the characteristics the assistant anticipated. Equation 2.1 relates diameter, an easily available measurement, to volume. Observe that volume increases as diameter does, the only reasonable situation. Likewise, Equation 2.2 does have harvest costs increasing slightly for larger trees. This reflects the fact that more area has to be cleared to work around a larger tree, it requires more time to cut down, more time to prepare for hauling, more effort to get to the mill, etc. The final equation shows that the logs have a basic wholesale value of nine cents per board foot, but this value increases slightly (.03 cents) as larger boards can be sawed from the logs.

As we watch the assistant, he has these thoughts: "We can afford to harvest any tree whose sale value exceeds its harvest cost. Or

$$s > c \tag{2.4}$$

or

$$9 + .03\ v > 1.2 + .01\ v\ . \tag{2.5}$$

Hey, wait a minute. That says that we can afford to harvest <u>any</u> tree!" Why? Whenever the first inequality, Equation 2.4, is true, a profit occurs, but Equations 2.4 and 2.5 give equivalent representations of the same fact. The latter inequality always is satisfied because it reduces to $7.8 + .02\ v > 0$ which is true for all $v > 0$, the only relevant values. "This result does not make sense because small trees have no value for sawmill logs. I'd better check dimensions on all quantities:

$$c\left(\frac{\$}{\text{tree}}\right) = 1.20\left(\frac{\$}{\text{tree}}\right) + .01\left(\frac{\$}{\text{bd. ft.}}\right) \times v\left(\frac{\text{bd. ft.}}{\text{tree}}\right) \tag{2.6}$$

$$s\left(\frac{¢}{\text{bd. ft.}}\right) = 9\left(\frac{¢}{\text{bd. ft.}}\right) + .03\left(\frac{¢\ \text{tree}}{\text{bd. ft.}}\right) \times v\left(\frac{\text{bd. ft.}}{\text{tree}}\right) \tag{2.7}$$

(The dimensions on the last constant (.03) may seem odd, but they result from (cents per board foot) per (board foot per tree). The product, $.03 \times v$, must have dimensions of cents per board foot before it can be added to another quantity with the same dimensions.) When cost and sale value have different dimensions, they cannot be compared as you cannot add eggs and apples or pounds and ounces without conversion to a common scale. If we multiply the cost equation by 100 (cents per dollar) and the sale value equation by volume (board feet per tree), both will represent value per tree in cents:

$$sv > c$$

or

$$9\ v + .03\ v^2 > 120 + v \tag{2.8}$$

or

$$.03\ v^2 + 8\ v - 120 > 0$$

Small values of v will not satisfy this inequality. What values of v will? If we set the left hand side equal to zero, we can use the quadratic formula to solve for the value of v separating the solution from the inadmissible values:

$$.03\ v^2 + 8\ v - 120 = 0$$

or (2.9)

$$v = \frac{-8 \pm \sqrt{64 - 4\ (.03)(-120)}}{2(.03)}$$

$$v \doteq -281 \text{ or } v \doteq 14.2 \text{ board feet per tree.}$$

(The notation $\doteq$ signifies approximate equality.) Only the positive solution interests us because our relationship has no meaning for negative v. Given this minimum economic volume, we can use Equation 2.1 to establish the minimum economic diameter:

$$14.2 = 10 + .007\ (d - 5)^3$$

or

$$(d - 5)^3 = \frac{14.2 - 10}{.007} = 600$$

so

$$d = 5 + \sqrt[3]{\frac{14.2 - 10}{.007}} \doteq 13.4 \text{ inches.} \tag{2.10}$$

The assistant, rather proud of himself, showed his results to the chief forester. He took pains to explain his approach, background Equations 2.1-2.3, the resultant Equation 2.8 and the minimum diameter in Equation 2.10. The following exchange then occurred.

Chief: "That can't be right!"

Assistant: "It has to be right. Equations don't lie."

Chief: "If all else fails, THINK! Right now we are harvesting a stand of mostly 14" trees, but we hardly are breaking even. We can't possibly break even on 14" trees in the new area because it lies much further from the mill."

Where did the assistant go astray? Critically examine his reasoning. His first thought was correct: The firm can afford to harvest any tree whose sale value exceeds its total cost. However, the relation

$$c = 1.20 + 0.01 v$$

includes only the cost of harvesting the tree, not the tree's cost of 0.02 v dollars per tree. When the harvest cost of 120 + v cents is replaced by the total cost of 120 + v + 2 v cents, we have

$$9 v + .03 v^2 > 120 + 3 v$$

or

$$.03 v^2 + 6 v - 120 > 0 .$$

Again, the minimum economic volume results from finding the values of v which satisfied

$$.03 v^2 + 6 v - 120 = 0 , \tag{2.11}$$

$$v = \frac{-6 \pm \sqrt{36 - 4 (.03)(-120)}}{2 (.03)}$$

or

$$v \doteq 18.3 \text{ or } -218 \text{ board feet.}$$

Again, only the positive solution has meaning for establishing the minimum economic diameter:

$$18.3 = 10 + .007\ (d - 5)^3$$

or

$$d = 5 + \sqrt[3]{\frac{18.3 - 10}{.007}} \doteq 15.6 \text{ inches}$$

Two final things should be said about this example. It may appear fairly detailed, but a forestry cost accountant could identify costs more precisely than Equations 2.1-2.3. For example, the $1.20 constant in Equation 2.2 represents all fixed costs of equipment for harvest and transportation, plus all costs of getting lumberjacks to the tree and setting it up for felling, etc. Secondly, the figures used here typify the situation of harvesting ponderosa pine in the more remote areas of the central Rocky Mountains about 1970. Since costs and prices can significantly depend on local conditions and change through time, the numbers used here might be irrelevant to a particular problem of interest to you. Nevertheless, the approach applies not only to ponderosa pine, but to other species of trees. The general approach applies to a wide spectrum of harvest, and even production processes outside of forestry.

2.2 DIAGNOSING THE PROBLEM

You cannot hope to solve a problem, textbook or otherwise, until you understand it. Thus, do not expect to successfully execute any other steps in the problem solving process until you resolve this one. In the process of solving the problem you should expect to make false starts, but they should provide a reason for recycling to a particular point. Remember the assistant in the forestry example. On his second attempt he still left out the cost of the trees, so he recycled to that point.

You need clear answers to three questions: What do I know? What do I want to determine? What restrictions must my solution satisfy? In the forestry example, the question was simple: What diameter of tree can be harvested economically? The available information was scant other than the location at which the harvest was to proceed. The restriction on the solution was simple: It had to be an economically feasible solution.

If any of the above questions do not seem to have reasonably clear answers, then the problem must be reexamined to identify features ignored or erroneous assumptions made. This process should become clearer as we present more solved examples. Counterexamples provide a useful means for understanding what a problem is by learning what it is not. Again in the forestry example, an answer of 16 inches would make sense while an answer of 19 board feet or a forest 30 miles away would not. As you study the examples here, keep the above questions in mind.

2.3 ESTABLISHING RELATIONSHIPS

Many problems involve relations among important variables. The relationship itself may be the object of interest; in this case you can bypass the next two steps in the problem solving process. Frequently, however, an important step in analyzing the nature of a relation occurs in the next two steps. In a more formal mathematical environment, special emphasis is placed on solving a different kind of problem from the ones we stress here. For example in geometry, proofs and construction of geometric figures occupy a central role. In the life sciences, however, interest usually centers on how variables relate to each other.

Biological relations often are expressed mathematically by using functions. Because a substantial discussion of functions and relations appears in a later chapter, we will not be very precise about them here; for now regard them as relationships which can be expressed in a mathematical manner, often as an equation.

Two strategies exist for establishing relationships. First you can look at established facts. Basic laws of chemistry and physics often are used in the development of biological relationships. For example, the product of the pressure and volume of a gas always equals a constant, provided the gas remains at a constant temperature. At a less formal level, many established empirical relationships exist within the life sciences. For example, consider the cost and volume relationships set out in Equations 2.1-2.3. They came from information established within the context of the problem. Established relationships can be found by direct research or by reference to research documented in textbooks, journals, or various other publications. In practice you will have to seek them out or establish them yourself. We generally will have to give you such relationships because you presently have no effective way for finding them.

In sorting out and finding the available information, whether in textbook or real problems, take care to distinguish between relevant and irrelevant information. Thus, textbook problems can legitimately include extra, irrelevant, or confusing information because the real world does. You cannot make a final evaluation of the relevance of all information at the point of establishing a relationship, but you certainly can identify some information as not particularly useful. The final decision on the relevance of information must await your tentative solution to check its compatibility with all aspects of the problem. For example, in the forestry cost analysis, lumberjacks' salaries would constitute irrelevant information because they were incorporated into Equation 2.2.

The other major source of information for helping establish relationships comes from mathematics itself. Within a problem, many mathematical relations are intrinsically present. For example, in many problems which involve length or distances, the Pythagorean Theorem can be very useful. It relates the lengths of the two sides of a right triangle to the length of its hypotenuse. If we symbolize lengths of two sides and the hypotenuse by respectively a, b, c as in Figure 2.1, this theorem says

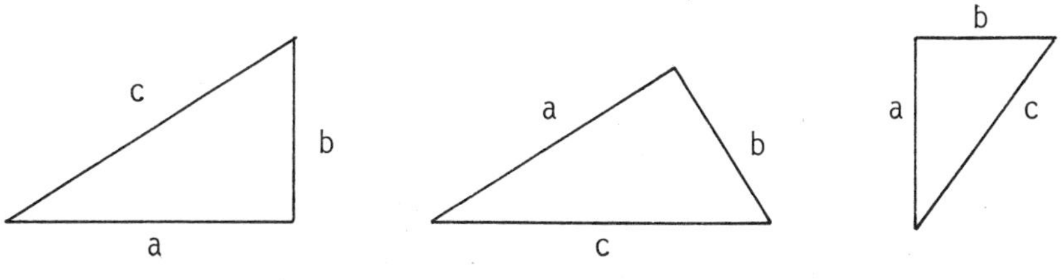

Figure 2.1

$a^2 + b^2 = c^2$, or $c = \sqrt{a^2 + b^2}$, or $a = \sqrt{c^2 - b^2}$, or $b = \sqrt{c^2 - a^2}$.

Similarly, facts from Euclidean or analytic geometry, trigonometry and algebra can be used. We will alert you to some of these at appropriate points. When you have to develop a new relationship from features of a particular problem setting, you should usually do some or all of the following things:

1. Draw relevant figures, including auxiliary components.
2. Introduce suitable and suggestive notation.
3. Think of similar or related problems you have solved before.
4. Create and solve a restricted version of your problem if you cannot at first solve the whole problem.

These strategies will help you attack the following two examples.

EXAMPLE 2.1: Suppose that as a fisheries biologist you decide to set a gill net on your next trip to a remote mountain lake. You have determined exactly where you want the net, between the points designated as A and B in Figure 2.2. Since you must backpack to the lake you do not want to carry in more net than necessary. How do you use what you now have at the lake, namely you, to determine the length of gill net?

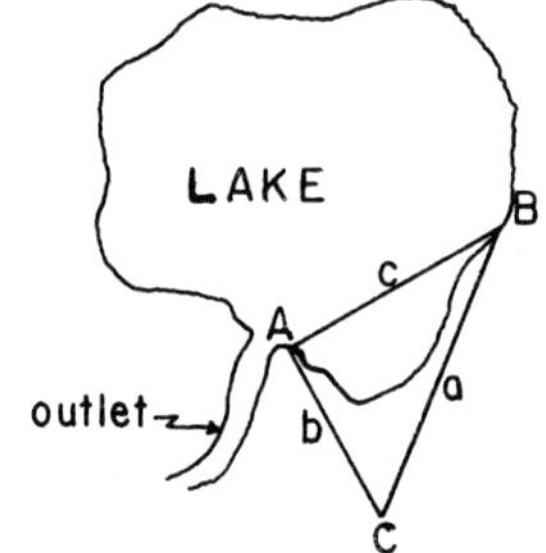

Figure 2.2

SOLUTION: Figure 2.2 provides a working drawing in which you seek the distance between A and B, labeled as c. If you pick an accessible point C such that the triangle ABC has a right angle at A, the Pythagorean theorem gives:

$$c = \sqrt{a^2 - b^2} .$$

The distances a and b could be paced off or measured because they are accessible. Upon substitution into the above equation, you get the length of gill net needed.

The original description of the problem may have appeared to convey little, if any, information needed for a solution. However, after drawing a figure and recalling a past relationship, you had enough information to find a solution. ∎

<u>EXAMPLE 2.2</u>: Consider a population of N_0 animals in a well defined region. Suppose that a certain proportion of this population, say r, give birth to one individual per year with the rest giving none. Develop a relationship which will give the number of individuals in the population after n years, assuming no losses from the population.

First we have the following:

N_0 = number of individuals in the population initially

r = proportion of individuals giving birth to one additional individual.

We must conclude that $0 \leq r \leq 1$; r = 1 means the entire population is giving birth while r = 0 implies that the population receives no new recruits. For present simplicity, assume that r remains constant from year to year. By calculating the size of the population for a few years, we may see a pattern developing. After the first year the population has $N_0 + N_0 r$ individuals because $N_0 r$ new individuals were born into the population. After the second year,

$$N = N_0 (1 + r) + N_0 (1 + r) r$$

$$= N_0 (1 + r)(1 + r)$$

$$= N_0 (1 + r)^2 .$$

The term $N_0 (1 + r) r$ gives the number of new births in the population already of size $N_0 (1 + r)$. After the third year,

$$N = N_0 (1 + r)^2 + N_0 (1 + r)^2 r$$

$$= N_0 (1 + r)^3 .$$

The above can be summarized this way:

Year	0	1	2	3	4
Number	N_0	$N_0 (1 + r)$	$N_0 (1 + r)^2$	$N_0 (1 + r)^3$	$N_0 (1 + r)^4$

This table suggests that the general relationship giving the number of individuals in the population after the n-th year, when no individuals were leaving the population, is

$$N = N_0 (1 + r)^n \tag{2.12}$$

A proof that this equation holds for all n = 1, 2, 3, ··· utilizes the principle of mathematical induction, a technique you probably have seen. A problem at the end of this chapter removes the restriction that no individuals may leave the population. ▌

Once you have a tentative relationship, make a preliminary check on its reasonableness. Have you used all relevant information in its development? Will it submit to mathematical analysis in the next step? Are all of your equations dimensionally conformable? See the discussion in the timber cost accounting example near Equations 2.6 and 2.7 for an illustration of this point. Most variables have dimensions involving basic units like length, weight, time and cost. In multiplication or division they multiply and divide like algebraic variables; dimensions must be identical before quantities can be added or subtracted.

2.4 PICKING USEFUL MATHEMATICAL TOOLS

Students naturally try to use techniques of a section on exercises at its end. Real problems rarely come so neatly packaged. Thus, a few of the problems in the exercises will require tools described at an earlier point in the book than where they occur. Many problems in the life sciences, nevertheless, can be categorized as dealing with (i) likelihood, (ii) behavior of functions, (iii) rates of change (minimization or maximization), and (iv) accumulation.

In problems which deal with likelihood of occurrences, probability almost certainly will be involved. In using probability, an essential technique involves the careful definition of what constitutes outcomes of interest. This will be illustrated in the next chapter.

To investigate the behavior of a specified function, look for two general kinds of behavior: How does the function behave at the limits of its permissible domain? What does it look like toward the middle of its domain? The tools of limits and differentiation, and later, partial differentiation apply to such problems.

Rates of change of functions occupy an important place in biology. Derivatives illuminate how a function is changing across its domain. In certain instances when the rates of change become zero, the function has achieved a minimum or a maximum.

When accumulation or area appears, the tools of integral calculus often prove useful. It may be accumulated work, accumulated area under a curve, total incident radiation during a specified time period, or any such response.

As you progress through this book, the task of picking the right tool will become easier. You will learn to identify relevant tools; for now do not worry about specific examples or specific types of tools.

2.5 APPLYING THE SELECTED TOOL

This particular step does not lend itself to close examination here because the use of a selected tool depends heavily on the tool. We offer only this suggestion: Regardless of the tool selected, be alert for its limitations. Do not apply it indiscriminately under circumstances for which it was not designed.

2.6 INTERPRETING THE CONSEQUENCES

First, after you think you have a solution, check that it satisfies the problem. To examine a tentative solution, you might ask these questions: Does it make sense mathematically? Does it seem practically sensible? Has it used the available relevant information? If these questions have been answered with yes, you should turn to drawing practical implications from your solution. Otherwise, you probably need to again seek a solution, correcting the deficiencies you have found.

To illustrate "Does the solution make sense mathematically?", recall the cost relationship of Equation 2.11:

$$.03\ v^2 + 6\ v - 120 = 0$$

Look at this equation. The term .03 v^2 adds a positive amount to its left hand side, regardless of the value of v. Without this term the equation is satisfied by v = 20. Thus, the solution to the entire equation, including .03 v^2, has to be smaller than v = 20. This little example illustrates how you might check a solution for plausibility.

The first solution that the assistant took to his chief simply illustrates an answer which was not practically sensible. By omitting one essential fact, the basic cost of the lumber before harvest, an answer of 13.4 inches resulted. From the chief forester's experience this was not a sensible answer. This illustrates another method for checking a solution, namely it should appear compatible with past experience.

Once you have satisfied yourself that the answer can be defended mathematically and practically, you should draw practical implications from it. In actual settings this means that mathematical results should be translated into management policy and implications such as recommendations for action. Of course, these kinds of uses cannot be fully simulated in a textbook environment. Again using the timber problem, the management policy would be to cut all trees having a diameter of at least 15.6 inches.

As you use mathematics in your career, always remember that mathematics is a tool to be used, not a lord to be worshiped. Mathematical answers have no practical value until they are turned into usable conclusions.

2.7 SUMMARY

This summary consists of a completely solved example which illustrates the points brought out in the previous sections. As you read through the example, see if you can visualize an alternate method which would lead to the solution of the problem.

EXAMPLE 2.3: A wildlife researcher is designing methodology for studying one seed juniper, a species important for browsing mule deer. He needs to measure the volume of individual trees and the surface area available for browsing. He decides that he can satisfactorily approximate these quantities from measurements of the thickness of each tree's canopy along four axes located every two feet above the ground. To measure growth he must be able to return and evaluate the same measurements. Thus, he wants to permanently locate eight points, equally spaced around a circle and 10 feet from the tree (see Figure 2.3). He plans to pick one point adjacent to the shallowest point of tree's canopy and 10 feet from the trunk. How should he locate the remaining seven points?

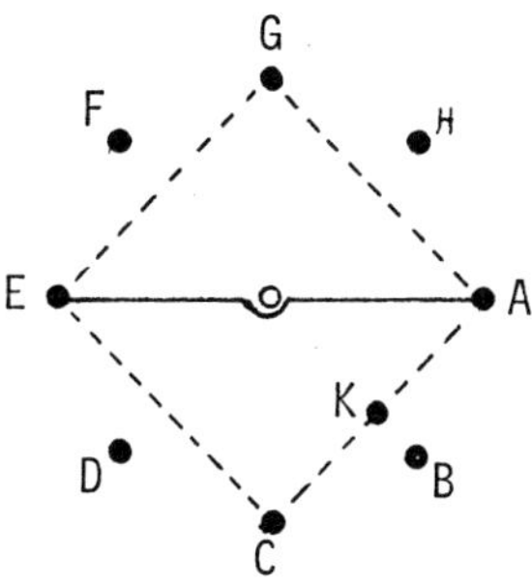

Figure 2.3

The wildlife researcher faces these restrictions: The strategy must be easy to use in the field; he cannot draw a circle around the tree because of thick foliage near the ground; and he will have along a 20 foot pole, shaped as shown in Figure 2.3, for the subsequent measuring process.

AN APPROACH: This problem is rather simple to diagnose. We must locate eight points (A-H, Figure 2.3) equally spaced around the trunk of a tree; the process must be easy to use in the field. We will assume that the 10 feet is measured from the center of the trunk of the tree.

We now need to establish relationships which will locate the points. In Figure 2.4 the distance $\overline{OA} = \overline{OC} = 10$ feet, and the triangle AOC is a right triangle. We can calculate the length $\overline{AC}$ using the Pythagorean Theorem: $\overline{AC} = \sqrt{10^2 + 10^2} = \sqrt{200} \doteq 14.1$ feet. Now he should take along a 28.2' cord. By marking its midpoint and attaching it to the ends of the pole, he can locate the point C as shown in Figure 2.4. If the cord also is marked at 7.05 feet, he can locate the

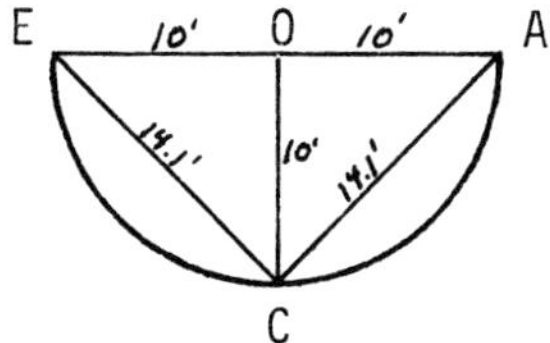

Figure 2.4

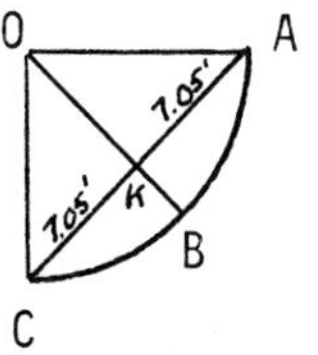

Figure 2.5

point labeled K in Figure 2.5 at the same time he locates the point C. If he puts a stake at K and rotates the pole over it, the points B and F lie at the end of the pole. By repeating this process on the other side of the tree, he can locate the points D and H.

This approach satisfies the restrictions. Reread the example and see if you can identify the steps in the problem solving process outlined at the beginning of the chapter. Try to devise another solution to this problem and identify each step as you progress through the problem. ▌

EXERCISES

1. Temperature is recorded on four common scales: Fahrenheit (F), Centigrade (C), Rankine (R) and Kelvin (K). Under standard conditions, water freezes at 32°F and 0°C while it boils at 212°F and 100°C. The other two scales measure temperature relative to absolute zero, -273°C, the lowest possible temperature. A 1°F change in temperature produces a change of 1°R and likewise for 1°C relative to 1°K.
   a. Establish twelve equations relating temperature on any scale to any other scale.
   b. What is 77°F on the other scales? 198°K? 70°C? 561°R?

2. Refer back to Example 2.2. Find a relationship giving the number of individuals in the population if the proportion q of individuals leave the population; still assume that the proportion p of the remaining individuals reproduce as before. What restrictions must p and q satisfy?

3. Given the following information about a solution containing sodium pentabarbital (an anesthetic): A bottle contains 100 cc. (cubic centimeters) of liquid; 1 cc. of liquid from the bottle contains 64.8 mg. (milligrams) of sodium pentabarbital. In order to anesthetize a rat the amount of sodium pentabarbital needed to be injected intraperitoneally is 4.2 mg. per 100 grams of body weight of the rat. How many cc.'s of the initial solution should be given

3. Continued
   to a 350 gram rat to anesthetize him?

4. The radium isotope ^{228}Ra decreases in its intensity of radiation approximately 9.8% every year. If I_0 denotes the original intensity, what is the intensity after 1 year, 2 years, 4 years and n years?

5. The average diameter of human capillaries is 0.01 mm. (millimeter) and the linear velocity in the capillary is approximately 0.6 mm./second. Calculate the volume of outflow from the capillary per second and per hour.

6. The average diameter of an adult's pulmonary air-cell is 0.2 mm. while that of a newborn infant is 0.07 mm. If we assume that the cells are spherical in shape and the volume of air-cells in an adult lung and infant lung are 1600 cm^3 (centimeters) and 68 cm^3, respectively, find the approximate number of air-cells and their total surface area in the adult and the infant. ($A = 4\pi r^2$, $V = 4\pi r^3/3$).

7. Water in contact with most soil materials will rise to a height of $h = \frac{2\sigma}{rpg}$ where h is the height, r is the radius of the capillary in cm., σ = 73 dynes/cm^3, p = 1 gram/cm. and g = 980 dynes/gram. If r = 5μ (microns) = 0.0005 cm., calculate h. What are the units of h?

8. In the human body blood flow is regulated largely by changes in the radius of arterioles, small blood vessels which have muscle fibers in their walls. When the muscles contract, the blood vessels narrow, resistance increases, and flow decreases. Poiseville's formula for the nonturbulent flow of a liquid through a tube approximately describes blood flow. It states:

$$f = \frac{r\pi p}{8L\eta}$$

   when η symbolizes viscosity measured in poises (one poise = 1 dyne-second/cm^2); r and L symbolize the radius and length of the tube; p = pressure difference between ends of the tube in mm. of

8. Continued

mercury; f = flow in cm^3/sec. Under certain conditions these are valid assumptions: Mean arterial blood pressure is 100 mm. of mercury while in the great veins entering the heart it equals zero; the heart pumps approximately 5 liters of blood per minute against a resistance of .04 poise. For conversion purposes, 1 atmosphere of pressure = 760 mm. of mercury = 1.013×10^6 dyne/cm^2, 1 liter = 1000 cm^3.

a. If blood flows an average of one meter in a complete circuit from the heart through peripheral tissues and back, a tube of what constant radius would give the same flow?
b. By what percent would the radius have to be increased to double flow?
c. Cholesterol sometimes is deposited inside arteries. What decrease in flow would result from a 10% reduction in the diameter of the artery with no pressure change? How much would the pressure have to increase to maintain the original flow?
d. How much would the pressure have to increase to maintain the flow of 5 liters/min.? Assume the conditions of part c.

Chapter 3

WILL IT HAPPEN?

Will *it* happen? Define *it* however you want! More precisely, you might ask if the result of a particular happening belongs to a specified set of possible outcomes. The first three sections of this chapter present some elementary set theory and a brief review of the real number system. Some of this material may be familiar to you from high school or other college courses while some may be completely new. We suggest that you pay particular attention to the definitions because you use sets in your everyday life; you just do not call them by that name.

Will it *happen?* Death and taxes are sure; some uncertainty attaches to almost all other occurrences. Probability provides a way for quantifying the degree of uncertainty. This idea, to which the latter part of the chapter is devoted, occupies a central place in biology. The fourth section introduces the idea of probability and considers some simple cases while the last two sections concern computation of probability and several important discrete probability distributions.

3.1 COLLECTING THINGS

People, animals and even plants collect things. People collect stamps, coins, matchbooks, antiques, etc. Animals collect nesting material, food or even, like packrats, just anything! Plants collect nutrients from the soil and carbon dioxide from the air; some, like the mesquite bush of the desert or lichens in the alpine areas, even collect soil. Collectors tend to organize things. Taxonomy provides a striking illustration of man's attempt to organize biology.

The terminology of set theory is a tool we can use to make precise statements about collections. As collections have individual components, sets have elements; as collections have parts, sets have subsets. Thus, consider the following definitions:

- ELEMENTS are basic objects of interest.

  SETS are collections of elements with a clear criteria for the inclusion of an element in the set.

- Notation for membership: If a is an element of the set A we write $a \in A$; $a \notin A$ symbolizes a does not belong to A.

- A set B is a SUBSET of the set A, symbolized $B \subseteq A$, if all of the elements in B belong to the set A. In terms of memberships, $B \subseteq A$ provided if $b \in B$, then $b \in A$.

- A set B is a PROPER SUBSET of the set A, symbolized $B \subset A$, when $b \in B$, implies $b \in A$, but there exists at least one element $a \in A$ such that $a \notin B$.

- The NULL or EMPTY SET has no elements. It is denoted by $\emptyset$. The empty set is a subset of every other set.

The essential feature in the above definitions involves membership or belonging. The ideas set forth above are illustrated in the following examples.

<u>EXAMPLE 3.1</u>: Here are some examples of different sets: The students enrolled in the course Mathematics for Biologists; the fish in Lake Mead; the animals, plants or insects belonging to a particular taxonomic phylum; all acid forming bacteria; the trees common to New York; the big game animals of Colorado. ▌

<u>EXAMPLE 3.2</u>: Consider one of the sets from Example 3.1 in more detail. The set of big game animals in Colorado has elements of elk, deer, pronghorn (antelope), bighorn sheep, mountain goat, black bear, mountain lion and wild turkey. A subset of this could be described as animals with antlers. This subset consists of only elk and deer. If we let B symbolize the set of big game animals in Colorado, A symbolize the set

† Some writers choose not to distinguish between subsets and proper subsets. In this case $\subset$ symbolizes both.

of antlered big game animals in Colorado and H symbolize the hooved big game animals in Colorado, then

B = {deer, elk, pronghorn, bighorn sheep, mountain goat, black bear, mountain lion, wild turkey}

A = {deer, elk}

H = {deer, elk, pronghorn, bighorn sheep, mountain goat}

and $A \subset B$, $A \subset H$, $H \subset B$ so we could write $A \subset H \subset B$. If x symbolizes (is another name for) bighorn sheep, then $x \in B$, $x \in H$, but $x \notin A$. ▌

EXAMPLE 3.3: Let G represent the gamma ray emitting radioisotopes.

$$G = \{{}^{139}\text{Ba},\ {}^{140}\text{Ba},\ {}^{60}\text{Co},\ \cdots\}$$

If $x = {}^{14}\text{C}$, then $x \notin G$ because ${}^{14}\text{C}$ is a pure beta emitter, not a gamma emitter. ▌

A set can be specified in two ways: By listing its elements or by giving the criteria for an element to belong to the set. Example 3.2 gave a verbal description of the sets B, A and H and then listed the elements in each. To specify more complex sets we need a uniform scheme for describing a set by its defining criteria. Sets have different sizes, a fact already apparent from the above example. We will need a consistent way for describing this size. Thus we have the following definitions:

- The SET BUILDER NOTATION $A = \{x | x \text{ satisfies stated property}\}$ specifies A as comprising all elements x satisfying the stated property.
- The SIZE of a set A, symbolized $S(A)$, gives the number of distinct elements in A if A is finite.
- A FINITE SET has elements which can be arranged and counted one by one until the last element is encountered. The null set is considered to be a finite set; it satisfies $S(\phi) = 0$.
- An INFINITE SET is any set which is not a finite set. $S(A) = \infty$ symbolizes the size of an infinite set A.

The vertical slash in the set builder notation separates the symbol for the elements in the set from the specifying property. The specifying property can be any clear criteria. It may include equalities, inequalities, other set memberships, or verbal descriptions.

EXAMPLE 3.2 (continued): The sets B, A, H could have been defined by

$B = \{x|x$ is a big game animal in Colorado$\}$

$A = \{x|x$ is an antlered big game animal in Colorado$\}$

$H = \{x|x$ is a hooved big game animal in Colorado$\}$

Then $S(B) = 8$, $S(A) = 2$, $S(H) = 5$. Notice that $A \subset H \subset B$ and $S(A) < S(H) < S(B)$. ▌

When a set B is a subset of the set A, you might feel that A is larger than B. In a sense it is because when $B \subseteq A$, $S(B) \leq S(A)$. If $B \subset A$, $S(B) < S(A)$, provided B is a finite set; when $S(B) = \infty$, then because A has at least as many elements as B, $S(A) = \infty = S(B)$. Intuition sometimes fails in comparing sets of infinite size because the elements of an infinite set may lie very close to each other, or very far apart.

EXAMPLE 3.4: Consider the sets

$A = \{x|x = 1, 2, 3, \cdots\}$ and $B = \{y|y = 10, 20, 30, \cdots\}$.

Clearly, if $y \in B$, then $y \in A$ so $B \subset A$. Yet these sets have the same size. To verify this, suppose you have elements in both sets lined up in increasing order. Start through each set, counting an element at a time. As you pass $13 \in A$ you pass $130 \in B$; with $277 \in A$ you find $2770 \in B$ and so on. Thus A and B have the same size because each element of A has a matching element in B, and vice versa. When B contains only every tenth element of A, intuitively it seems like B is only one-tenth as large as A. This apparent paradox has a simple solution: The statement $S(A) = \infty$, describes something no number can exceed. In this sense it is bigger than any number, and so is one-tenth of it. ▌

EXERCISES

1. Give an example of a finite set and an example of infinite set in your own area of study.

2. Classify the following sets as either finite or infinite.
   a. The set of fresh water species of fish.
   b. The set of positive integers.
   c. The set of nitrogen molecules in the atmosphere.
   d. The number of scales on a particular species of fish.
   e. The number of cells in a cancerous tumor.
   f. The set of rare earth elements.
   g. The number of evergreen trees in the United States.

3. Describe each of the following sets by using set builder notation.
   a. The set of people with blood type A, Rh positive.
   b. The set of gaseous radioactive elements.
   c. The set of warm water fish.
   d. The set of acid forming bacteria.
   e. The set of cereal grains.

3.2 CATEGORIZING YOUR THINGS

The previous section introduced some basic ideas about sets and gave biological illustrations. Sets can be used to describe characteristics of biological objects. Objects in different sets are related to one another through their characteristics. To develop operations on sets, we offer these definitions:

- The UNIVERSAL SET is the totality of elements under consideration and usually is denoted by U.
- EQUALITY of two sets A and B, symbolized as $A = B$, means that every element in A is also an element in B, and conversely. Symbolically $a \in A$ implies $a \in B$, and $b \in B$ implies that $b \in A$.

- The INTERSECTION of two sets A and B, symbolized by $A \cap B$, is the set which contains all elements common to both A and B. $C = A \cap B$ requires that the elements $c \in C$ simultaneously satisfy $c \in A$ and $c \in B$. The key word is "and".
- Two sets A and B are DISJOINT or MUTUALLY EXCLUSIVE when they have no common elements. Disjoint sets satisfy $A \cap B = \phi$.
- The UNION of two sets A and B, symbolized by $A \cup B$, is that set containing all elements in either A or B. $C = A \cup B$ means that $c \in C$ if $c \in A$ or $c \in B$ or $c \in A \cap B$. The key word is "or".

The universal set plays the obvious role of defining the reference background. The intersection of two sets usually contains fewer elements than either of the individual sets. Some elements in the first set may not be in the second, and vice versa; only the elements in both A and B belong to the intersection. Disjoint sets describe entirely different parts of the same universe; disjoint sets share no elements. The union of two non-empty sets defines a set at least as large as either of its individual components because the union includes all elements in either set, or in both. These ideas are illustrated in the ensuing examples.

VENN DIAGRAMS provide a convenient means for visualizing operations on sets. A Venn diagram consists of a schematic figure depicting the universal set within which the sets under discussion are sketched to show their relation to one another. Various schemes of shading show the particular part of interest. The shapes used to represent the universal set and other sets have no significance. Initially we will use rectangles for the universal set and circles of varying sizes for sets therein.

The Venn diagram in Figure 3.1 shows U, the universal set, the two sets A and B, as well as $A \cap B$ and $A \cup B$. This figure delineates the set A with horizontal lines and the set B with vertical lines. The area containing both vertical and horizontal lines represents $A \cap B$ while an area

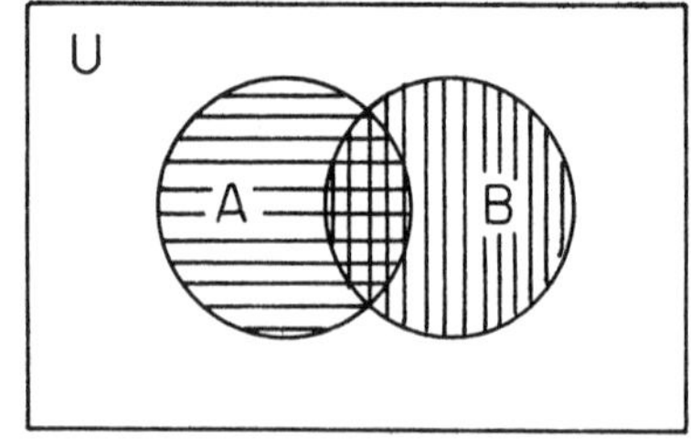

Figure 3.1

which has either vertical lines, horizontal lines, or both, represents $A \cup B$. Notice that

$$S(A \cup B) = S(A) + S(B) - S(A \cap B)$$

because the addition of $S(B)$ to $S(A)$ counts $A \cap B$ a second time.

EXAMPLE 3.5: Let U, the universe of interest, consist of aquatic vertebrate animals and

$C = \{x | x$ names a fish which can be caught in fresh water$\}$

$S = \{y | y$ names a fish which can be caught in salt water$\}$

$F = \{z | z$ is the name of a fish$\}$.

The Venn diagram in Figure 3.2 shows relations among these three sets and their universe. Notice first $C \cap S \neq \phi$ because some fish spend at least part of their life in fresh water and part of it in salt water; salmon provide a clearcut illustration. Some fish spend their entire lives in fresh water; trout, crappie and catfish provide familiar examples. Notice that $F = C \cup S$ because all fish spend most of their lives in either fresh water or salt water. F does not exhaust the universe because not all aquatic vertebrates are fish; you know that whales and dolphins are mammals. About 17,000 species of fish have been identified (as many as 40,000 may exist), so $S(F) > 17,000$. ▌

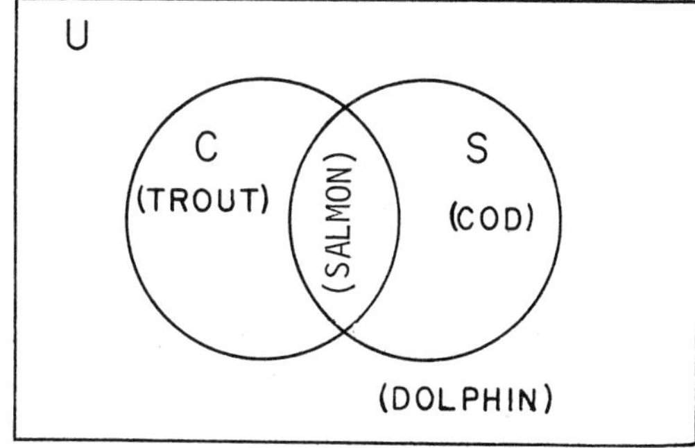

Figure 3.2

EXAMPLE 3.6: Let B be the set of registered Brown Swiss dairy cows in Wisconsin and J the set of registered Jersey dairy cows in Wisconsin. These sets are disjoint, $B \cap J = \phi$ because no cow can be registered as both Brown Swiss and Jersey. ▌

<u>EXAMPLE 3.7</u>: Figure 3.3 represents a typical mammalian cell. From the structure of the cell we can identify various sets and subsets. Let us define the following sets:

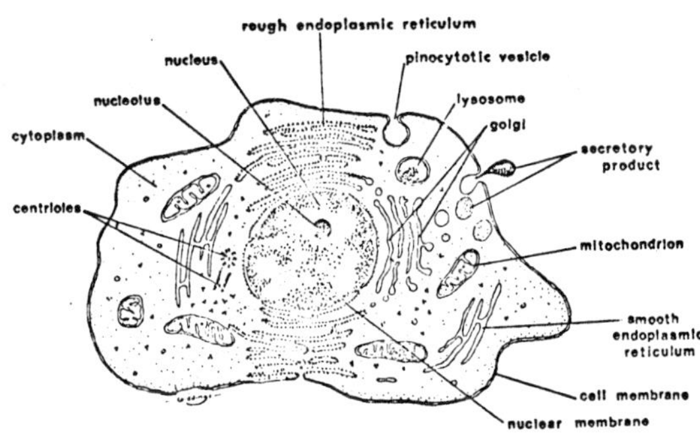

Figure 3.3

$C = \{x | x \text{ is an element of the cytoplasm}\}$,

$N = \{y | y \text{ is an element of the nucleus}\}$,

$A = \{z | z \text{ is the nucleolus}\}$, and

$B = \{b | b \text{ is a lysosome}\}$.

The set C includes the elements lysosomes, golgi bodies, centrosomes, endoplasmic reticulum, and mitochondria. The set N includes reticulum, karyolymph, nucleolus, and nuclear membrane. Clearly $A \cap B = \phi$ because a nucleolus and a lysosome differ. $A \subset N$ and $B \subset C$. Also $C \cap B = B$ as only a lysosome is common to both sets. $C \cup B = C$ and $N \cup A = N$.

Even though the mammalian cell represented in Figure 3.3 is a very complex unit, it has structures which definitely relate to each other and which play a role in its life. Similar relations exist in plant cells. ▌

<u>EXAMPLE 3.8</u>: Figure 3.4 shows the distribution of soil types over an area along the eastern foothills of the Rocky Mountains in Colorado. Notice that we could define several types of soils. For instance we could speak of the set of loam soils, or the set of sandy loam soils in the area. We could speak of the area consisting of Vona sandy loam, or Ascalon sandy loam soil, or clay soils. What might possibly be a universal set for these various sets?

We could construct a similar figure to describe the plant communities existing in this same area; it would show certain types of plants as being predominant in some areas but sparse in others. ▌

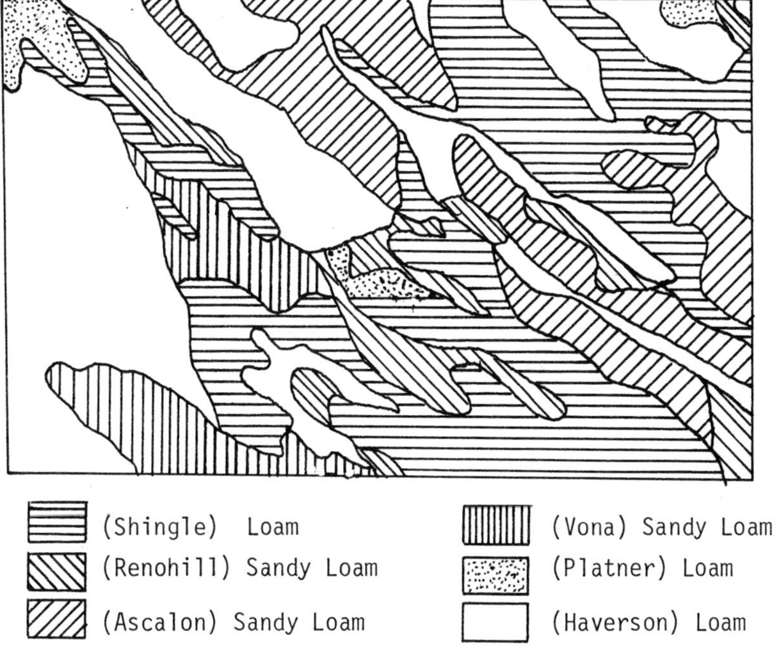

(Shingle) Loam
(Renohill) Sandy Loam
(Ascalon) Sandy Loam
(Vona) Sandy Loam
(Platner) Loam
(Haverson) Loam

Figure 3.4

In looking at Figure 3.2 notice that a set outside of $C \cup S$ does exist, as do sets outside of C and outside of S. This idea of a set outside another set is associated with the idea of a complement.

- The COMPLEMENT of a set with respect to some stated or understood universe consists of the set of elements in the universe which are not in the set. This operation usually is denoted by the symbol '. $A' = \{x | x \notin A\}$.

EXAMPLE 3.6 (continued): Recall that this example had the sets B and J of registered Brown Swiss and Jersey dairy cows in Wisconsin. Consider the universe to be all dairy cows in Wisconsin. Then B' symbolizes all cows in Wisconsin other than Brown Swiss. This includes all non-registered cows, all Jerseys, all Holsteins, etc. Likewise J' denotes all non-Jersey cows in Wisconsin. Earlier we noted that $B \cap J = \phi$ so $(B \cap J)' = U$. ▌

EXAMPLE 3.5 (continued): Recall that C, S, F respectively symbolized the sets of sometimes fresh water fish, sometimes salt water fish, and all fish. The vertically ruled region in Figure 3.5 denotes S' while the horizontally ruled region indicates C'. Furthermore, note that $F = C \cup S$ and $F' = C' \cap S' = (C \cup S)'$. This is denoted by the doubly ruled area in Figure 3.5 and stands for all aquatic creatures other than fish. Likewise, the unruled region stands for $C \cap S$, the set of fish which spend some of their life in fresh water and some in salt water. Note that $(C \cap S)' = C' \cup S'$, the region without any ruling. This illustrates one of DeMorgan's laws, the next topic. ▮

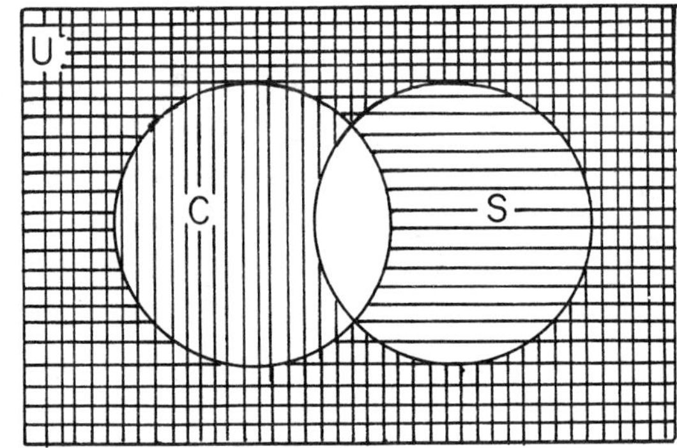

Figure 3.5

Now that complements have been defined and illustrated, we will need to know how complements of intersections and unions relate to the component sets. These relationships are called DeMorgan's laws:

- DeMorgan's laws state that

$$(A \cup B)' = A' \cap B'$$

$$(A \cap B)' = A' \cup B'$$

DeMorgan's laws give rules needed to distribute complementation over union and over intersection. Quantities in parentheses are treated the same way for sets as in algebra. In other words, the statement $(A \cap B)'$ indicates two operations; first find the intersection of A and B, then take the complement of that set. The first of DeMorgan's laws is proved below but the other is left as a problem. The previous two examples illustrated them.

PROOF: This proof relies on a simple fact you should prove for yourself: If two sets C and D satisfy $C \subseteq D \subseteq C$, then $C = D$.

We begin by showing that $(A \cup B)' \subseteq (A' \cap B')$.

If $x \in (A \cup B)'$, then $x \notin A \cup B$;

if $x \notin A \cup B$, then $x \notin A$ and $x \notin B$;

if $x \notin A$ and $x \notin B$, then $x \in A' \cap B'$;

or equivalently, $(A \cup B)' \subseteq (A' \cap B')$.

The first step relies on the definition of a complement; the second uses the definition of a union; verify the third from a Venn diagram; the last follows from the definition of subset.

Now we establish that $(A' \cap B') \subseteq (A \cup B)'$:

If $x \in A' \cap B'$, then $x \in A'$ and $x \in B'$;

if $x \in A'$ and $x \in B'$, then $x \notin A$ and $x \notin B$;

if $x \notin A$ and $x \notin B$, then $x \notin A \cup B$;

if $x \notin A \cup B$, then $x \in (A \cup B)'$;

or equivalently, $(A' \cap B') \subseteq (A \cup B)'$.

Now that we have established

$(A \cup B)' \subseteq A' \cap B' \subseteq (A \cup B)'$,

it follows that $(A \cup B)' = A' \cap B'$. ▼

Thus far we have dealt with only two sets. Many practical problems involve more. Certain cautions must be used in working with more than two sets; these cautions parallel the need for doing addition and multiplication in the right order. For example, if you were asked to draw a Venn diagram of $A \cup B \cap C$, you might not be sure whether this means $(A \cup B) \cap C$ or $A \cup (B \cap C)$. That these two mean very different things should be clear from the two Venn diagrams in Figure 3.6.

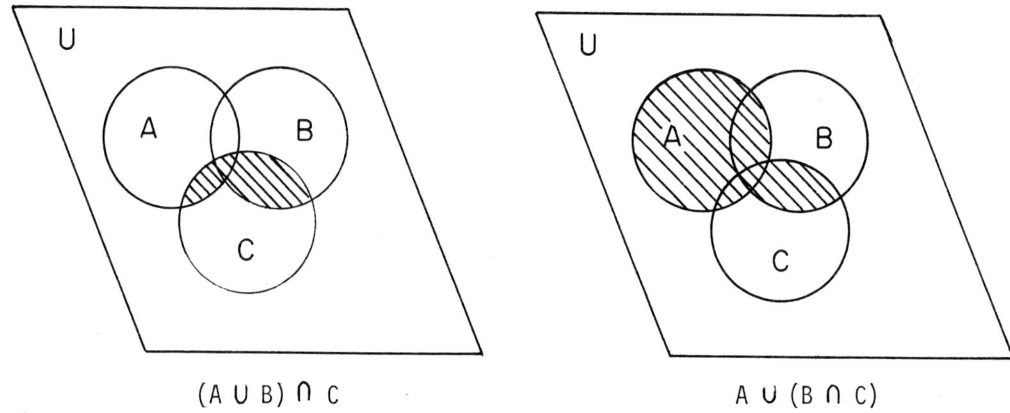

$(A \cup B) \cap C$ $\qquad$ $A \cup (B \cap C)$

Figure 3.6

EXAMPLE 3.9: A wildlife conservation officer questioned 60 duck hunters about the composition of their day's bag limit of ducks with these results: 8 hunters had bagged mallards only; 8 hunters had bagged teal only; 10 had bagged mallards and pintails; 6 had bagged teal and pintails but not mallards; 30 had bagged teal; 6 had bagged mallard and pintails but not teal; and 25 had bagged pintails. The officer has to compile these numbers before he files a report containing the answers to the following questions:

1. How many hunters bagged none of the three species?
2. How many hunters bagged all three species?
3. How many hunters bagged at least two different types of ducks?
4. How many hunters bagged pintails only?
5. How many hunters bagged only a single species of duck?

To answer these, he needs a Venn diagram reflecting his information. The definitions of union, intersection, and complementation provide the basis for answering these questions about the universe of 60 duck hunters. Let

M = $\{m|m$ is a hunter that bagged a mallard duck$\}$

P = $\{p|p$ is a hunter that bagged a pintail duck$\}$

T = $\{t|t$ is a hunter that bagged a teal duck$\}$

We must first translate the given information into sizes involving sets. From the fact that 8 hunters bagged mallards only, $S(M \cap P' \cap T') = 8$ because these hunters got mallards, but not pintails or teals. Continuing in this fashion, the initial information can be summarized as:

S- given	$M \cap P' \cap T'$	$T \cap P' \cap M'$	$M \cap P$	$T \cap P \cap M'$	T	$M \cap P \cap T'$	P
S- ordered	$M \cap P' \cap T'$	$M' \cap P' \cap T$	$M \cap P$	$M' \cap P \cap T$	T	$M \cap P \cap T'$	P
S (S)	8	8	10	6	30	6	25

The second row merely specifies all sets in the order M, P, T.

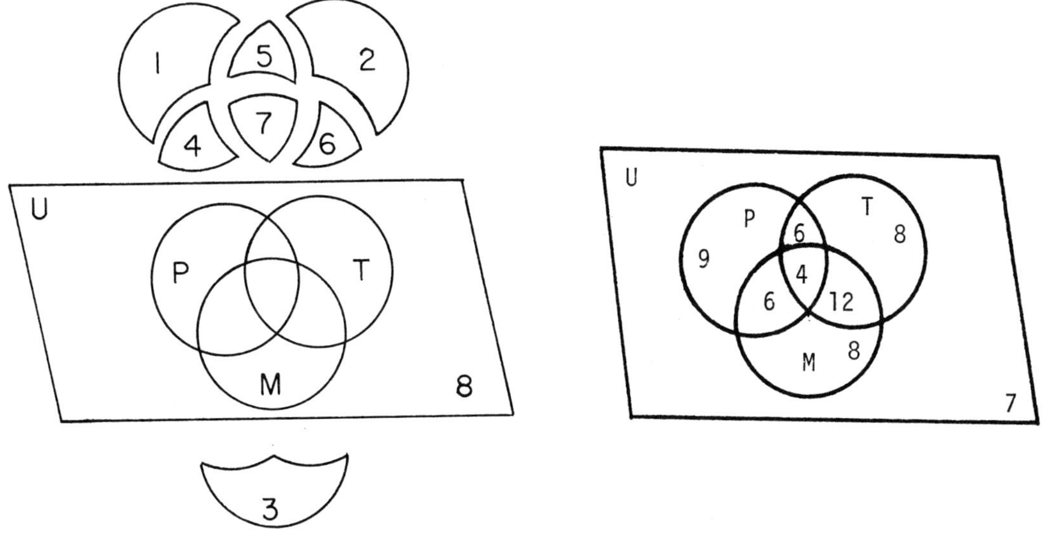

Figure 3.7

Figure 3.8

Now we can begin filling in the exploded Venn diagram shown in Figure 3.7 where the eight basic subsets are numbered. From $S(M \cap P) = 10$ and $S(M \cap P \cap T)$ is found this way: No element can be in both T and T', so $M \cap P$ (areas 4 and 7) can be decomposed into the disjoint sets $M \cap P \cap T$ (area 7) and $M \cap P \cap T'$ (area 4). Thus we have

$$S(M \cap P) = S(M \cap P \cap T) + S(M \cap P \cap T')$$

or

$$S(M \cap P \cap T) = S(M \cap P) - S(M \cap P \cap T') = 10 - 6 = 4 .$$

To next determine the number of hunters that bagged mallards and teals but not pintails, we can decompose $M \cap T$ (areas 6 and 7) into two disjoint sets, $M \cap P \cap T$ (area 7) and $M \cap P' \cap T$ (area 6). Now we must approach this portion differently from the previous part as we have different available information. From

$$30 = S(T) = S(M \cap P' \cap T) + S(M \cap P \cap T) + S(M' \cap P \cap T) + S(M' \cap P' \cap T)$$

we have

$$S(M \cap P' \cap T) = S(T) - S(M \cap P \cap T) - S(M' \cap P \cap T) - S(M' \cap P' \cap T)$$

$$= 30 - 4 - 6 - 8 = 12 .$$

Thus 12 hunters bagged mallards and teals but not pintails. Continuing in a similar fashion we can deduce the size of each subset. Verify the numbers given in Figure 3.8. The questions now have these answers; (1) 7; (2) 4; (3) 28; (4) 9; (5) 25. ▮

EXAMPLE 3.10: In Chapter 1 the relationship that exists between mathematics, biology and statistics was mentioned. We now offer the following Venn diagram for further clarification of this relationship. Write a verbal description of the 8 distinct areas of the Venn diagram.

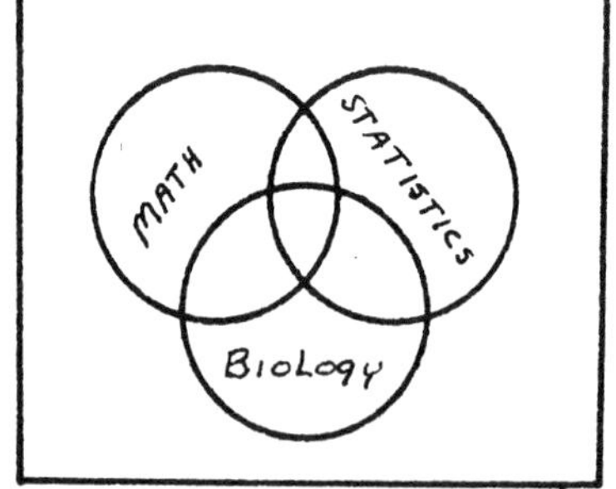

Figure 3.9

EXERCISES

1. Let P and Q be two sets. Draw Venn diagrams showing $P \cap Q$, $P \cap Q'$, $P' \cap Q$, $P' \cap Q'$.

2. Identify each of the areas 1-8 in the following Venn diagram by using the operations of union, intersection, and complementation different from the notation used in Example 3.9.

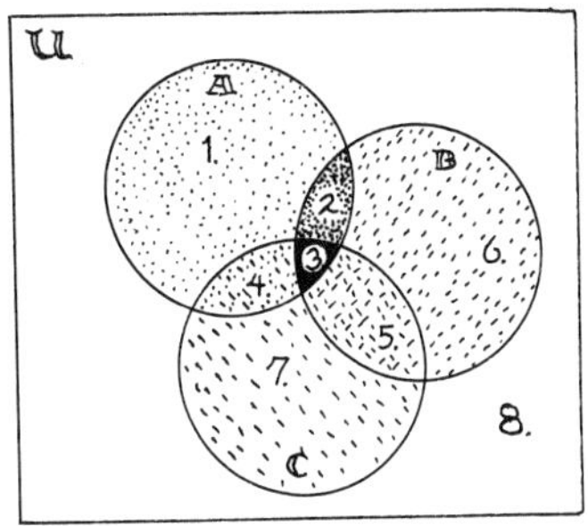

3. Show by use of a Venn diagram that the set $A \cap B'$ is the set of elements in A but not contained in B. This may also be denoted by $A-B$ and sometimes is referred to as the DIFFERENCE SET.

4. Let A = {elk, mule deer, black bear, bighorn sheep, mountain goats} and B = {white tail deer, antelope, elk}. List the elements in $A \cap B$, $A \cup B$, and $A-B$. (See exercise 3.)

5. Describe $A \cap B$ if
   a. A and B are disjoint;
   b. $B \subset A$;
   c. $B \subseteq A$;
   d. $A = B$.

6. By use of Venn diagram, show that $A \cap (B \cup C) = (A \cap B) \cup (A \cap C)$.

7. Is the following true or false: $A \cup (B \cap C) = (A \cup B) \cap (A \cup C)$? Justify your answer.

8. Verify or refute this assertion: $(A \cap B) \cup C = A \cap (B \cup C)$.

9. Show that $A = B$ if $(A \cap B) = (A \cup B)$, and conversely.

10. Show that $(A \cap B)' = A' \cup B'$. (Second of DeMorgan's Law.)

11. In a survey of 100 students, the numbers studying various courses were found to be: physiology, 28; botany, 30; zoology, 42; physiology and botany, 8; physiology and zoology, 10; botany and zoology, 5; all three courses, 3.
    a. How many students were studying none of the above?
    b. How many students had zoology as their only course?
    c. How many students had botany as their only course?

12. In a later survey of the 100 students (see exercise 11), the number of students in the various courses were found to be: botany only, 18; botany but not physiology, 23; botany and zoology, 8; botany, 26; zoology, 48; zoology and physiology, 8; no course, 24.
    a. How many students took physiology?
    b. How many students took botany and physiology but no zoology?

13. A timber inspector was supposed to take particularly careful notes on the oak, hickory and ash trees being brought out of a wood lot. In inspecting 100 trucks, he reported 5 trucks had all three kinds of wood; 10 had oak and hickory; 8 had hickory and ash; 20 had oak and ash; 30 had hickory; 23 had oak and 50 had ash. The inspector turning in this report was fired. Why?

14. In determining blood types, the factors considered are the presence or absence of three antigens. The antigens are antigen A, antigen B, and antigen Rh. Draw the appropriate Venn diagram for the following sets. Let $A = \{x|x \text{ has antigen A}\}$, $B = \{x|x \text{ has antigen B}\}$ and $R = \{x|x \text{ has antigen Rh}^+\}$.
    a. $A \cap B \cap R = \{x|x \text{ has type (AB, Rh}^+)\}$
    b. $A \cap B \cap R' = \{x|x \text{ has type (AB, Rh}^-)\}$
    c. $A \cap B' \cap R = \{x|x \text{ has type (A, Rh}^+)\}$
    d. $A \cap B' \cap R' = \{x|x \text{ has type (A, Rh}^-)\}$
    e. $A' \cap B \cap R = \{x|x \text{ has type (B, Rh}^+)\}$
    f. $A' \cap B \cap R' = \{x|x \text{ has type (B, Rh}^-)\}$
    g. $A' \cap B' \cap R = \{x|x \text{ has type (0, Rh}^+)\}$
    h. $A' \cap B' \cap R' = \{x|x \text{ has type (0, Rh}^-)\}$

15. Are two families from the same taxonomic phylum disjoint? Justify your answer.

16. In a study of ABO blood groups, 5000 people were tested and 1325 had only antigen A, 1642 had only antigen B and 1437 had neither antigen A nor B. How many individuals had both antigens?

17. A herd of 100 goats is examined for two types of parasites. Fifty goats are found to have type A parasites and 18 goats have type B parasites. Among these, 6 goats had both types A and B parasites. How many goats had neither of these types of parasites?

18. In a study of genetic characteristics of the garden pea, 800 plants were examined for the following traits: Tall or dwarf plant, round or wrinkled pod shape, and green or yellow pea color. Let D represent the set of dwarf plants, R the set of plants having round pod shape, and G the set of plants with green pea color. The following observations were made: There were 258 dwarf plants, 433 plants having a round pod shape, 431 plants with green pea color, 105 dwarf plants with round pod shape, 183 plants with round pods and green pea color, 209 dwarf plants with green pea color, and 67 plants had all three traits. How many tall plants had yellow pea color and wrinkled pods? How many tall plants had green pea color and wrinkled pods?

3.3 SETS FULL OF NUMBERS

NUMERICAL SETS are sets whose elements are numbers. Such sets occupy a prominent role throughout the sciences, the life sciences included. Numbers result almost any time a biologist evaluates a response. You have known some things about numbers for many years; other features of numbers you have learned about more recently. This section begins by describing some familiar characterizations of numbers in terms of sets. (We will use these sets in the rest of the book to specify the kinds of numbers for which certain results hold because some results will hold only for certain subsets of numbers.)

The real number system as we know it today had a very simple beginning in numbers for counting. Over the span of several thousand years it has been expanded to meet new needs, approximately in the order given below.

1. The NATURAL NUMBERS 1, 2, 3, 4 $\cdots$ (three dots mean and so on) were introduced for counting. In mathematics they often are called the POSITIVE INTEGERS. Here the positive integers will be denoted by the symbol N. The sum or product of two (or more) positive integers is again a positive integer.

2. The positive fractions, ordinarily called the POSITIVE RATIONAL NUMBERS, were needed to divide a quantity of something into several parts. For example how could three merchants divide up five sacks of grain? The positive rational numbers include numbers like 2/3, 8/5, 169/17. The positive rationals include the positive integers because 3/1 is the integer 3.

3. The remaining positive numbers are called the POSITIVE IRRATIONAL NUMBERS, and include numbers such as $\sqrt{2}$, $\sqrt{17/7}$, π. This extension was required to allow solution of equations such as $x^2 = 2$, and to allow the calculation of the area of a circle: $A = \pi r^2$.

4. The number ZERO, symbolized 0, represented a great step forward. It first was used as a position holder to distinguish the integers like 101 and 11 from each other. As $101 = 1(100) + 0(10) + 1(1)$, the zero in 101 really means no tens. Thus it was natural to use the same symbol for 6 - 6, 121 - 121, $\sqrt{13} - \sqrt{13}$.

5. NEGATIVE INTEGERS, NEGATIVE RATIONAL NUMBERS, and NEGATIVE IRRATIONAL NUMBERS such as -3, -1/6, and $1 - \sqrt{2}$, respectively, expanded the number system to allow operations such as 2 - 5, (1/3) - 1/2, and $1 - \sqrt{2}$.

   In the absence of a sign, the positive sign is understood. Thus 7 and +7, 7/8 and +7/8, 5 and +5 mean, respectively, the same things. Zero can be regarded as either an integer or a rational number, but it has no sign.

Let J denote the integers, positive, negative and zero. Use Q to denote the rational numbers. The irrational numbers, H, comprise the rest of the numbers introduced here.

The REAL NUMBER SYSTEM, denoted by R, consists of all positive and negative rational and irrational numbers, together with zero. Be careful of the name real: These numbers are not real in the sense that you can grab hold of them to either love them or do violence to them. The word real is used in contrast to those numbers involving $\sqrt{-1}$, often called IMAGINARY numbers. Such numbers prove useful in higher mathematics and in some of the physical sciences, particularily those dealing with the theory of electricity. This book will not need the imaginary numbers, except for an explanation at the end of Section 13.11. Thus here, number will imply a real number unless specifically noted to the contrary.

EXAMPLE 3.11: The sets N, J, Q, H, and R are defined above. Verify the following relations: $N \subset J \subset Q \subset R$, $H = R \cap Q' = R - Q$. Put these relations into words to see that they are true. Further, notice that $S(N) = \infty$. Why? What does this say about the sizes of J, Q, H, and R? ▌

As a biologist you are accustomed to seeing things: Plants, animals, cells, and so on. Some features of mathematics can be visualized. Thus we need to introduce a visualization of the reals so later we can progress to graphs and other figures. Consider the REAL NUMBER LINE. Every number has a corresponding position on the line, and every point on the line corresponds to exactly one real number. We construct the real number line in the following manner: Draw a line and pick some intermediate point to represent zero (0), a point often called the ORIGIN. Select some unit of measure. Go to the right this distance, make a mark, and label this point 1. Now repeat this to get a second point to label 2; continuing in this way, establish points for 3, 4, 5, ⋯. Now repeat this process, but go to the left, starting at the zero point or origin. In this manner you will establish points for -1, -2, -3, ⋯. The result thus far appears in Figure 3.10.

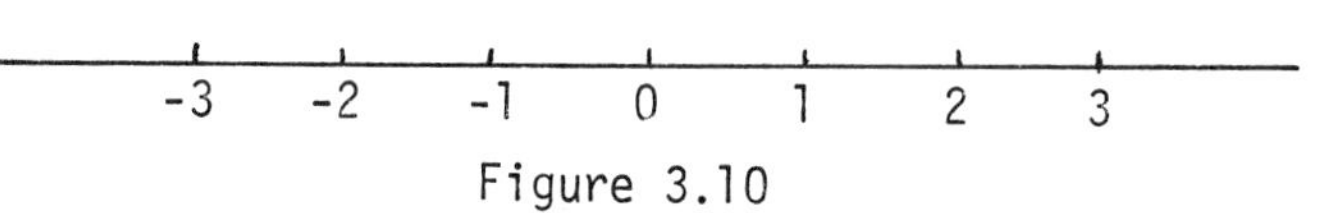

Figure 3.10

Now insert the rationals. Put the number 1/2 half the distance between 1 and 2, represent -8/3 by the point 2/3 of the way from -2 to -3, and so on. After we have included the rationals, the real number line is full in a sense: An infinite number of rationals lie within any small distance of any point now on the real number line. The integers lacked this denseness because no integer lies closer than one unit to another integer. The denseness of the rationals is like a fine filter: It looks solid, yet it contains many invisible holes. By continuing and finally inserting the irrational numbers, we completely fill the real number line so no holes remain.

The position of a number on the real number line establishes an order. Consider the point corresponding to the number a. If another number b lies to the right of a, then we say that b is larger than a, but if b lies to the left of a, then b is less than a. Now you are ready to consider these:

- If a, b, $c \in R$, then:

  $a < b$, read a is less than b, or
  $a = b$, read a is equal to b, or
  $a > b$, read a is greater than b.

- If $a < b$ and $b < c$, then $a < c$.
- If $a < b$, then $ac < bc$ if $c > 0$, but $ac > bc$ if $c < 0$.
- The ABSOLUTE VALUE of a real number $a \in R$ is given by

$$|a| = \begin{cases} a & a > 0 \\ 0 & a = 0 \\ -a & a < 0 \end{cases}$$

- The absolute value satisfies these properties, a, $b \in R$:

$$|a \times b| = |a| \times |b|$$
$$|a + b| \leq |a| + |b| ,$$

  where $\leq$ is read, less than or equal to.

You probably have seen these relations before. The first statement says that of two numbers either one is bigger, or they are equal. The next statement affirms the truth of all statements like: If $-2 < 1$ and $1 < 7$, then $-2 < 7$. The subsequent statement applies to multiplying both sides of an inequality by the same number. For a positive multiplier, the sense (direction) of the inequality stays the same, but a negative multiplier reverses the sense. The absolute value of a real number measures how far it lies from zero, disregarding signs. For example $|-3| = -(-3) = 3 = |3|$, reflecting the fact that both -3 and 3 lie three units away from zero. The statement $|a \times b| = |a| \times |b|$ says that the absolute value of a product may be evaluated by either finding the product and then taking its absolute value, or by first finding the absolute value of each component of the product before multiplying them. If a and b have the same sign, then $|a + b| \leq |a| + |b|$. The inequality applies when the numbers have opposite signs:

$$|-3 + -5| = |-8| = 8 = |-3| + |-5|, \text{ but } |-3 + 5| = |2| = 2 < |-3| + |5| = 8.$$

EXAMPLE 3.12: If two different aspects of a problem place restrictions on a solution, you need to find what, if any, values satisfy both sets of restrictions. For example suppose you need to find what values of x satisfy $|x| \leq 5$ and $0 \leq |x-4| \leq 2$. If $A = \{x| \ |x| \leq 5\}$, and $B = \{x| \ |x-4| \leq 2\}$, then you need $A \cap B$. If x satisfies $|x| \leq 5$, it must also satisfy $-5 \leq x \leq 5$. (Check other values if this seems obscure.) Similarly, $|x-4| \leq 2$ is equivalent to $-2 \leq x-4 \leq 2$ or $2 \leq x \leq 6$. Thus now, $A = \{x| -5 \leq x \leq 5\}$ and $B = \{x| \ 2 \leq x \leq 6\}$. These sets overlap from 2 to 5 so $A \cap B = \{x| \ 2 \leq x \leq 5\}$. ▮

The operation $S(A)$ gives the number of elements in the set A, but it tells us nothing about their magnitude. Thus we need:

- AN UPPER BOUND of a nonempty numerical set S is a real number u such that $s \leq u$ for every $s \in S$.
- The LEAST UPPER BOUND (sometimes abbreviated l.u.b.) of a nonempty numerical set S is an upper bound for S such that no smaller upper bound for S exists.
- A LOWER BOUND of a nonempty numerical set S is a real number l such that $l \leq s$ for every $s \in S$.
- The GREATEST LOWER BOUND (sometimes abbreivated g.l.b.) of a nonempty set S is a lower bound for S such that no larger lower bound for S exists.

The names of these concepts suitably describe their meaning. The idea of a least upper bound introduces an element of uniqueness; a set may have many upper bounds, but only one least upper bound. Similar comments apply on the lower side.

EXAMPLE 3.13: Consider driving an automobile on an interstate highway near a major city. Speeds frequently have been posted as a minimum of 45 miles per hour and a maximum of 70 MPH until recently. $S = \{s \mid 45 \leq s \leq 70\}$ gives the permissible speeds. Speeds of 17 MPH and 98 MPH provide lower and upper bounds, respectively, but not very useful ones because you may get fined if you drive them. Thus, the g.l.b. and l.u.b. of 45 MPH and 70 MPH have a very direct meaning. ▌

EXERCISES

1. Draw a line, choose a unit of length, and pick a point for zero. Mark down the points located at -3, 4, 7/2, -15/5, $\sqrt{7/2}$, π, ($\pi = 3.14159\cdots$), $\pi/2$.

2. Find the fallacy. If a = 2 and b = 4; then $a < b$

b.	Multiply by a	$a^2 < ab$
c.	Subtract b^2	$a^2 - b^2 < ab - b^2$
d.	Factor	$(a - b)(a + b) < b(a - b)$
e.	Divide	$a + b < b$
f.	Substitute a = 2, b = 4	$6 < 4$

3. Find the values of X for which

   a. $X^2 - 9X + 20 = 0$

   b. $X^2 - 9X + 20 > 0$

   c. $X^2 - 9X + 20 < 0$

4. Classify each of the following numbers according to the categories: real number, positive integer, negative integer, rational number, irrational number, and integer.

   -4, 4/5, 3π, 2, -.25, 7.8, $\sqrt{3}$, $\sqrt{5}$, .3782, $\sqrt{9}$, -18/7, 0, 29, $\sqrt[3]{64}$

   Locate them on a real number line.

5. Show that $a^2 + b^2 > 2ab$ if a and b are unequal real numbers.

6. A normal erythrocyte (red blood cell) is a biconcave disk, varying from 6.7 - 8.0 microns in diameter while a microcyte is a very small erythrocyte of size less than 6.7 microns. A macrocyte is an abnormally large erythrocyte of size from 8 to 12 microns. For each of the above give, if one exists, a lower bound, an upper bound, a greatest lower bound, and a least upper bound.

7. A normal person has 14-16 grams of hemoglobin per 100 cc. (cubic centimeters) of blood and 5-10 thousand white blood cells per cubic millimeter. A person with pernicious anemia has 5-8 grams of hemoglobin per 100 cc. and 3-4 thousand white blood cells per cubic millimeter. Give a least upper bound and greatest lower bound for hemoglobin and white blood cells if a person has pernicious anemia.

8. Give an example from your own area of an application of an upper bound, least upper bound, lower bound, and greatest lower bound.

9. Consider the task of placing postage on a package, assuming a price of 10¢ per whole or fractional ounce. Establish bound relationships in these two situations:
   a. You have $1.25. What can you mail?
   b. You have a 2 pound package. How much money do you need?

10. Give a least upper bound and a greatest lower bound for the areas of the following:
    a. A triangle with height 2 and a base of 3 units.
    b. A circle of radius 2 units.
    c. A trapezoid with a height of 2 units and bases of length 3 and 5 units. (A trapezoid is a 4 sided configuration with two sides (bases) being parallel and the distance between these bases is its height.)

3.4 MEASURING ITS CHANCE OF HAPPENING

Probability occurs in biology for two different reasons. Biology contains truly random mechanisms: Genetic assortment offers a primary example. Because most organisms are subject to genetic assortment, we can be confident that random mechanisms underlie many aspects of biological material. On the other hand, many apparently uncertain phenomena have causes which we do not yet understand. We can attribute their variation to random mechanisms, but seek causes for the variation. Continuing research may eventually provide a basis for precisely predicting this variation from many, many factors.

Biology abounds with examples of happenings whose specific results cannot be known in advance, even though the possible results can be enumerated: A wheat field may be infected with rust, but then again, it may escape infection; a female field mouse may give birth to 0, 1, 2, $\cdots$, 8 or more, live young during the month of May, but we do not know how many; a sterilized male screwworm fly will move an unknown distance from its point of release; a fertilized egg may hatch or it may not; the fertilization of a particular ovum of a higher animal will produce either a male or female individual, if any. Certain words have special meanings in the discussion of the likelihood or probability of the occurrence of such happenings:

- A TRIAL is one happening of an occurrence whose outcome is not known in advance.
- An OUTCOME is a distinct result of a trial.
- An EVENT consists of a set of outcomes.

These ideas are closely related to each other and to the ideas of sets. An *outcome* merely is an element of a set while an *event* is a subset of the possible outcomes. *ITS* of the title of this section really means an event. A *trial* expands our view of sets because it embodies the concept of uncertainty with which we will soon associate probability. The following examples illustrate the diversity of situations to which these definitions apply:

<u>EXAMPLE 3.14</u>: Many plants have seed pods with variable numbers of seeds per pod. Some varieties of green beans have no more than ten seeds per pod. If we select a pod from such a bean plant and count its number

of seeds, then the trial amounts to the selection of the pod and its subsequent examination. An outcome could be the number of seeds per pod, that is, one of 0, 1, ···, 10, so that the universal set is given by u = {0, 1, ···, 10}. In this setting an event could consist of pods having more than four seeds, that is, E = {4, 5, ···, 10}. ∎

EXAMPLE 3.15: A beehive has three kinds of bees: a queen, drones, and workers. If we select a bee from the hive, it could be any one of these three. An event could be getting a female, which technically consists of getting either a queen or a worker. ∎

EXAMPLE 3.16: Commercial roses sometimes are graded as 1, 2, 3 or culls. A trial could consist of selecting a rose from a rose nursery. An outcome could be the grade of that rose and an event could be a marketable rose, which would mean that the rose cannot be a cull. A grower usually will try to minimize the number of culls through his management practices. ∎

EXAMPLE 3.17: Deer can have several offspring. A trial could be a birth and our evaluation of its size. An event could be a multiple birth, that is, a doe having 2, 3 or, rarely, more than 3 fawns. The age of the doe, her condition and the condition of her range (where she eats) can greatly influence the frequency of multiple births. Eventually we can consider how probability depends on certain external features like these. ∎

EXAMPLE 3.18: Lakes sometimes have an algae bloom making them rather undesirable as recreational sites. Thus, a trial could consist of evaluating the algae condition of a lake by classifying it as not in bloom or in bloom; the absence of a bloom would be an event of interest here. ∎

Probability measures the degree of uncertainty associated with uncertain events. The title of this section involves the word *MEASURING* in this sense. We use the word probability in our everyday life without being particularly concerned about its technical meaning. At this level it appears fruitful to regard probability as a primitive concept which can be characterized from three different perspectives. You need a feeling for what probability means; each view will broaden your understanding. No one view adequately describes probability. At an advanced level some aspects of probability can be defined very precisely, but the same ideas emerge. Each of the views has its proponents, so presently no one precise definition encompasses all features of probability used by biologists. Thus, these three views:

- SUBJECTIVE VIEW OF PROBABILITY, symbolized by $P_s(E)$, describes our degree of belief in the likelihood of the occurrence of the event E.
- FREQUENCY VIEW OF PROBABILITY, symbolized by $P_f(E)$, regards probability as an eventual relative frequency evaluated over many trials. If the event E occurs r times in t trials, then $P_f(E)$ is the eventual value of r/t as t continues to increase.
- A PRIORI VIEW OF PROBABILITY, symbolized by $P_a(E)$, concerns a trial with $n = S(U)$ equally likely and disjoint outcomes of which $n_E = S(E)$ manifest the event E; $P_a(E) = n_E/n$.

The subjective view of probability describes how we act in the face of one trial of an uncertain event when we already have the probability of various outcomes. The second view, the frequency view, concerns how we talk about the idea of probability when pressed for an interpretation. The third view is used in calculating probability. Note the apparent circularity of the *a priori* view of probability. Unless probability has been defined, how can we know what constitutes equally likely outcomes? They must be outcomes having the same probability of occurrence. This ambiguity usually is avoided by picking a description of outcomes so that the supposition of equally likely events appears plausible. Our examples will illustrate the importance of this.

The word RANDOM occurs in many discussions involving probability. It basically conveys the idea of no pattern. Random devices often are used to pick elements from the POPULATION, namely, the set of elements available for selection, in order to get a fair description of that set. Under random selection each element in the population has an equal opportunity of being the one selected. Thus under random selection the *a priori* view of probability relates the likelihood with which we see various outcomes to the frequency with which that outcome exists in the population. Several random devices exist. Most statistical textbooks and tables contain randomly ordered sets of digits. If these are used in certain systematic ways, they will produce nearly random selections from a population. In other settings, you can generate your own random device by assigning numbers to objects in the population, putting these numbers on plastic disks or something similar, dropping them into a container and mixing them. When a disk is selected after a thorough mixing, you have a legitimate basis for regarding it as a randomly selected object. A random selection from the population results from the selection of the object having the same number as the disk.

Certain ideas about games of chance can provide convenient illustrations of probability unencumbered by other considerations. For example suppose you hear the odds associated with a sporting event or election stated as 3:2, this means that someone's subjective probability of that event's occurrence equals $\frac{2}{5} = \frac{2}{(2 + 3)}$. Weather reports can be viewed as games of chance if you regard the decision to take an umbrella relative to probability. The probabilities given in weather forecasts are area-weighted relative frequencies developed from many years' data under similar meteorological conditions. You can use the probability of precipitation given in a weather report as your subjective probability of precipitation. If you might need an umbrella, you can use this probability for making a decision, or you can always ignore it and play a game of possibly getting rained on.

Probability has three related, but reasonably apparent properties. If U denotes the set of all possible outcomes, then $P(U) = 1$. Further, if A' is the complement of A, then $P(A) + P(A') = 1$, or $P(A') = 1 - P(A)$ because together A and A' exhaust the elements in U. If $A = U$, then $A' = \phi$ so we can conclude that $1 = P(U) = P(A) = 1 - P(A')$, or that $P(\phi) = 0$.

EXAMPLE 3.19: Early in the fall term most college phone systems suffer from the lack of a directory. Some male students have used several variations of this caper: They select and call any possible phone number. If a person of the opposite sex answers, they ask for a date to get a coke, go bowling, or some such activity.

Suppose that the numbers from 4600 through 8399 belong to phones in rooms having two occupants, but they have no known pattern relative to the sex of the rooms' occupants (coed dorms, perhaps). Suppose a fellow is trying to find a date where 60% of the rooms belong to girls and in 35% of the girls' rooms at least one of the girls would accept such an offer. What is the probability the fellow will contact a girl on any particular completed call? That she will accept?

If the fellow randomly selects a number between 4600 and 8399, using a deck of cards for example, then each of the 3800 (8399 - 4599) numbers has an equal chance of being called. Because 60% of these correspond to girls rooms, (2280 or them), P_a(contacting girl) = 2280/3800 = 0.60. Here a trial consists of selecting a number between 4600 and 8399; the random selection gives each an equal chance of being selected; the event E consists of selecting a girl's room, which occurs in 2280 of the outcomes.

Now only 35% of the 2280 girls' rooms will produce a favorable response, 798 of them, so P_a(date) = 798/3800 = 0.21. The probability of an unfavorable response is 1 - 0.21 = 0.79. ▮

EXAMPLE 3.20: The sex of a particular fertilization has a substantial element of randomness to it. The outcome will be either male or female. If these two outcomes are equally likely then P(male) = P(female) = $\frac{1}{2}$. This comes reasonably close to the actual frequencies in many higher animals, but for some animals it can be off drastically. When it fails, it does so because the two events are not equally likely. Reproductive mechanisms apparently produce male and female gametes with equal frequency, but thereafter some gametes have a reduced chance of surviving through fertilization into a developing embryo. For example, dogs produce nearly 60% males, humans about 52%, and chickens as low as 47%.

These figures represent reasonably typical situations. Extreme environmental conditions and wider selection of species can produce substantially more variation. ▌

The simultaneous occurrence of two events may relate to the probability of their individual occurrence, but it may not. Look back at Example 3.19. Note that the final probability of 0.21 can be obtained as $(0.60 \times 0.35) = 0.21$. Sometimes the probability of simultaneous occurrences (get a girls' room <u>and</u> get a favorable response) can be calculated from the probabilities of the individual occurrences. To identify these simplifying situations, we introduce:

- Two events, A and B, are called INDEPENDENT (probabilistically) if the occurrence or nonoccurrence of either event gives no information about the probability of occurrence or nonoccurrence of the other.
- For two independent events, A and B, $P(A \cap B) = P(A)\ P(B)$. The probability of the events A and B occurring simultaneously equals the product of their individual probabilities.
- For any two events, A and B, $P(A \cap B) = P(A) + P(B) - P(A \cap B)$.
- For two disjoint or mutually exclusive events, A and B, $A \cap B = \phi$ and $P(A \cap B) = P(A) + P(B)$.

Independent events are most easily comprehended as totally unrelated aspects of the same situation, like the day of the week on which you were born and your hair color. It applies to a larger class of situations depicted by Figure 3.11, which can be viewed as a

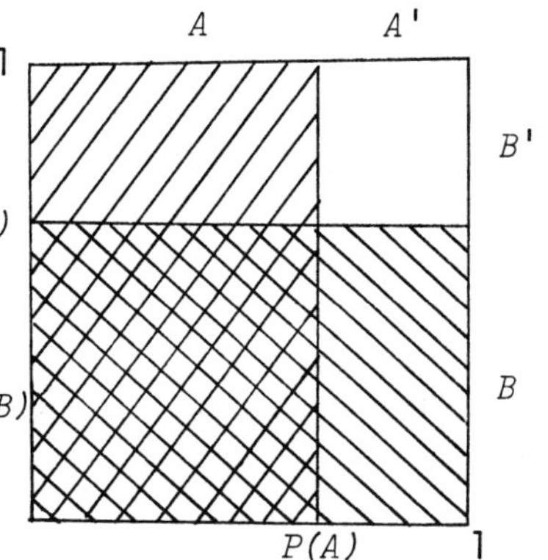

Figure 3.11

Venn diagram. The first part of the horizontal axis represents the occurrence of A and has a length of $P(A)$; the remaining distance to 1 corresponds to A' and its probability. Together $P(A) + P(A') = 1 = P(U)$. The vertical axis treats B analogously. This figure should convey the idea that the elements of A and B are distributed throughout U so B takes its proportionate share of elements in A and A' without reference to where they are. If the dividing lines were slanted A and B would not be independent; for example, if the line dividing B' from B sloped upward, B would be getting more and more of U as it went from A to A'.

The formula for $P(A \cup B)$ occurs naturally from the earlier observation that $S(A \cup B) = S(A) + S(B) - S(A \cap B)$. Figure 3.12 should confirm this. Part (a) shows $A \cap B$ and indicates $A \cup B$ as the shaded area. Part (b) demonstrates that $P(A)$ and $P(B)$ both contain the contribution of $P(A \cap B)$, so it must be subtracted off once. If $A \cap B = \phi$, as in part (c), then $S(A \cap B) = 0$ so $P(A \cap B) = 0$ and $P(A \cup B) = P(A) + P(B)$.

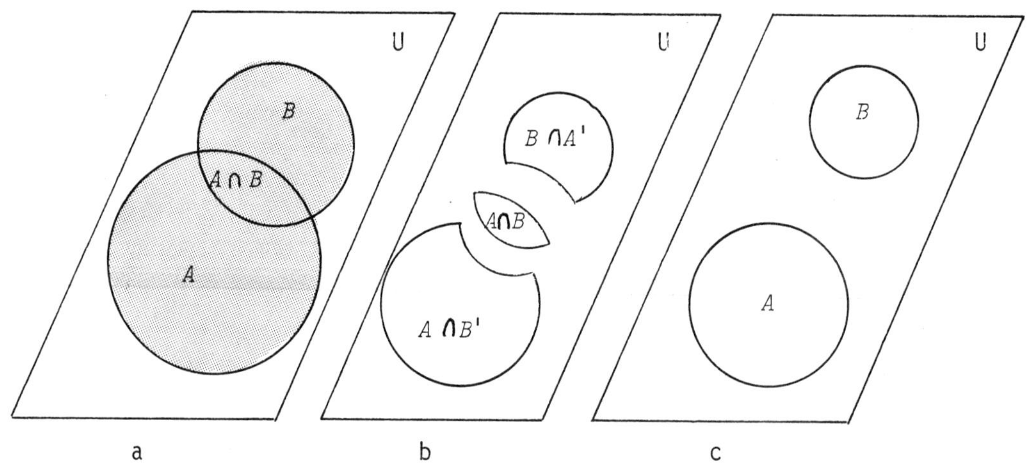

Figure 3.12

<u>EXAMPLE 3.21</u>: Annual and biannual higher plants commonly have two growth stages, often called vegetative and reproductive, with some possible overlap. The plant establishes roots, stems, leaves, and stores food during the first stage while during the second it puts up seed stalks,

flowers and produces fruit, often consuming stored food. Humans consume the vegetative parts of produce such as lettuce, onions, and radishes. When such a plant BOLTS, it enters the reproductive stage and usually becomes unpalatable. You might hear a commercial lettuce grower say, "Only 1% of my lettuce bolted last year, but it looks like nearly 5% did thid year." He is estimating the P_f (that a randomly selected lettuce plant bolts in the specified year). Notice that it changes from year to year in response to changing meteorological conditions. ▌

EXAMPLE 3.22: Consider now the sex of human twins. Twins occur in two ways: Monozygotic or identical twins result from a fertilized ovum splitting into two zygotes; dizygotic or fraternal twins result when two ova are fertilized by two separate sperm. Let E = {first born twin is female} and F = {second born twin is female}.

Now if the twins are monozygotic, these events are not independent because if one occurs, the other will also. Thus, $P(E) = P(F) = P(E \cap F) = \frac{1}{2}$. Likewise, $P(E \cup F) = P(E) + P(F) - P(E \cap F) = \frac{1}{2}$.

If, instead, twins are dizygotic, they result from separate sperm. Since the sperm have been generated by a genetic mechanism involving a random assortment, their sex determining properties should be independent so E and F will be independent. Consequently, two girls occur with probability $P(E \cap F) = P(E)\, P(F) = \left[\frac{1}{2}\right]\left[\frac{1}{2}\right] = \frac{1}{4}$. Thus dizygotic twins both will be girls with only half the probability that monozygotic twins will be. Similarly, $P(E' \cap F') = P(E')\, P(F') = \left[\frac{1}{2}\right] \frac{1}{2} = \frac{1}{4}$ gives the probability of two boys. The remaining probability is associated with one boy and one girl, in either order:

$$P[(E \cap F') \cup (E' \cap F)] = P(E \cap F') + P(E' \cap F) = \frac{1}{4} + \frac{1}{4} = \frac{1}{2}.$$

Note: All of these calculations provide approximations to the extent that $\frac{1}{2}$ is not quite correct. ▌

The examples above all involve finite universal sets, but infinite ones do exist. For example, suppose you are concerned with mutations; you might ask, "How many trials will it take to find 30 mutations?" It could take any number of trials so you cannot know an upper limit on the number required. This involves a universal set of infinite size, a problem which must await our consideration of limits in Chapter 5. As we progress further, you will see there is another order of infinity concerned with all the numbers on the real line. Many characteristics such as length and weight of organisms can be regarded as continuous variables because they assume every value between some definite limits. Before we can work with probability of such responses we need integral calculus, a topic which we will take up in Chapter 8.

EXERCISES

1. Clearly specify the trial and its outcomes, an event and the universal set for these Examples:
   a. 3.15 b. 3.16 c. 3.17

2. Describe a familiar occurrence to which probability can be applied. Illustrate the meaning of the subjective, frequency and *a priori* views of probability in your setting.

3. Reconsider the phoning situation of Example 3.19. Let the event A correspond to the determination of the sex of the person who answers the phone; let B correspond to the answer to the request for a data, assuming caller and answerer of opposite sexes. Is A independent of B? Defend.

4. *The Statistical Abstract of the United States* (1970 edition, p. 48) reports these figures:

Year	Male Births Per Female Births	Live Births (thousands)
1950	1.054	3,632
1955	1.051	4,104
1960	1.049	4,258
1963	1.053	4,098
1964	1.047	4,027
1965	1.051	3,760
1966	1.049	3,606
1967	1.050	3,521
1968	1.053	3,502

4. Continued
   Do you think the sex of a newborn should be regarded as a random event? If so, what is the probability of this random event? Explain.

5. For discussion: A life insurance company does not know how long a particular one of its clients will live. Is it gambling or prudent because it can evaluate what will happen to many of its clients?

6. If A, B, C are events, extend $P(A \cup B) = P(A) + P(B) - P(A \cap B)$ to show that $P(A \cup B \cup C) = P(A) + P(B) + P(C) - P(A \cap B) - P(A \cap C) - P(B \cap C) + P(A \cap B \cap C)$.

7. Suppose one digit is chosen from {1, 2, 3, 4, 5}, and then a second is selected from those remaining. List all 20 possible outcomes. Assuming that these outcomes have the same probability, find the probability that an even digit (2 or 4) will be selected
   a. the first time,
   b. the second time,
   c. both times.

   Repeat for the odd digits (1, 3, 5).

*3.5 COUNTING SCHEMES

In calculating probabilities using the *a priori* view of probability, you need to evaluate $n_E = S(E)$ and $n = S(U)$. Straightforward schemes exist for counting the elements in certain kinds of sets. The following deal with numbers of arrangements of distinguishable objects, like ones of different colors or uniquely numbered ones, as well as indistinguishable ones. Here we will introduce, explain, and illustrate several counting schemes.

When objects are selected from a set, they may be selected and returned, that is, selected WITH REPLACEMENT. Or objects may be kept out once they are selected, called selection WITHOUT REPLACEMENT.

- An ARRANGEMENT of r elements from the set S identifies an order for those objects.

- When an arrangement is selected WITH REPLACEMENT, any element of S can appear in any position in the arrangement.
- There are n^r distinct arrangements of r elements selected with replacement from n distinct elements.

An arrangement merely distinguishes a symbol configuration like NUTS from STUN. The distinction between selection with and without replacement has a parallel to reusable and destructable objects; more possible arrangements occur when an object can be used multiple times.

EXAMPLE 3.23: How many four-letter arrangements can be constructed from the three leters D, O, and T?

To get four-letter arrangements from three letters, we must use selection with replacement. Here, $S = \{D, O, T\}$, $r = 4$, and $n = 3$ so there are $3^4 = 81$ arrangements of these three letters. For example DDDD, DOOO, OOTT, DOTO are members of this 81 arrangements. The 3^4 occurs this way: Any one of three letters can appear in each of four positions. ▌

A different formula applies under selection without replacement:

- When an arrangement is selected WITHOUT REPLACEMENT, an element of S can appear in at most one position in the arrangement.
- The FACTORIAL notation,

$$n! = (n)(n - 1) \cdots (2)(1) = n(n - 1)! \;, \tag{3.1}$$

gives the number of distinct arrangements of n elements selected without replacement from n distinct elements. By definition $0! = 1! = 1$.

Selection without replacement allows no duplication so each element can be used once. The first object can be selected n ways. With one object removed, the second object can be selected n - 1 ways. The first two can be selected n(n - 1) ways. Continuing in this fashion, n! results.

EXAMPLE 3.24: How many three-letter arrangements can be selected from S = {D, O, T}? With n = 3, n! = (3)(2)(1) = 6. Four of these arrangements are DOT, DTO, TOD, TDO. What are the other two? ▌

Another formula applies when we select fewer than all available elements.

- The number of distinct arrangements of r elements selected without replacement from n distinct elements is given by

$$_nP_r = n(n - 1) \cdots (n - r + 1) = \frac{n!}{(n - r)!} \qquad (3.2)$$

- The name PERMUTATION sometimes describes a specific arrangement of r elements selected from n distinct elements.

If we select fewer than n objects from S, then we take only as many terms from (n)(n - 1)(n - 2) $\cdots$ (1) as we select objects. Thus we take r terms and leave off n - r. These last terms have a product of (n - r)! so if we divide n! by (n - r)! the formula for $_nP_r$ results.

EXAMPLE 3.25: How many two letter arrangements can be selected from S = {D, O, T, S} if we sample without replacement?

Here, n = 4 and r = 2 so $_4P_2 = \frac{4!}{2!} = \frac{(4)(3)(2)(1)}{(2)(1)} = 12$. These twelve include DO, DT, OD, and TD, but not DD, OO, TT, or SS since we are sampling without replacement. ▌

Finally, another formula applies if we ignore order, that is, if we regard DO and OD as the same. For example in discussing dizygotic twins, we noted that a boy and a girl occur with probability $\frac{1}{2}$ while two boys or two girls each occur with probability $\frac{1}{4}$. A mixed pair can occur as boy, girl or girl, boy. After birth, these two outcomes become indistinguishable. For such situations we need

- The number of distinct COMBINATIONS (unordered arrangements) of r elements selected without replacement from n distinct elements is given by

$$\binom{n}{r} = \frac{n!}{r!(n - r)!} = \binom{n}{n - r}, \qquad r = 0, 1, \cdots, n . \qquad (3.3)$$

Alternately, Equation 3.3 gives the number of ways r indistinguishable elements can be placed in n positions; thus, it also gives the number of ways to arrange r elements of one kind among n - r of another kind.

When distinguishable elements are selected, combinations differ from permutations in one major way: Combinations ignore order within the arrangement while permutations do not. For example, the one combination of letters DOT has the 6 = 3! permutations DOT, DTO, ODT, OTD, TDO, TOD. Note that these three letters have $3^3 = 27$ arrangements under selection with replacement. These include DDD and DDT; you should list all 27 arrangements with replacement.

Some of the examples involve number balls, not because they should interest you; instead, they illustrate the basic ideas devoid of distracting details of more interesting problems.

EXAMPLE 3.26: (a) The game of pool involves 15 balls numbered from 1 to 15; 7 of the balls are striped and 8 are solid colored. In how many ways can two balls be taken from the table? This act can be accomplished three ways: (1) One ball can be taken, but returned before the other is taken. (2) Remove one ball, then another without replacing the first one. (3) Remove both balls simultaneously. How many ways can these be done?

(1) $15^2 = 225$;

(2) ${}_{15}P_2 = 15!/13! = 15(14) = 210$;

(3) $\binom{15}{2} = \frac{15!}{2!13!} = 15(14)/2 = 105.$

The distinction between the last two involves this: The pair (ball 6, ball 3) differs from the pair (ball 3, ball 6) when done one ball at a time, but not when done together.

(b) In how many ways can 5 balls be taken from the table at one time? How many of these contain exactly two striped balls? At least two striped balls?

At first you might think that the five balls could be selected in ${}_{15}P_5 = \frac{15!}{10!} = (15)(14)(13)(12)(11) = 360{,}360$, but this relates to selecting the balls in an order, not to the selection of five balls at once. Any particular set of five balls can occur in 5! = 120 ways so that there are only 360,360/120 = 3003 distinct sets of five balls. Note that

$$3003 = \frac{15!}{10!} \times \frac{1}{5!} = \binom{15}{5} ,$$

a way to have obtained the result directly. Now for the next question: Two striped balls can be picked from the seven striped ones in

$$\binom{7}{2} = \frac{7!}{2!5!} = \frac{(7)(6)}{(1)(2)} = 21 \text{ ways;}$$

similarly the solid colored balls can be picked in $\binom{8}{3} = 56$ ways. Because there are 56 triplets of solid colored balls for each pair of striped balls, there are (21)(56) = 1176 sets of two striped and three solid balls. Applying parallel reasoning for any number of striped balls, we find that there are 56, 490, 1176, 980, 280, 21 sets of five balls containing 0, 1, $\cdots$, 5 striped balls. Note that these numbers sum to 3003 because they exhaust the possibilities. The last question requires the sum 1176 + 980 + 280 + 21 = 2457 to get the number of sets containing 2, 3, 4, or 5 striped balls.

Each of these counts can be turned into a probability if the five balls are chosen in such a way that each set of five has the same probability of selection, that is, with probability $1/3003 \doteq 0.0003$. Thus P(2 striped balls) = $1176/3003 \doteq 0.392$, and P(at least two striped balls) = $2457/3003 \doteq 0.818$. ▌

EXERCISES

1. A Wildlife Conservation Officer stops a fisherman with two undersized (illegal) fish in his catch of 10 fish. If the warden randomly selects two fish for measurement, what is the probability he will

1. Continued

   select neither of the undersized fish? Repeat, assuming that the officer selects three fish randomly.

2. A couple plan to have three children. What is the probability they will have exactly one boy? At least one boy? Assume boy and girl babies are equally frequent.

3. A nominating committee of an organization having 100 eligible members must select:

   a. One candidate for president, one for vice president, and one for secretary-treasurer. How many possible slates of candidates does the committee have to select among?

   b. Three delegates must be elected to go to an annual meeting. How many sets of three eligible members does the committee have to chose among?

4. A cage contains 5 apparently identical white rats, but some have been innoculated against a particular virus. Find the probability that you select the innoculated rat(s) if:

   a. You select one and one is innoculated.

   b. You select three and two were innoculated.

   c. You select two and two were innoculated.

   d. You select two, but you want them in the order they were innoculated; two were innoculated, one on Monday and one on Friday.

5. A patient has a reoccurring disorder which you can treat in any one of five ways. You can give the same or a different treatment on each occurrence. How many possible ways can you treat the patient if

   a. The patient has three occurrences of the disorder and you consider using the same treatment any number of times?

   b. The patient has four occurrences and you cannot reuse a treatment? (A patient can become allergic to a treatment.)

   c. The patient has six occurrences and you cannot reuse a treatment?

6. A store mixes its eggs left over from yesterday with new ones today. As the cartons are undated, you cannot tell the two ages of eggs apart. Suppose you randomly select four dozen eggs from a rack containing 40 new dozens and 20 old dozens. What is the probability of getting
   a. Four new dozens?
   b. Two new and two old dozens?
   c. Four old dozens?
7. Suppose you have purchased 4 blue spruce, 3 ponderosa pine, and 3 Douglas fir trees to plant along one edge of your yard. How many different planting patterns do you have if
   a. You can distinguish each of the trees from the others?
   b. You cannot differentiate among trees of the same species?
   c. You cannot differentiate among trees of the same species, but you have already chosen where to plant the blue spruce?

*3.6 COMMON PROBABILITY DISTRIBUTIONS

What is a probability distribution? When an uncertain event can have several distinct outcomes, each outcome has a probability of occurring. The distribution of probability across distinct outcomes is called the PROBABILITY DISTRIBUTION of the event. For example, dizygotic twins consist of 0, 1 or 2 girls with probability $\frac{1}{4}$, $\frac{1}{2}$, $\frac{1}{4}$. These numbers give the distribution of probability across the three outcomes 0, 1, and 2.

Many situations in biology have associated probability distributions. The probability distributions presented here deal with counts; each finds application throughout the life sciences. We will specify the conditions under which each applies, and give $P(x = k)$, the symbol for the probability that k counts will occur. The formulas are not derived from the conditions; these derivations and a much more comprehensive treatment can be found in many probability texts, like *Modern Probability Theory and Its Applications* by Emanuel Parzen (1960).

Biology abounds with situations having exactly two possible outcomes on a single trial, often called *success* and *failure*. Many involve the male-female (or female-male if you prefer) dicotomy; many others involve some form of alive and dead, with hatch or germination, fertilization or pollination, or survival to reproduction illustrating kinds of "alive". Other kinds of illustrations occur throughout economic biology, primarily the agricultural, horticultural and pharmaceutical industries, where a product may or may not be acceptable for marketing. The first probability distribution arises from a population of fixed size sampled without replacement.

- HYPERGEOMETRIC probability distribution
  1. Consider the event E of size $S(E)$ = M in the sample space U of size $S(U)$ = N.
  2. Randomly select n elements from U.
  3. Let x represent the number of these n elements which belong to E.

  The observable outcome x can assume any of the values 0, 1, ⋯, Min(M,n) where the upper limit denotes the lessor of M and n.

$$P(x = k) = \frac{\binom{M}{k}\binom{N - M}{n - k}}{\binom{N}{n}} \qquad k = 0, 1, \cdots, \mathrm{Min}(M,n) \qquad (3.4)$$

- An observable outcome, like x above, sometimes is called a RANDOM VARIABLE.

How does this occur? First, $P(x = k)$ should give the relative frequency of combinations (not permutations) which have exactly k elements from E and n - k from E'. As any combination of k elements from E will satisfy this condition, the k elements can be obtained in $\binom{M}{k}$ ways. For each, the remaining n - k elements can be obtained in $\binom{N - M}{n - k}$ ways because $S(E')$ = N - M. Thus, the product of these gives the number of ways to get k elements from E and n - k from E'. The denominator arises from the number of unordered ways to get n objects from U without replacement.

A random variable has this distinctive property: It describes something observable, whose observed value is governed by a probability distribution. The number of puppies in a litter is observable after their birth, but before, this number has a probability distribution depending on such things as the breed of the mother. Random variables are denoted by italic letters.

EXAMPLE 3.26 (continued): Previously we calculated the probability of getting two striped balls in a random selection of five balls. Although the earlier calculations began with basic principles, we can now calculate the probability directly from Equation 3.4. With N = 15, M = 7, N - M = 8, n = 5, k = 2 and n - k = 3, we get

$$P(x = 2) = \frac{\binom{N - M}{n - k}\binom{m}{k}}{\binom{N}{n}} = \frac{\binom{8}{3}\binom{7}{2}}{\binom{15}{5}} = \frac{(56)(21)}{3003} = 0.392 \ ,$$

as before.

Observe that the same results would have appeared if M = 8 and k = 3. This amounts to nothing more than counting solid colored balls, rather than striped ones. ▌

EXAMPLE 3.27: Effective management of a natural population requires knowledge of its size. Several methods for estimating population size involve capturing individuals, marking them, releasing them back into the population, and subsequent recapture. Under random dispersal of the marked individuals and random sampling for the recaptures, the number of marked recaptures follows the hypergeometric probability distribution. Specifically, N equals the population size, M equals the number marked, n equals the number of individuals in recapture sample, and x equals the number of marked individuals in the recapture sample.

A method sometimes called the Peterson estimate, or the Lincoln Index, estimates population size by $M \frac{n}{x}$. This method arose by observing that the fraction marked in the sample, $\frac{x}{n}$, should be close to $\frac{M}{N}$, the fraction actually marked in the population. If these are equal, then $\frac{M}{N} = \frac{x}{n}$, or $Mn = Nx$, or $N = n \frac{M}{x}$. This method has two kinds of difficulties: (i) the case $x = 0$; (ii) the sample fraction rarely equals the population fraction. The hypergeometric probability distribution has been used to investigate the impact of such matters. These investigations have shown that this estimation procedure gives answers with acceptable precision only when $\frac{M}{N}$ and $\frac{x}{n}$ have values of at least 0.01 or more. Because the marking of 1% of the population often proves prohibitively expensive, more complex estimation procedures also have been developed. ∎

Still consider trials with two identifiable outcomes, but now require independent trials with a probability of each occurrence not changing over trials.

- BINOMIAL probability distribution
    1. A trial can have exactly two outcomes, called *success* and *failure*.
    2. Prob(*success*) = P remains constant from trial to trial. (For a finite population, this requires selection with replacement.)
    3. Let x represent the number of *successes* in n independent trials. Then

$$P(x = k) = \binom{n}{k} p^k (1 - p)^{n-k} \qquad k = 0, 1, \cdots, n \qquad (3.5)$$

(Sometimes 1 - p is written as q.)

This results because the k *successes* can occur anywhere among the n - k *failures*. Thus,

$P(x = k)$ = (number of ways for k *successes* to be interspersed among n - k *failures*) × Prob(k *successes* and n - k *failures*).

Recognize the first part as the number of ways for k *successes* to occur in n trials, that is, $\binom{n}{k}$. Recall that $P(A \cup B) = P(A)\,P(B)$ when A and B are independent trials. Successively applied here, this gives

Prob(k *successes* followed by n - k *failures*) = $p^k (1 - p)^{n-k}$

These give the formula stated for the binomial probability distribution.

The situation described above includes not only sampling with replacement, but also many situations where nature does the sampling for us. Even under sampling without replacement, sampling with replacement provides a satisfactory approximation when small samples are selected from large populations.

EXAMPLE 3.22 (continued): Recall that this example dealt with sex of offspring. Suppose that you and your spouse want to have a "perfect" family, one boy and one girl. Let a *success* represent a girl and a *failure* a boy (or interchange this identification depending on your biases). The probability p = Prob(girl) remains constant from conception to conception and approximately equals $\frac{1}{2}$. Genetic assortment assures independence of sex in the n = 2 trials. Thus, if x = number of girls,

$$P(x = 1) = \binom{2}{1} p^1 (1 - p)^{2-1} = \frac{2!}{1!1!} \frac{1}{2} \times \frac{1}{2} = \frac{1}{2} .$$

Likewise, x = 0 represents two boys:

$$P(x = 0) = \binom{2}{0} p^0 (1 - p)^2 = \frac{2!}{0!2!} \left[\frac{1}{2}\right]^0 \left[\frac{1}{2}\right]^2 = \frac{1}{4}$$

where 0! = 1.

These calculations assume p = .50, but p = .48 provides a slightly better approximation to the real situation. Check in this case that $P(x = 0) = .2704$, $P(x = 1) = .4992$, $P(x = 2) = .2304$. Observe that $P(x = 0) + P(x = 1) + P(x = 2) = 1.0000$ as must occur because these outcomes exhaust the possibilities. ▌

EXAMPLE 3.28: Zygotes resulting from irradiated sperm or pollen usually have less chance of maturing than unirradiated ones. For example, when Cornish chicken hens are fertilized with semen collected from Leghorn roosters and irradiated at a dosage of 300 roentgens, about 60% of the eggs hatch compared to a normal egg hatch of over 80%. What is the probability of various numbers of such eggs hatching if n = 5 eggs are incubated? Check the conditions for applicability of the binomial distribution.

The probability of 3 eggs hatching is given by

$$\binom{5}{3}(0.6)^3(1 - 0.6)^{5-3} = \frac{5!}{3!2!} (0.216)(0.16) = .3456 .$$

Continuing similarly, the following table results:

k	0	1	2	3	4	5
$P(x = k)$	.01024	.07680	.23040	.34560	.25920	.07776

You probably expected for most of the results to occur at $x = .60 \times 5 = 3$, but note the substantial probability on either side of this value. This hints at the amount of variation present in natural phenomena even when the underlying features are fixed. ▌

EXAMPLE 3.29: The probability of anesthetizing a male rat for experimental purposes is about 0.8. What is the probability of getting 19 anesthetized rats out of 20 trys? Here n = 20, p = 0.80 and k = 19. Thus the required probability is

$$\binom{20}{19}(0.8)^{19}(1 - 0.8) = \frac{20!}{19!1}(0.8)^{19}(0.2)$$

$$= 20\ (0.8)^{19}(0.2) \doteq (20)(0.0144)(0.2) = 0.0576\ .$$

(The sign $\doteq$ means approximately equals.) ▌

Waiting time plagues us all; we wait for buses, trains, planes, stoplights, meals in a cafeteria, a shower in a dormitory, and so on! Biologists also wait for nature, for plants to mature, for insect populations to explode, for animals to die, and so on. Many probability distributions are associated with such phenomena, but one of the simplest of these closely associates with the situation of the binomial distribution. It involves the number of trials needed to get r *successes*.

- NEGATIVE BINOMIAL probability distribution
    1. A trial again can have exactly two outcomes, called *success* and *failure*.
    2. Prob(*success*) = p remains constant from trial to trial.
    3. Let x represent the number of *failures* encountered in the course of achieving r *successes*. Then

$$P(x = k) = \binom{r + k - 1}{k} p^r (1 - p)^k \qquad k = 0, 1, \cdots \tag{3.6}$$

Note that $x + r$ gives the number of trials needed to achieve r *successes*.

- The GEOMETRIC probability distribution is another name for the negative binomial probability distribution with r = 1.

$$P(x = k) = p(1 - p)^k \qquad k = 0, 1, \cdots$$

The distinction between the binomial and the negative binomial may seem small at first, but one major difference exists. Sampling under the negative binomial will go on and on until the r *successes* occur while under the binomial sampling quits with precisely n trials, regardless of the number of *successes*.

<u>EXAMPLE 3.22</u> (continued): You probably know a couple who kept having children until they got the number of boys or girls they wanted. Suppose you and your spouse decided to have children until you have two girls. Check the conditions on the negative binomial; let p = Prob(girl) = $\frac{1}{2}$. The x in the negative binomial represents the number of boys and r, the number of girls. Thus

$$P(x = 0) = \binom{2 + 0 - 1}{0}\left(\frac{1}{2}\right)^2\left(1 - \frac{1}{2}\right)^0 = \frac{1}{4}$$

$$P(x = 1) = \binom{2 + 1 - 1}{1}\left(\frac{1}{2}\right)^2\left(1 - \frac{1}{2}\right) = \frac{1}{4}$$

Continuing in this fashion, check that

k	0	1	2	3	4	5	...
$P(x = k)$	.250	.250	.188	.125	.078	.047	

These probabilities add to .938 indicating a probability of 1 - 0.938 = 0.062 that you will have to have more than 5 boys before you get 2 girls. Most parents would become discouraged by the time they got enough boys for a basketball team! ▌

<u>EXAMPLE 3.30</u>: State departments of game and fish stock some lakes with species of fish which do not naturally reproduce there. For example, the natural spawning habitat might regularly get covered with silt, killing

the eggs laid on the bottom of the lake. If a constant number of young fish are stocked into the lake each year and they undergo constant mortality (1 - p), then the age distribution of that species of fish is given by the geometric probability distribution. This means that a randomly selected fish has an age of 1 (survived mortality through age 0) with probability p, age 2 with p(1 - p), age 3 with $p(1 - p)^2$ and so on. ▌

EXAMPLE 3.29 (continued): Approximately how many trials would it take to anesthetize 19 rats if the probability of anesthetizing any particular rat is 0.8? Let x = number of rats anesthetized so

$$P(x = 0) = \binom{18}{0}(0.8)^{19}(0.2)^0 = (0.8)^{19} \doteq 0.0144$$

$$P(x = 19) = \binom{19}{1}(0.8)^{19}(0.2)^1 \doteq 0.0548$$

Continuing in this fashion we find that

k	0	1	2	3	4	5	6	7	8	9	10
$P(x = k)$	.0144	.0548	.1095	.1533	.1687	.1552	.1241	.0887	.0576	.0346	.0194

If we add up these probabilities, we find that with 29 trials (x = 10) the probability of getting 19 anesthetized rats is approximately 0.98. This does not say that we will get 19 anesthetized rats but that we have a very high probability of obtaining 19 anesthetized rats in 29 attempts. ▌

The final probability distribution dealing with counts involves this situation: The number of objects in a fixed unit is counted. We may be interested in fungus spores per leaf, deer per square mile, red blood cells per cubic centimeter, radioactive emissions per minute, or fish caught per hour. The following probability distribution applies to these situations, among others:

• POISSON probability distribution - Within a small fractional unit of material or time,

1. Prob(exactly one count) $\doteq \lambda \times$ (size of the small unit),
2. Prob(no counts) $\doteq 1 - \lambda \times$ (size of the small unit),
3. Prob(two or more counts) $\doteq 0$.
4. Let x represent the number of counts in a whole unit. Then

$$P(x = \mathrm{k}) = \frac{\lambda^{\mathrm{k}}}{\mathrm{k}!} e^{-\lambda} \qquad \mathrm{k} = 0, 1, \cdots \tag{3.7}$$

λ gives the average count rate per unit of material.

This probability distribution has had a long and sometimes glamorous history. It is named after its originator, S. D. Poisson (1781-1840). For many years it was regarded as more of a curiosity than as a powerful tool, being applied to suicides of women and deaths of enlisted men in the Prussian army who were killed by kicks from a mule. More recently, it has found extensive use in areas associated with blood cells, bacteria, radioactive decay, population dynamics, nerve impulses, and problems in the telephone industry, to name a few.

The Poisson distribution applies when discrete events occur in time or space so that:

1. The conditions of experimentation remain constant during a useful period of observation. Thus, for example, the probability of an event occurring during a specific time interval depends on the length of the interval rather than the specific times.

2. Events occur independently of each other.

Conversely the Poisson distribution does not apply if these assumptions fail. A comparison of data with the appropriate Poisson distribution provides a basis for validating these assumptions. Specifically, failure of the Poisson to apply often suggests nonindependent events.

The individual terms in the Poisson probability distribution depend on both the count rate λ and the specific number of counts, x. When $0 < \lambda < 1$, $P(x = 0) > P(x = \mathrm{k})$ for any other k, but for $\lambda > 1$, larger values of x produce the maximum probability. This agrees with common sense because a large average count rate should produce a large count. The component $e^{-\lambda}$ serves exclusively as a standardization factor so the sum of the probabilities equal one. (In case you have forgotten, the number e = 2.71818 ···, the base for natural logarithms, arises throughout mathematics.)

The following examples illustrate a few kinds of situations to which the Poisson probability distribution is applied. The exercises and problems provide more illustrations, but many other applications remain unmentioned.

EXAMPLE 3.31: In evaluating the amount of a virus in a medium, a virologist may introduce a fixed amount of medium into a solution and then dilute the mixture. After thorough mixing the solution may be regarded as consisting of many one cubic centimeter (1 cc.) units of volume. Even with thorough mixing some of the 1 cc. units will have no virus particles, others will have exactly one virus particle, still others will contain exactly two virus particles, and so on. When many particles are involved, and the volumes examined are small relative to the volume available, the assumptions underlying the Poisson probability distribution are satisfied approximately. Further, experience shows that the number of particles per small volume does follow the Poisson distribution. ▌

EXAMPLE 3.32: Plant ecologists use the Poisson probability distribution in some applications. Some species of plants appear to be randomly distributed on the ground while others display various non-random patterns. These patterns often are studied by randomly locating a number of quadrats, that is, small square areas. If the plant species of interest is small relative to the quadrat, and is randomly distributed, then the number of that plant species in a quadrat should follow a Poisson probability distribution.

Shimwell (1971, p. 25 and 27) gives two relevant sets of data. The first deals with a study of *Senecio jacobaea*, a shrub whose relatives are found throughout the world in the dryer climates; the data set appears in the second row of Table 3.1.

Individuals Per Quadrat	0	1	2	3	
Frequency in 100 quadrats	46	36	15	3	
Poisson probability, $\lambda = 0.75$	0.4736	0.3552	0.1326	0.0317	...
Expected number	47	36	13	3	

Table 3.1

The average number of plants per quadrat is calculated by $(0 \times 46 + 1 \times 36 + 2 \times 15 + 3 \times 3)/100 = 0.75$. The Poisson probabilities are calculated using this as the parameter value. Thus if x = number of plants in a randomly selected quadrat,

$$P(x = 0) = \frac{(0.75)^0}{0!} e^{-0.75} = e^{-0.75} = 0.4735$$

can be looked up directly in Table III for e^x. Similarly,

$$P(x = 1) = \frac{(0.75)^1}{1!} e^{-0.75} = \frac{0.75}{1} (0.4735) = 0.3552$$

$$P(x = 2) = \frac{(0.75)^2}{2!} e^{-0.75} = \frac{0.56}{2} (0.4735) = 0.1326$$

$$P(x = 3) = \frac{(0.75)^3}{3!} e^{-0.75} = \frac{0.422}{6} (0.4735) = 0.0317 ,$$

as given in the third row of the table. Note that these probabilities sum to 0.9941, leaving a probability of only 0.0059 on four or more occurrences. The expected number results from multiplying the probabilities by 100 for the number of quadrats, and rounding. This data set shows close agreement with its fitted Poisson distribution, thus validating the suggestion that these plants have a random distribution on the ground.

Shimwell's second set of data concerns *Geranium robertianum*, a plant related to the household geranium, and is summarized in Table 3.2.

Individuals Per Quadrat	0	1	2	3	4	5	≥6
Frequency in 100 quadrats	36	9	5	10	14	11	15
Poisson probability $\lambda = 2.5$	0.0821	0.2053	0.2566	0.2138	0.1336	0.0668	0.0418
Expected number	8	21	26	21	13	7	4

Table 3.2

The computations have been left for you to verify. These observed and expected frequencies disagree; the observed results clearly indicate some sort of clumping.

The comparison of observed with expected has been strictly intuitive here. Precise methods exist and would be found in most statistics textbooks under the heading of chi-square. This topic lies outside of the intent of this book. ▌

EXAMPLE 3.33: Accidents occur in many situations. With an extended period of observation we will find that some individuals will incur several accidents even if accidents are rare. Multiple accidents may occur simply by chance, but they also can indicate a predisposition toward accidents. If accidents occur simply by chance in an independent fashion, then the number of accidents often follows the Poisson probability distribution; departure from the Poisson frequently indicates nonindependent accidents, that is, a proneness for accidents. ▌

EXERCISES

1. Under sampling from the hypergeometric probability distribution, suppose N = 12, M = 4, n = 5. Evaluate $P(x = 3)$. Describe a situation from your discipline to which this might apply; check each of the assumptions.

2. Suppose n = 4 and $p = \frac{1}{3}$ in a binomial probability distribution. Evaluate $P(x = k)$ for all values of k; verify that these probabilities sum to unity. Describe a situation from your discipline which might follow this probability distribution; check the assumptions.

3. Establish a relation between $P(x = k + 1)$ and $P(x = k)$ for the negative binomial probability distribution. Observe that once you have evaluated $P(x = 0)$, then $P(x = 1)$ can be evaluated from $P(x = 0)$ and so on.

4. For $r = 2$ and $p = \frac{1}{3}$, evaluate $P(x = k)$ for $k = 0, 1, 2, 3$ for the negative binomial probability distribution. Illustrate its applicability with a situation from your discipline; check assumptions.

5. Establish a relation between $P(x = k + 1)$ and $P(x = k)$ for the Poisson probability distribution.

6. Evaluate $P(x = k)$ for $k = 0, 1, 2, 3, 4$ in a Poisson probability distribution having $\lambda = \frac{1}{2}$; repeat for $\lambda = 2$ and $\lambda = 4$. How does this distribution respond to changing values of the parameter λ? Give an example from your discipline which should follow one of these distributions; check assumptions.

7. Assume that the probability of a girl birth is 0.5 and independent births. Compute the probability that exactly 6 out of 10 babies born will be girls.

8. Ten per cent of the people contracting a certain disease do not recover. What is the probability that all of three people contracting the disease do not recover? What is the probability that 2 out of 3 of the people will recover? What is the probability that at least 2 out of the 3 will survive?

9. Suppose there is an average of one bacteria per 2 cubic centimeters of water in a swimming pool. What is the probability that a selected cubic centimeter of water would have no bacteria? What is the probability that none of four independently selected cubic centimeters of water would have any bacteria?

10. Referring to Example 3.28, what is the probability of getting 16 out of 20 rats anesthetized?

11. Suppose 10,000 bacteria are moving independently and randomly in a volume of 20,000 cubic centimeters of medium. What is the probability that a selected cubic centimeter would contain no bacteria?

11. Continued

What is the probability that a given cubic centimeter will contain 3 or more bacteria?

12. If the probability of the birth of a boy and a girl are equally likely, how frequently will families of 5 children (with no multiple births) have

a. no girls? b. 1 girl? c. 3 girls? d. 5 girls?

13. The cure rate of a certain type tumor is about 15%. What is the probability that out of 7 separate tumors of this type 4 of them are cured?

14. Suppose that under certain specific conditions, radioactive particles are reaching a Geiger counter at an average rate of 5 per second. Assuming a Poisson distribution of numbers of particles, what is the probability function for x, the number of counts, in a randomly chosen 10-second interval? In three independently selected such intervals? (Hint:

$$P(x = k) = \frac{(\lambda I)^k e^{-\lambda I}}{k!}$$

where I is time, x is the number of counts, λ is the average count per unit time.)

REFERENCES

Shimwell, D. W. (1971). *The Description and Classification of Vegetation.* U. Washington Press, Seattle, 322 pp.

Chapter 4

DESCRIBING RELATIONS

Relations come in many styles and varieties; for example they can be permanent or transient, strong or weak, or clear or diffused. A relation describes an association of elements in one set with those in another set. Relations occur throughout biology, frequently involving numbers. Thus here we will focus particular attention on relations between numerical sets. For example, how does an insect's activity relate to its body temperature, and in turn, how does its body temperature relate to ambient temperature? Before you can deal with such matters you must learn how to describe and examine relations.

This chapter introduces relations. The first section presents relations in somewhat general terms while the second associates numerical relations with graphs. Because straight lines occupy an important place in biology, the next section discusses straight lines from different viewpoints. The remainder of the chapter presents and illustrates specific relations, called functions, which occur frequently in biology. The final section relates the contents of this chapter to material in the remainder of the book.

It seems important to emphasize that functions, the basic content of this chapter, have an important place in the use of mathematics in biology. The material in the subsequent chapters will explain how to develop and further examine functions.

4.1 WHAT IS A RELATION?

You know about quantitative relations like less than, greater than, and equal. Although biologists use many relations between numerical sets, relations exist for other kinds of sets. Such situations illustrate relations with a minimum of distracting details. Before we proceed to definitions, the central ideas are illustrated in the following examples.

EXAMPLE 4.1: Blood types must be monitored carefully when blood is transfused from a donor to a recipient. Blood is typed into four primary groups labeled A, B, AB and O. A person with type O blood is called a universal donor because a person with type A, B, AB, or O can receive type O blood. A person with type AB blood is referred to as a universal recipient because he can accept blood from a person having any of the other types. These relationships are sketched in Figure 4.1. A transfusion begins with a donor (the domain in the definitions to follow) and goes to a recipient (the range defined below). The relation describes who may be a donor to whom.

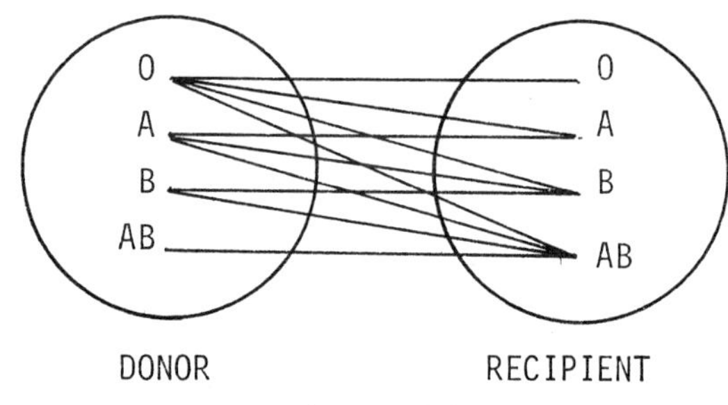

Figure 4.1

Blood is typed according to several different factors. This example describes one of the primary factors. Another is called the Rh factor. (See Problem 14 page 50 .) Thus the above applies to people having the same Rh factor, and compatability in other factors. ▌

EXAMPLE 4.2: Biology abounds with examples of dominance. For example birds display a phenomenon called the pecking order. It is especially noticeable among domestic fowl, like chickens, when several birds occupy a pen or cage. It goes this way: The dominant bird pecks every other bird, but is immune to being pecked. The next dominant bird pecks all others but the dominant one. The last bird in the dominance order pecks no bird, but is pecked by every other bird.

Suppose a cage contains birds labeled A, M, and S. (A,S) could specify that A pecks S. A whole pecking order could be (M,S), (A,S), (M,A); this indicates the dominance order M, A, S. Here the relation states who (the domain) can peck whom (the range). ▌

The following definitions generalize the ideas illustrated above:

- A RELATION between two sets A and B specifies a correspondence between elements of A and elements of B such that for each element $a \in A$, the set B contains at least one element $b \in B$ which corresponds to a.
- The set A from which the elements are selected is called the DOMAIN. The set B in which the corresponding elements appear is called the RANGE.
- A FUNCTION is a relation for which each element in the domain has exactly one corresponding element in the range.
- A set of ORDERED PAIRS, such as (a,b), can be used to state a relation's correspondences. The first element in the parenthesis comes from the domain while the second gives its corresponding element in the range.

Relations concern the correspondence between elements in two sets. We begin by picking elements from the set called the domain; we look for the related elements in the set called the range.

EXAMPLE 4.1 (continued): The blood donor-recipient association does form a relation, but not a function. To see this, let the domain be the possible donors, and the range be the possible recipients:

$$D = \{x | x \text{ identifies a person who can serve as a donor}\}$$

$$R = \{y | y \text{ identifies a person who can serve as a recipient}\}.$$

Most healthy people would belong to both R and D, but practical interest would center on the members of R in need of a transfusion. Both D and R have subsets of people of each blood type. For example let D_{AB} denote the possible AB donors; define other subsets analogously.

If $x \in D_0$, then x relates to any $y \in R$, because type 0 is the universal donor. If instead, $x \in D_{\text{B}}$, then x relates to any $y \in (R_{\text{B}} \cup R_{\text{AB}})$. The donor-recipient association is a relation, because for each donor there exists at least one recipient. It is not a function because each donor can give to more than one recipient. ∎

EXAMPLE 4.2 (continued): Recall the pecking order involved birds labeled A, M, S. Let D = {A,M} symbolize the domain and R = {A,S} the range. The relation is "pecks". The dominance order is M, A, S, because M dominates (pecks) both A and S, and A dominates S. This relation can be stated by using the three ordered pairs {(M,A), (M,S), (A,S)}. In each part the first element gives an element from the domain while the second gives a corresponding element from the range. This relation is not a function because to the element M in the domain there corresponds more than one element in the range.

Animal dominance certainly occurs in birds, but it also appears throughout most animal behavior work. Similarly in plants, an associated idea called species dominance occurs both within a particular stand of plants and during successional stages as plant communities change. ▌

EXAMPLE 4.3: Consider the relation "is the mate of" for a particular species. The domain could be taken to be the set of males and the range the set of females. For a monogamous species which pair off, such as ducks, geese, foxes, or humans in most societies, this relationship is a function because with each male there is associated precisely one female, his mate. Here the domain and range share no elements in contrast to the previous two examples.

Other species such as elk establish harems where one male will breed a number of females. If the domain were males and the range females, then the set of ordered pairs specifying each male and each female in his harem would not be a function because some males have more than one associated female, a violation of the definition of a function. ▌

A relation can be stated either by: i) Specifying the relation verbally or as an equation; or ii) by listing all of the corresponding pairs of elements as ordered pairs (a,b) as illustrated in Example 4.2. Functions occupy a special place among relations: They display an important kind of uniqueness. To each element in the domain, there corresponds exactly one element in the range, a condition not required of all

relations. Figure 4.2 depicts this schematically. The left part describes

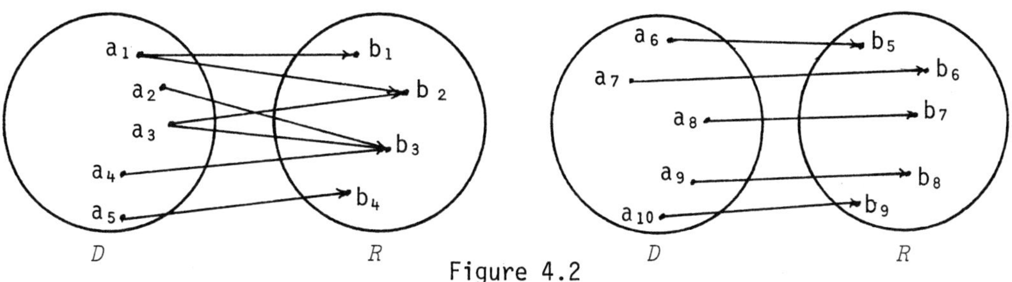

Figure 4.2

a relation because some points in the domain have several corresponding points in the range, like a_3 to which both b_2 and b_3 correspond. On the right this does not occur: Each of a_6 - a_{10} has exactly one corresponding point. The fact that both a_7 and a_8 have b_6 as their corresponding point causes no problem. A function may allow the same point in the range to correspond to several points in the domain.

Now let us turn attention to relations and functions relating elements in numerical sets.

EXAMPLE 4.4: The yield of most crops increases as the number of seeds planted per acre increases, up to some limit. Past this limit, production falls off because the plants compete for limited resources such as sunlight, CO_2, water and soil nutrients. For example, consider the yield of brewers barley relative to seeds

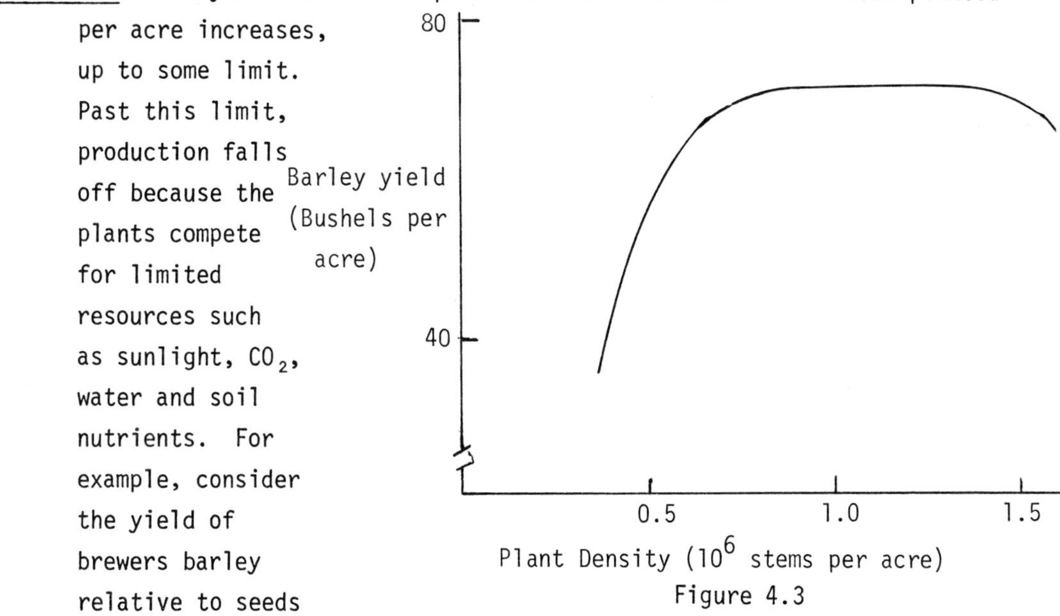

Figure 4.3

per acre. The curve in Figure 4.3 depicts the expected yield under a specific set of conditions for various numbers of seeds per acre. Notice how yield falls off at high plant densities.

Here, to each point in the domain

$$D = \{x|\ 0 \leq x \leq 2.0 \times 10^6 \text{ stems per acre}\}\ ,$$

the mean yield function relates one point in the range

$$R = \{y|\ 0 \leq y \leq 100 \text{ bushels per acre}\}\ .$$

Figure 4.3 illustrates a function between two numerical sets. It has a domain of plant density and a range of expected yield. It is a function because each plant density has exactly one associated expected yield. Again this illustrates that more than one point in the domain may give the same point in the range. Specifically, densities from 0.65×10^6 to 1.5×10^6 stems per acre both give expected yields of about 70 bushels per acre. ▮

A relation may be specified either by stating all pairs of correspondences, or by describing the nature of the relation; this applies in particular to the kind of relations called functions. Special notation is used to denote functions. The function, namely, all of the correspondences, is denoted by a single letter like f, g, or h. You frequently will see a symbolism like $D \overset{f}{\rightarrow} R$ used to say that the function f specifies a functional relation between the domain D and the range R. Suppose f relates a ε D to b ε R. This was written as (a,b), but now we will also write b = f(a) or (a,f(a)). While f denotes the entire set of correspondences, (a,f(a)) denotes one specific correspondence. In Example 4.4, the entire function between plant density and expected yield is shown by the curve. We could denote this by g (for growth, say). We noticed that g(0.65) = 70 = g(1.50), that is, densities of 0.65 and 1.50 both gave expected yields of 70; check to see that g(1.00) $\doteq$ 76.

A function behaves much like a machine or a transformer. Many machines take an input, operate on it to produce an output, that is, they transform the input into an output. Consider a scale for example. You have an object (an element in the domain) and when you place it on the scale (the function), a number results which is the weight of the object (an element in the range). So long as you use the same object you get the same weight.

A new object has a weight which may differ from that of the object before, or it may have the same weight.

This view of a function clearly distinguishes the function (the machine) from its values (its output) as is illustrated in Figure 4.4. This view also makes it clear that a specific function can relate only certain kinds of domains to certain kinds of ranges. For example, the domain for $f(x) = \sqrt{x}$ can include only nonnegative numbers ($x \geq 0$) in our context. By contrast, $g(x) = x^2$ can have any real numbers in its domain. Reflect on these machines: A microscope, an electronic calculator, a drill press or lathe, or a letter addressing machine as they might relate to functions.

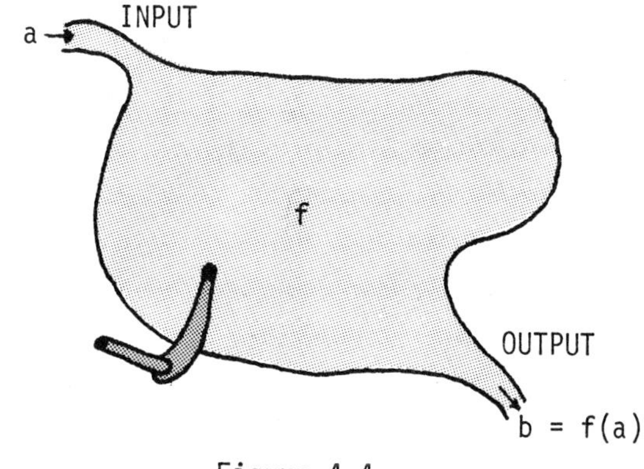

Figure 4.4

This discussion of relations and functions has dealt with the technical meaning of these terms in mathematics. Many terms in the mathematical sciences have related, but less precise meanings in general usage, and in other disciplines. The name function does not escape this problem. In biology it frequently means that two biological characteristics depend on each other; this usage comes closer to being a relation as described above, but it may even fall short of that. In other situations, a biologist may speak of the function of an organ when he really is describing its task or role. These comments do not criticize biologists; they merely alert you to a troublesome fact: The meaning of a word depends on its user.

Functions involving numerical sets appear throughout biology. We have seen that graphs provide a powerful way to visualize the function. The next section focuses directly on using graphs to describe functions relating numerical sets.

4.2 NUMERICAL RELATIONS

Many biological problems involve relations between numerical sets. You should have used the rectangular coordinate system in your previous algebra, so we will review the system only briefly. The system begins with two perpendicular lines called axes intersecting at a point called the origin as depicted in Figure 4.5. Customarily one axis is horizontal so that the other is vertical. Frequently, the horizontal axis is called the x-axis and the vertical axis is called the y-axis, but a problem may suggest more meaningful labels for axes. For example in a problem involving population size, it would make sense to refer to a t-axis (t for time) and a P-axis (P for population), or S-axis (S for size).

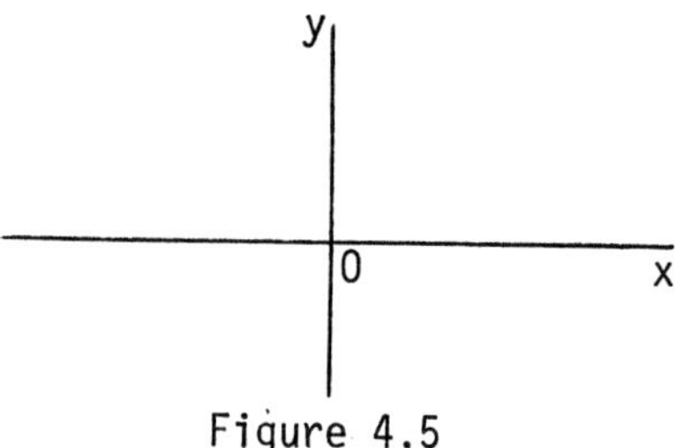

Figure 4.5

Along these axes, scales are marked off to represent units. Frequently, as shown in Figure 4.6, equal lengths are marked off along the two axes for the scale, but the nature of a problem may dictate otherwise. For instance, it would be quite inconvenient to use equal length scales if at $x = 5$ you needed to plot $y = 25{,}372$.

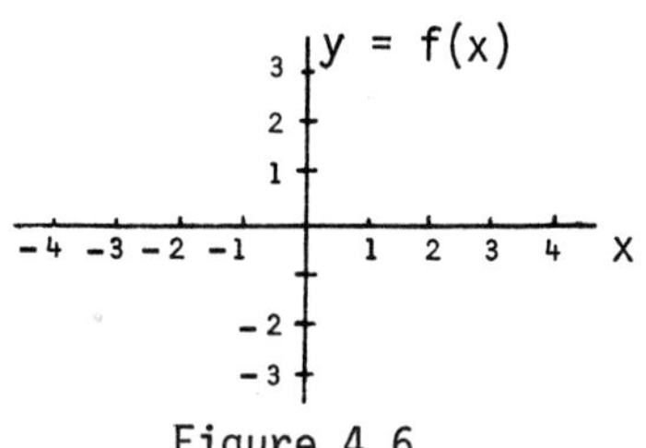

Figure 4.6

An ordered pair (x_0,y_0) can be plotted in this coordinate system by regarding the values of x_0 and y_0 as directed distances as shown in Figure 4.7. Specifically, x_0 gives a length along the horizontal axis, measured to the right for positive x, but measured to the left if x is negative; y is related to the vertical axis in the same manner. The point located x_0 units to the right (or left) of the origin and y_0 units above (or below) the origin uniquely represents the ordered pair (x_0,y_0).

The horizontal coordinate of a point often is referred to as its ABSCISSA, while its vertical coordinate frequently is called its ORDINATE. The axes divide the plane into four QUADRANTS, numbered I, II, III, IV counterclockwise from the positive x axis as indicated in Figure 4.7. This numbering convention provides a convenient way for discussing the behavior of a function on a graph.

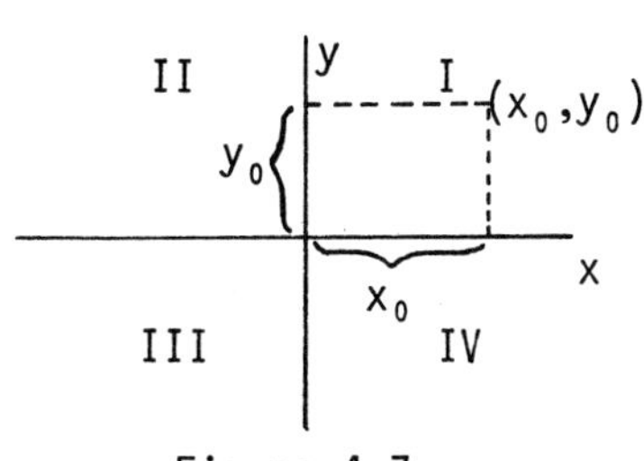

Figure 4.7

EXAMPLE 4.5: Plot the ordered pairs (4,3) and (-2,-3).

To plot the point (4,3), erect a perpendicular to the x-axis at x = 4 and go 3 units upward to the point (4,3). The same procedure plots (-2,-3) because the perpendicular is erected at x = -2 and then 3 units are measured off in the negative or downward direction. The point (4,3) has an abscissa of 4, an ordinate of 3, and lies in the first quadrant. The other point lies in the third quadrant. ▮

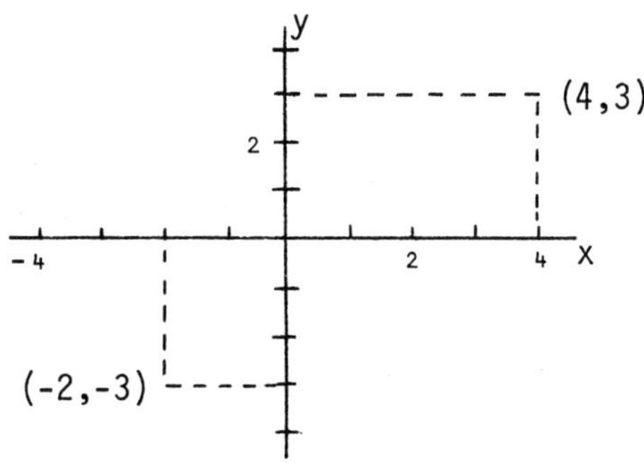

Graphs serve an important role in describing a relation or function. At this point we have a coordinate system in which we can plot ordered pairs, and functions and relations defined in terms of ordered pairs. Thus it should seem natural to combine the two. The result is a graph such as we used in Figure 4.4 to display how barley yield changed with plant density. The following terms often are used in discussing graphs:

- An INDEPENDENT VARIABLE is a general element from the domain set.
- An independent variable sometimes is called a function's ARGUMENT.

- A DEPENDENT VARIABLE is a general element from the range set.
- The SOLUTION SET of a relation is the set of ordered pairs satisfying the relation.
- The GRAPH of a relation presents the solution set of the relation in a coordinate system.

When a relation is graphed, the horizontal axis usually represents the independent variable (domain) and the vertical axis represents the dependent variable (range). Again be careful about the meaning of words; independent does not imply lack of association. View it this way for a function: You can pick a value of the independent variable, the input, and drop it into a function machine; the machine transforms your input into the appropriate output, the resulting value of the dependent variable. The function creates the dependence by how it transforms domain points into range points. A function's argument is merely a widely used synonym for independent variable.

When a dependent variable is related to an independent variable through an equation such as $y = x^2 + 3$, y really is being regarded as a function of x and could be more carefully written as y(x). The (x) often is dropped when it is not a focus of interest. Nevertheless, it is present. The solution set merely describes those points in the coordinate system which behave according to a specified relation.

<u>EXAMPLE 4.6</u>: Consider $y(x) = x^2 + 3$. The independent variable x can be any real number while the range consists of the real numbers greater than or equal to 3. In set notation the solution set of $y = x^2 + 3$ can be written as $\{(x,y) | y = x^2 + 3\}$, and is shown as a graph in Figure 4.8. This depicts a function

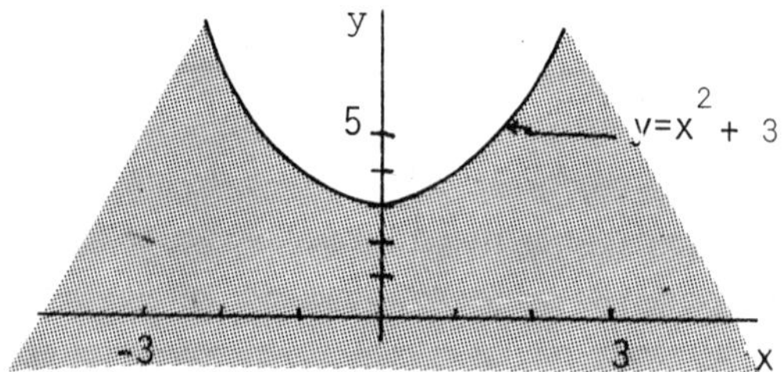

Figure 4.8

because the curve has exactly one value of y for each value of x.

Sometimes we will need to represent solution sets for inequalities as well as equalities. The shaded region in Figure 4.8 represents $\{(x,y) | y < x^2 + 3\}$. ▌

EXAMPLE 4.7: A glucose tolerance test illustrates solution sets in biology. A glucose tolerance test operates this way: A patient abstains from food for 12 hours, gives samples of blood and urine and then drinks a single dose of glucose (a sugar). Blood and urine samples then are obtained 0.5, 1, 2, and 3 hours later for sugar determinations. These determinations provide a basis for detecting hyperglycemia unresponsiveness which may be due to either lack of insulin or insulin resistance. Normally the venous blood sugar reaches a peak (120-160 mg.) between 30 minutes and 1 hour after the glucose dose and returns to near the fasting level by 2 hours as shown in Figure 4.9. Increased glucose tolerance (flat or inverted curve) is noted in hypoinsulinism and other diseases. Decreased glucose tolerance (a high or prolonged curve) occurs in diabetes and other diseases. Thus comparison to a typical glucose tolerance curve indicates whether a person has a normal or abnormal glucose curve, and possibly even detects the nature of the disorder.

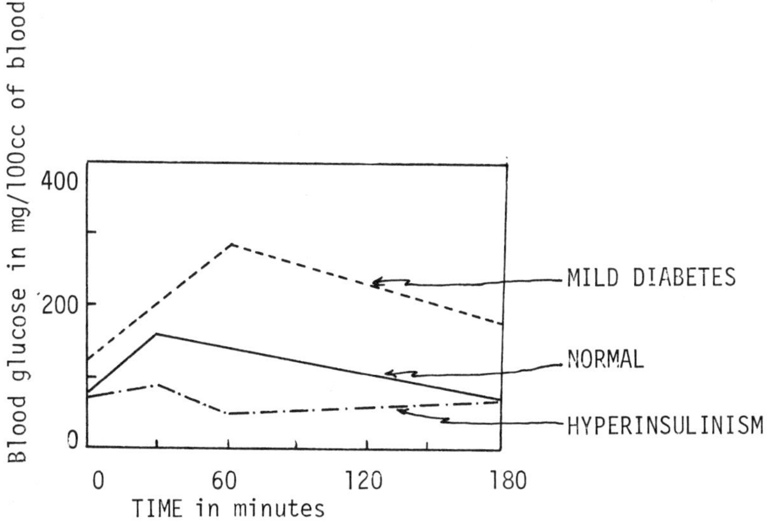

Figure 4.9

The three curves in Figure 4.9 could be labeled f_D, f_N and f_H for mild diabetes, normal and hyperinsulinism. Each is a function. Why? Note that at the critical time t = 60, $f_D(60) \doteq 280$ mg. ($\doteq$ means approximately equals), $f_N(60) \doteq 150$ mg. and $f_H(60) \doteq 50$ mg., so $f_D(60) > f_N(60) > f_H(60)$. In fact if the domain is restricted to $0 \leq t \leq 180$ minutes, $f_D(t) > f_N(t) > f_H(t)$ for each value of t. ▮

EXAMPLE 4.8: An example involving a relationship between night temperature in °C and growth rate of 48 day old pepper plants is shown in Figure 4.10. The graph indicates that as the night temperature increases from 10°C to about 20°C the growth increases and then as the night temperature increases up to 30°C the rate of growth declines. The optimum temperature for growth is about 20°C.

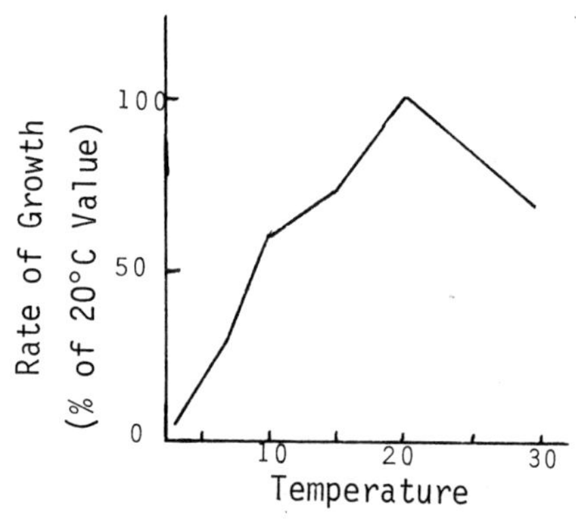

Figure 4.10

Here time is the independent variable and rate of growth is the dependent variable. The curve specifies a solution set which is a function because it has exactly one value for rate of growth for each relevant temperature. Describe the domain and range for this function. ▮

A function can be stated in two essentially different ways. It can be described as a solution set for some sort of equation, or all points of interest may be listed. To do the latter, of course there must be only a finite number of points for graphing. This brings us to note that either the range or the domain can be finite sets consisting of discrete entities such as the integers.

Discrete sets closely related to the integers occur throughout biology, but many of the relationships involving discrete sets can be conveniently and satisfactorily approximated by smooth functions. For example, a smooth function may provide a very convenient and satisfactory approximation to population size, a variable which definitely is discrete. An operation exists to translate any function into a discrete one. Its notation, often called "square bracket notation", appears thus: $[\![x]\!]$. This notation means to take the largest integer value which is less than or equal to x as is illustrated by $[\![2.3]\!] = 2$, $[\![15.93]\!] = 15$, $[\![-14.7]\!] = -15$. Note that if x is an integer then $[\![x]\!] = x$ and in all other cases the value of the $[\![x]\!]$ is found by taking the next integer to the left of x on the real number line.

EXAMPLE 4.9: Consider the growth of a bacterial population through time. Suppose f(t) is a smooth function which gives at various times the number of cells expected from a single cell. For example if a strain of bacteria divides every 43 minutes, f(0) = 1, f(43) = 2, f(86) = 4, f(129) = 8, $\cdots$. If a population has an initial size of N_0, then the subsequent population size can be denoted by $N(t) = N_0f(t)$. Because population size must be discrete, this really means $N = [\![N_0f(t)]\!]$. Under most conditions $N_0f(t)$ provides a close approximation to $[\![N_0f(t)]\!]$ because the population increases by one cell at a time. The smooth curve in Figure 4.11a depicts $N_0f(t)$. It remains close to the disconnected horizontal line segments giving actual population size.

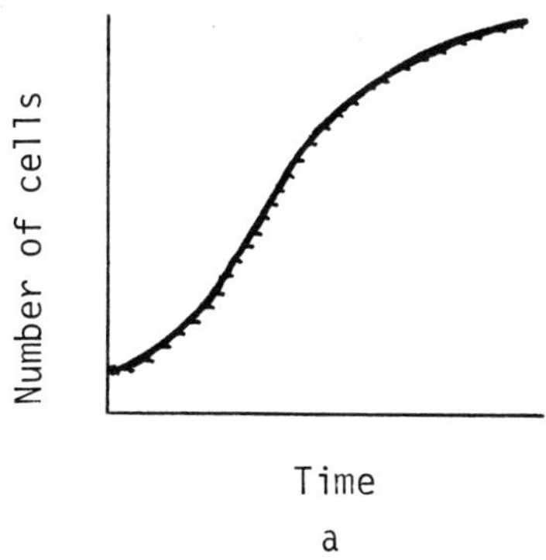

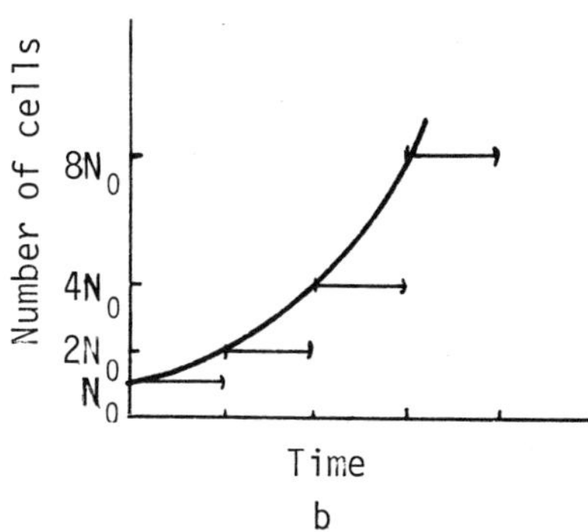

Figure 4.11

By contrast work with tissue cultures may seem similar, but their size cannot be satisfactorily approximated by a smooth function. Cells in a tissue culture can be synchronized so that they all divide at very nearly the same time. Figure 4.11b shows a smooth approximation to such a population size function. As the real size function jumps from N_0 to $2N_0$, then on to $4N_0$, and so on, the actual function cannot be well approximated by a smooth curve.

The distinction between these two situations lies in magnitude of the step relative to the response size. In the first case it would be one in thousands or even millions while in the second case it has a magnitude similar to the response size. ▌

One final point should be made about graphs. The interesting values along an axis may be far removed from zero. For example an animal may be injected with a material which should have no effect for about 24 hours, but then the reaction needs to be monitored closely. This needs an axis which starts at about 20 hours and continues as long as the reaction is of interest. Likewise, in talking about the size of mature organisms, small individuals are not expected. It would be reasonable to consider then only individuals about some minimal size. Two notations are commonly used to indicate a broken axis. They consist of either putting a kink in the axis shortly past the zero point, or drawing a pair of slanted lines through the axis where it is broken. In Figure 4.12 one convention is used on the horizontal axis while the other is used on the vertical axis. In practice only one of the notations would be used on one graph.

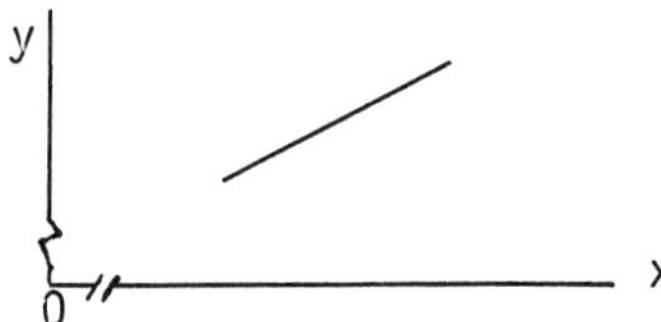

Figure 4.12

EXERCISES

1. If f is the function with domain R, the reals, defined by $f(x) = 5x^4 + 4x^3 + 3x^2 + 1$ find
   a. $f(0)$; b. $f(1)$; c. $f(-1)$;
   d. $f(2)$; e. $f(-2)$.

2. If f is the function with domain R defined by $f(x) = 3x^2 - x + 2$, and $a, h \in R$ find

   a. $f(a)$; b. $f(-a)$; c. $-f(a)$; d. $f(a + h)$;

   e. $f(a) + f(h)$; f. $\frac{f(a + h) - f(a)}{h}$

3. Suppose that a cell divides every hour so that the number of cells, N, present at a given time is $N = 2^{[\![t]\!]}$ where $t \in R$. Thus N is a function of t: $N(t) = 2^{[\![t]\!]}$.

   a. How many cells are present at time $t = \frac{1}{2}$, 1, 2, 3, 6, 7.3, 8?
   b. Graph the function on an N,t-plane.
   c. What is the domain of N?
   d. What is the range of N?

4. Let $f(x) = [\![x]\!]$ for $x > 0$, where $[\![x]\!]$ is the greatest integer less than or equal to x.

   a. Draw the graph of f(x).
   b. Give a practical example of such a function.
   c. Draw the graph of $g(x) = [\![x + 6]\!]$ for $x \geq 0$.

5. On an x,y-plane, sketch the graph of the solution sets for the following.

   a. $A = \{(x,y) \mid y \leq x^2 + 1\}$
   b. $B = \{(x,y) \mid y > 2x^2 + 3x + 1\}$
   c. $C = \{(x,f(x)) \mid f(x) < 3x + 1\}$
   d. $D = \{(x,y) \mid y = mx + 2, m = 1\}$

6. Does the following graph describe a function? If not, explain why not.

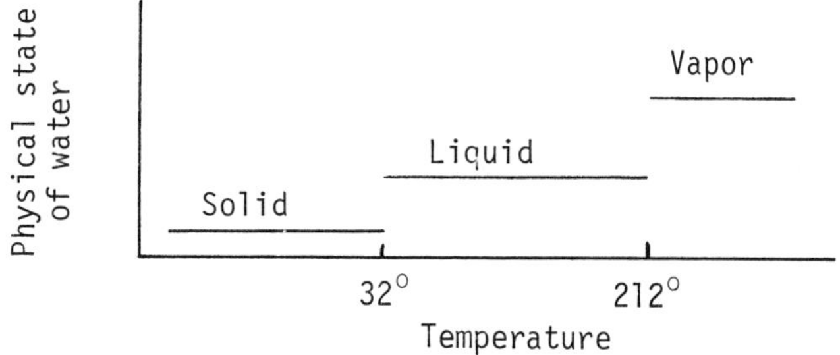

7. The following graphs represent relations from R to R.

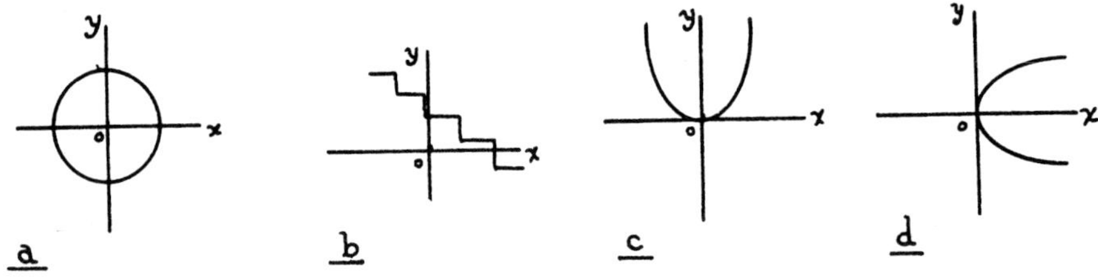

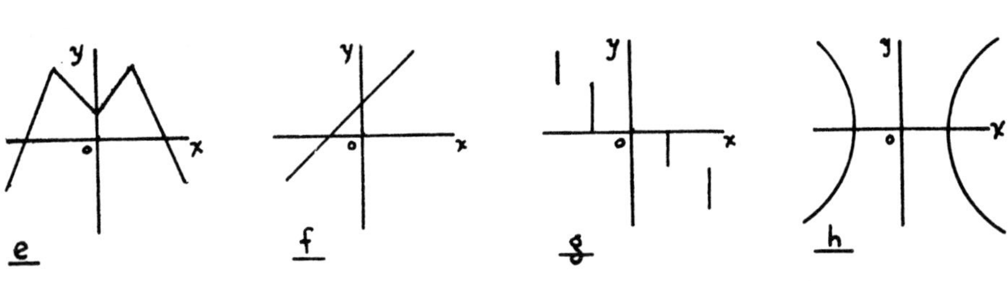

a. In which of these relations is y a function of x?

b. In which of these relations is x a function of y?

In exercises 8-15 determine which of the given relations from R to R are functions.

8. $A = \{(x,y) \mid y = 4x + 3\}$

9. $A = \{(x,y) \mid x^2 + y^2 = 4\}$

10. $A = \{(x,y) \mid x = 4\}$

11. $A = \{(x,y) \mid x + y = 0\}$

12. $A = \{(x,y) \mid y^2 = x^2 + 2\}$

13. $A = \{(x,y) \mid y^2 = x^2\}$

14. $A = \{(x,y) \mid y = 1\}$

15. $A = \{(x,y) \mid xy = 0\}$

16. Express the perimeter of an equilateral triangle (all three sides are the same length) as a function of its height.

17. The surface area of a sphere is given by $S_A = 4\pi r^2$ and the volume is given by $V = \frac{4}{3}\pi r^3$. Express the volume as a function of the surface area. Is the word function used properly here?

18. A rectangular trough without a top is to be made from a long piece of metal which has a width of 12 inches. The trough is to be made by turning up the edges of the metal. Express the cross-sectional area of the trough in terms of the height.

19. Water is being pumped into a cylindrical tank which has a diameter of 15 feet and a height of 30 feet. Express the volume of the water in the tank in terms of the depth and determine the domain and range of the function.

20. A rectangle is inscribed in a circle of radius 20 inches. Express the area of the rectangle in terms of the length of the longest side of the rectangle.

21. An open topped box is formed from a rectangular piece of cardboard by cutting out equal squares from each of the four corners and then bending up the sides. If the original dimensions of the cardboard were 16 inches by 20 inches, express the volume of the box as a value of a function whose independent variable is the length x of an edge of the squares cut out.

4.3 STRAIGHT LINES

Many kinds of relations occur in biology, but straight lines are used widely. In using a straight line to describe a relation, the biologist usually is not asserting that the true relation behaves exactly as a straight line over all relevant values of the dependent and independent varibles; instead he is saying that a straight line provides a satisfactory approximation to the true relation over a limited range of interest. On the other hand, biology does provide numerous examples of nearly linear relations. This section begins by considering straight lines as entities; applications appear later in the section. We begin with this definition:

- A STRAIGHT LINE is the graph of the solution set specified by the function $y = mx + b$, where m and b are constants.

In this general form the parameter m represents the slope of the line and b is the value of y at which the line intersects the ordinate axis. The slope of a nonvertical line in the x,y-plane is given by the ratio of the change in the vertical coordinate to the change in the horizontal coordinate. The slope also can be described as the rise divided by the run. The change in the ordinate direction frequently is designated by Δy and in the abscissa direction by Δx. Thus, the slope is given by $m = \Delta y/\Delta x$. Figure 4.13 shows the graph of the line $y = mx + b$ passing through the points (x_1,y_1) and (x_2,y_2). The slope is

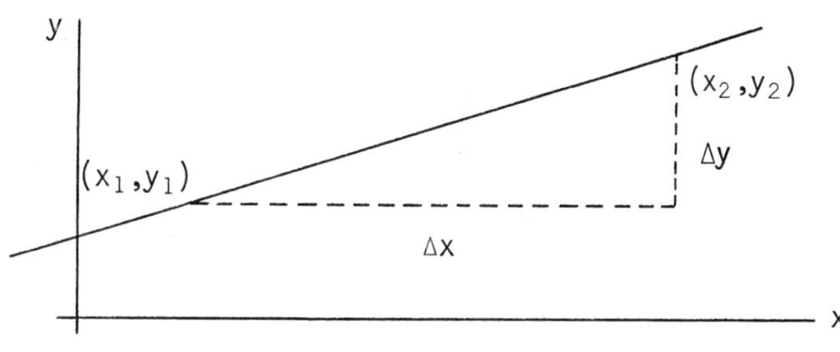

Figure 4.13

$$m = \Delta y/\Delta x = (y_2 - y_1)/(x_2 - x_1)$$

because $\Delta y = y_2 - y_1$ and $\Delta x = x_2 - x_1$.

Observe that the same slope occurs between any two pairs of points on a straight line. This fact characterizes straight lines, but does not apply to curves generally. When $m > 0$ the line has a positive slope and it slopes upward to the right, that is, as x increases y also increases. Conversely, when $m < 0$ the line slopes downward to the right so y decreases as x increases. Of course, the line with $m = 0$ is horizontal and parallel to the abscissa axis.

EXAMPLE 4.10: Figure 4.14 shows the graph of $y = 2x + 3$, $y = -x + 1$, and $y = -x$. Note the effect of positive and negative slopes, and the role b plays in the positioning of the line. ▮

The specification of a straight line as $y = mx + b$ gives a completely general form called the SLOPE-INTERCEPT form; m represents the slope and b the intercept. A line can be specified also

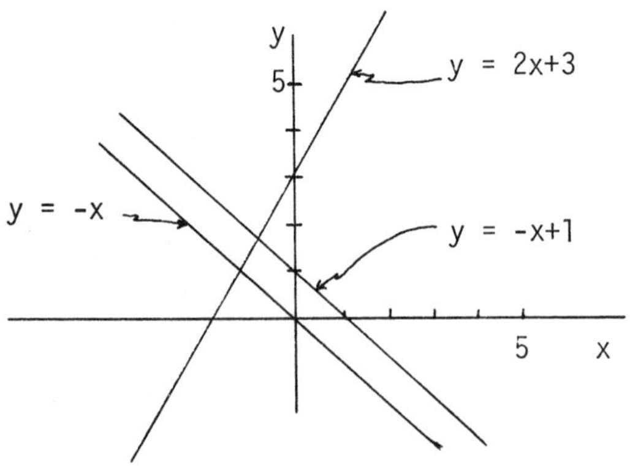

Figure 4.14

by its slope and any point it passes through, or by two points it passes through. Only a small amount of algebraic effort lies between either of these specifications and the standard slope-intercept form.

EXAMPLE 4.11: Find the equation of the line with a slope of 3 which also passes through the point (4,2).

We only need to find its intercept, because its slope is given. The line $y = 3x + b$ has a slope of 3; to make it pass through (4,2) it must satisfy $2 = 3(4) + b$, or $b = -10$. Thus we have $y = 3x - 10$. This line is plotted in Figure 4.15. To check our arithmetic, the point (4,2) should satisfy $y = 3x - 10$; upon substitution we have $2 = 3(4) - 10$, truly an equality. ▌

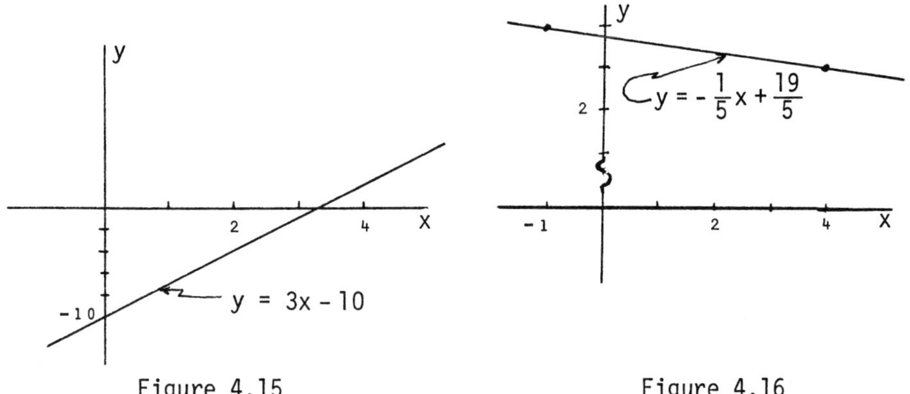

Figure 4.15

Figure 4.16

EXAMPLE 4.12: Establish that the equation of a line passing through the two points (-1,4) and (4,3) is $y = -1/5x + 19/5$.

First find the slope, then proceed as in the previous example. In going from $(x_1,y_1) = (-1,4)$ to $(x_2,y_2) = (4,3)$, $\Delta x = x_2 - x_1 = 4 - (-1) = 5$ and $\Delta y = y_2 - y_1 = 3 - 4 = -1$ and thus $m = \Delta y/\Delta x = -1/5$. The line through (4,3) with slope $m = -1/5$ must satisfy $y = -x/5 + b$, or upon substitution of the point (4,3) we have $3 = -4/5 + b$ so $b = 19/5$. As a check, verify that the resulting line, $y = -x/5 + 19/5$ passes through the other point, (-1,4). The line is plotted in Figure 4.16. ▌

Several other terms appear in discussion of straight lines. Three or more points are COLLINEAR when they lie on the same straight line. Two lines, defined by $y = m_1x + b_1$ and $y = m_2x + b_2$, are PARALLEL when $m_1 = m_2$. Two such lines are PERPENDICULAR when $m_1 = -1/m_2$. To check for collinearity you merely need to be assured that all of the points lie on the same straight line. This occurs when the slopes between each pair of points are all equal. For this test to work, you must check every pair, not only two pairs. To see this, take two points on one line and two more on a parallel line.

EXAMPLE 4.13: From Example 4.12, we found that $y = (-1/5)x + 19/5$ passed through (-1,4) and (4,3). The line $y = (-1/5)x + 0$ shown in Figure 4.17 is parallel to the first line because both have a slope of -1/5. The points (-1,4), (1,18/5), and (4,3) all lie on the same line. (Find the slope between each pair.) The line $y = 5x + 4$ is perpendicular to both $y = (-1/5)x + 19/5$ and $y = -x/5$ because these parallel lines have a slope of -1/5, namely, the negative reciprocal of the slope of 5 in $y = 5x + 4$. ▌

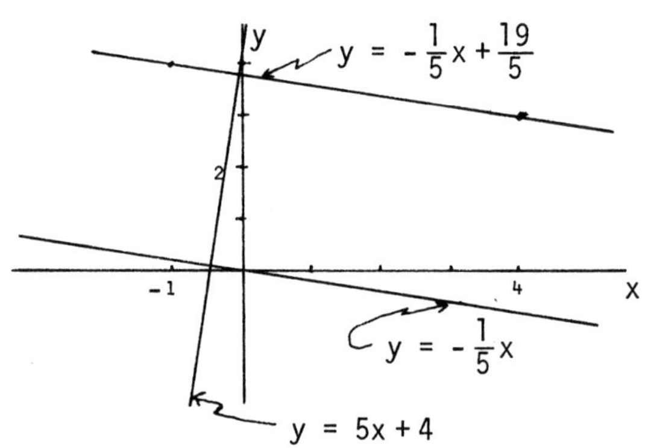

Figure 4.17

Now that we have discussed some mathematical aspects of straight lines, we will give several biological examples.

EXAMPLE 4.14: Figure 4.18 shows the twig diameter-twig length relation of bitter brush, a species mule deer browse on. Here, length is the dependent variable and twig diameter is the independent variable. They are related through the equation length = 1.25 + 89.83 (diameter). ▌

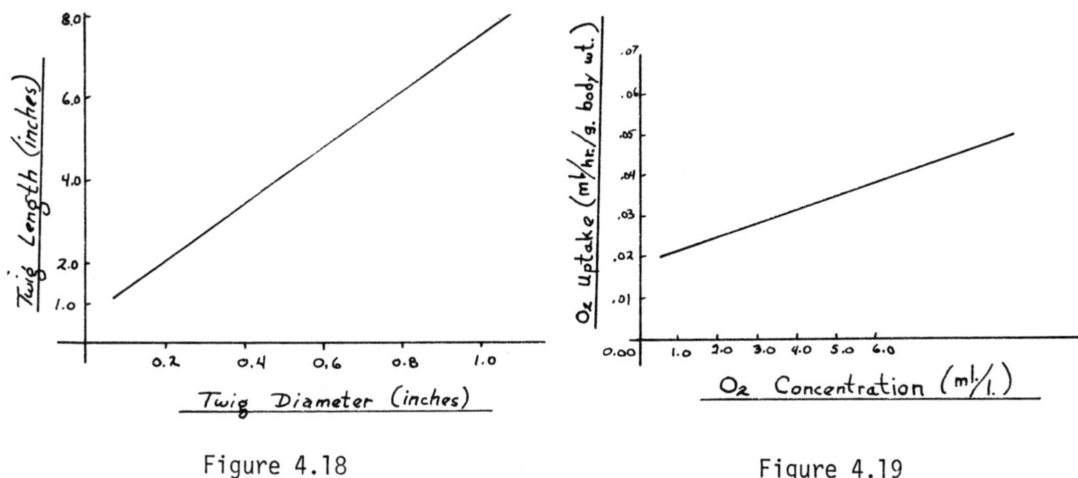

Figure 4.18

Figure 4.19

EXAMPLE 4.15: Lobster, a salt water crustaceon, has an oxygen consumption which depends on the oxygen concentration in his surrounding environment as depicted in Figure 4.19. ▮

EXAMPLE 4.16: In certain instances the rate at which nerve impulses are conduct-d along a nerve fiber depends upon the nerve diameter. Figure 4.20 shows the relation between the independent variable, diameter, and the dependent variable, conduction velocity along myelineated nerve fibers, in meters per second. ▮

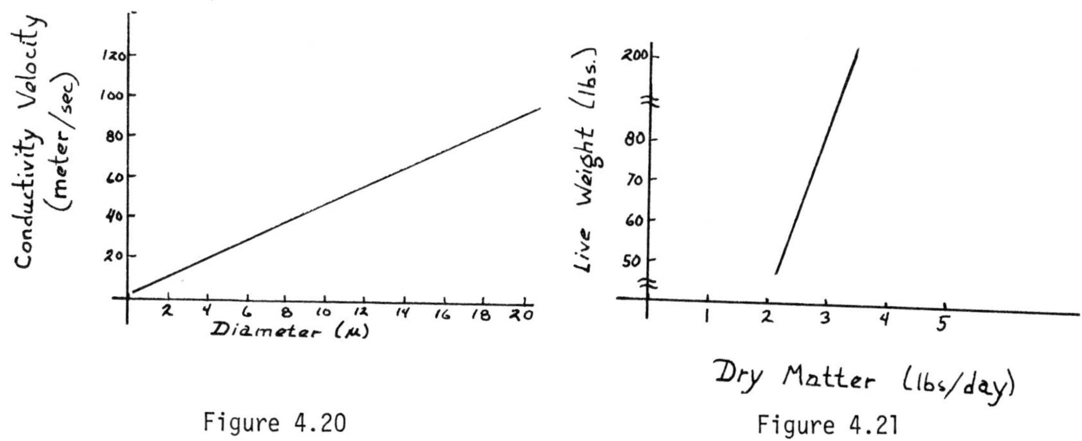

Figure 4.20

Figure 4.21

EXAMPLE 4.17: The food requirements of most animals depend upon their size. For example, the relation between live weight and dry matter maintenance requirements for sheep of different weights appears in Figure 4.21. Note that the broken scale on this graph omits live weights below the range of interest. ▌

EXAMPLE 4.18: The size of an antler on a deer depends substantially upon the age of the animal carrying the antler. Figure 4.22 displays the relation between antler weight during October and the age of mule deer in the Cache la Poudre deer herd in Northern Colorado. ▌

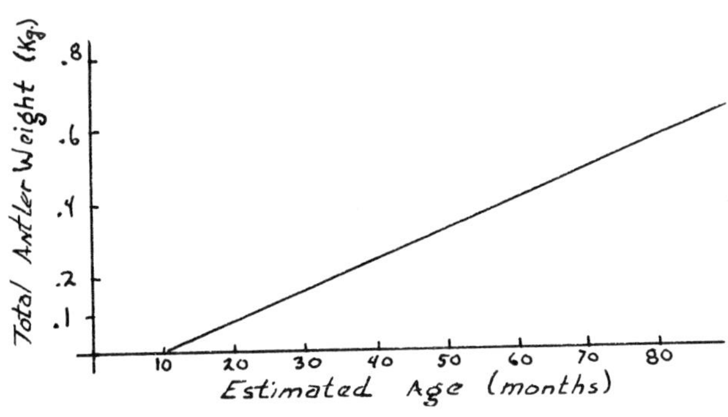

Figure 4.22

EXERCISES

Give the y-intercept and the slope of each of the following:

1. $y = 2x - 1$ 2. $x - 3y + 2 = 0$ 3. $x = 5y + 10$

4. Given the points A(-1,-2) and B(3,4)
   a. Write the equation of a line which contains C(-1,4) and is parallel to the line containing A and B.
   b. Write the equation of the line which contains C(-1,4) and is perpendicular to the line containing A and B.

5. Find the equation of the straight line through
   a. The two points (1,2) and (3,4)
   b. The two points (-1,-1) and (2,1)
   c. The two points (3,4) and (3,7)
   d. The two points (-5,4) and (5,4)
   e. The two points (0,1) and (2,0)
   f. A y-intercept of 2 with a slope of 3
   g. A y-intercept of -1 and the point (2,2)

Four points A, B, C, D are specified for each of the following problems. Prove that they form the vertices (corners) of a rectangle. (Hint: Let appropriate pairs of points define lines, and then check for parallelism and perpendicularity.)

6. A(4,8), B(8,6), C(6,2), D(2,4). Is this a square?

7. A(3,5), B(1,1), C(-2,1), D(-1,3).

8. A(-1,-2), B(-2,-4), C(2,-3), D(0,-2).

9. A(-3,-4), B(-2,5), C(4,2), D(-5,3).

Determine whether the following points are collinear.

10. (2,3), (3,5) and (1,1).

11. (0,3), (1,6) and (-1,0).

12. (2.0,-1.5), (5.0,3.0) and (12.5,14.25).

13. Prove that (-2,1), (-1,-2), (2,2) and (3,-1) are vertices of a parallelogram. (Hint: A parallelogram has two pairs of parallel sides.)

14. The maximum amount of a certain substance, A grams, that can be dissolved in 100 milliliters of water at a temperature of T degrees centigrade was found to be as shown in the following table:

A	76	80	90	100	105	115
T	5	15	25	35	45	55

Plot A versus T and draw what you consider to be the best fitting straight line. Assuming the relationship between A and T to be of the form $A = cT + b$ where b and c are constants, estimate the values of b and c from the line you drew through the data points.

15. Plot the function $y = 9.2x - 20$ using a unit on the y-axis ten times that of the unit on the x-axis; repeat using equal units.

16. Plot the function $y = 0.32x + 25$ with equal units on the x and y-axis but with an "interrupted" y-axis in order to save space.

17. A copper rod is exposed to different temperatures. Its length L changes almost linearily with temperature T if T < 150°C. Find the equation relating L and T if you obtain the following measurements: (20°,76.467 cm.) and (90°,76.547 cm.).

18. At sea level water freezes at 32°F and 0°C while it boils at 212°F and 100°C. Find °C as a function of °F, and °F as a function of °C. What is the reading in °C when the temperature is 68°F? 98.6°F?

19. When bacteria are put into a nutrient broth, a period of adaptation usually occurs (a period without a change in the number of bacteria); afterwards the number of bacteria increases for a period of time, and finally decreases. During the period of adaptation the bacteria form the enzymes necessary for them to use the food in the broth. After the period of adaptation they reproduce until they consume all the available food and begin to die. In an idealized experiment the following results were obtained:

Time (hr.)	Number of Bacteria	Time (hr.)	Number of Bacteria
0	10^3	5.5	3.2×10^7
0.5	10^3	6.0	10^8
1.0	10^3	6.5	3.2×10^8
1.5	3.2×10^3	7.0	10^9
2.0	10^4	7.5	10^9
2.5	3.2×10^4	8.0	10^9
3.0	10^5	8.5	10^9
3.5	3.2×10^5	9.0	7.5×10^8
4.0	10^6	9.5	5.6×10^8
4.5	3.2×10^6	10.0	2.1×10^8
5.0	10^7	10.5	1.6×10^8

Graph the number of bacteria as a function of time using appropriate scales. Identify the period of adaptation and the period of growth.

20. It is known that a deficiency of oxygen and an excess of carbon dioxide influences the EEG (electroencephalogram), which is the electrical potential obtained from electrodes attached to the surface

20. Continued

of the head. This potential (EEG) is an indicator of the over-all activity of the cortex.

The following three EEG traces are typical of recordings one might obtain (1) during sleep, (2) during alert wakefulness, and (3) during drowsiness.

(1) During sleep (synchronized):

(2) During alert wakefulness (desynchronized):

(3) During drowsiness (intermediate):

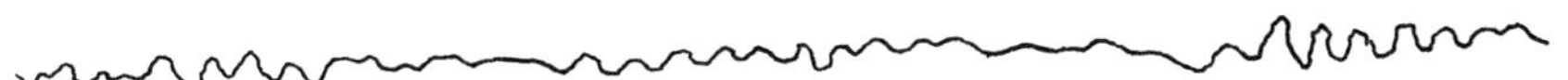

During sleep the electrical potential is large in amplitude and low in frequency. It has been hypothesized that this type of record is caused by the synchronization of sinusoidal electrical potential changes in individual cortical neurons. During alert wakefulness the pattern is broken due to the desynchronization of the electrical potentials of individual neurons. Because of the cancellation of unsynchronized potentials a rather flat record is obtained. During drowsiness an alternation of the synchronized and desynchronized waveforms is apparent.

It was found that the composition of inspired air affected the EEG of a sleeping cat; it could change from (1) synchronized to (2) desynchronized or (3) intermediate. Research was carried out to determine which was the influencing factor: The low concentration of oxygen or the high concentration of carbon dioxide.

20. Continued

These were the results:

AIR MIXTURE (partial pressure of the gas in mm. of mercury)		
O_2	CO_2	
84	15	synchronized
120	20	intermediate
140	18	intermediate
155	14	synchronized
100	25	desynchronized
140	19	intermediate
90	27	desynchronized
145	22	desynchronized
140	16	synchronized
94	15	synchronized
85	18	intermediate
105	22	desynchronized
120	11	synchronized
105	20	intermediate
120	29	desynchronized
110	18	intermediate
89	12	synchronized

Draw a graph in which the x-axis represents the concentration of carbon dioxide and the y-axis the oxygen concentration. Label a point "d" if the EEG was desynchronized, "i" if it was intermediate, and "s" if it was synchronized.

If the EEG was synchronized, it was interpreted that the cat continued sleeping undisturbed. If it became desynchronized or intermediate, it was interpreted that the cat had become disturbed and awakened by a change in its breathing. Using the table, decide for yourself which affects the EEG, the oxygen or the carbon dioxide (in the range of concentration with which we are dealing).

21. You wish to estimate the thickness of a coating which covers half of a plane glass slide by focusing the microscope at the stations 1-8 which are equally spaced as shown in the figure below and measuring the vertical displacement of the microscope for each station. Note that the surface is not horizontal. Estimate the thickness of the coating on the slide from the data below. Explain the difficulties you had in estimating the thickness of the coating.

Station	1	2	3	4	5	6	7	8
Adjustment (10⁻ cm.)	0	2	4	9	28	31	37	37

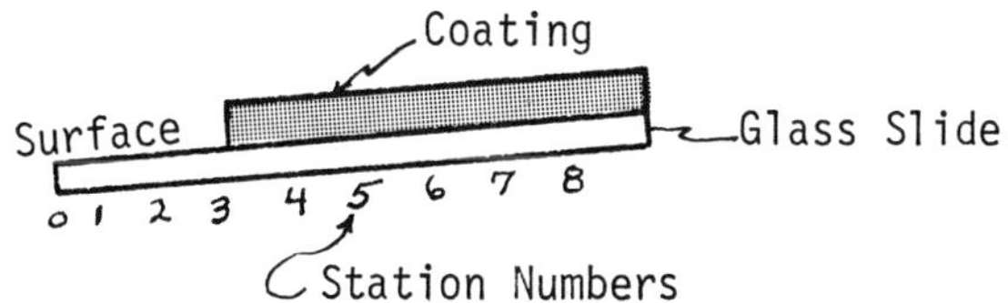

22. Let $A(x_1,y_1)$ and $B(x_2,y_2)$ be two points in the x,y-plane and let $C(\overline{x},\overline{y})$ be the coordinate of the midpoint of the segment $\overline{AB}$. Show that $\overline{x} = \frac{x_1 + x_2}{2}$ and $\overline{y} = \frac{y_1 + y_2}{2}$.

23. Using the results of Problem 5, find $C(\overline{x},\overline{y})$ where
    a. A(2,3), B(-2,-1)
    b. A(2,5), B(-4,3)
    c. A(0,1), B(1,0)

4.4 CURVACEOUS FUNCTIONS

The previous section noted how straight lines frequently can serve as satisfactory approximations to biological relations. Straight lines frequently are useful, but situations do occur in which they dismally fail to describe a biological relationship. These cases require curves of some sort. From theoretical considerations you may know the exact mathematical form that a curve should take; if not you may still know enough about its general behavior to approximate it with the right kind of smooth curve.

Here we will introduce and discuss three important and widely used curved relations, one at a time. Each is immediately illustrated by an algebraic example. Biological examples appear toward the end of the section.

- The standard form of the equation for a PARABOLA can be stated in either of two ways:

  (i) a parabola with its vertex at (h,k) and opening either up or down is given by the solution set of the equation $y - k = a(x - h)^2$ or

  (ii) $y = ax^2 + bx + c$

Look at Figure 4.23 to see the shape of a parabola. The VERTEX at (h,k) is where the curve changes direction, much as the sides of a triangle change direction at its vertices or corners. Of course if $(h,k) = (0,0)$, then the parabola becomes $y = ax^2$ and goes through the origin.

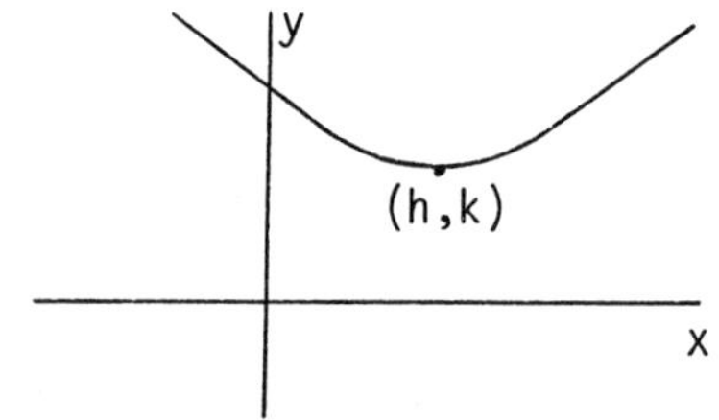

Figure 4.23

A parabola is a curve which opens upward and continues to go upward from its minimum, or opens downward and continues to go downward from its maximum. In either standard form, $a > 0$ produces a parabola which opens upward and has its minimum at its vertex while $a < 0$ produces a parabola which opens downward and has its maximum at its vertex. Relations between the constants in the two forms of the parabola are developed in Example 4.28. By interchanging the roles of x and y, a parabola opening either to the right or to the left could be generated, but these would not be functions.

<u>EXAMPLE 4.19</u>: The solid curve in Figure 4.24 shows the graph of $y - 1 = (x - 2)^2$, or equivalently of the function $y = 1 + (x - 2)^2$. Note that for $a = 1 > 0$, the parabola opens upward. This curve could also be described by the equation $y(x) = x^2 - 4x + 5$, a fact you

should establish by squaring and collecting terms. The dashed curve represents the related function $y = 1 - (x - 2)^2$. Since $a = -1 < 0$, the parabola opens downward. The alternate form for this function is $y = -x^2 + 4x - 3$. ▌

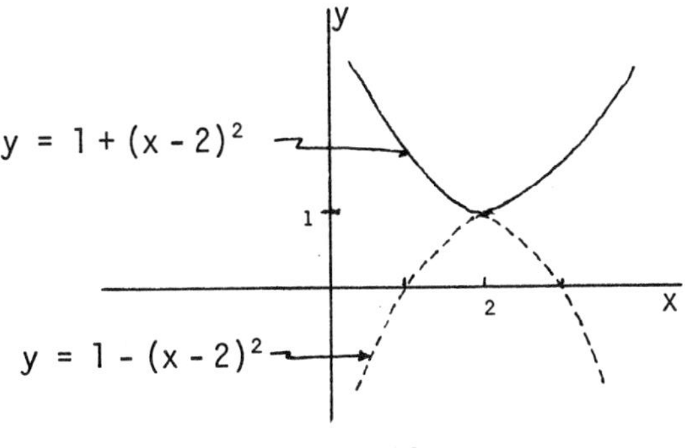

Figure 4.24

Now consider another curve:

- A standard form of the RECTANGULAR HYPERBOLA is given by the solution set of $y = \frac{k}{x}$ where k is a specified constant.

Figure 4.25 displays a rectangular hyperbola with a general k. At $x = k$, $y = \frac{k}{k} = 1$ and at $x = 1$, $y = \frac{k}{1} = k$. This particular sketch has $k > 1$, but $0 < k < 1$ gives the same general shape. The curves change from the first and third quadrants to the second and fourth where $k < 0$. The rectangular hyperbola gets closer and closer to the axes as x moves away from the origin, or as x gets very close to the origin. When this occurs we say that the curve becomes ASYMPTOTIC to the axis. Asymptotes are presented in more detail in the next chapter. That branch of the rectangular hyperbola which lies in the first quadrant interests biologists because the curve continues to decrease, a property which appears many places in biology. It also turns up in the generalized version of Boyles gas law which relates the volume, temperature and pressure of an ideal gas by

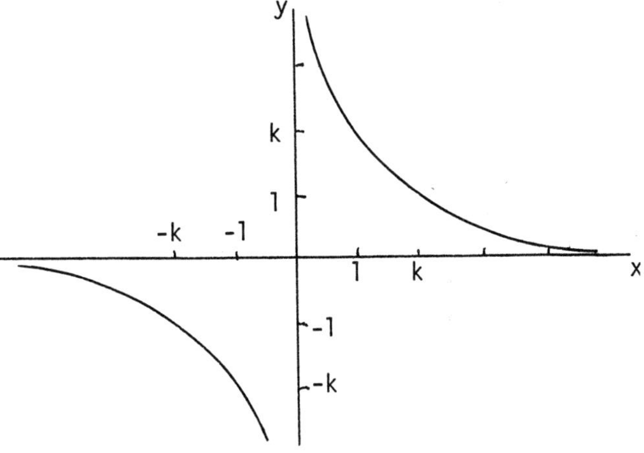

Figure 4.25

PV = C where P is pressure, V is volume and C is a constant and that gas remains at a constant temperature.

EXAMPLE 4.20: Figure 4.26 presents the graph of the hyperbola specified by the function $y = 4/x$, or equivalently by $xy = 4$. Although this hyperbola has a branch in both the first and third quadrants, interest usually centers on the positive values of x, which produce the curve in only the first quadrant. ▮

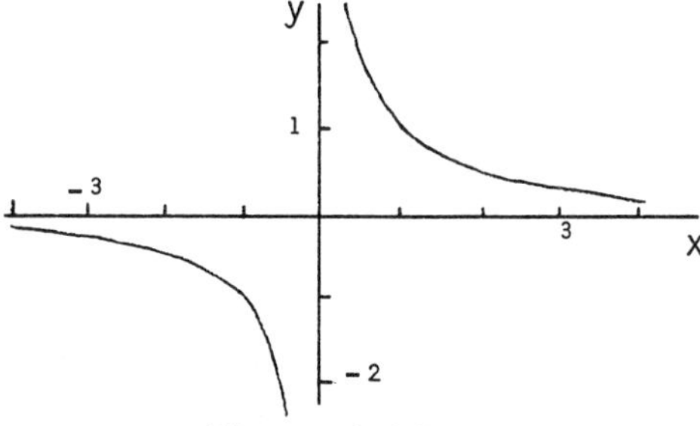

Figure 4.26

Finally consider a third curve:

- The standard form of a CIRCLE with its center at (h,k) and a radius of r is given by the solution set of $(x - h)^2 + (y - k)^2 = r^2$.

Figure 4.27 shows a circle centered at (h,k) where $h < 0$ and $k > 0$. The circle is a familiar curve which satisfactorily represents certain relations in biology such as distance an animal moves away from a point in a fixed time. A segment of this curve also approximates other relations.

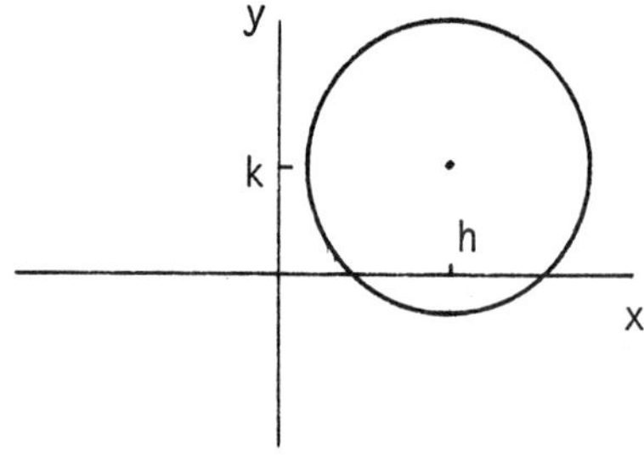

Figure 4.27

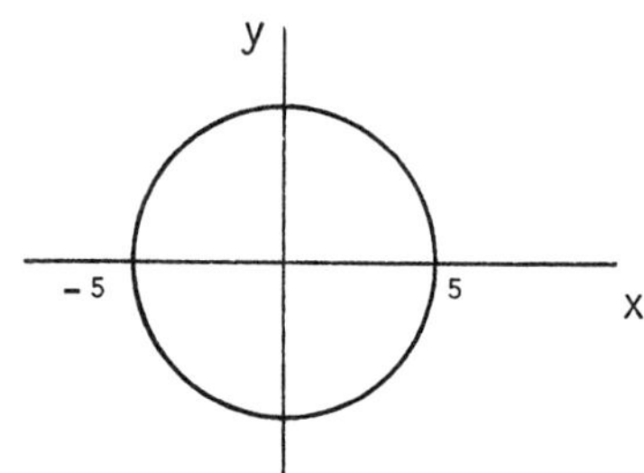

Figure 4.28

EXAMPLE 4.21: Figure 4.28 displays the graph of the circle defined by $x^2 + y^2 = 25$. This also could be specified by the two functions $y_1(x) = \sqrt{25 - x^2}$ and $y_2(x) = -\sqrt{25 - x^2}$ because $x^2 + y^2 = 25$ implies that $y^2 = 25 - x^2$. ▮

Now consider all of these curves together. Both the parabola which opens up or down, and the hyperbola define functions because each value of x has one associated value of y. The circle describes a relation which can be represented by two functions; one for its upper half, another for its lower half. The following examples illustrate these ideas in real situations.

EXAMPLE 4.22: The quantity of an antigen which precipitates during an antigen-antibody reaction depends on the quantity of the antigen present. For a constant quantity of antibody, Figure 4.29 depicts the relation between antigen quantity and the rapidity of the antigen-antibody precipitation. Note that this has the shape of a parabola with $a > 0$. ▌

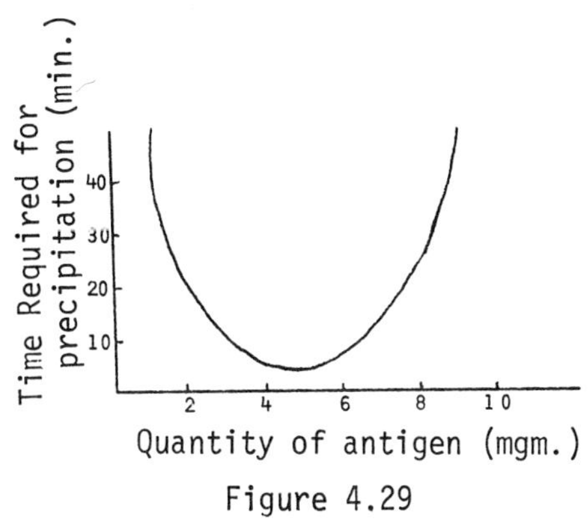

Figure 4.29

EXAMPLE 4.23: Various animals reproduce over a period of time with a peak sometime in the middle. Figure 4.30 shows the propagation of the prairie grouse, obviously not exactly following any of the curves we have discussed. Nevertheless, a parabola opening downward and centered around June 11 will closely approximate this function over most times of interest. ▌

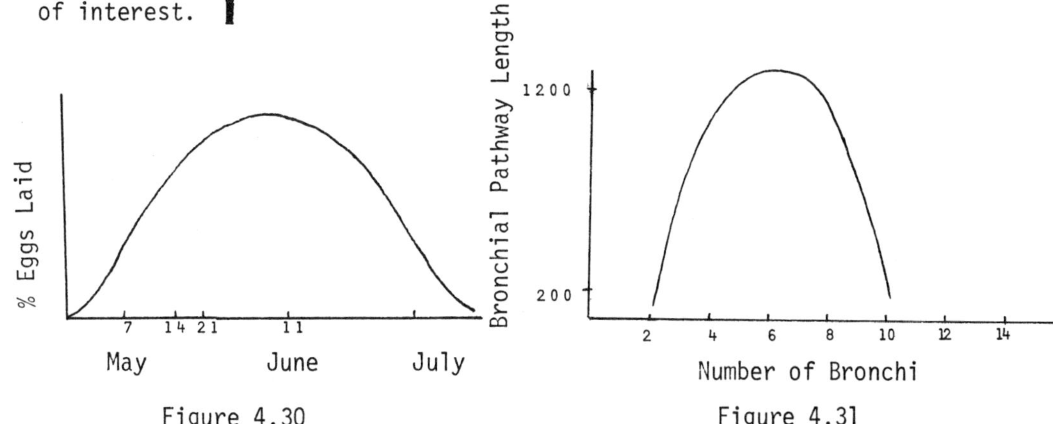

Figure 4.30

Figure 4.31

EXAMPLE 4.24: The number of bronchi depend on the length of the associated bronchial pathway. Specifically, Figure 4.31 displays the distribution of the terminal bronchi in the left lung of a dog relative to the bronchial pathway length. Again note that this has the shape of a parabola with $a < 0$. ▌

EXAMPLE 4.25: Boyles Law states that the pressure and volume of a gas have a constant product, namely, $PV = C$. Figure 4.32 shows this relationship in the positive quadrant because it makes sense to talk only of positive pressure and volume. This relationship can also be represented by the function $P(V) = C/V$, a hyperbola. ▌

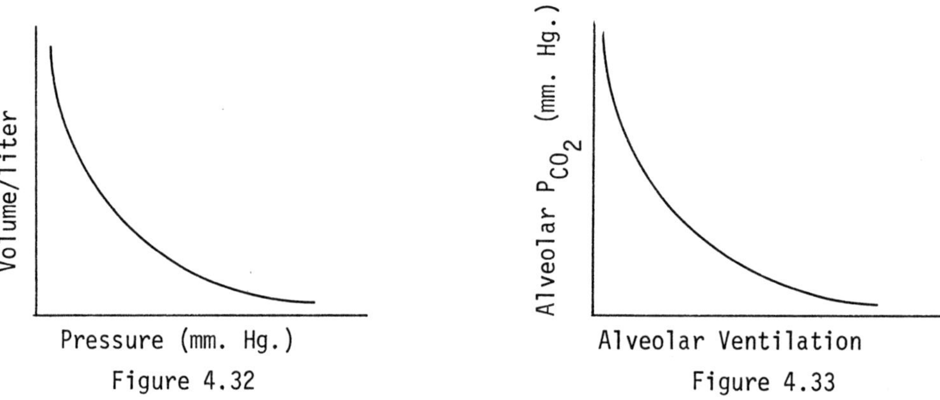

Figure 4.32

Figure 4.33

EXAMPLE 4.26: In respiratory physiology, alveolar ventilation relates to carbon dioxide pressure in the following way: $O_2 \times P_{CO_2} = k$. This relationship is depicted in Figure 4.33. Again this has the shape of a hyperbola with all values being positive and thus the graph lies in the first quadrant. ▌

The parabola was represented both in a standard form involving a deviation from its vertex, and as a second-degree polynomial. More generally, functions of the sort $y = a_0 + a_1x + \cdots + a_kx^k$, where k is a positive integer and the a's ($a_k \neq 0$) are constants, are called POLYNOMIALS of power or degree k. They find various applications in biology; polynomials of degree $k = 3$ having particular relation to volumes of objects. For example, in the cost analysis of timber in Chapter 2, Equation 2.1 is a polynomial of this sort.

When a parabola or circle is specified by a polynomial we can complete the square to locate the center of the circle or the vertex of the parabola. This technique is illustrated in the next two examples.

<u>EXAMPLE 4.27</u>: Give the coordinates of the center of the circle and its radius if $x^2 + 4x + y^2 - 6y - 3 = 0$.

For this relation we must complete the square in both x and y. Because the coefficients of the squared terms already are one, we begin by squaring one half of the coefficient of the linear terms, and add and subtract this quantity to obtain

$$(x^2 + 4x + 4 - 4) + (y^2 - 6y + 9 - 9) - 3 = 0 \ .$$

If the coefficients of the squared terms were different than one, then you must divide both sides of the equation by that quantity which will make the coefficients one. Upon collecting terms we have

$$(x^2 + 4x + 4) + (y^2 - 6y + 9) - 16 = 0$$

or

$$(x + 2)^2 + (y - 3)^2 = 16$$

Recall that the general equation of a circle with its center at (h,k) and a radius of r is $(x - h)^2 + (y - k)^2 = r^2$. We have $h = -2$, $k = 3$ and $r = 4$. By completing the square, the task of graphing the circle becomes much easier. ∎

<u>EXAMPLE 4.28</u>: Locate the vertex of a general parabola by completing the square in the polynomial $y = Ax^2 + Bx + C$.

To make the coefficient of the squared term one, factor an A out of the right side of the equation:

$$y = A\left[x^2 + \frac{B}{A}x + \frac{C}{A}\right] .$$

Next take 1/2 of the coefficient of the linear term in x, square that result, and add and subtract this square:

$$y = A\left[x^2 + \frac{B}{A}x + {\frac{B}{2A}}^{2} - {\frac{B}{2A}}^{2} + \frac{C}{A}\right] .$$

$$y = A\left[\left[x + \frac{B}{2A}\right]^2 - \left[\frac{B}{2A}\right]^2 + \frac{C}{A}\right] = A\left[\left[x + \frac{B}{2A}\right]^2 + \frac{4AC - B^2}{4A^2}\right]$$

or

$$y = A\left[x + \frac{B}{2A}\right]^2 + \frac{4AC - B^2}{4A}$$

and finally

$$y - \frac{4AC - B^2}{4A} = A\left[x + \frac{B}{2A}\right]^2 .$$

Recognizing that the original equation describes a parabola, we can now locate the vertex. In the standard form $y - k = a(x - h)^2$, the vertex lies at (h,k) so $h = -\frac{B}{2A}$ and $k = \frac{4AC - B^2}{4A}$. This illustrates completing the square on a quadratic for the case of a parabola. ▌

EXERCISES

Graph the following curves. If the curve is a parabola give its vertex; if the curve is a circle, give its center and radius; if the equation describes a straight line, give the slope and intercept.

1. $y = x^2 + 2$
2. $xy = 4;\ x > 0$
3. $y = x^2 + 6x + 11$
4. $3x + 2y = 7$
5. $y = x^2 - 6x - 9$
6. $y = x^{1/2};\ x \geq 0$
7. $y = 3x^2 + 5x - 4$
8. $2x^2 + 2y^2 + 8x = 0$
9. $x^2 + 10x + y^2 - 10y = 100$
10. $xy = -6;\ x \neq 0$
11. $-2y^2 + 3y - x + 4 = 0$

12. The following set of data was gathered in a study of how a particular gas behaves at a constant pressure. The readings give temperature in °K and volume in ml.

T (°K)	274	137	548	685	412
V (ml.)	500	250	1000	1250	750

Plot volume versus temperature. Then try to deduce an equation which descirbes the behavior of the gas.

13. The set of data given below represent the temperature of a compost pile on various days after building the pile and turning it. Plot temperature versus time, using time as the independent variable, and connect the points with lines. Does the graph define a function? If not, explain why.

Temperature (°F)	130	151	178	145	163	157	139	120
Time (Days)	1	3	5	7	9	11	13	17

14. The following set of data was gathered in a study of how a particular gas behaves at a constant temperature. The readings give the pressure P in atmospheres and volume V in ml.

P(atmospheres)	6	12	3	2.4	4	2	8
V (ml.)	500	250	1000	1250	750	1500	375

Plot volume versus pressure. Then try to deduce an equation which describes the behavior of the gas.

15. Find the equation of the circle which has its center at (1,6) and which is tangent to (just touching) the line $x - y = 1$.

16. Find the equation of a circle with:
   a. Its center at (-2,5) and tangent to the line $x = 8$.
   b. Its center at (-3,-5) and tangent to the line $12x + 5y - 4 = 0$.

4.5 POWERFUL FUNCTIONS

Two important functions in biology involve the independent variable with an exponent. These functions could have been discussed in the last section because they define curves, but they have a special relation to logarithms which we will pursue in the next section. The first subsection below deals with the kind of powerful function where the independent variable is raised to a constant power while the second subsection involves the independent variable as the exponent.

4.5a POWER FUNCTIONS

- A POWER FUNCTION is given by an equation of the type $y = bx^k$ where b and k are constants.

This family of functions takes on a variety of shapes depending upon the constant power k. All go through the origin for $k > 0$, but are asymptotic to the x and y-axis for $k < 0$. Figure 4.34 shows sketches of several of these curves with differing values of k. Notice that $k = 1$ produces a straight line through the origin ($y = bx$) while $k = 0$ produces the horizontal line $y = b$. The values of $k > 0$ produces curves which rise as x increases, while $k < 0$ results in the opposite; $0 < k < 1$ produces curves which increase more and more slowly as x increases while the $k > 1$ produces curves which go up faster and faster as x increases.

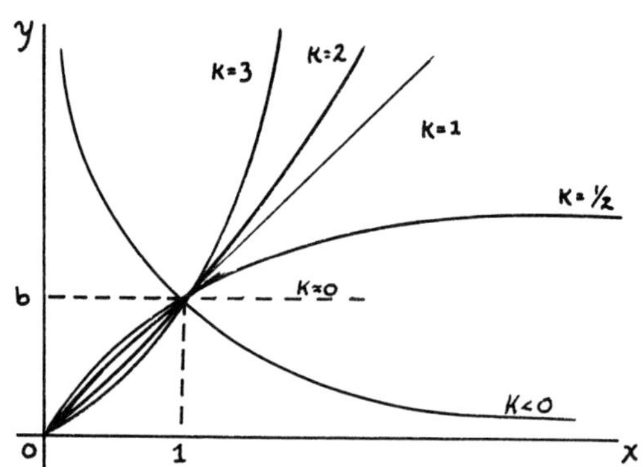

Figure 4.34

These curves, sometimes called ALLOMETRIC GROWTH CURVES, occur throughout biology. Various organisms grow at a rapid rate for a period of time and then slow down as the organism ages. Thus, an independent variable of age and a dependent variable of size often are related by a power curve with $0 < k < 1$. Such an example appears in Example 4.29 below.

These power functions have a special relation to functions we considered in the last section. If a polynomial has all constant equal to zero except the one associated with the highest power, then the polynomial produces a power type curve, except that the power must be an integer. Thus, a power curve with an integer as its constant power also is a polynomial.

EXAMPLE 4.29: Figure 4.35 shows the relationship between the volume of a fir tree and its age. The general shape of the curve is similar to the curves with $k > 1$ in Figure 4.34. ▮

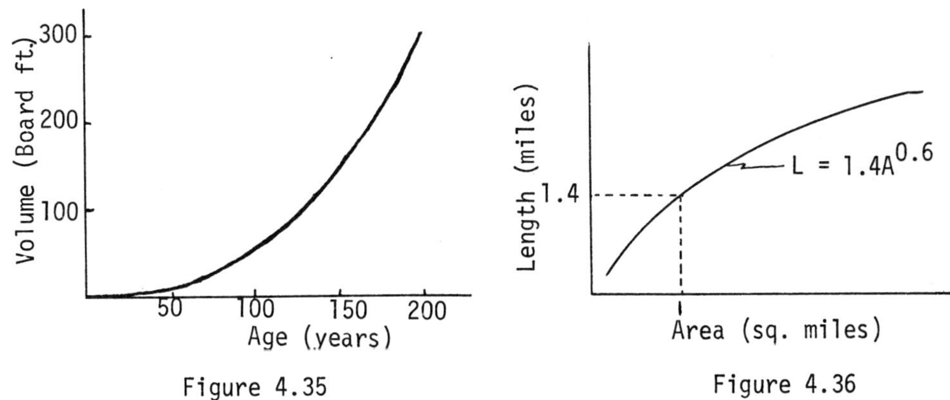

Figure 4.35

Figure 4.36

EXAMPLE 4.30: The length (L) of a stream channel has a fairly clear relation to the area (A) which it drains. In particular, $L = 1.4\ A^{0.6}$ is one such relation. This power function specifies the way in which length has to increase in order to drain an increasing area. Of course, this relation represents the average behavior of many stream channels over many areas because any one particular stream channel drains a specified area. This should be regarded as an approximation to a true relation. A specific stream channel may manifest modest deviation from it because of local geological phenomena. Nevertheless, if the data points are plotted on the graph, they would lie both above and below the curve, and show the general trend displayed in Figure 4.36. ▮

EXAMPLE 4.31: A nerve operates by creating electrical impulses in response to the specific stimulus which excites it. For example, the amount that a pen is depressed on a monkey's leg (or your arm) is translated into electrical impulses so he (or you) can perceive the sensation. The data which give rise to Figure 4.37 resulted from the study of a single afferent fiber of the saphenous nerve in the thigh of a monkey. Thirty skin indentations were delivered at each specified intensity, one every three seconds. The number of impulses in a specified period of time was recorded. The various curves give the average number of impulses per time interval. This phenomenon is governed by a relation which can be represented by the allometric equation $r = qs^n$ where r = response number and s = stimulus intensity in microns, with different constants applying to each electrical frequency. ▮

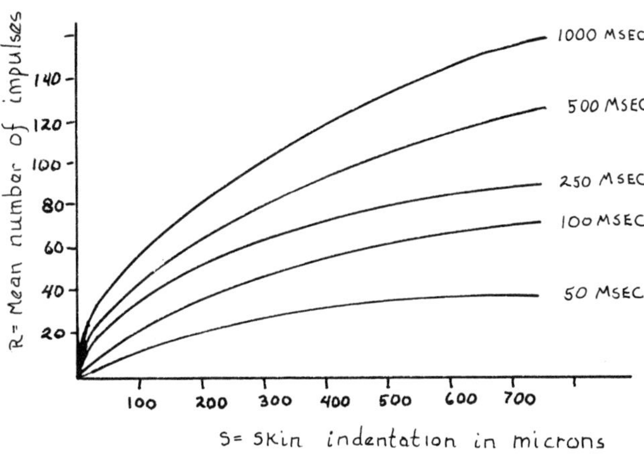

Figure 4.37

4.5b EXPONENTIAL FUNCTIONS

The family of exponential functions occurs in the description of many biological phenomena such as the growth or death of a population, the buildup or decay of daughter products of radioactive isotopes, and more generally, any process where the rate of change of a response is proportional to the current size of the response. This family is specified by:

- An EXPONENTIAL FUNCTION is given by an equation of the type $y = bc^{kx}$ where b, c and k are constants. Often c = e where e is the base of natural logarithms.
- The EXPONENTIAL NOTATION specifies $e^{f(x)}$ as $\exp[f(x)]$.

Figure 4.38 sketches this family of curves for various values of the parameter k. Notice how the sign of the multiplier in the exponent dominates the nature of the function. A negative coefficient puts the term with the power in the denominator while a positive coefficient leaves it in the numerator. Thus, when the coefficient of x is positive ($k > 0$), the curve increases as x increases while the curve for $k < 0$ decays to 0.

The form of $y = be^{kx}$ is common in most applications. (Recall that $e = 2.71828\cdots$.) As we progress, real problems will lead to complicated exponents on e. The exp notation makes these simplier to read and write.

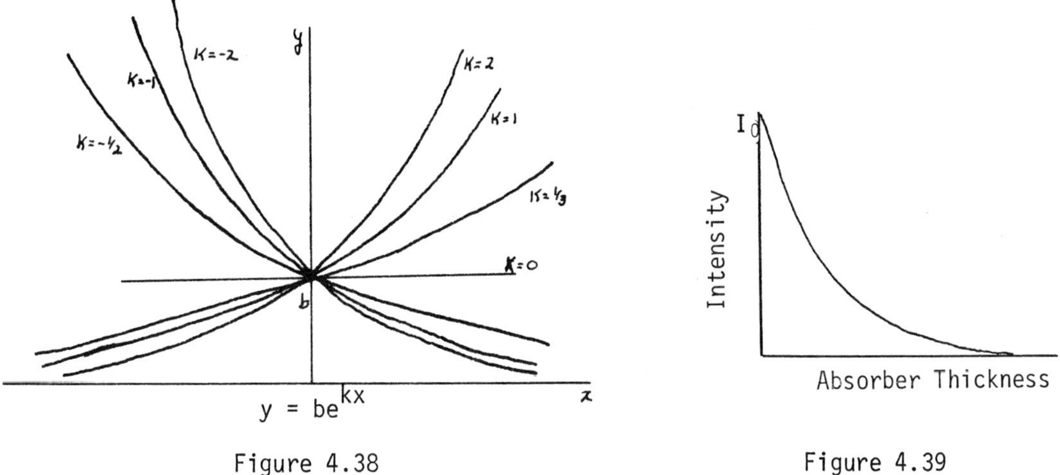

Figure 4.38

Figure 4.39

When you compare exponential curves in Figure 4.38 to the power curves in Figure 4.34, they may appear similar. Both families curve upward for some parameter values, but downward for others. Note three distinct differences: Because $e^0 = 1$, all of the exponential curves pass through (0,b) while the power curves go through (1,b); the exponential curves exist for all x while most power curves exist only for $x > 0$; and all of the upward curving exponential curves ($k > 0$) increase faster as x increases while for $0 < k < 1$ the power curves flatten out.

Section 4.7 will deal exclusively with growth, competition and survival curves. There you will see various examples of exponential curves so that the examples here will be more brief than usual. When the independent variable is time, the x given in the definition above is replaced by t for obvious notational meaning.

<u>EXAMPLE 4.32</u>: The exponential function describes the absorption of photons in an absorber. If a photon beam of initial intensity I_0 passes through an absorber of thickness x, then the residual intensity satisfies $I = I_0e^{-\mu x} = I_0 \exp[-\mu x]$. The constant μ changes with the material

used as an absorber. A graph of one such function appears in Figure 4.39. It shows only $x > 0$, because thickness cannot be negative. ▌

EXAMPLE 4.33: A drug is eliminated from the circulatory system by organs such as the liver, and by consumption by other organs that use it. This sort of decay frequently follows an exponential curve of the form $A = A_0e^{kt}$ with $k < 0$. Consider the following set of data, and its plot in Figure 4.40. Note that $A = A_0$ at $t = 0$ ($A_0 \doteq 1$ here).

t	.1	.3	.6	.8	1.4	1.6	2.3	3.0	3.5	4.2	4.8	5.3	5.9
A	.97	.90	.83	.77	.69	.60	.48	.41	.34	.32	.24	.19	.17

At this point we have not explained how to find the equation which describes the data given. We will return to this example later and calculate the values of A_0 and k. Note also that this is the graph of real data and the points DO NOT FOLLOW PRECISELY the exponential function due to the errors common to such experiments. ▌

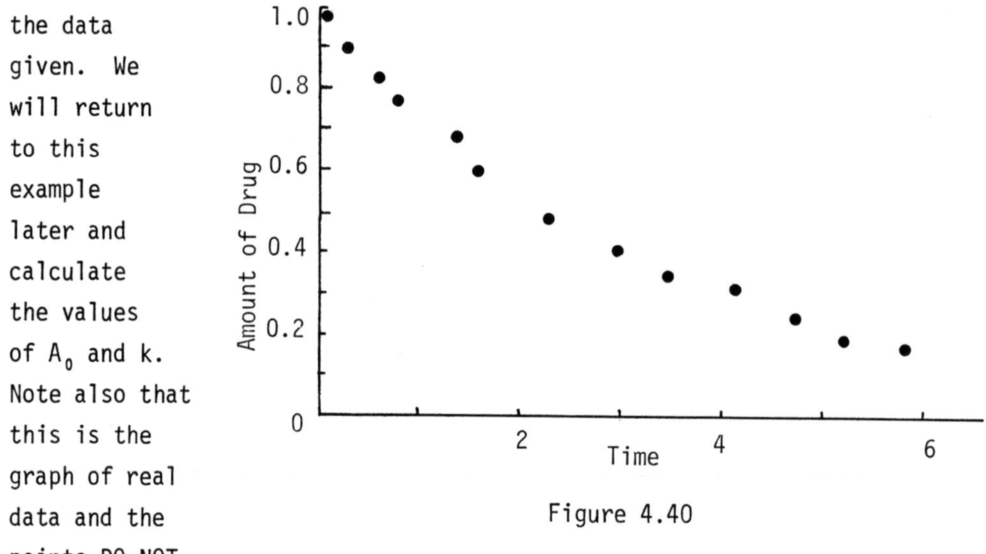

Figure 4.40

4.6 TAKING THE POWER OUT OF POWERFUL FUNCTIONS

The power and exponential functions both involve exponents or powers. The logarithmic transformation changes an exponent on the regular scale to a multiplicative coefficient on the logarithmic scale. This section will explain how logarithms can be used to simplify examination of the power and exponential functions. (If you feel a need for review on logarithms, study the first part of Appendix I before you go any further in this section.) Special graph papers have been created so that the exponential and power functions appear as straight lines on them; we will explain their use later in this section. They provide a convenient way to estimate the parameters in known functions from data, and an effective way to choose a curve for a set of data.

Recall two properties of logarithms using any base b. First, a logarithm of a product equals the sum of the logarithms of the multipliers in the product:

$$\log_b(MN) = \log_b M + \log_b N \; ; \tag{4.1}$$

and secondly the logarithm of a number raised to a power equals the power times the logarithm of that number:

$$\log_b(M^p) = p \log_b M \; . \tag{4.2}$$

For this section and most applications, the base is taken as either 10 or $e = 2.71828\cdots$ for which the notations log and ln, respectively, replace $\log_b$. The notation log refers to common logarithms and ln refers to natural logarithms.

Now, if we take the common log of both sides of a power function, $y = bx^k$, and use the relations given above in Equations 4.1 and 4.2 we have:

$$\begin{aligned} \log y &= \log bx^k \\ &= \log b + \log x^k \\ &= \log b + k \log x \; . \end{aligned} \tag{4.3}$$

If we let $b^* = \log b$, $y^* = \log y$ and $x^* = \log x$ then Equation 4.3 can be rewritten as

$$y^* = kx^* + b^* \; ,$$

the equation of a straight line involving x^* and y^* as variables. Figure 4.41 shows a power function plotted on an ordinary scale on the left and on a log-log scale on the right. Note the use of terms here:

The familiar, equally spaced scale, is called a linear scale in contrast to the log side.

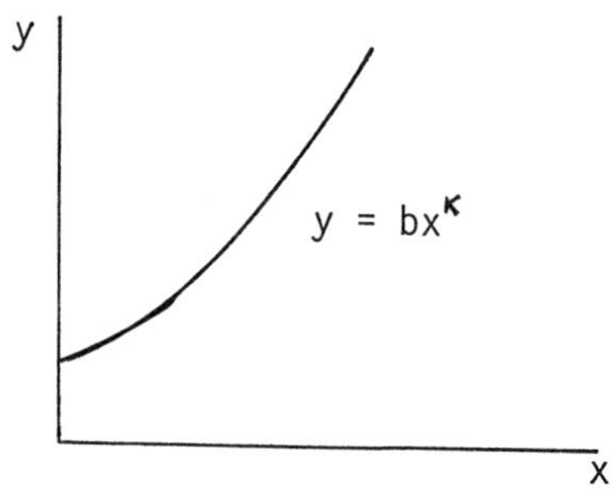

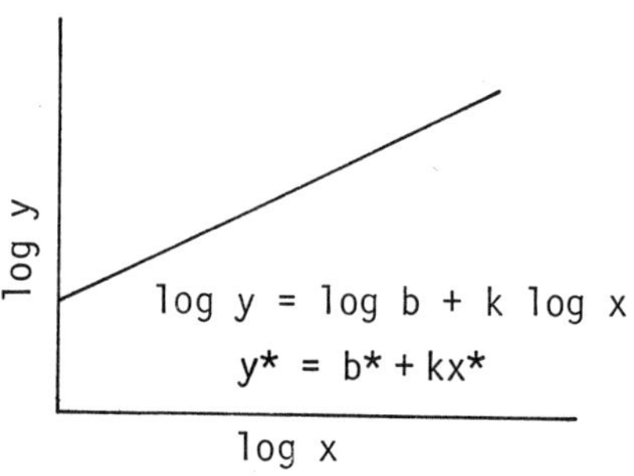

Figure 4.41

The intercept and slope on the logarithmic scale have well defined relations to the two parameters in a power function. The slope of the line on the log scale equals the power k of x in the power function and the intercept on the log scale equals the logarithm of the multiplier b in the power function.

EXAMPLE 4.30 (continued): The length (L) of a stream channel relates to the area (A) which it drains through the power function $L = 1.4\ A^{0.6}$. Equivalently, by taking the common logarithm of both sides of the equation we have:

$$\log L = 0.6 \log A + \log 1.4$$

$$= 0.6 \log A + 0.146$$

or

$$L^* = 0.6\ A^* + 0.146\ , \qquad (4.4)$$

where $L^* = \log L$, $A^* = \log A$ and $\log 1.4 = 0.146$. From this relationship we can establish a table of values of A and log A, which in turn give log L and then L.

A	1	5	10	20	50
log A	0.000	0.699	1.000	1.301	1.699
log L	0.146	0.576	0.746	0.927	1.165
L	1.400	3.677	5.573	8.448	14.639

The entries in this table can be calculated directly using Equation 4.4. For example, when A = 10, A* = log A = 1, so L* = log L = 0.6 x 1 + 0.146 = 0.746, and by looking up the antilogarithm we obtain L = 5.573. We can draw the line given by Equation 4.4 from its slope and intercept or also from values in the table. Either method we choose, Figure 4.42 shows the results. On the left is a linear plot while on the right is the logarithmic plot. ▌

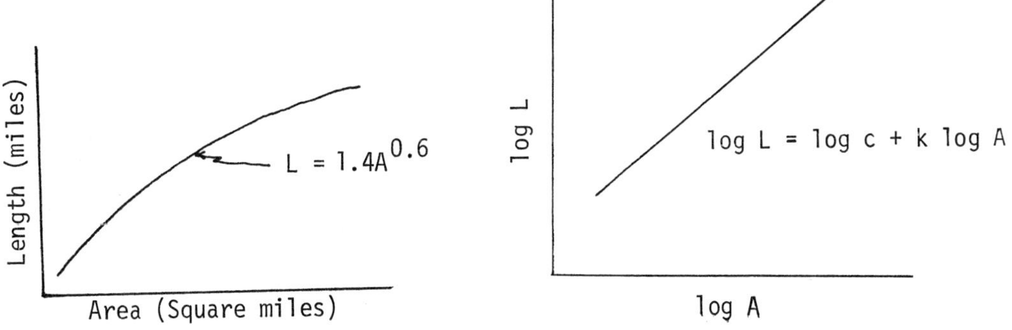

Figure 4.42

The line on the log-log plot can be drawn easily when the coefficients of the power function are known, but if you want to plot data on that scale, you would have to evaluate log x and log y for each point. A scheme has been developed to avoid this effort. The idea revolves around this reasoning: Each x has a unique log x and vice versa, so a correspondence like the one indicated in Figure 4.43 can be set up.

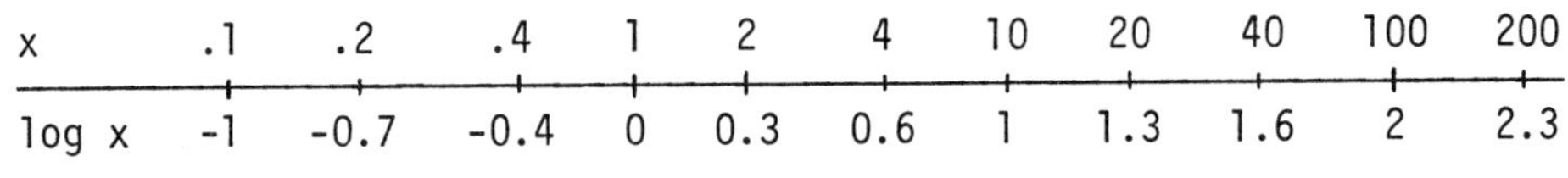

Figure 4.43

This has log x equally spaced on the lower side of the line with the corresponding value of x above. Observe that on the left part of the scale, which corresponds to small values of x (between 0 and 1), the scale has been stretched, but for x > 1 it has been compressed. If a scale of x is drawn and labeled like the upper line, then the points can be plotted on a scale of log x without having to obtain the log of each value of x. The scale of y can be treated in exactly the same way to allow you to make a log-log plot without taking logarithms. The paper created for this purpose is called logarithmic paper and will be illustrated in the following example.

EXAMPLE 4.30 (continued): In an earlier consideration of this example we established the stream length and area drained for several different points. These data points are plotted below in Figure 4.44.

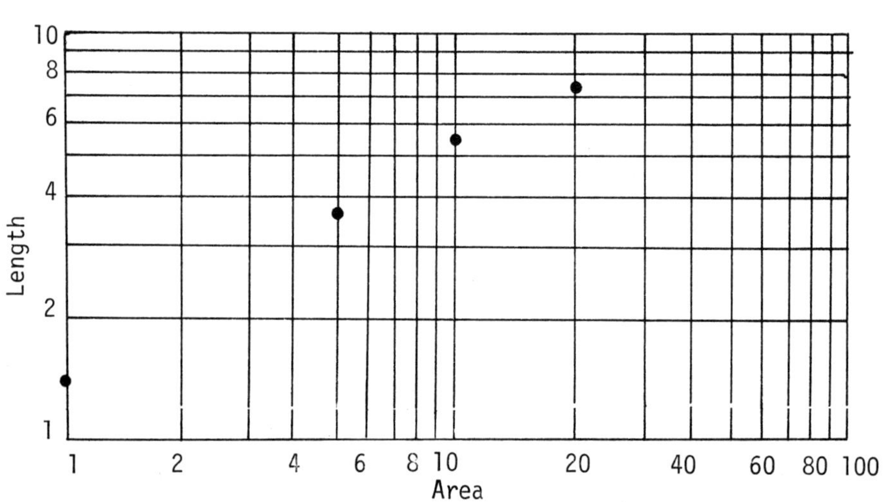

Figure 4.44

For example to plot the point (5,3.677), we plot the point A = 5 on the abscissa and L = 3.677 on the ordinate axis. The construction of the logarithmic paper makes the location of this point equivalent to the location of the logarithms of the numbers, plotted on ordinary scales. Thus even though we plotted the number (5,3.677), the paper gives us the result of plotting (log 5,log 3.677) on an ordinary scale. Observe that if we had just plotted the data in Figure 4.44 but did not have the relation we could obtain it from the plotted data. The equation of a straight line on a log-log plot would be $\log A = \log C + k \log L$. Now log C would be the intercept and k would be the slope of the line drawn on the graph. ▮

We can apply almost identical reasoning to the exponential function. It was expressed as $y = bc^{ax}$ and as $y = be^{kx}$. Using common logarithms on the first form gives an equation

$$\log y = \log b + ax\log c \ . \qquad (4.5)$$

If we let $y^* = \log y$, $b^* = \log b$, and $m = a\log c$ then Equation 4.5 becomes

$$y^* = mx + b^* \ ,$$

again the equation of a straight line. In most biological applications the power function translates into a straight line when both x and y are transformed to the log scale, but the exponential function becomes linear in log y and x, not log x. The log-log graph paper used to linearize power functions is replaced by paper called semi-log graph paper in order to handle the exponential function. The semi-log graph paper has x on the linear scale but y is labeled to correspond to a log scale.

EXAMPLE 4.33 (continued): Recall that this example concerned the disappearance of a drug from the circulatory system. A symbolized the fraction of the drug in a given blood sample at time t hours. The data is reproduced below:

t	0.1	0.3	0.6	0.8	1.4	1.6	2.3	3.0	3.5	4.1	4.7	5.3	5.9
A	.97	.90	.83	.77	.69	.60	.48	.41	.34	.32	.24	.19	.18

This set of data was plotted on an ordinary linear scale in Figure 4.40, but it is plotted on semi-log scales in Figure 4.45. As was pointed out previously, the data do not fall precisely on the line due to the errors encountered in the experiment. However in order to illustrate a

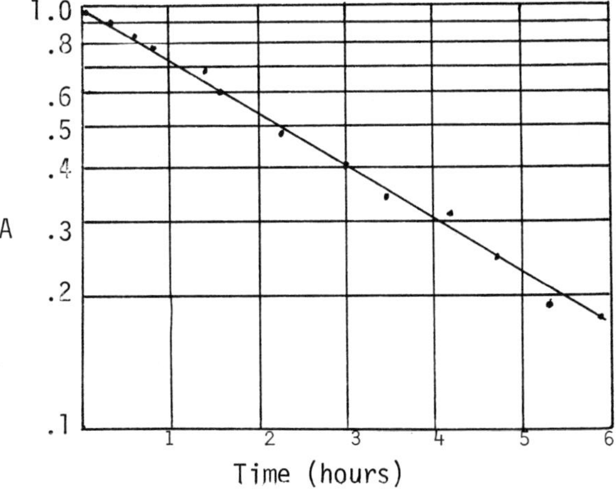

Figure 4.45

process in order to estimate the constants A_0 and k, we must first draw a line through the data points which appears to represent the trend of the data. (Precise statistical methods exist for obtaining this line but cannot be discussed here. So for our purposes, the free hand drawn line will suffice.) Because $A = A_0e^{kt}$, the equation of the line drawn through the points represents either

$$\log A = \log A_0 + (k\log e)t$$

or

$$\ln A = \ln A_0 + kt$$

depending upon whether data was plotted as log A versus t or ln A versus t. We will calculate the values of A_0 and k using both kinds of logarithms.

We can pick two ordered pairs (t,A) on our fitted line to calculate the slope and A_0. To pick the two points for our calculations, we must first plot the data and then draw the line which best fits the data as was pointed out previously. We cannot use actual data points because suppose we picked the two actual data points (1.4,0.69) and (1.6,0.60). By drawing a line through these two points on Figure 4.45 you will observe that this line DOES NOT describe the general trend of the data. Suppose we pick pairs (1.6,0.62) and (3.0,0.41) from the line. Notice that the A value at t = 1.6 on the line differs from the value given in the data table, because the data point lies off the line. Using common logarithms with these pairs we can calculate the slope (m) of the line as

$$m = \frac{\log 0.62 - \log 0.41}{1.6 - 3.0} = \frac{(0.79239 - 10) - (9.61278 - 10)}{-1.4} = -0.12829 \ .$$

Because klog e is the slope, we have

$$k\log e = -0.12829 \text{ and } k = \frac{-0.12829}{\log e} = \frac{-0.12829}{0.4343} = -0.2954 \ .$$

To estimate A_0 we have

$$\log A_0 = \log A - (k\log e)t \ .$$

Any ordered pair (t,A) that lies on the line satisfies the equation, so we can use the point (1.6,0.62) to get

$$\log A_0 = \log 0.62 - (-0.12829)(1.6)$$
$$= 9.79239 - 10 + 0.205264$$
$$= 9.99765 - 10 \ .$$

from which $A_0 = 0.9946$. Both A_0 and the slope can be observed from the graph to give us confidence that our answers are in the appropriate range. However, calculation of the values as illustrated above is always more precise than trying to read them directly from the graph.

Calculating these values using natural logarithms we have

$$k = \frac{\ln 0.62 - \ln 0.41}{1.6 - 3.0} = \frac{-0.47804 - (-0.89160)}{-1.4} = \frac{0.41356}{-1.4} = -0.2954 \ .$$

To estimate A_0 we have

$$\ln A_0 = \ln A - kt \ .$$

Proceeding as above, we can again use the point (1.6,0.62) to give

$$\ln A_0 = \ln 0.62 - (-0.2954)(1.6)$$
$$= -0.47804 + 0.47264$$
$$= -0.0054 \ .$$

Upon taking the antilogarithm, this gives $A_0 = 0.995$.

The two calculations agree within rounding error giving the equation

$$y = 0.995\ e^{-0.2954t} \ .$$

Note that had we drawn the line a little differently we would obtain slightly different values for A_0 and k. In fact, the data lack the precision indicated by the number of decimals carried in the above calculations. The digits were carried throughout so you could follow the details of the process. Practically we have found that

$$y = e^{-0.3t} \ . \ \blacksquare$$

We have suggested that data can be plotted and a straight line drawn through it without recourse to precise statistical techniques. But we just noted that different lines would give slightly different estimates of the parameters. So long as the data fits the specified relation reasonably well, this produces no serious problems. If, however, the data has substantial scatter, then drawing a good straight line becomes troublesome. Biologists tend to pick up a transparent ruler and place it through the data so that approximately half the data is above and half below the ruler throughout the entire range of the data. This turns out to be a bit tricky, so if you try it you may find yourself a bit frustrated. The statistical topic of regression, which lies outside the scope of this book, deals with this matter. Thus, if you find that you need to fit a curve to data with substantial variation you will need to learn something about least squares concepts and associated topics in regression.

So far we have proceeded supposing that you knew which curve applied to your data. When it comes time to pick a curve to fit a set of data, an alternate strategy can be invoked. The data can be plotted on the regular arithmetic scale and if it appears to have a curved shape, then the data can be plotted on both log-log and semi-log graph paper. If the data appears to follow a straight line on the log-log paper, the power function should be seriously considered, but if it follows a straight line on the semi-log graph paper, an exponential function should be considered. If the data approximately follow an underlying relationship, this technique will cause no confusion because only one of these graphs should display a straight line. If the data contains a substantial amount of variation away from the true relation, this approach may not always yield a clear answer because each of the graphs could have a somewhat linear appearance. Precise statistical techniques for handling these problems go far beyond the scope of this book, but techniques do exist and would be considered in a comprehensive course on regression analysis.

One feature of the graph papers having logarithmic scales should be noted. Most of the graph papers correspond to the base 10 logarithms with the result that one cycle on the logarithmic scale ranges from 1 to 10, another cycle from 10 to 100, a further cycle from 100 to 1000, and so on.

Each cycle need not correspond to these particular multiples of ten, but in picking graph paper you should have as many cycles along each axis as you have changes in the power of ten. In other words, if you want to plot data which ranges from 1.5 to 1200 you will need four cycles along that axis with the first cycle ranging from 1 to 10, the second from 10 to 100, the third from 100 to 1000 and the final one from 1000 to 10,000. The 1.5 will be plotted in the first cycle while the 1200 would be plotted in the fourth cycle. You should look at commercially available log-log and semi-log graph paper to appreciate the number of different scale sizes which can be obtained. They differ primarily in the number of cycles available on a graph paper, and this closely relates to the range of values that you want to be able to plot.

EXERCISES

1. A bacteriocidal (antibacterial) agent was added to a nutrient broth containing 10^7 viable bacteria per milliliter. The following table displays the number of viable bacteria remaining at various times thereafter:

time in minutes	0	10	20	30	40	50	60	70
number of bacteria	10^7	10^6	10^5	10^4	10^3	10^2	10^1	1

   a. Plot the number of bacteria as a function of time.
   b. Plot the logarithm of the number of bacteria as a function of time, or plot the data on the appropriate log type paper.
   c. Establish an equation relating the number of bacteria, N, to the time, t.

2. For the equation $y = 0.5\ x^2$,
   a. Plot y versus x, $0 \leq x \leq 5$.
   b. What transformation would linearize this equation?
   c. Plot the curve on that transformed scale which you established in part b.

3. Given the equation $y = 2e^{-t}$:

t	1	1.5	2	2.5	3
y	.74	.45	.27	.16	.10

a. Plot y versus t.

b. Plot ln y versus t.

c. From your plot in part b, establish the equation relating y and t. Does it give back the original equation?

4. Bacteria, placed in a nutrient broth, undergo a period of adaptation during which they do not reproduce but instead form enzymes necessary for them to use the food in the broth. After the period of adaptation they begin reproducing until they have depleted the food supply, after which they begin to die off. An idealized experiment was carried out and the following results were obtained (t = time in hours and N = number of bacteria):

t	N	t	N	t	N
0	1000	3.5	3.2×10^5	7	10^9
0.5	1000	4	10^6	7.5	10^9
1	1000	4.5	3.2×10^6	8	10^9
1.5	3200	5	10^7	8.5	10^9
2	10^4	5.5	3.2×10^7	9	7.5×10^8
2.5	3.2×10^4	6	10^8	9.5	5.6×10^8
3	10^5	6.5	3.2×10^8	10	2.1×10^8
				10.5	1.6×10^8

a. Plot the number of bacteria versus time.

b. Plot the logarithm of the number of bacteria as a function of time.

c. In which of the curves can you clearly distinguish the period of adaptation?

d. In which of the two curves can you measure the duration of time during which the bacteria reproduce exponentially? (Hint: What does an exponential function look like when plotted on semi-logarithmic paper?)

4. Continued
   e. Establish an equation relating the number (N) of bacteria to time during the exponential phase of reproduction.
   f. Over what values of t can you use the equation found in part e? Why must there be restrictions?

5. How would you estimate the constants q and n in Example 4.31?

4.7 FUNCTIONS IN REALITY

Individual organisms and populations both grow, but a population can also decline. This section presents functions which describe the size of a population. The exponential function occupies a central role in such situations. The exact population size has only a probabilistic relation to time so we really will consider the average or expected value of this variable. This approach satisfactorily demonstrates the general behavior of large populations without undue attention to probabilistic details.

Growth occurs in all kinds of populations, including such diverse ones as dodder (an orange parasitic vine) or aphids (a kind of insect which sucks juices from a plant) invading an alfalfa field, algae in a lake, tribolium (a flour weavil), malarial mosquitos, polio virus, wild mallard ducks, or humans. They share the characteristic that the population will grow from its present size depending upon the environmental conditions.

One general growth function is given by an exponential function of the form $N = N_0e^{kt}$ where k is a constant, t represents time and N_0 symbolizes the number present at $t = 0$. Many populations have a fixed proportion of organisms which reproduce in some fairly consistent manner. Thus, the growth of a population relates directly to the size of the reproducing part and so growth changes in proportion to the current population size. In the absence of external constraints, growth in proportion to current size leads to the exponential function. You will see how to step from the growth condition to the size function once the needed tools have been developed. The form set out above uses e, the base of natural logarithms because it arises naturally in this context. However, other forms also are used. For example, microbiologists frequently write this relation as $N = N_0 2^{t/t_0}$. Why would they do this? It simply involves a basic phenomenon of microorganisms: A microorganism reproduces by dividing into two daughter organisms. If t_0 represents the time from cell division to cell division, then t/t_0 gives the number of time periods so $2^{t/t_0}$ gives the number of cells which have resulted from each individual in the original population.

The exponential function was discussed in the last section, but its origin in population studies has an interesting and noteworthy aspect. Thomas Robert Malthus (1776-1834) published a pamphlet in 1798 in which he pointed out the geometric or multiplicative nature of unrestrained population growth. Although he was not the first to perceive this phenomenon, he succeeded in publicizing it. In honor of this, exponential growth frequently is described as Malthusian growth and the parameter k sometimes is called the Malthusian growth parameter. Review Figure 4.38 which shows exponential functions for various values of the parameter k. Note that they all curve upward at an increasing rate for $k > 0$. Thus a population experiencing exponential growth eventually will overrun its habitat. This was Malthus' basic point.

Natural populations do grow, but as they approach the carrying capacity of their habitat the growth rate usually slows down. This is why previously the statement "growth in proportion to current size" was qualified by "in the absence of external constraints". A population overrunning its habitat becomes subject to numerous external forces. These reductions in growth have been recognized for many years and have led to the introduction of several other population growth functions:

- The LOGISTIC GROWTH CURVE is specified by the equation

$$N = N_0 \frac{1 + b}{1 + be^{-kt}} , \tag{4.6}$$

where b, k and N are positive constants.

- The GOMPERTZ GROWTH CURVE is specified by the equation

$$N = ae^{-be^{-kt}} , \tag{4.7}$$

where a, b, and k are positive constants.

- The VON BERTALANFFY GROWTH CURVE is specified by the equation

$$N = N_0(1 - be^{-kt})^3 , \tag{4.8}$$

where N_0, b and k are positive constants, and $b < 1$.

- The MONOMOLECULAR GROWTH CURVE is specified by the equation

$$N = \frac{N_0}{1 - b}(1 - be^{-kt}) , \tag{4.9}$$

where N_0, b and k are positive constants and $b < 1$.

All four of these curves have a period of accelerating growth from the initial population, and then level off to the maximum population supportable by the environment. For each curve the several parameters specify the curve's shape, the rate at which it approaches its maximum, and the population size supportable by the environment. The curves differ in how fast these transitions occur. We will examine them further as we develop mathematical tools in the next three chapters. Figure 4.46 shows the general shape of these curves. This sketch begins with an accelerating portion and then flattens out. For some parameter values the accelerating part may be absent, essentially occurring before $t = 0$. Note that all of these forms contain a term of the sort e^{-kt} while for exponential growth it was e^{kt}. This apparent change of notation

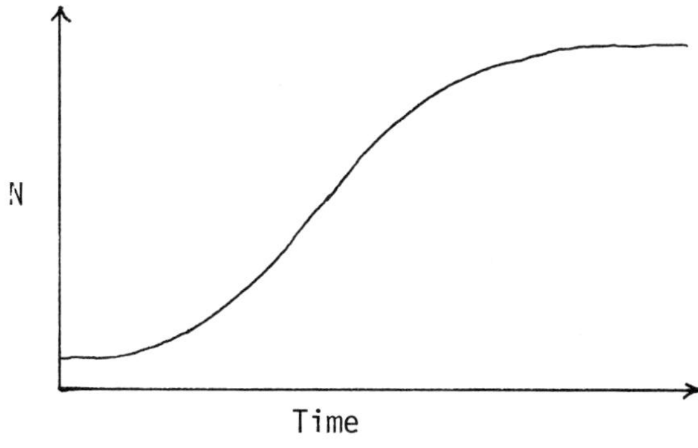

Figure 4.46

has a simple explanation. The exponent can be either positive or negative for the exponential, but it must be negative for these latter curves to make sense for otherwise they go to zero or become negative as time continues. The stated negative in the exponent emphasizes this fact.

The decline of a population or the decay of materials occupies as important a place in biology as does growth. Let us first consider decay. Of the several ways to describe decay, a very common form uses the time necessary for half the material to decay, often called its half life. Frequently decay follows an exponential function having a negative coefficient. If we let $t_{\frac{1}{2}}$ symbolize the half life of the substance, namely, for the time required for the original amount (A_0) to reach $\frac{A_0}{2}$, then

$$\frac{A_0}{2} = A_0 \exp[kt_{\frac{1}{2}}]$$

This expression gives $k = \frac{-\ln 2}{t_{\frac{1}{2}}} = \frac{-0.693}{t_{\frac{1}{2}}}$. Since the half lives of many materials, particularly radioactive substances, are well established, this equation provides a simple way to determine k. A use of this process is illustrated in the next example.

<u>EXAMPLE 4.34</u>: CARBON DATING. Archaeologists often want to age (establish the age of) artifacts or relics which they uncover. If the artifact was once part of a living thing, the method of carbon dating has proved quite helpful in establishing its age. Carbon dating rests on the assumption that the radioactive substance carbon 14, an isotope of carbon denoted by ^{14}C, has been present in the atmosphere at an essentially constant level for eons. Since any radioactive substance continually disintegrates, there must be some replenishment occurring to justify this assumption of an equilibrium level. New carbon 14 constantly is being produced by cosmic bombardment of nitrogen in the upper atmosphere, apparently at a sufficient rate to balance the disintegration taking place.

All living objects contain a large amount of carbon of which then a small fixed percentage would have been drawn in as ^{14}C because plants do not distinguish between ^{12}C and ^{14}C. If the level of ^{14}C has remained constant in the atmosphere, it seems reasonable to assume that an artifact being examined, say a cubic centimeter of wood, contained the same amount of ^{14}C when it was living as does a comparable piece of the same material now. When it died, the ^{14}C disintegration continued, but no new ^{14}C was taken in. When the artifact is counted with a geiger counter to establish its present rate of ^{14}C disintegration, a rate less than the original rate will occur. By associating the present rate with the correct point on the decay curve, the age of the artifact can be estimated.

Suppose that an artifact has ^{14}C disintegrating at a rate of 30% of the rate of a corresponding piece of material living today. This implies that the decay has depleted 70% of the original ^{14}C. We need to establish the time required for the original ^{14}C to decay this much. The half life of ^{14}C has been experimentally found to be approximately 5568 years. An exponential decay function with this half life would have $k = \frac{-0.693}{5568} = -0.000124$. The constant c in the exponential decay function $y = ce^{kt}$ gives the emission rate at time $t = 0$. Since the artifact radiates at 30% of this rate, we must solve the following equation for t:

$$0.30c = ce^{-0.000124t} .$$

Taking the natural logarithm of both sides of this equation we have

$$\ln 0.30 = -0.000124t ,$$

and so

$$t = \frac{\ln 0.30}{-0.000124} = \frac{-1.204}{-0.000124} = 9710 \text{ years}$$

To reiterate, this says that it has taken approximately 9710 years for the original ^{14}C to decay to 30% of its original emission rate. Thus, the archaeologist would conclude that the artifact is approximately 9700 years old. ▌

The survival of organisms in a declining population has some similarity to the decay of particles considered above: When a particle decays, it ceases to exist as it was before the decay. Similarly, when an organism dies, it ceases to exist as it was when it was living. Survival describes a very important aspect of a declining population and finds application throughout biology. The interpretation of the constants in a survival curve will vary across areas of biology, but all of them concern the probability that a randomly selected individual will survive. Broadly speaking survival curves can be divided into two classes, exponential and sigmoid. (Sigmoid curves are discussed below.) By now you have seen enough of the exponential functions that we need not explain them much further, other than to observe that in considering survival only the variables on the axis change from our previous examples. Exposure or stress becomes the independent variable. It may represent time, a concentration of a drug or a radioactive dosage, temperature, stress, or any of a multitude of other things which could influence the survival of organisms. The dependent variable represents the fraction of the population surviving that exposure. Under most circumstances this fraction decreases smoothly as exposure increases, in conformity with the biological observation that more individuals die as stress increases. The graphs in Figure 4.47 give both a linear and semi-log plot of the exponential decay curve. On the linear scale the curve can be regarded as describing the probability that a randomly selected individual will survive if it is subjected to the associated exposure. The semi-log plot linearizes this function and provides an easy way to examine it.

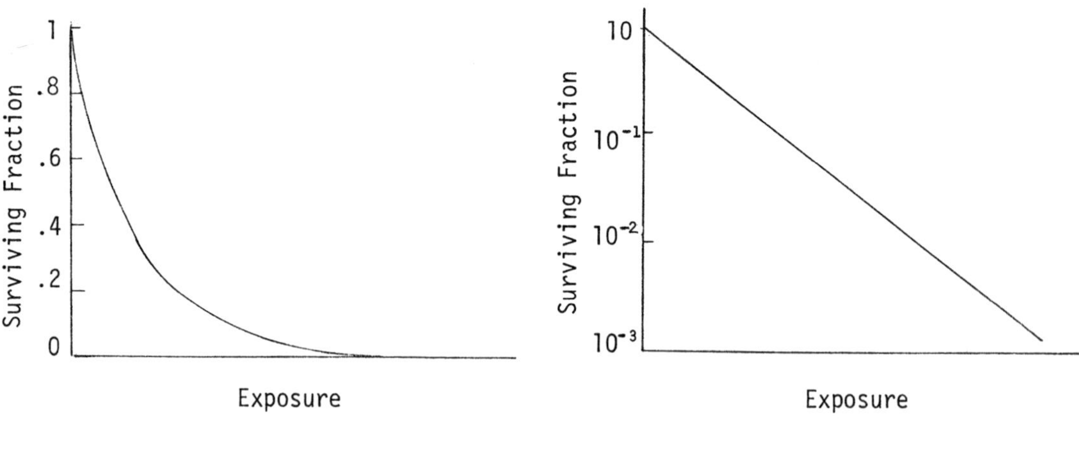

Figure 4.47

EXAMPLE 4.35: Type A spermatogonia result during the production of gametes for reproduction and have varying sensitivity to x rays. The semi-log plot in Figure 4.48 shows how the survival curve of type A spermatogonia relates to radiation with x rays. From data the parameters of the exponential curve were estimated as $k = -0.00263$ and $A = 1.12$. ∎

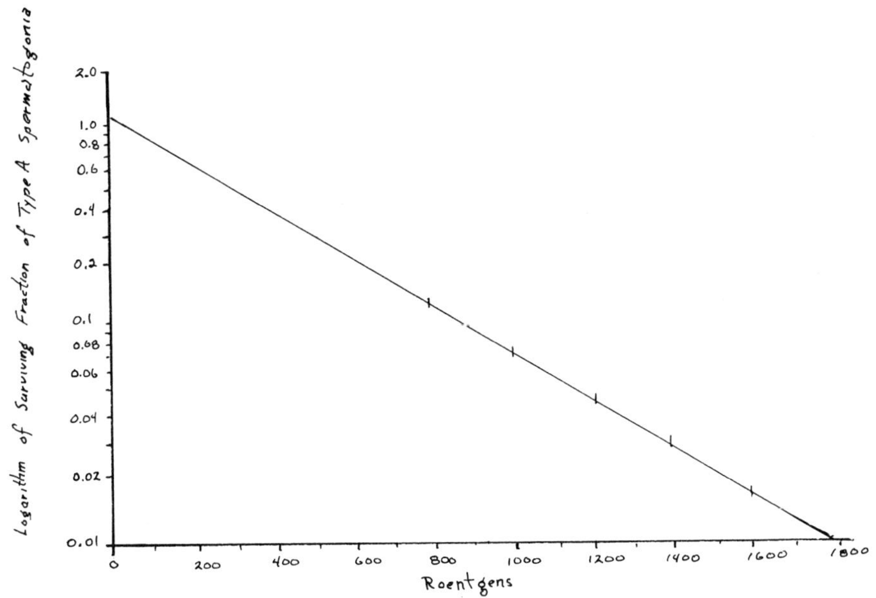

Figure 4.48

One characteristic of an exponential survival curve invalidates it for some responses. The surviving fraction (number of cells surviving a given dose or stress divided by the number of cells in an untreated situation) decreases rather markedly at low exposures, but for some kinds of exposure a small dose causes virtually no damage. This suggests that there should also be survival curves which have a shoulder before entering their sharply declining portion. This leads us to introduce another survival curve which was developed from a probabilistic point of view.

- The SIGMOID SURVIVAL FUNCTION is specified by the equation

$$s = 1 - (1 - e^{-kD})^n , \qquad (4.10)$$

where k is a positive constant and n a positive integer which together determine the curve and thus the surviving fraction; D is a variable describing the dose or stress applied to the organism under study.

A sigmoid survival curve with a value of $n > 1$ is plotted on a linear axis to the left in Figure 4.49 and to the right on a semi-log scale. Looking at these curves and comparing them to the earlier exponential survival curves presented in Figure 4.47, you should observe the following:

1. Both of the semi-log plots have greatly extended axes on the surviving fraction, compared to the original scales.
2. The essential difference between the exponential and sigmoid appears at low dosages. The shoulder of the sigmoid curve is absent from the exponential curve.
3. Greater ranges of survival are demonstrable with the semi-log plot for either of these functions than with the linear plot.

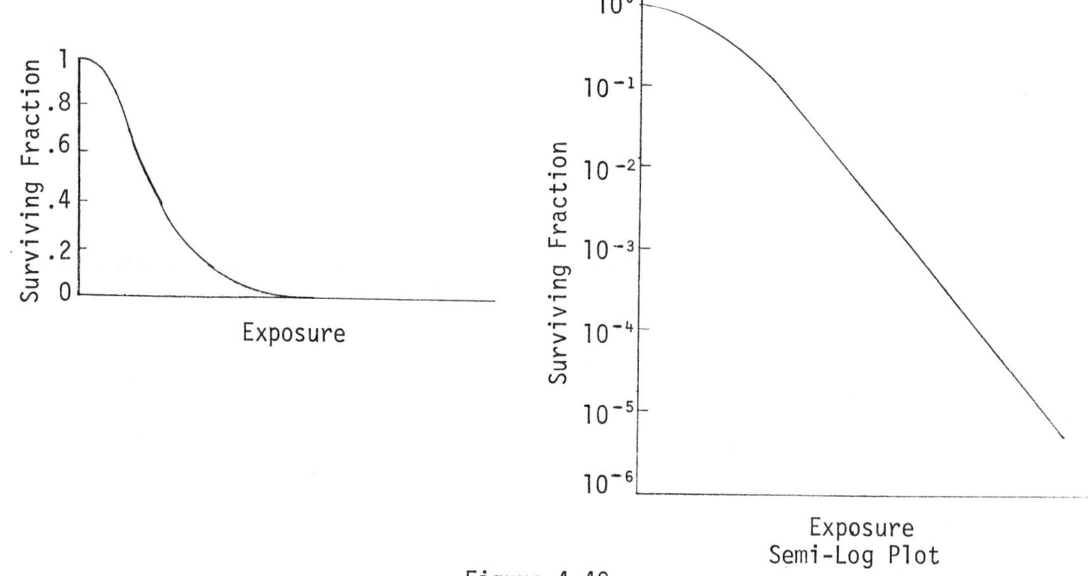

Figure 4.49

Another figure will demonstrate a further aspect of the sigmoid survival curves. Figure 4.50 displays several different survival curves, plotted for varying values of n. For survival rates below 0.1 each of them can be reasonably well approximated by a straight line so that extrapolation of the linear portion of the survival curve back to a stress of 0 (ordinate intercept) yields an estimate of n. Notice that the shoulder or threshold region on each of the curves becomes more pronounced as n increases.

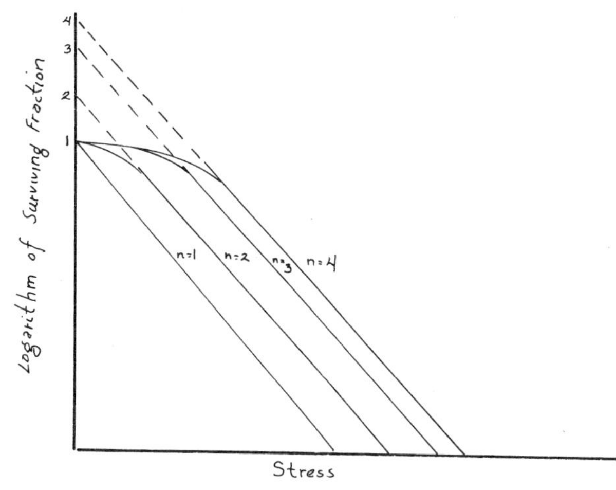

Figure 4.50

How would you estimate n and k for the sigmoid survival curve? Using the binomial expansion on the right side of the equation, we obtain, if we assume n to be an integer:

$$s = 1 - (1 - e^{-kD})^n$$

$$= 1-[1-n(1)^{n-1}(e^{-kD}) + \frac{n(n-1)(1)^{n-2}(e^{-kD})^2}{2!} - \frac{n(n-1)(n-2)(1)^{n-}(e^{-kD})^3}{3!} + \cdots]$$

$$= ne^{-kD} - \frac{n(n-1)\ e^{-2kD}}{2!} + \frac{n(n-1)(n-2)\ e^{-3kD}}{3!} - \cdots .$$

When D is sufficiently large, the terms e^{-2kD}, e^{-3kD}, $\cdots$ will be very small compared to e^{-kD} so that we can approximate s by the first term, so

$$s \doteq ne^{-kD} ,$$

where $\doteq$ is the symbol used for approximated by or approximately equals. The survival curve now has the form of an exponential equation so that the parameters n and k can be estimated by taking natural logarithms of both sides of the approximated survival equation to obtain

$$\ln s \doteq \ln ne^{-kD} = \ln n + \ln e^{-kD} = \ln n - kD$$

We have now reduced the approximated survival equation to an equation which is linear in logarithms so that by plotting the survival data as ln s versus D we can obtain a straight line with an ordinate intercept of ln n and a slope of -k. In reality we are estimating n and k from the straight portion of the survival curve, that is, from data associated with the large values of D, because this is where the approximation applies. By examining the graph of the survival curves in Figure 4.50, sufficiently large values of D must mean those values that get beyond the shoulder of the curve and onto the portion of the survival curve that appears to be straight. Thus, for different survival curves, these sufficiently large values of D change. The determination of a sufficiently large value of D poses no practical problem because an experimenter usually is working with surviving fractions that are small enough to insure that D is sufficiently large.

EXAMPLE 4.36: The effect of γ-rays on a strain of potato virus is shown in the table below and in Figure 4.51. Assume that the survival is approximately exponential, $s = ne^{-kD}$, and estimate the value of n and k.

$D \times 10^5$ rads	3	5	6	8	10	12	14
Surviving fraction	0.40	0.25	0.175	0.10	0.05	0.03	0.015

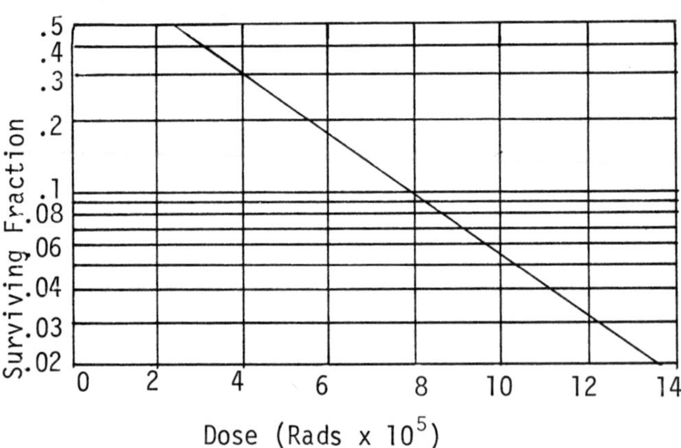

Figure 4.51

The points $(3\times10^5, 0.43)$ and $(6\times10^5, 0.175)$ lie on the line. Then because $\ln s = \ln n - kD$,

$$k = -\frac{\ln 0.175 - \ln 0.43}{6\times10^5 - 3\times10^5} = -\frac{-1.74297 - (-0.84397)}{3\times10^5} = \frac{0.89900}{3\times10^5}$$

$$= 0.2996 \times 10^{-5} .$$

Now to estimate n, use the point $(3\times10^5, 0.43)$ in $\ln n = \ln s + kD$:

$$\ln n = \ln 0.43 + (0.2996\times10^{-5})(3\times10^5)$$

$$= -0.84397 + 0.889$$

$$= 0.04503$$

and $n \doteq 1.05$. Thus the survival equation describing how a potato virus responds to γ-rays is approximately

$$s = 1.05 \exp[-0.3\times10^{-5} D] . \quad \blacksquare$$

EXERCISES

1. In a radioactive decay process, if one half of the atoms have decayed in one day (24 hours), what fraction is left after 10 hours?

2. Radium-228 has a half life of 6.7 years. How much time is required for a reduction to 20% of the original amount?

3. Assuming exponential growth, find the law governing the growth of 10^5 bacteria if at t = 24 hours there are 3×10^7 bacteria.

4. The population of deer in a well defined region has been growing at a rate proportional to its own size. If the population now contains 40,000 deer while twenty-five years ago it had 15,000, find the anticipated population ten years from now provided management practices do not change and its environment does not become limiting.

5. From a skeleton recently uncovered on a highway project, a 1 gram sample of bones radiated ^{14}C at 32% of the living rate. How old are the bones?

6. The data below illustrate the removal of labeled ferric globulinate from the plasma of man. Time (t) is measured in minutes and the amount of iron (A) is measured in milligrams per milliliter of plasma.

t	0	60	120	180	240	360
A	0.26	0.14	0.11	0.09	0.043	0.020

   a. Demonstrate that the data is exponential.

   b. Assuming the form of the curve to be represented by $A = A_0e^{-kt}$, estimate k and A_0.

7. Suppose a 1 gram piece of wood found in a cave gives 6.2 ^{14}C disintegrations per minute while 1 gram of the same type of living wood yields 15.3 disintegrations per minute. Approximately what age does the ^{14}C dating method give for this piece of wood?

8. A colony of bacteria grows in proportion to its size (it grows exponentially) until limiting factors set in. If initially a colony contains 2000 bacteria and one day later it has 5000 bacteria, what will be the size of the population of the colony after five days after it begins growing if it grows without restriction?

9. One hundred fruit flys are placed in a large glass jar which will support a maximum fly population of 2000. If after 50 days the population has grown to 500 flies, when will the population reach 50% of the capacity of the jar?

10. If the rate of growth of a certain population of animals follows the exponential growth law and if the size of the population doubles every 60 years, estimate the number in the population in the year 2000 if there were 60,000 individuals in the population in 1950.

11. Assume a culture of bacteria grows at a rate which is proportional to the number of bacteria present. At a given time there are 1000 bacteria present and 10 hours later there are 7000 bacteria present. How many bacteria will be present in t hours after the first count? 20 hours?

12. A colony of bacteria has 3×10^6 individuals initially and 9×10^6 individuals 2 hours later. Find the growth laws for $N = N_0e^{kt}$ and $N = N_02^{ht}$ utilizing the above information. How long does it take the colony to double its size?

13. By subjecting a virus to x rays, certain parts of the virus were inactivated and reproduction by the virus ceased. If the number (N) of viruses surviving a dose r is governed by $N = N_0e^{-kr}$ where N_0 is the initial number of viruses and k a constant, what dose is required to inactivate 80% of the virus?

14. The number of bacteria in a certain culture grows from 100 to 300 in 12 hours. How many bacteria were present after 6 hours, assuming ideal exponential growing conditions?

15. Potassium-42, ^{42}K, is used as a tracer in labeling experiments and has a half life of approximately 12.5 hours. If you have a sample which originally contained N_0 atoms, how many atoms would you have after 30 hours? How long would it take the sample to decay to $\frac{1}{1024}N_0$?

16. Potassium-42, ^{42}K, is a radioactive tracer. A cell is injected with ^{42}K and after 1 hour the amount of ^{42}K is 85% of its original value. Has any of the ^{42}K been removed from the cell? (Half life of ^{42}K = 12.5 hours.)

17. A biochemical solution was incubated with adenosine triphosphate (ATP) in which the terminal phosphate group contained radioactive phosphorous-32, ^{32}P. After 18 hours of incubation free radioactive phosphate was found to be present. What correction has to be applied in order to get an accurate estimate of the ATP that has broken down? (Half life of $^{32}P \doteq 14.2$ days.)

18. The following experimental data represents the absorption of x rays by lead plates of varying thicknesses and should follow the exponential law $I = I_0e^{kt}$.

t (thickness)	1/8"	1/4"	1/2"	3/4"	1"
I (counts per minute)	774	688	462	272	168

a. Plot the logarithm of counts per minute versus absorber thickness and estimate I_0 and k, where t is the thickness of the lead absorber in inches.

b. Why will your estimates be crude?

19. Suppose that bacteria are growing in a certain culture and have a rate of growth which follows the exponential function $N = N_0e^{kt}$ where N is the number of bacterial colonies at any time t, N_0 is the number of bacterial colonies at time 0, and k is constant to be estimated. The following information becomes available: Between the second and fourth day (a 48 hours period) the number of colonies increased by 1675, while between the fifth and seventh days (a 48 hour period) the number of colonies increased by 2785.

a. Estimate N_0 and k from the above data.

b. Give an expression for the increase of the number of colonies during the rth day after the culture was started.

c. Evaluate the population increase during the 10th day after the culture was started.

d. Discuss the effect N_0 and k have on the general shape and location of the curve.

20. A veterinarian about to perform surgery on a dog estimates that it will take him 45 minutes to complete the procedure. The dog weighs 20 kilograms; 20 mg. of sodium pentobarbitol per kilogram of body weight is required for the bare maintenance of surgical anesthesia; and the half life of sodium pentobarbitol in dogs is five hours.

a. How much anesthesia should the veterinarian initially administer to maintain anesthesia over the estimated duration of the operation?

b. At the end of the 45-minute period the veterinarian realizes that the dog is beginning to emerge from anesthesia and that the surgery will take an additional 20 minutes to complete. How much additional sodium pentobarbital should be administered in order to maintain anesthesia for the rest of surgery?

21. The following data approximates the results obtained from subjecting HeLa S cells to 250 kVp x rays:

Dose	0	100	150	200	250	300	400	450	500	550	600
Fraction Surviving	1.0	.5	.4	.3	.17	.13	.03	.02	.01	.006	.005

Determine the values of n and k for the survival curve described by the above data.

4.8 HINDSIGHT FOR FORESIGHT

What have we accomplished? What remains to be done? So far our efforts have concentrated on two aspects of functions, namely, what they are and why a biologist should know about them. The first chapter dealt with the role of mathematics, and more particularly of functions, in biology while the second chapter dealt with developing mathematical counterparts for real problems. The third and fourth chapters have laid the foundations and begun detailed considerations of functions. The latter part of this chapter has set out a number of functions which have central roles in biology. Their appearance should not frighten you, but should indicate the work that we have left to do, because you will have a lot to learn about those and other functions.

Now that you have some insight into what functions are and where they fit into biology, two major issues remain unresolved. How does a specified function behave? How should you reason from biologically defensible assumptions to the associated functions? The first question serves as the focus of the next three chapters. Chapter 5 deals with how functions behave at specified points, including points where they may behave irregularly, and at the edges of their domain. This topic relates to the mathematical concept of a limit. The subsequent two chapters deal with the mathematical concept of derivatives. This tool proves exceedingly useful in examining how the functions accelerate, or in other words, how fast they are increasing or decreasing. For example we noted that an exponential function with a positive power increases at an ever faster rate. The topic of derivatives will lend precision to such ideas and allow us to pin down other important characteristics of functions. Together these topics will provide a very effective way for examining a function to learn what it does at all points of interest.

The other topic, how to arrive at a function, has already received passing attention. In several of the immediately preceding sections we pointed out the shapes of various curves when plotted on various graph papers. These shapes can hint at what function underlies a particular response, but it does not provide a means for reasoning from your understanding of the biological phenomena to a mathematical function. Many ways of describing the mathematical behavior of functions in biology will involve derivatives that are introduced in Chapter 6 and pursued in Chapter 7. The topic of integration reverses the process of differentiation. Thus, it provides a means of reasoning from how a function changes to the function itself. Chapter 8 introduces integration.

The remainder of the book further develops derivatives, integrals and allied topics as we go to more complex situations, particularly to those simultaneously involving several independent variables.

This brief sketch of our progress should give you a little idea why the topics ahead have the organization they have. Our progress to this point should show some of the places that mathematics can become involved in biology.

REVIEW EXERCISES

1. When a population is declining in an exponential fashion, what are the advantages of plotting the data on semi-logarithmic paper?

2. Are the growth and decay relations given in this chapter functions?

3. A bacterial population is growing exponentially according to $N = N_0e^{kt}$ and a second population of bacteria is growing exponentially also according to the law $N = 0.5\ N_0e^{2kt}$ at time t. How do the populations differ in initial size and growth rate? Calculate the size of each population at $t = \frac{1}{4k}$, $\frac{1}{2k}$, and $\frac{1}{k}$. Find the time at which the populations will be equal in size.

4. Anthropologists have found that in the United States the half life of an affair between a man and a woman is approximately 7 months. Assuming that affairs follow the exponential decay, how many couples were having an affair initially if after 1.5 years there were still 342 couples?

5. Two bacterial populations are growing exponentially and have size $A \exp[t/t_0]$ and $\frac{A}{3} \exp[2t/t_0]$, respectively, at time t. At what time will the populations be of equal size?

6. A piece of wood found in an Egyptian tomb gave 7.04 counts/minute/gram more than normal background (every where present) radiation. A modern piece of the same type of wood yielded 12.5 counts/minute/gram above background. How old is the tomb? What have you assumed?

7. A bacterial culture which is growing exponentially increases from 2×10^6 cells to 3×10^8 cells in 4 hours. What is the time between successive fissions if there is no mortality? Assume the fission time is $T = \frac{1}{k}$ where k is the constant in the exponent of an exponential function.

8. Technetium-99m, ^{99m}Tc, has a half life of 6 hours. A known quantity is injected into the brain. After 1 hour the amount of ^{99m}Tc has been reduced to 85.2% of its original value. Has any of the radioactive material been removed from the brain tissue?

9. The following set of data represents the number of white pine trees per acre as a function of average diameter at breast height in inches. Plot the data on log-log scales to find the equation relating the number of trees per acre to diameter.

N = number of trees per acre
Dbh = diameter at breat height

N	4000	2700	1300	710	450	310	230	185	150	120	100
Dbh	1	1.4	2	3	4	5	6	7	8	9	10

10. The following set of data represents the x ray survival of HeLa S3-91V cells 0 hours and 13 hours irradiated during the division cycle measured from mitosis (0 hours).

Dose rads	0	50	100	150	200	250	300	350	400	450
Surviving fraction at 0 hours	1	.9	.4	.28	.14	.08	.042	.02	.006	
Surviving fraction at 13 hours	1	.8	.67	.58	.42	.28	.18	.11	.09	.06

Determine the values of n and k for the survival curves described by the above data.

Chapter 5

WHAT IS YOUR LIMIT?

Expressions like *pushed into the corner* or *boxed up* describe situations of stress. You may remain cool in such a situation while another person may explode; in other settings you may not at all like the way you behave. Functions also have varying behavior at different points in their domain; different functions have varying behavior at the same point.

Some functions have a fairly obvious behavior throughout their domain, but many useful and significant functions have obscure behavior at important points in their domain. Thus, we will introduce the idea of a limit and some of its basic properties in the next section, then look at the continuity, or the smoothness, of functions and conclude by considering some troublesome types of limits. The concept of a limit lies at the foundation of calculus. This chapter shares with the next one a study of basic mathematical facts. Consequently, we will place far more emphasis on specified algebraic functions than biological examples. A few interesting examples will appear in this chapter, but the really interesting applications involving limits mainly await Chapter 7.

Before we get into a close examination of limits and continuity, you might wonder, "Really what is a limit?" In one sense it is a sneaky concept. If you want to look at a gorge or a canyon, you do not run off its edge and look at it as you fall; you cautiously walk toward the edge and look over, and then inch a little closer for a further look. Similarly, to see a bird you do not jump to where he is perched. Instead you creep toward him without getting there. To find the limit of a function at a point, we sneak toward the point of interest, simultaneously investigating the associated values of the function. Often this limiting value equals the value of the function at the point, but occasionally the limit may not exist. For example, what altitude should you give for a point on the edge of a vertical cliff? The top? The bottom? Or some point in between?

Limits are closely associated with continuity of functions. To be continuous at a point, the value of the function at a point must equal its limit there. A continuous function is like a rolling prairie: It contains no sharp dropoffs like a gorge. The intent of this chapter then will be to give you an intuitive feeling for what a limit is.

5.1 WHAT HAPPENS WHEN YOU GET REAL CLOSE?

This section does not present a mathematically rigorous definition of a limit. Instead its examples should help you see what a limit is and what it is not.

EXAMPLE 5.1: Consider how the function defined by $f(x) = x^2 + 1$ behaves near $x = 2$:

x	f(x)	x	f(x)
1	2	3	10
1.5	3.25	2.5	7.25
1.9	4.61	2.2	5.84
1.95	4.8025	2.1	5.41
1.98	4.9204	2.05	5.2025
1.99	4.9601	2.01	5.0401
1.999	4.996001	2.001	5.004001
1.9999	4.99960001	2.0001	5.00040001

The left hand side shows that f(x) approaches 5 as x approaches 2 from below. Likewise the right side shows that f(x) gets close to 5 as x drops from 3 down toward 2. Notice that $x = 2$ does not appear on either side. A limit describes how a function behaves near a point, not at the point. ▌

To progress further we need a usable definition of a limit. A formal rigorous definition can be stated only using precise mathematical language. Nevertheless, the essence of the concept embodied in that rigorous definition appears below:

- A function f has a LIMIT L as x approaches the point x_0 if f(x) can be made arbitrarily close to L for all values of x in its domain sufficiently near x_0. This usually is symbolized as

$$\lim_{x \to x_0} f(x) = L$$

Note that x_0 must lie either in the domain of f or on its edge. For example, if $D = \{x|\ 3 < x < 5\}$, then $3 \leq x_0 \leq 5$.

After having seen this definition and Example 5.1, you may wonder why bother with values of x other than x_0? In that example we found f(2) = 5; this also appeared to be the limit of the function at x = 2. We cannot always get the limit of a function by evaluating it at the point in question. A function may have a limit at a point in its domain which differs from the value of the function evaluated there. A function may fail to have a limit at a particular point in its domain even though the function could be well defined there. Conversely a function may have a limit at a point even if the function is not defined there.

EXAMPLE 5.2: Consider the function defined by $f(x) = 2^{-1/x^2}$ near x = 0. We cannot evaluate f(0) because 1/0 has no numerical value, but we can still examine the limit by looking at values of x very close to 0:

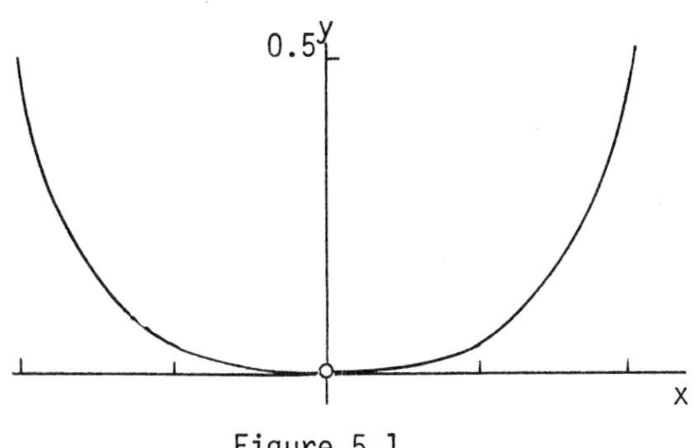

Figure 5.1

x	±1	±0.5	±0.25	±0.1	±.01
f(x)	1/2	0.0625	$\frac{1}{2^{16}}$	$\frac{1}{2^{100}}$	$\frac{1}{2^{10000}}$

An examination of this table should convince you that values of x close to 0 make the function nearly 0, a fact which prompts us to conclude that

$$\lim_{x \to 0} 2^{-1/x^2} = 0 .$$

This function cannot be evaluated at the point in question, $x = 0$, even though it has a limit of 0 as $x \to 0$. Its graph appears in Figure 5.1; the hole at (0,0) indicates that the function is not defined there. ▌

To further examine and use limits, we need some further facts about them. The theorems stated below are not proven here because their complete proof requires the rigorous definition of a limit which we avoided. Nevertheless, the theorems state results which should seem apparent and which can be proved.

<u>THEOREM 5.1</u>: $\lim_{x \to x_0} c = c$, where c is a constant not depending on x.

<u>THEOREM 5.2</u>: $\lim_{x \to x_0} x = x_0$.

<u>THEOREM 5.3</u>: $\lim_{x \to x_0} (mx + b) = mx_0 + b$.

The remaining theorems all involve at least one function; for notational simplicity consider two functions f and g, where

$$\lim_{x \to x_0} f(x) = B \qquad \text{and} \qquad \lim_{x \to x_0} g(x) = C.$$

<u>THEOREM 5.4</u>: $\lim_{x \to x_0} [f(x) + g(x)] = \lim_{x \to x_0} f(x) + \lim_{x \to x_0} g(x) = B + C$.

This theorem also applies with the plus sign replaced by a minus sign.

<u>THEOREM 5.5</u>: $\lim_{x \to x_0} kf(x) = k \lim_{x \to x_0} f(x) = kB$ where k is a constant.

<u>THEOREM 5.6</u>: $\lim_{x \to x_0} [f(x)g(x)] = \left[\lim_{x \to x_0} f(x)\right]\left[\lim_{x \to x_0} g(x)\right] = B \times C$.

<u>THEOREM 5.7</u>: $\lim_{x \to x_0} \frac{f(x)}{g(x)} = \frac{\lim_{x \to x_0} f(x)}{\lim_{x \to x_0} g(x)} = \frac{B}{C}$, provided $C \neq 0$.

THEOREM 5.8: $\lim_{x \to x_0} [f(x)]^m = \left[\lim_{x \to x_0} f(x)\right]^m = B^m$ where m is any real number.

These theorems state results which should seem reasonable in this sense: The limit of the sum equals the sum of the respective limits; the limit of a constant times a function equals that constant times the limit of the function; the limit of a product of two functions equals the product of their respective limits; the limit of the ratio of two functions equals the ratios of their respective limits, provided the denominator does not go to zero; the limit of any power of a function equals the same power of the limit of that function. The next example illustrates most of these theorems.

EXAMPLE 5.3: Evaluate $\lim_{x \to 3} 7 \frac{x^2 \sqrt{x^2 + 2}}{2x^2 - 17}$.

$$\lim_{x \to 3} 7 \frac{x^2 \sqrt{x^2 + 2}}{2x^2 - 17} = 7 \lim_{x \to 3} \frac{x^2 \sqrt{x^2 + 2}}{2x^2 - 17} \qquad \text{Theorem 5.5}$$

$$= 7 \frac{\lim_{x \to 3} (x^2 \sqrt{x^2 + 2})}{\lim_{x \to 3} (2x^2 - 17)} . \qquad \text{Theorem 5.7 (denominator} \neq 0)$$

Now consider the numerator and denominator separately.

$$\lim_{x \to 3} x^2 \sqrt{x^2 + 2} = \left[\lim_{x \to 3} x^2\right]\left[\lim_{x \to 3} (x^2 + 2)^{1/2}\right] \qquad \text{Theorem 5.6}$$

$$= \left[\lim_{x \to 3} x\right]^2 \left[\lim_{x \to 3} (x^2 + 2)\right]^{1/2} \qquad \text{Theorem 5.8}$$

$$= (3)^2 \left[(\lim_{x \to 3} x^2) + 2\right]^{1/2} \qquad \text{Theorem 5.2, Theorem 5.4}$$

$$= 9\sqrt{11} \qquad \text{Theorem 5.8, Theorem 5.2}$$

Similarly, $\lim_{x \to 3} (2x^2 - 17) = 1$ so

$$\lim_{x \to 3} 7 \frac{x^2 \sqrt{x^2 + 2}}{2x^2 - 17} = 7 \frac{9\sqrt{11}}{1} = 63\sqrt{11} . \quad \blacksquare$$

EXAMPLE 5.4: $\lim_{x\to 2} [(x - 2)^2 + 2(x - 2) + 3] = 3$. In your verification of this make sure you check what theorems were actually used, and that they apply. ▌

EXAMPLE 5.5: Evaluate $\lim_{h\to 0} \frac{5h^2}{2h^2 + 3h}$.

If $f(h) = \frac{5h^2}{2h^2 + 3h}$, evaluation of $\lim_{h\to 0} f(h)$ encounters difficulty because the denominator goes to 0 at $h\to 0$. Even though none of our theorems provide a basis for directly evaluating this limit, some algebraic manipulation will change it into a form which we can evaluate. Factor h out of the numerator and denominator:

$$\lim_{h\to 0} \frac{5h^2}{2h^2 + 3h} = \lim_{h\to 0} \frac{h}{h}\left[\frac{5h}{2h + 3}\right] = \lim_{h\to 0} \frac{5h}{2h + 3} .$$

From Theorem 5.7 this equals 0 because the numerator approaches 0 while the denominator does not. The fraction h/h always equals 1, provided $h \neq 0$. Here, $h \neq 0$ because we only get close to $h = 0$ when $h\to 0$. ▌

EXAMPLE 5.6: If $f(x) = x^2 + 3$, we can investigate the limit $\lim_{h\to 0} \frac{f(x + h) - f(x)}{h}$ as a function of h; that is x remains unspecified but fixed while h varies:

$$\begin{aligned}\frac{f(x + h) - f(x)}{h} &= \frac{[(x + h)^2 + 3] - [x^2 + 3]}{h} \\ &= \frac{[x^2 + 2xh + h^2 + 3] - [x^2 + 3]}{h} \\ &= \frac{2xh + h^2}{h} ,\end{aligned}$$

then

$$\lim_{h\to 0} \frac{f(x + h) - f(x)}{h} = \lim_{h\to 0} \frac{2xh + h^2}{h} = \lim_{h\to 0} (2x + h) = 2x .$$

This illustrates how a limit can produce a function, as well as a numerical value. ▌

EXERCISES

Evaluate the following limits by use of the theorems on limits:

1. $\lim_{x\to 2} [(x - 2)^2 + 3]$

2. $\lim_{h\to 3} (3h^2 - 5h)$

3. $\lim_{x\to 3} 4\sqrt{2x + 3}$

4. $\lim_{x\to 4} \frac{x^2 - 6x + 9}{x^2 + 3x}$

5. $\lim_{x\to -3} \frac{x^2 + 6x + 9}{x^2 + x}$

6. $\lim_{h\to 0} \frac{4h^2 + 3h + 7}{3h + 2}$

7. $\lim_{h\to 0} \frac{(x + h)^2 - x^2}{h}$

8. $\lim_{x\to 0} \frac{x}{3}$

9. $\lim_{x\to 1} \frac{3x - 1}{\sqrt[3]{26x^2 + 2x - 1}}$

10. $\lim_{x\to 0} \frac{e^x + e^{-x}}{2}$

11. $\lim_{x\to 0} \frac{e^x - e^{-x}}{e^x + e^{-x}}$

12. $\lim_{x\to 3} [x + 1]$

13. Does $\lim_{x\to 4} \frac{10 + x}{10 - x}$ exist? Justify your answer.

14. Does $\lim_{h\to 0} \frac{(x + h)^3 - x^3}{h}$ exist?

15. Assume the growth of a population described by:
    a. The LOGISTIC GROWTH CURVE
    b. The GOMPERTZ GROWTH CURVE
    c. The VON BERTALANFFY GROWTH CURVE
    d. The MONOMOLECULAR GROWTH CURVE

    Evaluate the limit of each of the above curves as $t\to 0$. Does the limit of each exist at $t = 0$? Justify.

16. For the relation $f(x) = \frac{1}{x}$, set up a table as was done in Example 5.1 and try to determine what the limit of $f(x)$ will be as x approaches zero from the positive side.

5.2 WHAT HAPPENS WHEN YOU GET THERE?

You have observed that $\lim_{x \to x_0} f(x)$ sometimes equals $f(x_0)$. Such a function is well behaved in a sense that the function is CONTINUOUS. This section introduces and explores the concept of continuous functions and related topics. Specifically, a function may lack continuity at a point so its behavior must be investigated from either side. The previous section on limits answered the question, "What happens when you get real close?" This section deals with what happens at the point relative to what happens close by.

The limits in the last section were easily obtained because the functions involved were well behaved. In order to examine functions which are not as well behaved as those above we need the following:

- A function f has a RIGHT HAND LIMIT L^+ at a point x_0 if f(x) can be made arbitrarily close to L^+ for all values of x in its domain, where $x > x_0$ but close enough to x_0. This is usually sumbolized by
$$\lim_{x \to x_0^+} f(x) = L^+ .$$
- A function f has a LEFT HAND LIMIT of L^- at a point x_0 if f(x) can be made arbitrarily close to L^- for all values of x in its domain, where $x < x_0$ but close enough to x_0. This is usually symbolized by
$$\lim_{x \to x_0^-} f(x) = L^- .$$

The above definitions are merely formal ways for saying that we will look at the limiting behavior of the function near x_0 exclusively from the right or exclusively from the left. An immediate consequence of these definitions is given in the following theorem:

THEOREM 5.9: The limit of a function f as $x \to x_0$ exists if and only if its right hand limit at x_0 equals its left hand limit there and both limits are finite.

Symbolically we express Theorem 5.9 as

$$\lim_{x \to x_0} f(x) = L \quad \text{if and only if} \quad \lim_{x \to x_0^+} f(x) = L = \lim_{x \to x_0^-} f(x)$$

where L is a finite number.

You have probably observed that the limit of a function at a point in its domain can often be found simply by computing the value of the function at that point. For example

$$\lim_{x \to 2} (2x^2 - 3x + 4) = 2(2)^2 - 3(2) + 4 = 6$$

Not all functions are so well behaved. From Example 5.2 recall the behavior of the function $y = 2^{-1/x^2}$ as $x \to 0$. We were not able to evaluate f(0) but could evaluate the limit of this not so well behaved function as $x \to 0$.

Using the definitions for RIGHT and LEFT HAND limits and Theorem 5.9 we are now able to define what we mean by a well behaved function.

- A function f is called CONTINUOUS at a point x_0 in its domain when all of the following are satisfied:

  (i) $\lim_{x \to x_0} f(x) = L$ exists

  (ii) $f(x_0)$ exists

  (iii) $L = f(x_0)$

- A function f is called DISCONTINUOUS at a point x_0 in its domain if it is not continuous at x_0.

In common usage the word continuous means unbroken. The definition of a function which is continuous at x_0 merely lends some precision to this common idea. A function which is continuous at a point has no break in its graph there.

The following examples illustrate the above theorems and definitions.

EXAMPLE 5.7: In working with bacteria and tissue cultures, techniques have been developed to synchronize the divisions of individual cells. Consider a situation where a synchronized population of bacteria is allowed to double every hour. Figure 5.2 shows the growth of this population starting at time $t = 0$ with N_0 bacteria and lasting four hours.

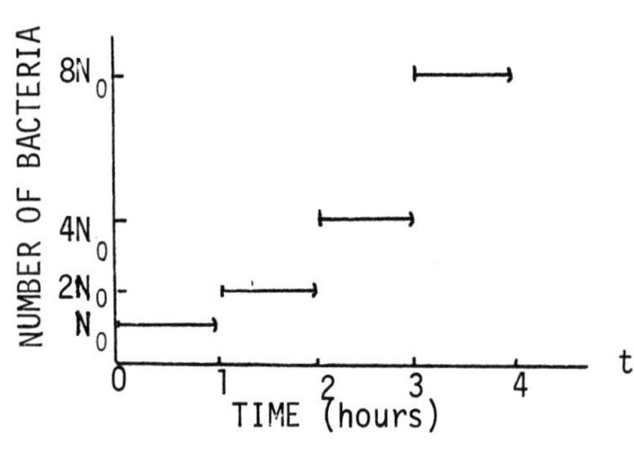

Figure 5.2

This growth function has a value of $2N_0$ slightly past one hour, but a value of N_0 just before one hour. The rounded right ends of the line segments in Figure 5.2 indicate that the end points are not included; the straight lines on the other ends indicate inclusion of the end points. The function which describes the growth of the bacteria could be specified by $N = N_0\, 2^{[\![t]\!]}$. If we examine some representative left and right hand limits we have:

$$\lim_{t \to 1^-} N(t) = N_0 \qquad \text{while} \qquad \lim_{t \to 1^+} N(t) = 2N_0; \text{ and}$$

$$\lim_{t \to 3^-} N(t) = 4N_0 \qquad \text{while} \qquad \lim_{t \to 3^+} N(t) = 8N_0.$$

Thus we see that the $\lim_{t \to 1} N(t)$ and $\lim_{t \to 3} N(t)$ do not exist because the left and right hand limits of $N(t)$ are not equal at either $x = 1$ or $x = 3$. Consequently the function also is discontinuous at these points. ▌

EXAMPLE 5.8: For a bit of variation consider a nonnumerical relation defined on a numerical set. This example will demonstrate the limitations of our definitions. Specifically, consider a relation $s(x)$ which describes the physical state of water. Define

$$s(x) = \begin{cases} \text{solid if } x \leq 32°\text{F} \\ \text{liquid if } 32°\text{F} \leq x \leq 212°\text{F} \\ \text{vapor if } x \geq 212°\text{F} \end{cases}$$

A graph of this relation appears in Figure 5.3. Observe that the relation has fairly obvious discontinuities at 32°F and 212°F. Note, however, because our definition really applies to numerical functions, we cannot carefully examine the continuity of this relation. Such problems have led mathematicians to conceive more general ideas of limits and continuity, a topic which you should realize can be handled.

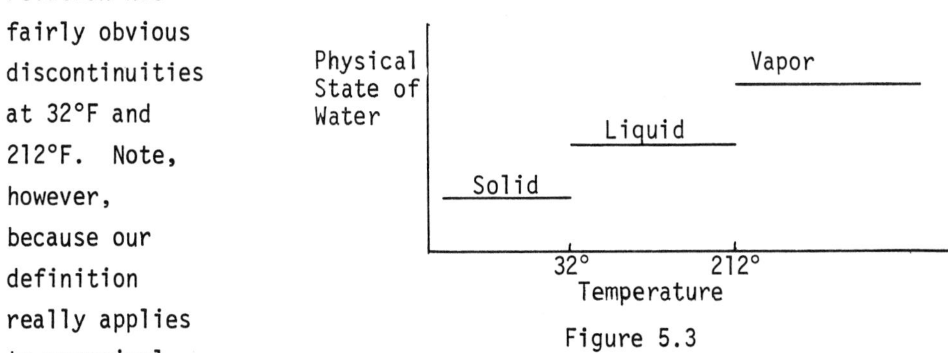

Figure 5.3

When you look at s(x) notice that the relation has two values at 32°F and 212°F. Water can exist in either of two states at each of these two points depending upon the energy it has. Further, observe that the relation has different left and right hand limits at these two points, but at all other points the relation clearly has a limit. ▌

To progress further into some interesting and important illustrations we need the following:

<u>THEOREM 5.10</u>: A sum, difference, product, or quotient of two functions is continuous at every point where each of its components is continuous, provided the denominator of the quotient does not go to zero.

A proof of this theorem simply requires the utilization of rules for working with limits; rather than prove it we illustrate how useful it is in some situations:

<u>EXAMPLE 5.9</u>: Consider a polynomial function defined by

$$p(x) = a_m x^m + a_{m-1} x^{m-1} + \cdots + a_1 x + a_0$$

where m and a_1 are constants and i = 0, 1, 2, $\cdots$, m. We will use results already developed to establish its continuity. Theorem 5.10 implies that the continuity of each additive component in this polynomial guarantees the continuity of p(x). Thus to establish the continuity of p(x), it suffices to demonstrate that each term of the sort cx^d is continuous.

$$\lim_{x\to x_0} cx^d = c \lim_{x\to x_0} x^d \qquad \text{by Theorem 5.5}$$

$$= c\left[\lim_{x\to x_0} x\right]^d \qquad \text{by Theorem 5.8 with } m = d \text{ and } n = 1$$

$$= cx_0^d \qquad \text{by Theorem 5.2}$$

Now check the definition of continuity. The term under consideration has a limit at the point x_0, the term can be evaluated at that point, and the limit equals the evaluated term. Now that we have established the continuity of each additive term in p(x), it follows that p(x) is continuous. ▌

We have discussed the general ideas of a limit, continuity and discontinuity. Now we need to explore the idea of discontinuous functions which may be slightly modified to make them continuous.

- A function f has a REMOVABLE DISCONTINUITY at a point x_0 in its domain if it is discontinuous there but $\lim_{x\to x_0} f(x) = L$ does exist so the function f^* defined as

$$f^*(x) = \begin{cases} f(x) \text{ if } x \neq x_0 \\ L, \; x = x_0 \end{cases}$$

is continuous at x_0.

EXAMPLE 5.2 (continued): We found that the function $f(x) = 2^{-1/x^2}$ has a peculiar behavior near x = 0, so first we will consider its behavior elsewhere. At any other point, for example at x = 3, the function has a limit and the limit does equal the value of the function. Thus, f(x) is continuous at every other point other than x = 0. We saw that $\lim_{x\to 0} f(x) = 0$, but that the function could not be evaluated at x = 0.

This function has a discontinuity at $x = 0$, but a removable one because we can define f^* as

$$f^*(x) = \begin{cases} 2^{-1/x^2}, & x \neq 0 \\ 0, & x = 0. \end{cases}$$

If we now check f^* for continuity we find:

(i) $\lim_{x \to 0} f^*(x) = 0$ does exist,

(ii) $f^*(x) = 0$ does exist, and

(iii) $\lim_{x \to 0} f^*(x) = f^*(0)$.

Thus we find that f^* is a continuous function. Even though f is discontinuous at $x = 0$, it has a removable discontinuity there. ▮

Examples 5.7 and 5.8 illustrate functions with discontinuities which cannot be removed by defining some function f^*. Why?

EXAMPLE 5.5 (continued): Recall that this example concerned the function $f(h) = \frac{5h^2}{2h^2 + 3h}$. The function is not continuous because f(0) is not defined. If we define

$$f^*(h) = \begin{cases} \frac{5h^2}{2h^2 + 3h}, & h \neq 0 \\ 0, & h = 0 \end{cases}$$

the we have removed the discontinuity at $h = 0$. ▮

EXERCISES

1. Find all values of x, if any, at which the following functions are discontinuous. For each discontinuity, state which of the three continuity conditions are violated.

   a. $\frac{4x}{x^2 + 4}$ b. $\frac{x^2 - 25}{x + 5}$ c. $5x^2 + 6x - 7$ d. $mx \pm b$

2. If the inventory in the stockroom of a microbiology laboratory is replenished once a week and the inventory declines steadily (in a

2. Continued
linear fashion) during the week, what is the appearance of the graph of the stock on hand? Is this graph continuous?

3. If $f(x) = \sqrt{x}$, is f continuous at 0? Give reasons for your answer.

4. The domain of a function f consists of all numbers x such that $-1 \leq x \leq 3$ and

$$f(x) = \begin{cases} -2 \text{ if } x \text{ is an integer} \\ 2 \text{ if } x \text{ is not an integer} \end{cases}$$

   a. Draw a graph of f.
   b. What is f(1)? What is $\lim_{x\to 1} f(x)$? Is f(x) continuous at x = 1?
   c. What is $f(\frac{3}{2})$? What is $\lim_{x\to \frac{3}{2}} f(x)$? Is f(x) continuous at $x = \frac{3}{2}$?

5. Does the $\lim_{x\to 3} \frac{x^3 - 27}{x^2 - 9}$ exist? If so, what is the limit; if not, why not?

6. Show that $\frac{x^3 - 27}{x^2 - 9}$ has a removable discontinuity at x = 3.

7. Test $f(x) = \frac{2x^2 - 18}{x - 3}$ for continuity at x = 3. If there are points of discontinuity, are they removable?

8. Test the following for continuity:

$$f(x) = \begin{cases} \frac{x^2 - 25}{x + 5} \text{ if } x \neq -5 \\ -10 \text{ if } x = -5. \end{cases}$$

   If there are points of discontinuity, are they removable?

9. Test $f(x) = \frac{1}{x - 2}$ for continuity. If discontinuous, state which rule for continuity was violated and if it is a removable discontinuity.

5.3 WAY OUT LIMITS AND LIMITS WAY OUT

A function may get large without bound either near some value of its independent variable or as its independent variable keeps getting larger without bound. On the other hand, a function may reach a plateau as the independent variable gets very large. For example, contrast the objectives expressed by population control groups such as Zero Population Growth with the unlimited Malthusian (exponential) growth portrayed in the last chapter. A population which ceases to grow would possibly reach a plateau or could even decline, while a population following Malthusian growth gets larger and larger without bound. You need to be able to investigate such functions. We begin with the following example:

EXAMPLE 5.10: A definite relationship exists among the volume, temperature and pressure of a gas in a container. Here pressure means total pressure, including atmospheric pressure. For a gas at a constant temperature the pressure, P, and volume, V, relate through PV = k where k is a positive constant. This could be written as $V = \frac{k}{P}$. What happens to the volume as the pressure gets smaller and smaller? Specifically, what is $\lim_{P\to 0} \frac{k}{P}$? Clearly we must be talking about $\lim_{P\to 0^+} \frac{k}{P}$ because P < 0 does not exist. (A vacuum merely reduces pressure below the atmospheric level.)

A few values of P near zero should suggest that the value of the function becomes very large.

P	1	0.5	0.25	0.1	0.01	0.001	0.0001	....
$\frac{k}{P}$	k	2k	4k	10k	100k	1000k	10000k	

Thus the closer we approach zero from the right, the larger positively $\frac{k}{P}$ becomes. Thus we write this observation as $\lim_{P\to 0} \frac{k}{P} = \infty$. Thus under these conditions we have an infinitely large volume as $P\to 0$. Physically this says that to continually reduce the pressure on a fixed amount of gas, it must be allowed to occupy an increasingly large volume, without bounds. ▌

EXAMPLE 5.11: Figure 5.4 shows the graph of $f(x) = \frac{1}{(x + 1)(x - 3)}$. Now $\lim_{x\to 3} f(x)$ and $\lim_{x\to -1} f(x)$ do not exist for two reasons:

$\lim_{x\to 3^+} f(x) \neq \lim_{x\to 3^-} f(x)$, and neither limit is finite. The same reasoning holds at $x = 1$. Thus this function has discontinuities at $x = -1$ and at $x = 3$. These discontinuities are not removable. ▌

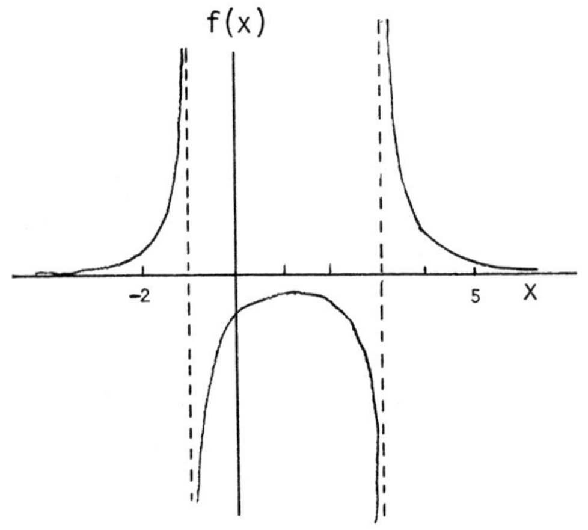

Figure 5.4

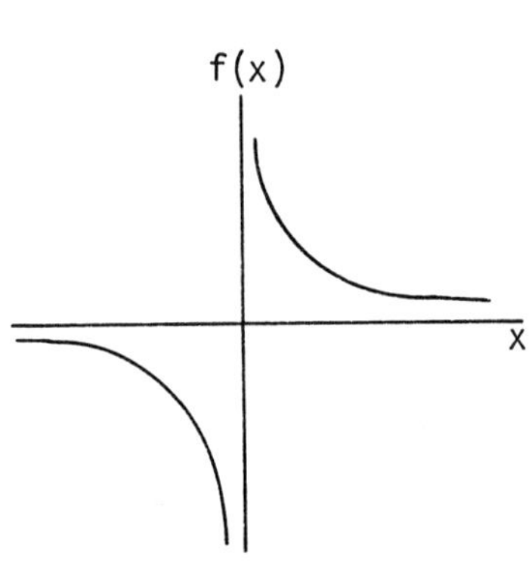

Figure 5.5

EXAMPLE 5.12: Consider the function defined by $f(x) = \frac{1}{x}$ which clearly exists for all $x \neq 0$. The graph of this function, a hyperbola, appears in Figure 5.5. Recalling from Example 5.10 that $\lim_{P\to 0^+} \frac{k}{P} = \infty$, we can infer that $\lim_{x\to 0^+} \frac{1}{x} = \infty$ also. On the other hand when we examine $\lim_{x\to 0^-} \frac{1}{x}$, negative x near zero produce very large negative numbers as is illustrated by:

x	-1	-0.5	-0.25	-0.1	-0.01	-0.001	-0.0001	...
f(x)	+1	-2	-4	-10	-100	-1000	-10000	

Intuitively we can see that as $x\to 0^-$, $f(x)$ gets very large negatively and thus $\lim_{x\to 0^-} f(x) = -\infty$.

This example also illustrates Theorem 5.9 in that $\lim_{x\to 0} \frac{1}{x}$ does not exist because $\lim_{x\to 0^-} \frac{1}{x} \neq \lim_{x\to 0^+} \frac{1}{x}$. Why is $f(x)$ not continuous at $x = 0$? When $\lim_{x\to 0^+} \frac{1}{x}$ is not finite, the function cannot be continuous there, regardless of the value of $\lim_{x\to 0^-} \frac{1}{x}$. ▌

The three previous examples have illustrated functions whose values become very large, either positively or negatively, at some finite point in the domain. We may however be interested in examining the behavior of the function as the independent variable becomes very large.

EXAMPLE 5.13: Examine the limit of the Logistic Growth Curve as $t \to 0$ and as $t \to \infty$.

The logistic growth curve is specified by $N = N_0 \dfrac{(1 + b)}{1 + be^{-kt}}$ and

$$\lim_{t \to 0} \frac{N_0(1 + b)}{1 + be^{-kt}} = N_0(1 + b) \lim_{t \to 0} \frac{1}{1 + be^{-kt}} = \frac{N_0(1 + b)}{1 + b} = N_0 .$$

$$\lim_{t \to \infty} \frac{N_0(1 + b)}{1 + be^{-kt}} = N_0(1 + b) \lim_{t \to \infty} \frac{1}{1 + be^{-kt}} = N_0(1 + b) .$$

Thus we see that N_0 is the initial value of the population at time zero and $N_0(1 + b)$ is the value which the growth curve approaches as $t \to \infty$. ∎

Limits of the sort $N_0(1 + b)$ have names:

- A function f has a VERTICAL ASYMPTOTE at a point x_0 if $\lim_{x \to x_0^-} f(x) = \infty$ (or $-\infty$) or if $\lim_{x \to x_0^+} f(x) = \infty$ (or $-\infty$).
- A function f has a HORIZONTAL ASYMPTOTE of H if $\lim_{x \to \infty} f(x) = H$ or if $\lim_{x \to -\infty} f(x) = H$.

EXAMPLE 5.13 (continued): The value of $N_0(1 + b)$ represents a horizontal asymptote. Figure 5.6 shows the graph of the Logistic Growth Curve starting at the initial value of N_0 growing to the asymptotic value of $N_0(1 + b)$.

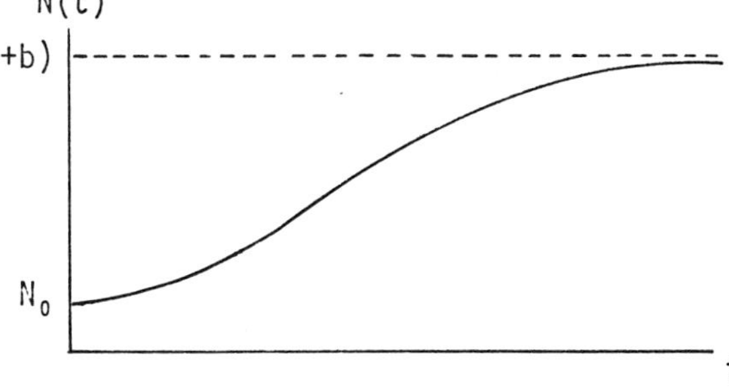

Figure 5.6

EXAMPLE 5.11 (continued): Reexamination of Figure 5.4 shows that the function f has vertical asymptotes at x = -1 and x = 3 because $\lim_{x\to -1^-} f(x) = \infty$, $\lim_{x\to -1^+} f(x) = -\infty$, $\lim_{x\to 3^-} f(x) = -\infty$ and $\lim_{x\to 3^+} f(x) = \infty$.

Thus in order to locate a vertical asymptote we must locate those values of x_0 for which $\lim_{x\to x_0} f(x) = \pm\infty$. ▮

There are certain functions which produce a result such as $\lim_{t\to\infty} f(x) = \infty$. This tells us that the limit becomes large without bound as $t\to\infty$. This is illustrated in the next example.

EXAMPLE 5.14: If cells divide by simple division so that each cell produces two daughter cells with no loss in cell numbers, the population size can be approximated by $N(t) = N_0 2^{t/t_0}$ where N(t) is the total number of cells present in the Nth generation, N_0 is the total number of cells present at time t = 0, t is the elapsed time and t_0 is the average time for one generation. If we assume that there is unrestricted growth, then we would be interested in how many cells would be present after a long period of time.

$$\lim_{t\to\infty} N_0 2^{t/t_0} = \infty ,$$

which is the result you would expect intuitively. Notice also that $\lim_{t\to 0} N_0 2^{t/t_0} = N_0$ which represents the number of cells present at time t = 0.

If $2^{t/t_0}$ were replaced by e^{kt}, $k > 0$, N(t) would represent the Malthusian growth described by the exponential function. It has the same behavior. ▮

5.4 INDETERMINATE FORMS

The evaluation of some limits cannot be completed because they produce quantities like 0/0 or ∞/∞. Yet, sometimes such limits still can be evaluated. Thus, consider the following definitions:

- The function f is called INDETERMINATE at the point x_0 if it can be written as $f(x) = g(x)/h(x)$ and either

  (i) $\lim_{x \to x_0} g(x) = 0 = \lim_{x \to x_0} h(x)$ or

  (ii) $\lim_{x \to x_0} g(x) = \infty = \lim_{x \to x_0} h(x)$ (or $-\infty$).

We already have operated on an indeterminate form in Example 5.5. The kinds of operations illustrated here will become central to use of limits in the next chapter where these sorts of limits appear in the limit definition of a derivative.

EXAMPLE 5.15: $\lim_{x \to 0} \frac{5x^2 + 6x}{2x^2 + 3x}$ cannot be evaluated by taking the ratio of the limit of the numerator to the limit of the denominator because the denominator has a limit of 0, in violation of Theorem 5.7. Because the numerator also goes to zero, the limit of the ratio assumes the indeterminate form 0/0; the limit may still have a value. Divide both the numerator and denominator by x to get

$$\lim_{x \to 0} \frac{5x^2 + 6x}{2x^2 + 3x} = \lim_{x \to 0} \frac{x}{x}\left[\frac{5x + 6}{2x + 3}\right] = \left[\lim_{x \to 0} \frac{x}{x}\right]\left[\lim_{x \to 0} \frac{5x + 6}{2x + 3}\right]$$

$$= \lim_{x \to 0} \frac{5x + 6}{2x + 3} = 2 \text{ .} \quad \blacksquare$$

Observe that by changing the 6 and 3 in the function of Example 5.15 to other values, the limit could be changed to any value. Thus, the intiial appearance of 0/0 tells us nothing about the limit's value, neither will ∞/∞. On the other hand, the appearance of the form $k/0$, $k \neq 0$, generally indicates a function which increases without bound while the form k/∞, $k \neq \infty$, indicates a limit which tends to zero.

EXAMPLE 5.16: Consider $f(x) = x^2 + 2x$ and evaluate $\lim_{h \to 0} \frac{f(x + h) - f(x)}{h}$ at x = 3. At x = 3,

$$\frac{f(x + h) - f(x)}{h} = \frac{(3 + h)^2 + 2(3 + h) - (3^2 + 6)}{h}$$

$$= \frac{9 + 6h + h^2 + 6 + 2h - 9 - 6}{h} = \frac{8h + h^2}{h} ,$$

and

$$\lim_{h \to 0} \frac{f(3 + h) - f(3)}{h} = \lim_{h \to 0} \frac{8h + h^2}{h} .$$

This has the indeterminate form of 0/0, but it can be resolved by
$\lim_{h \to 0} \frac{f(3 + h) - f(3)}{h} = \lim_{h \to 0} (\frac{h}{h})(8 + h) = \left[\lim_{h \to 0} \frac{h}{h}\right]\left[\lim_{h \to 0} (8 + h)\right] = 8$. ▌

EXAMPLE 5.17: For $f(x) = x^3 + 2x + 7$, evaluate $\lim_{h \to 0} \frac{f(x + h) - f(x)}{h}$, allowing x to have any value.

If x has any unspecified value which remains constant while we take the limit, then we can write

$$\frac{f(x + h) - f(x)}{h} = \frac{(x + h)^3 + 2(x + h) + 7 - (x^3 + 2x + 7)}{h}$$

$$= \frac{3x^2h + 3xh^2 + h^3 + 2h}{h} ,$$

so

$$\lim_{h \to 0} \frac{f(x + h) - f(x)}{h} = \lim_{h \to 0} \frac{h(3x^2 + 3xh + h^2 + 2)}{h}$$

$$= \lim_{h \to 0} (3x^2 + 3xh + h^2 + 2) = 3x^2 + 2 ,$$

for any value of x. Where did the indeterminate form appear? How was it resolved? ▌

EXAMPLE 5.18: Certain physiological processes are well described by

$$f(t) = \frac{be^{kt}}{a + e^{kt}} .$$

This function is continuous for all t, including t = 0, the starting time, for

$$\lim_{t\to 0} f(t) = \frac{\lim\limits_{t\to 0} be^{kt}}{\lim\limits_{t\to 0} (a + e^{kt})} = \frac{b}{a + 1} = f(0) .$$

As $t\to\infty$, problems occur when $k > 0$ because the indeterminate form ∞/∞ appears. Upon dividing both the numerator and denominator by e^{kt},

$$\lim_{t\to\infty} \frac{be^{kt}}{a + e^{kt}} = \lim_{t\to\infty} \frac{b}{\left[\frac{a}{e^{kt}} + 1\right]}$$

$$= \frac{\lim\limits_{t\to\infty} b}{\lim\limits_{t\to\infty} \left[\frac{a}{e^{kt}} + 1\right]} = b .$$

This happens because $\lim\limits_{t\to\infty} e^{kt} = \infty$ implies that $\lim\limits_{t\to\infty} a/e^{kt} = 0$. The strict inequality $k > 0$ is required here because $f(t) = b/(a + 1)$ when $k = 0$, independent of t. On the other hand no indeterminate form even arises. To see this, let $m = -k$, a positive quantity. From this, $k = -m$ so $e^{kt} = e^{-mt} = 1/e^{mt}$ and $\lim\limits_{t\to\infty} e^{kt} = \lim\limits_{t\to\infty} \frac{1}{e^{mt}} = 0$. ▮

A slightly different approach is applied to indeterminate forms having radicals in the denominator. The numerator often must be factored before the troublesome factor can be eliminated or you may have to multiply the numerator and denominator by an appropriate quantity. This is illustrated in the following example.

EXAMPLE 5.19: Investigate the continuity of $f(x) = \dfrac{x - 4}{\sqrt{x} - 2}$.

First it has no definition for $x < 0$ because $\sqrt{x}$ has meaning only for $x \geq 0$. As $\sqrt{x}$ denotes the positive square root, $\sqrt{x} - 2 \neq 0$ for all $x > 0$, except at $x = 4$, so f is continuous at all $x > 0$, except $x = 4$. Check the conditions for continuity! Problems appear at $x = 4$ because $\sqrt{x} - 2 = 0$ in the denominator. Nevertheless, $x - 4$ can be viewed as $(\sqrt{x} - 2)(\sqrt{x} + 2)$:

$$\lim_{x\to 4} \frac{x - 4}{\sqrt{x} - 2} = \lim_{x\to 4} \frac{(\sqrt{x} - 2)(\sqrt{x} + 2)}{\sqrt{x} - 2} = \lim_{x\to 4} \sqrt{x} + 2 = 4 .$$

Even though $\lim_{x\to 4} f(x) = 4$, the function is not continuous there because $f(4)$ is not defined. However, it does have a removable discontinuity there.

An alternate approach to evaluating the limit would be to multiply the numerator and denominator by $\sqrt{x} + 2$

$$\lim_{x\to 4} \frac{x - 4}{\sqrt{x} - 2} = \lim_{x\to 4} \frac{(x - 4)(\sqrt{x} + 2)}{(\sqrt{x} - 2)(\sqrt{x} + 2)} = \lim_{x\to 4} \frac{(x - 4)(\sqrt{x} + 2)}{x - 4}$$

$$= \lim_{x\to 4} \sqrt{x} + 2 = 4 .$$

This is precisely the same answer as we obtained by the previous method.

The quantity $\sqrt{x} + 2$ is called the conjugate of the denominator. The method of multiplying a fraction by either the conjugate of the numerator or denominator is a tool which is sometimes helpful in evaluating certain indeterminate forms involving radials. The conjugate of $a + b$ is $a - b$ and the conjugate of $a - b$ is $a + b$. ▮

EXERCISES

1. Assume that the growth of a population is described by
   a. The LOGISTIC GROWTH CURVE
   b. The GOMPERTZ GROWTH CURVE
   c. The VON BERTALANFFY GROWTH CURVE
   d. The MONOMOLECULAR GROWTH CURVE
   e. The EXPONENTIAL GROWTH CURVE

   Evaluate the limit of each of the above curves as $t\to\infty$.

Evaluate the following limits:

2. $\lim_{x\to 2^-} \frac{3x^2 + x - 10}{x^2 - 4}$

3. $\lim_{x\to 3^+} \frac{2x^2 - 7x + 3}{2x - 6}$

4. $\lim_{x\to -3} \frac{x^3 + 27}{2x - 6}$

5. $\lim_{x\to 4} \frac{3\sqrt{x} - 6}{x - 4}$; work two ways.

6. $\lim_{x\to\infty} \frac{3x + 7}{x - 3}$

7. $\lim_{x\to\infty} \frac{4x^2 + 5x + 1}{9x^3 - 8x^2 + 7x + 6}$

8. $\lim_{x\to-\infty} \frac{\sqrt{9x^2 + 16}}{2x - 10}$

9. $\lim_{x\to\infty} \frac{4x^3 - 6x^2 + 11}{7x(x^2 + 9)}$

10. $\lim_{x\to 10^+} \frac{10 + x}{10 - x}$

11. $\lim_{x\to 10^-} \frac{10 + x}{10 - x}$

12. Does $\lim_{x\to 10} \frac{10 + x}{10 - x}$ exist?

13. Find the horizontal asymptotes, if they exist, of the growth functions given in problem 1.

14. Evaluate $\lim_{x\to b} \frac{\sqrt{x + 1} - \sqrt{b + 1}}{x - b}$. (Hint: Multiply the numerator and denominator by the conjugate of the numerator; $\sqrt{x + 1} + \sqrt{b + 1}$.)

15. Evaluate $\lim_{x\to\infty} \frac{e^x + e^{-x}}{2}$.

16. Evaluate $\lim_{x\to\infty} \frac{e^x - e^{-x}}{e^x + e^{-x}}$.

17. Wiess' law is based on the finding that the intensity of an electric current required to excite a living tissue (threshold strength) depends on the duration of the current through $i = \frac{a}{t} + b$, where i is the current strength, t is time, a and b are positive constants. Describe the behavior of i when $t\to 0$ and when $t\to\infty$.

18. A colony of bacteria grows according to the Logistic Growth curve $N = \frac{N_0(1 + b)}{1 + be^{-kt}}$ with a limiting density of 5×10^9 cells per milliliter. When the density of the population is low, the number of bacteria doubles every 40 minutes. What will the population density be after 2 hours if initially the population size was

a. 10^9 cells?

b. 10^8 cells?

**5.5 A SPECIAL LIMIT

This brief section introduces the limit definition of the number $e = 2.71828\cdots$, the base of natural logarithms. It provides the basis for an important relation between the Binomial and Poisson probability distributions. This section will not contain any exercises.

Recall from Section 3.6 that

$$P_B(x) = \binom{n}{x} p^x (1 - p)^{n-x} \qquad \text{Binomial} \qquad (3.5)$$

and

$$P_P(x) = \frac{\lambda^x}{x} e^{-\lambda} \qquad \text{Poisson} \qquad (3.7)$$

We will show that when $np = \lambda$

$$\lim_{n\to\infty} P_B(x) = P_P(x) .$$

The product np gives the theoretical average number of successes in n trials. For this to remain constant as n increases, the condition for a success must become rarer, like quadrat size (a square area of ground) shrinking. In turn this transforms the conditions for the binomial into those for the Poisson. As we investigate this limit, we will encounter an unfamiliar limit closely related to e.

When $np = \lambda$, $p = \lambda/n$ so

$$P_B(x) = \binom{n}{x} p^x (1 - p)^{n-x} = \frac{n!}{x!(n - x)!} p^x (1 - p)^{n-x}$$

$$= \frac{n(n - 1) \cdots (n - x + 1)}{1(2) \cdots (x)} \left(\frac{\lambda}{n}\right)^x \left(1 - \frac{\lambda}{n}\right)^{n-x}$$

$$= \frac{\lambda^x}{x!} \left(\frac{n}{n}\right)\left(\frac{n - 1}{n}\right) \cdots \left(\frac{n - x + 1}{n}\right) \frac{\left(1 - \frac{\lambda}{n}\right)^n}{\left(1 - \frac{\lambda}{n}\right)^x} .$$

Thus as $n\to\infty$,

$$\lim_{n\to\infty} P_B(x) = \frac{\lambda^x}{x!} \left[\lim_{n\to\infty} \frac{n}{n}\right]\left[\lim_{n\to\infty} \frac{n - 1}{n}\right] \cdots \left[\lim_{n\to\infty} \frac{n - x + 1}{n}\right] \frac{\lim_{n\to\infty} \left(1 - \frac{\lambda}{n}\right)^n}{\lim_{n\to\infty} \left(1 - \frac{\lambda}{n}\right)^x}$$

First consider the limits of the form $\lim_{n\to\infty} \frac{n-k}{n}$, $k = 0, 1, \cdots, x - 1$. These limits all equal 1 because for any fixed k,

$$\lim_{n\to\infty} \frac{n-k}{n} = \lim_{n\to\infty} \left(1 - \frac{k}{n}\right) = 1 .$$

Likewise the limit in the denominator tends to 1 as $n\to\infty$ because x does not change with the limiting process. Thus

$$\lim_{n\to\infty} P_B(x) = \frac{\lambda^x}{x!} \lim_{n\to\infty} \left(1 - \frac{\lambda}{n}\right)^n .$$

This last limit does not tend to 1 as you might guess at first. The term $\frac{\lambda}{n}$ gets small as n increases, making $1 - \frac{\lambda}{n}$ get close to 1, but simultaneously the power on $1 - \frac{\lambda}{n}$ increases. Thus, we need to examine it directly. If $n = -\lambda m$, then

$$\left(1 - \frac{\lambda}{n}\right)^n = \left(1 + \frac{1}{m}\right)^{-\lambda m} = \left(1 + \frac{1}{m}\right)^{m\ -\lambda}$$

or

$$\lim_{n\to\infty} \left(1 - \frac{\lambda}{n}\right)^n = \lim_{m\to\infty} \left[\left(1 + \frac{1}{m}\right)^m\right]^{-\lambda} = \left[\lim_{m\to\infty} \left(1 + \frac{1}{m}\right)^m\right]^{-\lambda} ,$$

because λ remains constant during the limiting process. The following table gives values of $\left(1 + \frac{1}{m}\right)^m$:

m	1	2	3	4	5	10	20	30	40	50	
$\left(1+\frac{1}{m}\right)^m$	$(2)^1$	$\left(\frac{3}{2}\right)^2$	$\left(\frac{4}{3}\right)^3$	$\left(\frac{5}{4}\right)^4$	$\left(\frac{6}{5}\right)^5$	$\left(\frac{11}{10}\right)^{10}$	$\left(\frac{21}{20}\right)^{20}$	$\left(\frac{31}{30}\right)^{30}$	$\left(\frac{41}{40}\right)^{40}$	$\left(\frac{51}{50}\right)^{50}$	...
	2	2.25	2.37	2.44	2.49	2.56	2.66	2.68	2.69	2.70	

The values in the bottom row of this table appear to be approaching a limit slightly greater than 2.70. Although we cannot prove it here, these values do have a limit of $e = 2.71828\cdots$. Thus,

$$\lim_{m\to\infty} \left(1 + \frac{1}{m}\right)^m = e = 2.71828\cdots .$$

This gives the special limit of this section. The preceeding table lends credence to the asymptotic value. Now to complete the limiting process on the binomial, we collect the developed results:

$$\lim_{n\to\infty} P_B(x) = \lim_{n\to\infty} \binom{n}{x} p^x (1 - p)^{n-x}$$

$$= \frac{\lambda^x}{x!} \left[\lim_{n\to\infty} \left(1 - \frac{\lambda}{n}\right)^n\right]$$

$$= \frac{\lambda^x}{x!} \left[\lim_{m\to\infty} \left(1 + \frac{1}{m}\right)^m\right]^{-\lambda}$$

$$= \frac{\lambda^x}{x!} (e)^{-\lambda} = \frac{\lambda^x}{x} e^{-\lambda} = P_P(x) \ .$$

Chapter 6

RATES OF CHANGE

Almost every variable changes under the influence of intrinsic and extrinsic factors. A rate of change usually varies with the values of contributory variables. For example, the rate at which a plant grows depends on temperature, availability of nutrients, and length of day, to mention a few. For it to grow faster some of these independent variables have to change. In general we need to describe how biological entities respond to changes in causitive variables; this applies equally to plants or animals, individuals or populations, or simple or complex structures.

We have been telling you that calculus lay ahead. Now you have arrived at the gate! Calculus concerns two basic topics: differentiation and integration. This chapter explores differentiation as a mathematical concept while the next is devoted entirely to applications of derivatives. The subsequent chapter then introduces integration, the other basic concept. For now we restrict attention to how a dependent variable changes with its independent variable. In later chapters we will study how several independent variables simultaneously influence a dependent variable.

This chapter has a simple structure. It begins by introducing the rate of change of a function between two points. Limits are used to pull those two points together to a common point producing the rate of change of the function at the common point, namely, the derivative of the function there. The next section presents the tools of differentiation in seven subsections while the final one presents higher derivatives, namely, derivatives of derivatives. Because this chapter mainly concerns mathematical tools, the examples are algebraic or are very simple biological examples. The next chapter is devoted entirely to the uses, usually the biological uses, of derivatives.

6.1 AVERAGE RATE OF CHANGE

The weight of any biological object, such as an organ in a higher plant or animal, changes through time; it can be denoted by a function $W(t)$. At a later time $(t + \Delta t)$ the weight of the organ would be $W(t + \Delta t)$ so that from time t to $(t + \Delta t)$ its weight would change from $W(t)$ to $W(t + \Delta t)$. We will denote this change in weight as:

$$\Delta W = W(t + \Delta t) - W(t) .$$

The symbol Δ is the Greek capital letter delta which is usually used in mathematical applications to denote a small change. The average rate of change (growth) in the weight of the organ during the time interval $(t, t + \Delta t)$ results when ΔW is divided by Δt, the length of the time interval:

- The AVERAGE RATE OF CHANGE of a function f over some interval t to $t + \Delta t$ in its domain is

$$\text{A.R.C.} = \text{Average rate of change} = \frac{f(t + \Delta t) - f(t)}{\Delta t}$$

A graph provides a clear representation of an average rate of change. In Figure 6.1, suppose f represents the function of interest. Over a time interval $(t, t + \Delta t)$, the function f changes by $f(t + \Delta t) - f(t)$ in the ordinate direction during a time change of Δt. From Chapter 4 we know that the change in the ordinate divided by the corresponding change in the abscissa gives the slope of a straight line. Thus the average rate of change merely gives the slope of the line connecting the two points on the curve $(t, f(t))$ and $(t + \Delta t, f(t + \Delta t))$. This line usually is called the SECANT LINE joining the two points.

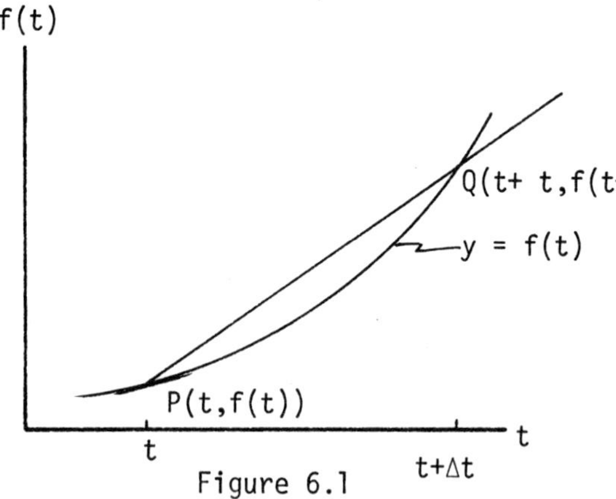

Figure 6.1

EXAMPLE 6.1: The growth in weight of a cancerous tumor sometimes can be described by a function as simple as $W(t) = bt^2$ where b is some constant and t (time) is the independent variable. The average rate of change in the weight of the tumor over some time interval of length Δt is given by

$$\text{A.R.C.} = \frac{\Delta W(t)}{\Delta t} = \frac{W(t + \Delta t) - W(t)}{(t + \Delta t) - t}$$

$$= \frac{b(t + \Delta t)^2 - bt^2}{\Delta t}$$

$$= \frac{bt^2 + 2bt\ \Delta t + b(\Delta t)^2 - bt^2}{\Delta t} .$$

After simplification this becomes

$$\frac{\Delta W(t)}{\Delta t} = 2bt + b\Delta t .$$

This represents the slope of the secant line between the points $(t,W(t))$ and $(t + \Delta t,W(t + \Delta t))$. This quantity has the units of weight per unit of time.

Suppose we want to know the average rate of change in the weight of the tumor between t = 100 days and 101 days so Δt = 1 day. The expression A.R.C. = $2bt + b\Delta t$ gives the average rate of growth during this time interval as A.R.C. = $2b(100) + b(1) = 200b + b = 201b$ units of weight per day.

If, on the other hand, we were to calculate the average rate of change of the weight between t = 50 days and t = 51 days, A.R.C. = 101b units of weight per day. This illustrates that the average rate of change usually changes with different choices of time intervals. Draw a graph of the function and examine the slope of the secant line for the above points. Note that the slope of the secant line changes as its end points change. ▮

EXAMPLE 6.2: Consider the progress of a red blood cell as it moves through a particular capillary bed. When a single cell or a small group of cells is identified with a radioactive tag, its position can be observed at various times after a reference time taken as t = 0. Its position translates directly into distance traveled (D) in various amounts of time:

time elapsed (t)	distance traveled (D)
0	
0.2 seconds	0.16 millimeters
0.4 seconds	0.32 millimeters
0.6 seconds	0.48 millimeters

This gives specific values of distance traveled as a function of time; the graph shown in Figure 6.2 connects these points with a staight line. The ratio $\Delta D/\Delta t$ gives the average speed of the red blood cell for any time interval during which the process remains stable. Here

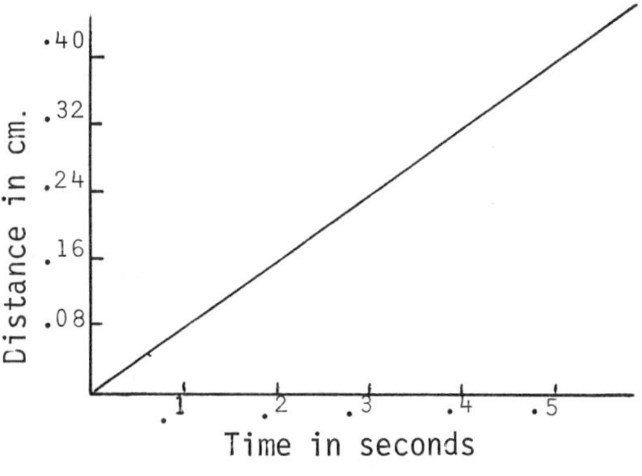

Figure 6.2

$$\frac{\Delta D}{\Delta t} = \frac{0.16 \text{ mm.}}{0.2 \text{ sec.}} = 0.8 \text{ mm./sec.} \quad \blacksquare$$

The above example holds only for brief periods of time and for a red blood cell well into the capillary bed. On the arterial side of a capillary bed, blood still has a pulsed flow which cannot be linear, but the pulsing vanishes in the capillaries.

6.2 THE RATE OF CHANGE AT A POINT

The average rate of change introduced in the previous section applies over an entire interval. The pertinent question for the biologist concerns the rate of change at a particular point. An average rate of change evaluated for progressively smaller increments of the independent variable tends to the instantaneous rate of change of the function, namely its derivative. Thus we now give this limit definition:

• The DERIVATIVE or INSTANTANEOUS RATE OF CHANGE of a function f at a point x in its domain is defined as

$$\lim_{\Delta x \to 0} \frac{f(x + \Delta x) - f(x)}{\Delta x}.$$

Graphically this can be represented as in Figure 6.3. The line through the two points $P(x,f(x))$ and $Q(x + \Delta x, f(x + \Delta x))$ has a slope equal to the average rate of change. As Δx becomes smaller and smaller, the secant line, PQ, tends toward the tangent to the curve at the point P. Thus, the slope of the line PL equals the instantaneous rate of change of the function f at the point x.

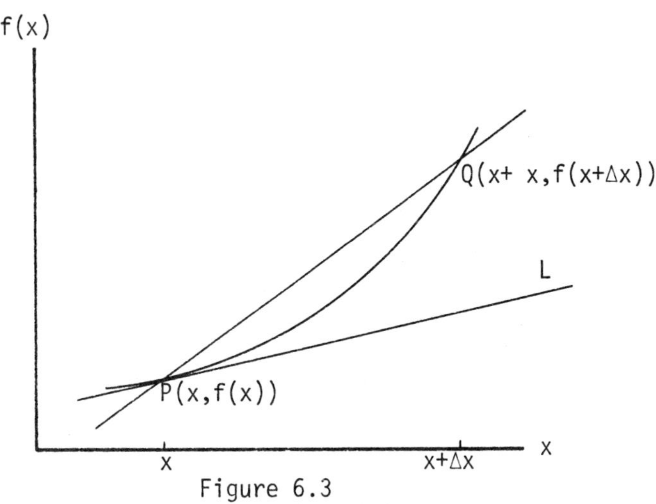

Figure 6.3

The symbol Δx is used widely to denote an incremental change in x, but any other symbol will work equally well. Here x symbolizes the independent variable. In a specific situation a more suggestive letter can be used, like our use of t in Examples 6.1 and 6.2.

When $y = f(x)$, the derivative of the function is symbolized by the following notations in various contexts:

$$\frac{dy}{dx}, \frac{df(x)}{dx}, y', f'(x), D_x f(x), Dy, \text{ or } \dot{y}.$$

The first two emphasize that the derivative is with respect to x, but differ in their emphasis of the functional dependence on x. The next two indicate the derivative by ', but lose an indication of the variable relative to which the derivative is being taken; this poses no problem so long as the independent variable has an obvious contextual meaning, otherwise it can become troublesome. The next two use the capital letter D to indicate the derivative, and again differ in their emphasis on the involvement of the independent variable. The final notation survives from its original introduction by Newton (1642-1727). Even though it cannot identify the independent variable, you may see it used occasionally. We usually will use only the first four with the choice resting on clarity in each situation. You should recognize the equivalence of the other forms.

A point on nomenclature: The word derivative is a noun for which the word differentiate is the corresponding verb. For example when you differentiate a function, you get a derivative. When a function has a derivative, it sometimes is described as being differentiable.

EXAMPLE 6.2 (continued): This example concerned the rate of progress of a red blood cell through a capillary bed. The original situation did not apply on the arterial edge of the capillary bed because the pulsed blood flow still appears there. To work on this edge, we need to use a shorter time interval to detect the variation in velocity.

The following set of observations could arise if we measured the progress of the cell every 0.1 second:

Time	Distance (mm.)	ΔD	$\frac{\Delta D}{\Delta t}$
0	0.000		
		0.048	0.48
0.1	0.048		
		0.096	0.96
0.2	0.144		
		0.144	1.44
0.3	0.288		
		0.096	0.96
0.4	0.384		
		0.048	0.48
0.5	0.432		
⋮	⋮	⋮	⋮

Notice now that the average speed of the blood particle differs greatly from interval to interval. A still smaller time increment, of say $t = 0.05$ second, might yield the following:

Time	Distance (mm.)	ΔD	$\frac{\Delta D}{\Delta t}$
0.00	0.000		
		0.016	0.32
0.05	0.016		
		0.032	0.64
0.10	0.048		
		0.040	0.80
0.15	0.088		
		0.056	1.12
0.20	0.144		
⋮	⋮	⋮	⋮

Concentrate on the particle's movement time near t = 0.10 second. The A.R.C. $\frac{\Delta D}{\Delta t}$ was 0.48 and 0.96 in the incremental intervals adjoining t = 0.1 second in the first set of measurements, but in the second set it changed to 0.64 and 0.80. These latter values differ by much less than the former ones. If we evaluate $\frac{\Delta D}{\Delta t}$ for progressively shorter time intervals, the larger value of $\frac{\Delta D}{\Delta t}$ would get smaller while the smaller value would get larger, until eventually they would be practically indistinguishable. The common value of $\frac{\Delta D}{\Delta t}$ as $\Delta t \to 0$ is its derivative, denoted as $\frac{dD}{dt}$. ▌

EXAMPLE 6.3: Find the derivative of f(x) = c, where c is a constant.

Its derivative, evaluated from the definition, is

$$\frac{df(x)}{dx} = \lim_{\Delta x \to 0} \frac{f(x + \Delta x) - f(x)}{\Delta x}$$

$$= \lim_{\Delta x \to 0} \frac{c - c}{\Delta x} = 0 .$$

This says that the constant function does not change with x. This fact should be apparent when you recall that the graph of f(x) = c is a straight line parallel to the x-axis and has a slope of zero. ▌

EXAMPLE 6.4: Find the derivative of the function describing a straight line, y = mx + b.

We must revert to the limit definition of the derivative. The definition had $\Delta x \to 0$, but any symbol can be used for the incremental quantity, provided it is used throughout, as was noted after the definition. The symbol h sometimes is used in place of Δx:

$$\frac{dy}{dx} = \lim_{h \to 0} \frac{y(x + h) - y(x)}{h}$$

$$= \lim_{h \to 0} \frac{m(x + h) + b - (mx + b)}{h}$$

$$= \lim_{h \to 0} \frac{mx + mh + b - mx - b}{h}$$

$$= \lim_{h \to 0} \frac{mh}{h} = \lim_{h \to 0} m = m \ .$$

Thus, the derivative of a function describing a straight line equals the slope of the line. Note that this result agrees with Example 6.3, for if $m = 0$, $dy/dx = m = 0$, the same result is there. This merely confirms the fact that the constant function can be represented as a horizontal line, that is, a line with a slope of zero. ▌

EXAMPLE 6.5: Find the instantaneous rate of change of the function $y(x) = x^3 - x + 1$ at $x = 2$.

From the limit definition of the derivative we have

$$\frac{dy}{dx} = \lim_{\Delta x \to 0} \frac{y(x + \Delta x) - y(x)}{\Delta x}$$

$$= \lim_{\Delta x \to 0} \frac{[(x + \Delta x)^3 - (x + \Delta x) + 1] - (x^3 - x + 1)}{\Delta x}$$

$$= \lim_{\Delta x \to 0} \frac{x^3 + 3x^2\Delta x + 3x(\Delta x)^2 + (\Delta x)^3 - x - \Delta x + 1 - x^3 + x - 1}{\Delta x}$$

$$= \lim_{\Delta x \to 0} \frac{3x^2\Delta x + 3x(\Delta x)^2 + (\Delta x)^3 - \Delta x}{\Delta x}$$

$$= \lim_{\Delta x \to 0} (3x^2 + 3x\Delta x + (\Delta x)^2 - 1) = 3x^2 - 1 \ .$$

Thus the instantaneous rate of change of $y = x^3 - x + 1$ equals $3x^2 - 1$ for any value of x.

Again, $3x^2 - 1$ represents the slope of the line tangent to the curve $y = x^3 - x + 1$ at any x. To evaluate this derivative at x = 2, insert x = 2 into the expression for the derivative. This frequently is denoted by

$$\left.\frac{dy}{dx}\right|_{x = 2} = 3(2)^2 - 1 = 11 .$$

This derivative tells us that the function is increasing at a rate of 11 units in the ordinate direction for every unit change in the abscissa direction at x = 2. (The value of the derivative is different at other values of x.) Thus, the line tangent to the curve $y = x^3 - x + 1$ at x = 2 has a slope of 11. Now that we have the slope, we can determine the equation of the tangent line by finding a point which is on the tangent line. The tangent line must pass through the point of tangency: At x = 2, y = 7 so the equation of the line is $y - 7 = 11(x - 2)$ or $y = 11x - 15$. ∎

The differentiability and continuity of a function at a point have a noteworthy relation: When a function has a derivative at a point, it must be continuous there. To see this, first observe that the derivative of a function f at x_0 cannot be defined unless $f(x_0)$ exists. For $(f(x_0 + \Delta x) - f(x_0))/\Delta x$ to remain defined as $\Delta x \to 0$, $\lim_{\Delta x \to 0} f(x_0 + \Delta x)$ must equal $f(x_0)$; otherwise the numerator would remain nonzero so the limit of the ratio would become undefined. Thus the conditions for continuity are satisfied. (The definition of continuity in Section 5.2 dealt with $\lim_{x \to x_0} f(x)$; the same limit is being considered here although this may not seem immediately obvious. Observe that $x_0 + \Delta x \to x_0$ as $\Delta x \to 0$.)

EXAMPLE 6.5 (continued): The polynomial function

$$y(x) = x^3 - x + 1 \qquad \text{has} \qquad y'(x) = 3x^2 - 1 .$$

Because $y'(x)$ exists for all x, $y(x)$ is continuous at all x. This result was established using a different argument in Example 5.9 where the continuity of a general polynomial was investigated. ∎

Above we demonstrated that when a function has a derivative at a point, it must be continuous there. The reverse does not hold; a function can be continuous at a point without being differentiable there. A continuous function can approach one of its points of continuity with a positive slope from one side, but with a negative slope from the other side. The derivative cannot be defined at such a point because $[f(x_0 + \Delta x) - f(x_0))/\Delta x]$ has a positive right hand limit, but a negative left hand limit. Such a point of continuity often is called a CORNER POINT because the function has to turn sharply there.

EXAMPLE 6.6: The function defined by $f(x) = |x - 4| + 2$ has the graph shown in Figure 6.4. Even though f(x) is continuous for all x, in particular at x = 4, it has no derivative at x = 4. To see this, recall the definition of an absolute value. Specifically, for x < 4,

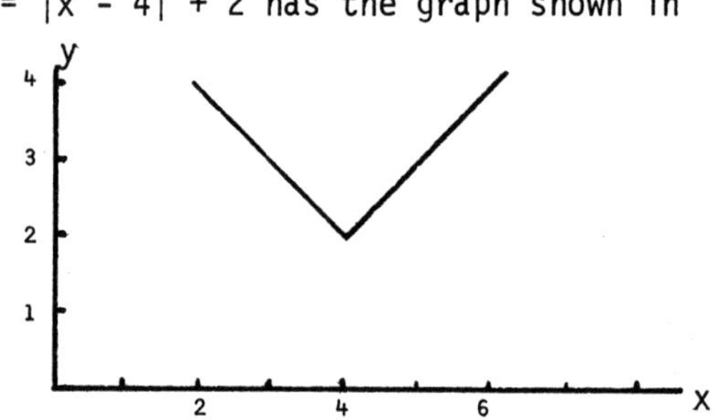

Figure 6.4

$$f(x) = |x - 4| + 2$$
$$= -(x - 4) + 2 = 6 - x$$

so rewrite f as

$$f(x) = \begin{cases} x - 2 & \text{for } x \geq 4 \\ 6 - x & \text{for } x < 4 \,. \end{cases}$$

This does satisfy $\lim_{x \to 4} f(x) = f(4)$. For $x \geq 4$

$$\lim_{\Delta x \to 0^+} \frac{f(4 + \Delta x) - f(4)}{\Delta x} = \lim_{\Delta x \to 0^+} \frac{(4 + \Delta x - 2) - (4 - 2)}{\Delta x} = 1 \,,$$

but for x < 4,

$$\lim_{\Delta x \to 0^-} \frac{f(4 + \Delta x) - f(4)}{\Delta x} = \frac{(6 - 4 - \Delta x) - (6 - 4)}{\Delta x} = -1 \,.$$

Thus from Theorem 5.9, the limit of the difference quotient does not exist at x = 4 so f(x) has no derivative there. ∎

EXERCISES

1. a. Find the average rate of change in $y = (x + 2)^2$ over the interval $(x, x + \Delta x)$.
   b. Find the instantaneous rate of change of y for any value of x.
   c. Find the average rate of change as x goes from 3 to 3.1, 4 to 4.1, and 5 to 4.8.
   d. Sketch the curve and show the rates of change in parts a—c.

For problems 2-7, find the average rate of change of the function over the interval $(x, x + \Delta x)$.

2. $y = 2x + 3$
3. $f(x) = \frac{2}{x}$
4. $y = \frac{2}{2x + 3}$
5. $y = \frac{1}{x^3}$
6. $y = \sqrt{3x}$
7. $y = 3x^3 - 2x + 4$
8. Find f' in problem 3.
9. Find y' in problem 7.
10. Consider a population in which the number of individuals in the population can be described by $N(t) = N_0 \frac{t^2}{1 + t^2}$ where N_0 is a constant.
    a. Find the expression for the average rate of change between t and $t + \Delta t$.
    b. Find the expression for the instantaneous rate of change for any value of t.
    c. Find the average rate of change in the population during the interval of t = 5 units to t = 5.2 units.
    d. Find the instantaneous rate of change in the size of the population when t = 5 units.
    e. Explain what the result obtained in part (d) means to you.

11. Assume that a population has an initial size of N_0 and grows according to the law $N(t) = N_0 + kt^2$ where k is a positive constant and t is time measured in days. Find the average rates of growth in the time intervals from

    a. $t = 0$ to $t = 3$ b. $t = 3$ to $t = 10$ c. $t = 0$ to $t = 10$.

12. Find the instantaneous rate of growth in problem 11 when

    a. $t = 0$ b. $t = 3$ c. $t = 10$.

13. Explain the difference between average rate of growth and instantaneous rate of growth assuming you are talking to a biologist with less training in mathematics than you have. (He knows nothing about limits.)

14. In what situation does it make no difference whether we speak of average rate of change or instantaneous rate of change?

15. Equation 2.1 of Chapter 2 relates the volume, V, in board feet of a tree to the diameter in inches, d, through the expression

$$V = 10 + 0.007(d - 5)^3 \quad \text{for} \quad d > 10.$$

    Find the instantaneous rate of change of the volume when $d = 15$ inches.

16. On his 12th birthday a child was measured as 105 cm. tall and on his 13th birthday his height was measured as 120 cm. If we assume a constant monthly growth rate, what is this average monthly growth rate?

17. At the start of a bacterial experiment a culture was found to contain 10^4 individuals. The growth was observed and it was found that $N(t) = 10^4(1 + 0.75t + t^2)$ described the number, N, of bacteria at any time t. If t is expressed in hours

    a. Find the expression for the average rate of growth of the population between times t and $t + \Delta t$.
    b. Find the expression for $\frac{dN}{dt}$.
    c. Evaluate part (b) when $t = 1/2$ hour and also when $t = 3$ hours.

18. Consider the length L of a metal bar as a function of the temperature, T. If a change in length from L = 1 meter to 1.000013 meters occurs when the temperature goes from 0°C to 0.1°C, what is the average rate of change of the length with respect to the temperature?

19. Suppose that a right-circular cylinder has a constant height of 12 inches. If V is the volume of the cylinder in cubic inches, and r is the radius of the base in inches, find the average rate of change of V with respect to r as r changes from

    a. 4 to 4.4 inches b. 4 to 4.1 inches c. 4 to 4.01 inches

    d. Find the instantaneous rate of change of V with respect to r when r = 4 inches.

6.3 THE TOOLS OF DIFFERENTIATION

The derivative of a function is defined in terms of the limiting average rate of change. This definition was used to evaluate derivatives in Example 6.3, 6.4 and 6.5, but its application was tedious, even for simple functions. The limit definition of a derivative can be applied to general kinds of functions to produce rules of differentiation. Once established, these rules can be used to directly get most derivatives. Regard the rules we will develop as only tools to simplify the process of getting derivatives; the limit definition remains central to their interpretation.

This section is divided into seven subsections, each dealing with a separate tool. Each subsection begins with its central result. Examples then illustrate the use of the result; a mathematical development of the tool concludes most subsections. The use of a tool, mathematical or otherwise, may not demand an understanding of how it works. Nevertheless, the more you understand about a tool the better you usually can use it. The proofs in the following sections were included for this reason. They were deferred to the end of the subsection so you could see how each tool worked before you encountered its proof.

6.3a THE DERIVATIVE OF ax^n

First consider the derivative of the general form $y = ax^n$:

For any constants a and n

- $$\frac{d}{dx}(ax^n) = anx^{n-1} , \qquad (6.1)$$

- $$\frac{d}{dx}(a) = 0 ,$$

- $$\frac{d}{dx}(ax + b) = a .$$

EXAMPLE 6.7: If $y = 3x^{10}$ find $\frac{dy}{dx}$.

$$\frac{dy}{dx} = \frac{d}{dx} 3x^{10} = 3(10)x^9 = 30x^9 . \quad \blacksquare$$

EXAMPLE 6.8: If $y = 4x^{-3}$, find $\frac{dy}{dx}$.

$$\frac{dy}{dx} = \frac{d}{dx} 4x^{-3} = 4(-3)x^{-4} = -12x^{-4} . \quad \blacksquare$$

Now we will develop the proof of Equation 6.1.

PROOF OF THE POWER RULE: For the moment we will verify Equation 6.1 only for positive integer values of n. Apply the limit definition of a derivative to $y(x) = ax^n$:

$$\frac{dy}{dx} = \lim_{\Delta x \to 0} \frac{y(x + \Delta x) - y(x)}{\Delta x} = \lim_{\Delta x \to 0} \frac{a(x + \Delta x)^n - ax^n}{\Delta x}$$

$$= a \lim_{\Delta x \to 0} \frac{(x + \Delta x)^n - x^n}{\Delta x} \qquad (6.2)$$

Recall the binomical expansion:

$$(r + s)^n = r^n + nr^{n-1}s + \frac{n(n - 1)}{2!} r^{n-2}s^2 + \cdots + s^n$$

When this is applied to Equation 6.2 with $r = x$ and $s = \Delta x$, we get

$$\frac{dy}{dx} = a \lim_{\Delta x \to 0} \frac{\left[x^n + nx^{n-1}(\Delta x) + \frac{n(n-1)}{2} x^{n-2}(\Delta x)^2 + \cdots + (\Delta x)^n\right] - x^n}{\Delta x}$$

$$= a \lim_{\Delta x \to 0} \left[\frac{nx^{n-1}(\Delta x) + \frac{n(n-1)}{2} x^{n-2}(\Delta x)^2 + \cdots + (\Delta x)^n}{\Delta x}\right]$$

Because the numerator and denominator have a common factor of Δx, this becomes

$$\frac{dy}{dx} = a \lim_{\Delta x \to 0} \left[nx^{n-1} + \frac{n(n-1)}{2} x^{n-2}(\Delta x) + \cdots + (\Delta x)^{n-1}\right]$$

$$= a \left[nx^{n-1} + \frac{n(n-1)}{2} x^{n-2} (\lim_{\Delta x \to 0} \Delta x) + \cdots + (\lim_{\Delta x \to 0} \Delta x)^{n-1}\right]$$

$$= anx^{n-1} .$$

The other two results stated in the above box were established earlier, in Examples 6.3 and 6.4. ▼

6.3b THE DERIVATIVE OF A SUM OF FUNCTIONS

The derivative of a sum of differentiable functions equals the sum of the derivatives of the individual functions:

- If the functions f and g have derivatives at a common point x in their respective domains, then so does $y(x) = f(x) + g(x)$ with

$$\frac{dy(x)}{dx} = \frac{df(x)}{dx} + \frac{dg(x)}{dx} \qquad \text{or} \qquad y'(x) = f'(x) + g'(x) \qquad (6.3)$$

Repeated application of Equation 6.3 extends the result to the sum of any number of differentiable functions. For example,

$$\frac{d}{dx}\left[f(x) + g(x) + h(x)\right] = f'(x) + g'(x) + h'(x) .$$

EXAMPLE 6.9: Evaluate the derivative of $y = 2x^3 + 6x^2 + 2x + 3$.

Recall that functions like this one (polynomials) are defined and continuous for all real numbers x. To apply Equation 6.3 as it stands, we must break y up into the sum of exactly two functions:

$$y = (2x^3 + 6x^2) + (2x + 3) .$$

Thus if we let $f(x) = 2x^3 + 6x^2$, and $g(x) = (2x + 3)$, then

$$\frac{dy(x)}{dx} = \frac{df(x)}{dx} + \frac{dg(x)}{dx} = \frac{d}{dx}(2x^3 + 6x^2) + \frac{d}{dx}(2x + 3) .$$

Again apply Equation 6.3 to each of these parts, and then use $\frac{d}{dx}(ax^n) = anx^{n-1}$ three times with $\frac{d}{dx}(3) = 0$:

$$\frac{dy(x)}{dx} = \frac{d}{dx}(2x^3) + \frac{d}{dx}(6x^2) + \frac{d}{dx}(2x) + \frac{d}{dx}(3)$$

$$= (2)(3)x^2 + (6)(2)x + 2 + 0 = 6x^2 + 12x + 2 .$$

The same result occurs if we use the extended version of Equation 6.3. ▮

A straight forward extension of the approach used in this example shows that all polynomials are differentiable at all values of their independent variable. A function is differentiated with respect to its independent variable, regardless of the symbol for the independent variable.

EXAMPLE 6.10: Find the derivative of $m(w) = (2w + 6)^3$ with respect to w.

The derivative of $m(w) = (2w + 6)^3$ can be evaluated by recourse to the binomial expansion as

$$\frac{d}{dw} m(w) = \frac{d}{dw}(2w + 6)^3$$

$$= \frac{d}{dw}(8w^3 + 72w + 216w + 216)$$

$$= \frac{d}{dw} 8w^3 + \frac{d}{dw} 72w^2 + \frac{d}{dw} 216w + \frac{d}{dw} 216$$

$$= 24w^2 + 144w + 216 .$$ ▮

An alternate method of calculating this derivative will be given in the next section.

We now return to prove Equation 6.3.

PROOF OF THE SUM RULE: Wherever $y(x) = f(x) + g(x)$ exists, the definition of a derivative gives

$$\frac{dy(x)}{dx} = \lim_{\Delta x \to 0} \frac{y(x + \Delta x) - y(x)}{\Delta x}$$

$$= \lim_{\Delta x \to 0} \frac{[f(x + \Delta x) + g(x + \Delta x) - (f(x) + g(x))]}{\Delta x} .$$

By rearranging the terms, this becomes

$$\frac{dy(x)}{dx} = \lim_{\Delta x \to 0} \left[\frac{f(x + \Delta x) - f(x)}{\Delta x} + \frac{g(x + \Delta x) - g(x)}{\Delta x}\right] .$$

Because the limit of a sum equals the sum of the individual limits, this becomes

$$\frac{dy(x)}{dx} = \lim_{\Delta x \to 0} \frac{f(x + \Delta x) - f(x)}{\Delta x} + \lim_{\Delta x \to 0} \frac{g(x + \Delta x) - g(x)}{\Delta x}$$

$$= \frac{df(x)}{dx} + \frac{dg(x)}{dx}$$

$$= f'(x) + g'(x) . \blacktriangledown$$

This proof extends to more than the sum of two functions simply by including more terms. Thus to find the derivative of a polynomial, simply take the derivative of the polynomial term by term.

6.3c THE CHAIN RULE FOR DERIVATIVES

You will use the chain rule given below almost every time you evaluate a derivative. You need to completely master it for your work with derivatives. It states:

- When $y = f(u)$ and $u = g(x)$, where f and g are differentiable functions, then

$$\frac{dy}{dx} = \left[\frac{dy}{du}\right]\left[\frac{du}{dx}\right] . \tag{6.4}$$

This result is known as the CHAIN RULE for derivatives.

Of course it applies for only those values of x where the functions are defined and the constituent derivatives exist. The next example illustrates its use before we discuss it.

EXAMPLE 6.10 (continued): Recall that if $m(w) = (2w + 6)^3$, $m'(w) = 24w^2 + 144w + 216$. The function m has two essentially different parts, namely its cubic structure, and what is cubed. If we write $y = f(u) = u^3$ and $u = g(w) = 2w + 6$, then the functions m and y have the same values at all w, but they have a different emphasis on the involvement of w. Upon application of the chain rule,

$$\frac{dm(w)}{dw} = \frac{dy}{dw} = \frac{dy}{du}\frac{du}{dw}$$

$$= \left[\frac{d}{du}(u^3)\right]\left[\frac{d}{dw}(2w + 6)\right] = (3u^2)(2) .$$

To complete this, we must substitute $u = g(w) = (2w + 6)$ to get

$$\frac{dm(w)}{dw} = (3u^2)(2) = 6(2w + 6)^2 .$$

This is the same result as we obtained before because $6(2w + 6)^2 = 24w^2 + 144w + 216$.

This example does not fully illustrate the utility of the chain rule because m(w) can be expanded into a polynomial easily. For a function like $r(x) = (3x^4 + 12x + 1)^{12}$, the algebraic labor makes expansion almost prohibitive; the chain rule gives the derivative of r(x) simply as

$$r'(x) = 12(3x^4 + 12x + 3)^{11}\frac{d}{dx}(3x^4 + 12x + 1)$$

$$= 12(3x^4 + 12x + 3)^{11}(12x^3 + 12)$$

$$= 144(x^3 + 1)(3x^4 + 12x + 3)^{11} . \quad \blacksquare$$

The chain rule allows you to decompose a function into more elementary parts. When $y = f(u)$ and $u = g(x)$, we really mean $y(x) = f(g(x))$, namely, y is composed of two basic patterns described by the separate functions f and g. Some mathematicians prefer to emphasize the separate roles of f and g by writing Equation 6.4 as

$$\frac{dy(x)}{dx} = \frac{df(u)}{du}\frac{dg(x)}{dx}$$

This form clearly indicates that the first derivative, $\frac{dy}{du}$ or $\frac{df(u)}{du}$ views the primary function y as a function of u, not of x; the second derivative, denoted by either $\frac{du}{dx}$ or $\frac{dg(x)}{dx}$, concentrates on how $u = g(x)$, the argument of f, changes with x. The form of the chain rule stated immediately above does precisely symbolize its sense, but most students find the form $\frac{dy}{dx} = \frac{dy}{du}\frac{du}{dx}$ more suggestive to remember.

The chain rule is used with many functions. For example it provides the basis for writing

$$\frac{du^n}{dx} = nu^{n-1}\frac{du}{dx} \tag{6.5}$$

when we set $f(u) = u^n$ and let $u = g(x)$ be any differentiable function of x. Note that if $u = x$, the above application of the chain rule reduces to Equation 6.1 as it should:

$$\frac{dx^n}{dx} = nx^{n-1}\frac{dx}{dx} = nx^{n-1}, \qquad \text{because} \qquad \frac{dx}{dx} = 1 .$$

<u>EXAMPLE 6.11</u>: If $y = (x^3 + 1)^{1/3}$, find $\frac{dy}{dx}$.

Use of the chain rule with $y = f(u) = u^{1/3}$ and $u = g(x) = (x^3 + 1)$, gives

$$\frac{dy}{dx} = \frac{d}{du}(u^{1/3})\frac{d}{dx}(x^3 + 1) = \frac{1}{3}(u^{-2/3})\,3x^2 = x^2(x^3 + 1)^{-2/3}$$

upon substitution of $u = g(x) = (x^3 + 1)$. ▌

<u>EXAMPLE 6.12</u>: Our proof of $\frac{dx^n}{dx} = nx^{n-1}$ in section 6.3a assumed that n was a positive integer. Now suppose it is a positive rational number, that is, $n = p/q$ where both p and q are positive integers. Apply the chain rule to $y = x^{p/q}$ or equivalently to $y^q = x^p$:

$$\frac{d}{dx}(y^q) = \frac{d}{dx}(x^p) \qquad \text{or} \qquad qy^{q-1}\frac{dy}{dx} = px^{p-1} .$$

Now solve the last expression for the derivative of interest.

$$\frac{dy}{dx} = \frac{px^{p-1}}{qy^{q-1}} .$$

To complete this, recall that $y = x^{p/q}$:

$$\frac{dy}{dx} = \frac{p}{q}\,\frac{x^{p-1}}{(x^{p/q})^{q-1}} = \frac{p}{q}\,x^{(p/q)-1} = nx^{n-1} .$$

This shows that the result given in section 6.3a holds even if n is not an integer. It will be left as a later exercise for you to show that the result also holds for negative values for n. ▌

To complete this section we need to develop Equation 6.4.

PROOF OF THE CHAIN RULE: From the definition of a derivative,

$$\frac{dy}{dx} = \lim_{\Delta x \to 0} \frac{\Delta y}{\Delta x}$$

where Δy is the incremental change in y for an incremental change Δx in x. However, an incremental change in x also produces an incremental change Δu in u which will go to 0 as Δx does. Provided $\Delta u \neq 0$, we can write

$$\frac{dy}{dx} = \lim_{\Delta x \to 0} \frac{\Delta y}{\Delta x} = \lim_{\Delta x \to 0} \left[\frac{\Delta y}{\Delta u}\,\frac{\Delta u}{\Delta x}\right] = \left[\lim_{\Delta x \to 0} \frac{\Delta y}{\Delta u}\right]\left[\lim_{\Delta x \to 0} \frac{\Delta u}{\Delta x}\right]$$

$$= \left[\lim_{\Delta u \to 0} \frac{\Delta y}{\Delta u}\right]\left[\lim_{\Delta x \to 0} \frac{\Delta u}{\Delta x}\right] = \left[\frac{dy}{du}\right]\left[\frac{du}{dx}\right] \qquad (6.6)$$

A careful handling of the situation when $\Delta u = 0$ formalizes this observation: $\Delta u = 0$ corresponds to u being constant near x, this implies both that y does not change (because the argument of f does not change) and that $\frac{du}{dx} = 0$ near x. Thus the chain rule gives the correct result, namely $\frac{dy}{dx} = 0$.

This proof may look like mere symbolic hand waving. The introduction of $\frac{\Delta u}{\Delta u} = 1$ into Equation 6.6 and the subsequent association of its denominator with Δy, and its numerator with Δx, has substantial content. It only formalizes an observation we have already made: The change in y arises partly from the structure of f and partly from how its independent variable, g(x), depends on x. ▼

EXERCISES

Find the derivative of the following expressions with respect to their independent variable.

1. $y = 2x^2$
2. $y = (3v + 4)^5$
3. $y = 4t^2 + 3t^3 - t + 3$
4. $y = (3x - 1)^{1/2}$
5. $y = 1/x^3$
6. $y = \dfrac{12x^4 - 3x^3 + 2x^2 + x - 7}{x}$
7. $y = (3x^4 - 2x^{-3} + 4x)^4$
8. $y = (2 - 7x)^{11}$
9. $y = (4 - w^2)^{1/2}$
10. $y = Ax^3 + Bx^2 + Cx + D$
11. $y = \sqrt{4x - 1}$
12. $y = (2x^2 + 3)(x + 2)$
13. $y = x^2 - 2x - 1/x^2$
14. $y = \sqrt[5]{3x^2 + 2x + 1}$
15. $y = x\sqrt{x + 1}$
16. Find the derivative of $(3x^2 + 4)^2$ by using the chain rule, and then by expanding the quantity before you take the derivative. Are the results the same?
17. Assume that a population grows from an initial size of N_0 according to $N = N_0 + kt^2$, where k is a positive constant and t is time measured in days. Find the instantaneous rate of growth when
    a. $t = 0$ b. $t = 3$ days c. $t = 10$ days.
18. Equation 2.1 of Chapter 2 relates the volume, V, in board feet of a tree to the diameter, D, in inches by the expression

    $$V = 10 + 0.007(D - 5)^3 \text{ for } D > 10.$$

    Use the chain rule to find the instantaneous rate of change of the volume; evaluate it at D = 15 inches.
19. Use the limit definition of a derivative and show that
    $\frac{d}{dx} kf(x) = k \frac{d}{dx} f(x)$ where k is a constant.

20. The specific heat of wood for temperatures between $0°C \leq T \leq 106°C$ is given by $S = 0.266 + 0.00116T$ where S is the specific heat and T is the temperature of wood in degrees centigrade. Find S' and evaluate this derivative at T = 20°C. Interpret your answer.

*21. Find y' if $y = \frac{x}{x - 1}$. (Hint: Divide the denominator into the numerator.)

6.3d THE DERIVATIVE OF A PRODUCT OF TWO FUNCTIONS

In Section 6.3b we discovered how to take the derivative of a sum of differentiable functions. This section presents the derivative of a product of two or more differentiable functions:

- When $y(x) = f(x)g(x)$,

$$\frac{dy}{dx} = f(x)\frac{dg(x)}{dx} + g(x)\frac{df(x)}{dx}$$

$$= f(x)g'(x) + g(x)f'(x) . \qquad (6.7)$$

In words Equation 6.7 states that the derivative of a product of two functions equals the first function times the derivative of the second function plus the second function times the derivative of the first function.

EXAMPLE 6.13: Find the derivative of $y = (2x^2 + 3)(4x + 1)^2$.

Regard this as the product of $f(x) = 2x^2 + 3$ and $g(x) = (4x + 1)^2$. Application of Equation 6.7 gives

$$\frac{dy}{dx} = f(x)g'(x) + g(x)f'(x)$$

$$= (2x^2 + 3)\frac{d}{dx}(4x + 1)^2 + (4x + 1)^2\frac{d}{dx}(2x^2 + 3) .$$

Now applying the power rule and then the chain rule:

$$\frac{dy}{dx} = (2x^2 + 3)(2)(4x + 1)\frac{d}{dx}(4x + 1) + (4x + 1)^2(4x) .$$

$$= (2x^2 + 3)(2)(4x + 1)(4) + (4x + 1)^2(4x) .$$

Finally factor (4x + 1) out of the above expression and collect terms to obtain:

$$\frac{dy}{dx} = (4x + 1)[8(2x^2 + 3) + 4x(4x + 1)]$$

$$= 4(4x + 1)(8x^2 + x + 6) \ . \quad \blacksquare$$

EXAMPLE 6.14: Find the derivative of $y = f(x)g(x)h(x)$ where f, g and h are differentiable functions.

Equation 6.7 tells us how to find the derivative of the product of two functions, but not of three. If however we write y as

$$y = [f(x)g(x)]\ h(x) \ ,$$

we have the product of two functions whose derivative is given by

$$\frac{dy}{dx} = [f(x)g(x)]\frac{d}{dx}h(x) + h(x)\frac{d}{dx}[f(x)g(x)] \tag{6.8}$$

Now apply Equation 6.7 to $\frac{d}{dx}[f(x)g(x)]$:

$$\frac{d}{dx}[f(x)g(x)] = f(x)\frac{d}{dx}g(x) + g(x)\frac{d}{dx}f(x) \ .$$

When this is substituted into Equation 6.8, we find that

$$\frac{dy}{dx} = f(x)g(x)\frac{d}{dx}h(x) + h(x)[f(x)\frac{d}{dx}g(x) + g(x)\frac{d}{dx}f(x)]$$

$$= f(x)g(x)h'(x) + f(x)h(x)g'(x) + g(x)h(x)f'(x) \ . \tag{6.9}$$

This type of procedure produces the derivative of the product of any number of desired terms. The result always has the same form as Equation 6.9. The derivative has an many additive terms as the product has functions; each term consists of the derivative of one of the functions times the product of the other functions. $\blacksquare$

We now give the proof of Equation 6.7.

PROOF OF THE PRODUCT RULE: From the definition of a derivative, we have

$$\frac{dy}{dx} = \lim_{\Delta x \to 0} \frac{y(x + \Delta x) - y(x)}{\Delta x} = \lim_{\Delta x \to 0} \left[\frac{f(x + \Delta x)g(x + \Delta x) - f(x)g(x)}{\Delta x}\right] .$$

By adding and subtracting $f(x + \Delta x)g(x)$ in the numerator, we have

$$\frac{dy}{dx} = \lim_{\Delta x \to 0} \left[\frac{f(x + \Delta x)g(x + \Delta x) - f(x + \Delta x)g(x) + f(x + \Delta x)g(x) - f(x)g(x)}{\Delta x}\right]$$

$$= \lim_{\Delta x \to 0} \left[\frac{f(x + \Delta x)[g(x + \Delta x) - g(x)]}{\Delta x} + \frac{g(x)[f(x + \Delta x) - f(x)]}{\Delta x}\right]$$

upon rearrangement of terms. The rules governing the use of limits give

$$\frac{dy}{dx} = \lim_{\Delta x \to 0} f(x + \Delta x) \lim_{\Delta x \to 0} \frac{g(x + \Delta x) - g(x)}{\Delta x} + \lim_{\Delta x \to 0} g(x) \lim_{\Delta x \to 0} \frac{f(x + \Delta x) - f(x)}{\Delta x} .$$

Look at the above to see that it contains

$$\lim_{\Delta x \to 0} \frac{f(x + \Delta x) - f(x)}{\Delta x} = \frac{df(x)}{dx} \quad \text{and} \quad \lim_{\Delta x \to 0} \frac{g(x + \Delta x) - g(x)}{\Delta x} = \frac{dg(x)}{dx}$$

so

$$\frac{dy}{dx} = f(x) \frac{dg(x)}{dx} + g(x) \frac{df(x)}{dx} .$$

Thus, the derivative of a product of two functions equals the first function times the derivative of the second, plus the second function times the derivative of the first. ▼

6.3e THE DERIVATIVE OF A QUOTIENT OF TWO FUNCTIONS

We have the following concerning a quotient:

- When $y(x) = \frac{f(x)}{g(x)}$ with f and g each being differentiable, then

$$\frac{dy(x)}{dx} = \frac{d}{dx}\frac{f(x)}{g(x)} = \frac{g(x)\frac{df(x)}{dx} - f(x)\frac{dg(x)}{dx}}{g(x)^2}$$

$$= \frac{g(x)f'(x) - f(x)g'(x)}{g(x)^2} \qquad (6.10)$$

In words Equation 6.10 states that the derivative of a quotient equals the denominator times the derivative of the numerator minus the numerator times the derivative of the denominator, all divided by the denominator squared. Again Equation 6.10 applies only at those values of x where f and g are both defined, differentiable, and $g(x) \neq 0$.

EXAMPLE 6.15: If $y = \frac{x^2}{(x + 1)}$, find y'.

Application of Equation 6.10 to $f(x) = x^2$ and $g(x) = (x + 1)$ gives

$$y' = \frac{dy}{dx} = \frac{(x + 1)\frac{d}{dx}(x^2) - x^2 \frac{d}{dx}(x + 1)}{(x + 1)^2} .$$

Upon computing the indicated derivatives, we find

$$y' = \frac{dy}{dx} = \frac{(x + 1)2x - x^2}{(x + 1)^2} = \frac{x(x + 2)}{(x + 1)^2} .$$

An alternate method for finding the derivative would be to treat the quotient as the product because

$$y = \frac{x^2}{x + 1} = x^2(x + 1)^{-1} .$$

The rule for finding the derivative of a product gives

$$y' = x^2 \frac{d}{dx}(x + 1)^{-1} + (x + 1)^{-1}\frac{d}{dx}(x^2)$$

and

$$y' = x^2(-1)(x + 1)^{-2} + (x + 1)^{-1}(2x) .$$

Obtaining a common denominator and simplifying again gives $y' = \frac{x(x + 2)}{(x + 1)^2}$. This illustrates that the derivative of a quotient does not depend on whether you view it as a quotient or as a product. ▌

EXAMPLE 6.16: Find the derivative of $y = \frac{(4x + 3)^2(x - 1)^3}{(3x^2 + 4)}$.

This quotient has a product in the numerator. Initially treating y as a quotient, we find that

$$\frac{dy}{dx} = \frac{(3x^2 + 4)\frac{d}{dx}[(4x + 3)^2(x - 1)^3] - (4x + 3)^2(x - 1)^3 \frac{d}{dx}(3x^2 + 4)}{(3x^2 + 4)^2} .$$

Applying Equation 6.7 to the product and using the chain rule, we find

$$\frac{dy}{dx} = \frac{(3x^2 + 4)[(4x + 3)^2 3(x-1)^2 + (x-1)^3 2(4x + 3)(4)] - (4x + 3)^2(x-1)^3 6x}{(3x^2 + 4)^2} .$$

Check to see that after simplification this becomes

$$\frac{dy}{dx} = \frac{(4x + 3)(x - 1)^2(36x^3 + 9x^2 + 98x + 4)}{(3x^2 + 4)^2} .$$

This resulted by proceeding step by step beginning with the function's basic form as a quotient. As you learn, begin by going step by step; write down each step in its entirety until the final result is reached. After some practice, you may be able to do some steps mentally. However, as a word of caution, do not try to do the problems mentally until you have thoroughly mastered a particular technique. ▌

We will now prove the statement of Equation 6.10.

PROOF OF THE QUOTIENT RULE: We will judiviously use the derivative of a product, Equation 6.7, to establish the derivative of a quotient. When $y(x) = f(x)/g(x)$, multiplication of both sides of this equation by $g(x)$ gives $f(x) = y(x)g(x)$. Now applying the derivative of a product to this,

$$\begin{aligned} f'(x) &= g(x)y'(x) + y(x)g'(x) \\ &= g(x)y'(x) + \frac{f(x)}{g(x)} g'(x) \end{aligned} \tag{6.11}$$

because $y(x) = \frac{f(x)}{g(x)}$. This can be solved for $y'(x)$, the quantity of interest, by first subtracting the last term in Equation 6.11 from both sides and then dividing both sides of the equation by $g(x)$ to give

$$y' = \frac{dy}{dx} = \frac{f'(x) - \frac{f(x)}{g(x)} g'(x)}{g(x)} .$$

After obtaining a common denominator and simplifying, this becomes

$$y' = \frac{g(x)f'(x) - f(x)g'(x)}{g(x)^2} . \quad ▼$$

Again, the derivative of a quotient of two differentiable functions equals the denominator times the derivative of the numerator minus the numerator times the derivative of the denominator, all divided by the denominator squared.

EXERCISES

Use the properties of derivatives developed thus far to evaluate $\frac{dy}{dx}$ for these functions:

1. $y = (2x^2 + 3)(x + 2)$

2. $y = \frac{x + 1}{x - 1}$; do two ways

3. $y = x\sqrt{x + 1}$

4. $y = (x - 1)^{1/2} (2x + 3)^5$

5. $y = x(x + 2)(x - 1)$

6. $y = \frac{2x^2 - 5x + 4}{x}$

7. $y = \frac{x}{x - 2}$; do three ways

8. $y = \frac{2x^2 + 1}{x - 1}$

9. $y = \frac{2x^2 - 3x + 4}{x^2 - 2x + 3}$

10. $y = (2x + 3)^4(3x - 2)^{7/3}$

11. $y = \frac{\sqrt{x - 2}}{\sqrt[3]{x + 2}}$

12. $y = \frac{\sqrt{x^3 - 1}}{(1 + 4x)^2}$

13. $y = \frac{(x^2 + 2)^2(x - 1)}{(3x^2 + 2x + 1)^3}$

14. $y = \frac{x^3 + 1}{x^2 - 3} (x^4 - 2x^{-3} + 1)$

15. $y = \left[\frac{3x^2 - 2}{2x + 1}\right]^7$

16. $y = 1/x^3$

17. $y = 1/g(x)$ where g is differentiable

18. $y = \frac{x^2 + 2}{(x - 1)(2x^3 - 1)}$

19. $y = (x^2 + 3)(x^2 - 3)^{-1}$

20. $y = \frac{(x + 2)^3(x - 1)^4}{(x - 5)^2 x^3}$

21. Suppose a biological phenomenon follows the law $y = \frac{1}{3} x^3 - x^2 - 3x + 3$, locate the highest and lowest points on the curve.

22. According to Stefan's law, a body emits radiant energy according to the formula $R = KT^4$ where R is the rate of emission of radiant energy per unit area, T is the Kelvin Temperature of the surface and K is a physical constant.
    a. Find $\frac{dR}{dT}$.
    b. At what rate does R increase with respect to T if T = 300°K?

23. Light-response curves (photosynthetic rate of the individual leaves as a function of light intensity) may be described by $P = \frac{P_{max} I}{I + K} - R$ where P_{max} is the asymptotic rate of photosynthesis, I is the light intensity, R is the dark respiration; R, P_{max} and K are constants for each species.

    a. Find $\frac{dP}{dI}$. b. Find $\lim_{I \to 0} P$. c. Sketch the curve.

24. Show that the rate of change of the area of a circle with respect to its radius is equal to its circumference.

25. Use the limit definition of a derivative to establish the formula for the derivative of a quotient.

*26. Show that Equation 6.1 also holds for

    a. an integer n less than zero.

    b. a rational number less than zero.

    (Hint: If n is a negative integer, m = -n is positive and $x^n = \frac{1}{x^m}$.)

6.3f THE DERIVATIVES OF LOGARITHMS

Logarithms have an important place in biology; Chapter 4 presented several important illustrations. To take the derivative of the logarithmic functions, we need:

- When u is a positive differentiable function of x,

$$\frac{d}{dx}(\ln u) = \frac{1}{u}\frac{du}{dx} \tag{6.12}$$

$$\frac{d}{dx}(\log_b u) = \frac{\log_b e}{u}\frac{du}{dx} \tag{6.13}$$

for any base b > 0.

We will use Equation 6.12 without proof. Its use is illustrated in the next examples.

EXAMPLE 6.17: If $y = \ln 2x^3$, find y'.

Let $u = 2x^3$ so we have

$$y' = \frac{d}{dx}(\ln 2x^3) = \frac{1}{2x^3} 6x^2 = \frac{3}{x}. \quad \blacksquare$$

EXAMPLE 6.18: $\frac{d}{dx}(\ln (x^2 + x - 1)) = \frac{2x + 1}{x^2 + x - 1}$. ∎

EXAMPLE 6.19: If $y = \log \sqrt{x}$, find y'.

$$\frac{dy}{dx} = \frac{d}{dx} \log \sqrt{x} = \frac{\log (e)}{\sqrt{x}} \frac{d}{dx} \sqrt{x}.$$

When $\sqrt{x}$ is treated as $x^{1/2}$, this becomes

$$\frac{dy}{dx} = \frac{\log (e)}{\sqrt{x}} \left[\frac{1}{2} x^{-1/2}\right] = \frac{\log e}{2x}.$$

The properties of logarithms could have reduced our work this way:

$$y = \log \sqrt{x} = \log x^{1/2} = \frac{1}{2} \log x,$$

$$\frac{dy}{dx} = \frac{1}{2} \frac{d}{dx} \log x = \frac{\log e}{2x}. \quad \blacksquare$$

6.3g THE DERIVATIVES OF EXPONENTIALS

The derivative of the exponential function, e^u, occurs frequently in biology:

- When u is a differentiable function of x, then

$$\frac{d}{dx}(e^u) = e^u \frac{du}{dx}. \qquad (6.14)$$

The function e^x occupies an exclusive niche among differentiable functions; it equals its derivative. This strange and sometimes wonderful property will resurface in several ways as we progress. It relates intimately to the main result in the last section, namely, $\ln x$ has a derivative of 1/x. This relation can be explained when we get to integration in Chapter 8, but a glimpse will emerge in the proof of Equation 6.14. (A reminder: e^u sometimes is written as exp[u].)

EXAMPLE 6.20: Find the derivative with respect to x of $e^{(x-m)^2}$.

Letting $(x-m)^2 = u$ in Equation 6.14, we have

$$\frac{d}{dx} e^{(x-m)^2} = e^{(x-m)^2} \frac{d}{dx} (x-m)^2 = 2(x-m)\, e^{(x-m)^2} \ . \quad \blacksquare$$

EXAMPLE 6.21: Find y' if $y = e^{x^{1/3}} = \exp[x^{1/3}]$.

Let $u = x^{1/3}$, and apply Equation 6.14:

$$y' = \frac{d}{dx} \exp[x^{1/3}] = \exp[x^{1/3}] \frac{d}{dx} (x^{1/3}) = \frac{x^{-2/3}}{3} \exp[x^{1/3}] \ . \quad \blacksquare$$

The proof of Equation 6.14 relies on the properties of logarithms and on knowing the derivative of ln u. We proceed as follows:

PROOF OF THE EXPONENTIAL RULE: Let $y = e^u$ where u is a differentiable function of x. From the properties of logarithms we have ln y = u. The derivative of both sides of ln y = u gives

$$\frac{du}{dx} = \frac{1}{y} \frac{dy}{dx} \ .$$

To solve for $\frac{dy}{dx}$, multiply both sides of the equation by y and substitute $y = e^u$:

$$\frac{dy}{dx} = y \frac{du}{dx} = e^u \frac{du}{dx} \ . \quad \blacktriangledown$$

EXERCISES

Find the derivatives with respect to x of the following expressions.

1. $y = e^{3x}$
2. $y = e^{3x^2}$
3. $y = xe^x$
4. $y = e^{1/x}$
5. $y = e^{x \ln x}$
6. $y = e^x \log x^2$

7. $y = e^{\sqrt{2 + x}}$

8. $y = \frac{e^x - e^{-x}}{2}$

9. $y = \frac{e^x - e^{-x}}{e^x + e^{-x}}$

10. $y = x^2 e^{x^3}$

11. $y = e^x e^{-x}$

12. $y = \frac{\ln (2x^2 + 3)}{e^x}$

13. $y = I_0 e^{\mu x}$

14. $y = I_0 e^{\mu x} - N_0 e^{-kx}$

15. $y = \ln (x + 3)^4$

16. $y = \ln \frac{\sqrt{x - 1}}{x^3}$

17. $y = \ln (\sqrt{x} - \sqrt{x + 2})$

18. $y = \ln\left[\frac{\sqrt{x^2 + 9}}{\sqrt{(x + 2)^2}}\right]$

19. $y = x^2 e^x \ln x$

20. $y = \frac{e^x + e^{-x}}{2}$

21. Modify the procedure used in proving Equation 6.14 to find $\frac{d}{dx} a^x$ where a is a positive constant.

22. If $W = \frac{W_0 (1 + b)}{1 + be^{-kt}}$, the logistic growth curve, where b, k and W_0 are constants, find $\frac{dW}{dt}$ and explain its meaning.

23. If $N = a\exp[-be^{-kt}]$, the Gompertz growth curve, where a, b, and k are positive constants, find $\frac{dN}{dt}$.

24. If $N = N_0(1 - be^{-kt})^3$, the Von Bertalanffy growth curve, where N_0, b, and k are positive constants, find $\frac{dN}{dt}$.

6.4 HIGHER DERIVATIVES

To investigate functions, you will need the rate of change of their rate of change as well as their rate of change. As the derivative of a function gives its rate of change, the derivative of its derivative, usually called its second derivative, gives the rate of change of the rate of change.

You have experienced these in at least one situation. When you take a trip (automobile, boat, plane, train), the distance, D, from your departure point changes with time. Your velocity equals your rate of change of distance relative to time, that is, $\frac{dD}{dt}$. The derivative of $\frac{dD}{dt}$, the second derivative of D, gives the rate of change of velocity and usually is called acceleration. At a constant velocity, acceleration equals zero because velocity is not changing. Positive acceleration occurs when the vehicle picks up speed and pushes you back against your seat; negative acceleration, associated with slowing down, does the opposite.

Let f symbolize a function of the independent variable x, and f' its derivative. The derivative of f' if it exists is called its SECOND DERIVATIVE and is variously denoted as f", $\frac{d^2f}{dx^2}$, or D^2f. Thus,

$$f''(x) = \frac{d}{dx}\left[\frac{d}{dx} f(x)\right] = \frac{d^2}{dx^2} f(x) = D^2f(x) \;. \tag{6.15}$$

The 2's in the symbol $\frac{d^2}{dx^2}$ should not be read as squares or second powers. The entire expression should be read as the derivative taken twice with respect to x.

This process can be repeated again and again to produce higher derivatives. The nth derivative of a function results from taking n successive derivatives of the function. Such derivatives commonly are symbolized by $f^{(n)}$, $\frac{d^nf}{dx^n}$ or D^nf.

EXAMPLE 6.22: If $y = x^4 + 3x^3 + x^2 - 5$, then

$$y' = \frac{dy}{dx} = 4x^3 + 9x^2 + 2x$$

$$y'' = \frac{d^2y}{dx^2} = \frac{dy'}{dx} = 12x^2 + 18x + 2$$

$$y''' = \frac{d^3y}{dx^3} = \frac{dy''}{dx} = 24x + 18$$

$$y^{(4)} = \frac{d^4y}{dx^4} = 24, \qquad y^{(n)} = \frac{d^ny}{dx^n} = 0 \quad \text{for} \quad n = 5, 6, \cdots. \quad \blacksquare$$

We previously discussed the geometric interpretation of the first derivative as the slope of the line tangent to the curve at any point. In a like manner the second derivative represents the instantaneous rate of change of the first derivative, and the nth derivative represents, in general, the instantaneous rate of change of the (n-1)st derivative.

6.5 SUMMARY

We collect the major results given in the preceding seven subsections for your convenience in study and reference. Consider functions f, g, and u defined over an independent variable x. Assume that the functions are all differentiable relative to their independent variable. Further, suppose that k, n, and the a_i symbolize constants. Then at those points where the derivatives exist,

POWERS:
$$\frac{du^n}{dx} = nu^{n-1}\frac{du}{dx} \qquad (6.1)$$

POLYNOMIALS:
$$\frac{d}{dx}(a_n x^n + a_{n-1}x^{n-1} + \cdots + a_2x^2 + a_1x + a_0)$$
$$= na_n x^{n-1} + (n-1)a_{n-1}x^{n-2} + \cdots + 2a_2x + a_1$$

CHAIN RULE:
$$\frac{df}{dx} = \left[\frac{df}{du}\right]\left[\frac{du}{dx}\right] \qquad (6.4)$$

SUM:
$$\frac{d}{dx}(f(x) + g(x)) = \frac{df(x)}{dx} + \frac{dg(x)}{dx} = f'(x) + g'(x) \qquad (6.3)$$

PRODUCT:
$$\frac{d}{dx}(f(x)g(x)) = f(x)\frac{dg(x)}{dx} + g(x)\frac{df(x)}{dx}$$
$$= f(x)g'(x) + g(x)f'(x) \qquad (6.7)$$

QUOTIENT:
$$\frac{d}{dx}\left[\frac{f(x)}{g(x)}\right] = \frac{g(x)\frac{df(x)}{dx} - f(x)\frac{dg(x)}{dx}}{[g(x)]^2}$$
$$= \frac{g(x)f'(x) - f(x)g'(x)}{[g(x)]^2} \qquad (6.10)$$

LOGARITHM: $$\frac{d}{dx}(\ln u) = \frac{1}{u}\frac{du}{dx} \quad \text{and} \tag{6.12}$$

$$\frac{d}{dx}(\log_b u) = \frac{\log_b e}{u}\frac{du}{dx} \tag{6.13}$$

EXPONENTIAL: $$\frac{d}{dx}e^u = e^u \frac{du}{dx}\ . \tag{6.14}$$

EXERCISES

1. The following graph represents the graph of the function $y = f(x)$. Sketch a graph of the first and second derivatives using your knowledge of derivatives, clearly indicating what happens at each of the points A - G.

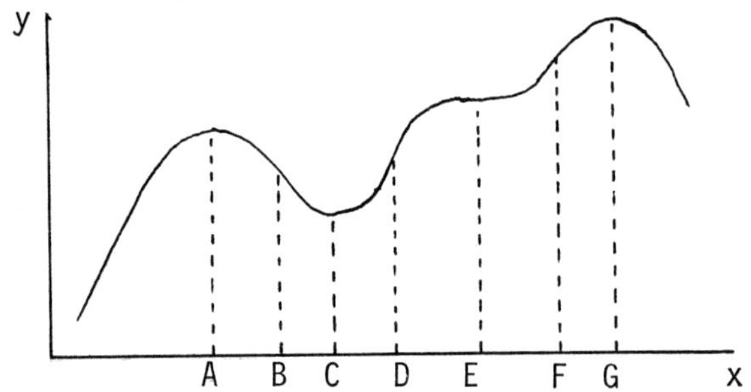

2. If $y = 3x^2 + 2x + 3$, find y' and y'''.

3. If $y = x^{-2}$, find y'''.

4. If $w = \dfrac{W_0(1 + b)}{1 + be^{-kt}}$ where W_0, b, k are constants, find $\dfrac{d^2W}{dt^2}$.

5. If $y = x \ln x$, find y''.

6. If $y = \log(3x^2 - 5)$, find y''.

7. Find y'' given $y = e^{-x} \ln x$.

8. The equilibrium constant K of a balanced chemical reaction changes with the absolute temperature T according to the law given by

$$K = K_0 \exp[(- cT_0(T - T_0)/2T]$$

where K_0, c and T_0 are constants. Find the rate of change of K with respect to T.

9. Evaluate the second derivative of the logistic growth curve,

$$W = \frac{W_0(1 + b)}{1 + be^{-kt}}, \text{ at } t = \frac{\ln b}{K}.$$

10. According to Stefan's law, a body emits radiant energy according to the formula $R = KT^4$, where R is the rate of emission of radiant energy per unit area, T is the Kelvin temperature of the surface and K is a physical constant. Find $\frac{dR}{dT}$. Does this represent a first or second derivative?

Chapter 7

USING THE TOOLS OF DIFFERENTIATION

This chapter relates closely to the last one. Chapter 6 handed you a new tool to put in your tool box. As a set of wrenches or measuring cups come in several sizes, you saw rules for evaluating the derivatives of several classes of functions. Here we will demonstrate some ways in which those tools can fit into biological situations.

The first section presents situations where rates of change serve as the primary focus. The next two sections describe how derivatives can be used to describe the behavior of specified functions. They focus mainly on algebraic functions, but the next two sections emphasize similar matters for biological situations. The remaining sections cover other applications including differentiation of complicated functions, impacts of small measurement errors, series approximation and a tool for evaluating indeterminate forms.

7.1 RELATED RATES OF CHANGE

Two variables sometimes are related by a known equation with each variable being a differentiable function of an independent variable, say t. Differentiation of both sides of the equation with respect to the independent variable, t, produces a relation between the derivatives of the dependent variable and the independent variable, t. When t happens to represent time, we obtain a relation between time-rates of change. The next examples illustrate this idea.

EXAMPLE 7.1: This example typifies problems which arise in using radio telemetry to study the movements of animals.

Suppose you want to study a white-tail deer to monitor its daily travels and habits. Good information of this sort can help in habitat management. Suppose you trap the deer and attach a transmitter to its neck. Using sophisticated electronic equipment you and an assistant determine that the deer is moving east at the rate of 0.2 meters/second and south at the rate of 0.3 meters/second. At what rate is the distance between you and the deer changing as the deer passes the point located at (45 meters, 108 meters), as indicated in Figure 7.1?

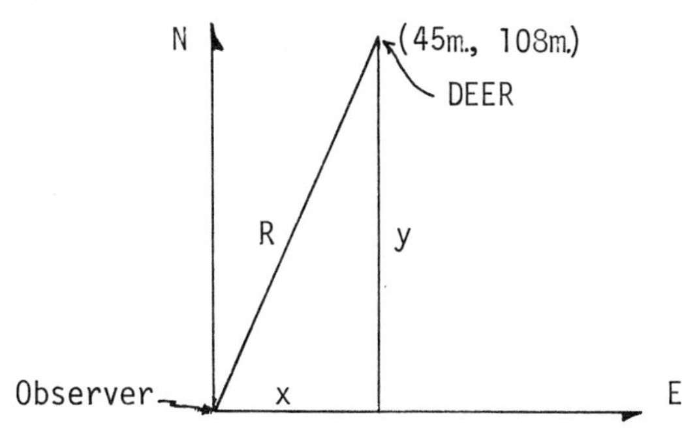

Figure 7.1

Figure 7.1 shows the physical set-up of the problem and introduces appropriate notation. The given information translates into x = 45 meters, y = 108 meters, $\frac{dx}{dt}$ = 0.2 meters/second and $\frac{dy}{dt}$ = -0.3 meters/second. Now $\frac{dx}{dt}$ = 0.2 meters/second from the fact that the deer is going east (away from the observer) at the rate of 0.2 meters/second, but $\frac{dy}{dt}$ = -0.3 meters/second because the deer is going south at this rate so the north-south distance between the observer and the deer is decreasing. To evaluate the quantity of interest, $\frac{dR}{dt}$, we must find a relationship between R and known quantities. From the right triangle in Figure 7.1 we know that $R^2 = x^2 + y^2$. Because R, x and y all depend upon t, differentiation of both sides of this equation with respect to t gives

$$2R \frac{dR}{dt} = 2x \frac{dx}{dt} + 2y \frac{dy}{dt}$$

and solving for $\frac{dR}{dt}$ we have

$$\frac{dR}{dt} = \frac{x \frac{dx}{dt} + y \frac{dy}{dt}}{R} . \tag{7.1}$$

We know the value of every quantity on the right-hand side of Equation 7.1 except R, but it can be calculated from $R = \sqrt{x^2 + y^2} = \sqrt{45^2 + 108^2} = \sqrt{13689} \doteq 117$ meters. Substituting these quantities into Equation 7.1 gives

$$\frac{dR}{dt} = \frac{(45 \text{ meters})(0.2 \text{ meters/second}) + (108 \text{ meters})(-0.3 \text{ meters/second})}{117 \text{ meters}}$$

$$= \frac{-23.4 \text{ meters /second}}{117 \text{ meters}} = -0.2 \text{ meters/second.}$$

Observe that the units must be carried along in order to determine whether or not we have used the proper conversion factors. This result tells us that at the instant in question the distance R is decreasing at the rate of 0.2 meters/second. It was unnecessary in this example to solve for R and then differentiate. However, we would obtain exactly the same answer if we had solved for R and then differentiated. You should verify this fact for yourself. ▌

EXAMPLE 7.2: As an employee of Sawdust Logging Company you must present some figures to the management concerning the rate of growth of the company's sawdust pile. The pile is created by the sawdust falling from a tube onto a conical pile, where the radius of the base equals one and a half times the height. The sawdust is added to the pile at the rate of 60π cubic feet per hour. This question needs an answer, "How fast is the pile of sawdust growing when it has a height of 90 feet?"

In one sense the question has already been answered because the 60π cubic feet per hour gives a growth rate. Thus some other aspect of growth apparently is needed. The rate at which the base is expanding relates directly to how soon a new pile will have to be started. The radius of the base provides one reasonable measure of size because it can be translated into basal area. Figure 7.2 represents a cross section of the pile and introduces suitable notation.

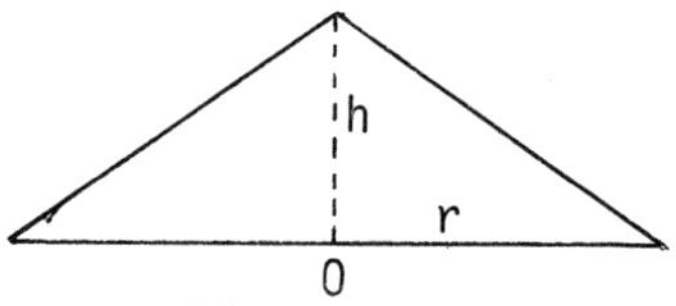

h
r
0

Figure 7.2

It should be evident that if h represents the height and r the radius of the pile of sawdust at its base, both of these change with time, t. The volume, V, of a cone relates to its height, h, and its radius, r, through $V = \frac{\pi r^2 h}{3}$. The 60π cubic feet/hour must refer to the rate of change of the sawdust pile since it has units of cubic feet per hour so $\frac{dV}{dt} = 60\pi$ feet^3/hour. We need to find how fast the radius of the base, $\frac{dr}{dt}$, is changing when the sawdust pile has a height of 90 feet. To evaluate $\frac{dr}{dt}$ from our knowledge of $\frac{dV}{dt}$, we need to express the volume as a function of the radius of the base: Because $r = \frac{3}{2}h$ or $h = \frac{2}{3}r$, substitution into the volume equation gives

$$V = \frac{\pi r^2 h}{3} = \frac{\pi r^2 \left(\frac{2}{3} r\right)}{3} = \frac{2\pi r^3}{9}.$$

Since V and r both are functions of t, we can take the derivative of this equation with respect to t:

$$\frac{dV}{dt} = \frac{6\pi r^2}{9}\frac{dr}{dt} = \frac{2}{3}\pi r^2 \frac{dr}{dt}.$$

Upon solving this for $\frac{dr}{dt}$,

$$\frac{dr}{dt} = \frac{3\frac{dV}{dt}}{2\pi r^2}. \qquad (7.2)$$

At h = 90 feet and $r = \frac{3}{2}h = \frac{3}{2}$ (90 feet) = 135 feet, substitution of the known values into Equation 7.2 gives

$$\frac{dr}{dt} = \frac{(3)\ 60\pi \text{ ft.}^3\text{/hr.}}{2\pi(135 \text{ ft.})^2} = \frac{(3)\ 30 \text{ ft.}^3\text{/hr.}}{(135 \text{ ft.})^2} \doteq 0.0049 \text{ feet/hour}$$

Again check the units. They are admissible since $\frac{dr}{dt}$ should be expressed in units of length per unit of time. The answer of 0.0049 feet/hour gives the rate of change of the radius of the pile of sawdust when the height of the pile is 90 feet. At other heights the rate of change of the radius would be different. Intuitively you should be able to see that as the height increases, the rate of change of the radius will become smaller; when the pile first started, the radius was changing somewhat more rapidly than later. ▌

The problem you encounter in working related rate problems is in trying to relate what is known to what you want to find. In Example 7.2 we sought information about the radius, knowing only the height and the rate at which the volume of the pile was changing. Recall the discussion of establishing relations in Section 2.3. This task often will be even more difficult on real biological problems than it is in a textbook environment. In most cases, however, you usually will be able to find the appropriate relationships.

EXERCISES

1. It has been established experimentally that the specific heat of ethyl alcohol can be approximated over the temperature range from 0°C to 60°C by this equation:

$$S(T) = 5.068 \times 10^{-1} + 2.86 \times 10^{-3}T + 5.4 \times 10^{-6}T^2$$

where T is given in °C. What is the rate of change of specific heat at

a. 0°C? b. 10°C?

2. A small weather balloon is released from ground level 75 feet from an observer, also at ground level. If the balloon goes straight up at a rate of 2.5 feet per second, how rapidly will the balloon be receding from the observer 40 seconds later?

3. The adiabatic law for the expansion of air is $PV^{1.4} = C$ where P denotes the pressure, V the volume of the air in the chamber and C is a constant. At a certain instant the volume of the chamber is 12 feet^3 and the pressure is 60 pounds per square inch. If the volume decreases at the rate of 1 foot^3 per second, how rapidly is the pressure changing? Is the pressure increasing or decreasing?

4. The surface area of a cube is increasing at the rate of 5 square inches per hour. How fast is the volume of the cube changing when the surface area is equal to 30 square inches?

5. A rainbow trout comes out of the water after a fly. This causes concentric rings to be produced in the otherwise calm lake. If the radius of the ring is increasing at the rate of 1.5 feet per second, find the rate of increase of the circumference of the ring and the area of the ring when the radius of the ring is

a. 3 feet b. 10 feet.

6. The pressure, P, and volume, V, of a given quantity of a gas at a constant temperature satisfy the relation $PV = C$ where C is a constant. At the instant when the pressure equals 5 pounds per square inch and the volume equals 100 cubic feet, the volume is decreasing by 2 cubic feet per minute. Find the rate of change of P at this instant.

7. A metal cylinder is shrinking as it cools. The radius and the height are decreasing at the rate of 0.001 and 0.01 inches per minute respectively. When the altitude is 100 inches and the radius is 5 inches, find the rate of decrease of

   a. the curved surface area   b. the volume.

8. Reconsider the situation in Example 7.2. Suppose the management rejected this answer by saying, "We need to know how fast the area covered by the sawdust pile is growing." What would be your answer?

9. You have bacteria growing in circular colonies. If the radius of a colony is increasing at the rate of 2 millimeters per day, how fast is the area changing at the end of the fourth day? What assumptions are you making?

10. Two concentric circles are expanding, the outer radius at a rate of 3 feet per second and the inner one at 5 feet per second. At a certain instant when the outer radius is 10 feet and the inner radius is 3 feet, is the area between the circles increasing or decreasing? How fast?

11. An elk is running horizontally at 520 feet per minute directly away from a man on a ridge which is 100 feet above the elk. How fast is the distance between the elk and the man changing when the elk is 250 feet from the man?

12. Wheat is pouring from a spout at 3 cubic feet per minute onto a conical pile whose diameter at the base always equals 3 times its altitude. At what rate is the altitude increasing when the altitude is 4 feet?

13. A blower at a sawmill is depositing sawdust in a conical pile whose radius is always three-fourths of its height. The sawdust is being deposited at a rate of 2 cubic feet per minute. How fast is the altitude of the pile increasing when the base of the pile is 20 feet in diameter?

14. A pill is composed of a right circular cylinder with a hemisphere on each end. The length of the cylindrical portion is twice the radius of the cylinder. Find the rate of change of the volume of the solid with respect to the radius of the cylinder.

15. A certain type of spherical pill dissolves at the rate of 0.64π cubic centimeters per minute.
    a. At what rate is the radius of the pill changing when $r = 0.2$ centimeters?
    b. How soon will the pill be completely dissolved if r equals 1 centimeter initially?

16. In the construction of a snowman one afternoon, a spherical snowball was being rolled for part of the body. The volume of the ball was increasing at a rate of 8 cubic feet per minute. Find the rate at which the radius is increasing when the snowball is 4 feet in diameter if the height of the snowman is to be 8 feet.

17. A funnel in the shape of a cone is 10 inches across the top and 8 inches deep. A fluid is flowing into the funnel at the rate of 12 cubic inches per minute and out at the rate of 4 cubic inches per minute. How fast is the surface of the fluid rising when it is 5 inches deep?

18. A patient is receiving a liquid intravenously. The reservoir holding the liquid is conical with the vertex down. The fluid flows at the rate of 946 cubic centimeters per hour and the reservoir is 30 centimeters across the top and 10 centimeters deep.
    a. At what rate is the fluid level dropping when the fluid is 5 centimeters deep?
    b. How is the radius changing when the fluid is 5 centimeters deep?

19. For a perfect gas, the volume V, pressure P, and absolute temperature T relate through PV = RT where R is a constant. It is found that P = 20 pounds per square inch and V = 200 cubic inches when T = 400°K. Find T and its time rate of change at an instant when P = 15 pounds per square inch and V = 300 cubic inches if P is decreasing at the rate of 0.4 pounds per square inch per second while V is increasing at the rate of 0.3 cubic inches per second.

20. An elk has a cancerous tumor on its cheek of approximately spherical shape (volume = $\frac{4}{3}\pi r^3$ and the surface area = $4\pi r^2$, where r is the radius of the tumor). The size of the tumor is increasing at the rate of 10 cubic millimeters per week. How fast is the radius of the tumor changing when the surface area is 16π square millimeters?

21. Water is leaking out of a conical tank (vertex down) at the rate of 0.5 cubic feet per minute. The tank has a diameter 30 feet across the top and a depth of 10 feet. When the water is 8 feet deep, it is rising at the rate of 1.5 feet per minute. At what rate is water being poured into the tank?

22. The shape of a water reservoir can be approximated by a cone with its vertex down. The depth of the reservoir is 30 feet and the diameter is 180 feet. Assume that water is flowing in at the rate of 60π cubic feet per hour and that there is approximately 2π cubic feet of silt in this water per 24 hours.
    a. How fast, in general, is the reservoir silting up if we assume all silt coming into the reservoir is deposited there?
    b. How is the volume of the silt changing when the silt is 4 feet deep?
    c. How fast is the radius of the silted portion changing when its height is 4 feet?

23. If a rod shaped (cylindrical) bacterium has a radius of r and length L, how does its surface area change with changes in its radius and length?

24. A rod shaped bacterium of length 100μ and a diameter of 20μ swells upon being subjected to a medium of different ionic concentration

24. Continued
than its normal environment. Its radius increases at the rate of 0.1μ/hour and its length increases at the rate of 0.25μ/hour. At what rate is the volume of the cell changing?

25. A solution is passing through a conical filter 24 inches deep and 16 inches across the top into a cylindrical vessel of diameter 12 inches. At what rate is the level of the solution in the cylinder rising when the depth of the solution in the filter is 12 inches and is falling at the rate of 1 inch/minute?

26. A conical icicle, whose height is always 12 times the radius of the base, is being formed by dripping water. If the volume is increasing at the rate of 1 cubic centimeter per hour, at what rate is the height increasing when the height is 8 centimeters?

*27. A cylindrical drum containing a tissue fixative solution is lying on its side in an anatomy laboratory and is filled half full. Determine a formula for the rate at which the upper area of the solution changes with respect to its depth in the drum as the fixative is removed from the drum.

7.2 HIGHS AND LOWS

You might encounter this situation: You have a function you need to study, probably in relation to a particular biological response. You may have found the relation by the graphic means discussed near the end of Chapter 4; you may have derived it by precise means set out in subsequent chapters; or another investigator may have suggested it. However you got the function, a graph can provide great assistance in understanding it. Limits, discussed in Chapter 5, can identify ultimate behavior at the edges of a function's domain, but how can you find out about its behavior elsewhere? You can sketch the middle of a graph by evaluating the function at judiciously chosen points; if you choose points poorly, you may miss very important features of the graph. This and the next section concern how to pick points at which to examine the function.

The location and magnitude of maxima and minima of a function are a great help in understanding the behavior of a function. They come in two varieties, absolute and relative:

> Consider a function f defined over some domain, D, and a point $x_0 \in D$.
>
> - f has an ABSOLUTE MAXIMUM of $f(x_0)$ at x_0 if $f(x_0) \geq f(x)$ for all $x \in D$.
> - f has an ABSOLUTE MINIMUM of $f(x_0)$ at x_0 if $f(x_0) \leq f(x)$ for all $x \in D$.
> - f has a RELATIVE MAXIMUM at x_0 provided $f(x_0) \geq f(x)$ for all $x \in D$ which lie sufficiently close to x_0.
> - f has a RELATIVE MINIMUM at x_0 provided $f(x_0) \leq f(x)$ for all $x \in D$ which lie sufficiently close to x_0.
> - The word EXTREMUM describes either a relative maximum or a relative minimum.

A function has an absolute maximum at x_0 if its value there is the largest of its values. By contrast, a relative maximum describes a value which is the largest among all nearby values. From the top of a hill you can go only down (a relative maximum), even though another hill may be higher. The highest hill in the region of interest such as a county (the domain) corresponds to the absolute maximum. Absolute and relative minima similarly deal with smallest values; from a minimum you can go only up, like being at the bottom of a dry lake or other depression. Any local high or low point is called an extremum while two or more local high or low points are called extrema.

Some functions have an absolute maximum or an absolute minimum on an interval, but others do not. Consider the behavior of the function $f(x) = 2x^3$ over the closed interval $[0,2]$, namely restrict x to $0 \leq x \leq 2$. Over this interval f rises smoothly from $f(0) = 0$ to $f(2) = 16$. As $f(0) = 0 \leq f(x)$ for any x in this interval, the function has an absolute minimum of 0 at $x_0 = 0$. Similarly, the function has an absolute maximum of 16 at $x_0 = 2$. Now consider the open interval $(0,2)$, namely $0 < x < 2$ (excluding the endpoints). Without these endpoints, no point in the interval gives either the largest or smallest value of the function.

Frequently, however, a function will have local extrema, its humps and dips, as well as the possibility of absolute extremum. A continuous function has both an absolute maximum and an absolute minimum on a closed interval. This fact should become intuitively apparent if you do this: Think of placing your pencil on the graph of a function at the left boundary of the interval. Now follow the graph of the function across to the other boundary. In doing this you will trace out every permissible value of the function, including its largest and smallest.

EXAMPLE 7.3: Consider how the function sketched in Figure 7.3 behaves over the closed interval [a,d]. It has a relative minimum at x = a, a relative maximum at x = b, a relative and absolute minimum at x = c, and a relative and absolute maximum at x = d. The function has extrema at x = a, b, c, and d.

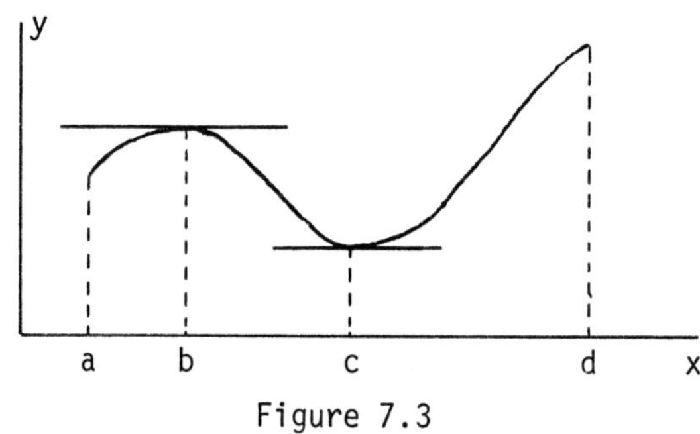

Figure 7.3

The graph has an important feature at x = b and x = c, namely, at those extrema which occur away from the boundary: The tangent to the curve becomes horizontal (has a slope of zero) there. These two tangents are shown in Figure 7.3. ▮

This last observation suggests a relation between the appearance of extrema away from the boundaries and the condition $f'(x_0) = 0$:

THEOREM 7.1: If a function f has a relative extremum at an interior point $x = x_0$ in its domain, and if the derivative $f'(x_0)$ exists, the derivative is zero.

This theorem states a necessary condition for a relative extremum to exist: The function's first derivative, if it exists, must equal zero there. However, a vanishing first derivative does not assure a relative extremum at that point. The next example illustrates this point.

EXAMPLE 7.4: The first derivative of the function sketched in Figure 7.4 equals zero at A, B, C and D; the tangent to the curve has a slope of zero at those points. But it has extrema at A, B, D and E. Neither a relative maximum nor a relative minimum occurs at C even though the derivative there equals zero. This illustrates that not all points having f'(x) = 0 give extrema. The special kind of point illustrated by C will be discussed in the next section. The point E illustrates another nuisance situation, namely, a point at which the derivative does not exist, yet where it has an extrema. Theorem 7.1 does not apply here because one of its conditions fail, namely, f'(E) does not exist. ▌

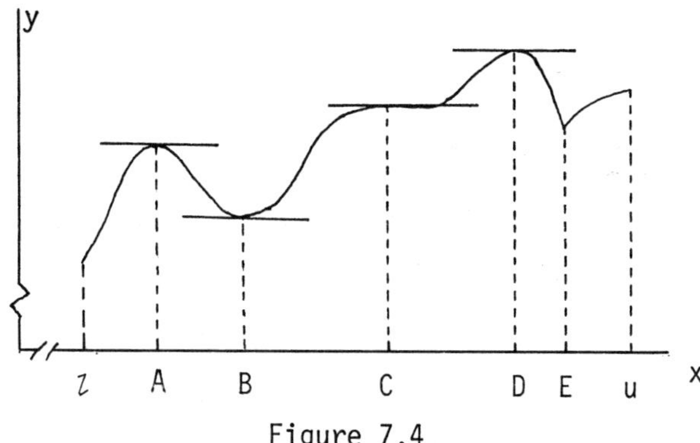

Figure 7.4

These examples suggest that we need a scheme for isolating all of the points A, B, C, D, and E in Figure 7.4.

- The extreme behaviors of a function f occur for $l \leq x \leq u$ at values of x called the CRITICAL VALUES, which are:

    a. values of x where $\frac{df(x)}{dx} = 0$,

    b. values of x where $\frac{df(x)}{dx}$ does not exist,

    c. the boundary points, l and u.

- If x_c is a critical value, then $(x_c, f(x_c))$ is called a CRITICAL POINT of the function.

Usually our primary interest focuses on those critical values where $f'(x) = 0$, but we should not forget the others, such as point E in Figure 7.4. The importance of the boundary points lies mainly in finding the absolute maximum and minimum in which case each critical point must be investigated.

EXAMPLE 7.5: Find the extrema of $f(x) = x^3 - 3x^2 - 9x$.

This function is defined and differentiable for all x because it is a polynomial and

$$f'(x) = 3x^2 - 6x - 9 = 3(x + 1)(x - 3) \;.$$

At $x = -1$ and $x = 3$, $f'(x) = 0$ so these critical values need further investigation. A graph of the function appears in Figure 7.5. The function has a relative maximum at the critical point (-1,5) and a relative minimum at the critical point (3,-27).

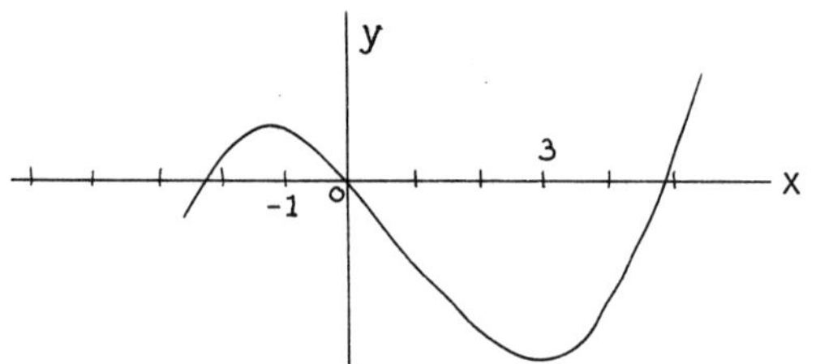

Figure 7.5

If f had been defined only for $0 \leq x \leq 4$, then its absolute maximum of 0 would have occurred at $x = 0$ while both its relative and absolute minimum of -27 would still have occurred at $x = 3$. ∎

EXAMPLE 7.6: Investigate the extreme behavior of $y = 1 + (x - 1)^{2/3}$.

We need the derivative

$$y' = \frac{2}{3}(x - 1)^{-1/3} = \frac{2}{3}\frac{1}{\sqrt[3]{x - 1}} \;.$$

This derivative does not go to zero for any x; it does not exist at $x = 1$. Figure 7.6 shows the graph of this function on the left, and its derivative on its right. The point $(1,1) = (x,f(x))$ is called a CORNER. The function definitely turns a corner there; its derivative changes from large negative to large positive values. A relative minimum occurs at the point (1,1) even though $f'(1)$ does not exist. ∎

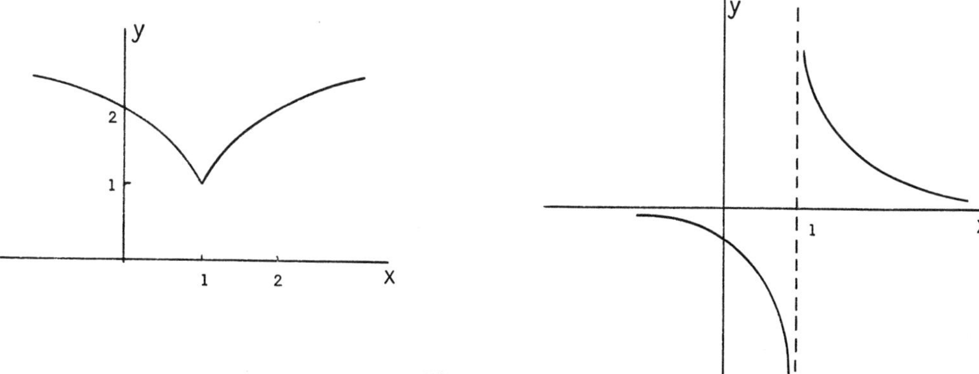

Figure 7.6

In these last two examples we have evaluated the functions' behavior at the critical points by sketching the graph, but this defeats our purpose. We need an easy way to see how the graph behaves. Look back at the relative minimum and relative maximum in Example 7.5. The derivative of the function changes in a special way as we pass $x = -1$ and $x = 3$ from left to right. Provided $f'(x_0)$ exists, we have at $x = x_0$:

- At a relative minimum, $f'(x_0) = 0$ and $f'(x)$ changes from a negative (-) to a positive (+) quantity as we cross $x = x_0$. At a relative maximum $f'(x_0) = 0$ and $f'(x)$ changes from a positive (+) to a negative (-) quantity as we cross $x = x_0$.

These criteria can be applied at any x_0 which does not lie on the boundary (edge) of the domain; at a boundary the function is not defined on one side of x_0.

Why do these criteria work? Suppose f has a relative minimum at $x = x_0$. Its graph looks like the curve segment on the left in Figure 7.7. A line tangent to the curve at $(x_0,f(x_0))$ has a slope of zero, but for $x < x_0$ it slopes downward (negative) while for $x > x_0$ it slopes upward (positive). A relative maximum at $(x_0,f(x_0))$ is illustrated on the right in Figure 7.7. As the slope is examined from left to right, the slope changes from positive through zero to negative.

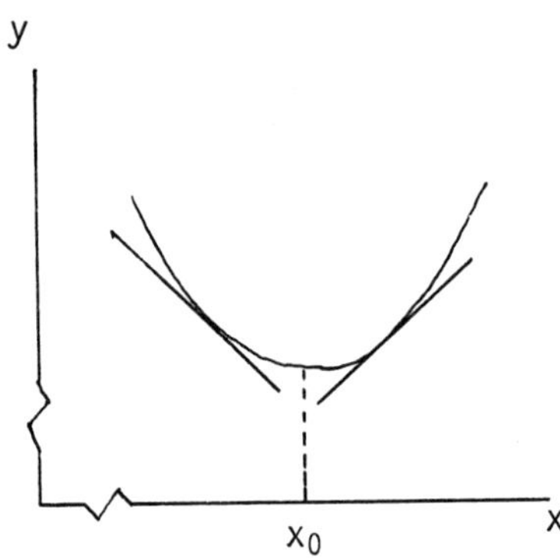

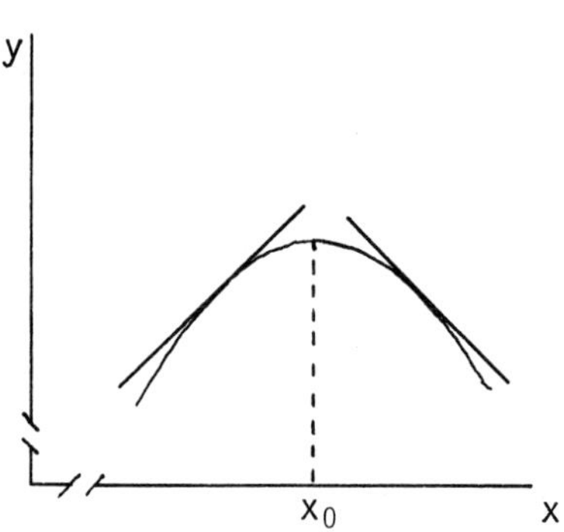

Figure 7.7

EXAMPLE 7.5 (continued): When $y = x^3 - 3x^2 - 9x$, $f'(x) = 3(x + 1)(x - 3)$ so $f'(x) = 0$ at $x = -1$ and $x = 3$. According to the criteria, we should examine $f'(x)$ at points on either side of $x = -1$ and of $x = 3$ to establish the behavior of the function there. At $x = -2$, $x = 0$, $x = 2$ and $x = 4$, $f'(x)$ has the values 15, -9, -9, and 15, respectively. Near $x_0 = -1$, the slope of the tangent line changes from 15 at $x = -2$ to -9 at $x = 0$; this change from positive to negative across $x = -1$ implies that f has a relative maximum at $x = -1$. Similarly near $x = 3$, $f'(x)$ changes from -9 at $x = 2$ to 15 at $x = 4$, implying a relative minimum there. Figure 7.8 shows f and f'; note that f' is zero where f has relative extrema.

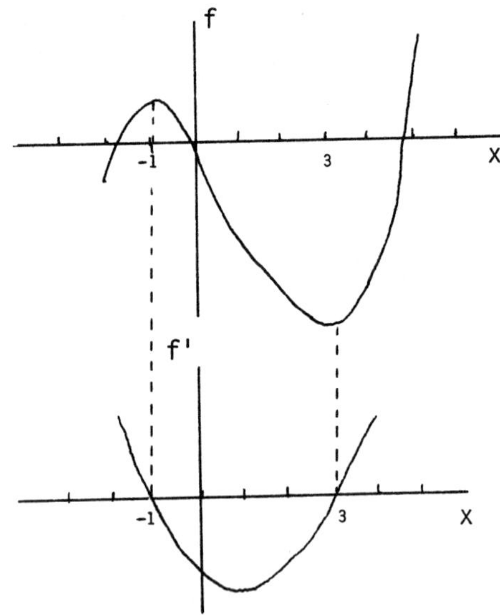

Figure 7.8

We examined the behavior of $f'(x)$ near $x = -1$ by evaluating $f'(-2)$ and $f'(0)$. In reality we need to know only the sign of these derivatives. The points $x = -2$ and $x = 0$ were selected simply for arithmetic convenience; any values slightly below and above $x = -1$ would work equally well. For example, both factors of $f'(x) = 3(x + 1)(x - 3)$ are negative for any $x < -1$, so their product is positive. Slightly above $x = -1$, one of the factors is negative while the other is positive, giving a negative product. Thus the sign change (+ to -) can be detected without actually evaluating f' at various points. ▌

<u>EXAMPLE 7.7</u>: Examine the function $y = x^3$ for relative maxima or minima.

From $y' = 3x^2$, $y' = 0$ at $x = 0$, its single critical point. Notice from the graph in Figure 7.9 that the tangent to the curve at $x = 0$ has a slope of zero, but neither a relative maximum nor relative minimum occurs there. This fact is born out by examining f' at points on either side of $x = 0$, say $x = -1$ and $x = 1$, $f'(-1) = 3$ and $f'(1) = 3$. The first derivative does not change sign across $x = 0$ so neither a relative maximum nor minimum occurs there.

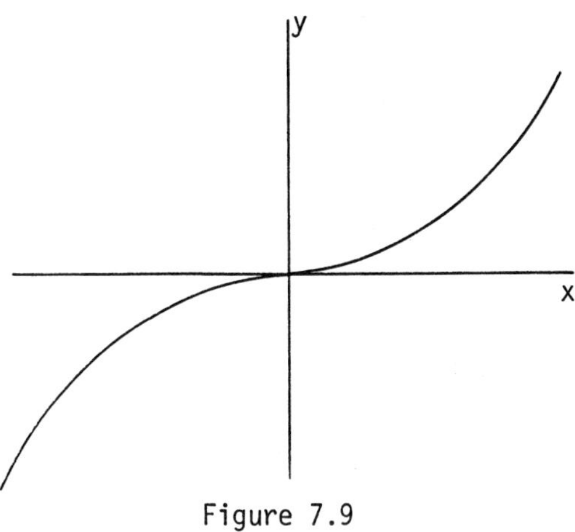

Figure 7.9

The first derivative cannot become negative; hence the function cannot have a relative maximum or minimum. We will discuss this important point in the next section. ▌

The next theorem gives another, usually easier, method for determining whether a maximum or minimum occurs at a critical point.

<u>THEOREM 7.2</u>: A continuous function f, which has continuous first and second derivatives near and at $x = x_0$, has a relative maximum at $x = x_0$ if $f'(x_0) = 0$ and $f''(x_0) < 0$, or a relative minimum if $f'(x_0) = 0$ and $f''(x_0) > 0$.

EXAMPLE 7.5 (continued): The results of Theorem 7.2 can be applied to $y = x^3 - 3x^2 - 9x$ which has $y' = 3(x^2 - 2x - 3)$ and $y'' = 6(x - 1)$. The points $x = -1$ and $x = 3$ previously were identified as critical points because $y' = 0$ there. The second derivative evaluated at these points yields $f''(-1) = -12 < 0$; thus, according to the criteria of Theorem 7.2 a relative maximum occurs at $x = -1$. Similarly, because $f''(3) > 0$, a relative minimum occurs at $x = 3$. These results agree with our earlier examination of this function. ▌

Why does Theorem 7.2 work? We will use Figure 7.10 to help you see why. The top part of the figure shows a function f, while the middle and bottom parts of the figure display the associated f' and f". All of f, f', and f" are continuous. The function does special things at a, b, $\cdots$, g. A relative maximum occurs at $x = a$ so f' passes through zero from positive to negative; the graph of f' shows this near $x = a$. Now follow f from f(a) to f(c). It drops and then flattens off to a minimum; the graph of f' dips down and comes back to zero. Now you relate the rest of the graph of f' to the graph of f. Because f" is the derivative of f', the graph of f" relates to the graph of f' just as the graph of f' related to the graph of f. A relative maximum occurs at $x = a$; $f'(a) = 0$ and $f''(a) < 0$.

In summary, we have found how to isolate points at which the extreme behaviors of a function occur, and how to examine the nature of that extreme behavior. First, isolate the critical points. Look at f" at those points where $f' = 0$; $f'' < 0$ implies a relative maximum, $f'' > 0$ implies a relative minimum, while the case $f'' = 0$ remains to be discussed. If f' does not exist at a critical point, investigate it on either side of that point. If f" does not exist where $f' = 0$, compare the behavior of f' from the left to the right. A change from + to - indicates a relative maximum while the opposite indicates a relative minimum. To find absolute extrema compare the function's value at interior critical points to each other and to the boundary points.

7.3 A SPECIAL KIND OF POINT

Look at Figure 7.10 again. Specifically notice that the curve opens downward for values to the left of $x = b$, but it opens upward from $x = b$

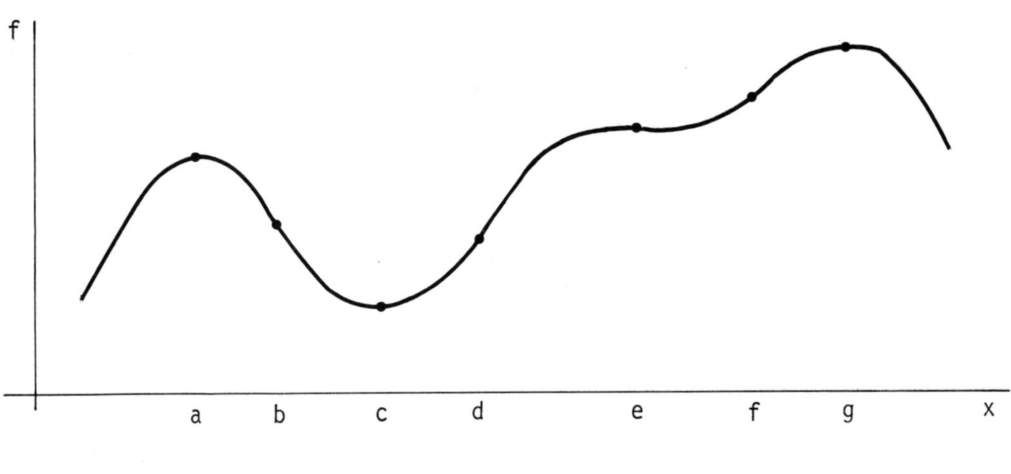

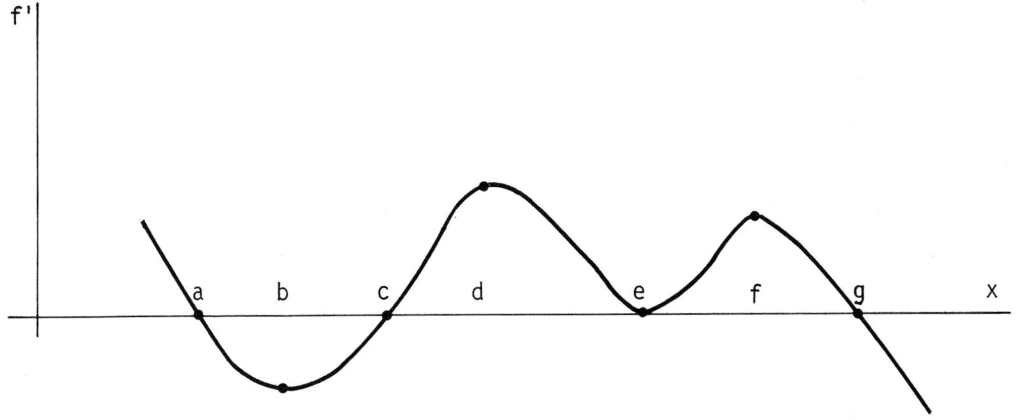

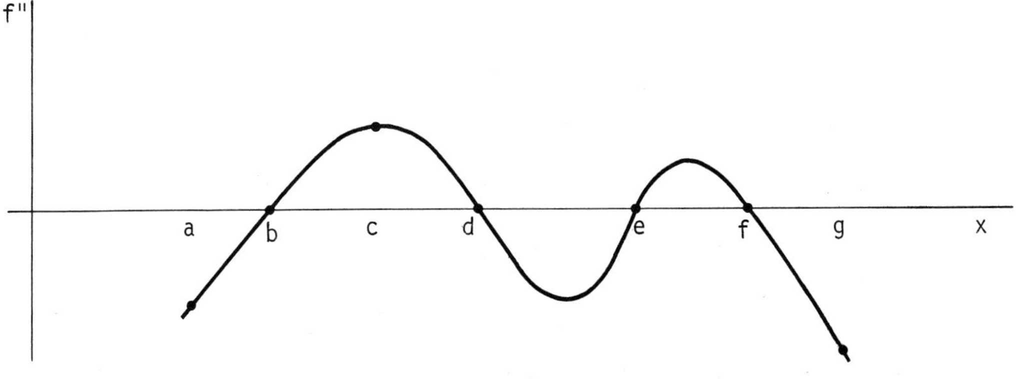

Figure 7.10

to $x = d$. The first part of the curve is described as being CONCAVE DOWNWARD, while the second part is CONCAVE UPWARD. Continuing across the graph, it is concave downward from $x = d$ to $x = e$, upward on to $x = f$, and downward thereafter. The concavity changes at the points $x = b$, d, e, and f. Such points are called INFLECTION POINTS. Their location can help in graphing a function because the function is shaped differently on opposite sides of inflection points. Now look at the middle graph in Figure 7.10: The derivative, f', has extrema at the same values of x where the function has inflection points. Then because the second derivative is the derivative of the derivative, extrema of the derivative occur where the second derivative of the function equals zero. This observation is summed up in the next theorem.

THEOREM 7.3: When a function f has a second derivative, $f''(x_0)$, at $x = x_0$ and the graph of $y = f(x)$ has an inflection point at $x = x_0$, then $f''(x_0) = 0$.

Notice the similarity of this theorem to Theorem 7.1. Both tell you where to look for particular features of a graph, neither assures you that the feature will appear at each candidate point. For a candidate point to be an extremum, the first derivative had to change sign across the critical value. At an inflection point the second derivative must similarly change sign.

EXAMPLE 7.8: Locate the relative maxima, minima, and points of inflection, if they exist, of $y = x^3 - 3x^2 - 9x + 5$. If an inflection point exists, find the equation of the line tangent to the curve at the inflection point.

We need to calculate the first two derivatives:

$$y' = 3(x^2 - 2x - 3) \qquad \text{and} \qquad y'' = 6(x - 1) .$$

Now $y' = 0$ when $x = -1$ or $x = 3$. Applying the second derivative test, $f''(-1) = -12$ and $f''(3) = 12$ which establish that f has a relative maximum at $x = -1$ and a relative minimum at $x = 3$. Because $y'' = 0$ at $x = 1$, check for a possible inflection point at $x = 1$. As x passes 1, y'' changes from negative to positive, so an inflection point does occur there. To calculate the equation of the tangent line at the inflection point, we need its slope. This is found by evaluating $f'(1) = -12$. A line with this slope passing through the inflection point $(1,-6)$ has an equation of $y + 6 = -12(x - 1)$ or $y = -12x + 6$. The function is graphed in Figure 7.11, using what we have found. Again, to evaluate the nature of an extremum point, we need only the sign of the second derivative, not its value. ▌

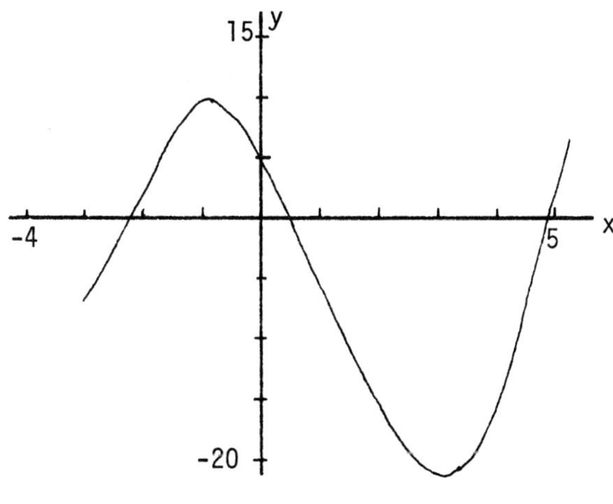

Figure 7.11

<u>EXAMPLE 7.9</u>: Does the graph of $y = x^4$ have an inflection point?

From $y' = 4x^3$ and $y'' = 12x^2$, the second derivative equals zero at $x = 0$, but it never will be negative. Thus no inflection point exists even though the second derivative goes to zero. Figure 7.12 indicates the similarity of this curve with the quadratic; near the origin it flattens more than the quadratic and for $|x| > 1$ the quartic $(y = x^4)$ goes up faster. ▌

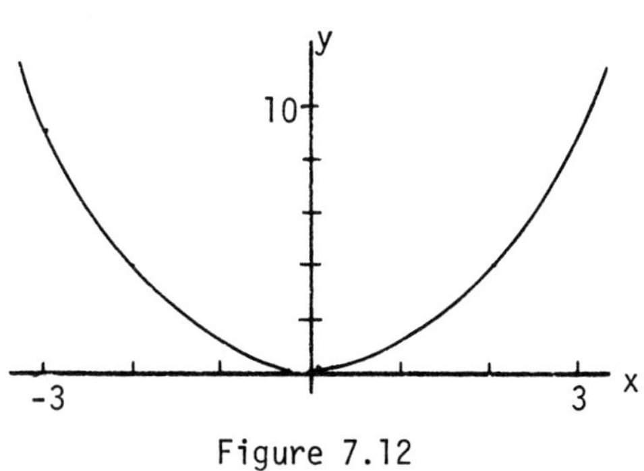

Figure 7.12

In conclusion, a graph of a function f has an inflection point at $x = x_0$ in its domain when the graph is concave in one sense for $x < x_0$, but in the opposite sense for $x > x_0$, for all x sufficiently close to x_0.

EXERCISES

Locate and investigate all critical points of the following functions. Evaluate the relative and absolute extrema. Sketch each function indicating its behavior over its whole domain. When an inflection point exists, obtain the equation of the line tangent to the curve there. Take the domain, D, as $D = \{x| -\infty < x < \infty\}$ unless specified otherwise.

1. $y = (x - 2)^2, |x| \leq 6$
2. $y - 2 = x^2 + 4x$
3. $y = -2x^2 + 4$
4. $y = -x^3 + 3x^2 + 24x + 7$
5. $y = x^3 + 9x^2 + 27x + 5$
6. $y = (x - 2)^3, |x| \leq 10$
7. $y = 6x^4$
8. $y = x^3 - 6x^2 + 9x + 15$
9. $y = x^2 \ln \frac{1}{x}$
10. $N = N_0e^{kt}$ where $N_0 > 0$ and $k > 0$
11. $N = \dfrac{N_0(1 + b)}{1 + be^{-kt}}$ where $N_0 > 0$, $b > 1$, and $k > 0$
12. $y = \dfrac{e^x + e^{-x}}{2}$
13. $y = (x + 2)^2(x - 3)^3$
14. $y = (x - 2)^3(x + 1)^2$
15. $y = x^3 + x^2 - x + 1$, $-2 \leq x \leq \frac{1}{2}$

7.4 OPTIMIZATION

Maximization and minimization of functions have numerous practical counterparts. Most practical problems have a resource (like monetary) constraint on their solution. You may need to maximize production or a similar quantity within a fixed resource base. Or you may need to accomplish a task for minimum expenditure of resources. Resources should be viewed broadly, not only as money. They may include materials, time (yours and/or other persons'), timber wasted in a logging operation, water used to irrigate greenhouse or field crops, lettuce left to rot in a field, etc. The problem of maximization or minimization also occurs in biology outside of resource allocation. The next example illustrates a resource allocation problem.

EXAMPLE 7.10: You are assigned to build a fence around a rectangular field plot. One side is bounded by a river which runs straight along that edge. Because you must conserve money, you need to fence off a given amount of land, A, using as little fence as possible. Assume no fence is needed along the river.

First, introduce appropriate notation as in Figure 7.13. Let x symbolize the length of the plot to be enclosed and y its width. The area enclosed, $A = xy$, is fixed by x and y. The fence has a length of $L = x + 2y$, but because $y = \frac{A}{x}$, it also satisfies $L = x + \frac{2A}{x}$. Now we need to minimize this expression of total length of fence relative to x. Upon taking the derivative of $L = x + \frac{2A}{x}$ with respect to x, we obtain

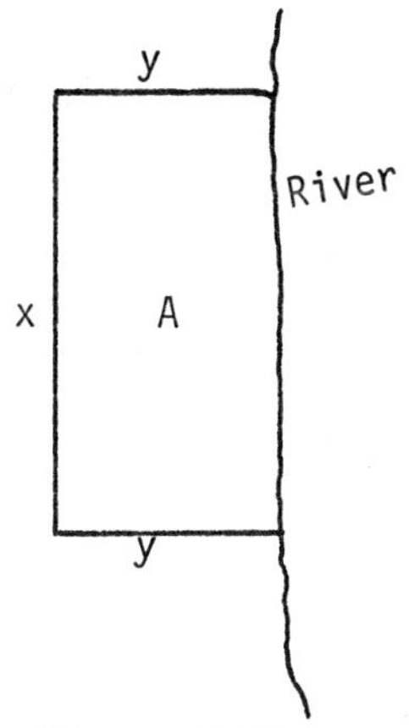

$$\frac{dL}{dx} = 1 - \frac{2A}{x^2} .$$

Figure 7.13

Set this last expression equal to zero to find critical points:

$$\frac{dL}{dx} = 1 - \frac{2A}{x^2} = 0 \qquad \text{or} \qquad x = \sqrt{2A}$$

The positive root was used because we are dealing with length. Because

$$\frac{d^2L}{dx^2} = \frac{4A}{x^3} ,$$

will be greater than zero for $x > 0$, $x = \sqrt{2A}$ yields a minimum. From $x = \sqrt{2A}$ and $y = \frac{A}{x}$, we get $y = \frac{\sqrt{2A}}{2}$. The rectangular plot will have a length twice its width. ▮

EXAMPLE 7.11: Suppose you need to construct cylindrical soil sampling cans of volume V. You need the dimensions of the cylindrical can which will minimize the amount of material used in making the can; assume it needs no top. These dimensions will then be given to a fabricator for the actual construction.

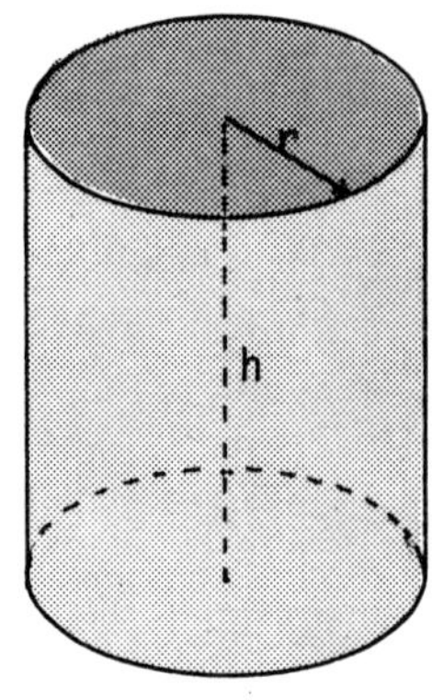

Figure 7.14

Let r denote the radius of the can, h its height, and A the amount of material in the can. (See Figure 7.14.) $A = \pi r^2 + 2\pi rh$; the term πr^2 comes from the area of the bottom and $2\pi rh$ from the material in the can's side. The volume of the cylindrical can, V, is given by $V = \pi r^2 h$. As you seek the dimensions of a can which will minimize the amount of material for a fixed volume, you must minimize the surface area of the can.

Observe that the equation for the surface area contains two independent variables, r and h. You need to express the surface area as a function of a single variable. Because the volume of the can is known, h depends on r according to $h = \dfrac{V}{\pi r^2}$. When this is substituted into the surface area equation, $A = \pi r^2 + \dfrac{2\pi rV}{\pi r^2} = \pi r^2 + \dfrac{2V}{r}$. Now that the amount of material in the can has been expressed strictly in terms of constants and r, this area needs to be minimized. Upon differentiation, $\dfrac{dA}{dr} = 2\pi r - \dfrac{2V}{r^2}$ so $\dfrac{dA}{dr} = 0$ when $r = \sqrt[3]{\dfrac{V}{\pi}}$. Because $\dfrac{d^2A}{dr^2} = 2\pi + \dfrac{4V}{r^3}$ and $r > 0$, the second derivative remains positive. Thus cans of radius $r = \sqrt[3]{\dfrac{V}{\pi}}$ will have minimum surface area. To calculate the corresponding value of h, substitute $r = \sqrt[3]{\dfrac{V}{\pi}}$ into $h = \dfrac{V}{\pi r^2}$ and obtain $h = \dfrac{V}{\pi\left(\dfrac{V}{\pi}\right)^{2/3}} = \sqrt[3]{\dfrac{V}{\pi}}$. The can having the desired volume which also minimizes the amount of material needed will have a height equal to its radius.

We also would obtain the same answer by expressing the surface area, A, as a function of h by eliminating r from the volume equation. The height was eliminated because it was easiest. You should verify that the same answer results from either approach; this should give a comforting feeling because this means you do not need to worry about what variable to use. ▌

*<u>EXAMPLE 7.12</u>: Derivatives are used in the development of various statistical procedures. We will illustrate one such use with the binomial probability distribution introduced in Section 3.6:

$$P(x) = \binom{n}{x} p^x (1 - p)^{n-x} , \qquad x = 0, 1, \cdots, n .$$

Here p gives the probability of a success on a single trial, and x represents the number of successes in n trials.

The probability of a success on any single trial, symbolized by p, is an unknown parameter characterizing the population. Suppose $x = 6$ in $n = 15$ trials. How should we estimate p? The principle of MAXIMUM LIKELIHOOD proceeds this way: The probability of the observed result changes with the parameter value. Estimate the parameter as that value which maximizes the probability of the observed result. This maximization frequently uses derivatives.

Here, for a fixed n and x, what value of p maximizes

$$P(x) = \binom{n}{x} p^x (1 - p)^{n-x} ?$$

Upon taking derivatives (suppose $x \neq 0$ and $x \neq n$),

$$\frac{dP}{dp} = \binom{n}{x} [xp^{x-1}(1 - p)^{n-x} - (n - x)p^x(1 - p)^{n-x-1}]$$

$$= \binom{n}{x} p^{x-1}(1 - p)^{n-x-1} [x(1 - p) - (n - x)p]$$

$$= \binom{n}{x} p^{x-1}(1 - p)^{n-x-1}(x - np) .$$

To find the critical point, set $\frac{dP}{dp} = 0$:

$$\binom{n}{x} p^{x-1}(1 - p)^{n-x-1}(x - np) = 0 ,$$

$$x - np = 0 \qquad \text{or} \qquad p = \frac{x}{n} .$$

The other two solutions to $\frac{dP}{dp} = 0$, namely $p = 0$ and $p = 1$, here correspond to meaningless solutions. They describe an event which cannot occur, or one which always occurs; neither of these situations has any meaning here because we have 6 successes in 15 trials, not 0 or 15. Evaluate the second derivative to see that a maximum, not a minimum occurs at $p = \frac{x}{n}$.

To emphasize that this is an estimate, not the true value, statisticians frequently write $\hat{p} = \frac{x}{n}$. For our data this becomes $\hat{p} = \frac{6}{15} = 0.4$. This estimate is quite reasonable because it actually equals the average number of successes per trial (x successes in n trials). ▮

The key to solving optimization problems lies in relating the known information to what you want to find, as was also the case in the related rate problems.

EXERCISES

1. The specific weight of water, S, at a temperature t°C is closely approximated by the equation $S = 1 + at + bt^2 + ct^3$ for $0° < t < 100°C$ where $a = 5.3 \times 10^{-5}$, $b = -6.53 \times 10^{-6}$, $c = 1.4 \times 10^{-8}$. At what temperature will water have a maximum specific weight?

2. Find the dimensions of the rectangular pasture of greatest area which can be enclosed with 1000 feet of wire fencing.

3. An animal scientist has 640 feet of fencing. He plans to enclose a rectangular area and divide it into 5 pens with fences parallel to the short end of the rectangle. What dimensions of the enclosure make its area a maximum?

4. An agronomist wishes to select 10 acres of his experimental land along a river and divide it into 8 small plots by means of a fence running parallel to the river and 9 fences perpendicular to it. Show that the total amount of fencing to be used is a minimum if the length of the 9 cross fences equals the length of the fence parallel to the river.

5. You want to build a rectangular box which will hold a maximum amount of fluid and have a surface area of 1200 square inches. What dimensions should the box have for a square base and an open top?

6. What would be the dimensions in Problem 5 if the box has a lid?

7. You are assigned to build some open top boxes from 10" x 16" pieces of tin by folding up the edges after cutting equal small squares from each corner. What dimensions will result in boxes having the largest possible volume?

8. The rate of a certain auto-catalytic reaction is given by this expression: $V = kx(a - x)$, where x represents the amount of the product, a is the initial amount of the substance being catalyzed and k is a positive constant. The rate of reaction depends on the concentration of both the remaining substance $(a - x)$ and the product (x). At what concentration x will the rate be maximal?

9. In water and in solutions, the product of the concentrations of the hydrogen ions $[H^+]$, and the hydroxyl ions $[OH^-]$ is very close to 10^{-14} mole. Find the ratio $[H^+]/[OH^-]$ that minimizes the sum of the concentrations.

10. You are at a bridge crossing a river which runs due south. You need to reach a cabin located 1.5 miles south and 3 miles east of the bridge. You can travel 6 miles per hour in an available boat, but in the rugged local terrain you can only make 2 miles per hour on land. How should you go in order to reach the cabin in a minimum amount of time? (Hint: Distance = rate × time.)

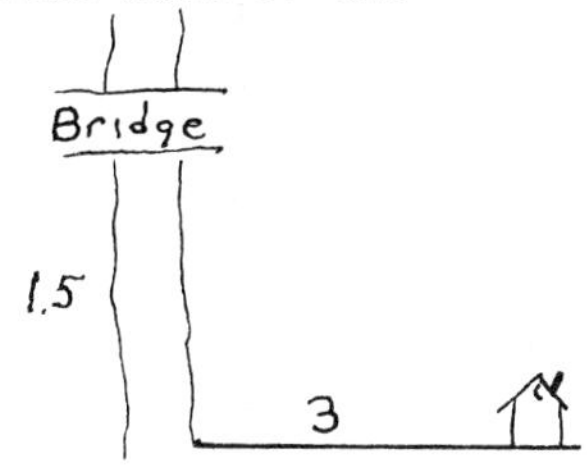

11. Late in the afternoon a hunter is at a point which is 1 mile from the nearest road and his car is 3 miles straight on down the road. If he can walk 2 miles per hour in the forest and 4 miles per hour on the road, how should he plan his course in order to reach his car in the minimal amount of time?

12. Suppose you work for the Bureau of Sport Fisheries in Yellowstone National Park. Every day you need to go from a reasearch cabin on the shore of the lake to a point across the lake 9 miles north and 6 miles west of the cabin. You can paddle your canoe at 3 miles per hour, but you can walk at 5 miles per hour. How far should you walk if you are interested in getting to your destination in the shortest time? Assume that you would not encounter any obstacles to either mode of transportation other than needing to cross the lake.

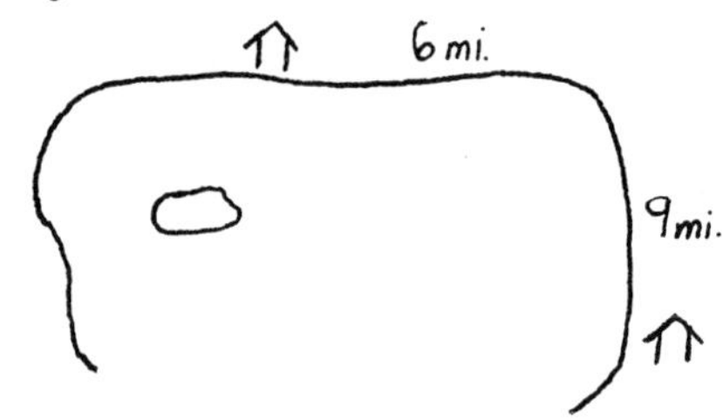

13. A man lives in a cabin on an island out in a lake while his girl-friend lives in a cabin on the shore. In this area, the lake has an essentially straight shore with the island 1.5 miles from the shore. If he rows straight to the shore he has to walk 3 miles to get to her house. How can this anxious fellow get to her home most quickly? How much time will he save, over the route straight to the shore, and over rowing all the way, assuming he can row at 4 miles per hour and walk at 5 miles per hour?

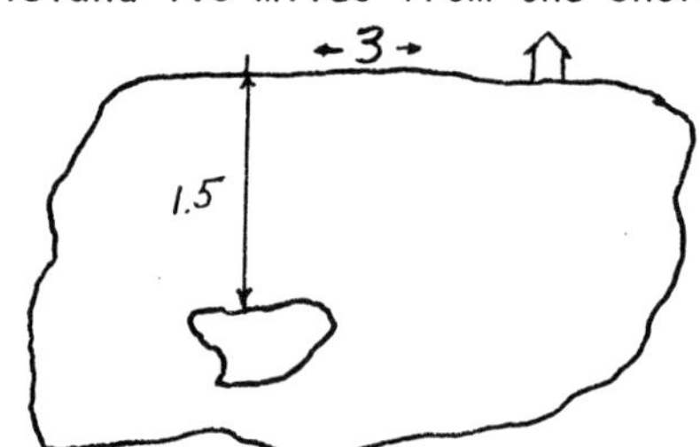

14. An island is at a point A, 6 miles off shore from the nearest point B on a straight beach. A girl lives on the beach at point C, 7 miles down the beach from B. If the man on the island can row a boat at 4 miles per hour and walk at the rate of 5 miles per hour, where should he land in order to get from the island to the girl's house in the least possible amount of time.

15. An open irrigation ditch of given cross-sectional area is to be dug and lined with concrete to prevent seepage. If the two equal sides are perpendicular to the flat bottom, find the relative dimensions of the ditch which require the least amount of concrete to build.

16. Consider a cylindrically shaped cell of length L and radius R. If the volume is fixed, what ratio of L to R will give the minimum total surface area?

17. It has been shown clinically that when a person coughs, the diameter of the trachea and bronchi shrink. It is assumed that the pressure which serves to empty the lungs during exhalation has an unwanted side effect of compressing the bronchi and windpipe. The simplest assumption states that constriction of the tubes increases in direct proportion to the pressure. Let r_P represent the radius at a particular point under a pressure of P, where the pressure on the lungs is measured as the excess above atmospheric pressure and let a be a constant of proportionality. If r_0 is the radius in the absence of pressure on the lungs, $r_0 - r_P = aP$. This formula provides a satisfactory approximation only when P lies between 0 and $r_0/2a$. For P greater than this, the resistance of the tube to compression becomes greater. If it were not for this we would suffocate whenever we cough. Poiseville's law states that the resistance to the flow of air through a tube increases in proportion to the reciprocal of r_P^4. Thus, $R = k/r_P^4$ where R is the resistance and k is a constant of proportionality. The flow equals the pressure divided by the resistance:

$$\text{flow} = \frac{P}{R} = \frac{Pr_P^4}{k} = \frac{(r_0 - r_P)\, r_P^4}{ak} .$$

The velocity is given by the flow divided by the area of a cross section of the tube. Thus,

$$V(r_P) = \frac{\text{flow}}{\pi r_P^2}$$

Find the radius at which the flow will be maximum and the radius at which the velocity will be a maximum.

18. A greenhouse plot is to contain 500 square inches and have a weed free border of 6 inches at the top and bottom, and 3 inches on each side. Seeds will be planted inside the border. What dimensions of the plot will yield the largest area for seeding?

19. The strength of a beam of rectangular cross-section varies directly as the product of its width and the square of its depth. What dimensions will make the strongest beam that can be cut from a log 16 inches in diameter?

20. An apple orchard now has 40 trees per acre and the average yield per tree is 500 apples. For each additional tree planted per acre, the yield per tree is reduced by about 1 dozen apples. How many trees per acre should be planted in order to get the largest crop of apples?

7.5 ANOTHER LOOK AT FUNCTIONS IN REALITY

This section presents no new material, neither has it any examples because it can be regarded as several related examples. Section 4.7 dealt with functions in reality with particular emphasis on population growth curves and survival curves. Here we will apply all of the tools we have developed to the growth curves and leave the survival functions for you to examine.

7.5a EXPONENTIAL GROWTH CURVE

The exponential function, symbolized here as $W = W_0e^{kt}$, has numerous associations with growth, both of individuals and of populations. Look at its derivative

$$\frac{dW}{dt} = kW_0e^{kt} = kW \ . \qquad (7.3)$$

It grows in proportion to its present size. Much biological growth does behave this way until limiting factors set in.

Observe that $W' > 0$ when $k > 0$. This implies that W grows with time. But as W increases, Equation 7.3 shows that W' also increases. Thus the exponential function with a positive parameter increases at an ever increasing rate. Higher derivatives tell us nothing further because for each n

$$\frac{d^nW}{dt^n} = k^nW$$

remains always positive or always negative when $W > 0$, the only relevant values.

When $k < 0$, $W' < 0$. This situation describes a decreasing response. The size of an individual rarely behaves this way, but a population subject to predation, disease or extermination pressures can. Certain cleansing organs such as the liver and kidneys remove metabolic waste products from the blood stream and also some materials introduced intraveneously. Once a material has reached an elevated level, its removal often nearly follows an exponential curve with $k < 0$ because the cleansing organ succeeds in capturing the material in proportion to its presence.

Although the exponential function is defined for all t, interest usually is restricted to $t \geq 0$. Because the function is continuous throughout its domain, $\lim_{t \to 0} W(t) = W(0) = W_0 e^0 = W_0$. Thus, W_0 gives the initial value of the exponential function. In population considerations W_0 frequently is replaced by N_0, and sometimes W is also replaced by N to emphasize the involvement of numbers. As t continues onward, $\lim_{t \to \infty} W(t) = \infty$ when $k > 0$, the situation of concern to Malthus. This shows that although an exponential function may supply a very good approximation for some values of t, it cannot apply to all t because real responses have upper bounds past which the system breaks down. When $k < 0$, $\lim_{t \to \infty} W(t) = 0$. This properly describes many decreasing biological functions and related decay functions.

The graph of any exponential function looks like one of the two curves in Figure 7.15. The upward trending curve corresponds to $k > 0$ while the other curves downward in response to the negative parameter. They behave nearly as reciprocals of each other because $e^{-b} = 1/e^b$. This graph was drawn assuming that the same W_0 applied for both signs on k. Exponential functions associated with various situations will usually differ in W_0, the initial value. The derivative, Equation 7.3

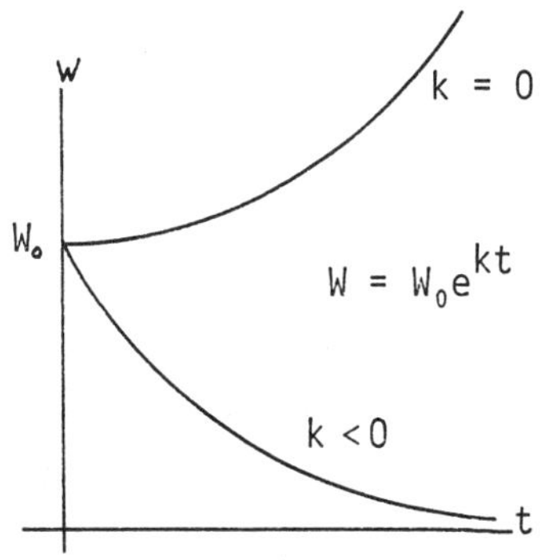

W

k = 0

W_o

$W = W_0 e^{kt}$

k <0

t

Figure 7.15

shows that changes in the magnitude of k alter how fast the function increases. Otherwise, the basic shape remains unchanged.

Remember that the estimation of an exponential function's parameters was discussed in Section 4.6.

7.5b THE LOGISTIC GROWTH CURVE

The Logistic growth curve is specified by the function

$$W = W_0 \frac{1 + b}{1 + be^{-kt}}$$

where b, k and W_0 are positive constants. Notice that $W = W_0$ at $t = 0$ so W_0 is the amount, quantity or number of individuals present at time zero.

What does this curve look like? To sketch it, we will examine its first and second derivatives and two of its limits. First, because of W's continuity, we have already established that

$$\lim_{t \to 0} W_0 \frac{1 + b}{1 + be^{-kt}} = W_0 \ .$$

Quantities following the Logistic growth curve will continue to grow from W_0 if the system remains undisturbed for a long time. Thus, we need to examine

$$\lim_{t \to \infty} W_0 \frac{1 + b}{1 + be^{-kt}}$$

This limit equals $W_0(1 + b)$ because $e^{-kt} \to 0$ as $t \to \infty$.

After establishing the limiting behavior of this function we need to examine how it behaves in between. Let us examine its rate of growth.

$$\frac{dW}{dt} = \frac{W_0(1 + b)bke^{-kt}}{(1 + be^{-kt})^2} > 0 \qquad (7.4)$$

for $t > 0$, because b and k both are positive constants. What have we found out? Because the derivative remains positive, the response grows from its initial value of W_0 toward its upper bound of $W_0(1 + b)$. The presence of the horizontal asymptote agrees with the fact that $\lim_{t \to \infty} W'(t) = 0$.

With the first derivative always a positive, either of the two curves in Figure 7.16 could describe the graph of W. Both have initial values of W_0, both are asymptotic to $W_0(1 + b)$ and both have positive slopes. How then can we distinguish between them? The first one has an inflection point in the region $t > 0$ while the second does not; we will find that curve 2 has an inflection point, but it occurs before the process starts, namely for $t < 0$. The second derivative of W can be used to investigate inflection points and thereby distinguish between the curves:

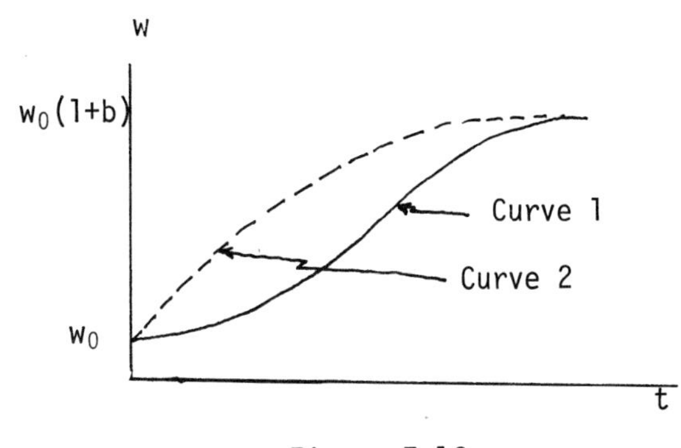

Figure 7.16

$$\frac{d^2W}{dt^2} = \frac{(1 + b)bk^2W_0e^{-kt}(be^{-kt} - 1)}{(1 + be^{-kt})^3}$$

The second derivative will be zero either when $e^{-kt} = 0$ or when $be^{-kt} - 1 = 0$, for all other terms in the numerator are constants. The term e^{-kt} remains positive for any t; it goes to zero only as $t \to \infty$, a case already considered. Thus we must determine what values of t make $be^{-kt} - 1$ equal zero:

$$be^{-kt} - 1 = 0$$

or

$$e^{-kt} = \frac{1}{b}$$

and by taking the natural logarithm of both sides, the solution for t gives

$$t = -\frac{1}{k} \ln \frac{1}{b} = \frac{\ln b}{k} .$$

This will occur during the time of interest for $k > 0$ only when $b > 1$. By examining the second derivative we are able to show that this is indeed the inflection point. Thus, we must conclude that curve 1 describes the

Logistic growth curve for $b > 1$; but curve 2 applies for $b \leq 1$. Recall that the extreme behaviors of the first derivative occur where the second derivative equals zero. (See the discussion of Figure 7.10 at the end of Section 7.2.) Thus the maximum or minimum values of the derivative (rates of growth) occur at the inflection points. Here, the inflection point appears at $t = \frac{\ln b}{k}$; here $W = \frac{W_0}{2}(1 + b)$ so the maximum rate of growth occurs when W equals 1/2 of its final value. Why is this a maximum rather than a minimum? The graph of the Logistic curve appears in Figure 7.17, drawn for $b > 1$; it really applies for $0 < b \leq 1$, but the vertical axis at $t = 0$ lies to the right of the inflection point in this case.

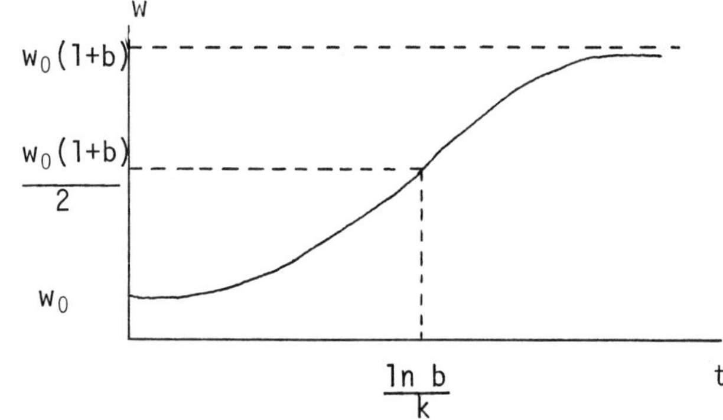

Figure 7.17

Functions closely related to the Logistic growth curve appear in diverse areas of biology, far afield from population size. The change of variable $u = e^{kt}$ has this effect

$$y = W_0 \frac{1 + b}{1 + be^{-kt}} = W_0 \frac{1 + b}{1 + \frac{b}{u}} = \frac{W_0(1 + b)u}{u + b} .$$

By taking $c = W_0(1 + b)$, this becomes

$$y = \frac{cu}{b + u} . \qquad (7.5)$$

This curve has the same shape as the Logistic, except with the horizontal axis stretched by the transformation $u = e^{kt}$. This produces a particularly interesting variation from the Logistic near $u = 0$, where Equation 7.5 goes to zero.

Several forms of equations equivalent to Equation 7.5 have appeared for example in theoretical considerations of photosynthesis. The photosynthetic production of a leaf varies with substrate conditions.

In particular, available solar radiation and carbon dioxide level have been used as variables symbolized by u when other variables were held constant.

7.5c THE GOMPERTZ GROWTH CURVE

We will examine this curve in the same manner as we have the previous ones. It is specified by the function

$$W = ae^{-ce^{-kt}}$$

where a, c and k are positive constants. It has the limits

$$\lim_{t \to 0} W = ae^{-c} \qquad \text{and} \qquad \lim_{t \to \infty} W = a \ ,$$

and the derivatives

$$W' = ae^{-ce^{-kt}}(-ce^{-kt})(-k) = ack\ e^{-kt-ce^{-kt}} \tag{7.6}$$

$$W'' = \frac{dW'}{dt} = ack\ e^{-kt-ce^{-kt}}(-k + kce^{-kt})$$

$$= ack^2(ce^{-kt} - 1)e^{-kt-ce^{-kt}} \tag{7.7}$$

Be sure to verify these using $\frac{de^u}{dt} = e^u \frac{du}{dt}$, taking $u = -ce^{-kt}$ for W' and $u = -kt - ce^{-kt}$ for W". When a, c and k are all positive constants, $W' > 0$ and W" will be equal to zero when either $e^{-kt-ce^{-kt}} = 0$ or $ce^{-kt} - 1 = 0$. No value of t satisfies the first equation. The second equation gives

$$ce^{-kt} - 1 = 0 \qquad \text{or} \qquad ce^{-kt} = 1 \ .$$

Upon taking the natural logarithm of both sides and solving for t, $t = \frac{1}{k} \ln c$. Because $k > 0$, an inflection point occurs at a positive time only when $c > 1$, as was the case with the Logistic. By checking values on either side of $t = \frac{1}{k} \ln c$ in the second derivative, we see this value of t is indeed an inflection point. At the inflection point, $W = ae^{-1}$.

Again $\lim_{t\to\infty} W' = 0$ which is in agreement with the horizontal asymptote at $W = a$.

The Logistic and Gompertz curves have the same general shapes as gleaned from their limits and derivatives. How do they differ? To compare them let us get them to agree in initial values and asymptotes. Then we can compare them for intermediate values of t.

The Gompertz curve involves its parameters in a more complex manner than the Logistic, so solve for the Logistic's parameters in terms of the Gompertz's parameters. At t = 0 and $t\to\infty$ the two curves should be equal so

$$W_0 = ae^{-c} \qquad \text{and} \qquad W_0(1 + b) = a$$

or

$$W_0 = ae^{-c} \qquad \text{and} \qquad b = e^c - 1 \ .$$

Substituting these values into the function for the Logistic, (denoted by W_ℓ), gives

$$W_\ell = ae^{-c}\,\frac{1 + (e^c - 1)}{1 + (e^c - 1)e^{-kt}} = \frac{a}{1 + (e^c - 1)e^{-kt}} \tag{7.8}$$

Now how do the heights of these curves compare? We have shown that the Gompertz curve has a height of ae^{-1} at its inflection point. The Logistic curve has the slightly larger value of a/2 at its inflection point. Each of the curves has a third parameter, k. If these parameters are equal, then the two curves have their inflection points at the same value of t with the result appearing in Figure 7.18.

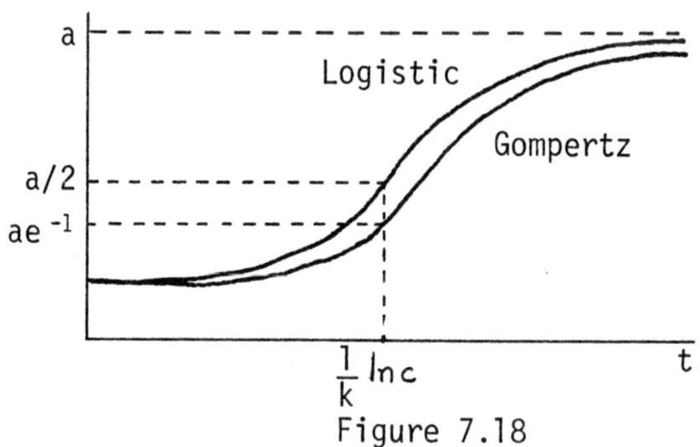

Figure 7.18

As the k for the Gompertz increases, the inflection point will occur at smaller values of t so eventually the curves will intersect twice.

7.5d THE VON BERTALANFFY GROWTH CURVE

The von Bertalanffy growth curve is specified by the function

$$W = W_0(1 - be^{-kt})^3,$$

where W_0 and k are positive and $0 < b < 1$. It has limits of

$$\lim_{t\to 0} W = W_0(1 - b)^3 \qquad \text{and} \qquad \lim_{t\to\infty} W = W_0$$

Its first and second derivatives are

$$\frac{dW}{dt} = 3W_0bke^{-kt}(1 - be^{-kt})^2$$

and

$$\frac{d^2W}{dt^2} = 3W_0bk^2e^{-kt}(1 - be^{-kt})(3be^{-kt} - 1) .$$

Even though one value of t makes the first derivative zero, neither a relative maximum nor minimum occurs there because the first derivative can never become negative.

The second derivative equals zero for $t = \frac{\ln b}{k}$ and $t = \frac{1}{3}\left[\frac{\ln b}{k}\right]$ both of which are negative for $0 < b < 1$ and $k > 0$, their permissible values. Consequently, the von Bertalanffy function curves upward as depicted in Figure 7.19.

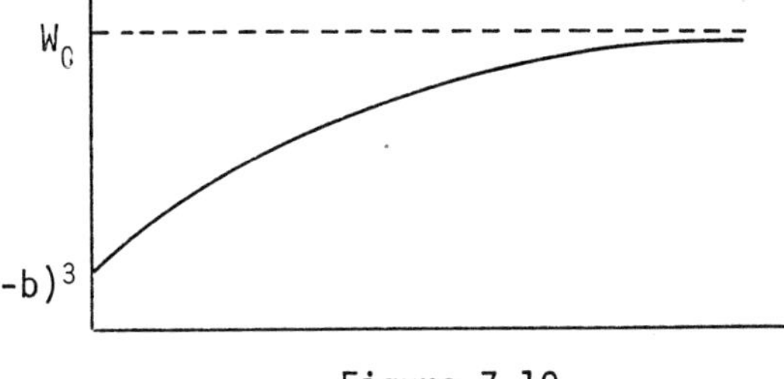

Figure 7.19

EXERCISES

1. Show $\lim_{t\to\infty} W' = 0$ in Equation 7.4.

2. Examine the behavior of Equation 7.5 by looking at the first and second derivatives, obtaining asymptotes, if any, locating the inflection point, if any, and finally sketch the function.

3. Carefully explain why $\lim_{t \to \infty} W' = 0$ in Equation 7.6.

4. Show that the von Bertalanffy growth curve does not have a relative maximum or minimum. What are its absolute maximum and minimum?

5. Show that the survival curve $S = 1 - (1 - e^{-kD})^n$ has no relative maximum or minimum.

6. Examine the limit of the first derivative of the equation given in Problem 7 as $D \to 0$ and as $D \to \infty$.

7. The Gompertz growth curve can be written as a survival curve if $k < 0$. The equation can be rewritten as $N = ae^{-ce^{\ell t}}$ where a, c, and $\ell = -k$ are positive constants. Examine the first and second derivatives for relative maxima, minima, and inflection points, if they exist. Locate any asymptotes and sketch the curve.

8. When a body is surrounded by a cooling liquid which is held at a constant temperature T_0, the body dissipates heat to the liquid. The body temperature, T, decreases approximately according to the formula $T = T_0 + ae^{-kt}$ where a and k are positive constants. Locate all asymptotes, relative maxima and minima, inflection points and sketch the graph. If a body of temperature $T_1 > T_0$ is immersed at $t = 0$, what value does the constant a have?

9. The function $C = k(e^{-at} - e^{-bt})$ represents the concentration-time relationship of a drug or dye in the blood after injection where k, a and b are positive constants and $b > a$. Locate all relative maxima and minima, inflection points, asymptotes and sketch the graph.

10. Let x be the amount of fertilizer applied to a certain crop. The total yield, y, from the crop cannot be raised indefinitely by applying more and more fertilizer. A restricted growth equation is given by

$$y = c(1 - e^{-kx})$$

where $k > 0$. It provides a good approximation to how yield changes

10. Continued
with fertilizer level (x). Graph this function by locating relative maxima and minima, inflection points and asymptotes.

11. A concrete aquaduct is to be constructed. In cross-section it will have a horizontal bottom, two vertical sides, and a semicircular top. If the cross-sectional area must be 72 square feet, what dimensions will require the least concrete, assuming all walls are equally thick?

12. A farmer is about to plant apple trees in a region of favorable climate and soil. He plans on planting 25 trees per acre and the extension agent tells him he will get about 370 apples per tree with that tree density. However, he also tells the farmer that for each additional tree planted, the yield per tree will be decreased by 10 apples per tree. What is the optimum number of trees to plant per acre?

13. In a certain locality and type of soil, it is found that 20 apple trees per acre will yield 500 apples per tree and that the yield will be decreased by 15 apples per tree for each additional tree per acre.
    a. What number of trees per acre should be planted to obtain the largest crop with all other factors being ignored except for density per acre?
    b. Assuming you have only one acre to plant, how many trees would you plant? Explain your answer.

14. A funnel of a specific volume is to be in the shape of a right circular cone. Find the ratio of the height to the basal radius if the least amount of material is to be used in its manufacture. (Hint: The curved surface area of a right circular cone whose altitude is h and whose basal radius is r is given by $A = \pi r\sqrt{r^2 + h^2}$.)

15. A rumor is spreading among students at a university. The rumor spreads at the rate of $r = kp(1 - p)$ where p is the ratio of students who have heard the rumor and k is some constant. For what value of p does the rumor spread fastest?

16. Water is leaking from an old oaken ice-cream bucket in such a manner that during a certain time interval the volume (cubic inches) of water in the bucket is given by $V = 1000 - 20t + t^2$ where t is time in minutes. The water causes the wood to swell, thus gradually stopping the leak. When will the leak be stopped? How much water will remain in the bucket?

*17. If the strength of a rectangular beam is proportional to the width times the square of the depth, what relative dimensions should be chosen, when cutting a beam from a circular log, to achieve maximum strength?

*18. Does the survival equation given in Problem 7 possess an inflection point?

7.6 STRATEGIES FOR DIFFERENTIATING COMPLICATED FUNCTIONS

Real problems do not always produce nice functions, in fact, messy ones seem more the rule than the exception. This section presents two techniques for slightly easing the pain. We have used both before, but without specific mention.

7.6a IMPLICIT DIFFERENTIATION

Most relations we have encountered were defined by an EXPLICIT equation of the type $y = f(x)$ in which $f(x)$ was an expression involving only one variable.

An equation like $x^2 + y^2 = 4$ defines a relation between x and y IMPLICITLY rather than explicitly. Sometimes these are called IMPLICIT RELATIONS. The equation $x^2 + y^2 = 4$ really defines two functions of x, one given by $y = \sqrt{4 - x^2}$ and the other by $y = -\sqrt{4 - x^2}$. Relations like $x^2 + y^2 = 4$ can be solved for $y = f(x)$, but an equation like $2y^5 - 3y^4 + 4y^3 - 10y^2 + 5y + x^5 - 3x^2 = 0$ cannot be solved for y in terms of x. This equation also defines an implicit relation between x and y.

A method called IMPLICIT DIFFERENTIATION can be used to find a derivative from an implicit relation; it avoids solving the defining equation for the dependent variable in terms of the independent variable. It merely consists of taking the derivative of a whole equation and then solving for the derivative of interest. It is illustrated in the following examples.

EXAMPLE 7.13: If $x^3 + 2x^2y + 4y^4 = 0$, find $\frac{dy}{dx}$.

To find $\frac{dy}{dx}$, we know that y should be regarded as the dependent variable and x as the independent variable. Differentiate the entire equation with respect to x:

$$\frac{d}{dx}(x^3 + 2x^2y + 4y^4) = \frac{d}{dx}0$$

$$\frac{dx^3}{dx} + \frac{d}{dx}2x^2y + \frac{d}{dx}4y^4 = 0$$

Using the already developed rules for taking derivatives,

$$3x^2 + 2x^2\frac{dy}{dx} + y\frac{d}{dx}2x^2 + 16y^3\frac{dy}{dx} = 0$$

or

$$3x^2 + 2x^2\frac{dy}{dx} + 4xy + 16y^3\frac{dy}{dx} = 0 .$$

When this is solved for $\frac{dy}{dx}$, we get

$$\frac{dy}{dx} = \frac{-3x^2 - 4xy}{2x^2 + 16y^3}$$

Thus, without expressing y as a function of x, we found an expression for $\frac{dy}{dx}$ by means of implicit differentiation. ▮

EXAMPLE 7.14: Find the equation of the tangent to the circle $x^2 + y^2 = 4$ at the point $(1,\sqrt{3})$.

We have the point so we need only the slope to determine the equation of the line. Implicit differentiation gives

$$\frac{d}{dx}(x^2 + y^2) = \frac{d}{dx}4$$

$$2x + 2y\frac{dy}{dx} = 0$$

and so $\frac{dy}{dx} = -\frac{x}{y}$. This gives $-\frac{\sqrt{3}}{3}$ as the slope of the tangent line at $(1,\sqrt{3})$. To have this slope and pass through the point $(1,\sqrt{3})$, the line must satisfy $\frac{y - \sqrt{3}}{x - 1} = -\frac{\sqrt{3}}{3}$, or $y = -\frac{\sqrt{3}}{3}x + \frac{4\sqrt{3}}{3}$. ▮

EXAMPLE 7.15: If $y^2 = 3x^4$ find $\frac{d^2y}{dx^2}$.

We need to find $\frac{dy}{dx}$ and then differentiate it to obtain the second derivative. The derivative of the defining equation gives

$$\frac{d}{dx}(y^2) = \frac{d}{dx}(3x^4)$$

$$2y\frac{dy}{dx} = 12x^3 \qquad \text{so} \qquad \frac{dy}{dx} = 6\frac{x^3}{y}.$$

Now differentiating again,

$$\frac{d^2y}{dx^2} = \frac{d}{dx}\left(\frac{dy}{dx}\right) = \frac{d}{dx}\left(\frac{6x^3}{y}\right) = \frac{y(18x^2) - 6x^3\frac{dy}{dx}}{y^2}.$$

Upon substituting $6\frac{x^3}{y}$ for $\frac{dy}{dx}$ this yields

$$\frac{d^2y}{dx^2} = \frac{18x^2y - 6x^3\left(\frac{6x^3}{y}\right)}{y^2} = \frac{18x^2y^2 - 36x^6}{y^3}$$

Thus, without solving for y in terms of x we obtained the second derivative of y with respect to x.

This example illustrates more than technique. When $y^2 = 3x^4$, $y = \sqrt{3}\,x^2$ so $y' = 2\sqrt{3}\,x$ and $y'' = 2\sqrt{3}$. With some algebraic manipulation y' and y'' given earlier reduce to these simple results. Note this: Many problems have several analytic approaches to their solution, with one often easier than the others. Use implicit differentiation when the defining relation cannot be solved for $y = f(x)$ with reasonable effort; otherwise, establish the function and differentiate as usual. ▮

EXERCISES

1. Find y' by implicit differentiation; determine the equation of the line tangent to the graph of the given equation at the specified point P.

   a. $x^2 + y^2 = 25$; P(-4,-3)

   b. $x^3 - 9x^2 - y^2 + 27x - 2y - 28 = 0$; P(4,-2)

   c. $x^3 - 2x^2y = x + 2y$

   d. $y = 3x - 3y + x^3 - yx^2$

   e. $\sqrt{x} + \sqrt{y} = 9$

   f. $x^2y^3 = x^4 - y^4$

2. If $Ax^2 + Bxy + Cy^2 + Dx + Ey + F = 0$, find y'.

3. Find $\frac{d^2y}{dx^2}$ by implicit differentiation if $x^3 + y^3 = 9$.

4. Find the equation of the tangent line to the curve $x^3 + y^3 = 9$ at the point (1,2).

5. For the adiabatic expansion of a gas (expansion without loss of heat) we have $PV^a = c$ where P is pressure, V is volume, a and c are constants. Find $\frac{dP}{dV}$ by implicit and explicit differentiation.

6. The number of ions I in a solution satisfies $\frac{A + I}{A - I} = e^{2At}$ where A is a constant and t is time. Use implicit differentiation to compute $\frac{dI}{dt}$ in terms of A and I.

7. Given $y = \frac{u^2 - 1}{u^2 + 1}$ and $u = \sqrt[3]{x^2 + 2}$, find y' without solving for y in terms of x.

8. A point moves along the curve $y = x^3 - 3x + 5$ so that $x = \frac{\sqrt{t}}{2} + 3$ where t is time. At what rate is y changing when t = 4?

7.6b LOGARITHMIC DIFFERENTIATION

The labor involved in taking certain derivatives, particularly ones involving quantities raised to a power, or products or quotients, can be reduced greatly by taking the natural logarithm of the expression, using the properties of logarithms and then differentiating, implicitly if necessary. This technique is called LOGARITHMIC DIFFERENTIATION. We used it in establishing the derivative of e^u. These examples further illustrate its use.

EXAMPLE 7.16: If $y = \frac{\sqrt{4 - x^2}}{(x + 2)^{3/2}}$ find y'.

By taking the natural logarithm of this equation, we get

$$\ln y = \frac{1}{2} \ln(4 - x^2) - \frac{3}{2} \ln(x + 2) \ ,$$

Upon differentiating both sides of the equation

$$\frac{1}{y}\frac{dy}{dx} = \frac{1}{2}\left[\frac{-2x}{4 - x^2}\right] - \frac{3}{2}\left[\frac{1}{x + 2}\right] = \frac{x - 6}{2(4 - x^2)}$$

and

$$\frac{dy}{dx} = \frac{y(x - 6)}{2(4 - x^2)}$$

Now substitute y into the last equation:

$$\frac{dy}{dx} = \frac{\sqrt{4 - x^2}\ (x - 6)}{(x + 2)^{3/2}\ 2(4 - x^2)} = \frac{x - 6}{2(x + 2)^{3/2}\ (4 - x^2)^{1/2}}$$

Had we been interested in evaluating the derivative at some point x_0, then it would have not been necessary to substitute for y. We could have found and substituted the corresponding value of y. For example to evaluate y' at x = 0,

$$y = \frac{\sqrt{4 - 0}}{(0 + 2)^{3/2}} = \frac{2}{2^{3/2}} = 2^{-1/2}$$

$$y' = \frac{2^{-1/2}\ (0 - 6)}{2(4 - 0)} = \frac{-6}{8\sqrt{2}} = \frac{-3\sqrt{2}}{8}$$

We would arrive at the same result if we evaluated y' as

$$y' = \frac{0 - 6}{2(0 + 2)^{3/2}\sqrt{4 - 0}} = \frac{-6}{2 \times 2^{3/2} \times 2} = \frac{-6}{2^{7/2}} = \frac{-3\sqrt{2}}{8} . \quad \blacksquare$$

EXAMPLE 7.17: The Gompertz growth curve, given by the function $y = ae^{-ce^{-kt}}$, proved troublesome to differentiate. Logarithmic differentiation gives

$$\ln y = \ln a - ce^{-kt} \qquad \text{so} \qquad \frac{1}{y}\frac{dy}{dt} = cke^{-kt} .$$

Solution of this gives a meaningful form involving y, as well as the complete solution.

$$\frac{dy}{dx} = cke^{-kt}\ y = ack\ e^{-kt-ce^{-kt}} . \quad \blacksquare$$

EXAMPLE 7.18: To this point, we had no way to differentiate $y = u^u$, but logarithmic differentiation provides the tool:

$$\ln y = u \ln u \qquad \text{or} \qquad \frac{1}{y}\frac{dy}{dx} = \frac{u}{u}\frac{du}{dx} + \frac{du}{dx}\ln u$$

so

$$\frac{dy}{dx} = y(1 + \ln u)\frac{du}{dx} = u^u(1 + \ln u)\frac{du}{dx} . \quad \blacksquare$$

EXERCISES

Find the derivatives of the following, using both logarithmic differentiation, and the usual methods. In each case which is easier?

1. $y = \sqrt{\dfrac{x + 2}{x - 2}}$

2. $y = \dfrac{\sqrt{3x + 2}}{(x + 2)^3}$

3. $y = \dfrac{x + 3}{(x^2 - 5)^2}$

4. $y = \dfrac{\sqrt{x + 9}}{\sqrt[3]{x - 3}}$

5. $y = \dfrac{(x^2 + 3)(x^3 - 4)}{(x + 7)^2}$

6. $y = x \ln x$

7. $y = \dfrac{x\sqrt{x + 2}}{\sqrt[3]{x - 2}}$

8. $\dfrac{3x}{\sqrt{(x + 1)(x^2 - 2)}}$

9. $y = (x^2+3x-1)(x+2)^2(x^3-3x+2)$

10. $y = x^2 \ln(x + 1)$

11. $y = \frac{\ln(x + 1)}{\ln(x - 1)}$

12. $y = e^x \ln x^2$

13. $y = 4x \ln\sqrt{x + 1}$

14. $y = \frac{\ln(x + b)}{x + b}$

15. $y = e^{x^2} + 2x \ln x$

16. $y = (e^x)^x$

17. $y = x^{e^x}$

18. $y = x^{\ln x}$

19. $y = 2^{6x}\, 3^{4x^2}$

20. Use the method of logarithmic differentiation to find $\frac{d}{dx} a^u$ where u is a differentiable function of x.

21. Would logarithmic differentiation be useful in calculating the first derivative of $S = 1 - (1 - e^{-kD})^n$? State your reasons.

7.7 DIFFERENTIALS AND SMALL ERRORS

So far we have used derivatives directly. They reflect the rate at which one variable changes relative to another. A slight change of view will allow us to evaluate the comparative change of the dependent variable to the independent variable. This will provide a basis for approximating the change in the dependent variable produced by a small change in the independent variable. Specifically, a small imprecision in measurement can be magnified greatly so that the measurement of interest may be far less precise than you expect.

7.7a DIFFERENTIALS

A geometrical interpretation of the derivative of a function was advanced in the last chapter. We will follow the same reasoning in introducing the differential. Consider a function $y = f(x)$ from x to $x + \Delta x$, and let $\Delta y = f(x + \Delta x) - f(x)$ denote the change in y induced by the change in x. Recall that

$$f'(x) = \lim_{\Delta x \to 0} \frac{f(x + \Delta x) - f(x)}{\Delta x} = \lim_{\Delta x \to 0} \frac{\Delta y}{\Delta x},$$

and thus

$$\lim_{\Delta x \to 0} \left[\frac{\Delta y}{\Delta x} - f'(x) \right] = \lim_{\Delta x \to 0} \frac{\Delta y - f'(x)\Delta x}{\Delta x} = 0 \ .$$

For this limit to go to zero,

$$\Delta y \doteq f'(x)\Delta x$$

when Δx gets close to zero. Now define the DIFFERENTIAL of x, dx, to be the number Δx. Define the differential of y (with respect to dx at x) by the equation

$$dy = f'(x)dx \ .$$

From $\Delta y \doteq f'(x)\Delta x$ above, it follows that dy approximates Δy. As $\Delta x = dx \to 0$, dy becomes a better approximation to Δy.

Figure 7.20 provides a basis for comparing Δy and dy. There P = (x,y) and Q = $(x + \Delta x,\ y + \Delta y)$ satisfy the function; T corresponds to the intersection of the vertical line through $x + \Delta x$ with the line tangent to the curve at P. The slope of the tangent line equals $f'(x)$ so $dy = f'(x)dx = f'(x)\Delta x$ gives the change in the height of the tangent line from x to $x + \Delta x$. The quantity $\Delta y - dy$, the distance from the tangent to the curve, gives the amount by which dy differs from Δy. This difference generally becomes smaller as Δx decreases. Of course how fast f' changes near x also influences the closeness of this approximation.

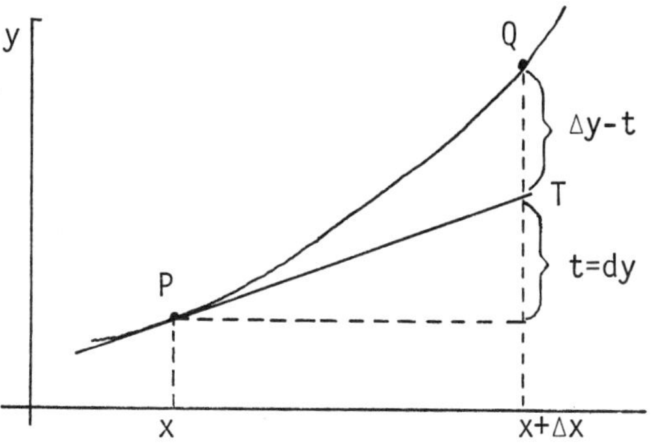

Figure 7.20

The reason for differentials is this: It is much easier to calculate dy than Δy, and in most cases the differential provides a satisfactory approximation of Δy.

<u>EXAMPLE 7.19</u>: When $y = 4x^3 + 4x^2 - 3x + 3$, find dy.

From

$$dy = f'(x)dx \qquad \text{and} \qquad f'(x) = 12x^2 + 8x - 3 \ ,$$

then

$$dy = (12x^2 + 8x - 3)dx \ . \quad \blacksquare$$

EXAMPLE 7.20: If $y = x^2 + 2x$, find both dy and Δy as x goes from $x = 1$ to $x = 1.5$.

$$\Delta y = f(x + \Delta x) - f(x) = (x + \Delta x)^2 + 2(x + \Delta x) - x^2 - 2x$$

$$= 2x\Delta x + (\Delta x)^2 + 2\Delta x \ .$$

When $\Delta x = 0.5$ and $x = 1$, $\Delta y = 2.25$, but $dy = f'(x)dx = (2x + 2)dx = 2$ when $dx = 0.5$ and $x = 1$. The two results, Δy and dy, differ by 0.25. If instead we consider $x = 1$ and $x = 1.01$, then $\Delta x = 0.01$ and $\Delta y = 0.0401$ while $dy = 0.04$. These differ by far less than the earlier pair because dx is much smaller in the latter case. $\blacksquare$

EXAMPLE 7.21: Approximate $\sqrt[5]{33}$ using differentials.

Select the function $y = f(x) = x^{1/5}$ because a distinctive characteristic of $\sqrt[5]{33}$ is its power of 1/5. Note that we seek $f(33)$. The results on differentials imply that $y + \Delta y \doteq y + dy$. In particular consider $x = 32$ and $dx = 1$; the associated value of the function is $f(32) = (32)^{1/5} = 2$. Check that $dy = \frac{1}{5} x^{-4/5} dx$; thus $dy = \frac{1}{5} (32)^{-4/5} (1) = \left(\frac{1}{5}\right)\left(\frac{1}{10}\right) = \frac{1}{80} = 0.0125$. Thus $(33)^{1/5} = y + \Delta y \doteq y + dy = 2 + 0.0125 = 2.0125$. More precisely, $\sqrt[5]{33} \doteq 2.01235$; the approximate result differs from this by less than 0.01%. $\blacksquare$

EXAMPLE 7.22: Approximate the volume of the cellular membrane of a spherical cell of radius 4μ (μ = microns) when the cellular membrane is $1/16\mu$ thick.

Let r and V symbolize the radius and volume of a sphere and recall that $V = \frac{4}{3}\pi r^3$. We seek ΔV when $r = 4\mu$ and $\Delta r = 1/16\mu$. Using differentials, we find $\Delta V \doteq dV = 4\pi r^2 dr = 4\pi(4\mu)^2\left(\frac{1}{16}\mu\right) = 4\pi\mu^3$. Now proceeding directly, $V = \frac{4}{3}\pi(4\mu)^3 = \frac{256\pi}{3}\mu^3$ gives the volume of the interior of the cell, while $V + \Delta V = \frac{4}{3}\pi\left(4\frac{1}{16}\mu\right)^3$. By subtraction then,

$\Delta V = \frac{12.481\pi}{3.072}\mu^3 = 4.06\pi\mu^3$; the approximate value of $4\pi\mu^3$ compares favorably with the exact value. ▌

EXERCISES

1. Approximate the volume of a spherical bacterium as the radius increases from 2μ to 2.25μ. Solve by differentials and then calculate ΔV.

2. A surface coat of thickness t inches was applied evenly to all faces of a cube whose edges were a inches long. Approximately how many cubic inches of coating were used?

3. Approximate the square root of 65 by the use of differentials.

4. Approximate $(37)^{1/5}$ by the use of differentials.

5. Approximate the increase in surface area of a spherical cancerous tumor if the radius increases from 3 centimeters to 3.025 centimeters.

6. A cubical metal box has sides each 0.25 inches thick and an interior volume 40 cubic inches. Use differentials to approximate the volume of metal used to make the box.

7. Is there another interpretation and solution to Example 7.22?

8. Let $y = x^3$ and assume x takes on a small increment Δx. Calculate Δy, the actual change in y, and dy, the approximate change. Show the corresponding parts on a drawing.

9. The surface area of a spherical cell is $S = 4\pi r^2$ and the volume is $V = \frac{4}{3}\pi r^3$. How does a small change in r affect S and V?

10. From a table entry, ln 4 = 1.3863. Approximate ln 4.01 and ln 3.99.

11. Ten grams of a certain substance is being dissolved and the number of grams, Q, left undissolved at the end of t seconds is given by $Q = 10\,e^{-0.01t}$. Find approximately the change in Q when t increases from 200 to 202 seconds.

12. If we consider a blood vessel of length ℓ and of radius r, then Poiseville's law states that the resistance R of the blood vessel is

12. Continued
given by $R = \frac{k\ell}{r^4}$ where k is a constant of proportionality. How does a small change in r affect the resistance R if ℓ remains fixed?

13. When a muscle contracts against a force F, the speed of shortening V decreases with an increasing force. Specifically $(F + a)(V + b) = c$ where a, b and c are constants. How is V affected by a small change in F?

14. The size of a slow growing bacterial population is given by $N = N_0 + 46t + 2t^2$ where t is time in hours and N_0 is the initial number of bacteria.
    a. Find the approximate change in the number of bacteria corresponding to a small change of Δt in time;
    b. Find the actual change in the number of bacteria due to a small change of Δt in time.

15. In a metabolic experiment the mass M of glucose decreased according to $M = M_0 - (0.31)t^2$ where t is time in hours and M_0 is the initial amount of glucose in the system.
    a. Calculate the approximate change in the mass of glucose as t goes from 2 hours to 2 hours 10 minutes.
    b. What is the actual change in the glucose mass in part a?

16. With a certain amount of a gas at a constant temperature, the pressure P and volume V satisfy the relation $PV = C$, where C is a constant. If $P = 20$ pounds per square inch when $V = 60\pi$ cubic inches, find
    a. the approximate change in V if P increases by 0.4 pounds/in.^{2};
    b. the approximate change in V if P decreases by 0.4 pounds/in.^{2};
    c. the actual change in V if P increases by 0.4 pounds/in.^{2};
    d. the actual change in V if P decreases by 0.4 pounds/in.^{2}.

17. A bone of cylindrical shape is measured by an anatomist. He then wants to calculate the curved surface area of the bone. Approximately how much will this area change if he makes an error of Δr in the radius and ΔL in the length? (Hint: Treat the differential of r^2L in a fashion similar to the derivative of a product.)

*18. For what values of x may $\sqrt[5]{x}$ be used in place of $\sqrt[5]{x+1}$, if an allowable error must be less than 0.001?

7.7b IMPACT

Suppose that $y = f(x)$ where x will be evaluated experimentally and y computed from it. Let (x_0, y_0) be the true values of the variables at some point, but suppose x is measured as $x = x_0 + dx$. This produces an error in y because $f(x + dx) = y_0 + \quad y \neq y_0$. To evaluate the impact of such errors we need the following definitions:

- The RELATIVE ERROR in y is the error Δy divided by the true value y.
  R.E. $= \dfrac{\Delta y}{y_0}$.
- The PERCENTAGE ERROR is the relative error multiplied by 100.
- If $|\Delta y| \leq b$, then the MAXIMUM RELATIVE ERROR is $\dfrac{b}{|y_0|}$ so the relative error satisfies

$$\left|\frac{\Delta y}{y_0}\right| \leq \frac{b}{|y_0|}$$

- The MAXIMUM PERCENTAGE ERROR is the maximum relative error multiplied by 100.

How much will an error dx in measuring x change the value of y? Each of the above definitions gives *an* answer. The quantity $\Delta y = f(x + dx) - f(x)$, measures the error in y induced by an error, dx, in x. The importance of Δy equaling 1 cubic centimeter depends heavily on whether y_0 = 100 liters, or 1 liter or 100 cubic centimeters. The relative error recognizes this because it equals $1/(100 \times 1000) = 0.00001$, $1/(1 \times 1000) = 0.001$, or $1/100 = 0.01$, respectively. The percentage error merely expresses the same information on a percentage scale, becoming 0.001%, 0.1% or 1%, respectively. Even when the actual error may remain unknown, it may be possible to bound it. For example if a length is recorded as 103 centimeters, to the nearest centimeter, the actual length lies between 102.5 and 103.5 centimeters. For this, the maximum relative error is $0.5/103 = 0.0049$ or 0.49%, but it may be less, perhaps much less.

This discussion of the impact of small errors accompanies differentials because $\frac{\Delta y}{y_0} \doteq \frac{dy}{y_0}$ and $dy = f'(x)dx$ so $\frac{\Delta y}{y_0} \doteq \frac{f'(x_0)dx}{y_0}$.

EXAMPLE 7.23: Suppose we have a spherical cell of radius 100μ, which was measured with an error of 1% giving a measured radius of 101μ. What is the percentage error in the volume of the cell?

A sphere has a volume of $V = \frac{4}{3}\pi r^3$ and $dV = 4\pi r^2 dr$. The relative error of the volume would be $\frac{dV}{V} = \frac{4\pi r^2 dr}{\frac{4}{3}\pi r^3}$ or $\frac{dV}{V} = \frac{3dr}{r}$. Now $\frac{dr}{r}$, the relative error in measuring of the radius, equals 0.01 here because the error in the radius was 1%. So $\frac{dV}{V} = 3(0.01) = 0.03$, or a 1% error in measuring the radius would create a 3% error in the volume.

The above gives the actual error if the radius is measured with an error of exactly 1%. Ordinarily, however, a 1% possible error in the radius means $|\frac{dr}{r}| \leq 0.01$. Thus $|\frac{dV}{V}| = 3\,|\frac{dr}{r}| \leq 0.03$ gives the MAXIMUM relative error and 3% the MAXIMUM percentage error. ▌

EXAMPLE 7.24: Consider a cone, 24 inches high, having a basal radius measured as 4 inches with an error of at most 2%. Approximate the maximum possible error and the percentage error in the computed volume.

A cone of height h and radius r has a volume of $V = \frac{\pi r^2 h}{3}$. Here $h = 24$ so $V = 8\pi r^2$ and $dV = 16\pi r dr$.

The error in V caused by an error in r will be $dV = 16\pi(4)dr = 64\pi dr$. From the maximum error in r of 2%, $|dr| \leq 4(0.02) = 0.08$ inches, so in turn, $|dV| \leq 64\pi(0.08) = 16.1$ cubic inches. Thus, the error in volume will not exceed 16.1 cubic inches due to a 2% maximum error in the measurement of the radius. Consequently the volume can be too large or too small by as much as 16.1 cubic inches.

The relative error in V due to an error dr in r is given by $\frac{dV}{V} = \frac{16\pi r dr}{8\pi r^2} = \frac{2dr}{r}$. Because the relative error in r is bounded as

$|\frac{dr}{r}| \leq 0.02$, $|\frac{dV}{V}| = 2\ |\frac{dr}{r}| \leq 2(0.02) = 0.04$. The maximum percentage error in V is approximately 4%. ▌

<u>EXAMPLE 7.25</u>: Game biologists encounter this situation: As a deer herd grows in an area, it requires more and more food until finally the animals begin to damage their range. At this point the range damage becomes very apparent visually and animals' growth patterns change noticeably. Two problems arise here. First, how should range damage be measured? Second, how does increasing range damage influence the physical condition of animals in the deer herd?

Antlers are shed each winter and regrow each summer. The yearling bucks still are growing during their second summer while their first sizeable antlers are growing. Thus, it seems reasonable to expect that the volume of antlers on the yearling age class should provide a reasonably precise index of range quality. Notice that this solves one of the frequently troublesome problems of biology, namely, what measurable response really reflects on the biological concern? Even though this response may not measure all aspects of range quality, it does index those aspects of range quality which impinge on deer growth, the response of actual concern.

Now why has this example been brought up in a section on small errors? Volume, V, of antlers cannot be measured very efficiently at a check station, but the diameter, D, of antlers at a standard point can be. Thus, diameter can also be taken as an index of range quality since it increases closely with volume. This last observation includes an interpretive hazard closely associated with differentials and small errors.

Suppose that mean yearling antler diameter, D, in a particular locality changed by 10% from one year to the next. Has range quality changed 10%? No, to the extent that volume indexes range condition, range quality possibly has changed more nearly 30%! Why? The cross section area of an antler closely relates to D^2 while its total length changes with D. Thus, $V = kD^3$ provides a volume-diameter relation; it proves to be a remarkably good approximation in view of the variation in shapes of antlers, but the restriction to yearling age classes helps here. From $dV = 3kD^2 dD$, $dV/V = 3dD/D$ so a change of 10% in D produces approximately a 30% change in V. ∎

EXERCISES

1. A square has sides of length x inches, but it was measured as $(x + dx)$ inches, where the error is $dx > 0$. Draw a figure showing the exact error in the computed area and its differential approximation.

2. If we would assume a red blood cell to be spherical in shape, what would be the percentage error in estimating its volume and surface area if an error of 0.1μ is made in measuring the diameter as 7.5μ?

3. A right circular cylinder has an altitude of 10 inches. If the radius of the base changes from 2 inches to 2.06 inches, use differentials to approximate the corresponding change in the volume of the cylinder. By what percentage has the volume changed?

4. If the diameter of a spherical cell is measured and found to be 20μ with a maximum possible error of 0.05μ, use differentials to approximate the maximum possible error in computing the surface area of the cell. What is the maximum relative error? What is the maximum percentage error?

5. The diameter of a sphere is measured and the result is used to compute the volume of the sphere. If the measurement of the diameter has a maximum possible error of 0.02 inches and a maximum acceptable error in computing the volume is 3 cubic inches, approximate the diameter of the largest sphere to which the process can be applied.

6. The diameter of a spherical cell is to be measured and its volume computed. If the diameter can be measured within an accuracy of 0.1%, what is the maximum percentage error in the determination of the volume? (Caution: The accuracy of the measurement of the radius is not 0.005%. Express the volume in terms of the diameter.)

7. The area of a square greenhouse plot is computed by using 11 feet as the length of a side. Suppose the plot yielded 20 bushels of tomatoes and that this was to be converted to bushels per acre equivalent.

7. Continued

   What is the approximate error, relative error and percentage error in the computed area if the true length of a side was 10.98 feet? In per acre yield? (1 acre = 43,560 ft^2 = 10×66^2 ft^2)

8. In order to compute the volume of a cubical box with an error of at most 2%, with what relative accuracy should the edge of the box be measured?

9. The side of a square is measured as 10 inches, with an error of at most 0.3 inches, and then the area is computed. Find approximately the maximum relative error and the percentage error in calculating the area A.

10. With what largest percentage error, approximately, is it necessary to measure the radius of a sphere in order to compute its surface area with less than 4% error?

11. Find an approximate formula for the area of a circular ring, with an outer radius r and a small width of dr.

12. An orange with a diameter of 3 inches has a skin which is 1/8 inch thick.

    a. What is the approximate volume of the skin?

    b. What percentage of the total volume is skin?

13. The altitude of a right circular cone is twice the radius of the base. The altitude is measured as 12 inches with a maximum possible error of 0.005 inches. Find the approximate error in the calculated volume of the cone.

14. Stefan's law for the emission of radiant energy from the surface of a body is given by $R = KT^4$ where R is the rate of emission per unit area, T is the temperature in °Kelvin and K is a constant. A 1% change in T corresponds to what change in R? A relative error of at most 0.03 in T results in what size of a relative error in R?

15. If the radius of the base of a right-circular cone is half its altitude and if the radius of the base is measured as 2 inches with a possible error of ±0.01 inch, approximate the maximum possible error in calculating the volume of the cone.

16. The radius of a circle was measured and found to be 19.73 centimeters with a maximum possible error of 0.05 centimeters. What is the maximum possible error in calculating the area of the circle?

17. The volume of a sphere is to be measured by submerging it in water and then calculating the radius from the amount of water the sphere displaces. If the volume was measured as 512 ± 5 cubic inches, what is the radius and the relative error in the radius?

18. A cylindrical shaped bacterium is 10μ long. If its radius grows from 2μ to 2.06μ, what is the percentage change in the volume of the bacterium?

19. In measuring the acceleration of gravity, g, by means of a pendulum, the formula $g = \frac{4\pi^2 \ell}{T}$ is used where ℓ is the length of the pendulum in inches and T is the period of time in seconds. Assuming the error in measuring ℓ is negligible, express the error dg in terms of the error dT. Express the relative error of g in terms of the relative error of T.

*7.7c APPROXIMATING ROOTS OF AN EQUATION

If f is known, how should you find those values of x for which $f(x) = 0$? This problem can arise in several ways. For example, suppose f_1 describes how a process responds to the variable x_1 relying on one feature of the process. A different feature could lead to a related function f_2. Specifically f_1 might describe the carrying capacity of a changing habitat while f_2 described population size. For what x does $f_1(x) = f_2(x)$? This is equivalent to seeking the zeros of $f(x) = f_1(x) - f_2(x)$.

The Newton method for approximating the roots of an equation uses reasoning similar to that used in defining differentials. Let the curve in Figure 7.21 represent the behavior of the function f in the neighborhood of its root r where $f(r) = 0$. We seek a method for locating r starting from an initial approximation x_0. The line tangent to the curve at $P_0 = (x_0, y_0) = (x_0, f(x_0))$ will intersect the x-axis at a point labeled x_1, provided $f'(x_0) \neq 0$. In the figure, x_1 lies closer to r than x_0 was;

this generally happens provided f' changes slowly between x_0 and r. To find x_1, the tangent through P_0 has a slope of $f'(x_0)$ and passes through $(x_1,0)$: Thus

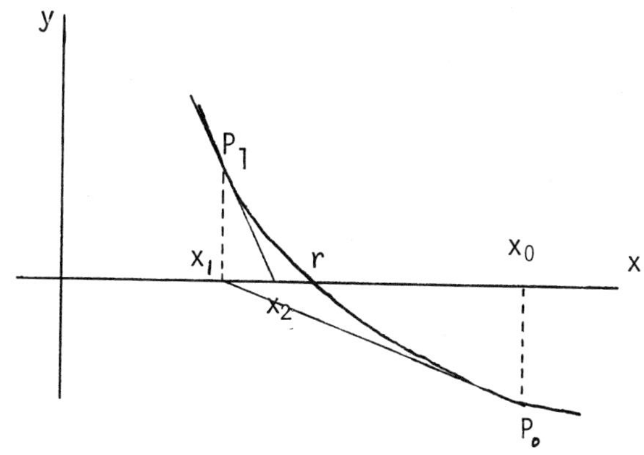

Figure 7.21

$\frac{y_0 - 0}{x_0 - x_1} = f'(x_0)$, or

$y_0 = (x_0 - x_1)f'(x_0)$, so

$x_1 = x_0 - \frac{y_0}{f'(x_0)}$. Finally using $y_0 = f(x_0)$, we get

$$x_1 = x_0 - \frac{f(x_0)}{f'(x_0)} .$$

By repeating the above reasoning at $P_1 = (x_1,y_1) = (x_1,f(x_1))$, we find x_2:

$$\frac{y_1 - 0}{x_1 - x_2} = f'(x_1) \qquad \text{or} \qquad x_2 = x_1 - \frac{f(x_1)}{f'(x_1)} .$$

In fact this approach will take us from any approximation x_{n-1} to a successive one x_n through

$$x_n = x_{n-1} - \frac{f(x_{n-1})}{f'(x_{n-1})} . \tag{7.9}$$

As $x_n \to r$, $f(x_n) \to 0$ so the successive subtractive factors decrease in size.

In practice, we start with an initial approximation x_0 to r. We repeatedly apply Equation 7.9 until x_{n-1} and x_n become equal within our desired computational precision. The sequence $x_0, x_1, \cdots, x_n, \cdots$ approachs r provided the function does not have maxima, minima, or inflection points between either x_0 and r, or x_1 and r. This merely places restrictions on x_0. For an f whose derivative changes very slowly near r, x_0 can be a very poor approximation indeed; for a function having a critical value near r, x_0 needs to be a fairly good approximation.

EXAMPLE 7.26: Use Newton's method to approximate the root between 0 and 1 of $x^3 - 3x + 1 = 0$, to one decimal digit of accuracy. Figure 7.22 shows the graph of this function in the region of $x = 0$ and $x = 1$. A root lies between $x = 0$ and $x = 1$ because the polynomial drops from 1 to -1 over this interval. Thus an initial approximation to the root could be $x_0 = 0.5$. The first approximation is given by

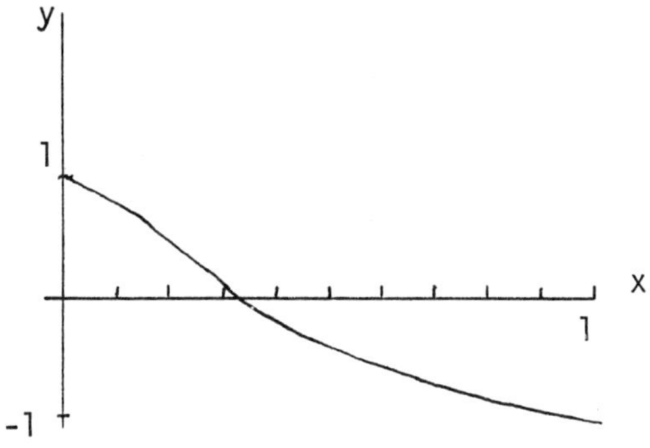

Figure 7.22

$$x_1 = x_0 - \frac{f(x_0)}{f'(x_1)} \qquad \text{where} \qquad f'(x) = 3x^2 - 3$$

or

$$x_1 = 0.5 - \frac{f(0.5)}{f'(0.5)} = 0.5 - \frac{-0.375}{-2.25} = 0.5 - 0.1667 = 0.3333 .$$

The second approximation is then given by

$$x_2 = 0.3333 - \frac{f(0.3333)}{f'(0.3333)} = 0.3333 - \frac{0.037125}{-2.666733}$$

$$= -0.3333 + 0.01392 = 0.34722 .$$

As x_1 and x_2 have the same first decimal digit, we have satisfied this problem's criterion. Thus, $r = 0.3$ would be accurate to one decimal place. Of course, repeated application of the process will give more digits of accuracy. ▮

EXERCISES

1. Draw a figure to show how Newton's method can produce an x_1 further away from the true root, r, than x_0 was.
2. In view of your answer in Problem 1, what general guidelines should be followed when using Newton's method? What are some of the pitfalls of the process?

3. Use Newton's method to find the two roots of $x^2 - 6x + 6 = 0$, accurate to 2 decimal places.

4. Check the result you obtained in Problem 3 by using the quadratic formula.

5. Use Newton's method to find the root of $x^3 + 3x - 5 = 0$ between 1 and 2.

6. How do you know a root exists between 1 and 2 in Problem 5?

*7.8 A TOOL FOR APPROXIMATING FUNCTIONS

Think back to Chapter 1. There we pointed out that biologists get functional relations from several places. In particular, they sometimes are able to develop functions by understanding the biological process and associated physical and chemical processes. Other times they have to approximate the exact function by another function which possesses the essential properties of the exact function. Approximation often is necessary because some part of the essential biological, chemical, or physical processes is not clearly understood, although the final result of the biological process can be seen.

Within mathematics, another kind of approximation appears. How can a given function be approximated by another function? In fact, what does this question mean? Think back to Section 7.5c where we compared the Logistic and Gompertz growth curves. These two curves had the same shape; by appropriately choosing parameters, the Gompertz could be forced to have the same initial value (at $t = 0$), the same limiting value (at $4 = \infty$), and the same inflection point (at $t = \frac{\ln c}{k}$). In a sense then the Gompertz would display some of the distinctive properties of the Logistic, that is, the Gompertz could be regarded as an approximation to the Logistic.

Polynomial functions occupy an important place in mathematical approximation. Their form,

$$a_0 + a_1x + a_2x^2 + \cdots a_nx^n$$

allows them to assume a variety of shapes. Over a restricted range, any continuous function can be approximated to any desired accuracy by polynomial, provided enough terms are taken. In many cases only a few terms will give a surprisingly good approximation. This section introduces one of the main tools used to get from a known function to its polynomial approximation. It introduces two approximations which differ in the point about which they are expanded, namely, whether the approximating polynomials involve powers of x or of x - a. This chapter presents only an introduction to series expansion. It ignores certain technical points, like whether a particular function has a series expansion. Chapter 16 deals with such topics in substantially more detail.

*7.8a MACLAURIN SERIES EXPANSION

When a function can be written as

$$f(x) = a_0 + a_1x + a_2x^2 + \cdots + a_nx^n + \cdots,$$

how do we determine the coefficients $a_0, a_1, a_2, \cdots, a_n, \cdots$ in the expansion? When f has continuous derivatives of all orders, take a few of the derivatives to get

$$f'(x) = a_1 + 2a_2x + 3a_3x^2 + \cdots + na_nx^{n-1} + \cdots$$

$$f''(x) = 2a_2 + 6a_3x + \cdots + n(n-1)a_nx^{n-2} + \cdots$$

$$f'''(x) = 6a_3 + \cdots + n(n-1)(n-2)a_nx^{n-3} + \cdots$$

In general, $f^{(n)}(x) = n!a_n + \cdots$, but if $x = 0$ then all of the subsequent terms involving x are zero so $a_0 = f(0)$, $a_1 = f'(0)$, $a_2 = f''(0)/2!$, $a_3 = f'''(0)/3!, \cdots, a_n = f^{(n)}(0)/n! \cdots$. This last term sometimes is called the GENERAL TERM because it will work for any n. Substituting these values into the original expression, we have:

- The MACLAURIN SERIES EXPANSION of the function f with continuous derivatives of all orders is given by

$$f(x) = f(0) + \frac{f'(0)x}{1!} + \frac{f''(0)x^2}{2!} + \frac{f'''(0)x^3}{3!} + \cdots + \frac{f^{(n)}(0)x^n}{n!} + \cdots$$

This expansion provides the basis for numerical values of various mathematical quantities found in many tables. Series expansions have become widely used in computer computation. In sophisticated computing environments the right name sends the computer off to evaluate almost any standard mathematical function, but the computation often is done by series expansion. In less sophisticated environments, or for unusual functions, you may need to develop your own series expansions.

You may wonder why computers do not use tables. Accurately evaluated tables require large amounts of storage, one of the scarce resources in a computer, but a few lines of computer program can evaluate a series expansion in less time than you can blink an eye.

EXAMPLE 7.27: Find the Maclaurin expansion for e^x.

Remember the special property of $f(x) = e^x$? Because $f^{(n)}(x) = e^x$, $f^{(n)}(0) = 1$ so the Maclaurin series gives

$$e^x = f(0) + \frac{f'(0)x}{1!} + \frac{f''(0)x^2}{2!} + \frac{f'''(0)x^3}{3!} + \cdots$$

$$= 1 + x + \frac{x^2}{2!} + \frac{x^3}{3!} + \frac{x^4}{4!} + \cdots$$

Specifically, if $x = 2$,

$$e^2 = 1 + 2 + \frac{2^2}{2} + \frac{2^3}{3!} + \frac{2^4}{4!} + \frac{2^5}{5!} + \frac{2^6}{6!} + \cdots$$

$$\doteq 1 + 2 + 2 + \frac{8}{6} + \frac{16}{24} + \frac{32}{120} + \frac{64}{720} = 7.356 .$$

This does not differ very far from $e^2 = 7.389$ because the omitted terms are getting smaller quickly; more terms would improve the approximation. ▌

7.8b TAYLOR SERIES EXPANSION

When a function possesses derivatives of all orders at $x = a$, we can write

$$f(x) = a_0 + a_1(x - a) + a_2(x - a)^2 + \cdots + a_n(x - a)^n + \cdots .$$

To determine the constants $a_0, a_1, a_2, \cdots, a_n, \cdots$, proceed as for the Maclaurin expansion by taking derivatives, but then let $x = a$ in each derivative. In so doing we obtain

- The TAYLOR SERIES EXPANSION about the point a of the function f with continuous derivatives of all orders is given by

$$f(x) = f(a) + \frac{f'(a)(x - a)}{1!} + \frac{f''(a)(x - a)^2}{2!} + \frac{f'''(a)(x - a)^3}{3!} + \cdots + \frac{f^{(n)}(a)(x - a)^n}{n!} + \cdots$$

The Maclaurin series expansion is a special case of the Taylor series expansion with $a = 0$. Presently we will not use the Taylor series to evaluate any series; we need this expansion to develop a method for evaluating limits of indeterminate forms.

EXERCISES

Obtain the first 4 non-zero terms in the Maclaurin series of each of the following:

1. $f(x) = 1 + x + x^2$
2. $f(x) = (1 + x)^3$
3. $f(x) = (1 - x)^{-1/3}$
4. $f(x) = e^{-x}$
5. $f(x) = 1 + 2x + x^2 + 4x^3$
6. $f(x) = (4 + x^3)^{1/2}$
7. $f(x) = a^x$
8. $f(x) = e^x - e^{-x}$
9. Why can e^{-x} be approximated by $1 - x$ for small values of x?
10. Evaluate $e^{0.1}$ accurate to three decimal places.
11. Evaluate $10^{0.1}$ accurate to 3 decimals by two different methods.
12. Write a Maclaurin series for $f(x) = \frac{a^x}{e^x}$ and obtain the form for the general term.

7.9 L'HOSPITAL'S RULE FOR EVALUATING INDETERMINATE FORMS

Suppose we need to evaluate $\lim_{x \to a} \frac{f(x)}{g(x)}$, but that $\lim_{x \to a} f(x) = 0$ and $\lim_{x \to a} g(x) = 0$, or $\lim_{x \to a} f(x) = \pm\infty$ and $\lim_{x \to a} g(x) = \pm\infty$. In Section 5.4 we demonstrated that such ratios could assume any value, depending on the nature of f and g. Here we add L'Hospital's Rule to the tools developed there. L'Hospital's Rule states

$$\lim_{x \to a} \frac{f(x)}{g(x)} = \lim_{x \to a} \frac{f'(x)}{g'(x)}$$

The following development of this rule applies to the first of these indeterminate forms (0/0). When f(x) and g(x) can be expressed in Taylor series about a, the conditions f(a) = 0 and g(a) = 0 give

$$f(x) = f'(a)(x - a) + \frac{f''(a)(x - a)^2}{2!} + \frac{f'''(a)(x - a)^3}{3!} + \cdots$$

and

$$g(x) = g'(a)(x - a) + \frac{g''(a)(x - a)^2}{2!} + \frac{g'''(a)(x - a)^3}{3!} + \cdots$$

so

$$\lim_{x \to a} \frac{f(x)}{g(x)} = \lim_{x \to a} \frac{f'(a) + \frac{f''(a)(x - a)}{2!} + \frac{f'''(a)(x - a)^2}{3!} + \cdots}{g'(a) + \frac{g''(a)(x - a)}{2!} + \frac{g'''(a)(x - a)^2}{3!} + \cdots}$$

after dividing the numerator and denominator by the common quantity (x - a). It follows that

$$\lim_{x \to a} \frac{f(x)}{g(x)} = \lim_{x \to a} \frac{f'(x)}{g'(x)}$$

If f'(a) and g'(a) both equal zero, another indeterminate form occurs and we obtain:

$$\lim_{x \to a} \frac{f(x)}{g(x)} = \lim_{x \to a} \frac{f''(a) + \frac{1}{3}(x - a)\, f'''(a) + \cdots}{g''(a) + \frac{1}{3}(x - a)\, g'''(a) + \cdots}$$

$$= \lim_{x \to a} \frac{f''(x)}{g''(x)}$$

This process can be repeated as necessary until a form which is no longer indeterminate results. Basically, L'Hospital's Rule states that if an indeterminate form is obtained, take the limit of the derivative of the numerator divided by the derivative of the denominator. This process can be carried out as many times as is needed until the limit can be evaluated.

<u>EXAMPLE 7.28</u>: Evaluate $\lim\limits_{t \to 2} \dfrac{\ln(t - 1)}{t - 2}$.

As $t \to 2$ $\ln(t - 1) \to \ln 1 = 0$ and $t - 2 \to 0$. Hence we have the indeterminate form 0/0. Having satisfied this condition, and differentiability, apply L'Hospital's rule:

$$\lim_{t \to 2} \frac{\ln(t - 1)}{t - 2} = \lim_{t \to 2} \frac{\frac{1}{t - 1}}{1} = \lim_{t \to 2} \frac{1}{t - 1} = 1 \ . \quad \blacksquare$$

<u>EXAMPLE 7.29</u>: Evaluate $\lim\limits_{x \to \infty} \dfrac{x^3}{e^{2X}}$.

As $x \to \infty$, $x^3 \to \infty$ and $e^{2X} \to \infty$ also so we have the indeterminate form ∞/∞. L'Hospital's rule also works in this case:

$$\lim_{x \to \infty} \frac{x^3}{e^{2X}} = \lim_{x} \frac{3x^2}{2e^{2X}} \ .$$

However this remains indeterminate, of the form ∞/∞; applying L'Hospital's rule once again we have

$$\lim_{x \to \infty} \frac{x^3}{e^{2X}} = \lim_{x \to \infty} \frac{3x^2}{2e^{2X}} = \lim_{x \to \infty} \frac{6x}{4e^{2X}} \ .$$

Again an indeterminate form results, but a third application of the rule gives

$$\lim_{x \to \infty} \frac{x^3}{e^{2X}} = \lim_{x \to \infty} \frac{3x^2}{2e^{2X}} = \lim_{x \to \infty} \frac{6x}{4e^{2X}} = \lim_{x \to \infty} \frac{6}{8e^{2X}} = 0 \ . \quad \blacksquare$$

L'Hospital's rule usually provides an easy method for evaluating indeterminate limits. However, any of the techniques developed earlier, such as factoring and rationalizing also may work. You must pick the tool which is the most appropriate for your current problem.

EXERCISES

Evaluate the following limits.

1. $\lim_{x\to 1} \frac{x^4 + 3x^2 - 6x + 2}{x^3 - 2x^2 + 5x - 4}$

2. $\lim_{x\to 2} \frac{x^3 - 3x^2 + 4}{x^3 - 5x^2 + 8x - 4}$

3. $\lim_{x\to -3} \frac{x^2 - 9}{x + 3}$

4. $\lim_{x\to -3} \frac{x^2 - 9}{x - 3}$

5. $\lim_{x\to -3^+} \frac{x^2 + 9}{x + 3}$

6. $\lim_{x\to \infty} \frac{x - 2x^3}{x^2}$

7. $\lim_{h\to 0} \frac{x^2h + 3xh + h^3}{2xh + 5h^2}$

8. $\lim_{k\to 0} \frac{(2z + 3k)^3 - 4k^2z}{2z(2z - k)^2}$

9. $\lim_{x\to 2} \frac{x^2 + x - 6}{x^2 - 4}$

10. $\lim_{x\to 2} (3x^3 - 4x^2 + 2)$

11. $\lim_{x\to -2} (5x^3 - 2x + 3)$

12. $\lim_{x\to \infty} \frac{e^{-\frac{1}{x}}}{x}$

13. $\lim_{x\to \infty} \frac{\ln x}{x - 1}$

14. $\lim_{x\to 1} \frac{\ln x}{x - 1}$

15. $\lim_{x\to 1} \frac{\ln(2x - 1)}{1 - x^2}$

16. $\lim_{x\to \infty} \frac{e^x}{x^3}$

17. $\lim_{x\to \infty} \frac{e^{-x}}{x^7}$

18. $\lim_{x\to 0} \frac{2^x - 3^x}{x}$

19. $\lim_{x\to \infty} \frac{e^x}{\ln x}$

20. $\lim_{x\to \infty} \frac{x}{\ln x}$

21. $\lim_{x\to 0} \frac{\ln x}{10^x}$

22. $\lim_{x\to \infty} \frac{(\ln x)^2}{e^x}$

23. $\lim_{x\to 0^+} \frac{\sqrt{5 + x^2}}{x}$

24. $\lim_{x\to 0^-} \frac{\sqrt{5 + x^2}}{x}$

25. Does the $\lim_{x \to 0} \frac{\sqrt{5 + x^2}}{x}$ exist? See Problems 23 and 24.

26. $\lim_{x \to -3} \frac{x^3 + 27}{2x + 6}$

27. $\lim_{x \to 4} \frac{3\sqrt{x} - 6}{x - 4}$, work two ways.

28. $\lim_{x \to 3} \frac{\sqrt{3 + 2x} - 3}{x - 3}$, work two ways.

29. $\lim_{x \to -2^+} \frac{3x^2 + x - 10}{x^2 - 4}$

30. $\lim_{x \to -2} \frac{3x^2 + 2x - 10}{x^2 - 4}$

Chapter 8

INTRODUCTION TO INTEGRATION

We sometimes use the phrase *putting it all together* to describe the assembling of many pieces of a problem or perspective so we can grasp the whole thing. This process of accumulation has its counterpart in calculus where integration can be viewed as the summing many small pieces. Properly interpreted, integrals represent such things as leaf areas, accumulated incident solar radiation (the light which makes plants grow), total energy output or work of a muscle, etc. The first part of this chapter presents a geometric introduction to integration.

Areas have application in biology, but the biologists' greatest interest in integration lies elsewhere. A basic theorem, often called the fundamental theorem of calculus, relates integration to differentiation. It provides the basis for viewing integration as the reverse of differentiation; it opens the bridge from a derivative back to its function. Many biological phenomena can be described by how they change, or equivalently by how the derivative of a related function behaves. These convert simply into differential equations as shown in Section 8.9. Integration occupies a central role in solving these equations for the underlying function; we will present the simplest of these in the latter part of this chapter.

8.1 AREA

Should a biologist be interested in areas? Biologists regularly use not only the areas of regular geometric figures, but also areas under curves. For example some physiological processes change fast enough that a measurement at one time cannot adequately describe them.

EXAMPLE 8.1: Consider blood pressure in an artery. The curve in Figure 8.1 shows how pressure changes over a very short time for a typical healthy young adult human. Within the span of one second it drops from the systolic pressure of more than 120 mm. to the diastolic pressure of about 80 mm. Blood pressure changes remarkably with age, activity, and ailments.

The area under the curve during one full cycle provides a summary of the aggregate pressure during that cycle. This area will change with variations in the systolic and diastolic pressures and with the shape of the pressure curve. The area under the curve divided by the cycle length gives an average or mean pressure which is used in some studies of blood pressure. The area inside the rectangle in Figure 8.1 equals the area under the curve during the same time span; the area under the curve but above the rectangle (singly lined) equals the area (cross hatched) inside the rectangle, but above the curve. ▌

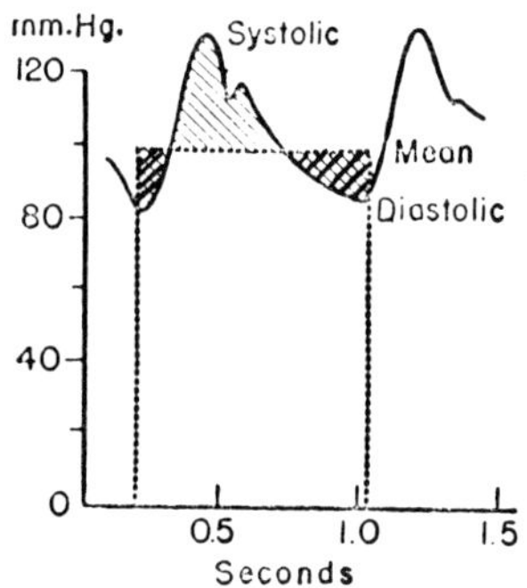

Figure 8.1

Every triangle and every rectangle has an associated number called its area. A triangle has an area equal to one-half of the product of the length of its base and height; the area of a rectangle equals the product of the lengths of its base and height. Other figures such as parallelograms and polygons have easily calculated areas. However, how can we find the area under a portion of a curve?

To describe what we mean by such an area, let f be a function continuous on a closed interval $[a,b]$; initially assume that $f(x) > 0$ for all $x \varepsilon [a,b]$. Consider the shaded region in Figure 8.2 bounded by the graph of $y = f(x)$, the lines $x = a$ and $x = b$, and the x-axis. This shaded region identifies the area we want to find.

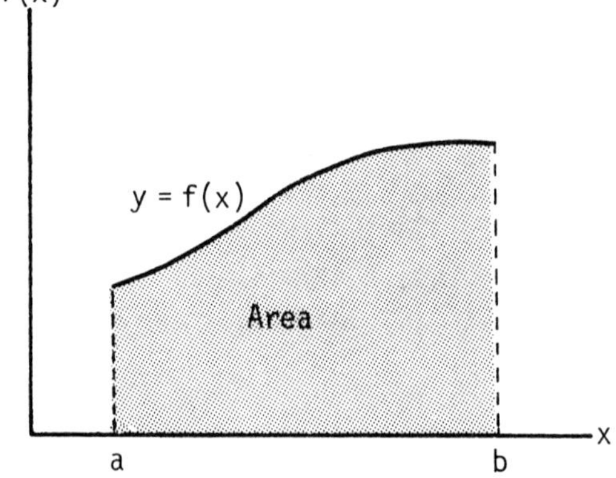

Figure 8.2

This area exists but we do not yet know how to calculate it exactly, so we begin by approximating it. An approximation results from dividing [a,b] into subintervals so that the area over each can be approximated by that of a rectangle. Later the size of each subinterval can be shrunk, using limits, to give the exact area. Thus, if the closed interval [a,b] is divided into n subintervals by $a = x_0 < x_1 < \cdots < x_n = b$, we can denote the ith subinterval by t_i and its width by $\Delta_i x = x_i = x_i - x_{i-1}$. Select any point $u_i \in t_i$, one in each subinterval as indicated in Figure 8.3. The products $f(u_i)\Delta_i x$, $i = 1, 2, \cdots, n$, give the areas of rectangles of heights $f(u_i)$ and widths $\Delta_i x$; their sum approximates the area under the curve between a and b.

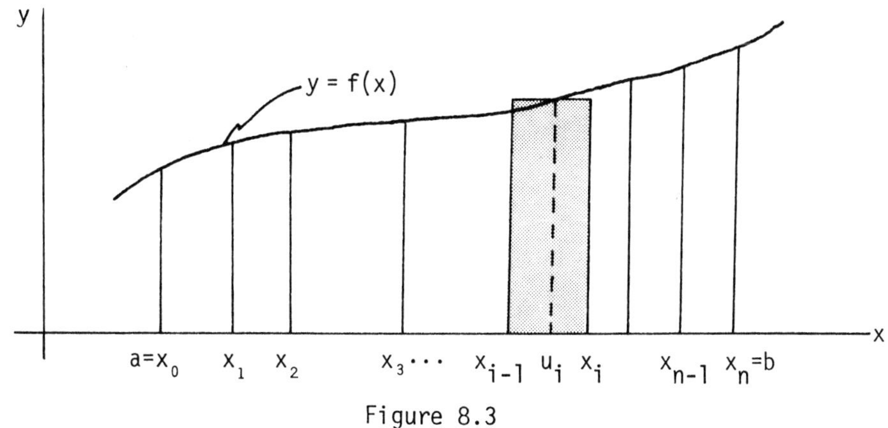

Figure 8.3

First consider choosing $u_i \in t_i$ so that $f(u_i)$ gives the absolute maximum of f on t_i. The product $f(u_i)\Delta_i x$ provides an upper bound for that part of the area under the curve which lies above t_i. Thus the sum of these products provides an upper bound for the actual area as illustrated in Figure 8.4.

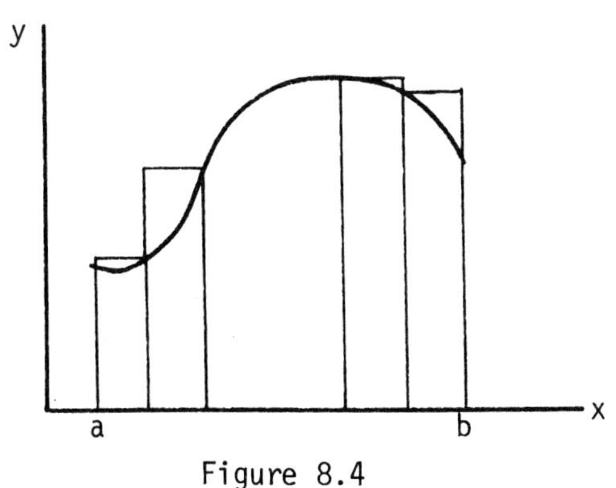

Figure 8.4

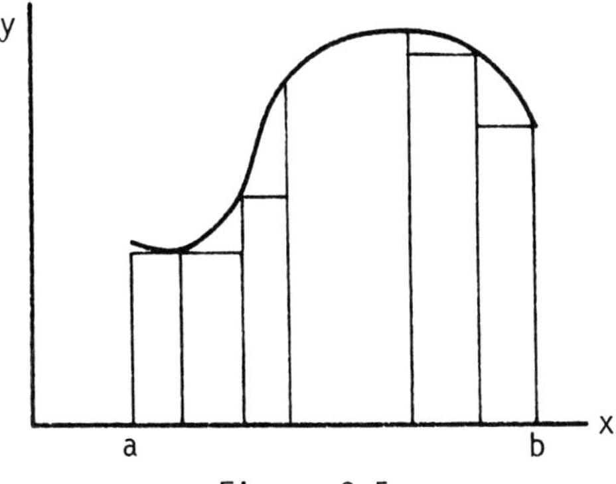

Figure 8.5

On the other hand if we choose $u_i \in t_i$ to minimize f on t_i, the sum of the products $f(u_i)\Delta_i x$ provides a lower bound for the actual area, as illustrated in Figure 8.5. If we let A_L and A_U symbolize the areas calculated by using the minimal and maximal values on each subinterval, then A_L and A_U provide lower and upper bounds for the actual area, A. When $A_U = A_L$, their common value gives A, the actual area under the curve.

Although $a = x_0 < x_1 < \cdots < x_n = b$ can be selected to subdivide [a,b] in any manner, it is convenient to make an equal subdivision so all $\Delta_i x = (b - a)/n$. This convenience introduces no loss of generality for the situations considered in this book. It will be used in the sequel unless specifically noted to the contrary.

EXAMPLE 8.2: To illustrate the above ideas, we take a simple and familiar area: That of a trapezoid. Consider the function $f(x) = 2x$ and let R be the region from $x = 2$ to $x = 4$, as shaded in Figure 8.6. You can calculate the area shaded directly as $A = 2\left[\frac{4 + 8}{2}\right] = 12$ square units since the area of a trapezoid equals $\frac{1}{2}$ of the height between the parallel bases times the sum of the length of the bases.

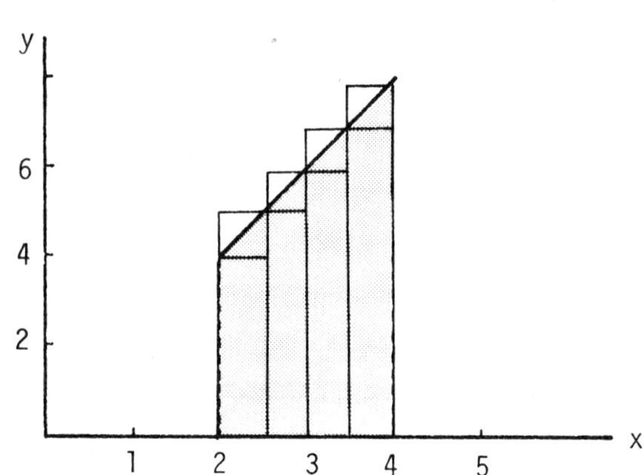

Figure 8.6

Find an upper bound and a lower bound for the area of R using: a) 4 equal length subregions, and b) 8 equal length subregions. When the interval [2,4] is divided into 4 subintervals of equal length, $\Delta_i x = \frac{4 - 2}{2} = \frac{1}{2}$. The maximal value of f on each subinterval occurs at the right edge of the subinterval. For example f ranges from $f(2) = 4$ to $f\left(\frac{5}{2}\right) = 5$ on $t_i = \left(2, \frac{5}{2}\right)$ so f has a maximal value of

5 on t_i. Thus choose the u_i to be the right-hand endpoint of their respective subintervals. The larger rectangles (Figure 8.5) over each subinterval have an area of $f(u_i)\Delta_i x = 2u_i\left[\frac{1}{2}\right] = u_i$. Now we can evaluate A_U:

$$A_U = (5)\left(\frac{1}{2}\right) + (6)\left(\frac{1}{2}\right) + (7)\left(\frac{1}{2}\right) + (8)\left(\frac{1}{2}\right) = 13 \text{ square units} .$$

To evaluate A_L, take u_i at the left edge of each subinterval:

$$A_L = (4)\left(\frac{1}{2}\right) + (5)\left(\frac{1}{2}\right) + (6)\left(\frac{1}{2}\right) + (7)\left(\frac{1}{2}\right) = 11 \text{ square units} .$$

When the number of subintervals is increased to 8, we have:

$$A_U = \left(\frac{9}{2}\right)\left(\frac{1}{4}\right) + \left(\frac{10}{2}\right)\left(\frac{1}{4}\right) + \left(\frac{11}{2}\right)\left(\frac{1}{4}\right) + \left(\frac{12}{2}\right)\left(\frac{1}{4}\right) + \left(\frac{13}{2}\right)\left(\frac{1}{4}\right) + \left(\frac{14}{2}\right)\left(\frac{1}{4}\right)$$
$$+ \left(\frac{15}{2}\right)\left(\frac{1}{4}\right) + \left(\frac{16}{2}\right)\left(\frac{1}{4}\right)$$
$$= \frac{100}{8} = 12.5 \text{ square units} .$$

Proceed similarly to find $A_L = 11.5$ square units.

Notice that the units attached to these bounds are the units associated with area, namely square units. Also notice that A_U and A_L approach each other as the number of subintervals is increased. The first approximation bounded A by $11 < A < 13$ while the second gave $11.5 < A < 12.5$. The smaller the intervals, the tighter the bounds.

The true area also can be calculated as the difference between the area of two triangles. The area of the larger triangle is $A_1 = (4)(8)/2 = 16$ square units; the area of the smaller triangle is $A_2 = (2)(4)/2 = 4$ square units. These give a total area of $A = A_1 - A_2 = 16 - 4 = 12$ square units. ▮

This discussion of area has assumed that $f(x) \geq 0$ for $x \in [a,b]$, but not all functions satisfy this requirement. For example, consider the function whose graph appears in Figure 8.7. An approximation to the area between the curve and the x-axis is given by $A_1 - A_2 - A_3 - A_4 + A_5$ if area is interpreted as a positive quantity. More generally, however, area can be regarded as a signed quantity with the negative being associated with areas lying below the x-axis. The negative signs occur for those areas having $f(u_i) < 0$ because $f(u_i)\Delta_i x$ gives a negative number. Thus, when f can become negative, the area between the curve $y = f(x)$ and the x-axis for $a \leq x \leq b$ means the net area. A negative net area then indicates more area below the x-axis than above it.

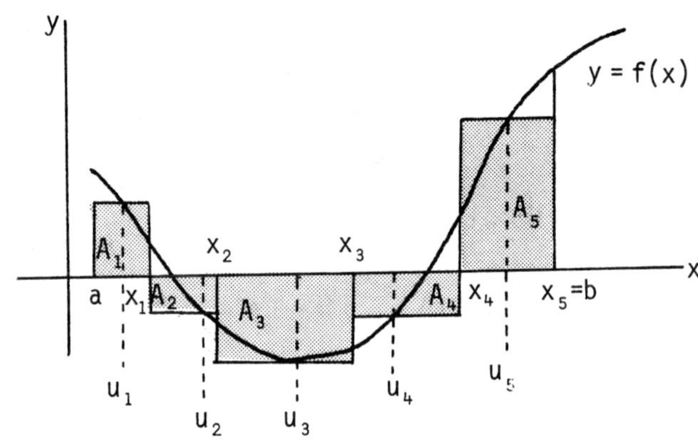

Figure 8.7

<u>EXAMPLE 8.3</u>: Green plants consume carbon dioxide, CO_2, in direct proportion to the rate at which photosynthetic products, such as sugars, are being produced. Certain physical factors, such as temperature, available light, and nutrients, govern the rate at which photosynthesis proceeds. If you were to measure net CO_2 uptake over a 24-hour period, a curve such as in Figure 8.8 would appear. Why should this curve ever dip below zero?

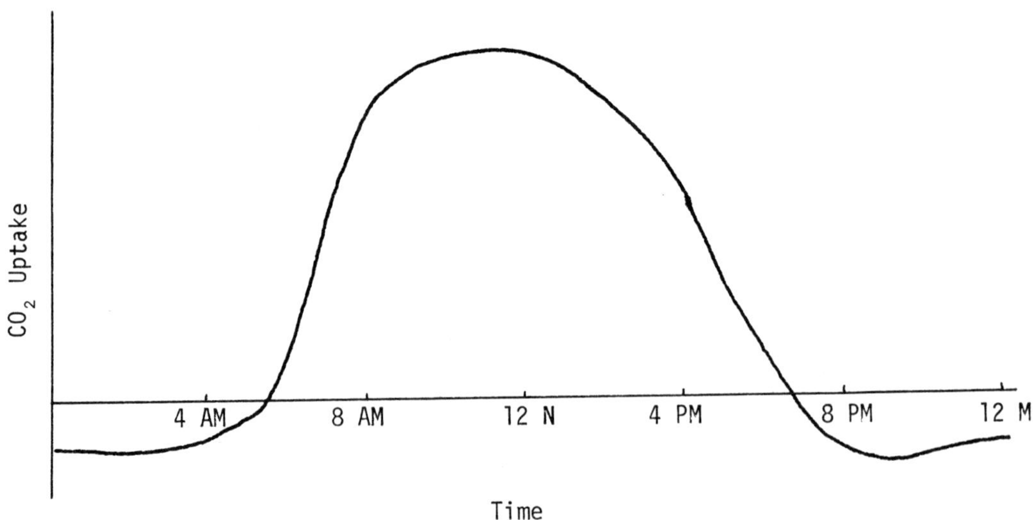

Figure 8.8

A plant oxidizes or burns sugars to produce energy, releasing CO_2 just as you do. Again this proceeds at a rate conditioned by factors such as temperature, but light has relatively little influence on it. Thus, even during the day net CO_2 uptake represents the excess of photosynthesis over respiration, namely, net energy conversion, but at night no photosynthesis offsets respiration.

The net area under the curve in Figure 8.8 represents the net CO_2 uptake, or equivalently energy conversion, during a 24-hour period. Because physical factors condition the rates of photosynthesis and respiration, negative net energy conversions can occur. For example, during a warm night followed by a cool heavily overcast day a plant could require more energy for its maintenance than it converts. ∎

EXERCISES

1. Sketch the region bounded by the x-axis and $y = x^2$ between $2 \leq x \leq 4$. Calculate an upper bound and lower bound for its area using
   a. 4 equal subintervals;
   b. 8 equal subintervals.

2. Calculate bounds for the area under $y = \frac{1}{x}$ between $\frac{1}{2} \leq x \leq 2$ using
   a. 6 equal subintervals;
   b. 12 equal subintervals.

3. If R is the region under the graph of $y = c$ for $0 \leq x \leq a$, show that $A_U = A_L$ for any set of subintervals you choose.

4. Calculate bounds (A_L and A_U) for the area under $y = x^2 - 4x + 3$ between $0 \leq x \leq 4$ using 8 equal subintervals.

5. Calculate bounds (A_L and A_U) for the area under $y = x$ between $-1 \leq x \leq 3$ using
   a. 4 equal subintervals;
   b. 8 equal subintervals.

6. Calculate the net area of the region defined in Problem 5.

8.2 SUMMATION NOTATION

To facilitate our discussion of summing the rectangular areas described in the previous section, we need a notation for summation:

- When a set of numbers has been ordered so its elements can be denoted as $a_1, a_2, a_3, \cdots$, then the sum of all the numbers from the mth number (a_m) through the nth number (a_n) is denoted by

$$\sum_{i=m}^{n} a_i = a_m + a_{m+1} + \cdots + a_n \qquad (8.1)$$

The symbolism on the left-hand side of Equation 8.1 compactly states a whole sentence. It says add up the numbers symbolized by a_i, beginning with the one numbered m and continuing through the one numbered n by increments of 1. Frequently m = 1, but other values do occur. The capital Greek letter sigma, $\sum$, corresponds to S for summation. The subscript i on the a's is called the INDEX OF SUMMATION. Several important properties of summation will be demonstrated in the next examples.

EXAMPLE 8.4: $\sum_{i=2}^{6} i(i - 1) = 2(1) + 3(2) + 4(3) + (5)4 + (6)5 = 70$.

Here the index of summation i has an initial value of 2 and is increased by 1 until 6 is reached. It also illustrates that the elements being added can be functions of the index of summation. ▌

EXAMPLE 8.5: $\sum_{i=1}^{6} a_i = a_1 + a_2 + a_3 + a_4 + a_5 + a_6 = \sum_{k=1}^{6} a_k$.

This example demonstrates that the name or symbol for the index of summation is immaterial. ▌

EXAMPLE 8.6: Suppose $a_i = c$, $i = 1, 2, \cdots, 8$. Then

$$\sum_{i=1}^{8} a_i = a_1 + a_2 + \cdots + a_8 = c + c + \cdots + c = 8c = \sum_{i=1}^{8} c$$

This illustrates that for a constant, the subscript is understood. Simply add up the constant the appropriate number of times. ▌

EXAMPLE 8.7: Summation of a function of the ordered elements poses no problem. Simply evaluate the function for each element and add up the results:

$$\sum_{k=2}^{5} kf(x_k) = kf(x_2) + kf(x_3) + kf(x_4) + kf(x_5)$$

$$= k\{f(x_2) + f(x_3) + f(x_4) + f(x_5)\}$$

$$= k \sum_{k=2}^{5} f(x_k)$$

This shows that the sum of a constant times a function equals the constant times the sum of the function. ▌

Equations 8.2 - 8.4 give useful summation formulas which are given here without proof. We will need them frequently in evaluating certain sums.

$$\sum_{i=1}^{n} i = 1 + 2 + \cdots + n = \frac{n(n+1)}{2} \tag{8.2}$$

$$\sum_{i=1}^{n} i^2 = 1^2 + 2^2 + \cdots + n^2 = \frac{n(n+1)(2n+1)}{6} \tag{8.3}$$

$$\sum_{i=1}^{n} i^3 = 1^3 + 2^3 + \cdots + n^3 = \left[\frac{n(n+1)}{2}\right]^2 \tag{8.4}$$

To gain a feeling for the above check them out with $n = 2$, 3, and 4. We also need this result which you can verify:

$$\sum_{i=1}^{n} (a_i + b_i) = \sum_{i=1}^{n} a_i + \sum_{i=1}^{n} b_i . \tag{8.5}$$

EXAMPLE 8.8: Evaluate $\sum_{i=1}^{15} 3i(2i - 5)$.

By multiplying the two quantities, $3i$ and $2i - 5$, and then applying Equation 8.5 we have

$$\sum_{i=1}^{15} 3i(2i - 5) = \sum_{i=1}^{15} (6i^2 - 15i) = \sum_{i=1}^{15} 6i^2 - \sum_{i=1}^{15} 15i .$$

Applying the result obtained in Example 8.7, we have

$$\sum_{i=1}^{15} 3i(2i - 5) = 6 \sum_{i=1}^{15} i^2 - 15 \sum_{i=1}^{15} i .$$

Now use Equations 8.2 and 8.3 with $n = 15$:

$$\sum_{i=1}^{15} 3i(2i - 5) = 6 \frac{(15)(16)(31)}{6} - 15 \frac{(15)(16)}{2} = 5640 .$$

Repeat this for an upper limit of 5 to get 105; also get this result by evaluating the five terms in the sum and adding them up. ▌

<u>EXAMPLE 8.9</u>: Evaluate $\sum_{i=1}^{n} (2 + i)(4 - i)$.

$$\sum_{i=1}^{n} (2 + i)(4 - i) = \sum_{i=1}^{n} (8 + 2i - i^2)$$

$$= \sum_{i=1}^{n} 8 + 2 \sum_{i=1}^{n} i - \sum_{i=1}^{n} i^2$$

$$= 8n + \frac{2n(n + 1)}{2} - \frac{n(n + 1)(2n + 1)}{6}$$

$$= \frac{n}{6} (53 + 3n - 2n^2) .$$

Thus for any value of n, we can evaluate this sum simply by substituting the appropriate value for n into this expression. ▌

<u>EXERCISES</u>

1. Show that $\sum_{i=1}^{n} (a_i + b_i) = \sum_{i=1}^{n} a_i + \sum_{i=1}^{n} b_i$

Evaluate the following sums two ways: (i) Write down the individual terms and add them up; (ii) use Equations 8.2 - 8.4 if possible.

2. $\sum_{i=1}^{7} (2i + 3)$ 3. $\sum_{i=3}^{6} \frac{i}{2i - 7}$ 4. $\sum_{k=5}^{10} (2k - 1)^2$

Evaluate the following sums.

5. $\sum_{i=1}^{n} i(i + 1)$

7. $\sum_{j=1}^{n} j(j + 1)(j + 2)$

6. $\sum_{k=1}^{n} (4k^2 - 4k + 1)$

8. $\sum_{j=m}^{n} j(j + 2)$ where $m < n$.

9. Evaluate $\sum_{i=1}^{8} f\left(\frac{i}{2}\right)\Delta_i x$ where $f(x) = x^2 - 4x + 3$ and $\Delta_i x = \frac{1}{2}$.

10. Evaluate $\sum_{i=1}^{4} f(2 + \frac{i}{2})\Delta_i x$ where $f(x) = x^2$ and $\Delta_i x = \frac{1}{2}$.

8.3 THE DEFINITE INTEGRAL

In this section we define the definite integral and discuss its properties. We still are concerned with a function f defined over the closed interval [a,b]. This interval is subdivided into n subintervals of equal length by $a = x_0 < x_1 < \cdots < x_n = b$ so that $\Delta_i x = x_i - x_{i-1} = (b - a)/n$ for $i = 1, 2, \cdots, n$. We now offer the following definitions:

- The DEFINITE INTEGRAL or RIEMANN INTEGRAL of the function f over the closed interval [a,b] is denoted $\int_a^b f(x)dx$ and is given by

$$\int_a^b f(x)dx = \lim_{n\to\infty} \sum_{i=1}^{n} f(x_i)\Delta_i x = \lim_{n\to\infty} \sum_{i=1}^{n} f(x_i)\,\frac{b - a}{n} \qquad (8.6)$$

provided this limit exists.

- If $\int_a^b f(x)dx$ exists, then f is called INTEGRATABLE over [a,b].

- The definite integral has this property:

$$\int_a^b f(x)dx = -\int_b^a f(x)dx \ .$$

- If a = b, then $\int_a^b f(x)dx = \int_a^a f(x)dx = 0$.

We could define the RIEMANN INTEGRAL or DEFINITE INTEGRAL as

$$\int_a^b f(x)dx = \lim_{n\to\infty} \sum_{i=1}^{n} f(x_i)\Delta_i x$$

where we subdivide the closed interval [a,b] into n subintervals of different lengths. Although we shall not do so, it can be proved that if the limit exists in Equation 8.6, the same limit exists for any sequence of subdivisions. To calculate the value of the integral using the definition, equal subintervals allow for easier calculations. The point where f is evaluated can lie anywhere in each subinterval; for calculating purposes it is often taken as the right-hand endpoint of the ith subinterval.

The above definition of the definite integral assumes that b > a but if a > b the convention

$$\int_a^b f(x)dx = -\int_b^a f(x)dx$$

resolves any problems because a > b implies subdivision from right to left giving negative $\Delta_i x$, and consequently a negative of the usual left to right definition.

The quantity

$$\sum_{i=1}^{n} f(x_i)\Delta_i x \ ,$$

is sometimes denoted by S_n and is called a RIEMANN SUM. The symbol for integration, $\int$, apparently resulted from an elongation of S from the first letter of sum. In the definition, f(x) is called the INTEGRAND, *a* the LOWER LIMIT of integration and *b* the UPPER LIMIT of integration.

Compare Equation 8.6 to the area approximations discussed in Section 8.1. The definite integral merely equals the limit of area approximations for a finer and finer subdivision along the x-axis. You might conveniently think of the limit in Equation 8.6 occurring by a repeated doubling of the number of subintervals. As the number of subintervals is doubled, the length of each is halved so $f(x_i)\Delta_i x$ gives an increasingly more precise approximation to a true area of each rectangle so their sum becomes a more accurate approximation to the true area. This process will be illustrated in the following examples.

EXAMPLE 8.10: Evaluate $\int_2^4 2x dx$.

Before you go any further, look back at Example 8.2; it concerned the same integrand and region. The graph of the function $y = 2x$ appears in Figure 8.9. We begin by dividing the closed interval [2,4] into n equal subintervals of length $\Delta_i x = \frac{4-2}{n} = \frac{2}{n}$.

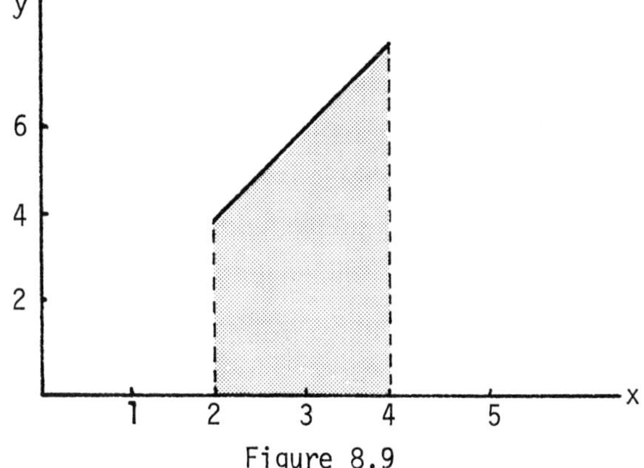

Figure 8.9

If we choose the arbitrary point, x_i, in each subinterval to be the right-hand endpoint of the interval and let $x_0 = 2$ and $x_n = 4$, we have $x_i = 2 + \frac{2i}{n}$ where i is the number of the subinterval. Using $f(x) = 2x$, then

$$f(x_i) = 2\left[2 + \frac{2i}{n}\right]$$

and

$$f(x_i)\Delta_i x = 2\left[2 + \frac{2i}{n}\right]\frac{2}{n} = \frac{8}{n} + \frac{8i}{n^2}.$$

Thus the Riemann sum becomes

$$S_n = \sum_{i=1}^{n} f(x_i)\Delta_i x = \sum_{i=1}^{n}\left[\frac{8}{n} + \frac{8i}{n^2}\right] = \sum_{i=1}^{n}\frac{8}{n} + \sum_{i=1}^{n}\frac{8i}{n^2}.$$

Now using the summation formulas given earlier for $\sum_{i=1}^{n} C$ and $\sum_{i=1}^{n} i$, we have

$$\sum_{i=1}^{n} f(x_i)\Delta_i x = 8 + \frac{8}{n^2}\sum_{i=1}^{n} i = 8 + \frac{8}{n^2}\,\frac{n(n+1)}{2} = 8 + 4\left[1 + \frac{1}{n}\right] = 12 + \frac{4}{n}\ .$$

Thus from

$$\int_2^4 2x dx = \lim_{n\to\infty} \sum_{i=1}^{n} f(x_i)\Delta_i x\ ,$$

we have

$$\int_2^4 2x dx = \lim_{n\to\infty} \left[12 + \frac{4}{n}\right] = 12 \text{ square units}\ .$$

You saw this same result when you looked back at Example 8.2. It was evaluated initially as the area of a trapezoid. Later in the same example it was evaluated as the difference in the areas of two triangles. The latter view will be generalized by the FUNDAMENTAL THEOREM OF THE CALCULUS, introduced in Section 8.5. ▌

The last example shows that a geometric approach yields the same area as the limit definition. Functions exist, however, which we cannot approach geometrically so we must fall back on the definition to evaluate their integrals. The next example illustrates this sort of function.

EXAMPLE 8.11: Find the area bounded by $y = x^2$ and the x-axis for $0 \le x \le 2$. The area is shown in Figure 8.10.

We begin by dividing $[0,2]$ into n equal subintervals each of a length $\Delta_i x = 2/n$ and if we choose x_i to be the right-hand endpoint, then $x_i = 2i/n$, $i = 0, 1, \cdots, n$. Now form the Riemann sum:

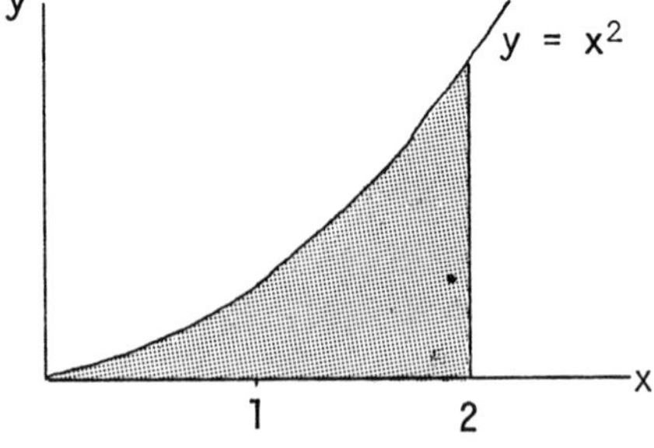

Figure 8.10

$$S_n = \sum_{i=1}^{n} f(x_i)\Delta_i x = \sum_{i=1}^{n} \left[\frac{2i}{n}\right]^2 \frac{2}{n} = \frac{8}{n^3}\sum_{i=1}^{n} i^2 = \frac{8}{n^3}\left[\frac{n(n+1)(2n+1)}{6}\right]$$

$$= \frac{4}{3}\left[2 + \frac{3}{n} + \frac{1}{n^2}\right] .$$

Upon taking the limit of the Riemann sum as all $\Delta_i x \to 0$, or equivalently as $n \to \infty$ we have

$$\int_0^2 x^2 dx = \lim_{n\to\infty} \sum_{i=1}^{n} f(x_i)\Delta_i x = \lim_{n\to\infty} \frac{4}{3}\left[2 + \frac{3}{n} + \frac{1}{n^2}\right] = \frac{8}{3} \text{ square units} .$$

In this situation no geometric method would give the exact area under the curve. ∎

The definition of an integral involves a limit; the integral exists whenever the limit does. The limit, and thus the integral, existed for all of the examples presented in this section. The integrals of most interesting functions do exist, but a few fail to exist. An integral may fail to exist for two distinctly different causes. If $\lim_{x \to x_0} f(x) = \pm\infty$ for $a \le x_0 \le b$, the integral may fail to exist. Difficulties of this sort are dealt with in Section 9.2. An integral also may fail to exist because its integrand fluctuates violently within the interval of integration. However, most of the functions we will consider will be well behaved functions.

EXERCISES

For the specified integrand, compute $S_n = \sum_{i=1}^{n} f(x_i)\Delta_i x$ accurate to two decimal places using the specified number of equal subintervals. Take x_i as indicated. Then obtain the exact value of the integral by using n equal subintervals and letting $n \to \infty$.

1. $\int_0^2 \left[1 + \frac{x^2}{2}\right]dx$; 4 subintervals and x_i = right-hand endpoint.

2. $\int_0^2 \left[1 + \frac{x^2}{2}\right]dx$; 8 subintervals

   a. x_i = midpoint,

   b. x_i = the left-hand endpoint,

   c. x_i = the right-hand endpoint.

3. Are the results obtained in Problems 1 and 2 what you would expect? Explain.

Find the exact value of the given definite integral.

4. $\int_1^3 (2x + 3)dx$ Hint: Let $x_i = 1 + \left[\frac{b-a}{n}\right]i = 1 + \left[\frac{3-1}{n}\right]i = 1 + \frac{2i}{n}$.

5. $\int_1^4 (x^2 + 1)dx$ Hint: Let $x_i = 1 + \frac{3i}{n}$.

6. $\int_0^3 (2x^2 + 3)dx$ 7. $\int_1^3 (x^3 - 1)dx$ 8. $\int_2^4 \frac{x^3 + 2}{3} dx$

8.4 PROPERTIES OF INTEGRALS

Some very useful basic properties of definite integrals are given in this section. Some of the properties will be stated while others will be proved. As you become familiar with these properties, notice their great similarity to analogous properties of derivatives and limits.

We have already given two properties of integrals in the form of definitions. These properties were

$$\int_a^b f(x)dx = -\int_b^a f(x)dx \tag{8.7}$$

and

$$\int_a^a f(x)dx = 0 . \tag{8.8}$$

You may wonder when does a function have an integral? The following theorem states one condition which assures an integrable function:

THEOREM 8.1: If f is continuous on the interval [a,b], then f(x) is integrable on [a,b], namely $\int_a^b f(x)dx$ exists.

We shall assume that any integrand hereafter is a continuous function unless otherwise stated. Theorem 8.1 then assures the existence of the integrals of such functions. For convenience in stating the following properties we shall assume that the lower limit of integration is less than the upper limit of integration. However, the properties are true for other limits of integration because of the two properties stated in Equations 8.7 and 8.8.

THEOREM 8.2: If k is any constant then $\int_a^b kf(x)dx = k\int_a^b f(x)dx$.

PROOF: By direct application of the limit definition of a definite integral, we have

$$\int_a^b kf(x)dx = \lim_{n\to\infty} \sum_{i=1}^{n} kf(x_i)\Delta_i x = \lim_{n\to\infty} k \sum_{i=1}^{n} f(x_i)\Delta_i x$$

because $\sum kf(x) = k \sum f(x)$; now using $\lim_{x\to x_0} kf(x) = k \lim_{x\to x_0} f(x)$,

$$\int_a^b kf(x)dx = k \lim_{n\to\infty} \sum_{i=1}^{n} f(x_i)\Delta_i x = k\int_a^b f(x)dx \;.\blacktriangledown$$

THEOREM 8.3: $\int_a^b [f(x) + g(x)]dx = \int_a^b f(x)dx + \int_a^b g(x)dx$.

PROOF: From the limit definition of an integral we have

$$\int_a^b [f(x) + g(x)]dx = \lim_{n\to\infty} \sum_{i=1}^{n} [f(x_i) + g(x_i)]\Delta_i x \;.$$

From the properties of summation we have

$$\sum_{i=1}^{n} [f(x_i) + g(x_i)]\Delta_i x = \sum_{i=1}^{n} f(x_i)\Delta_i x + \sum_{i=1}^{n} g(x_i)\Delta_i x \;.$$

Thus

$$\int_a^b [f(x) + g(x)]dx = \lim_{n\to\infty} \sum_{i=1}^{n} f(x_i)\Delta_i x + \lim_{n\to\infty} \sum_{i=1}^{n} g(x_i)\Delta_i x$$

because the limit of a sum equals the sum of the limits. The result now follows simply by noting that these sums define $\int_a^b f(x)dx$ and $\int_a^b g(x)dx$.▼

THEOREM 8.4: For any a, b, and c,

$$\int_a^b f(x)dx + \int_b^c f(x)dx = \int_a^c f(x)dx \;,$$

whenever each of the integrals exist. The proof of this statement is left as an exercise.

THEOREM 8.5: When the functions $\phi(x)$, $\theta(x)$ and $f(x)$ are defined on $[a,b]$, and $\phi(x) \leq f(x) \leq \theta(x)$, then

$$\int_a^b \phi(x)dx \leq \int_a^b f(x)dx \leq \int_a^b \theta(x)dx \ .$$

Biologists sometimes need only an upper or lower bound on an integral. This theorem provides the bounds without requiring an exact form for f, provided f can be bounded by known functions. Theorem 8.5 has several noteworthy special cases. You may legitimately use only one or the other of the bounds. In particular if $\phi(x) = 0 \leq f(x)$ over $[a,b]$, then $0 \leq \int_a^b f(x)dx$, namely, a non-negative function has a non-negative integral. The proof of this theorem is left as an exercise.

Because a function lies between its absolute minimum and maximum over an interval, we also have:

THEOREM 8.6: If m and M symbolize the absolute minimum and maximum of a function f over the interval $[a,b]$, then

$$m(b - a) \leq \int_a^b f(x)dx \leq M(b - a) \ .$$

This has the simple geometric interpretation shown in Figure 8.11. The integral $\int_a^b f(x)dx$ represents the area bounded by the curve $y = f(x)$, the x-axis and $a \leq x \leq b$. If m is the absolute minimum of $y = f(x)$, then $m(b - a)$ is the area of a rectangle of height m and width $b - a$, lying completely below f on $[a,b]$. If M is the absolute maximum of $y = f(x)$ then $M(b - a)$ is the area of a rectangle with a height of M and a width of $b - a$ completely including f. By examining

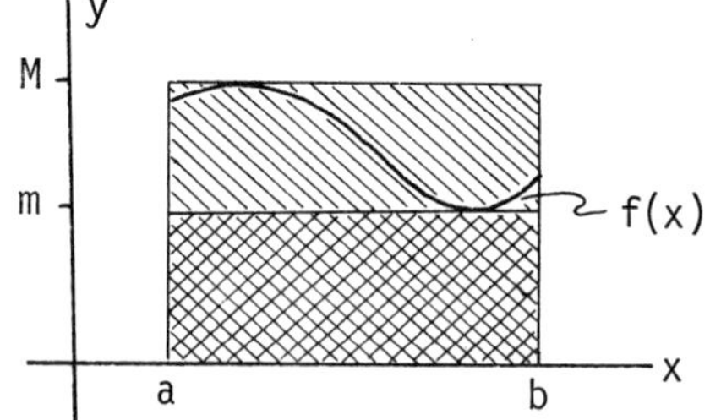

Figure 8.11

Figure 8.11 it should be clear that the area $\int_a^b f(x)dx$ certainly is bounded by $m(b - a)$ and $M(b - a)$.

EXERCISES

1. Give an intuitive argument using areas to prove Theorem 8.4.
2. Intuitively argue the proof of Theorem 8.5.
3. Show Theorem 8.4 is true for the case $a < b < c$.
4. Show Theorem 8.4 is true when a, b, c are distinct and in any order.
5. Show Theorem 8.4 is true when two of the three points, a, b, c are identical.
6. Show that $\int_0^1 \sqrt{1 + 2x^3}\,dx$ lies between 0 and $\frac{3}{2}$.

8.5 EVALUATION OF DEFINITE INTEGRALS

You have encountered a definition of the definite integral and some of its properties; you have used its interpretation as an area. As yet you have no easy way to evaluate an integral because the limit definition produces cumbersome algebra and the area approach works only for familiar geometric figures.

The result stated in this section, often called the FUNDAMENTAL THEOREM OF CALCULUS, unites differential calculus with integral calculus to form a single branch of mathematics, usually called calculus. We now offer the following result:

- The FUNDAMENTAL THEOREM OF CALCULUS states that if f is a continuous function on the interval [a,b] and F is any function such that $F'(x) = f(x)$, then

$$\int_a^b f(x)dx = F(b) - F(a) .$$

This theorem states a simple but powerful result. To evaluate the integral of f between a and b, merely find a function F, called the ANTIDERIVATIVE of f, which satisfies $F'(x) = f(x)$; evaluate this antiderivative at the upper limit b and subtract the value of the antiderivative evaluated at the lower limit a. In Example 8.10 we evaluated the area under $y = 2x$ for $2 \leq x \leq 4$ several ways. We could have evaluated the area by use of the Fundamental Theorem by letting $F(x) = x^2$ because $\frac{d}{dx} x^2 = 2x$. Then $F(4) - F(2) = 16 - 4 = 12$ square units as we obtained before.

A given function does not have a unique antiderivative because the derivative of a constant equals zero. Although x^2 may seem like the obvious antiderivative of 2x, $F_1(x) = x^2 + 2$, $F_2(x) = x^2 + 17$, $F_3(x) = x^2 - 31$ all are also antiderivatives of 2x because $F_1'(x) = F_2'(x) = F_3'(x) = 2x$. Thus, if F(x) is an antiderivative of f(x), then so is F(x) + C where C is any constant. This prompts us to state three basic antiderivative formulas:

- When $f(x) = x^n$, $n \neq -1$, then $F(x) = \int x^n dx = \frac{x^{n+1}}{n+1} + C$. (8.9)
- When $f(x) = \frac{1}{x}$, $x > 0$, then $F(x) = \int \frac{dx}{x} = \ln x + C$. (8.10)
- When $f(x) = e^x$, then $F(x) = \int e^x dx = e^x + C$. (8.11)

Verify that $F'(x) = f(x)$ in each of these cases because these antiderivatives serve as the basic building blocks for executing integration on the class of functions we have considered already.

In evaluating a definite integral, the value of C has no influence on the integral's value because

$$\int_a^b f(x)dx = [F(b) + C] - [F(a) + C] = F(b) - F(a) .$$

Thus without loss of generality we may take C = 0. Note, however, that this statement applies only to definite integrals. Later when we look at indefinite integrals and differential equations, we will sometimes have occasion to choose C's different from zero.

Another notation sometimes appears in the context of evaluating definite integrals. The middle expression in

$$\int_a^b f(x)dx = F(x)\Big]_a^b = F(b) - F(a) \qquad (8.12)$$

relates the antiderivative to the bounds of the integral without yet placing the numerical values for a and b in F(x).

EXAMPLE 8.12: Find an antiderivative of $f(x) = 3x^2$.

Begin by factoring out the constant: $\int f(x)dx = \int 3x^2dx = 3\int x^2dx$. Now apply Equation 8.9 with n = 3 to get $F(x) = 3\left[\frac{x^3}{3} + C\right] = x^3 + 3C$. In particular, as C varies, we find $F_1(x) = x^3$, $F_2(x) = x^3 + 3$, $F_3(x) = x^3 + \pi$, $F_4(x) = x^3 + 97$. Check each of these to see that they are antiderivatives of f(x). For purposes of evaluating definite integrals, we can let C = 0 and use $F_1(x) = x^3$. ▮

EXAMPLE 8.13: Find an antiderivative of $f(x) = x^{2/3}$.

Use Equation 8.9 with n = 2/3 to find

$$F(x) = \frac{x^{5/3}}{5/3} = \frac{3}{5}x^{5/3}.$$

Again you can check this result simply by showing $F'(x) = f(x)$. ▮

EXAMPLE 8.14: Evaluate $\int_1^3 (2x - 3)dx$.

The function $f(x) = 2x - 3$ has a general antiderivative of $F(x) = x^2 - 3x + C$. To find this antiderivative, Equation 8.9 was used twice, once with n = 1 for the term 2x and again with n = 0 for the term -3. Each application produced an arbitrary constant, but they may be combined since the sum of two constants is a constant. Now use F(x):

$$\int_1^3 (2x - 3)dx = [x^2 - 3x + C]_1^3 = (3^2 - 3(3) + C) - (1^2 - 3(1) + C)$$

$$= 9 - 9 + C - 1 + 3 - C = 2. \text{ ▮}$$

Notice that C, the constant of antidifferentiation, appears as C - C in the evaluation of the definite integral. In applying the fundamental theorem hereafter, we can let C = 0 when choosing a particular antiderivative of the integrand of a definite integral.

EXAMPLE 8.15: Evaluate $\int_{-1}^{3} (2x^2 + 3x - 1)dx$.

If $f(x) = 2x^2 + 3x - 1$, why does $F(x) = \frac{2}{3}x^3 + \frac{3}{2}x^2 - x$? Using this F, we get

$$\int_{-1}^{3} (2x^2 + 3x - 1)dx = \left[\frac{2}{3}x^3 + \frac{3}{2}x^2 - x\right]_{-1}^{3}$$

$$= \left[\frac{2}{3}(3)^3 + \frac{3}{2}(3)^2 - (3)\right] - \left[\frac{2}{3}(-1)^3 + \frac{3}{2}(-1)^2 - (-1)\right]$$

$$= \frac{80}{3} .$$ ∎

EXAMPLE 8.16: Evaluate $\int_{3}^{7} e^x dx$.

When $f(x) = e^x$, $F(x) = e^x$, so

$$\int_{3}^{7} e^x dx = e^x\Big]_{3}^{7} = e^7 - e^3 = 1096.6 - 20.1 = 1076.5 .$$ ∎

EXAMPLE 8.17: Evaluate $\int_{0.6}^{1.7} \frac{2}{x} dx$.

When $f(x) = \frac{1}{x}$, $F(x) = \ln(x)$ so

$$\int_{0.6}^{1.7} \frac{2}{x} dx = 2\int_{0.6}^{1.7} \frac{1}{x} dx = 2 \ln x\Big]_{0.6}^{1.7} = 2(\ln 1.7 - \ln 0.6)$$

$$= 2(0.53063 - (-0.51083)) = 2.08292 .$$

You can check that $\ln 1.7 = 0.53063$ in Appendix Table II . ∎

EXAMPLE 8.18: In Chapter 4 we saw how an exponential function can describe the decay of a radioactive substance, or the decline of a population. A curve such as shown in Figure 8.12 indicates the *rate* of decay, a rate which changes with time.

The area shaded between t_0 and t_1 indicates the accumulated or aggregate decay over this time span. This statement certainly deserves an explanation. Suppose that the rate of decay remained constant over $[t_0,t_1]$, like an automobile cruising along on an open highway. As such rates give the amount of something per unit of time, amount = rate × time. For example an automobile going 55 miles per hour for 2 hours would go 110 miles = (55 miles per hour) × (2 hours). This is the area under the rate curve (line) y = 55 M.P.H. over an interval of length $t_1 - t_0 = 2$ hr. The area under a changing rate curve generalizes this idea to rates (speeds) which vary. Those parts of the time interval where the rate is low (slow) contribute little to the area (distance traveled). Let us return to the decay rate (f) shown in Figure 8.12. The area under the curve, and thus the accumulated decay, is given by

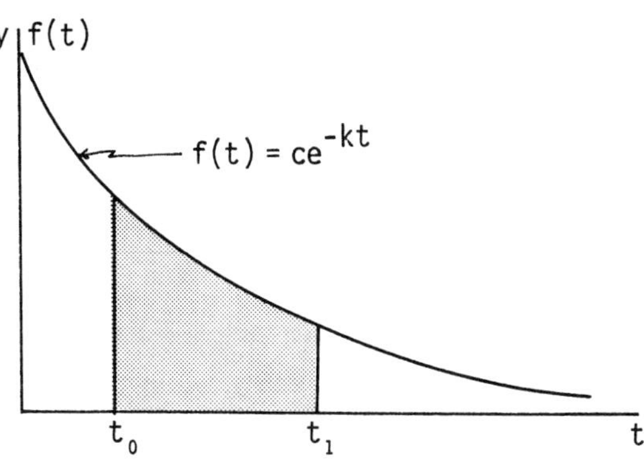

Figure 8.12

$$\int_{t_0}^{t_1} f(t)dt = \int_{t_0}^{t_1} ce^{-kt}dt = -\frac{c}{k}e^{-kt}\Big]_{t_0}^{t_1}$$

$$= \frac{c}{k}\left[e^{-kt_0} - e^{-kt_1}\right] .$$

Now what meaning does the constant c in f(t) have? The term e^{-kt} governs the shape of f(t), while c fixes its initial height. The role of c is particularily noticeable at t = 0 where f(0) = c. Look at the units on c: For f(t) to have units of amount per time (miles per hour, disintegrations per second), c must have the same units as f because the exponential function can have no units. Further, for kt to be unitless, k must have units of 1/time so c/k has units of (amount/time)/(1/time) = amount. In particular, if we seek relative rather than total amounts, we can consider an amount of one and let c = k.

Recall the discussion of carbon dating presented in Section 4.7, Example 4.33: As plants grow they extract carbon from the atmosphere, a small portion of it as the isotope ^{14}C. This isotope decays slowly so it still is disintegrating thousands of years later. By measuring the current rate of disintegration we can determine how long ago the plant was growing. Earlier we found that the exponential parameter $k = 0.000124$ for ^{14}C.

If a plant was growing 10,000 years ago, what proportion of the material has decayed during the last 2000 years? What proportion is left to decay? As these questions concern relative amounts rather than total amounts, we take $f(t) = ke^{-kt} = (0.000124)\ e^{-0.000124t}$. Thus the integral

$$\int_{8000}^{10000} (0.000124)\ e^{-0.000124t}\ dt$$

gives the decay in the last 2000 years. To evaluate this integral, observe that $F(t) = -e^{-0.000124t}$ satisfies

$F'(t) = f(t) = 0.000124t\ e^{-0.000124t}$, so

$$\int_{8000}^{10000} (0.000124)\ e^{-0.000124t}\ dt = -e^{-0.000124t}\Big]_{8000}^{10000}$$

$$= -\ e^{-1.24} - e^{-0.992}$$

$$= -[0.2893 - 0.3708] = 0.0815\ .$$

Thus only slightly more than 8% of the decay has occurred during the last 2000 years. Similarly, up until now,

$$\int_{0}^{10000} (0.000124)\ e^{-0.000124t}\ dt = -e^{-0.000124t}\Big]_{0}^{10000}$$

$$= -[0.2893 - 1.0000] = 0.7107$$

so about 71% of the decay has already occurred; or about 29% remains to decay. ▮

The value of a definite integral depends on its integrand, f(x), and its limits of integration, a and b; the letter used to denote the independent variable has no effect because the definition of an integral divides up the interval on this variable the same way, regardless of the symbol used. Thus,

$$\int_a^b f(x)dx = \int_a^b f(t)dt = \int_a^b f(u)du$$

so long as the same integrand and limits of integration are used throughout. The independent variable heretofore denoted by x or t is called a DUMMY VARIABLE because it can be replaced by any other symbol not already in use without affecting the value of the definite integral.

EXERCISES

Evaluate each of the following integrals by finding an antiderivative of the integrand.

1. $\int_1^3 2x dx$
2. $\int_{-3}^4 6 dx$
3. $\int_{-3}^0 3x^2 dx$
4. $\int_{-5}^5 x^3 dx$
5. $\int_{-4}^4 (4x^2 + 1)dx$
6. $\int_4^9 2(\sqrt{x} - x)dx$
7. $\int_1^5 \frac{1}{\sqrt{5z}} dz$
8. $\int_1^3 \left[1 + \frac{1}{x}\right]^2 dx$
9. $\int_0^t e^{-x} dx$
10. $\int_{N_0}^N \frac{dp}{p}$
11. $\int_0^{10} N_0 e^t dt$
12. $\int_3^4 \left[1 + \frac{1}{x}\right] dx$
13. Solve the integral equation $\int_{N_0}^N \frac{du}{u} = Kt$ for N.
14. Show $\int \frac{dx}{\sqrt{2x + 1}} = \sqrt{2x + 1} + C$ and then evaluate $\int_0^4 \frac{dx}{\sqrt{2x + 1}}$.
15. From $\frac{d}{dx}(1 + 3x^2)^{1/2} = \frac{3x}{\sqrt{1 + 3x^2}}$, evaluate $\int_0^2 \frac{xdx}{\sqrt{1 + 3x^2}}$.

*8.6 DERIVATION OF THE FUNDAMENTAL THEOREM

A proof of the fundamental theorem requires the development of several supporting ideas: (i) An integral viewed as a function of its upper limit; (ii) a mean value theorem for integrals; and (iii) the derivative of an integral. Each of these ideas has some merit in its own right, but we will not need them elsewhere in this book. We present these ideas in a starred section because they may be skipped without interrupting the continuity of this book. This section is intended for those desiring a deeper understanding of the Fundamental Theorem of Calculus.

The primary result presented in this section evaded eminent mathematicians from the Greeks well into the Renaissance. Newton, an Englishman, and Leibniz, a German, both contributed heavily to the birth and early growth of calculus. Newton apparently conceived the principle results about 1665, shortly after his graduation from Cambridge; our notation grew out of the work of Leibniz.

*8.6a INTEGRALS WITH A VARIABLE UPPER LIMIT

The definite integral $\int_a^b f(u)du$ has a unique value for any function f continuous on the closed interval [a,b]. At any intervening $x \in [a,b]$, $\int_a^x f(u)du$ also exists because continuity of f on [a,b] assures continuity of f on [a,x]. For any specified x, the definite integral $\int_a^x f(u)du$ has a unique value so it satisfies the properties of a function; we denote it by $H(x) = \int_a^x f(u)du$. This gives the area under f(x) between a and x.

<u>EXAMPLE 8.10</u> (continued): This example concerned the area under f(x) = 2x from x = 2 to x = 4. Recall that we obtained a geometrical evaluation of this area from the difference in areas of triangles with bases from 0 to 4 and 0 to 2. Again applying this strategy, but using the

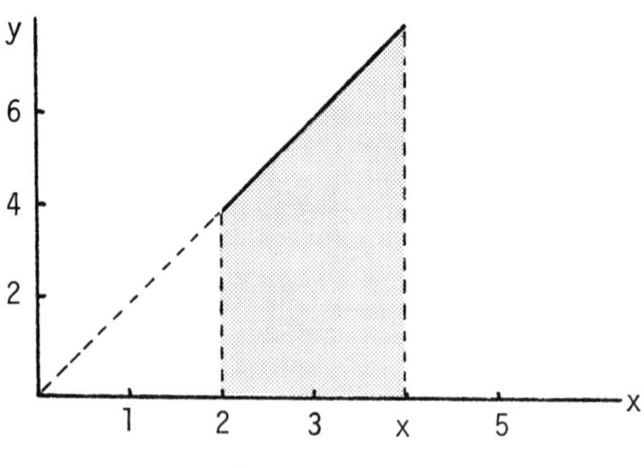

Figure 8.13

triangles shown in Figure 8.13,

$$H(x) = \int_2^x 2udu = \frac{1}{2}(2x^2) - \frac{1}{2}(4)(2) = (x^2 - 4) .$$

This gives, in particular, $H(4) = 12 = \int_2^4 2udu$, as before. ▌

EXAMPLE 8.19: Figure 8.14 shows a geometric interpretation of $H(x) = \int_a^x f(u)du$ where f represents an arbitrary continuous function on [a,b].

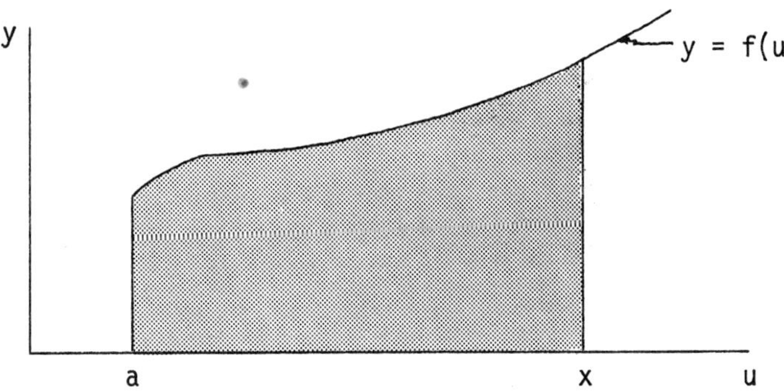

Figure 8.14

The shaded area represents the value of $H(x) = \int_a^x f(u)du$. It should be clear that the value of this integral depends upon x, the upper limit. ▌

*8.6b THE MEAN VALUE THEOREM FOR INTEGRALS

In evaluating the derivative of an integral we will need the following result:

THEOREM 8.7: If a function f is continuous on [a,b], there exists a number η, $a \leq \eta \leq b$, such that

$$\int_a^b f(u)du = (b - a) f(\eta) . \tag{8.13}$$

This theorem merely says that there exists a rectangle with a base of length b - a which has the same area as that under f between a and b, as is illustrated in Figure 8.15. The important feature of this theorem relates the height of this rectangle to values of f between a and b.

Specifically the height equals the value of the function at some point between a and b. In other words a height can be found so that the area inside the rectangle but above the curve exactly equals the area below the curve but above the rectangle.

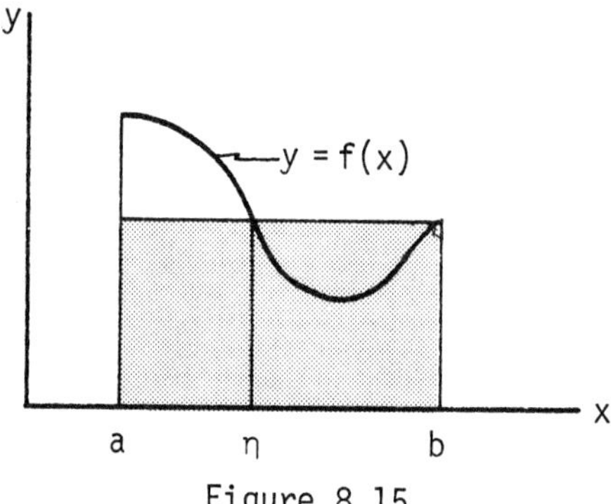

Figure 8.15

PROOF: Let m and M symbolize the absolute minimum and maximum values of f over [a,b]. From Theorem 8.6,

$$m(b - a) \leq \int_a^b f(u)du \leq M(b - a)$$

or

$$m \leq \mu = \frac{1}{b - a} \int_a^b f(u)du \leq M .$$

Because f is continuous, it assumes every value between m and M for some value of its independent variable; in particular, this implies the existence of a point η, $a \leq \eta \leq b$, at which f assumes the value μ:

$$f(\eta) = \mu = \frac{1}{b - a} \int_a^b f(u)du$$

Equation 8.13 follows immediately when the above equation is multiplied by b - a. ▼

*8.6c THE DERIVATIVE OF AN INTEGRAL

The derivative of an integral with a variable upper limit provides the next step on our path to the fundamental theorem:

<u>THEOREM 8.8</u>: Suppose that the function f is continuous on the interval [a,b]. If $H(x) = \int_a^x f(t)dt$, $a \leq x \leq b$, then H(x) has a derivative given by $H'(x) = f(x)$.

PROOF: For this first result concerning the derivative of an integral, we must revert to the limit definition of a derivative. To visualize this definition in terms of area, consider Figure 8.16 where $H(x) = \int_a^x f(t)dt$ is depicted by the horizontally ruled area. An increment, ΔH, in H results when x takes on a small increment Δx:

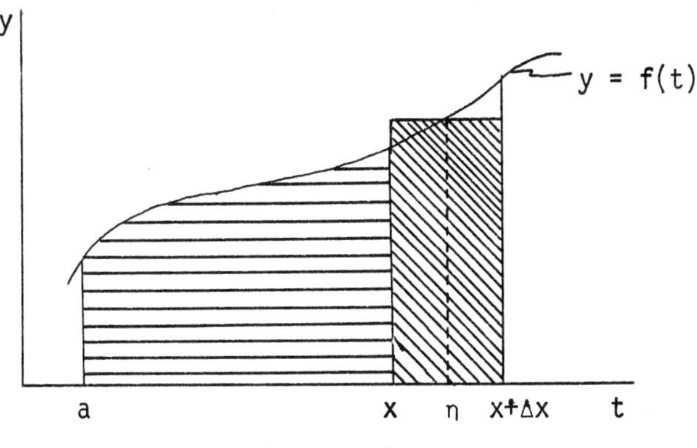

Figure 8.16

$$\Delta H = H(x + \Delta x) - H(x) = \int_a^{x+\Delta x} f(t)dt - \int_a^x f(t)dt$$

$$= \int_x^{x+\Delta x} f(t)dt$$

Thus ΔH is the area identified by the slanted rulings. The last equality utilizes Theorem 8.4. The mean value theorem assures the existence of an η between x and $x + \Delta x$, such that

$$\Delta H = H(x + \Delta x) - H(x) = \int_x^{x+\Delta x} f(t)dt = f(\eta)\Delta x \ .$$

To get H' divide both sides of this last equation by Δx and take its limit as $\Delta x \to 0$:

$$H'(x) = \lim_{\Delta x \to 0} \frac{\Delta H}{\Delta x} = \lim_{\Delta x \to 0} \frac{H(x + \Delta x) - H(x)}{\Delta x} = \lim_{\Delta x \to 0} \frac{\int_x^{x+\Delta x} f(t)dt}{\Delta x} = \lim_{\Delta x \to 0} f(\eta)$$

$$= f(x)$$

As $\Delta x \to 0$, $\eta \to x$ because η lies between x and $x + \Delta x$. Thus we arrive at the asserted result. ▼

EXAMPLE 8.10 (continued): In our last consideration of the function $f(x) = 2x$, $2 \leq x \leq 4$, we found $H(x) = \int_2^x f(t)dt = (x^2 - 4)$. Now we can either find $H'(x)$ directly or as the derivative of an integral:

$$H'(x) = \frac{d}{dx}(x^2 - 4) = 2x ,$$

and

$$H'(x) = \frac{d}{dx}\int_2^x f(t)dt = f(x) = 2x .$$

These two approaches give the same result as they must. ▌

<u>EXAMPLE 8.20</u>: The function $y = \int_{-2}^{x} (u^2 + e^u)du$ has a derivative given by

$$\frac{dy}{dx} = \frac{d}{dx}\int_{-2}^{x} (u^2 + e^u)du = x^2 + e^x .$$

Verify this another way: Evaluate the integral and take its derivative. ▌

In Theorem 8.8, x appeared as an upper limit. If the upper limit is some function of x, then we need this extension:

<u>THEOREM 8.9</u>: Let g be a function continuous on the closed interval [a,b], and let f be continuously differentiable on [a,b] and let

$$H(x) = \int_a^{f(x)} g(t)dt \qquad \text{where} \qquad a \le f(x) \le b .$$

H(x) has a derivative given by $H'(x) = g(f(x))f'(x)$.

This simply says that the derivative of an integral equals the integrand evaluated at the upper limit multiplied by the derivative of the upper limit. Theorem 8.8 presented the special case $f(x) = x$ because $f'(x) = 1$. The proof of this theorem uses the chain rule and the result established in the last theorem, so it will be left as an exercise.

<u>EXAMPLE 8.21</u>: If $y = \int_6^{x^2+2} (t^2 + 3)dt$, find $\frac{dy}{dx}$.

From the last theorem we have

$$\begin{aligned} y' &= ((x^2 + 2)^2 + 3)\frac{d}{dx}(x^2 + 2) \\ &= ((x^2 + 2)^2 + 3)(2x) . \end{aligned}$$

As an alternate method of evaluating the derivative of the integral, we first will evaluate the integral and then take its derivative.

$$\int_6^{x^2+2} (t^2 + 3)dt = \int_6^{x^2+2} t^2 dt + 3 \int_6^{x^2+2} dt$$

$$= \frac{t^3}{3}\Big]_6^{x^2+2} + 3t\Big]_6^{x^2+2}$$

$$= \frac{(x^2 + 2)^3}{3} - \frac{6^3}{3} + 3(x^2 + 2) - (3)(6)$$

$$= \frac{(x^2 + 2)^3}{3} + 3(x^2 + 2) - 90 .$$

Upon differentiating this function we find

$$\frac{d}{dx} \int_6^{x^2+2} (t^2 + 3)dt = \frac{d}{dx}\left[\frac{(x^2 + 2)^3}{3} + 3(x^2 + 2) - 90\right]$$

$$= (x^2 + 2)^2\, 2x + 3(2x) = ((x^2 + 2)^2 + 3)(2x) .$$

This same result was obtained above. Note how much easier Theorem 8.9 makes the differentiation. ▌

EXERCISES

1. If $H(x) = \int_a^x f(t)dt$, give a geometrical interpretation of $\frac{dH(x)}{dx}$.

2. If $H(x) = \int_a^x f(t)dt$, is it also true that $\int_a^x f(t)dt = H(x) + C$? Justify your statement.

3. Find the derivative of:

   a. $\int_{-1}^{t} x^2(1 - x^2)dx$, with respect to t.

   b. $\int_{x^2+2}^{102} (3z^2 + 4z - 1)dz$ where $|x| < 10$, with respect to x.

   c. $\int_{2x}^{3x+1} (t^2 + 2t - 3)dt$, with respect to x. (Hint: See Problem 4.)

4. Use the result of Theorem 8.9 to show that $\frac{d}{dx} \int_a^b g(t)dt = g(f_2(x))\frac{d}{dx} f_2(x) - g(f_1(x))\frac{d}{dx} f_1(x)$ where $b = f_2(x)$ and $a = f_1(x)$. (Hint: Break the interval $[f_1(x),f_2(x)]$ into $[f_1(x),c]$ and $[c,f_2(x)]$, where c is not a function of x.)

*8.6d THE FUNDAMENTAL THEOREM

We now have the tools to prove the Fundamental Theorem.

THEOREM 8.10: (THE FUNDAMENTAL THEOREM OF CALCULUS) If the function f is continuous on the closed interval [a,b] and if F is any function satisfying $F'(x) = f(x)$, then

$$\int_a^b f(x)dx = F(b) - F(a) .$$

PROOF: Let $H(x) = \int_a^x f(t)dt$. From Theorem 8.8, $H'(x) = f(x)$. Because the conditions of this theorem specify $F'(x) = f(x)$, $H'(x)$ must equal $F'(x)$. Functions having the same derivative can differ by at most a constant; if they differed by more than an additive constant, their derivatives could not be equal. Thus

$$F(x) + C = H(x) = \int_a^x f(t)dt .$$

To evaluate C, take $x = a$. From $\int_a^a f(t)dt = 0$, it follows that

$$0 = \int_a^a f(t)dt = F(a) + C ,$$

or $C = -F(a)$. Thus

$$\int_a^x f(t)dt = F(x) + C = F(x) - F(a) .$$

The theorem then follows by letting $x = b$.▼

8.7 SUBSTITUTION: A TECHNIQUE OF INTEGRATION

How would you evaluate $\int_0^5 xe^{-x^2}dx$? You already know how to evaluate $\int_a^b e^u du$; this provides the basis for evaluating this integral. The topic of substitution provides the tool needed to evaluate such an integral. We begin by looking at a definite integral as one member of a set of integrals all having the same form. This will allow us to focus attention on the process of finding antiderivatives without requiring that we get the specific value of a definite integral.

If $F'(x) = f(x)$, then we write all antiderivatives of f as

$$\int f(x)dx = f(x) + C , \tag{8.14}$$

where C is any constant, and call this the INDEFINITE INTEGRAL of f. Indefinite integrals share properties of definite integrals such as

$$\int kf(x)dx = k\int f(x)dx \qquad \text{and} \qquad \int [f(x) + g(x)]dx = \int f(x)dx + \int g(x)dx .$$

They do not share those properties which involved the limits of definite integration. Once an indefinite integral has been obtained, definite integrals with the same integrand can be evaluated easily:

$$\int_a^b f(x)dx = [\int f(x)dx]_a^b = [F(x) + C]_a^b = F(b) - F(a) .$$

When you have to evaluate $\int f(x)dx$, you may be able to use a familiar antiderivative or you may be lucky and find it in a table of integrals. (We will give an abbreviated table shortly.) Frequently, however, some sort of simple substitution is required before your integral conforms to a well known form or appears as an entry in an integral table. The critical skill needed to obtain an antiderivative lies in picking the substitution. It is one thing to know that

$$\int u^n du = \frac{u^{n+1}}{n+1} + C, \text{ for } n \neq -1$$

but another to realize that

$$\int \frac{(\ln x)^2}{x} dx$$

has this form. If we let $u = \ln x$ then $du = dx/x$ so

$$\int \frac{(\ln x)^2}{x} dx = \int (\ln x)^2 \frac{dx}{x} = \int u^2 du .$$

Now this last integral has a familiar form:

$$\int u^2 du = \frac{u^3}{3} + C = \frac{(\ln x)^3}{3} + C ,$$

again using $u = \ln x$.

This section began by asking how to evaluate $\int_0^5 xe^{-x^2}dx$. If we let $u = -x^2$ then $du = -2xdx$. Will this change $\int xe^{-x^2}dx$ into the known form

$\int e^u du$? Examine $\int e^u du$ closely; whatever we have for u, we must also have its perfect differential, du. When we rewrite the original integral as $\int e^{-x^2}(xdx)$ we do not quite get du = -2xdx. To get this perfect differential, and thus the form $\int e^u du$, we can multiply and divide the integrand by -2:

$$\int xe^{-x^2}dx = \int e^{-x^2}\frac{(-2xdx)}{-2} = -\frac{1}{2}\int e^{-x^2}(-2xdx) = -\frac{1}{2}\int e^u du \ .$$

Now that we have reduced the original integral to a standard form, we get

$$-\frac{1}{2}\int e^u du = -\frac{1}{2}e^u + C = -\frac{1}{2}e^{-x^2} + C$$

using $u = -x^2$. As a check we can take the derivative of the result. If we integrated properly the derivative should return the integrand:

$$\frac{d}{dx}\left[-\frac{1}{2}e^{-x^2} + C\right] = -\frac{1}{2}\left[-2xe^{-x^2}\right] = xe^{-x^2} \ .$$

Indeed we have integrated properly.

Again, the technique of substitution was used to accomplish this integration. The key in having success with substitutions is to make an intelligent guess about the general form the problem most closely resembles. As time goes on, you will become more adept in choosing your substitutions and manipulating integrands.

This process can be viewed another way. Again let $u = -x^2$ and then du = -2xdx. Upon examining the integrand observe that it already contains an xdx. Thus, when we solve du = -2xdx for xdx, we obtain $xdx = -\frac{du}{2}$. Now substitute this into the original integral:

$$\int xe^{-x^2}dx = \int e^{-x^2}(xdx) = \int e^u \frac{du}{-2} = -\frac{1}{2}\int e^u du \ .$$

We obtained this same expression earlier by multiplying and dividing the integrand by -2.

The original problem was to evaluate $\int_0^5 xe^{-x^2}dx$. Thus we need to evaluate this definite integral from the indefinite integral:

$$\int_0^5 xe^{-x^2}dx = \left[-\frac{1}{2}e^{-x^2} + C\right]_0^5 = -\frac{1}{2}[e^{-25} + C - e^{-0} - C] = \frac{1}{2}(1 - e^{-25}) \ .$$

An alternate method also exists for evaluating the definite integral without expressing the indefinite integral in terms of the original variable. Specifically, transform the limits as well as the integrand. We obtained

$$\int xe^{-x^2}dx = -\frac{e^u}{2} + C \; .$$

With $u = -x^2$, the upper limit of integration transforms from $x = 5$ into $u = -(5)^2 = -25$; $x = 0$ becomes $u = -(0)^2 = 0$:

$$\int_0^5 xe^{-x^2}dx = -\frac{1}{2}\int_0^{-25} e^u du = -\frac{1}{2}e^u\Big]_0^{-25} = -\frac{1}{2}[e^{-25} - e^0] = \frac{1}{2}(1 - e^{-25}) \; .$$

This same result was obtained before. Be aware of both methods. In some cases the first is the easier to apply while in other instances the second is easier.

A note of caution: Do not do the following:

$$\int_0^5 xe^{-x^2}dx = -\frac{1}{2}\int_0^5 e^u du = -\frac{e^u}{2}\Big]_0^5 = \frac{1}{2}(1 - e^{-5}) \; .$$

This incorrect result occurs because the limits on the second integral should apply to the dummy variable u, not to x. Always keep limits and variables of integration on the same scale.

In summary, then, it helps if you are familiar with a basic list of integrals, and if you know how to refer to tables of integrals for less familiar forms. But neither of these suffices; you have learned the kind of substitution which will reduce an integral to a standard or tabulated form. Often you will need to introduce constants as was illustrated above.

Next we give a brief list of indefinite integrals we have considered previously, and then illustrate how to adapt them in various situations:

$$\int u^n du = \frac{u^{n+1}}{n+1} + C \qquad n \neq -1 \tag{8.15}$$

$$\int \frac{du}{u} = \ln |u| + C \tag{8.16}$$

$$\int e^u du = e^u + C \tag{8.17}$$

$$\int a^u du = \frac{a^u}{\ln a} + C \tag{8.18}$$

In addition to the above we have also established the following general properties of integrals:

$$\int df(x) = f(x) + C \tag{8.19}$$

$$d\int f(x)dx = f(x)dx \tag{8.20}$$

$$\int 0 \, dx = C \tag{8.21}$$

$$\int kf(x)dx = k\int f(x)dx \tag{8.22}$$

$$\int (u \pm v)dx = \int udx \pm \int vdx \tag{8.23}$$

The formulas given in Equations 8.15 - 8.23 can be verified easily by taking the derivative of the right-hand side of the equation and showing that this equals the integrand on the left side of the equation.

The basic integration formulas given above cover many integrands by using the technique of substitution. This technique is illustrated in the following examples.

<u>EXAMPLE 8.22</u>: Evaluate $\int xe^{3x^2}dx$.

This appears to have the form $\int e^u du$. If we let $u = 3x^2$ then $du = 6xdx$. Multiplication and division of the integrand by 6 gives

$$\int xe^{3x^2}dx = \frac{1}{6}\int e^{3x^2}(6xdx) = \frac{1}{6}\int e^u du = \frac{1}{6}e^u + C \ .$$

Now substituting for u in terms of x, we obtain

$$\int xe^{3x^2}dx = \frac{1}{6}e^{3x^2} + C \ .$$

To check our answer, differentiate:

$$\frac{d}{dx}\left[\frac{1}{6}e^{3x^2} + C\right] = \frac{1}{6}(6xe^{3x^2} + 0) = xe^{3x^2} \ .$$

This equals the original integrand, affirming that we have integrated properly. ▮

EXAMPLE 8.23: Evaluate $\int \frac{e^x dx}{(1 + e^x)^4}$.

The term $(1 + e^x)^{-4}$ introduces most of the complexity into this integrand. Thus the form $\int u^n du$ provides the main hope for evaluating this integral. If we try $u = 1 + e^x$, $du = e^x dx$ and

$$\int \frac{e^x dx}{(1 + e^x)^4} = \int \frac{du}{u} = \int u^{-4} du = \frac{u^{-3}}{-3} + C .$$

Now substituting u into this we have

$$\int \frac{e^x dx}{(1 + e^x)^4} = - \frac{(1 + e^x)^{-3}}{3} + C . \quad \blacksquare$$

The most difficult step in using simple substitution is to decide what general form you want to work toward. Once you have decided upon the form, the substitution follows the straight-forward methods illustrated in the preceding two examples.

When using substitution in evaluating indefinite integrals, always leave the answer in terms of the original variable as illustrated in Examples 8.22 and 8.23. However, when evaluating a definite integral you also may change the limits to evaluate the integral in terms of the changed variable. This procedure is illustrated in the next example.

EXAMPLE 8.24: Evaluate $\int_0^2 \frac{(2x + 1)dx}{\sqrt{x^2 + x + 1}}$.

For the substitution $u = x^2 + x + 1$, $du = (2x + 1)dx$. When $x = 2$, $u = 7$ and when $x = 0$, $u = 1$. Thus we have

$$\int_0^2 \frac{(2x + 1)dx}{\sqrt{x^2 + x + 1}} = \int_1^7 \frac{du}{u^{\frac{1}{2}}} = \int_1^7 u^{-\frac{1}{2}} du$$

$$= 2u^{\frac{1}{2}}\Big]_1^7 = 2(\sqrt{7} - 1) .$$

If on the other hand we did not change the limits, we would have used $u = x^2 + x + 1$ and $du = (2x + 1)dx$ in this manner:

$$\int \frac{(2x + 1)dx}{\sqrt{x^2 + x + 1}} = \int u^{-\frac{1}{2}}du = 2u^{\frac{1}{2}} + C = 2(x^2 + x + 1)^{\frac{1}{2}} + C$$

$$\int_0^2 \frac{(2x + 1)dx}{\sqrt{x^2 + x + 1}} = 2(x^2 + x + 1)^{\frac{1}{2}}\Big]_0^2 = 2[(7)^{\frac{1}{2}} - 1] = 2(\sqrt{7} - 1) .$$

As the two answers are identical, you have to decide whether it is easier to change the limits or to simply resubstitute the original variables and then evaluate the integral. The problem usually dictates which is the best way to proceed. ▌

One final comment seems appropriate. Many textbooks, this one included, have some answers in the back. If you have evaluated an indefinite integral, it is comforting to find the same answer in the back. But suppose you find a different answer there. What should you conclude? You may be wrong, the book may be wrong, but there is a good chance that both you and the book are right. This can occur whenever several different paths exist for evaluating the indefinite integral.

<u>EXAMPLE 8.25</u>: Evaluate $\int(2x + 1)^3dx$.

First, we can expand the integrand using the binomial theorem to get

$$\begin{aligned}\int(2x + 1)^3dx &= \int(8x^3 + 12x^2 + 6x + 1)dx \\ &= 2x^4 + 4x^3 + 3x^2 + x + C_1 . \qquad (8.24)\end{aligned}$$

But directly using the substitution $u = 2x + 1$, we get

$$\int(2x + 1)^3dx = \frac{1}{2}\int u^3du = \frac{1}{2}\left[\frac{u^4}{4}\right] + C_2 = \frac{1}{8}(2x + 1)^4 + C_2 .$$

Even after expansion,

$$\int(2x + 1)^3dx = 2x^4 + 4x^3 + 3x^2 + x + \frac{1}{8} + C_2 , \qquad (8.25)$$

Equation 8.25 does not have exactly the same form as Equation 8.24 but C_1 and C_2 are arbitrary constants so $C_2 + \frac{1}{8} = C_1$ makes them equal. ▌

EXERCISES

Integrate each of the following. You may have to reduce some of the following integrals to one of the basic forms by the method of substitution, and then perform the integration. Assume the variables of integration are suitably restricted to make the integrands real and defined.

1. $\int \frac{dy}{\sqrt{1 + 3y}}$

2. $\int e^{x/2}dx$

3. $\int \frac{xdx}{(9 + 4x^2)^2}$

4. $\int xe^{x^2+3}dx$

5. $\int \frac{x^2dx}{x^3 + 1}$

6. $\int \frac{dx}{e^{2x}}$

7. $\int \left[x - \frac{1}{x}\right]^2 dx$

8. $\int \frac{dx}{x(1 + \ln x)^2}$

9. $\int \frac{e^x dx}{(3 + e^x)^3}$

10. $\int \frac{xdx}{(x^2 + 2)^2}$

11. $\int 3xe^{x^2}dx$

12. $\int \frac{1 + e^{2x}}{e^x} dx$

13. $\int x2^{x^2}dx$

14. $\int 2\sqrt{7t - 3}\, dt$

15. $\int (1 + \ln x)e^{x\ln x}dx$

16. $\int \frac{(5x - 1)dx}{5x^2 - 2x + 1}$

17. $\int \pi^{3x-1}dx$

18. $\int (e^{x/2} - e^{-x/2})dx$

19. $\int \frac{(3x^2 + 1)dx}{\sqrt{3x^3 + 3x - 4}}$

20. $\int \frac{e^{\sqrt{x}}2^{\sqrt{x}}}{\sqrt{x}} dx$

21. $\int_1^3 \frac{\sqrt[3]{\ln x}}{x} dx$

22. $\int_0^1 (2x - 5)^{17}dx$

23. $\int (x^3 - 2)^{\frac{1}{2}}x^2dx$

24. $\int \frac{2x}{\sqrt[4]{x} + 3} dx$

25. $\int 4x\sqrt{1 - 3x^2}\, dx$

26. $\int \frac{x^2 dx}{1 - 6x^3}$

27. $\int a^{2x} dx$

28. $\int (e^{2x} + 3)^2 e^{2x} dx$

29. $\int \left[x - \frac{x}{3} + \frac{4}{\sqrt{x}}\right] dx$

30. $\int_4^6 \frac{dx}{(x - 3)^3}$

31. $\int y^3\sqrt{1 + y^4}\, dy$

32. $\int \frac{3z}{\sqrt[3]{z^2 + 3}} dz$

33. $\int \frac{dx}{(a + bx)^{1/3}}$

34. $\int \frac{(1 + \sqrt{x})^2}{\sqrt{x}} dx$

35. $\int \frac{(x + 2)(x - 3)}{\sqrt{x}} dx$

36. $\int \frac{e^{1/x^2}}{x^3} dx$

37. $\int (e^x + 1)^2 dx$

38. $\int (e^{2x} + x^e) dx$

39. $\int \frac{(6x + 6)dx}{x^2 + 2x + 2}$

40. $\int \frac{x - 1}{x + 1} dx$

41. If ${}_AN_t$ represents the abundance of A-aged animals at the beginning of the year, the mean abundance of the animals throughout the year is given by ${}_A\overline{N}_t = \int_0^1 {}_AN_t\, e^{-(cf+m)t} dt$ where c, f and m are constants. Find the expression for ${}_A\overline{N}_t$.

8.8 APPLICATION OF DEFINITE INTEGRATION

Definite integrals are used in many areas of biology. We now present a few examples illustrating some applications. Other examples will appear in later chapters. Sometimes area is applied directly, but more often it is interpreted as the probability of an uncertain event. The next two subsections illustrate these applications.

8.8a AREAS BY INTEGRATION

Biologists sometimes need to evaluate the area which lies between two curves as is shown in Figure 8.17. You can calculate this area using available tools as follows:

$$\text{Area} = \int_a^b f_1(x)dx - \int_a^b f_2(x)dx .$$

This last expression consists of nothing more than the area under the curve $y = f_1(x)$ minus the area under the curve $y = f_2(x)$. This method can become troublesome when areas lie both above and below the x-axis. A simple method emerges if we briefly return to the definition of a definite integral.

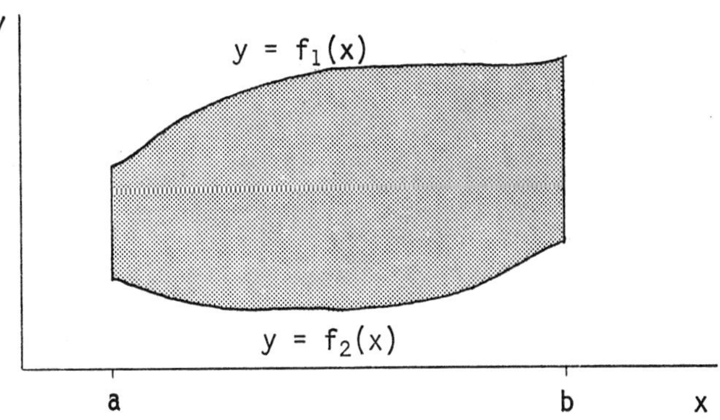

Figure 8.17

Consider a region as illustrated in Figure 8.18 and bounded by the lines $x = a$ and $x = b$, and the curves $y_1 = f(x)$ and $y_2 = g(x)$. Assume y_1 and y_2 are continuous functions on the interval $[a,b]$, where $y_2 \leq y_1$. First divide the interval $[a,b]$ into n subintervals with $a = x_0 < x_1 < \cdots < x_n = b$; in each choose an arbitrary point u_i. The ith subinterval has a

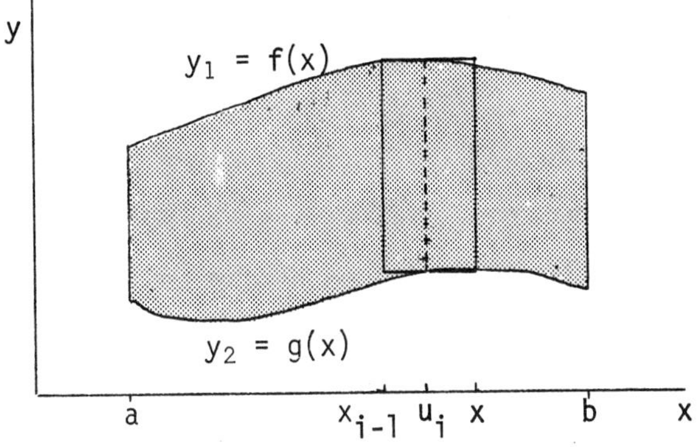

Figure 8.18

width of $\Delta_i x = x_i - x_{i-1}$. The line $x = u_i$ intersects the curves at $g(u_i)$ and $f(u_i)$, respectively. Construct a rectangle with a base of width $\Delta_i x$ and a height of $f(u_i) - g(u_i)$; it has an area of $\Delta_i A = [f(u_i) - g(u_i)]\Delta_i x$. Then, we define the *area* A of the region to be *the limit of the sum of all elements of area* $\Delta_i A$ as all $\Delta_i x \to 0$. Thus, we can evaluate the area in question as

$$A = \int_a^b [f(x) - g(x)]dx . \tag{8.26}$$

Examine Equation 8.26 to see that $f(x) - g(x)$ measures the height between the two curves and dx is the width of the representative rectangle as shown in Figure 8.18. The next two examples illustrate this technique.

EXAMPLE 8.26: Find the area between the parabolas $f_1(x) = x^2 - 4$ and $f_2(x) = -2x^2 + 2x + 1$.

Figure 8.19 shows the area in question. The limits of integration

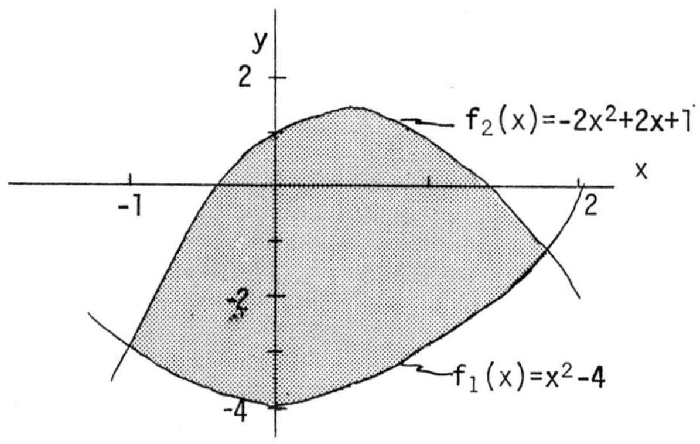

Figure 8.19

depend on where the curves intersect. This happens where $f_1(x) = f_2(x)$ or where

$$x^2 - 4 = -2x^2 + 2x + 1.$$

Solving this last expression for x gives

$$3x^2 - 2x - 5 = 0$$

or

$$x = \frac{-(-2) \pm \sqrt{4 - (4)(3)(-5)}}{6} = \frac{2 \pm \sqrt{4 + 60}}{6} = \frac{2 \pm 8}{6}$$

and $x = -1$ or $x = \frac{5}{3}$.

Thus the area in question equals

$$A = \int_a^b [f_2(x) - f_1(x)]dx = \int_{-1}^{5/3} [(-2x^2 + 2x + 1) - (x^2 - 4)]dx$$

$$= -\int_{-1}^{5/3} [3x^2 - 2x - 5]dx = -[x^3 - x^2 - 5x]_{-1}^{5/3}$$

$$= -\left[\left(\frac{5}{3}\right)^3 - \left(\frac{5}{3}\right)^2 - 5\left(\frac{5}{3}\right) - [(-1)^3 - (-1)^2 - 5(-1)]\right]$$

$$= \frac{256}{27} \text{ square units.}$$ ▮

EXAMPLE 8.27: Find the area of the region bounded by the parabolas $x = \frac{-y^2}{4}$ and $x = \frac{1}{4}(y^2 - 8)$.

The area in question appears in Figure 8.20. The curves intersect when $\frac{-y^2}{4} = \frac{1}{4}(y^2 - 8)$, namely, where $y = \pm 2$ or at $(-1,2)$ and $(-1,-2)$. We will encounter some difficulty if we integrate as before from the lower edge of the region to the upper edge. Specifically, we would need two integrals one for $-2 \leq x \leq -1$ and another for $-1 \leq x \leq 0$, because the equation defining the edge of the region changes at $x = -1$. By regarding y as the independent variable and x as the dependent, we can integrate from the left edge of the region to the right edge. This simplifies the integration because each edge is defined by one equation. A representative rectangular subregion is shaded heavily in the figure. (If this interchange of variables

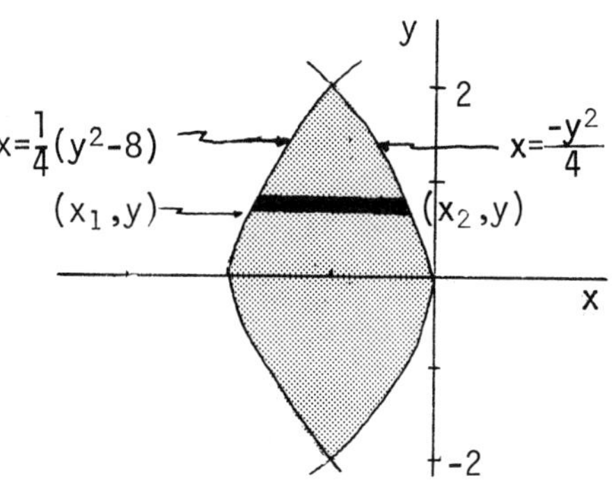

Figure 8.20

still bothers you, turn the book sideways so the negative x-axis points down; the typical area now is oriented as before.) The difference in the two functions defining the right and left edges is

$$x_2 - x_1 = \left(-\frac{1}{4}y^2\right) - \left(\frac{1}{4}(y^2 - 8)\right) = -\frac{y^2}{2} + 2 .$$

When this is incorporated into an integral, the area becomes

$$A = \int_a^b (x_2 - x_1)dy = \int_{-2}^{2}\left[\frac{-y^2}{2} + 2\right]dy = \left[\frac{-y^3}{6} + 2y\right]_{-2}^{2}$$

$$= \frac{-8}{6} + 4 - \left(\frac{8}{6} - 4\right) = \frac{-16}{6} + 8 = \frac{16}{3} \text{ square units.}$$ ∎

EXAMPLE 8.28: Leaves usually have small openings called stomata to allow passage of gases between their interior and exterior. Through them carbon dioxide passes in for capture by photosynthesis and the resulting oxygen passes out. Water vapor also escapes through the stomata, sometimes leading to dehydration. Thus plants have guard cells around the stomata to regulate their size. Action of the guard cells varies the shape of stomatal openings from a long narrow slit to nearly a circle. Throughout most of this variation the opening has approximately the shape of an ellipse with a constant length perimeter, typically about 35µ. How does the area of a stomatal opening change relative to its width?

First we need to establish an ellipse having the stated properties. Consider the notation introduced in Figure 8.21. An ellipse is specified by the ordered pairs (x,y) which satisfy the equation $\frac{x^2}{a^2} + \frac{y^2}{b^2} = 1$. (An ellipse is a generalization of a circle which allows a somewhat squashed

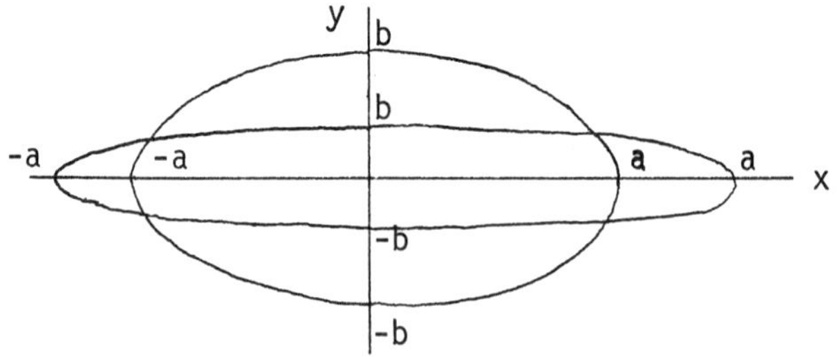

Figure 8.21

appearance. Figure 8.21 shows two ellipses. Each pair of values of a and b determines an ellipse.) An ellipse has a perimeter of approximately $2\pi\sqrt{\frac{a^2 + b^2}{2}}$; note the similarity of this to 2 r, the circumference of a circle. Thus we have

$$2\pi\sqrt{\frac{a^2 + b^2}{2}} = 35\mu \qquad \text{or} \qquad a^2 = \frac{2(35)^2}{(2\pi)^2} - b^2 \doteq 62 - b^2 \ .$$

because the stomata have a perimeter of 35μ. Thus the equation of the ellipse becomes

$$\frac{x^2}{62 - b^2} + \frac{y^2}{b^2} = 1 \ . \tag{8.27}$$

This ellipse becomes circular as $b^2 \to 31$ and approaches a slit as $b^2 \to 0$. We need to evaluate the area inside the ellipse specified by Equation 8.27 where $-\sqrt{62 - b^2} \leq x \leq \sqrt{62 - b^2} = a$. We can calculate the area of the ellipse in the first quadrant and multiply it by 4 to get the total area:

$$\text{Area} = A = 4\int_0^a ydx \ .$$

Now because $y = b\sqrt{1 - \frac{x^2}{a^2}} = \frac{b}{a}\sqrt{a^2 - x^2}$,

$$A = 4\int_0^a \frac{b}{a}\sqrt{a^2 - x^2}\, dx = \frac{4b}{a}\int_0^a \sqrt{a^2 - x^2}\, dx \ .$$

No simple substitution will change this integrand into any standard form we have yet encountered. Thus we must look for another method to evaluate it. As x ranges from 0 to a, $\sqrt{a^2 - x^2}$ traces out a quarter circle of radius a. (If this observation seems obscure, graph $x = \sqrt{a^2 - x^2}$ for $0 \leq x \leq a$.) Thus,

$$A = \frac{4b}{a}\int_0^a \sqrt{a^2 - x^2}\, dx = \frac{4b}{a}\left[\frac{\pi a^2}{4}\right] = \pi ab$$

where the divisor of 4 gives the area in the first quadrant of the circle. From $a = \sqrt{62 - b^2}$, we have $A = \pi ba = \pi b\sqrt{62 - b^2}$. This relates the area of the opening to the width of the opening. Notice also that quantities are dimensionally conformable: Area has units of μ^2 because both b and $\sqrt{62 - b^2}$ have units of μ. ▌

EXERCISES

1. Find the area bounded by the parabola $y^2 = 4x$ and the line $y = 2x - 4$.
2. Find the area enclosed within the parabolas $y = 6x - x^2$ and $y = x^2 - 2x$.
3. Find the area bounded by $y = 2 - x^2$ and $y = x$.
4. Find the area bounded by the x-axis, the curve $y = 10xe^{-2x^2}$ and the maximum ordinate on the curve.
5. Find the area of the region bounded by $y = 0$, $x = 9$, and $y = \sqrt{x}$.
6. Find the area of the region bounded by $y = 0$ and $y = x^2$ where $-2 \leq x \leq 1$.
7. Find the area of the region bounded by $y = x$ and $y = \sqrt{x}$.
8. Find the area of the region bounded by $y = x$ and $y = x^3 - 5x^2 + 7x$.

*8.8b PROBABILITY BY INTEGRATION

Our earlier consideration of probability focused on events displaying discrete outcomes, like the number of males in 50 births or the number of bacteria in a microscope field. Many biological responses assume a continuum of values. For example a plant or animal usually grows from a single fertilized ovum to maturity by assuming every intervening size. Probabilities associated with continuous responses cannot be described in the same way as for discrete responses. If we are counting the number of seeds which germinate (a discrete response), the count can be 2, or 3, or 4, but not 4.7 or 2.39. By contrast, a continuous response can assume any value in an interval. Thus probabilities are associated with intervals of possible outcomes, like $2 \leq x \leq 2.5$, rather than with specific values, like $x = 2$. How can we describe probabilities of such continuous variables? Consider a positive continuous function defined over the possible outcomes. The function will be low for those values of x which have only a slight chance of occurring; the function is higher over more likely values. This idea is analogous to describing decay by a rate curve, as was done in Example 8.18. If this example is not fresh in your memory, go back and look at it before continuing this section.

Integrals become associated with probabilities of continuous responses because they have two needed properties: $\int_a^a f(x)dx = 0$ and $\int_a^b f(x)dx \geq 0$ if $b > a$ and $f(x) \geq 0$. As the probability of any event cannot exceed one, not all functions can produce probabilities through integration. Suppose a response must lie between c and d, like human height lies between 0' and 10'. Then any function p satisfying $p(x) \geq 0$ for all $x \in [c,d]$ and $\int_c^d p(x)dx = 1$ can serve as a probability density function, sometimes abbreviated as *pdf*. Then the proportion of this unit area which lies between any bounds gives the probability of a value of the uncertain response lying within the same bounds:

$$P[a \leq x \leq b] = \int_a^b p(t)dt \qquad \text{where } c \leq a \leq b \leq d \,. \tag{8.28}$$

The x inside P[] represents a chance outcome, that is, an outcome which will have exactly one value. Equation 8.28 gives the probability that one random outcome will lie between a and b. By contrast t, the dummy variable of integration assumes every value between a and b.

EXAMPLE 8.29: Consider

$$p(x) = \begin{cases} \frac{4}{27}(3x^2 - x^3) & 0 \leq x \leq 3 \\ 0 & \text{otherwise} \end{cases}$$

As this function is nonzero only over [0,3], we need to check the integral only there:

$$\int_0^3 p(x)dx = \frac{4}{27}\int_0^3 (3x^2 - x^3)dx = \frac{4}{27}\left. x^3 - \frac{x^4}{4}\right]_0^3 = 1$$

Thus this function qualifies as a *pdf* because it also remains non-negative for all x. This *pdf* is graphed in Figure 8.22. The diagonally ruled area shows

$$P[0.5 \leq x \leq 1.0] = \int_{0.5}^{1.0} p(t)dt = \int_{0.5}^{1.0} \frac{4}{27}(3t^2 - t^3)dt = \left[\frac{4}{27}\right]\left[\frac{41}{64}\right] = 0.095 \,.$$

The horizontally ruled area gives

$$P[2.0 \leq x \leq 2.5] = \int_{2.0}^{2.5} \frac{4}{27}(3t^2 - t^3)dx = \frac{4}{27}\frac{119}{64} = 0.275 .$$

From the probabilities, an outcome has a higher probability of occurring in the interval $2.0 \leq x \leq 2.5$ than in the $0.5 \leq x \leq 1.0$. Notice from the graph that probability between the specified points simply equals the area under the probability density function between those two points. ∎

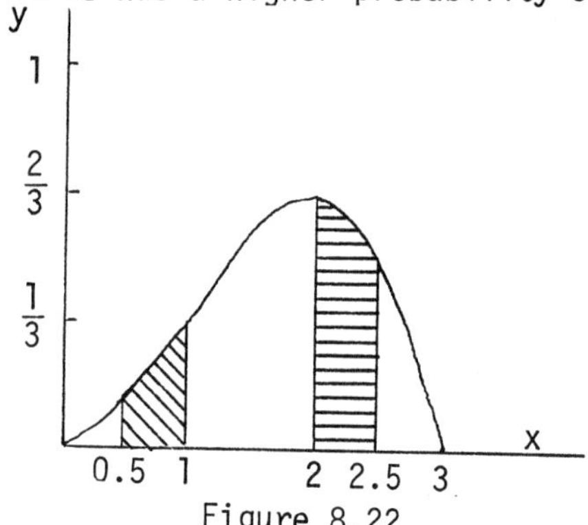

Figure 8.22

A large class of biological responses has a *pdf* of the form $p(x) = \lambda e^{-\lambda x}$, where λ is a parameter influencing the height of p(x). This function has the familiar shape appearing in Figure 8.23. Below we will verify that it is a *pdf*. This density function, called the EXPONENTIAL PROBABILITY DENSITY FUNCTION, describes diverse biological responses. It is used in studies of animal movement (from insect pests to large mammals). The distance an animal moves away from a release point often follows the exponential *pdf*. A large number of survival responses for physical, plant and animal characteristics follow the exponential distribution. Examples include the time until a particular radioactive particle decays, the lifetime of a light bulb, the length of time a seedling pine tree survives under natural conditions, and the length of time an insect lives in a cotton field. Of course, the constant λ in the exponential *pdf* changes from situation to situation. For example, the survival of an insect depends on the species of the insect and on the time of year.

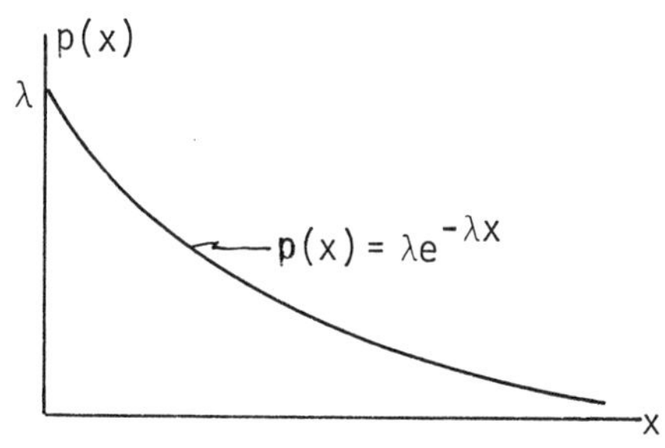

Figure 8.23

Now we will verify that p(x) really is a *pdf*. For a fixed interval [0,T],

$$\int_0^T \lambda e^{-\lambda t}dt = -\int_0^T e^{-\lambda t}(-\lambda dt) = -[e^{-\lambda t}]_0^T = [1 - e^{-\lambda T}] .$$

Usually T cannot be specified because there is a small, but positive, probability that the object survives for very long time. Thus, it seems appropriate to require that $T \to \infty$:

$$\lim_{T\to\infty} \int_0^T \lambda e^{-\lambda t}dt = \lim_{T\to\infty} [1 - e^{-\lambda T}] = 1 ,$$

because $\lim_{T\to\infty} e^{-\lambda T} = 0$.

First, suppose p(x) describes the time until the radioactive decay of a single particle. Then that value of T such that

$$\frac{1}{2} = \int_0^T \lambda e^{-\lambda t}dt = 1 - e^{-\lambda T}$$

or

$$e^{-\lambda T} = \frac{1}{2} \qquad \text{or} \qquad T = -\frac{1}{\lambda}\ln\frac{1}{2} = \frac{\ln 2}{\lambda} = t_{\frac{1}{2}} \tag{8.29}$$

gives the time by which there is a probability of 1/2 that the particle has decayed. This is referred to as the particle's half-life. A preliminary view of this concept was advanced in Section 4.7. Since we have associated the idea of the probability of continuous responses with calculus, we can precisely state the meaning of half-life: The half-life of a particle is that time, symbolized by $t_{\frac{1}{2}}$, at which $P[x < t_{\frac{1}{2}}] = \frac{1}{2} = P[x > t_{\frac{1}{2}}]$. Look back to Section 4.7 for illustrations of this concept. Example 4.31 concerns half-lives and carbon dating while the example thereafter deals with the half-life of an anesthetic.

EXAMPLE 8.30: Lygus bugs attack alfalfa and can seriously damage the growing parts of the plant. This results in stunted growth and reduced production. Under a certain set of laboratory conditions, lygus bugs have a half-life of 4 days. Evaluate the probability that a lygus bug survives only 2 days, and that one survives more than 10 days, if their survival follows the exponential density function. First, from Equation 8.29,

$$\frac{\ln 2}{\lambda} = t_{\frac{1}{2}}, \qquad \text{so} \qquad \lambda = \frac{\ln 2}{t_{\frac{1}{2}}} = \frac{0.693}{4} = 0.173 \ .$$

If x equals the length of life of a randomly selected lygus bug, then

$$P[x \le 2] = P[0 \le x \le 2] = \int_0^2 \lambda e^{-\lambda t} dt = -e^{-\lambda t}\Big]_0^2 = 1 - e^{-2\lambda}$$

$$= 1 - e^{-2(0.173)} = 1 - e^{-0.346} = 1 - 0.71 = 0.29 \ ,$$

while

$$P[x \ge 10] = P[10 \le x] = \lim_{T\to\infty} \int_{10}^{T} \lambda e^{-\lambda t} dt = \lim_{T\to\infty} (e^{-10\lambda} - e^{-\lambda T})$$

$$= e^{-10\lambda} = e^{-1.73} = 0.18 \ .$$

Consequently, the probability that a lygus bug survives between 2 and 10 days is given by

$$P[2 < x < 10] = 1 - P[2 \le x] - P[x \ge 10] = 1 - 0.29 - 0.18 = 0.53 \ . \quad \blacksquare$$

EXERCISES

1. Show that the following is a probability density function:

$$f(x) = \begin{cases} 0 & x < a \\ \dfrac{1}{b - a} & a \le x \le b \\ 0 & x > b \end{cases} \qquad \text{a and b are constants.}$$

   The above density function is called the *uniform density function*.

2. Calculate $P[1 \le x \le 3]$ if $f(x) = \frac{1}{10}$ and f(x) is a *pdf*, $-5 \le x \le 5$.

3. Consider an experiment whose outcomes are positive numbers. Given that the distribution of probability is described by a *pdf* which at any positive number x is e^{-x}, compute the probability that
   a. the outcome exceeds 2;
   b. the outcome is at most 1;
   c. the outcome lies between 1 and 2.

4. Show that the following is a probability density function:

$$f(x) = \begin{cases} \frac{x}{2} & \text{if } 0 \leq x \leq 2 \\ 0 & \text{elsewhere} \end{cases}$$

5. Calculate the following for the *pdf* given in problem 4:
   a. $P[1 \leq x \leq 1.5]$ b. $P[x > 1.5]$ c. $P[x < 1]$

8.9 SIMPLE DIFFERENTIAL EQUATIONS

Differential equations offer a compelling reason for a biologist to learn about integration. A differential equation is merely an equation involving a derivative. Such equations frequently appear because examination of biological phenomena can suggest how variables are related through either derivatives or differentials. The following example illustrates a situation where differential equations arise in biology.

EXAMPLE 8.31: Suppose we inject a given quantity A_0 of a dye or tracer into the blood supply at some time $t = 0$. Further suppose that 2% of the dye or tracer is removed from the system every minute. At the end of one minute we would have $A_0 - 0.02A_0$ or $0.98A_0$; at the end of two minutes we would have $0.9604A_0$ and so on. We can calculate the amount of dye or tracer in the system in this manner for any time interval. We have specified three quantities in the above description of the problem: The rule that governs the process over the short time interval of one minute; the assumption that the same rule applies from time interval to time interval and holds for a long period of time; and the initial amount of dye or tracer.

Now suppose we want to know how the system would behave if we lose 3%, 1% or some percentage other than 2%. In other words, we are changing the rule under which the system operates. For each of the desired percentages we would have to construct a table of values for each time interval. However, if we describe the system by a differential equation and find its general solution, then we have described the system through time for any rate of loss. The differential equation describing this type of a system is presented in Example 8.34. ▌

EXAMPLE 8.32: Obtain the function y whose graph passes through the point (3,4) and whose slope is given by $\frac{dy}{dx} = 6x^2 - 2x + 3$.

We need only rewrite the equation as

$$dy = (6x^2 - 2x + 3)dx \ .$$

Now, if we integrate both sides of this equation, the equality remains:

$$\int dy = \int (6x^2 - 2x + 3)dx \ .$$

Upon integration this becomes

$$y = 2x^3 - x^2 + 3x + C \ . \qquad (8.30)$$

Using the given conditions we can evaluate the constant C in Equation 8.30: When x = 3, y = 4 so

$$4 = 2(3)^3 - (3)^2 + 3(3) + C$$

or C = -50. Thus the equation which satisfies the original conditions is given by

$$y = 2x^3 - x^2 + 3x - 50 \ .$$

Only one constant of integration is needed in Equation 8.30 because $y + C_1 = 2x^3 - x^2 + 3x + C_2$ could be written as $y = 2x^3 - x^2 + 3x + C_2 - C_1$. Thus $C = C_2 - C_1$ embodies all of arbitraryness required.

Solutions of differential equations are very easy to check. Simply take the derivative of the result to arrive at the original differential equation. Upon differentiation of the result here, we indeed retrieve the original equation. ▌

EXAMPLE 8.33: Obtain an equation for y if y" = 12x - 8 subject to the conditions that y = 4 and y' = 3 at x = 2.

From $y'' = \frac{d^2y}{dx^2} = \frac{d}{dx}\left[\frac{dy}{dx}\right] = \frac{d}{dx} y'$,

$$dy' = (12x - 8)dx \ .$$

Upon integration of both sides we obtain

$$\int dy' = \int (12x - 8)dx \qquad \text{or} \qquad y' = 6x^2 - 8x + C_1 \ .$$

For y' to equal 3 at x = 2, C must satisfy

$$3 = 6(2)^2 - 8(2) + C_1 \qquad \text{or} \qquad C_1 = -5$$

and thus

$$y' = 6x^2 - 8x - 5 \ .$$

Repeating the process we have

$$dy = (6x^2 - 8x - 5)dx$$

and

$$y = \int dy = \int (6x^2 - 8x - 5)dx = 2x^3 - 4x^2 - 5x + C_2 \ .$$

Now use the initial condition y = 4 at x = 2 to evaluate C_2:

$$4 = 2(2)^3 - 4(2)^2 - 5(2) + C_2 \qquad \text{or} \qquad C_2 = 14 \ ,$$

so

$$y = 2x^3 - 4x^2 - 5x + 14 \ .$$

This function satisfies all of the stated conditions. It gives the unique solution satisfying the stated conditions. ▮

In some differential equations, we can separate the variables. These simple differential equations have the form

$$f(x)dx = g(y)dy \ .$$

The next examples in this section illustrate the technique of solving differential equations in which the variables may be separated. The next section then presents further illustrations from biology.

<u>EXAMPLE 8.34</u>: An idealized one-compartment dilution problem consists of a single continuously mixed chamber through which a fluid is flowing at a constant rate. It is described by the differential equation

$$\frac{dC(t)}{dt} = -\frac{F}{V} C(t) \ .$$

Here C(t) represents the concentration of the dye or tracer in the compartment at any time t, F is the flow rate and V is the volume of the compartment. Find the equation describing the concentration at any time t.

By separating the variables we can rewrite the equation as

$$\frac{dC(t)}{C(t)} = -\frac{F}{V}\,dt \ .$$

Upon integrating both sides of this equation we obtain

$$\ln C(t) = -\frac{F}{V}t + C_1 \ . \qquad (8.31)$$

Now rewrite this expression in exponential form:

$$C(t) = \exp\left[-\frac{Ft}{V} + C_1\right] = \exp[C_1]\exp\left[-\frac{Ft}{V}\right] .$$

The initial condition specifies a concentration of C(0) at t = 0. Thus we can evaluate the constant of integration as

$$C(0) = e^{C_1}e^{-0} = e^{C_1} \ .$$

Our equation then becomes

$$C(t) = C(0)e^{-Ft/V} \ .$$

This equation describes the concentration in the compartment at any time t provided we know C(0), F and V.

The constant of integration added in Equation 8.31 could have been chosen to be $\ln C_1$. This would lead to the form

$$\ln C(t) = -\frac{Ft}{V} + \ln C_1 \ .$$

Now rewriting the equation in exponential form, it becomes

$$\ln C(t) - \ln C_1 = -\frac{Ft}{V}$$

or

$$\ln\frac{C(t)}{C_1} = -\frac{Ft}{V} \qquad \text{and} \qquad \frac{C(t)}{C_1} = e^{-Ft/V} ,$$

and finally

$$C(t) = C_1e^{-Ft/V} \ .$$

Evaluating this last equation at the initial conditions, gives

$$C(0) = C_1e^0 \qquad \text{or} \qquad C_1 = C(0)$$

so we arrive at the same solution as before:

$$C(t) = C(0)e^{-Ft/V} \ .$$ ▌

EXAMPLE 8.35: Previously we discussed the topic of radioactive decay and carbon dating. The underlying assumption for radioactive decay is that the rate of decay is proportional to the amount present for decay. This can be stated as

$$\frac{dN}{dt} = -KN$$

where K is a constant of proportionality and N is the number of atoms present at any time; the negative appears because N is decreasing. To solve this differential equation, separate variables:

$$\frac{dN}{N} = -Kdt$$

and

$$\ln N = -Kt + \ln N_0$$

where $\ln N_0$ is our arbitrary constant of integration. Change this from the logarithmic to the exponential form:

$$\ln \frac{N}{N_0} = -Kt \qquad \text{or} \qquad N = N_0 e^{-Kt} \ .$$

Now referring back to Section 4.7, we can determine that $K = \frac{-0.693}{t_{1/2}}$ where $t_{1/2}$ represents the half-life of the radionuclide under investigation. This exponential form appears frequently when dealing with biological phenomena. ▌

EXAMPLE 8.36: The same mathematical form as was developed in Example 8.35 arises in the absorption of x rays passing through a homogeneous material. Figure 8.24 shows an absorber which is

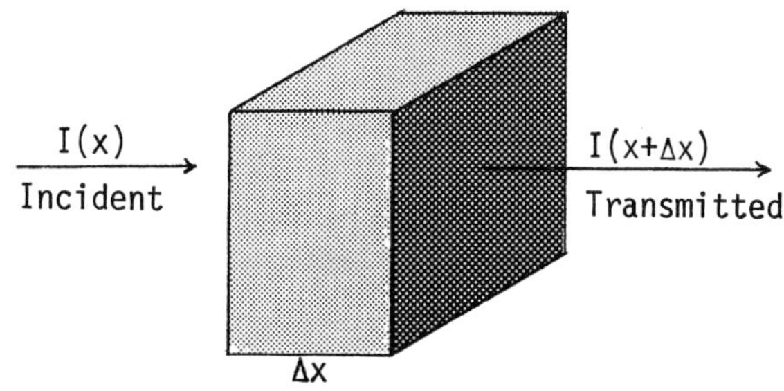

Figure 8.24

Δx units thick, with an incident ray of intensity $I(x)$ and a transmitted ray of intensity $I(x + \Delta x)$. The absorption is then $I(x) - I(x + \Delta x)$ and this is proportional to the intensity $I(x)$, the density of the medium and the thickness of the barrier. Thus

$$\Delta I = I(x + \Delta x) - I(x) = -DI(x)\Delta x \qquad (8.32)$$

where D is the density of the medium or barrier. The negative sign on the right side of Equation 8.32 is necessary because the x rays are being absorbed. Divide both sides of Equation 8.32 by Δx and take the limit as $\Delta x \to 0$:

$$\lim_{\Delta x \to 0} \frac{I(x + \Delta x) - I(x)}{\Delta x} = \lim_{\Delta x \to 0} (-DI(x))$$

or

$$\frac{dI(x)}{dx} = -DI(x) \ .$$

The solution of this differential equation is

$$I(x) = I(0)e^{-Dx} \ ,$$

where $I(0)$ is the intensity of the incident x rays. ▌

Additional problems involving simple differential equations solvable by the technique of separation of variables appear in the following exercises.

EXERCISES

Find the solutions to the following differential equations. If initial conditions are given, evaluate the constant of integration.

1. $2x(y^2 + 1)dy - ydy = 0$
2. $2x(7 + 1)dx - dy = 0$; where $y = 2$ when $x = 0$
3. $\frac{dy}{dx} = \frac{2y}{x}$; for $x > 0$ and $y > 0$
4. $V \frac{dV}{dx} = g$; when $x = x_0$, $V = V_0$
5. $\frac{dV}{dP} = -\frac{V}{P}$

6. $\frac{dr}{dt} = -4rt$; where $t = 0$ when $r = r_0$

7. $2xyy' = 1 + y^2$; when $x = 2$, $y = 3$

8. $e^{y^2}dx + x^2ydy = 0$

9. $(4 + x)y' = y^3$

10. Evaluate the constant in Example 8.34 if at $t = \frac{V}{F}$ the concentration is equal to 0.037 moles/liter.

11. Under ideal conditions, the rate of change of pressure above sea level is proportional to the pressure. If the pressure is 30 inches of mercury at sea level and 25 inches at 4000 feet, find the barometric pressure at 8000 feet.

12. Assuming that a population increases at a rate which is proportional to the number of individuals present, in how many years would the population double if it increases at the constant rate of 4% a year?

13. A bacterial population increases at a rate which is proportional to the number in the population. If this population doubles in size in one hour, in how many hours will it be 1000 times its original size? (Assume no factors limit growth.)

14. If a glucose solution is given intravenously, the concentration in the blood is increased at a constant rate R. However, at the same time, the glucose is converted and excreted at a rate which is proportional to the present concentration of glucose. Hence the rate of change in the glucose concentration, C, is given by $\frac{dC}{dt} = R - KC$ where K is a positive constant. Find an expression for the concentration at any time if $C = C_0$ at $t = 0$.

15. Assume that the rate $\frac{dN}{dt}$ at which radioactive atoms are formed is constant with time while the rate at which they decay is $-\lambda N$, where λ is the decay constant. Then the change in the number of radioactive atoms is given by $\frac{dN}{dt} = K - \lambda N$. Solve this expression for N if $N = 0$ when $t = 0$; $\lambda = \frac{0.693}{t_{1/2}}$. Sketch the graph of the resulting equation.

16. Newton's law of cooling states that the rate of cooling of a body is proportional to the difference between the temperature of the body and the room temperature. Find an expression for the temperature of the body if initially ($t = 0$) the temperature of the body equals T_0 and at time t the temperature is T_B.

17. If at this moment in time, the temperature of a certain body is 60° above room temperature and 20 minutes ago it was 70° above room temperature, what will the temperature difference be in 10 minutes? Two hours? When will the body temperature be 10° above room temperature? Use the results of Problem 16.

18. Use the results of Problem 16 to determine what the temperature of a body will be at the end of 2.5 hours if the centigrade temperature of the surrounding air remains 40° and the temperature of the body drops from 170° to 105° in 45 minutes.

19. Body tissues absorb additional nitrogen when air is inhaled under pressure greater than normal atmospheric pressure, as occurs in deep sea diving. Let P be the pressure of nitrogen in a tissue at any depth and let P_{max} be the nitrogen pressure in the tissue at the bottom of the sea. The rise in the nitrogen pressure in the tissue during descent depends on the pressure difference of $(P_{max} - P)$ and the area A of contact between the blood and the tissue. The following equation gives the rate of nitrogen absorption

$$\frac{dP}{dt} = kA(P_{max} - P)$$

where k is a constant which is characteristic of the tissue. Find an expression for P if we assume that $P = 0$ at $t = 0$. If $T_{\frac{1}{2}}$ represents the time for the tissue to become half saturated with nitrogen, obtain an expression for $T_{\frac{1}{2}}$.

20. In thermodynamics, the constant rate of reaction, k, in which one mole of a substance takes part depends upon the temperature, T, expressed in degrees Kelvin according to the law $\frac{d}{dT} \ln k = \frac{Q}{RT^2}$ where Q is the heat liberated or absorbed in the reaction and R is the gas

20. Continued
constant. Assume that R and Q do not change with a change in temperature. Find an expression for k.

21. Consider a simple reaction symbolized by $A + B \to x$ where two molecules react to give a third. The concentrations of A and B at time zero are a and b and the concentration of x at time zero is zero. The concentrations at time t of A and B are $a - x$ and $b - x$. The rate at which x is produced is given by $\frac{dx}{dt} = k(x - a)(x - b)$ where k is a constant. Find an expression for x in terms of t. (Hint: $1/[(x - a)(x - b)] = -1/[(b - a)(x - a)] + 1/[(b - a)(x - b)]$.)

22. An auto-catalytic reaction converts a substance into a new substance called the product in such a way that the product catalyzes its own formation. The rate of formation of the product is proportional to the amount, $a - x$, of unreacted substance and the amount x of the product. Thus $\frac{dx}{dt} = kx(a - x)$ where k is positive and a is the total initial amount of the product. Express x in terms of t. (Hint: $1/[x(a - x)] = 1/(ax) + 1/[a(a - x)]$.)

23. The table below presents the intensities in lead of a monochromatic photon beam. It was obtained to determine the absorption coefficient μ in the equation $I = I_0 e^{-\mu x}$. Using the data, determine the constants I_0 and μ. Then determine the thickness of a piece of lead in order for it to reduce the intensity to one-half of the original intensity; also to one-tenth the original intensity.

Absorber thickness x, cm:	1	2	3	4	5	6	7
Intensity I:	100	45	24	11	5.2	2.4	1.4

24. The table below presents the intensities in aluminum of a monochromatic photon beam. Using the data given, determine μ and I_0 in $I = I_0 e^{-\mu x}$. Then determine how thick a piece of aluminum will reduce the intensity to one-fourth of the original intensity.

Absorber thickness x, cm:	0	1	2	3	4	5	6	7	8
Intensity I:	140	105	73	50	33	28	17	15	9.2

25. Suppose a substance enters the circulatory system at time t = 0. A theoretical model of the system shows that the rate of increase of the substance is given by

$$\frac{dy}{dt} = k_1 e^{-\lambda_1 t} - k_2 e^{-\lambda_2 t} .$$

Here y symbolizes the amount of the substance in the circulatory system at time t and k_1, k_2, λ_1 and λ_2 are constants. Obtain an expression for the total amount of the substance in the circulation at any time t.

*26. A very simplified model for the spread of an epidemic assumes that if a population contains S susceptible and I infected individuals then $\frac{dS}{dt} = -kSI$ where k is a positive constant. A single infected individual is introduced into a population of S_0 susceptibles at time t = 0. If the population's size remains constant, explain why $S + I = S_0 + 1$ and then show that

$$S = \frac{S_0(S_0 + 1)}{S_0 + e^{(S_0+1)kt}} .$$

8.10 A FURTHER LOOK AT FUNCTIONS IN REALITY

Several growth curves were introduced in Chapter 4, their limits were examined in Chapter 5, and their derivatives were used in Chapter 7 to further examine their behavior. So far we have taken the growth curves as specified functions, considered their application, and examined their behavior. Some of the growth functions arose when an investigator needed a function which changed in a prescribed manner. In this section we will specify various conditions and then use the technique of differential equations to develop the growth curve from the behavior of the derivative.

First consider the case where a biological response grows in proportion to its current size. If W symbolizes the current size of a population, then at t = 0, we will assume that the population had an initial size of W_0. To say a population grows at a rate proportional to its current size, really means that

$$\frac{dW}{dt} = kW \tag{8.33}$$

where k is a constant of proportionality. To solve this equation we can separate the variables:

$$\frac{dW}{W} = kdt$$

Upon integration of both sides we obtain $\ln W = kt + \ln C$, where $\ln C$ is an arbitrary constant of integration. Changing this last equation from logarithmic form to exponential form gives

$$W = Ce^{kt} . \tag{8.34}$$

The initial condition specifies that the population has size W_0 at $t = 0$. Thus $W_0 = Ce^{k0}$ or $C = W_0$ and so

$$W = W_0e^{kt} . \tag{8.35}$$

This equation should be recognized as the equation describing the EXPONENTIAL GROWTH CURVE.

The exponential growth curve has no horizontal asymptote if $k > 0$; thus, it gets large without bound. Few, if any, biological responses behave like this. For example, any environment can only support a limited, even though perhaps very large, population. The number of individuals which the environment is capable of supporting is called the CARRYING CAPACITY OF THE ENVIRONMENT. Denote the carrying capacity or maximum population size by W_m. As a population grows toward this limit, the growth rate of the population must decrease. One possible way to express this fact is by the following equation:

$$\frac{dW}{dt} = kW\left[1 - \frac{W}{W_m}\right] . \tag{8.36}$$

Equation 8.36 states that the population grows at a rate almost in proportion to its size when W is small relative to W_m, but as W increases toward its maximum, the growth rate declines toward zero. This slight modification overcomes some of the criticisms of simple exponential growth.

To solve this differential equation, rewrite it as

$$\frac{dW}{W\left[1 - \frac{W}{W_m}\right]} = kdt \qquad \text{or} \qquad \frac{W_m dW}{W(W_m - W)} = kdt \;. \tag{8.37}$$

Equation 8.37 has no obvious solution in its present form. We must find a way to express the equation in a form which we can solve. The fraction $\frac{W_m}{W(W_m - W)}$ can be rewritten as $\frac{1}{W} + \frac{1}{W_m - W}$. (Check this last step by getting a common denominator.) Now we can rewrite Equation 8.37 as

$$\left[\frac{1}{W} + \frac{1}{W_m - W}\right]dW = kdt$$

You know how to integrate this:

$$\int \left[\frac{1}{W} + \frac{1}{W_m - W}\right]dW = \int kdt$$

or

$$\int \frac{dW}{W} + \int \frac{dW}{W_m - W} = \int kdt \;,$$

and

$$\ln W - \ln(W_m - W) = kt - \ln b$$

where ln b is the arbitrary constant of integration. Changing this from logarithmic to exponential form, we have

$$\ln \frac{bW}{(W_m - W)} = kt \qquad \text{or} \qquad \frac{bW}{(W_m - W)} = e^{kt} \;. \tag{8.38}$$

At $t = 0$, $W = W_0$ so we can solve for b:

$$b\,\frac{W_0}{(W_m - W_0)} = e^{k0} \qquad \text{or} \qquad b = \frac{W_m - W_0}{W_0}$$

This implies that $W_m = W_0(1 + b)$ so substituting these values back into Equation 8.38, we have

$$\frac{bW}{W_0(1 + b) - W} = e^{kt} \;.$$

To solve this for W, multiply to eliminate the denominator, then rearrange as follows:

$$bW = [W_0(1 + b) - W]e^{kt} = W_0(1 + b)e^{kt} - We^{kt}$$

$$bW + We^{kt} = W(b + e^{kt}) = W_0(1 + b)e^{kt}$$

and so

$$W = \frac{W_0(1 + b)e^{kt}}{b + e^{kt}} .$$

We arrive at the final result by dividing both numerator and denominator of the right-hand side of the last equation by e^{kt} to get

$$W = \frac{W_0(1 + b)}{1 + be^{-kt}} \qquad (8.39)$$

Recognize this as the LOGISTIC GROWTH CURVE.

The logistic growth curve came from making the assumption that $W' = kW\left[1 - \frac{W}{W_m}\right]$. It contains a multiplicative term which decreases the growth rate, but what can we do if we believe that the growth rate of this curve does not decrease fast enough? We could replace $1 - \frac{W}{W_m}$ by e^{-kt} to decrease the rate of growth more rapidly for large t, but this would reduce the growth rate only slightly for small t. Thus suppose we have a population whose growth can be characterized by

$$\frac{dW}{dt} = bWe^{-kt} . \qquad (8.40)$$

Again b and k are constants and W is the size of the population at time t. This equation allows more diversity than growth curves previously considered because curves satisfying Equation 8.40 can differ in either of the two constants b and k, not in only one as in Equations 8.33 and 8.36.

Separate the variables in Equation 8.40 and integrate as follows:

$$\frac{dW}{W} = be^{-kt}dt$$

and

$$\ln W = -\frac{b}{k}e^{-kt} + \ln C$$

where ln c is the arbitrary constant of integration. Next change this to exponential form:

$$\ln \frac{W}{c} = -\frac{b}{k} e^{-kt} \qquad \text{or} \qquad W = c \exp\left[-\frac{b}{k} e^{-kt}\right]$$

Recognize this as the GOMPERTZ GROWTH CURVE with the proper choice of the constants c and b.

The monomolecular growth equation arises when we assume that the growth rate can be described by

$$\frac{dW}{dt} = \frac{Wkbe^{-kt}}{1 - be^{-kt}}$$

where b and k are positive constants with $b > 1$. This equation says that the rate of growth is proportional to the number present and the quantity $\frac{e^{-kt}}{1 - be^{-kt}}$. This factor reduces the growth rate through time without regard to current size. For small values of t, this factor approximates $\frac{1}{1 - b}$ but for large values of t it becomes very small. Upon solving this last equation by separating the variables, we have

$$\frac{dW}{W} = \frac{kbe^{-kt}}{1 - be^{-kt}}$$

and upon integration gives

$$\ln W = \ln (1 - be^{-kt}) + \ln c$$

where ln c is the arbitrary constant of integration. Transforming this equation to exponential form, we have

$$W = c(1 - be^{-kt}) \ . \qquad (8.41)$$

Using the initial condition that $W = W_0$ at $t = 0$, we can solve for c:

$$W_0 = c(1 - be^{-k0}) \qquad \text{or} \qquad c = \frac{W_0}{1 - b} \ .$$

Substituting this into Equation 8.41, we arrive at the final form for the MONOMOLECULAR GROWTH CURVE of

$$W = \frac{W_0}{1 - b} (1 - be^{-kt}) \ .$$

In summary, some important growth curves have been defined in terms of differential equations. The technique of solution has been illustrated for these growth curves:

EXPONENTIAL GROWTH $$\frac{dW}{dt} = kW \rightarrow W = W_0 e^{kt} \qquad (8.42)$$

LOGISTIC GROWTH $$\frac{dW}{dt} = kW\left[1 - \frac{W}{W_m}\right] \rightarrow W = \frac{W_0(1 + b)}{1 + be^{-kt}} \qquad (8.43)$$

GOMPERTZ GROWTH $$\frac{dW}{dt} = bWe^{-kt} \rightarrow W = c \exp\left[-\frac{b}{k} e^{-kt}\right] \qquad (8.44)$$

MONOMOLECULAR GROWTH $$\frac{dW}{dt} = \frac{Wkbe^{-kt}}{1 - be^{-kt}} \rightarrow W = \frac{W_0}{1 - b}(1 - be^{-kt}) \qquad (8.45)$$

EXERCISES

1. Assume there exists a fixed upper bound for the size, N, of an individual, whether this individual is a cell, a tissue, a population or a crop. Let B be the upper bound toward which N tends asymptotically. This implies then that the rate of growth must approach zero as N approaches B. The following derivative has this property:

$$\frac{dN}{dt} = k(B - N)$$

where k is the constant of proportionality. If we assume that $N = N_0$ at $t = 0$, find a relation describing the size of an individual using this restricted growth model.

2. Bacteria reproduce by simple division so the growth rate of a bacteria population is proportional to the number of bacteria present. Write a differential equation for the rate of reproduction as a function of the number of bacteria. From the resulting equation, find out how many bacteria were present initially if at the end of three hours there were 10^4 bacteria and at the end of five hours there were 10^5 bacteria.

3. A substance in a chemical reaction is used up at a rate which is proportional to the amount of the substance present at any time.

3. Continued

If 0.9 of the substance is consumed in 4 hours, how much of the substance was left at the end of 1.25 hours?

4. A beam of light passing through a liquid loses intensity, I, as the distance, x, it has to move through the liquid increases. The rate of change of I is proportional to I itself, with a constant of proportionality, -k, which depends among other things on the liquid used. What is the value of k if $I = 1$ at $x = 0$ and $I = 0.01$ when $x = 10$ centimeters?

5. Newton's law of cooling states that the rate at which a body changes temperature is proportional to the difference between its temperature and that of the surrounding medium. If a body is in air at 35°F and the body cools from 120°F to 60°F in 40 minutes, find the temperature of the body after 100 minutes.

6. Consider the attack equation

$$\frac{dN_A}{dN} = aP^{1-b}(Pk - N_A)$$

where N_A is the number attacked, N is the number vulnerable to attack, P is the number of attackers and a, b and k are all constants. If $N_A = 0$ when $N = 0$, find an expression for N_A.

7. The rate of change of the number of fish in a particular area can be approximated by $\frac{dN}{dt} = -MN$ where M is the natural mortality of the fish. If the number of fish in the population at time t_p is R, find an expression for N at any time t.

8. After fingerling fish survive to an age of t_p', they are also capable of being caught by fisherman as well as being subjected to natural mortality. Thus $\frac{dN}{dt} = -(F + M)N$ where F is the fishing mortality, M is the natural mortality and N is the number of fish at any time t. If at time t_p' there are R' fish, find an expression for the number of fish at any time t.

9. In a predator-prey situation $\frac{dA}{dP} = -b\,\frac{A}{P}$ where A is the per capita effectiveness of predators, P is the number of predators, b is the

9. Continued
intraspecific competition among predators. Find the effectiveness per capita of the predators.

10. When predators are not stimulated to search very hard for prey we have

$$\frac{dN_A}{dN} = PB(Pk - N_A)$$

Where N_A is the number of prey attacked per unit area, N is the number of prey, P is the number of predators, B is the per capita effectiveness of predators, and k is the maximum per capita effectiveness of predators over a measured period. If $N_A = 0$ when $N = 0$, find an expression for N_A.

11. The rate of change in the concentration of an anesthetic in tissues satisfies $\frac{dC}{dt} = \frac{F}{\lambda V}(\lambda C_a - C)$ where F is blood flow in ml./min., V is the volume of the tissue, C_a is the concentration of the anesthetic in the arterial blood, C is the concentration of the anesthetic in the tissues and λ is a decay constant. Find an expression for C if at $t = 0$, $C = 0$ and at $t = T$, $C = C(T)$.

12. The rate of increase of a population of a certain city is proportional to the number present in the population. If the population increases from 40,000 to 60,000 people in 40 years, what will the population be in 70 years?

13. If a thermometer is taken from a room in which the temperature is 75° into the open where the temperature is 35°, and after 30 seconds the thermometer reads 65°, how long after it was moved from the room to the open air will the thermometer read 50°? (Hint: See Problem 16 of the last section.)

14. The relation which exists between the density of a prey population and the food ration available to each member of the predator population is given by $\frac{dr}{dP} = k(R - r)$ where P is the density of the population, r is the average size of the ration consumed by each predator, R is the maximal ration each predator would eat per unit time in the absence

14. Continued
of competitors and k is a constant. If r = 0 when P = 0, find an expression for r.

15. In a certain bacteria culture, the rate of growth of bacteria is proportional to the number present. If there are 10^3 bacteria present initially and the number doubles in 2 hours, how many bacteria will there be in 3½ hours?

16. The rate of growth of a population can be described by $\frac{dN}{dt} = (b - d)N$ where b is the birth rate, d is the death rate, N is the number of individuals in the population at some time t. Find an expression for N. Find the point in time (if any) where the rate of growth of the population is maximum. Sketch the curve.

*8.11 COMPUTER EVALUATION OF DEFINITE INTEGRALS

The length of this chapter may tempt you to think that you can integrate most functions of interest. This is not at all the case. Later we will introduce several more techniques of integration; some tables contain integrals of more than 500 specific forms. Yet, simple integrals like $\int_0^2 e^{-x^2} dx$ cannot be evaluated by using the fundamental theorem because their integrands have no antiderivatives. Numerical techniques exist to evaluate definite integrals. Although the techniques we will introduce can be used for hand calculation, you should expect to use them primarily on digital computers.

The solution to this problem may seem obvious: Let a computer evaluate $\sum_{i=1}^{n} f(x_i)\Delta_i x$ for $\Delta_i x$ chosen very small. Be careful! This approach contains a trap. The definition of an integral involves the limiting value of this sum as all $\Delta_i x \to 0$, but the computer will have to use some small, and usually equal, values for the $\Delta_i x$ so the sum will only approximate the integral. You may counter by observing that if the $\Delta_i x$ gets small enough, the error due to approximation will become negligible. You are right, but this gets to be a costly solution. Even more seriously it can produce nasty problems of computational precision arising from the addition of a very large number of very small quantities.

Cost closely relates to the number of times f gets evaluated. Thus, you really need an answer to this question: Given that f will be evaluated a fixed number of times, what scheme for combining these evaluations will give a good approximation to the integral? We will present three techniques which give reasonably accurate results.

We are going to approximate $\int_a^b f(x)dx$ by evaluating $f(x_i)$, $i = 0, 1, 2, \cdots, n$ at $x_i = a + \frac{i}{n}(b - a)$ and combine the evaluations according to

- A RIEMANN SUM

$$\int_a^b f(x)dx \doteq [f(x_1) + f(x_2) + \cdots + f(x_n)] \frac{b - a}{n}$$

$$= \frac{b - a}{n} \sum_{i=1}^{n} f(x_i) . \tag{8.46}$$

- The TRAPEZOIDAL RULE

$$\int_a^b f(x)dx \doteq [f(x_0) + 2f(x_1) + 2f(x_2) + \cdots + 2f(x_{n-1}) + f(x_n)] \frac{b - a}{2n}$$

$$= \frac{b - a}{2n} [f(x_0) + f(x_n) + 2 \sum_{i=1}^{n-1} f(x_i)] . \tag{8.47}$$

- SIMPSON'S RULE, or the PARABOLIC RULE, for n = 2m, an even integer,

$$\int_a^b f(x)dx = [f(x_0) + 4f(x_1) + 2f(x_2) + 4f(x_3) + \cdots$$

$$+ 2f(x_{n-2}) + 4f(x_{n-1}) + f(x_n)] \frac{b - a}{3n}$$

$$= \frac{b - a}{3n} [f(x_0) + f(x_n) + \sum_{j=1}^{m} (4f(x_{2j-1}) + 2f(x_{2j}))] . \tag{8.48}$$

To see why these work, look at three different approximations to the area under a segment of f as shown in Figure 8.25. On the left, f is approximated by a horizontal line of height $f(x_i)$ so the area over the ith interval is approximated by that of a rectangle. In the middle, f is

Figure 8.25

approximated by the straight line through the points $(x_{i-1}, f(x_{i-1}))$ and $(x_i, f(x_i))$ so the area over the ith interval is approximated by that of a trapezoid, namely by $\frac{\Delta_i x}{2}(f(x_{i-1}) + f(x_i))$. The addition of these terms produces the trapezoidal rule because each $f(x_i)$ except those at the end-points is added in twice, once as the left edge of a trapezoid, and once as the right edge.

Simpson's rule arises when f is approximated by a series of parabolas, a different one for each subinterval (x_{i-1}, x_{i+1}) where i is an odd integer. We do not prove Simpson's rule here. Should you desire to study such a proof consult a text dealing with elementary numerical analysis.

These three methods differ in how well they approximate the integral. To precisely evaluate their errors, we have to know the integral's exact value. We can, however, place an upper bound on their maximum error. For the three methods absolute errors are bounded, respectively by $\frac{(b-a)^2K}{2n}$, $\frac{(b-a)^3L}{12n^2}$, $\frac{(b-a)^5M}{180n^4}$ where K, L, M are the maximum values $|f'(x)|$, $|f''(x)|$ and $|f^{(4)}(x)|$ for $x \in [a,b]$. Each of these has the form $(b-a)\left[\frac{b-a}{n}\right]^j \frac{f^{(j)}(\xi_j)}{C_j} = (b-a)(\Delta_i x)^j \frac{f^{(j)}(\xi_j)}{C_j}$ where the $f^{(j)}(x)$ assume their maxima at $\xi_j \in [a,b]$. The term $(\Delta_i x)$ will tend to control the size of these errors provided n has been chosen reasonably large. For $\Delta_i x$ having typical values like 0.001 or smaller, the change in j from 1 to 2 to 4 can really reduce the value of $(\Delta_i x)^j$. The other variable term in these maximum errors, namely max $\frac{f^{(j)}(x)}{C_j}$, will decrease as j increases unless $f(x)$ behaves very irregularly over $[a,b]$. Thus Simpson's rule ordinarily gives substantially more precise approximations than either of the other two.

EXAMPLE 8.37: Approximate $\int_2^4 x^2dx$ with $n = 4$ using all three techniques given in this section.

In this case $f(x) = x^2$, $\Delta_i x = \frac{4-2}{4} = \frac{1}{2}$ and $x_i = a + \frac{i}{n}(b - a)$. Construct the following table:

i	0	1	2	3	4
x_i	2.0	2.5	3.0	3.5	4.0
$f(x_i)$	4.00	6.25	9.00	12.25	16.00

The Riemann Sum gives

$$\int_2^4 x^2dx \doteq \frac{b-a}{n} \sum_{i=1}^{n} f(x_i) = \frac{4-2}{4}(6.25 + 9.00 + 12.25 + 16.00) = 21.75.$$

The Trapezoidal Rule gives

$$\int_2^4 x^2dx \doteq \frac{b-a}{n}\left(f(x_0) + f(x_n) + 2\sum_{i=1}^{n-1} f(x_i)\right)$$

$$= \frac{4-2}{(2)(4)}(4.00 + 16.00 + 2(6.25 + 9.00 + 12.25)) = 18.75.$$

Simpson's Rule gives

$$\int_2^4 x^2dx \doteq \frac{b-a}{3n}[f(x_0) + 4f(x_1) + 2f(x_2) + 4f(x_3) + f(x_4)]$$

$$= \frac{4-2}{(3)(4)}[4.00 + 4(6.25) + 2(9.00) + 4(12.25) + 16.00] = 18\tfrac{2}{3}.$$

By direct evaluation,

$$\int_2^4 x^2dx = \frac{x^3}{3}\Big]_2^4 = \frac{64}{3} - \frac{8}{3} = 18\tfrac{2}{3}.$$

The three methods of approximation get increasingly close to the actual value of the integral. Simpson's Rule gives the exact value, but do not expect this usually to happen. It occurred here simply because we approximated $f(x) = x^2$ by a parabola which really involves no approximation at all! ▌

EXAMPLE 8.38: Approximate $\int_0^2 e^{-x^2}dx$ using n = 8 subdivisions. Get the approximation using all three methods.

From $f(x) = e^{-x^2}$, $\Delta_i x = \frac{2}{8} = \frac{1}{4}$ and $x_i = a + \frac{i}{n}(b - a)$, construct the following table:

i	0	1	2	3	4	5	6	7	8
x_i	0.0	0.25	0.50	0.75	1.00	1.25	1.50	1.75	2.00
$f(x_i)$	1.0000	0.9394	0.7788	0.5698	0.3679	0.2096	0.1054	0.0468	0.0183

Check that the methods give 0.7590, 0.8817, and 0.88207 as approximations to the exact value (to four digits) of 0.8821. Simpson's Rule gave nearly an exact approximation with only n = 8 subintervals. Even with n = 4, the three methods give 0.6352, 0.8806, 0.8818, respectively. ▌

So far this discussion has concerned mathematical techniques. Because a definite integral can be interpreted as an area, it seems appropriate to point out here that there exists a mechanical device called a planimeter which approximates areas of irregular figures. When the edge of an area is traced with the head of this device, it produces an approximation to the area through the vertical and horizontal movement of its attached arms. It is used to evaluate the area under various physiographic curves such as blood pressure curves. Its use on aerial photographs gives the areas of such things as wheat fields or lodge pole pine forests. It gives leaf areas for use in photosynthesis studies because leaf area closely relates to the photosynthetic capacity of the plant.

Biologists sometimes approximate irregularly shaped areas another way. They photograph the areas from a fixed height. Then they cut out the area with scissors and weigh it. Both the quality of the photographic paper, particularly the uniformity of its thickness, and the care used in the cutting influence the precision of this technique. Comparisons of such weights are equivalent to comparisons of the areas. Determination of actual areas require the establishment of the relationship between the weight of the areas' photographs and the actual areas. A few carefully established areas and their weights on photographic paper will give the proportionality constant.

EXERCISES

In the following problems, use the three approximate methods for evaluating the following integrals.

1. $\int_1^2 \frac{dx}{x}$; n = 2

2. $\int_1^2 \frac{dx}{x}$; n = 4

3. $\int_1^2 \frac{dx}{x}$; n = 10 Compare your answers obtained in Problems 1, 2 and 3 with ln 2.

4. $\int_0^2 e^{x^2} dx$; n = 4

5. $\int_0^2 e^{x^2} dx$; n = 8

6. Show that if f is a linear function the trapezoidal rule gives the exact value of $\int_a^b f(x)dx$ for all a and b and for all choices of the positive integer n.

7. The integral $S = \int_a^b V(t)dt$ gives the distance traveled (S) by an object moving at the velocity V(t) during the time from a to b. The following table records the velocity of a car at intervals of 0.1 hours. Use the trapezoidal rule to approximate the total distance traveled by the car during that hour.

t	0	0.1	0.2	0.3	0.4	0.5	0.6	0.7	0.8	0.9	1.0
V(t)	0	45	54	51	48	56	65	50	52	55	60

8. Find the errors involved in calculating the following by the three methods of this section.

   a. $\int_{-1}^{2} (x + 2x)dx$; $n = 3$

   b. $\int_{1}^{3} \frac{dx}{x}$; $n = 4$

   c. $\int_{5}^{9} \frac{dx}{x}$; $n = 4$

*Chapter 9

SPECIAL INTEGRALS

The presentation of integrals in the last chapter avoided several troublesome, but practically reasonable, conditions. Specifically, the integrands were assumed to be continuous and in most cases expressible as the derivative of another function; only closed intervals were considered for definite integration. This chapter will relax those conditions thereby slightly widening the class of integrals you can handle.

The first two sections discuss improper integrals, the first developing infinite limits of integration while the other presents techniques for dealing with ill-behaved integrands. The third section introduces the gamma function, a special function which arises in both biological and statistical situations. The last section illustrates uses of integrals arising from continuous probability distributions in statistics. The broadened view of integration presented in this chapter has substantial value to a biologist, but some people may judge it to be of less importance than several subsequent chapters. Thus, the entire chapter may be regarded as optional; little subsequent material demands this chapter as a prerequisite.

9.1 IMPROPER INTEGRALS: INFINITE LIMITS OF INTEGRATION

The definite integral originally was defined only over a closed interval $[a,b]$. Furthermore, for that definition to work, the integrand had to be finite throughout $[a,b]$; otherwise some of the products $(\Delta_i x)f(x_i)$ would not be defined so their sum could not be taken. For an integrand to remain finite throughout an interval, there must exist an upper bound M satisfying $|f(x)| \leq M$ for every $x \in [a,b]$. Such a function is called BOUNDED. Similarly, $[a,b]$ is called a BOUNDED INTERVAL in contrast to the interval which extends to infinity: $[a,\infty)$.

At times we need to relax each of these conditions of boundedness. The next section relaxes the boundedness on the integrand. This section deals with intervals which extend to infinity. This latter problem arises, for example, when we seek to determine the amount of radioactivity remaining in an organic sample. For this we need to integrate the appropriate decay rate function from the present (t_0) to infinity, namely over $[t_0,\infty)$. Such integrals are called IMPROPER INTEGRALS. To handle them we need the following definitions:

- Let a be a fixed number, and suppose that $f(x)$ is continuous on any domain $[a,h]$, $[k,b]$, or $[k,h]$, respectively. Then, we define the value of each of the following improper integrals as the limit on the right, provided that limit exists:

$$\int_a^\infty f(x)dx = \lim_{h\to\infty} \int_a^h f(x)dx \tag{9.1}$$

$$\int_{-\infty}^a f(x)dx = \lim_{k\to-\infty} \int_k^a f(x)dx \tag{9.2}$$

$$\int_{-\infty}^\infty f(x)dx = \int_{-\infty}^a f(x)dx + \int_a^\infty f(x)dx$$

$$= \lim_{k\to-\infty} \int_k^a f(x)dx + \lim_{h\to\infty} \int_a^h f(x)dx \tag{9.3}$$

If the limit on the right exists (is finite), the improper integral on the left exists or CONVERGES to the limit specified on the right. If the limit on the right fails to exist, the improper integral DIVERGES. This procedure is illustrated in the following examples.

<u>EXAMPLE 9.1</u>: Evaluate $\int_1^\infty \frac{dx}{x}$.

See Figure 9.1 for the area corresponding to this integral. Using the definition given above, we have

$$\int_1^\infty \frac{dx}{x} = \lim_{h\to\infty} \int_1^h \frac{dx}{x} = \lim_{h\to\infty} \ln x\Big]_1^h = \lim_{h\to\infty} [\ln h - \ln 1] = \infty .$$

Because the limit is not finite, the integral diverges. Thus an infinite area lies under the curve $f(x) = \frac{1}{x}$ to the right of 1. ∎

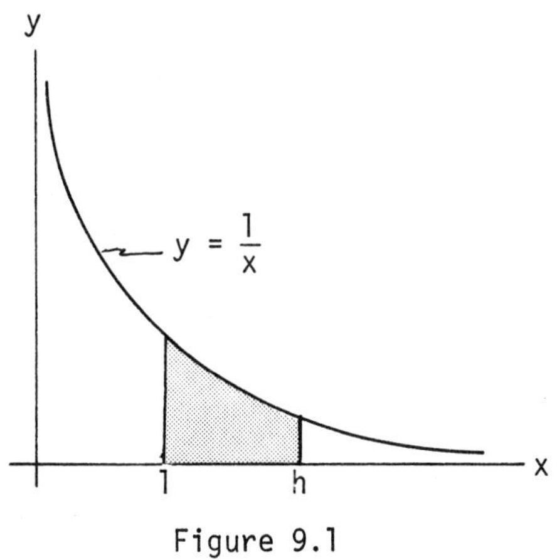

Figure 9.1

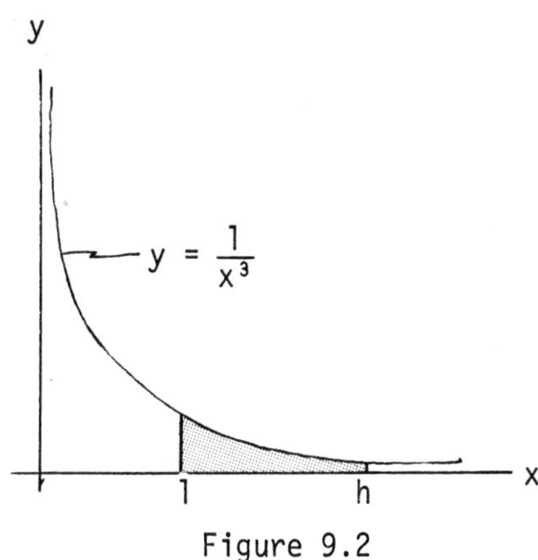

Figure 9.2

EXAMPLE 9.2: Evaluate $\int_1^\infty \frac{dx}{x^3}$.

The area corresponding to this integral is shaded in Figure 9.2. Recourse to the definition gives

$$\int_1^\infty \frac{dx}{x^3} = \lim_{h\to\infty} \int_1^h \frac{dx}{x^3} = \lim_{h\to\infty} \left. -\frac{1}{2x^2}\right]_1^h = -\lim_{h\to\infty}\left[\frac{1}{2h^2} - \frac{1}{2}\right] = \frac{1}{2} .$$

Thus, the integral converges to $\frac{1}{2}$. ∎

You might wonder why this integral converges while an apparently similar one in Example 9.1 diverged. The function $1/x^3$ approaches the x-axis much more rapidly than does $1/x$: The height $1/x^3$ approaches zero much faster than the width x approaches infinity in the sense that $\lim_{x\to\infty} x \frac{1}{x^3} = 0$. The same cannot be said for $1/x$ because $\lim_{x\to\infty} x \frac{1}{x} \neq 0$. In fact, more generally $\int_0^\infty x^n\, dx$ exists whenever $n + 1 < 0$ because

$$\int x^n dx = \begin{cases} \dfrac{x^{n+1}}{n+1} & n \neq -1, \text{ or } n + 1 \neq 0 \\[2ex] \ln x & n = -1, \text{ or } n + 1 = 0 , \end{cases}$$

and $\lim_{x\to\infty} \frac{x^{n+1}}{n+1} = 0$ for $n + 1 < 0$. The integral diverges for $n + 1 > 0$. The case $n + 1 = 0$ occupies a transitional role between the two previous cases. The integral also diverges for n + 1 = 0, but due to the properties of ln x, not of x^{n+1}.

EXAMPLE 9.3: Evaluate $\int_{-\infty}^{2} e^{3t}dt$.

Figure 9.3 shows the area corresponding to this integral.

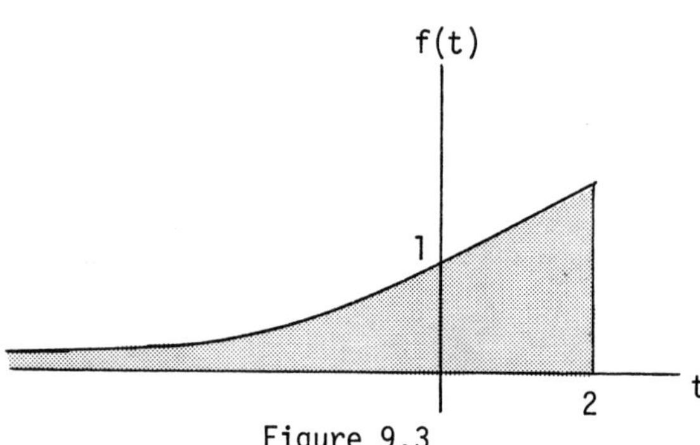

Figure 9.3

$$\int_{-\infty}^{2} e^{3t}dt = \lim_{h\to-\infty} \int_{h}^{2} e^{3t}dt = \lim_{h\to-\infty} \frac{1}{3}\left[e^{3t}\right]_{h}^{2} = \lim_{h\to-\infty} \frac{1}{3}\left[e^{6} - e^{3h}\right] = \frac{1}{3}e^{6}$$

so this integral converges. ▌

EXAMPLE 9.4: Evaluate $\int_{-\infty}^{\infty} x^5dx$.

From the definition given in Equation 9.3, we will break up the integral as:

$$\int_{-\infty}^{\infty} x^5dx = \int_{-\infty}^{0} x^5dx + \int_{0}^{\infty} x^5dx \ . \qquad (9.4)$$

Convergence of the original integral requires that both integrals on the right must converge.

$$\int_{0}^{\infty} x^5dx = \lim_{h\to\infty} \int_{0}^{h} x^5dx = \lim_{h\to\infty} \left.\frac{x^6}{6}\right]_{0}^{h} = \frac{1}{6}\lim_{h\to\infty}\left[h^6 - 0\right] = \infty \ .$$

The original integral diverges because at least one of the integrals on the right side of Equation 9.4 does. We could have broken the integral up as $\int_{-\infty}^{a} x^5dx + \int_{a}^{\infty} x^5dx$ where a is any number. However, for the integral to converge, both integrals must converge. In this case neither of the integrals converges. ▌

You might be *tempted* to evaluate Equation 9.4 by

$$\int_{-\infty}^{\infty} x^5 dx = \lim_{h\to\infty} \int_{-h}^{h} x^5 dx = \lim_{h\to\infty} \frac{x^6}{6}\Big]_{-h}^{h} = \lim_{h\to 0} 0 = 0 . \qquad (9.5)$$

This would say that the integral converges even though we have shown that it does not. This apparent contradiction appears because the limit in Equation 9.5 is taken so that the positive area exactly balances the negative area although both are increasing without bound, as shown in Figure 9.4.

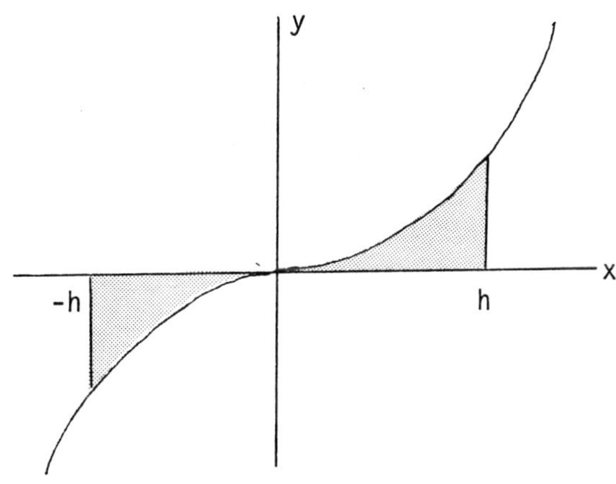

Figure 9.4

What are we actually doing in the above procedures? In evaluating $\int_a^{\infty} f(x)dx$ as $\lim_{h\to\infty} \int_a^h f(x)dx$, $\int_a^h f(x)dx$ is a definite proper integral whose value depends upon its upper limit h. Then we examine what happens to its value, the area under the curve, as $h\to\infty$. If a finite limit results, the integral converges but if the limit fails to exist, the integral diverges.

EXAMPLE 9.5: A radioactive nuclide decays according to the law $A = A_0e^{-kt}$, $k > 0$. Determine the total activity under the curve from $t = 1$ to $t = \infty$.

In Example 8.18 we showed that the total activity occurring between times t_0 and t_1 is given by

$$\int_{t_0}^{t_1} A_0e^{-kt} .$$

Here we need the area under the curve past t = 1 given by $\int_1^\infty A\, e^{-kt}dt$. This area is indicated in Figure 9.5. Now using the above definition we write

$$\int_1^\infty A_0 e^{-kt}dt = \lim_{h\to\infty} \int_1^h A_0 e^{-kt}dt = \lim_{h\to\infty} \frac{-A_0}{k} e^{-kt}\Big]_1^h$$

$$= \frac{-A_0}{k} \lim_{h\to\infty} [e^{-kh} - e^{k}] = \frac{A_0}{k} e^{-k} .$$

Here $\frac{-A_0}{k}[e^{-kh} - e^{-k}]$ represents the activity under the curve to any variable upper limit h as shown in Figure 9.5. Upon taking the limit, we find $\frac{A_0}{k} e^{-k}$ as the value of the integral. Thus, the integral converges. ∎

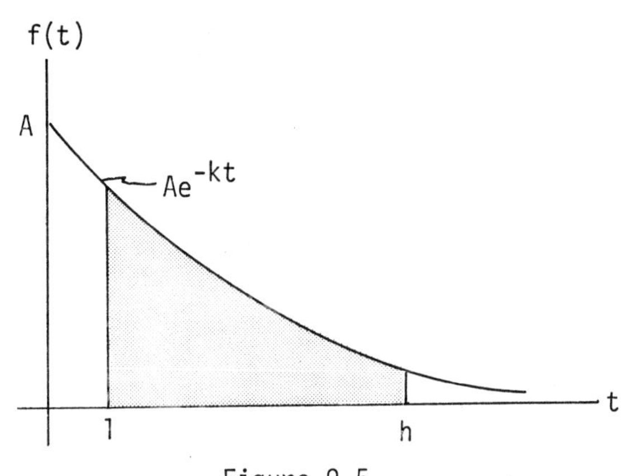

Figure 9.5

EXERCISES

Find the values of the following integrals, or demonstrate that they diverge.

1. $\int_2^\infty \frac{dx}{x^2}$

2. $\int_0^\infty \frac{dx}{(x + 3)^{3/2}}$

3. $\int_1^\infty x^{-p}dx;\ p > 1$

4. $\int_1^\infty x^{-p}dx;\ p = 1$

5. $\int_1^\infty x^{-p}dx;\ p < 1$

6. $\int_1^\infty e^{-x}dx$

7. $\int_{-\infty}^{0} e^{-x}dx$

8. $\int_{4}^{\infty} \frac{dx}{\sqrt[3]{x}}$

9. $\int_{0}^{\infty} xe^{-x^2}dx$

10. $\int_{5}^{\infty} \frac{dx}{(x - 4)^{3/2}}$

11. $\int_{-\infty}^{\infty} x^2e^{-x^3}dx$

12. $\int_{-\infty}^{\infty} x\sqrt{x^2 + 9}\, dx$

13. Find the area, if it exists, between the x-axis and the curve $y = e^x$ for $x \geq 0$.

9.2 IMPROPER INTEGRALS: ILL-BEHAVED INTEGRANDS

An integrand may have a discontinuity at an interior point or at either end of its interval of definite integration. A related problem occurs if the integrand has no antiderivative at a point.

CASE I: Discontinuity at an end-point.

Suppose that $f(x)$ is continuous when $a \leq x < b$ but that $f(x)$ is not defined at b or $|f(x)| \to \infty$ as $x \to b^-$. (Recall that $x \to b^-$ means that x approaches b from below.) Then $\int_a^b f(x)dx$ is an improper integral which can be defined in either of the following ways:

$$\int_a^b f(x)dx = \lim_{h \to b^-} \int_a^h f(x)dx = \lim_{\varepsilon \to 0^+} \int_a^{b-\varepsilon} f(x)dx \tag{9.6}$$

provided the limits on the right exist. Similarly, if $f(x)$ has a discontinuity at a, define the improper integral $\int_a^b f(x)dx$ as

$$\int_a^b f(x)dx = \lim_{h \to a^+} \int_h^b f(x)dx = \lim_{\varepsilon \to 0^+} \int_{a+\varepsilon}^{b} f(x)dx$$

Provided that the limits on the right exist. When the limits exist, the improper integral CONVERGES but otherwise it DIVERGES.

EXAMPLE 9.6: The integral $\int_1^2 \frac{dx}{\sqrt[3]{2 - x}}$ is improper because the denominator approaches 0 as $x \to 2$ so the integrand becomes undefined there. From the above discussion, we have

$$\int_1^2 \frac{dx}{\sqrt[3]{2 - x}} = \lim_{h\to 2^-} \int_1^h \frac{dx}{\sqrt[3]{2 - x}} = \lim_{h\to 2^-} \left[-\frac{3}{2} (2 - x)^{2/3} \right]_1^h$$

$$= -\frac{3}{2} \lim_{h\to 2^-} [(2 - h)^{2/3} - 1] = +\frac{3}{2} .$$

The graph of $y = \frac{1}{(2 - x)^{1/3}}$ appears in Figure 9.6 where the shaded area is bounded by the curve and the x-axis between $1 \leq x < h$. The value of this area will depend upon h. We are interested in how this area behaves as $h\to 2^-$. If the limit exists, then the area is finite; if the limit fails to exist, then the area is infinite. In this case the area has the finite value of $\frac{3}{2}$. ▌

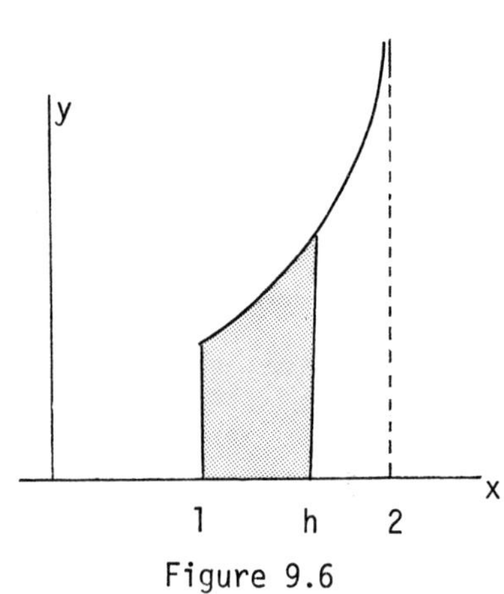

Figure 9.6

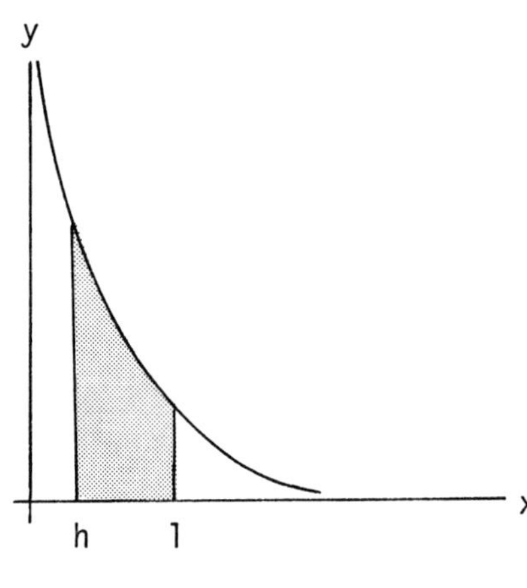

Figure 9.7

EXAMPLE 9.7: Evaluate $\int_0^1 \frac{dx}{x^2}$.

The integrand of $\int_0^1 \frac{dx}{x^2}$ has a discontinuity at $x = 0$ as shown in Figure 9.7:

$$\int_0^1 \frac{dx}{x^2} = \lim_{h\to 0^+} \int_h^1 x^{-2}dx = \lim_{h\to 0^+} \left[-\frac{1}{x} \right]_h^1 = -\lim_{h\to 0^+} \left[1 - \frac{1}{h} \right] = \infty .$$

This integral diverges because the limit on the right fails to exist. ▌

CASE II: Discontinuity at an Interior Point.

As we mentioned earlier, an integrand may be defined at the end-points but not at some interior point. In $\int_a^b f(x)dx$, suppose that f(x) is continuous at all points on the interval $a \leq x \leq b$ except at some point c, where $a < c < b$. This integral also is called an improper integral. It can be evaluated as the sum of two improper integrals provided they both exist:

$$\int_a^b f(x)dx = \int_a^c f(x)dx + \int_c^b f(x)dx \tag{9.7}$$

This procedure is illustrated in the following examples.

EXAMPLE 9.8: Evaluate $\int_1^4 \frac{dx}{(x-3)^2}$.

Evaluation of $\int_1^4 \frac{dx}{(x-3)^2}$ encounters difficulty at $x = 3$ because its integrand is undefined there. Thus, using Equation 9.7, we have

$$\int_1^4 \frac{dx}{(x-3)^2} = \lim_{h\to 3^-} \int_1^h \frac{dx}{(x-3)^2} + \lim_{h\to 3^+} \int_h^4 \frac{dx}{(x-3)^2}.$$

Now if the integral is to exist both limits on the right must also exist.

$$\lim_{h\to 3^-} \int_1^h \frac{dx}{(x-3)^2} = \lim_{h\to 3^-} \left[-\frac{1}{x-3}\right]_1^h = -\lim_{h\to 3^-}\left[\frac{1}{h-3} - \frac{1}{-2}\right] = \infty .$$

This limit fails to exist so the integral diverges. (It can be shown similarly that the second integral also diverges.) ▮

You would get an erroneous result if you integrated in the following manner:

$$\int_1^4 \frac{dx}{(x-3)^2} = -\frac{1}{x-3}\Big]_1^4 = -\left(1 - \frac{1}{-2}\right) = -\frac{3}{2}.$$

This approach is invalid because the fundamental theorem requires a continuous integrand. It cannot be used when its conditions fail.

Figure 9.8 gives a geometric interpretation of the process just illustrated. It shows an integrand undefined at the interior point c, but the shaded areas over [a,h] and [k,b] can be evaluated. When we write

$$\int_a^b f(x)dx = \int_a^c f(x)dx + \int_c^b f(x)dx = \lim_{k\to c^-} \int_a^k f(x)dx + \lim_{k\to c^+} \int_k^b f(x)dx ,$$

we are looking at the area under f between a and b, except for that portion close to c. The two limits shrink the size of the excluded slice to zero. If the limits exist, the area is finite, while if either of the limits fails to exist, the area is infinite so the integral diverges. One integral may converge while the other diverges. Their convergence depends upon how rapidly the curve becomes asymptotic to the line x = c.

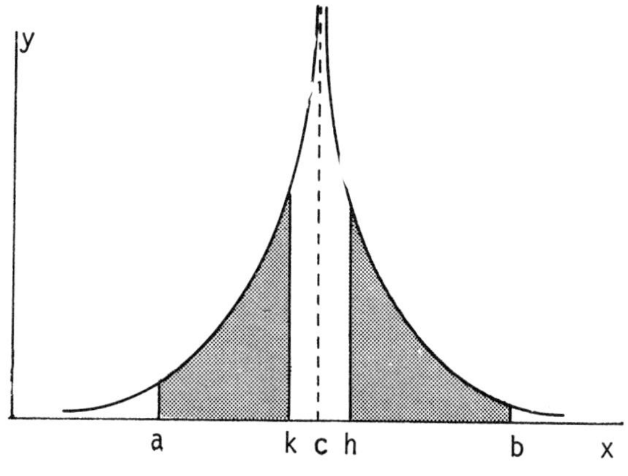

Figure 9.8

EXAMPLE 9.9: Evaluate $\int_0^2 \frac{dx}{(x-1)^{1/3}}$.

The point x = 1 needs special attention because the integrand is not defined at x = 1.

$$\int_0^2 \frac{dx}{(x-1)^{1/3}} = \lim_{k \to 1^-} \int_0^k \frac{dx}{(x-1)^{1/3}} + \lim_{h \to 1^+} \int_h^2 \frac{dx}{(x-1)^{1/3}}$$

$$= \lim_{k \to 1^-} \frac{3}{2} (x-1)^{2/3} \Big]_0^k + \lim_{h \to 1^+} \frac{3}{2} (x-1)^{2/3} \Big]_h^2$$

$$= \frac{3}{2} \lim_{k \to 1^-} \left[(k-1)^{2/3} + 1\right] + \frac{3}{2} \lim_{h \to 1^+} \left[1 - (h-1)^{2/3}\right]$$

$$= \frac{3}{2} + \frac{3}{2} = 3 .$$

Thus the improper integral, $\int_0^2 \frac{dx}{(x-1)^{1/3}}$, converges to 3. ▌

You may encounter an integral which combines the above properties as is illustrated next.

<u>EXAMPLE 9.10</u>: Evaluate $\int_1^\infty \frac{dx}{(x-2)^2}$.

The integral $\int_1^\infty \frac{dx}{(x-2)^2}$ is improper for two reasons. It has an undefined integrand at x = 2 as well as an infinite interval over which the integration occurs. It must be decomposed into separate integrals to handle the discontinuity at x = 2 and the infinite upper limit. Thus, we have

$$\int_1^\infty \frac{dx}{(x-2)^2} = \int_1^2 \frac{dx}{(x-2)^2} + \int_2^\infty \frac{dx}{(x-2)^2}$$

$$= \int_1^2 \frac{dx}{(x-2)^2} + \int_2^3 \frac{dx}{(x-2)^2} + \int_3^\infty \frac{dx}{(x-2)^2}$$

$$= \lim_{k\to 2^-} \int_1^k \frac{dx}{(x-2)^2} + \lim_{h\to 2^+} \int_h^3 \frac{dx}{(x-2)^2} + \lim_{m\to\infty} \int_3^m \frac{dx}{(x-2)^2}$$

$$= \lim_{k\to 2^-} \left[-\frac{1}{x-2}\right]_1^k + \lim_{h\to 2^+} \left[-\frac{1}{x-2}\right]_h^3 + \lim_{m\to\infty} \left[-\frac{1}{x-2}\right]_3^m$$

$$= -\lim_{k\to 2^-} \left[\frac{1}{k-2} + 1\right] - \lim_{h\to 2^+} \left[1 - \frac{1}{h-2}\right] - \lim_{m\to\infty} \left[\frac{1}{m-2} - 1\right]$$

$$= \infty + \infty + 1 = \infty .$$

The improper integral diverges because at least one of its constituent limits does not exist. ▮

These last three examples have illustrated discontinuities which resulted from a denominator going to zero, thereby producing an undefined integrand. Other kinds of discontinuities can occur. Any time a biological system receives a shock, it may respond in a discontinuous or nearly discontinuous manner. For example, when a manufacturing plant releases a pollutant, either intentionally or accidentally, into a body of water, the concentration of the pollutant in the water jumps quickly but then begins to decline through the combined efforts of dilution, decay and biological conversion. Likewise, growing individuals usually gain weight rather continuously, but when insect nymphs molt, their weight exhibits a measurable and discontinuous drop. The density of a migratory species changes abruptly upon the arrival of a new contingent of animals. We have previously discussed synchronization of cell reproduction in a tissue culture. When reproduction occurs in such a situation, population size has a large and practically discontinuous doubling.

A discontinuous yet bounded integrand still falls under this case. It is handled the same way as previous examples.

EXAMPLE 9.11: Evaluate $\int_4^6 [\![x]\!]\,dx$ where $[\![x]\!]$ symbolizes the greatest integer less than or equal to x.

This function has discontinuities at 5 and 6, as displayed in Figure 9.9. Consequently, we need to break the integral into two parts:

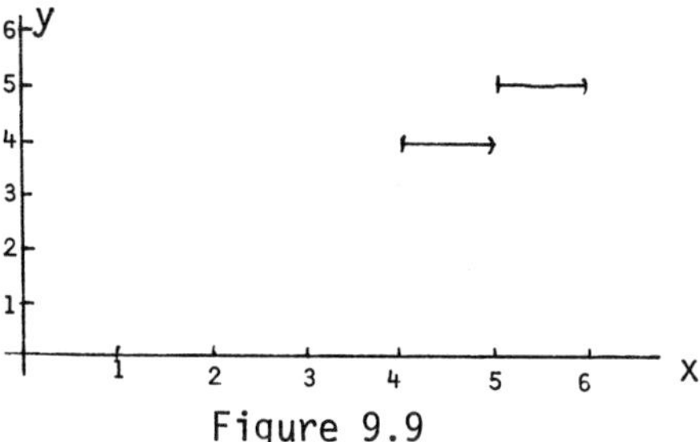

Figure 9.9

$$[\![x]\!] = \begin{cases} 4 & 4 \le x < 5 \\ 5 & 5 \le x < 6 \\ 6 & x = 6 \end{cases}$$

Again breaking the integral into parts, it becomes

$$\int_4^6 [\![x]\!]\,dx = \lim_{\varepsilon\to 0^+} \int_4^{5-\varepsilon} 4dx + \lim_{\varepsilon\to 0^+} \int_5^{6-\varepsilon} 5dx + \lim_{\varepsilon\to 0^+} \int_6^{6+\varepsilon} 6dx \ .$$

The last integral above covers only x = 6 so it contributes 0 to the sum. The other two give

$$\int_4^6 [\![x]\!]\,dx = \lim_{\varepsilon\to 0^+} [4x]_4^{5-\varepsilon} + \lim_{\varepsilon\to 0^+} [5x]_5^{6-\varepsilon} = 9 \ .$$

Check this answer by geometrically evaluating the area under $[\![x]\!]$ between 4 and 6. ▌

The same method applies when an integrand has a corner so its antiderivative has a discontinuity:

EXAMPLE 9.12: The function $f(x) = ke^{-a|x-\mu|}$ is symmetric about μ. (Its upper and lower tails match if the graph is folded about $x = \mu$.) It decreases quickly as x moves away from μ. To evaluate $\int_{-\infty}^{\infty} f(x)$, note that f(x) has a corner at $x = \mu$ as shown in Figure 9.10.

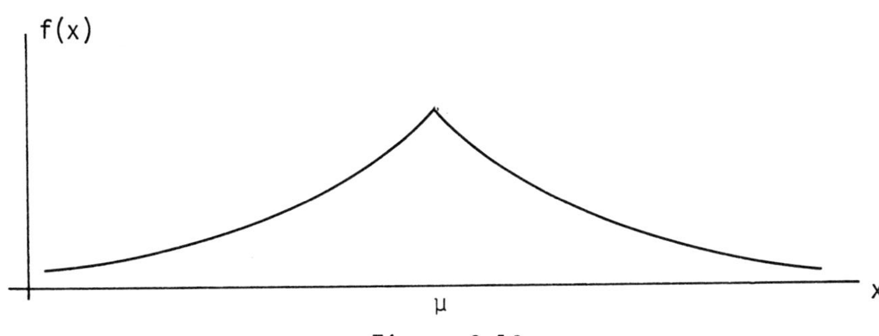

Figure 9.10

We need to evaluate

$$\int_{-\infty}^{\infty} ke^{-a|x-\mu|}dx = k\int_{-\infty}^{\mu} e^{-a|x-\mu|}dx + k\int_{\mu}^{\infty} e^{-a|x-\mu|}dx .$$

Because $x - \mu < 0$ for $x < \mu$, $|x - \mu| = \mu - x$ in the first integral on the right above, but in the second $|x - \mu| = x - \mu$ because $x > \mu$. Thus,

$$\int_{-\infty}^{\infty} ke^{-a|x-\mu|}dx = k\int_{-\infty}^{\mu} e^{-a(\mu-x)}dx + k\int_{\mu}^{\infty} e^{-a(x-\mu)}dx$$

$$= \lim_{r\to-\infty} \left[k \frac{1}{a} e^{-a(\mu-x)}\right]_r^{\mu} + \lim_{s\to\infty} \left[k - \frac{1}{a} e^{-a(x-\mu)}\right]_{\mu}^{s}$$

$$= k \frac{1}{a} (1 - 0) - k \frac{1}{a} (0 - 1) = 2 \frac{k}{a} . \quad \blacksquare$$

EXERCISES

Compute the value of the integral or show that it is divergent.

1. $\int_{-1}^{1} x^{-2}dx$
2. $\int_{0}^{1} x^{-p}dx;\ p < 1$
3. $\int_{0}^{1} x^{-p}dx;\ p > 1$
4. $\int_{0}^{1} \frac{dx}{\sqrt{1 - x}}$
5. $\int_{-8}^{1} x^{-2/3}dx$
6. $\int_{0}^{1} \frac{dx}{x}$

7. $\int_{-1}^{1} \frac{dx}{x^3}$

8. $\int_0^2 \frac{udu}{\sqrt{4 - u^2}}$

9. $\int_0^4 \frac{dx}{(3 - x)^2}$

10. $\int_0^\infty \frac{e^{-\sqrt{x}}}{\sqrt{x}}$

11. $\int_{-4}^{\infty} \frac{dx}{\sqrt{x + 4}}$

12. $\int_0^4 \frac{xdx}{\sqrt{16 - x^2}}$

*13. $\int_0^2 \frac{dx}{|x - 1|}$

*14. $\int_{-4}^{4} e^{-|x - 2|}dx$

**9.3 THE GAMMA FUNCTION

This section deals with an important improper integral, the gamma function. Biologists sometimes use the gamma function directly, but its greatest use lies in an associated probability distribution. This role will surface in the next section.

We begin with

- The GAMMA FUNCTION is defined by the improper integral

$$\Gamma(n) = \int_0^\infty x^{n-1} e^{-x}dx , \qquad n > 0 \tag{9.8}$$

This is a function: For each permissible value of n, Equation 9.8 has exactly one value. It is tedious, and not particularly instructive, to show that this improper integral converges for $n > 0$ so we will omit it. Figure 9.11 shows a graph of this function.

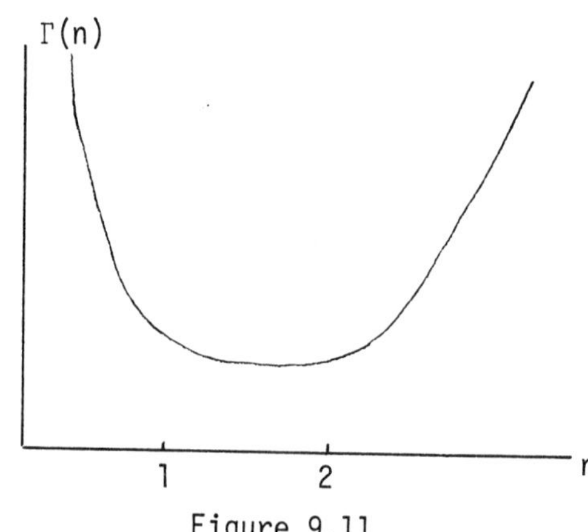

Figure 9.11

Note that the argument of Γ exceeds the power of x in the defining equation (Equation 9.8) by one. The reason for adopting this definition lies in a major property of the gamma function:

$$\Gamma(n) = (n - 1)\Gamma(n - 1) . \tag{9.9}$$

This property can be established using a technique called integration by parts which is discussed in Chapter 15.

EXAMPLE 9.13: Evaluate $\Gamma(1)$.

Using the definition above, we have

$$\Gamma(1) = \int_0^\infty e^{-x}dx = \lim_{h\to\infty} \int_0^h e^{-x}dx = \lim_{h\to\infty}\Big[- e^{-x}\Big]_0^h$$

$$= - \lim_{h\to\infty} (e^{-h} - 1) = 1 . \quad \blacksquare$$

EXAMPLE 9.14: Evaluate $\Gamma(2)$.

The property stated in Equation 9.9 gives

$$\Gamma(2) = \int_0^\infty xe^{-x}dx = (2 - 1)\Gamma(1) = (1)(1) = 1 . \quad \blacksquare$$

EXAMPLE 9.15: Evaluate $\int_0^\infty x^{2n+1}e^{-ax^2}dx$.

To evaluate this integral, let $u = ax^2$ and $\frac{du}{2a} = x\ dx$:

$$\int_0^\infty x^{2n}e^{-ax^2}x\ dx = \int_0^\infty \left(\frac{u}{a}\right)^n e^{-u}\ \frac{du}{2a} = \frac{1}{2a^{n+1}} \int_0^\infty u^{(n+1)-1}e^{-u}du = \frac{1}{2a^{n+1}}\ \Gamma(n+1). \quad \blacksquare$$

The recursive relation stated in Equation 9.9 allows us to obtain the following results if n is an integer:

$$\Gamma(n) = (n - 1)\Gamma(n - 1)$$

$$= (n - 1)(n - 2)\Gamma(n - 2)$$

$$= (n - 1)(n - 2) \cdots 3 \times 2 \times 1 \times \Gamma(1)$$

and finally that $\Gamma(n) = (n - 1)!$ for positive integer values of n. When n is not an integer, the value of $\Gamma(n)$ has to be determined by numerical integration of the sort discussed in Section 8.11. Tables of these results can be found in most mathematical handbooks. In particular, $\Gamma \frac{1}{2} = \sqrt{\pi}$.

The gamma function involves the product of e^{-x} and a power of x, integrated between 0 and ∞. This same integrand can have different limits. This has lead to defining the INCOMPLETE GAMMA FUNCTION by

$$I(x,n) = \int_0^x t^{n-1}e^{-t}dt \ .$$

Of course $I(x,n) < \Gamma(n)$ because the integrand remains positive over the interval of integration. Again, this function has been tabled extensively.

EXERCISES

Evaluate the following integrals:

1. $\int_0^\infty x^5e^{-x}dx$
2. $\int_0^\infty x^{3/2}e^{-x}dx$
3. $\int_0^\infty x^{5/2}e^{-x}dx$
4. $\int_0^\infty x^7e^{-x}dx$
5. $\int_0^\infty x^{9/2}e^{-x}dx$

**9.4 INTEGRALS ASSOCIATED WITH CONTINUOUS DISTRIBUTIONS IN STATISTICS

An important use of calculus in biology relates to probability distributions. Section 8.8b associated probability of continuous responses with areas under curves, and thus with integrals; here we will briefly review that topic. Two subsections then introduce common characterizations of probability distributions, and some important probability density functions (*pdf*'s). This section is intended primarily to show some applications of integrals as applied to statistics.

Sections 3.4 - 3.6 explained various features of probability. In particular, the idea of a probability distribution was introduced in Section 3.6. There we restricted attention to discrete responses, like counts. We avoided responses which could assume any value in an interval. Now that we have integration, we can consider a wider class of responses.

Mathematical variables were introduced in Chapter 4. They can assume any values in the domain and range sets that you choose to give them. Contrast this to the idea of a RANDOM VARIABLE. The latter, denoted by an *italic* letter, is associated with evaluating responses whose values occur in accordance with an underlying probability distribution.

Those values of a discrete random variable with high probability had the greatest chance of appearing; values with a low probability had only a slight chance of occurring.

Recall that a continuous response can assume any value in a range of possible values, like the time it takes you to answer an exam question. Ordinarily this could be any time between 0 and 60 minutes, like 27.2683··· minutes. Even though our evaluation has limited resolution (like to the nearest second), the real response can have any value. The density or frequency function may be interpreted this way: p(x)dx approximates the probability that the random variable x will lie between x and x + dx, as shaded in Figure 9.12. Notice that if x lies under the higher parts of p, then p(x)dx will be larger, for the same value of dx, than if x lies under a lower part of p. Thus, for intervals of fixed length, values of the random variable where p is high have a greater probability of occurring than do values of the random variable where p is low. In this sense p describes the density of probability, that is, the differential location of the probability.

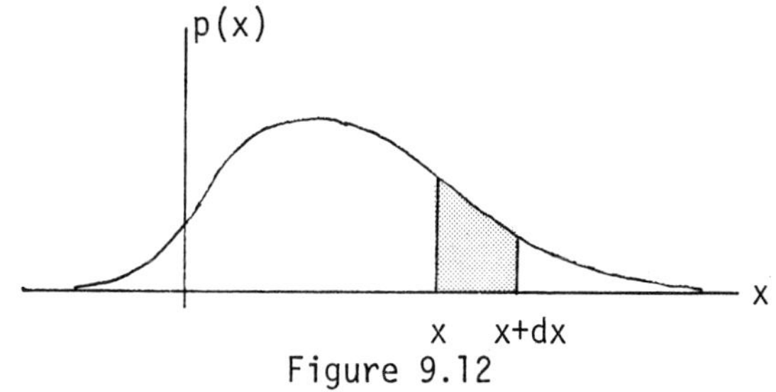

Figure 9.12

As dx is positive, $p(x) \geq 0$ for all values of x because probabilities are non-negative. When x is observed something will happen, namely with probability one some value will occur: $\int_{-\infty}^{\infty} p(x)dx = 1$. Of course, the values of x where $p(x) > 0$ may be a subset of the whole axis; at the remaining points $p(x) = 0$. Under these restrictions, the probability that the random variable x lies in [a,b] is given by

$$P[a \leq x \leq b] = \int_a^b p(x)dx \ .$$

Occasionally you may have a nonnegative function f possessing the right shape for a particular situation, but whose integral equals some finite value other than 1. In this case the function

$$p(x) = \frac{f(x)}{\int_{-\infty}^{\infty} f(x)dx} \qquad \text{satisfies } \int_{-\infty}^{\infty} p(x)dx = 1 \qquad (9.10)$$

and can be used as a probability density function.

EXAMPLE 9.16: For a purely operational example, consider $f(x) = 3x^2 - x^3$, $0 \leq x \leq 3$. Over this range, $f(x) \geq 0$; we can ignore the fact that $f(x) < 0$ for $x > 3$ because this lies outside the range of interest. Thus $f(x)$ satisfies one of the properties required of a probability density function. How about the other?

$$\int_0^3 f(x)dx = \int_0^3 (3x^2 - x^3)dx = \left[x^3 - \frac{x^4}{4}\right]_0^3 = \frac{27}{4}$$

so it is not a probability density function. However, the related function

$$p(x) = \begin{cases} \frac{4}{27}(3x^2 - x^3) & 0 \leq x \leq 3 \\ 0 & \text{otherwise} \end{cases}$$

does satisfy the requirements. This latter function was examined in Example 8.29; its graph appears there. Check to see that this $p(x)$ gives $P[\frac{1}{2} \leq x \leq 1] = 0.095$ and $P[2 \leq x \leq 2\frac{1}{2}] = 0.276$. Evaluate these probabilities by integration. If you have trouble, look back at Example 8.29. ∎

EXAMPLE 9.17: The exponential probability distribution was considered at length in Section 8.8b. Recall that it applies to a variety of observable responses like survival times. Let us begin with $f(x) = e^{-\lambda x}$, $x > 0$, and develop a probability density function. Thus a $p(x) = cf(x)$ must satisfy

$$1 = \int_0^{\infty} p(x)dx = \int_0^{\infty} cf(x)dx = c\int_0^{\infty} f(x)dx = c\int_0^{\infty} e^{-\lambda x}dx \, .$$

Now use the definition of an improper integral:

$$c \int_0^\infty e^{-\lambda x}dx = c \lim_{h\to\infty} \int_0^h e^{-\lambda x}dx = \frac{c}{\lambda} \lim_{h\to\infty} \int_0^h e^{-\lambda x}(\lambda dx)$$

$$= \frac{c}{\lambda} \lim_{h\to\infty} [-e^{-\lambda x}]_0^h = \frac{c}{\lambda} \lim_{h\to\infty} [1 - e^{-\lambda h}] = \frac{c}{\lambda} .$$

For $\frac{c}{\lambda}$ to equal 1, $c = \lambda$. Thus $p(x) = cf(x) = \lambda e^{-\lambda x}$, $x \geq 0$ has the properties required of a probability density function. If x has this distribution, show that $P[0 \leq x \leq 1] = 1 - e^{-1}$ and $P[x \geq 2] = e^{-2}$. The areas corresponding to these two probabilities are shaded in Figure 9.13. ▌

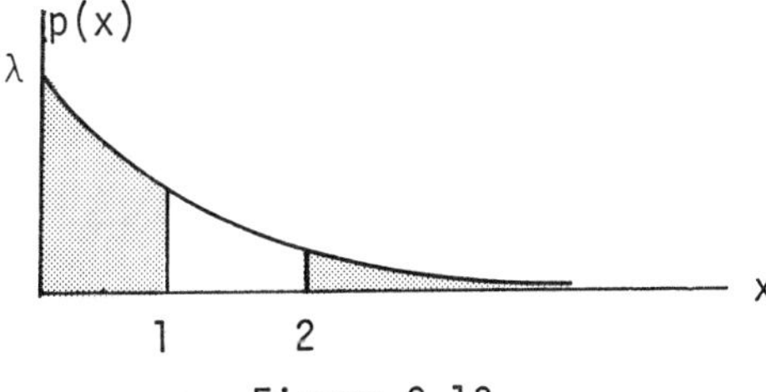

Figure 9.13

**9.4a THE MEAN AND VARIANCE OF A PROBABILITY DISTRIBUTION

Probability distributions have many shapes and forms. A direct comparison of the mathematical form of two probability density functions may not reveal how they are similar or how they differ. If we can define some important characteristics of probability distributions, then we can compare any two by comparing their important characteristics. Two important characteristics locate the center of a distribution and provide a measure of how concentrated the probability is about this center. Each of these characteristics can be defined in several ways, but here we give the most widely used two: The mean and variance.

You probably have encountered some of these terms: Average value, mean value, expected value, or expectation. You might not be able to precisely define them, but your attempts probably would touch on the idea of describing the general size of variable responses. For our purposes, these terms are near enough synonymous to regard them as describing one idea. Once this idea is expressed as an integral, it can be extended to a wider class of situations:

Suppose the random variable x has p(x) as its probability density function so

$$P[a \leq x \leq b] = \int_a^b p(x)dx .$$

- The EXPECTED VALUE or MEAN of x is given by

$$\mu = E(x) = \int_{-\infty}^{\infty} x\ p(x)dx . \qquad (9.11)$$

- The expected value of $f(x)$ is given by

$$E(f(x)) = \int_{-\infty}^{\infty} f(x)\ p(x)dx \qquad (9.12)$$

- The VARIANCE of x is obtained by taking $f(x) = (x - E(x))^2 = (x - \mu)$

$$\sigma^2 = var(x) = E(x - \mu)^2 = \int_{-\infty}^{\infty} (x - \mu)^2 p(x)dx . \qquad (9.13)$$

The average of a group of numbers is calculated by adding up the numbers and dividing this sum by the number of numbers summed. An expected value does the same thing for a random variable which follows a probability distribution. The values of x where p(x) is high contribute more to the expected value than do those values where p(x) is low; this corresponds to likely data values contributing more frequently to an average than unlikely values. An average involves division by n, but an expected value requires no division because the standardizatinn is built into p(x) with the condition $\int_{-\infty}^{\infty} p(x)dx = 1$. The two notations μ and $E(x)$ are used interchangeably to denote the expected value of a random variable, with μ often called its mean value. You might expect that

$$P[x \leq \mu] = \frac{1}{2} = P[x \geq \mu] ,$$

but this does not necessarily hold. This does hold for many probability distributions, in particular it holds for symmetric probability distributions, namely, ones satisfying $p(\mu - v) = p(\mu + v)$.

The idea of expectation extends to any function of a random variable. The variance is defined as the expected squared deviation of a random variable from its mean. Variance measures the dispersion of the probability distribution in this sense: A random variable whose probability is concentrated near its mean will have a smaller variance than a random variable whose probability is more spread out. To see this, consider random variables with probability densities as indicated in Figure 9.14, where both random variables have the same mean. In the integral $\sigma^2 = \int (x - \mu)^2 p(x)dx$, $(x - \mu)^2$ will be the same for both densities, but p(x) differs.

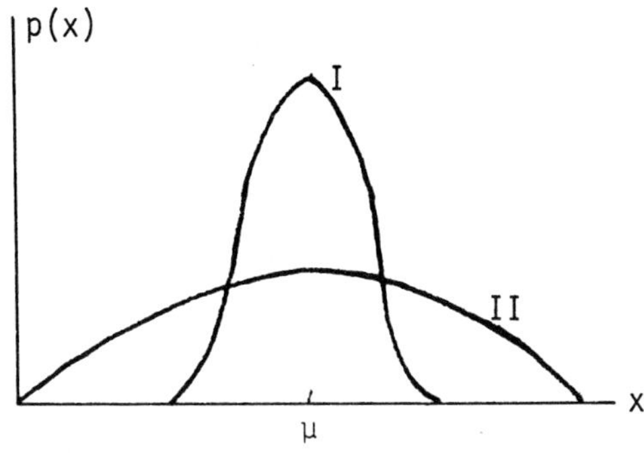

Figure 9.14

For small values of $(x - \mu)^2$, the first density is relatively much larger than for the second, but the reverse is true for large values of $(x - \mu)^2$. Thus $(x - \mu)^2\, p(x)$ remains of modest size for the first density, but becomes relatively larger for some values of x under the second. This implies that the first random variable has a smaller variance than the second. This argument does not really depend on the two random variables having the same mean because their respective means would be subtracted if they differed.

To evaluate variances, it will be convenient to know that

$$\sigma^2 = E(x - \mu)^2 = E(x^2) - [E(x)]^2 = E(x^2) - \mu^2 . \qquad (9.14)$$

To see this, expand $(x - \mu)^2$ and remember that

$$\int(f(x) + g(x))dx = \int f(x)dx + \int g(x)dx \qquad \text{and} \qquad \int kf(x)dx = k\int f(x)dx:$$

$$\sigma^2 = E(x - \mu)^2 = E(x^2 - 2x\mu + \mu^2)$$

$$= \int(x^2 - 2x\mu + \mu^2)p(x)dx = \int x^2 p(x)dx - 2\mu \int x\, p(x)dx + \mu^2 \int p(x)dx .$$

Now recall that $E(x^2) = \int x^2\, p(x)dx$, $\mu = E(x) = \int x\, p(x)dx$ and $\int p(x)dx = 1$ so

$$\sigma^2 = E(x^2) - 2\mu(\mu) + \mu^2(1) = E(x^2) - \mu^2 .$$

<u>EXAMPLE 9.16</u> (continued): Earlier we established that

$$p(x) = \begin{cases} \frac{4}{27}(3x^2 - x^3) & 0 \le x \le 3 \\ 0 & \text{otherwise} \end{cases}$$

is a probability density function. If a random variable x has this as its probability density function, Equation 9.11 gives

$$\mu = E(x) = \int_{-\infty}^{\infty} x\ p(x)\text{-}x = \frac{4}{27}\int_0^3 x(3x^3 - x^2)dx$$

$$= \frac{4}{27}\left[\frac{3x^4}{4} - \frac{x^5}{5}\right]_0^3 = \frac{4}{27}\left[\frac{(3)(81)}{4} - \frac{243}{5}\right] = 4\left[\frac{9}{4} - \frac{9}{5}\right] = \frac{9}{5}$$

The limits on the integral can be restricted to those values for which p(x) is positive. We can evaluate var(x) as either $E(x - \mu)^2$ or as $E(x^2) - \mu^2$. The first gives

$$\text{var}(x) = E\left[x - \frac{9}{5}\right]^2 = \frac{4}{27}\int_0^3 \left[x - \frac{9}{5}\right]^2(3x^2 - x^3)dx$$

$$= \frac{4}{27}\int_0^3 \left[x^2 - \frac{18}{5}x + \frac{81}{25}\right](3x^2 - x^3)dx$$

$$= \frac{4}{27}\int_0^3 \left[\frac{243}{25}x^2 - \frac{351}{25}x^3 + \frac{33}{5}x^4 - x^5\right]dx$$

$$= \frac{4}{27}\left[\frac{81}{25}x^3 - \frac{351}{100}x^4 + \frac{33}{25}x^5 - \frac{1}{6}x^6\right]_0^3$$

$$= \frac{4}{300}[(81)(12) - (351)(9) + (33)(108) - (50)(27)] = \frac{9}{25}.$$

To use the other approach, first evaluate $E(x^2)$:

$$E(x^2) = \frac{4}{27}\int_0^3 x^2\ (3x^2 - x^3)dx = \frac{4}{27}\left[\frac{3}{5}x^5 - \frac{1}{6}x^6\right]_0^3$$

$$= \frac{4}{27}\left[\frac{3}{5}3^5 - \frac{1}{6}3^6\right] = (4)(3^3)\left[\frac{1}{5} - \frac{1}{6}\right] = \frac{18}{5}$$

Now use $\text{var}(x) = E(x^2) - \mu^2$, to obtain

$$\text{var}(x) = \frac{18}{5} - \left[\frac{9}{5}\right]^2 = \frac{90 - 81}{25} = \frac{9}{25}. \quad ▮$$

EXAMPLE 9.17 (continued): We have established that $p(x) = \lambda e^{-\lambda x}$ is a probability density function. Find the expected value and variance of a random variable (x) which follows this exponential probability distribution.

$$\mu = E(x) = \int_0^\infty x\ p(x)dx = \int_0^\infty x\lambda e^{-\lambda x}dx$$

$$= \frac{1}{\lambda}\int_0^\infty (\lambda x)^{2-1}\ e^{-\lambda x}(\lambda dx) = \frac{1}{\lambda}\int_0^\infty u^{2-1}\ e^{-u}du = \Gamma(2)\ ,$$

where $u = \lambda x$. As $\Gamma(2) = 1\Gamma(1) = 1$, $\mu = E(x) = \frac{1}{\lambda}$. To find the variance of x, we first evaluate $E(x^2)$:

$$E(x^2) = \int_0^\infty x^2\ p(x)dx = \int_0^\infty x^2\ \lambda e^{-\lambda x}dx$$

$$= \frac{1}{\lambda^2}\int_0^\infty (\lambda x)^{3-1}\ e^{-\lambda x}(\lambda dx) = \frac{1}{\lambda^2}\int_0^\infty u^3\ e^{-u}du = \Gamma(3)\ .$$

Now as $\Gamma(3) = 2\Gamma(2) = (2)(1)\Gamma(1) = 2$, $E(x^2) = \frac{2}{\lambda^2}$ and so

$$\text{var}(x) = E(x^2) - \mu^2 = \frac{2}{\lambda^2} - \frac{1}{\lambda}^2 = \frac{1}{\lambda^2}\ .\ \blacksquare$$

The next subsection introduces three other important probability distributions and further illustrates the role of the expected value and the variance of a random variable. In this sense the entire subsection can be regarded as a continuing series of examples.

**9.4b IMPORTANT CONTINUOUS PROBABILITY DISTRIBUTIONS

1) UNIFORM DISTRIBUTION

The density function of the uniform probability distribution is given by

$$p(x) = \begin{cases} 0 & x < a \\ \frac{1}{b - a} & a \leq x \leq b \\ 0 & x > b \end{cases} \tag{9.15}$$

where a and b are constants satisfying $b > a$. The density function is shown in Figure 9.15.

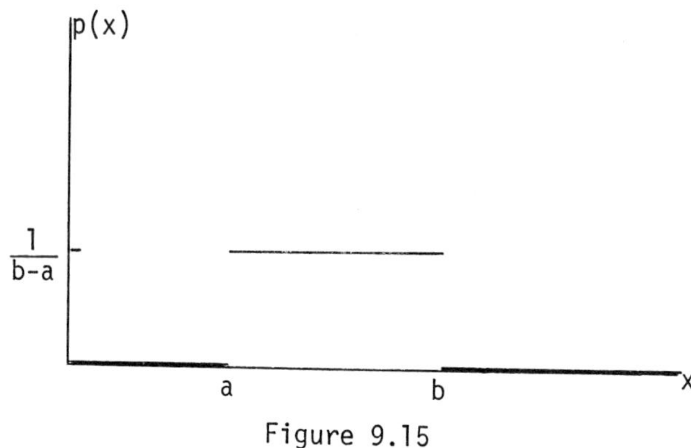

Figure 9.15

Does the above function satisfy the criteria necessary for a probability density function? It never becomes negative because b > a. Further,

$$\int_{-\infty}^{\infty} p(x)dx = \int_a^b \frac{dx}{b - a} = \frac{1}{b - a} x\Big]_a^b = \frac{1}{b - a} [b - a] = 1 ,$$

as required. Thus, the function given in Equation 9.15 can be regarded as a probability density function.

The uniform probability distribution assigns a probability of zero to any interval of values outside of [a,b]. Any interval inside [a,b] has a probability proportional to its length. Thus the uniform distribution is used whenever a random outcome has clearly known boundaries, but no pattern otherwise. Where does this occur in biology? It is used widely in studying breakage of objects such as chromosomes and DNA strands when subjected to radiation.

Now find the mean and variance of a random variable x following uniform probability distribution:

$$E(x) = \mu = \int_{-\infty}^{\infty} x\ p(x)dx = \int_a^b \frac{x}{b - a} dx$$

$$= \frac{1}{b - a} \frac{x^2}{2}\Big]_a^b = \frac{b^2 - a^2}{2(b - a)} = \frac{b + a}{2} .$$

Thus, $\mu = E(x)$ is located halfway between the ends of this symmetric distribution. The variance will be evaluated using both Equations 9.13 and 9.14. First use $\mu = \frac{(a + b)}{2}$:

$$\sigma^2 = \text{var}(x) = E\left[x - \frac{a + b}{2}\right]^2 = \int_a^b \left[x - \frac{a + b}{2}\right]^2 \frac{1}{b - a}\, dx$$

$$= \frac{1}{3(b - a)} \left[x - \frac{a + b}{2}\right]^3 \Bigg]_a^b$$

$$= \frac{1}{3(b - a)} \left[b - \frac{a + b}{2}\right]^3 - \left[a - \frac{a + b}{2}\right]^3 = \frac{(b - a)^2}{12} .$$

To use Equation 9.14 instead, we first need $E(x^2)$:

$$E(x^2) = \int_a^b x^2 \frac{1}{b - a}\, dx = \frac{1}{3(b - a)} x^3 \Big]_a^b = \frac{b^3 - a^3}{3(b - a)} = \frac{1}{3} (b^2 + ab + a^2) .$$

When this is combined with $E(x) = \frac{(a + b)}{2}$ we find

$$\sigma^2 = \text{var}(x) = E(x^2) - \mu^2 = \frac{b^2 + ab + a^2}{3} - \left[\frac{a + b}{2}\right]^2$$

$$= \frac{4(b^2 + ab + a^2) - 3(a^2 + 2ab + b^2)}{(3)(4)}$$

$$= \frac{b^2 - 2ab + a^2}{12} = \frac{(b - a)^2}{12} ,$$

the same result as the first approach gave.

This mean and variance simply illustrate what a mean and variance should do. The mean $\frac{a + b}{2}$ lies precisely in the middle of the interval: $a + \frac{b - a}{2} = \frac{a + b}{2}$. This tells nothing about the spread of the distribution for uniform distributions over [0,20], [5,15], [9,10] or [9.9,10.1] all have the same mean. The variance, $(b - a)^2/12$ clearly changes with the length of the interval. Thus, it reflects one aspect of dispersion.

2) EXPONENTIAL DISTRIBUTION

The probability density function

$$p(x) = \lambda e^{-\lambda x} , \qquad x > 0 , \tag{9.16}$$

$\lambda > 0$, specifies the exponential probability distribution. It was extensively illustrated in the earlier part of this section and in Section 8.8b. This probability distribution is applied to a wide class of survival problems. For instance Example 8.30 illustrated its use on the survival of lygus bugs.

We have already shown (Example 9.17) that if x follows the exponential probability distribution with parameter λ, then $E(x) = 1/\lambda$ and $\mathrm{var}(x) = 1/\lambda^2$.

3) GAMMA DISTRIBUTION

The gamma distribution is defined by the density function

$$p(x) = \begin{cases} 0 & x < 0 \\ \dfrac{x^{\alpha-1}e^{-x/\beta}}{\Gamma(\alpha)\beta^{\alpha}} & x \geq 0 \end{cases} \qquad (9.17)$$

The gamma distribution depends upon two parameters α and β; both must be positive for Equation 9.17 to represent a probability density function. Under these conditions, $p(x) \geq 0$ and

$$\int_{\infty}^{\infty} p(x)dx = \frac{1}{\Gamma(\alpha)\beta^{\alpha}} \int_0^{\infty} x^{\alpha-1}e^{-x/\beta}dx = 1 \ . \qquad (9.18)$$

The details of this verification are left for an exercise.

The gamma generalizes the exponential in this sense: $\alpha = 1$ and $\beta = 1/\lambda$ gives the exponential probability distribution previously considered. This appears in Figure 9.16 which shows the shape of the gamma probability density function for the values of $\alpha = \frac{1}{2}$, 1, 1½ and 2.

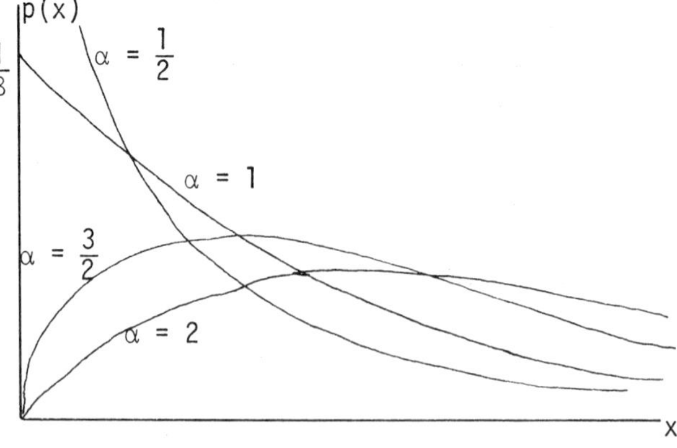

Figure 9.16

Observe that the shapes of the gamma probability distributions are similar to some of the general shapes of biological curves we have seen earlier. Thus, some biological data can be described by fitting a gamma probability distribution to it. This can be done using the method of least squares as mentioned earlier in regard to curve fitting.

The gamma probability distribution is important for several reasons. First, its shape ranges from more extreme than the exponential to the nearly normal distributions. The exponential decreases away from its maximum at its left boundary. The normal, to be discussed next, is symmetric about its mean. For most parameter values the gamma rises to its maximum fairly fast, then has a long tail to the right. The parameter α fixes this shape; the parameter β shrinks or stretches out the shape specified by α along the x-axis.

The gamma probability distribution often arises in studies of time until the occurrence of an event, like the hatching of larvae.

The evaluation of the mean and variance of a random variable following the gamma distribution is left as an exercise.

4) NORMAL DISTRIBUTION

The normal probability distribution is used widely by biologists. It applies to a variety of responses which measure some aspect of size such as weight, length, or yield from a plot. We will explore its mathematical properties shortly, but first look at Figure 9.17 to see its shape. This has a bell shape located over 0. Note that it is symmetric about $x = 0$: $p(x) = p(-x)$. This probability density function has most of its probability near zero with only small amounts of probability outside of $x = -3$ or 3.

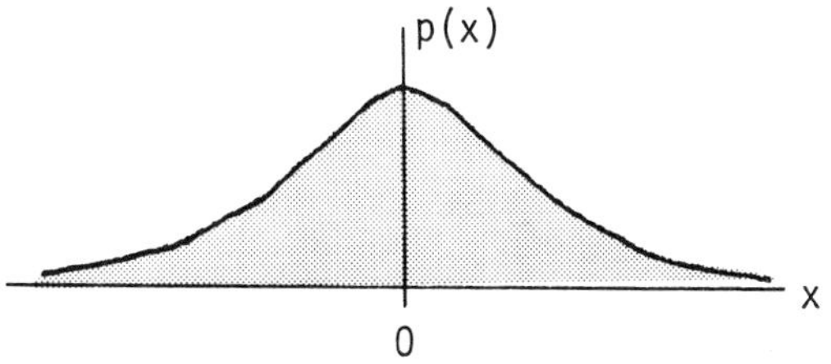

Figure 9.17

If x has the probability distribution shown in Figure 9.17, it satisfies $E(x) = 0$ and $\text{var}(x) = 1$. Many important responses have a mean and variance different from these values. Several such distributions appear in Figure 9.18. Changes in μ move the distribution along the x-axis. The parameter σ, or equivalently σ^2, fixes how closely the probability is concentrated about μ.

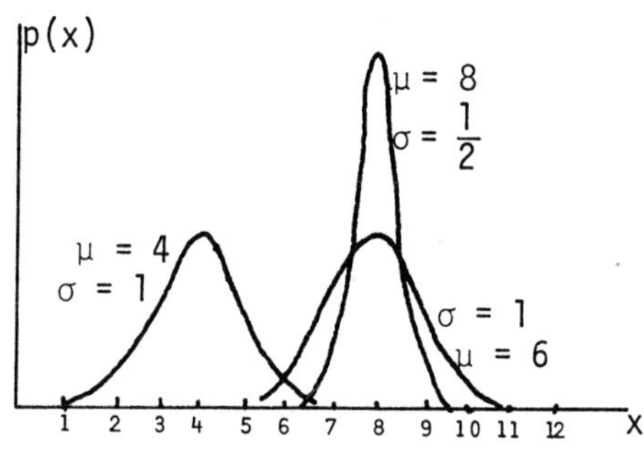

Figure 9.18

The density function of the normal probability distribution is given by

$$p(x) = \frac{1}{\sqrt{2\pi}\,\sigma} \exp - \frac{(x - \mu)^2}{2\sigma^2} \qquad \sigma > 0 \tag{9.19}$$

where μ and σ^2 are constants. The constant μ is the mean of the distribution and σ^2 is the variance of the distribution.

This function satisfies the properties of a probability density function: $p(x) > 0$, because e raised to any power is positive. Demonstration that $\int_{-\infty}^{\infty} p(x)dx = 1$ requires more effort. To show this, let $u = \frac{x - \mu}{\sqrt{2}\,\sigma}$ so $du = \frac{dx}{\sqrt{2}\,\sigma}$ or $dx = \sqrt{2}\,\sigma\, du$ and

$$\frac{1}{\sqrt{2\pi}\,\sigma} \int_{-\infty}^{\infty} \exp\left[- \frac{(x - \mu)^2}{2\sigma^2}\right] dx = \frac{1}{\sqrt{\pi}} \int_{-\infty}^{\infty} e^{-u^2} du = \frac{2}{\sqrt{\pi}} \int_0^{\infty} e^{-u^2} du$$

with the latter equality resulting because the integrand is symmetric about $u = 0$. Another change of variable will relate this to the gamma distribution: Let $v = u^2$ so $dv = 2u$ or $du = \frac{dv}{2\sqrt{v}}$ and

$$\frac{1}{\sqrt{2\pi}\,\sigma} \int_{-\infty}^{\infty} \exp - \frac{(x - \mu)^2}{2\sigma^2}\, dx = \frac{2}{\sqrt{\pi}} \int_0^{\infty} e^{-u^2} du = \frac{2}{\sqrt{\pi}} \frac{1}{2} \int_0^{\infty} v^{1/2-1} e^{-v} dv = \frac{1}{\sqrt{\pi}} \Gamma\left(\frac{1}{2}\right) = 1 ,$$

demonstrating the second property required of a probability density function.

We will demonstrate that $E(x) = \mu$, but leave $var(x) = \sigma^2$ for you. One approach uses the identity $x = (x - \mu) + \mu$:

$$E(x) = \int_{-\infty}^{\infty} x\, p(x)dx = \frac{1}{\sqrt{2\pi}} \int_{-\infty}^{\infty} [(x - \mu) + \mu] \exp\left[- \frac{(x - \mu)^2}{2\sigma^2}\right]dx$$

$$= \frac{1}{\sqrt{2\pi}} \int_{-\infty}^{\infty} (x - \mu) \exp\left[- \frac{(x - \mu)^2}{2\sigma^2}\right]dx + \mu \frac{1}{\sqrt{2\pi}} \int_{-\infty}^{\infty} \exp\left[- \frac{(x - \mu)^2}{2\sigma^2}\right]dx \; .$$

Use the substitution $u = x - \mu$ in the first integral and recall that $\int_{-\infty}^{\infty} p(x)dx = 1$ for the second:

$$E(x) = \frac{1}{\sqrt{2\pi}} \int_{-\infty}^{\infty} u \exp\left[- \frac{u^2}{2\sigma^2}\right]du + \mu \; .$$

The first integral equals zero because

$$\int_{-a}^{0} u \exp\left[- \frac{u^2}{2\sigma^2}\right]du = - \int_{0}^{a} u \exp\left[- \frac{u^2}{2\sigma^2}\right]du \; .$$

Check to see that this is true. Why does this imply that the first integral is zero? Thus the parameter μ in the normal probability distribution equals the expected value of a random variable which follows the normal probability distribution.

EXERCISES

1. Show that $var(x) = \sigma^2$ if x follows the normal probability distribution. You will find it easier to use $var(x) = E(x - \mu)^2$ than $var(x) = E(x^2) - \mu^2$.

2. Show that the gamma probability distribution satisfies $\int_{-\infty}^{\infty} p(x)dx = 1$.

3. Suppose x varies according to a gamma probability distribution. Find $E(x)$ and $var(x)$.

4. Can a random variable following a uniform probability distribution have an expected value of 0? If so, how? If not, why not?

5. Find the mean and variance for a random variable following a uniform probability distribution if

   a. $0 \leq x \leq 10$, b. $1 \leq x \leq 7$, c. $-2 \leq x \leq 2$.

6. Find the maximum and points of inflection for the normal distribution.

Chapter 10

MULTIDIMENSIONAL RELATIONS

Until now we have considered functions of only one independent variable. In reality, though, most interesting quantities depend on several independent variables. For example, fisheries biologists sometimes need to know how body weight of a particular species of fish depends on other identifiable variables. Length would be an important variable, but some of age, water temperature, availability of food, predatory pressure, etc., also could have an important role.

Our examination of functions of one variable began with the functions themselves (Chapter 4), introduced limits (Chapter 5), mainly as a prelude to derivatives (Chapters 6 and 7), and concluded with integrals (Chapter 8). We will follow a very similar route in dealing with functions of several variables. This chapter introduces the functions and presents visualizations of them as surfaces in three-dimensional space. The next (Chapter 11) will broaden the ideas of limits and derivatives, while the following one (Chapter 12) generalizes the idea of integration to several independent variables. These chapters merely extend familiar ideas so you should expect briefer explanations than before. As you begin each of the next three chapters, review the corresponding material in Chapters 4-8.

This chapter introduces functions of several independent variables with particular attention to the case of two independent variables. Graphs served a powerful role in providing a visualization of the behavior of a function of one independent variable. Here the generalization involves points, lines, curves, planes and surfaces in three-dimensional space. Finally, we will explain extensions to more than two independent variables, but not concentrate on them because no precise physical visualization of them exists within familiar frameworks.

10.1 WHAT ARE FUNCTIONS OF MORE THAN ONE VARIABLE?

Back in Chapter 4, we introduced the idea of a relation between elements in a *domain* set and a *range* set. A function was then defined as a relation having exactly one element in the range for each point in the domain. We observed that a set of *ordered pairs* can be used to state a function's correspondences. Do we really need a broader idea of a function? Nothing in this definition precludes elements in the domain set from being ordered pairs themselves. Thus we present the following definitions:

- A FUNCTION f of two independent variables x and y, sometimes denoted by $z = f(x,y)$, establishes a correspondence between elements in the domain set D and elements in the range set R such that each point $(x,y) \varepsilon D$ has exactly one corresponding point $z \varepsilon R$.
- A function of two independent variables can be specified by a set of ORDERED TRIPLETS $(x,y,z) = (x,y,f(x,y))$.

Figure 10.1 depicts a function of two independent variables where

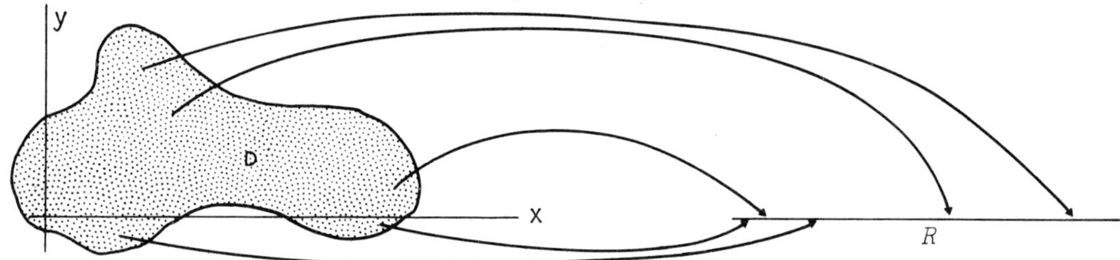

Figure 10.1

the domain set D consists of pairs of real numbers (x,y) and the range set R consists of a set of real numbers. This represents a common situation where the range and domain sets involve real numbers. Nothing in the definition, however, requires this. For example, the domain could consist of pairs of names of fathers and mothers. The function could give the name of their firstborn child, or it could give the child's weight. In the latter case it would be a numerical valued function of names. We will focus on numerical-valued functions of numerical arguments because differential and integral calculus can be extended readily to them; realize, nevertheless, that some applications require other kinds of functions.

We should comment briefly about notation. The letters x, y, and z have been used to denote the elements in the domain and range sets in these general discussions. As with functions of one independent variable, suggestive letters should be used when possible. Also note the change in the role of the letter y; before, it denoted the function or dependent variable while it now denotes one of the independent variables. This change of role, although fairly standard, confuses some students. The change arises from a common practice of labeling variables by consecutive letters with the dependent variable occurring last, unless more natural labels exist.

EXAMPLE 10.1: The area of a rectangle depends on its length and its width through

$$A = LW . \tag{10.1}$$

Here L and W suggestively symbolize length and width, the independent variables, and A symbolizes area, the dependent variable. We could write $A = f(L,W)$ where $f(L,W) = LW$. ∎

EXAMPLE 10.2: Another simple example involves the Fick method for directly measuring cardiac output. It operates from this observation: Carbon dioxide released by the lungs (x) can be measured, as can be the change (y) in carbon dioxide content of the blood leaving the lungs from when it entered. Because the blood flow through the lungs must equal cardiac output (C), we have

$$C = \frac{CO_2 \text{ absorbed per minute by the lungs}}{(\text{Venous } CO_2 \text{ per ml. of blood}) - (\text{Arterial } CO_2 \text{ per ml. of blood})}$$

$$= \frac{x}{y} . \tag{10.2}$$

For example, if venous blood coming to the lungs contained 58 cc. of carbon dioxide per 100 cc. of blood, but the arterial blood leaving the lungs had 52 cc., then 6 cc. of carbon dioxide per 100 cc. blood had to be given off by the blood as it passed through the lungs. A normal, resting person may exhale 300 cc. of carbon dioxide per minute, thus giving cardiac output of

$$C = \frac{x}{y} = \frac{300 \text{ cc. } CO_2/\text{minute}}{6 \text{ cc. } CO_2/100 \text{ cc. blood}} = 50(100 \text{ cc. blood})/\text{minute}$$

$$= 5000 \text{ cc. blood/minute}$$

$$= 5 \text{ liters blood/minute.}$$

Since the result x/y has units of 100 cc. blood/minute, Equation 10.2 commonly is written as

$$C = \frac{100x}{y} . \tag{10.3}$$

This equation specifies C as a function of the two independent variables x and y through the function $C = f(x,y) = 100x/y$. The domain set consists of all pairs (x,y) of permissible values for x and y. Under even unusual conditions, humans expel carbon dioxide, but strenuous activity by a large, vigorous athlete produces no more than 3000 cc. per minute; change in CO_2 concentration must lie between 0 and 100 because it is measured per 100 cc.. Thus,

$$D = \{(x,y)|0 < x < 3000,\ 0 < y < 100\},$$

and, in turn,

$$R = \{C|0 < C < \infty\} .$$

Even though this domain and range include all possible points, they include a very large number of points which cannot exist for a living individual. ▌

Here, x and y have been called independent variables, while C was called the dependent variable. These are used as other names for the FUNCTION'S ARGUMENTS and the FUNCTION'S VALUE at a point in its domain of definition. These names do not imply, nor should they suggest, mechanistic dependence or independence. In the actual biological situation, they all move together in response to activity levels, body size and characteristics, and ambient conditions. Two fairly accessible variables merely are being used to evaluate a third, inaccessible variable.

At the beginning of this section, we gave the definition of a function of two independent variables. We now give the definition of a function of several independent variables.

- If D is the domain in which $x_1, x_2, \cdots, x_n$ may vary independently, and, if a unique value of f is assigned to each point of this region, then we say that f is a function of the n independent variables $x_1, x_2, \cdots, x_n$ and denote the value of this function by $f(x_1, x_2, \cdots, x_n)$.

10.2 THE RECTANGULAR COORDINATE SYSTEM

We used graphs to provide a visualization of functions of one variable. A common and useful way to visualize functions of two variables uses three-dimensional representations, that is like the length-width-height space we live in. Earlier we drew a horizontal axis for the independent variable and, at right angles, a vertical axis for the function's value. Now, we need two axes for the two independent variables and a third for the dependent variable. These lie at right angles to each other in space.

In an effort to help you visualize three-dimensional space, we suggest that you find a corner, preferably with a floor covered with square tiles, and go sit in it. First, look at the floor; the domain will lie there. Look at where the wall on your right meets the floor. Take this as the positive x-axis, the junction of the other wall with the floor as the positive y-axis, and the intersection of the two walls as the z-axis. Look back to the floor; the joints between the tiles provide a grid system for locating points (x,y) on the floor. For example, (1,0) lies along the right wall, one tile out from the left wall, while to locate (3,4), come out to the third joint along the right wall and then out into the room four tile joints. Now, for each point (x,y), think of coming up a distance f(x,y). A surface results when all of these points are connected. For example, hold a sheet of paper by an edge. By holding it different ways and using various thicknesses of paper, you can make many different surfaces, each representing a function.

Now in this visualization, what does it mean to be a function? Does any surface represent a function? For a surface to describe a function, each point (x,y) in the domain must have exactly one point above it. This means that the surface cannot go straight up above any point on the floor, neither can the surface fold back on itself.

Remember that we numbered the four quadrants of our two-dimensional axis system I, II, III, IV counterclockwise. So far, we have concentrated on the extension of quadrant I, the positive OCTANT where $x > 0$, $y > 0$, $z > 0$. Each of the planes we described, the two walls and the floor, extend in all directions forever, at least conceptually. If you are sitting above a basement, those points below the floor have $z = f(x,y) < 0$. Although this cannot happen for the two functions given by Equations 10.1 and 10.2, we have seen negative functions of one independent variable and will see negative functions of two independent variables. Points through the wall to your right have $y < 0$, and through the wall to the left have $x < 0$.

Finally you need to understand how some features of a three-dimensional object can be communicated by a figure in two dimensions. A skillful artist can use many techniques to introduce an illusion of depth. Shading and perspective prove very effective, with the latter using diminishing size for objects meant to look further and further from the viewer. Successful utilization of such techniques requires skill and practice that few of us have. We will use a simpler

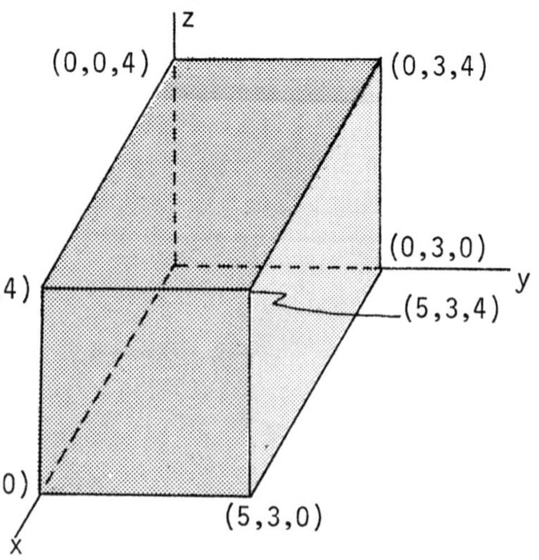

Figure 10.2

technique to communicate depth. It utilizes y and z axes perpendicular to each other as before, but the x-axis drawn obliquely as illustrated in Figure 10.2. Plotting of the point (5,3,4) is illustrated in Figure 10.2 along with some of its associated points. To plot the point (5,3,4), first recognize that we are being asked to plot the ordered triplet (x,y,z). We move along the x-axis five units, draw a line parallel to the y-axis and then move in positive y direction three units. At this point, we construct a line perpendicular to the x,y-plane and move up the line in the positive z direction four units. We have now plotted the point (5,3,4).

Another approach is used to communicate the three-dimensional character of a surface in only two dimensions. Think of slicing the three-dimensional surface with various planes. The surface will interesct a plane in a curve, often called the TRACE of the surface on the plane. In particular, intersections with the coordinate planes are called x,y-trace, x,z-trace, or y,z-trace, respectively. Now consider the trace of a surface on the plane $z = 1$, parallel to the (x,y) plane, but one unit higher. The plane will intersect the surface in a trace which can be drawn in two dimensions. If the surface is sliced again at $z = 2$, $z = 3$, $\cdots$, then the resulting traces can be drawn on the same two-dimensional graph as the one from $z = 1$. These curves are called CONTOURS of the surface. Cartographers use this technique in making maps, called contour maps, which show how the elevations vary in the area mapped. A flat, small area may have contours for every one-foot change in elevation while large or mountainous areas may have contours only for 100' changes in elevation. Similarly, the values of z for which contours of the function are drawn will depend on the function, and if clearly labeled, do not have to be equally spaced relative to z.

EXAMPLE 10.1 (continued): This example concerned the area (A) in a rectangle of length L and width W: $A = LW$. This function of the two variables L and W can be graphed as in Figure 10.3. The surface is described by the ordered triplet (L,W,LW) which lies a distance LW above the point (L,W) in the L,W-plane. In particular, the point (3,5,15) and associated points are indicated therein.

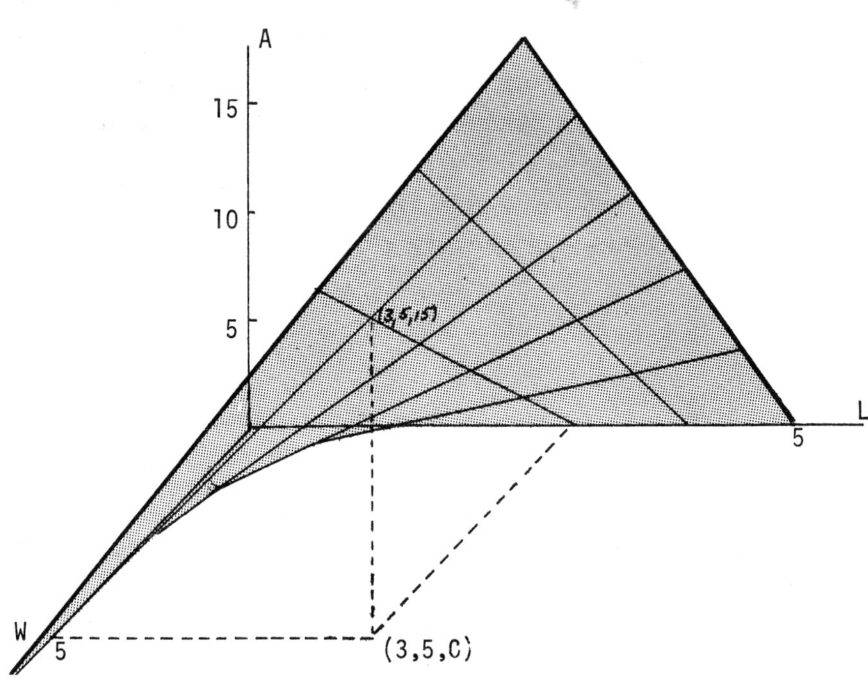

Figure 10.3

Slide a piece of paper into the corner of the room and then raise the corner of the paper farthest from the corner while holding the other corners down. This paper should give you some idea of what the surface looks like, even though this representation does not exactly mimic the surface. When you look at this construction, you cannot see the lowest part of the surface which lies in the corner. Likewise, in Figure 10.3, the low part near the corner lies behind the rising front of the surface.

To think of the surface another way, consider how it looks above selected lines in the L,W-plane. Part of the line L = 3 is drawn in the L,W-plane. Above this line, LW = 3W so the surface goes up as a straight line as indicated there. Similarly, the surface goes up as a straight line above lines such as (L,5,0) parallel to the L-axis. Even though the surface can be regarded as a collection of straight lines, it appears to curve. Why? Look at how the surface behaves above (L,L,0), the 45° line coming out from the corner in the

L,W-plane. Above it, the surface follows (L,L,L^2), a parabola. Now think back to the corner of the room. Suppose you had a wire shaped like half of a parabola. Place the vertex end into the corner and hold the other end at the appropriate height, equally distant from the two walls. Now take something straight like a pencil or a wooden dowel, put one end where the right wall and floor meet and lay it across the wire, keeping it parallel to the other wall. Many of these laid side by side approximate the surface.

The surface can also be described by a contour plot as appears in Figure 10.4. The planes A = 4, 8, 12, 16, intersect the surface in the traces or curves of Figure 10.4. Notice that the contours draw closer together as A increases, but this is most noticeable for large values of either variable. This implies that the surface goes up very steeply for large values of either L or W. ∎

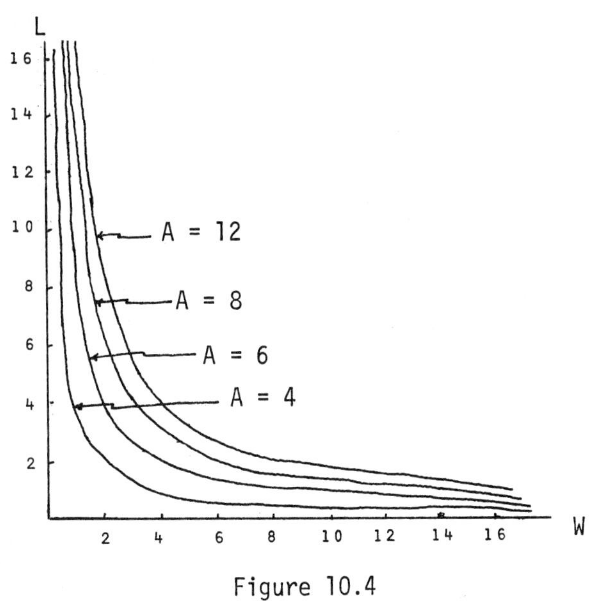

Figure 10.4

EXAMPLE 10.2 (Continued): This example related cardiac output (C) to carbon dioxide released by the lungs (x) and percent drop in carbon dioxide across the lungs (y) through C = 100x/y as given in Equation 10.3. First, the function's contours are drawn in Figure 10.5 and then these are stretched out vertically to give the three-dimensional representation shown in Figure 10.6. (The location of the x and y axes is reversed from before to emphasize the shape of the surface.)

The contours result from finding these combinations of x and y which give the same C. For example C = 4000 = 100x/y implies that

$40y = x$, a straight line from the origin. As y 0, that is, as we approach the C,x-plane, the height of the surface $C = 100x/y$ approaches infinity. Thus Figure 10.6 cannot show the surface for small values of y because they will not fit on the page. ▮

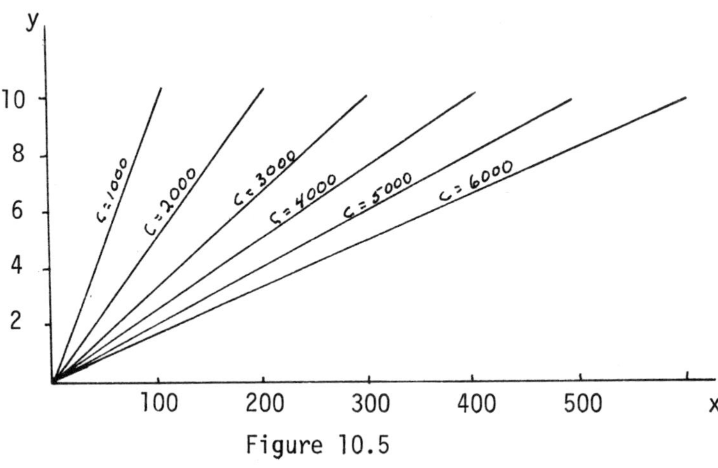

Figure 10.5

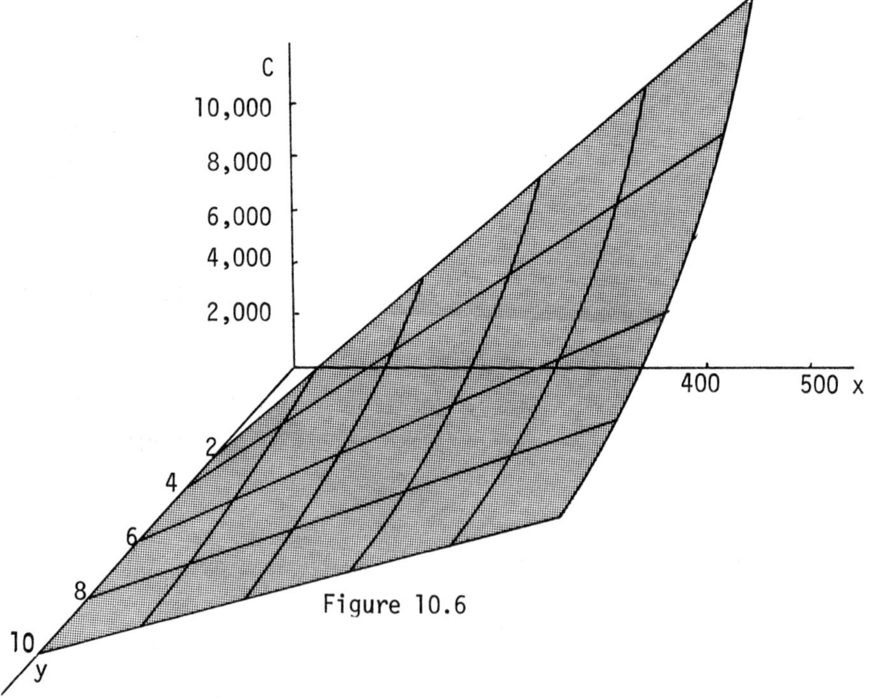

Figure 10.6

EXERCISES

1. Consider the equation $z = 16 - x^2 - y^2$.
   a. Describe it as a function specifying its domain and range.
   b. Describe the surface it defines by a contour plot.
   c. Represent it as a three-dimensional plot.
   d. Describe it in words. Describe it in terms of common objects. (What does it look like?)
2. Repeat Problem 1 using the equation $z = -40 - 5x^2 + 20x - y^2 + 12y$. How has the surface changed from Problem 1? (Try completing the square in both x and y.
3. Repeat Problem 1 using the equation $z = -52 - 5x^2 + 26x - y^2 + 14y - xy$. How has the surface changed from Problems 1 and 2? (Try completing the square in x and y, and in the cross product.)
4. Describe the surface specified by $z = e^{-x-2y}$.
5. If $z = 16 - x^2 - y^2$ defines a function, why doesn't $z^2 = 16 - x^2 - y^2$?
6. Does $z = [\![x + y]\!]$ define a function? Try describing the surface it specifies in terms of wooden blocks. (Recall that $[\![W]\!]$ gives the largest integer in W.)
7. Describe the surface defined by $y = e^{-x^2+2xy-y^2}$.
8. Describe the surface defined by $z = \frac{1}{x} e^{-y^2/x^2}$, $x > 0$.

10.3 DISTANCES, LINES AND PLANES IN SPACE

The Pythagorean Theorem provided the basis for finding the distance between two points in a plane; it has a simple generalization to more dimensions. The lines of two dimensions generalize to lines and planes in three dimensions.

The Pythagorean Theorem states that the length of the hypotenuse of a right triangle equals the square root of sum of squares of the lengths of its other two sides. Specifically this implies that the distance from the origin of a coordinate system (0,0) to a point (x,y) is $\sqrt{x^2 + y^2}$ as shown in Figure 10.7. Figure 10.8 will allow us to generalize this

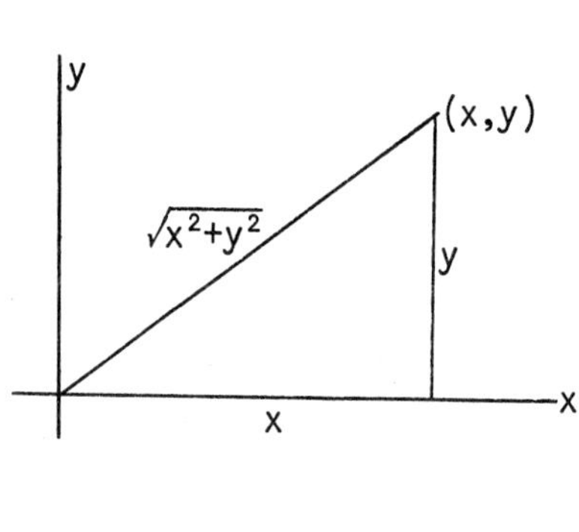

Figure 10.7

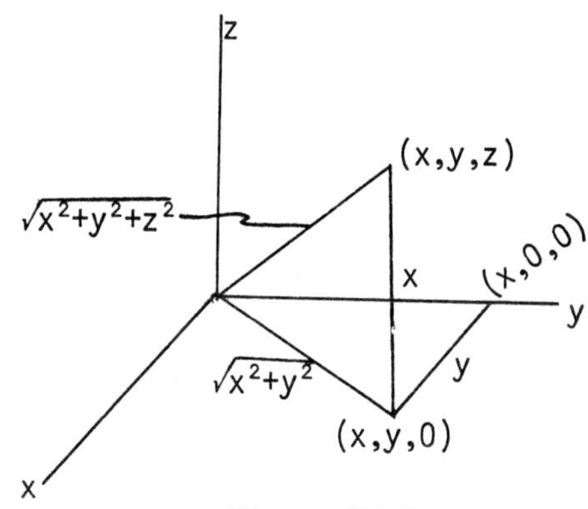

Figure 10.8

distance formula to three-dimensions. The Pythagorean Theorem applies to the right triangle formed by (0,0,0), (x,0,0), (x,y,0) giving the distance from (0,0,0) to (x,y,0) as $\sqrt{x^2 + y^2}$. But the corners (0,0,0), (x,y,0) and (x,y,z) also define a right triangle. A second application of the Pythagorean Theorem implies that the distance from (0,0,0) to (x,y,z) equals $\sqrt{(\sqrt{x^2 + y^2})^2 + z^2} = \sqrt{x^2 + y^2 + z^2}$. More generally, the distance from (x_0,y_0,z_0) to (x,y,z) is given by $\sqrt{(x - x_0)^2 + (y - y_0)^2 + (z - z_0)^2}$.

<u>EXAMPLE 10.3</u>: How far do the points (2,10,11) and (8,4,8) lie from the origin, and from each other?

The distance from the origin to (2,10,11) is

$$\sqrt{2^2 + 10^2 + 11^2} = \sqrt{4 + 100 + 121} = \sqrt{225} = 15 .$$

Similarly, from the origin to (8,4,8) is

$$\sqrt{8^2 + 4^2 + 8^2} = \sqrt{64 + 16 + 64} = \sqrt{144} = 12 .$$

The distance from (2,10,11) to (8,4,8) is

$$\sqrt{(2 - 8)^2 + (10 - 4)^2 + (11 - 8)^2} = \sqrt{(-6)^2 + 6^2 + 3^2} = \sqrt{36 + 36 + 9}$$
$$= \sqrt{81} = 9 .$$

Now what do these distances say about the shape of the triangle with corners at (0,0,0), (8,4,8) and (2,10,11)? How does this triangle lie in space? ∎

A PLANE consists of the set of points satisfying one equation of the form $cx + dy + ez = f$ where c, d, e, and f represent known constants. If $e \neq 0$ such an equation can be manipulated into a standard form to display its functional nature:

$$ez = f - cx - dy \qquad \text{or} \qquad z = (-\frac{c}{e})x + (-\frac{d}{e})y + \frac{f}{e} .$$

By letting $m = -\frac{c}{e}$, $n = -\frac{d}{e}$, $b = \frac{f}{e}$, this becomes

$$z = mx + ny + b . \tag{10.4}$$

This form looks much like the form $y = mx + b$ for a straight line in a two-dimensional coordinate system. The resemblance is intended and really present. The constant b gives the height of the plane above the origin where $x = y = 0$. The two coefficients m and n give the slopes of the traces of the plane in the x,z-plane and y,z-plane, respectively. In the x,z-plane, $y = 0$ so Equation 10.4 reduces to $z = mx + b$, the equation of a straight line of slope m. The other coefficient, n, has the same role relative to y as m has relative to x.

In introducing straight lines in Chapter 4, we gave the general slope-intercept form and then showed how to get that equation from a line specified by two points or by a point and a slope. A plane is fixed by three points, by a line and one point, or by two intersecting lines. Of course this statement assumes the absence of certain degeneracies. Coincident lines cannot determine a plane, nor can three points on the same line. Otherwise the general equation of a plane can be obtained from any defining conditions with a modest amount of algebraic effort. This will be illustrated shortly.

The intersection of two different planes in space defines a LINE. A line is a one-dimensional object in three-dimensional space much as a plane is a two-dimensional object in three-dimensional space. The algebraic specification of a line must restrict it to one dimension. A simple way adds a second linear equation involving only x and y to Equation 10.4, such as:

$$\begin{aligned} z &= mx + ny + b \\ x &= ey + f . \end{aligned} \tag{10.5}$$

The first of these specifies one plane while the second specifies a vertical plane so their intersection defines a line. Only those points which simultaneously satisfy two such equations lie on a line. More generally any pair of intersecting, distinct planes defines a line. A line also can be defined by any two distinct points on it.

The following comprehensive example illustrates these ideas about lines and planes.

<u>EXAMPLE 10.4</u>: Find the plane passing through the three points (1,3,13), (3,5,11) and (5,0,2).

To be a plane it must satisfy $z = mx + ny + b$; to pass through these three points its coefficients must satisfy

$$\begin{aligned} 13 &= m + 3n + b \\ 11 &= 3m + 5n + b \\ 2 &= 5m + b \end{aligned} \tag{10.6}$$

These three equations in three unknowns should define m, n and b. First, we can subtract the third equation from the first two equations to eliminate b:

$$\begin{aligned} 11 &= -4m + 3n \\ 9 &= -2m + 5n \ . \end{aligned}$$

We have now reduced the original system to two equations in two unknowns. Upon multiplication of the second equation by two, we have

$$\begin{aligned} 11 &= -4m + 3n \\ 18 &= -4m + 10n \end{aligned} \tag{10.7}$$

and by subtracting the first equation from the second, we have

$$10n - 3n = 18 - 11 \qquad \text{or} \qquad n = 1 \ .$$

Now we can substitute $n = 1$ into the first equation of Equation 10.7, $11 = -4m + 3n$, to obtain $11 = -4m + 3(1)$ or $m = -2$. Now substituting these values for m and n into the first equation of Equation 10.6, $13 = m + 3n + b$, we find $13 = -2 + 3 + b$ or $b = 12$. Thus the equation of the plane is specified by those points which satisfy the equation $z = -2x + y + 12$.

The plane appears in Figure 10.9. The three given points and the

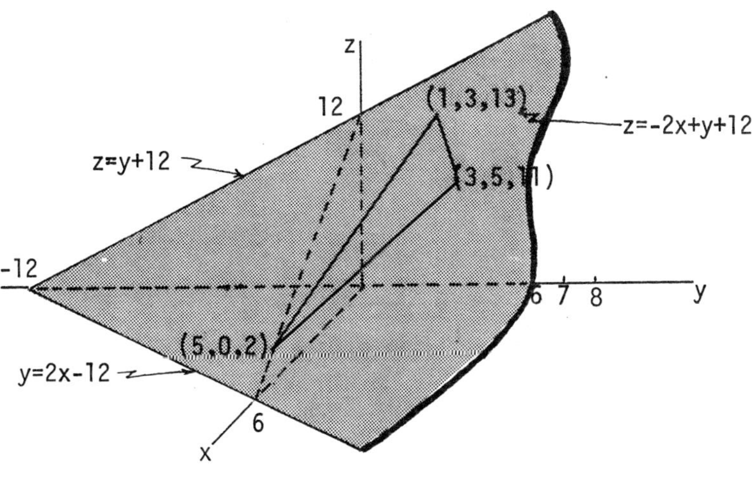

Figure 10.9

lines they determine are shown clearly. The x,z-trace of this plane is $z = -2x + 12$, obtained by taking $y = 0$ in the general equation for the plane. The corresponding y,z-trace and x,y-trace are $z = y + 12$ and $2x = y + 12$.

Only part of the plane is represented in Figure 10.9. The plane extends in all directions. The leading edge of this plane has been accented to emphasize that you are looking directly at an edge on this side, while it disappears through the coordinate planes on the other sides.

The line through (1,3,13) and (3,5,11) clearly lies on the plane $z = -2x + y + 12$ because both of these points do. The points (1,3,0) and (3,5,0) lie in the x,y-plane directly below the points under consideration. Check to see that these points both lie on $y = x + 2$. This specifies both a line in the x,y-plane and a vertical plane in the x,y,z-space. It intersects $z = -2x + y + 12$ in the straight line through (1,3,13) and (3,5,11). Check to see that both of these points lie on both planes. ∎

10.4 FUNCTIONS OF SEVERAL VARIABLES

A function can have any number of arguments or independent variables. Many important variables are influenced by many other variables. For example, the production of a plant depends heavily on the levels of such variables as light, nutrients, temperature, carbon dioxide and leaf surface area. Functions of three or more independent variables present only a slight conceptual extension from the two independent variable case, a modest notational change, but a gross alteration of visual representation. No uniformly suitable method exists for making a visual representation of a function of three or more independent variables because figures cannot be drawn in four or more dimensions. Even the extension of the idea of lines and planes encounters nuisance because, for example, two, three, four, five-dimensional planes exist in six-dimensional space.

A thorough treatment of these extensions leads to the idea of linear algebra, a topic worthy of whole books in its own right. When a biologist becomes serious about mathematics, he often learns about linear algebra after calculus.

To incorporate more variables the notation needs a simpler device than using more letters from the alphabet. Variables can be labeled as $x_1, x_2, \cdots, x_n$, using one symbol for the independent variables and a subscript to identify a dimension under discussion. The domain set consists of ordered n-tuples $\mathcal{D} = \{(x_1, x_2, \cdots, x_n)\}$. Individual components of the n-tuple may be restricted by the associated problem. A function has then one value belonging to its range $\mathcal{R}$ for each point in its domain. Thus, we write $y = f(x_1, x_2, \cdots, x_n)$ for $(x_1, x_2, \cdots, x_n) \in \mathcal{D}$ and $y \in \mathcal{R}$.

We will give one example to illustrate these extensions:

<u>EXAMPLE 10.5</u>: Theoretical studies of photosynthesis frequently assume that

$$P = \frac{bE}{1 + aE}$$

where P = gross rate of photosynthesis per unit leaf area,

E = incident radiation per unit leaf area,

b,a = constants depending upon plant material and environmental conditions.

Observe that $\lim_{E\to\infty} P = b/a$. In seeking interpretations of the constants a and b within ranges of relevant values of P and E, Chartier (1970) derived the relations

$$b = [C - F(r_a + r_s)]\left[\frac{1 - \frac{E + R}{\alpha E}}{F + R}\right] \tag{10.8}$$

and

$$-\frac{b}{a} = \frac{C - F(r_a + r_s)}{F + nR} \tag{10.9}$$

where C = concentration of carbon dioxide in the air,

r_a = diffusion resistance for CO_2 transfer in the boundary layer,

r_s = diffusion resistance for CO_2 transfer through stomata and cuticle,

F = net assimilation rate per unit leaf area,

R = respiration rate,

α = maximum efficiency of light energy conversion,

n = proportion of respiration occurring in nonphotosynthetic cells.

Equation 10.8 gives b as a function of the seven variables C, F, r_a, r_s, E, R and α. As the situation (plant species or environmental conditions) changes, some or all of these variables would change. The domain D consists of the set of admissible values for these variables and the range R, the resulting possible values for b. Note that Equation 10.9 could be solved for a as a function of b and another function of the listed variables.

Even though we could have used $x_1, x_2, \cdots, x_7$ to denote these variables, the notation used is fairly standard and somewhat suggestive. ▌

EXERCISES

1. Given the two general points (x_1,y_1,z_1) and (x_2,y_2,z_2), derive the equation representing the distance between the two points.

2. Plot the following points: A(6,-2,3) and B(8,-2,4). Find the distance from each to the origin and their perpendicular distances from the axes.

3. How far apart are the two points A(6,-2,3) and B(8,-2,4)?

4. Is the line through (6,8,16) and (0,11,10) perpendicular to the line through (6,8,16) and (10,0,8)? (Hint: Look back at the questions at the end of Example 10.3.)

5. What is the equation of the locus of a point which is always at a distance r from the point (x_0,y_0,z_0)? The equation just derived is the equation of a SPHERE with its origin at (x_0,y_0,z_0) and which has a radius of r.

6. Find the equation of a point (locus) which is always at a distance of r from the origin.

7. Give the equation of a sphere which has its center at (2,-2,3) and which is tangent to the x,y-plane.

8. Find the coordinates of the center and the radius of the sphere whose equation is $x^2 + y^2 + z^2 + 8x + 6y - 4z = -4$.

9. Calculate the perimeter of the triangle formed by connecting the points (4,6,1), (6,4,0) and (-2,3,4).

10. Determine the equation of the locus of a point the sum of whose distances from (0,2,0) and (0,-2,0) is 10.

11. Find the equation of the plane through (1,1,-1), (-2,-2,2) and (1,-1,2). Sketch the resulting plane.

12. Sketch the plane described by $2x + 3y + 6z = 12$.

13. Determine the equation of the plane which is
    a. Parallel to the x,y-plane and four units below it.
    b. Parallel to the y,z-plane and having an x-intercept of 4.
    c. Perpendicular to the z-axis at a point (0,0,6).

14. Find the equation of the plane which is parallel to the x,y-plane and passes through the point (3,-2,-4).

15. Develop the equation of the plane parallel to the z-axis, with a trace in the x,y-plane of $x + y - 2 = 0$.

16. Sketch the following planes, showing the intercepts and traces in each coordinate plane.

a.	$2x + 4y + 3z - 10 = 0$	d.	$2y - 5z = 6$
b.	$3x - 5y + 2z - 30 = 0$	e.	$2x - z = 0$
c.	$x + 2y = 6$	f.	$x - 7 = 0$

17. Write the equations of the planes whose intercepts are:
    a. (-3,0,0), (0,3,0) and (0,0,5)
    b. (4,0,0), (0,-1,0)
    c. (5,0,0)

REFERENCE

Chartier, P. (1970). A model of CO_2 assimilation in the leaf. PREDICTION AND MEASUREMENT OF PHOTOSYNTHETIC PRODUCTIVITY. Proceedings of the IBP/PP Technical Meeting, Trebon, 14-21 September, 1969. Edited by I. Setlik. Published by the Centre for Agricultural Publishing and Documentation, Wageninger, Netherlands.

Chapter 11

EXAMINING FUNCTIONS OF SEVERAL VARIABLES

Our examination of functions of one variable progressed from a consideration of the functions themselves through limits to derivatives and on to integrals. In following the same path for our study of functions of several variables, we began with the functions themselves in the last chapter and now will broaden the ideas of limits and derivatives.

This chapter begins by generalizing the idea of limits to functions of several variables; this leads to a natural extension of the idea of continuity. The next section generalizes the ideas about rates of change and derivatives to functions of several variables by introducing *partial derivatives*. Second, third and higher derivatives have natural extensions. This completes the generalizations of Chapter 6, so applications corresponding to the first part of Chapter 7 appear next. In particular, functions of two variables have maxima and minima, geometrically interpreted as domes and depressions or topographically as mountains and valleys. The final sections expand on those parts of Chapter 7 dealing with the differentiation of complicated functions and of differentials and small errors. These latter sections demonstrate complexities which arise from functions of several variables, but which were absent in our earlier work. This appears especially in the chain rule for partial derivatives and in total differentials.

Examples in the early sections mainly involve algebraic functions to illustrate mathematical operations. Biological situations appear in the examples in the latter sections. This allows you, as before, to grasp the new mathematical operations before trying to use them as tools.

11.1 LIMITS AND CONTINUITY

The basic concepts of limits and continuity do not change in going from one independent variable to several, but the definitions do get a little more complicated. The following definitions illustrate extensions of the definitions given in Chapter 5 to functions of more than one independent variable:

- The function g has the LIMIT L at the point (x_0,y_0) if $g(x,y)$ can be made arbitrarily close to L for all $(x,y) \in D$ which lie sufficiently close to (x_0,y_0). This usually is symbolized as

$$\lim_{(x,y)\to(x_0,y_0)} g(x,y) = L .$$

- The function g is called CONTINUOUS at $(x_0,y_0) \in D$ when all of the following are satisfied: (i) $g(x_0,y_0)$ exists, (ii) $\lim_{(x,y)\to(x_0,y_0)} g(x,y) = L$ exists, and (iii) $g(x_0,y_0) = L$.

Return to the definitions of the limit and continuity of a function of a single variable to see the similarity of the above with the earlier definitions. Again the idea of limits, whether for functions of one variable or of more than one variable, deals with the idea of closeness. The limit defined above involves a condition on x and y simultaneously. The values x and y must approach x_0 and y_0 simultaneously from any direction. Again, to be continuous at a point, a function cannot have a break or discontinuity there.

The limit theorems given in Chapter 5, Theorems 5.1-5.8, remain true if in them we replace $x\to x_0$ by $(x,y)\to(x_0,y_0)$. Specifically, limits of sums, products and ratios equal sums, products and ratios of limits of the constituent functions, provided all of the limits exist and in the case of ratios, the denominator does not equal zero. The notions of a limit and continuity for a function of two variables extend to functions of any other number of variables. The theorems (Theorems 5.1-5.8) presented in Chapter 5 extend to any number of independent variables.

<u>EXAMPLE 11.1</u>: Investigate the continuity of the function $f(x,y) = \frac{100x}{y}$, developed in Example 10.2.

The extension of Theorem 5.7, which concerns ratios, implies that $\lim_{(x,y)\to(x_0,y_0)} f(x,y) = 100x_0/y_0$ provided $y_0 \neq 0$. The function exists everywhere except where $y_0 = 0$, and $f(x_0,y_0) = \lim_{(x,y)\to(x_0,y_0)} f(x,y)$.

Thus, this function satisfies the properties of continuity except at the points on the line $y = 0$ in the x,y-plane. ▮

11.2 RATES OF CHANGE IN SPACE

We have used derivatives to examine how functions of one variable change. They proved powerful in helping us locate points in the domain where the function had some noteworthy behavior, like a relative maximum or minimum. We have extended the idea of representing a function of one variable by a curve to the representation of a function of two variables by a surface in space. Thus we may inquire, "How can we find the rate of change of a surface in space?"

A new complication quickly confronts us, however. It has this topographic equivalent: As we hike up a rounded hill, the climb may be very steep from one direction, but quite gentle from another. This suggests that our extension of a derivative needs a directional character. Thus, we are led to these definitions:

Consider a function of two arguments denoted as $z = f(x,y)$.

- The PARTIAL DERIVATIVE of f with respect to x is defined by

$$\frac{\partial z}{\partial x} = \lim_{\Delta x\to 0} \frac{f(x + \Delta x,y) - f(x,y)}{\Delta x} \tag{11.1}$$

  if the limit exists. It also is denoted by $\frac{\partial f}{\partial x}$, $\frac{\partial f(x,y)}{\partial x}$, and $f_x(x,y)$.

- The PARTIAL DERIVATIVE of f with respect to y is defined by

$$\frac{\partial z}{\partial y} = \lim_{\Delta y\to 0} \frac{f(x,y + \Delta y) - f(x,y)}{\Delta y} \tag{11.2}$$

  if the limit exists. It also is denoted by $\frac{\partial f}{\partial y}$, $\frac{\partial f(x,y)}{\partial y}$, and $f_y(x,y)$.

In discussing these, we will concentrate on the partial derivative of z with respect to x, but by interchanging the roles of x and y, identical statements apply to the partial derivative of z with respect to y. Equations 11.1 and 11.2 state formal definitions. What do they actually say? Each limit involves only one variable, not two as described in the last section. For example, the numerator of Equation 11.1 involves $f(x + x,y) - f(x,y)$; the change lies entirely in x: y does not change at all. For any fixed y, f(x,y) traces out a curve above y. This corresponds to walking across a hilly countryside going straight east, not northeast or southeast, assuming that x changes in the east-west direction. However, we could pick how far north or south we go (the fixed value of y) before we start eastward.

An important computational result follows from the fact that y remains constant in the limiting process. With y fixed, we really are dealing with a simple derivative like we examined in Chapters 6 and 7. Thus to evaluate a partial derivative, we merely regard f as a function of x alone by treating all other variables as constants.

<u>EXAMPLE 11.2</u>: Find $\frac{\partial z}{\partial x}$ and $\frac{\partial z}{\partial y}$ if $z = f(x,y) = 3x^2 + ye^{-x} + \frac{x}{y}$.

To find $\frac{\partial z}{\partial x}$, regard y as a constant and take derivatives as usual:

$$\frac{\partial z}{\partial x} = 6x - ye^{-x} + \frac{1}{y} .$$

Now proceed similarly to evaluate $\frac{\partial z}{\partial y}$, namely, regard x as a constant:

$$\frac{\partial z}{\partial y} = e^{-x} - \frac{x}{y^2} . \quad \blacksquare$$

Evaluation of a partial derivative at a particular point (x_0,y_0) utilizes notation similar to that of a simple derivative. The following notations are used:

$$\left.\frac{\partial z}{\partial x}\right|_{(x_0,y_0)} , \qquad f_x(x_0,y_0) , \qquad \frac{\partial f(x_0,y_0)}{\partial x} , \qquad \left.\frac{\partial f(x,y)}{\partial x}\right|_{(x_0,y_0)}$$

EXAMPLE 11.3: For symbolic variation, consider $g(u,v) = \frac{u^2}{v} + \frac{v^2}{u}$ and find $\frac{\partial g(1,2)}{\partial v}$. First we need

$$\frac{\partial g}{\partial v} = -\frac{u^2}{v^2} + \frac{2v}{u} .$$

Now evaluate this at u = 1 and v = 2:

$$\frac{\partial g(1,2)}{\partial v} = \left.\frac{\partial g}{\partial v}\right|_{(1,2)} = -\frac{1^2}{2^2} + \frac{2(2)}{1} = 3\,\frac{3}{4} . \quad \blacksquare$$

In three dimensions, one for the dependent variable and two for the independent variables, a partial derivative has a simple geometrical interpretation. When x and y vary through the domain of the definition of the function, z = f(x,y) will define a surface. (See Section 10.2.) Figure 11.1 shows such a general surface. Let P represent a general point on this surface. Cut this surface with planes through the point P, where these planes are parallel to the x,z-plane and y,z-plane. These cutting planes contain the points A,P,B and C,P,D as shown.

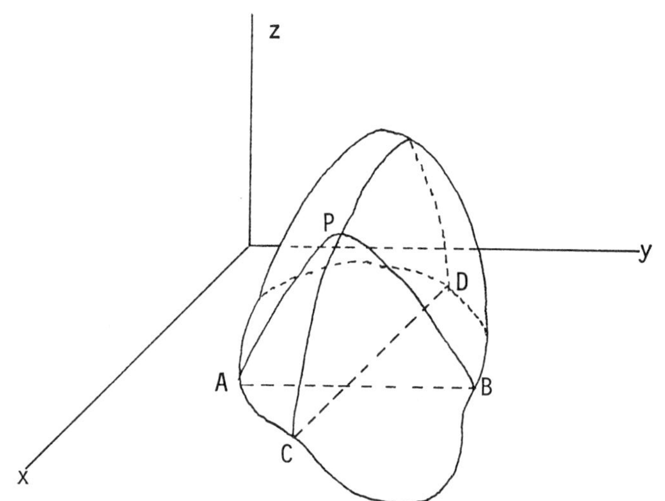

Figure 11.1

The cutting planes intersect the surface in curves displayed in Figures 11.2 and 11.3, namely the surface has these traces on the cutting planes. The curve in Figure 11.2 lies above the line $y = y_0$ in the

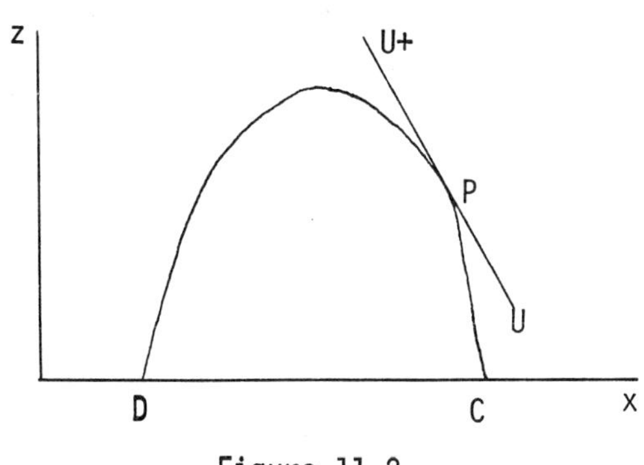

Figure 11.2

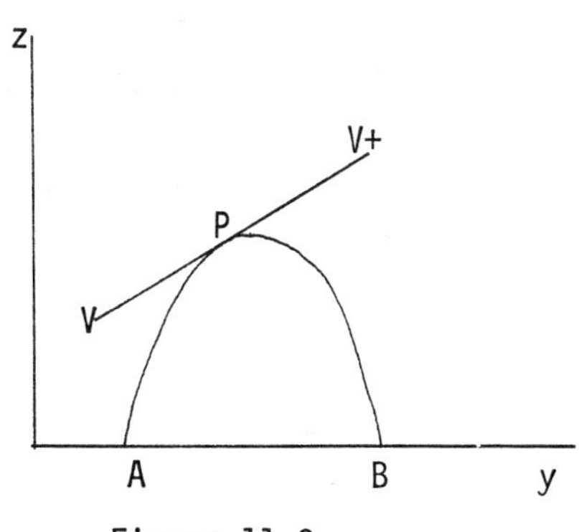

Figure 11.3

x,y-plane, effectively giving the value of the function for y held constant at $y = y_0$, but with x varying. The line UU^+ is tangent to the curve at the point P. Now if we return to the limit definition of the partial derivative of z with respect to x, we have

$$\frac{\partial z}{\partial x} = \lim_{\Delta x \to 0} \frac{f(x + \Delta x, y) - f(x,y)}{\Delta x} .$$

The numerator gives the change in z due to a small change in x while y remains constant; the denominator, Δx, measures the change in x. Thus the difference quotient simply gives the change in z divided by the change in x. This average rate of change of z in the x direction is analogous to the average rate of change for the function of a single variable presented in Chapter 6. The limit of the average rate of change of z in the x direction gives the slope of the line tangent to the curve at the point P, namely the slope of the line UPU^+. (A similar discussion relates $\frac{\partial z}{\partial y}$ to Figure 11.3.) Thus, the partial derivative of f with respect to x, evaluated at (x_0, y_0), gives the slope of the line tangent to the surface at P along the x direction. The lines VPV^+ and UPU^+ intersect at P. They determine a plane called the plane tangent to the surface at P.

<u>EXAMPLE 11.1</u> (continued): Recall that this continuing example concerns the relation between cardiac output (C) and carbon dioxide released by the

lungs (x) and the change (y) in carbon dioxide content of the blood across the lungs: $C = \frac{100x}{y}$. Find the plane tangent to the surface described by this function (see Figure 10.6) at x = 300 and y = 6.

First, at this point, $C = \frac{(100)(300)}{6} = 5000$ cc./minute. Next we need the slopes:

$$\left.\frac{\partial C}{\partial x}\right|_{(300,6)} = \left.\frac{100}{y}\right|_{(300,6)} = \frac{100}{6}$$

$$\left.\frac{\partial C}{\partial y}\right|_{(300,6)} = -\left.\frac{100x}{y^2}\right|_{(300,6)} = -\frac{5000}{6}.$$

This gives the respective slopes as m = 100/6 and n = -5000/6.

In our earlier work with planes (Section 10.3), we saw that $z = mx + nb + b$ describes a plane with slopes m and n in the x and y directions. Thus

$$C = \frac{100}{6}x - \frac{5000}{6}y + b$$

describes a plane with the required slopes. To force this plane through the point (x,y,C) = (300,6,5000), substitute into the above equation:

$$5000 = \frac{100}{6}(300) - \frac{5000}{6}6 + b$$

or b = 5000 - 5000 + 5000 = 5000 so the equation of the tangent plane becomes

$$C = \frac{50}{3}x - \frac{2500}{3}y + 5000 . \quad \blacksquare$$

The definition of a partial derivative given earlier explicitly involved two independent variables. It generalizes to any number of independent variables in the obvious way. The partial derivatives are calculated in the same way: Regard all independent variables except the one of interest as constant and take a simple derivative relative to the independent variable of interest. Rather than engage in more formalities of definitions, the next examples illustrate these operations:

EXAMPLE 11.4: If $z = 3x^2 + ye^{-x} + wx^3 + \frac{y}{v}$, find $\frac{\partial z}{\partial x}$, $\frac{\partial z}{\partial y}$, $\frac{\partial z}{\partial w}$ and $\frac{\partial z}{\partial v}$.

In finding $\frac{\partial z}{\partial x}$, we hold constant all of the independent variables except x and take the derivative as usual to obtain

$$\frac{\partial z}{\partial x} = 6x - ye^{-x} + 3wx^2 .$$

In a like manner

$$\frac{\partial z}{\partial y} = e^{-x} + \frac{1}{v}$$

$$\frac{\partial z}{\partial w} = x^3$$

$$\frac{\partial z}{\partial v} = -\frac{y}{v^2} .$$

Check these! ▌

EXAMPLE 11.5: If $f(x,y,z) = x^2y + y^2z + xz^2$, find $\frac{\partial f(2,-1,3)}{\partial x}$.

As

$$\frac{\partial f}{\partial x} = 2xy + z^2 ,$$

$$\frac{\partial f(2,-1,3)}{\partial x} = 2(2)(-1) + 3^2 = 5 . \quad ▌$$

EXERCISES

Find the partial derivatives of the given function with respect to each of the independent variables. (Regard the variables on the right hand side of the equality as the independent ones.)

1. $V = e^{2x} \ln y$
2. $g(x,y) = e^{3x^2y}$
3. $G(x,z) = ze^{-2x} + xe^{-3zx}$
4. $U = e^x \ln x + e^z \ln y$
5. $W = (x^2 + y^2)^3$
6. $U = \sqrt{x^2 + 4y^2}$
7. $f = (3x^2 - 2y + 3z)^{3/2}$
8. $g = \frac{10^{2x}}{4x + 3z}$
9. If $F(x,y) = 3x^2 + 7xy^3$, find $\frac{\partial F(3,1)}{\partial x}$, $\frac{\partial F(1,3)}{\partial y}$.

10. If $U = 3x^2y^2z$, find the rate of change of U in the y direction.

11. If $V = f(x,y,z)$, give the limit definition of $\frac{\partial V}{\partial x}$, $\frac{\partial V}{\partial y}$ and $\frac{\partial V}{\partial z}$.

12. If $z = mx + ny + b$, use partial differentiation to show that m and n represent the slope of the plane in the x and y direction, respectively.

13. Show that $z = \frac{\partial f(x_0,y_0)}{\partial x} x + \frac{\partial f(x_0,y_0)}{\partial y} y + b$ is parallel to the plane tangent to the surface $z = f(x,y)$ at (x_0,y_0).

14. Give a general formula for b in the equation in Problem 13 so that the plane described by the equation is tangent to the surface $Z = f(x_0,y_0)$.

15. Use the limit definition of a partial derivative to find $\frac{\partial z}{\partial x}$ and $\frac{\partial z}{\partial y}$ if $z = 3x^2 - 2xy + y^2$.

16. Find the slope of the line tangent to the trace of the surface $z = x^2 + y^2$ on the plane $y = 1$ at $(2,1,5)$.

17. Find the slope of the line tangent to the trace of the surface $36x^2 - 9y^2 + 4z^2 + 36 = 0$ on the plane $x = 1$ at $(1,\sqrt{12},-3)$.

18. The transfer of energy by convection from an animal results from a temperature difference between the animal's surface temperature and the surrounding air temperature. The convection coefficient is given by $h = \frac{kV^{1/3}}{D^{2/3}}$ where k is a constant, V is the wind velocity in cm./second and D is the diameter of the animal's body in centimeters. The units on h are calories/cm^2/minute/°C. What units does k have? Find $\frac{\partial h}{\partial V}$ and $\frac{\partial h}{\partial D}$.

19. The following equation gives an estimate of the oxygen consumption of a very well insulated mammal which is not sweating, but which is maintaining a body temperature of T_b°C in conditions at which its outside fur temperature is T_s°C and its weight is W: $m \doteq 2.5(T_b - T_s)\, W^{-0.67}$. Find $\frac{\partial m}{\partial T_b}$, $\frac{\partial m}{\partial T_s}$ and $\frac{\partial m}{\partial W}$. Explain briefly what each partial derivative represents.

20. If p represents the total density of leaves in a tree per unit of ground area and the leaves are uniformly distributed among n layers and r is the circular radius of the leaf, then $1 - \frac{p\pi r^2}{n}$ of the incident light gets through the first layer; of this proportion, only $1 - \frac{p\pi r^2}{n}$ gets through the second layer and so on. Thus $L = (1 - \frac{p\pi r^2}{n})^n$ is the amount of light penetrating all n layers. Find $\frac{\partial L}{\partial r}$, $\frac{\partial L}{\partial n}$ and $\frac{\partial L}{\partial p}$. What do each of these partials represent?

21. The amount of gas adsorbed per unit area of adsorbing surface σ_a, is related to the gas pressure, P, the temperature, T, and the heat of adsorption Q_a by the equation $\sigma_a = kP \exp\left[\frac{Q_a}{RT}\right]$ where R and k are positive constants. Discuss the behavior of σ_a in relationship to P and T.

22. The following equation gives the concentration of O_2 in plant roots:

$$C_R = C_P + \frac{QR^2}{2D} \ln \frac{R}{r}$$

where C_R is the concentration of the oxygen at the root surface (grams/cm.^{3}), C_P is the O_2 concentration at the liquid-gas interface in equilibrium with the partial pressure of O_2 in the gaseous phase (grams/cm.^{3}), Q is the O_2 consumption of the root (grams/(cm.^{3})(second)), R is the root radius (cm.) and D is the diffusion coefficient in the liquid-solid matrix around the root (cm.^{2}/second). Find $\frac{\partial C_R}{\partial C_P}$, $\frac{\partial C_R}{\partial Q}$, $\frac{\partial C_R}{\partial R}$ and $\frac{\partial C_R}{\partial r}$.

23. Stoke's law for the velocity of a particle falling in a fluid is given by $V = \frac{2}{9}(d_p - d)gr^2/n$ where V is the velocity of the fall in cm./second, g is the acceleration due to gravity, d_p is the density of the particle, d is the density of the liquid, r is the radius of the particle in cm. and n is the absolute viscosity of the liquid. Find $\frac{\partial V}{\partial d_p}$, $\frac{\partial V}{\partial d}$, $\frac{\partial V}{\partial r}$ and $\frac{\partial V}{\partial n}$.

24. The free water vapor diffusion coefficients D_V in soft woods is given by $D_V = 0.22 \left[\frac{T}{273}\right]^{1.75} \left[\frac{760}{P}\right]$ where T is the absolute temperature and P is the atmospheric pressure in mm. of Hg. Find $\frac{\partial V}{\partial P}$ and $\frac{\partial V}{\partial T}$ at T = 273°K and P = 6000 mm. of Hg.

25. The ability of a material to conduct heat as a result of transmitting molecular vibrations from one atom to another varies greatly, depending upon the chemical nature of the material and its gross external structure. Thermal conductivity can be evaluated from $k = \frac{Qx}{ATt}$. Here Q (calories) is the amount of heat flowing through a piece of material of thickness x (cm.) and cross-sectional area A (cm^2) during a time of t (seconds) when a temperature difference of T (°C) exists between the faces across which the heat is flowing. Find $\frac{\partial k}{\partial Q}$, $\frac{\partial k}{\partial x}$, $\frac{\partial k}{\partial A}$, $\frac{\partial k}{\partial T}$, and $\frac{\partial k}{\partial t}$. What units does k have?

26. The rate at which any chemical substance can be absorbed into a bacterium is proportional to the surface area A of the bacterium. Suppose that the substance has to be distributed throughout the whole volume (V) of the bacterium. The rate at which the substance can be delivered to any particular part presumably is proportional to A/V and thus the metabolism is m = kA/V where k is a constant. Assume you have a cylindrical shaped cell of length L and radius R which has a hemispherical cap on each end. Find out how m is affected by changes in the shape and size of the bacterium.

27. Repeat Problem 26 using a cylindrical cell with flat ends.

28. Determination of body surface area from actual measurements is at best a very crude and tedious process. The following formula relates the surface area A (m^2) to the weight W (kg.) and height H (m.) for humans:

$$A = 2.024W^{0.425}H^{0.725} .$$

Estimate your surface area. Calculate $\frac{\partial A}{\partial W}$ and $\frac{\partial A}{\partial H}$.

11.3 HIGHER PARTIAL DERIVATIVES

We may take partial derivatives of partial derivatives just as we took derivatives of derivatives for functions of a single variable. When only two successive partial differentials are taken, the result is called a SECOND ORDER PARTIAL DERIVATIVE. In a similar fashion, we can introduce derivatives of the third and higher orders. If $z = f(x,y)$, the following notation is used for derivatives of the second order:

$$\frac{\partial^2 z}{\partial x^2} = \frac{\partial}{\partial x}\left(\frac{\partial z}{\partial x}\right) = f_{xx} \;; \qquad \frac{\partial^2 z}{\partial y^2} = \frac{\partial}{\partial y}\left(\frac{\partial z}{\partial y}\right) = f_{yy} \;;$$

$$\frac{\partial^2 z}{\partial x \partial y} = \frac{\partial}{\partial x}\left(\frac{\partial z}{\partial y}\right) = f_{yx} \;; \qquad \frac{\partial^2 z}{\partial y \partial x} = \frac{\partial}{\partial y}\left(\frac{\partial z}{\partial x}\right) = f_{xy} \,.$$

The symbol $\frac{\partial^2 z}{\partial x \partial y}$ is read as the second partial of z with respect to y and then x while $\frac{\partial^2 z}{\partial y \partial x}$ is read as the second partial of z with respect to x and then y. The next example shows an interesting result related to these second order partial derivatives which we will state but not prove.

<u>EXAMPLE 11.6</u>: If $z = x^3 + 4x^3y - y^3$, find $\frac{\partial^2 z}{\partial x^2}$, $\frac{\partial^2 z}{\partial x \partial y}$ and $\frac{\partial^2 z}{\partial y \partial x}$.

$$\frac{\partial z}{\partial x} = 3x^2 + 12x^2y \qquad \text{and} \qquad \frac{\partial^2 z}{\partial x^2} = 6x + 24xy \,,$$

$$\frac{\partial z}{\partial y} = 4x^3 - 3y^2 \qquad \text{and} \qquad \frac{\partial^2 z}{\partial x \partial y} = 12x^2 \,,$$

and

$$\frac{\partial z}{\partial x} = 3x^2 + 12x^2y \qquad \text{and} \qquad \frac{\partial^2 z}{\partial y \partial x} = 12x^2 \,. \quad \blacksquare$$

Example 11.6 demonstrates that if all derivatives which we encounter are continuous and differentiable functions, then the order of partial differentiation is immaterial. This result holds for partial derivatives of all orders and will be taken without proof.

<u>EXAMPLE 11.7</u>: If $w = e^{3x+4y} + y^3$, find $\frac{\partial^3 w}{\partial y \partial x^2}$ using three different orders of partial differentiation.

$$\frac{\partial w}{\partial x} = 3e^{3x+4y} ; \qquad \frac{\partial^2 w}{\partial x^2} = 9e^{3x+4y} ; \qquad \frac{\partial^3 w}{\partial y \partial x^2} = 36e^{3x+4y} .$$

$$\frac{\partial w}{\partial x} = 3e^{3x+4y} \qquad \frac{\partial^2 w}{\partial y \partial x} = 12e^{3x+4y} ; \qquad \frac{\partial^3 w}{\partial x \partial y \partial x} = 36e^{3x+4y} .$$

$$\frac{\partial w}{\partial y} = 4e^{3x+4y} + 3y^2 ; \qquad \frac{\partial^2 w}{\partial x \partial y} = 12e^{3x+4y} ; \qquad \frac{\partial^3 w}{\partial x^2 \partial y} = 36e^{3x+4y} .$$

Notice that all of these orders of partial differentiation give the same answers. This helps verify our earlier statement that the order of partial differentiation makes no difference. ▌

EXERCISES

Find all partial derivatives of the second order for the given function for Problems 1-12.

1. $u = x^4y^3$

2. $w = r^3e^s \ln t$

3. $u = e^{3x^2+4y}$

4. $v = e^{x^2} \ln y$

5. $f(x,y) = x^4 - 2x^2y + y^3$

6. $g(x,y) \cdot \sqrt{x^2 - y^2}$

7. $f(x,y) = (x + 6x^3y^2)^2$

8. $u = e^x + e^y + e^z$

9. $z = e^{uvw^2}$

10. $f(x,y) = \frac{x^2}{y} - \frac{y}{x^2}$

11. $g(x,y) = 2x^3 - 3x^2y + xy^2$

12. $h(x\ y) = e^{-x/y} + \ln \frac{y}{x}$

13. If $z = x^2y^3 + 4xy^4$, find $\frac{\partial^3 z}{\partial x^2 \partial y}$ and $\frac{\partial^3 z}{\partial y^2 \partial x}$.

14. If $u = x^3 + x^2y$, show that $x\frac{\partial u}{\partial x} + y\frac{\partial u}{\partial y} = 3u$.

15. For $f(x,y,z) = ye^x + ze^y + e^z$, evaluate f_{xz}, f_{yz} .

16. When $f(r,s) = r^3s + r^2s^3 - rs^2$, calculate $\frac{\partial^3 f}{\partial r^2 \partial s}$, $\frac{\partial^3 f}{\partial s \partial^2 r}$.

17. If $g(r,s,t) = \ln(r^2 + s^2 + t^2)$, find $\frac{\partial^3 g}{\partial r \partial t \partial s}$, $\frac{\partial^3 g}{\partial r \partial^2 s}$.

18. Show that $u(x,y) = \ln(x^2 + y^2)$ satisfies the equation $\frac{\partial^2 u}{\partial x^2} + \frac{\partial^2 u}{\partial y^2} = 0$.

19. Consider a tube parallel to the x-axis containing fluid. Suppose that the tube contains other material in addition to the fluid. If the concentration of the material increases to the right, there will be a net flow of molecules to the left. The concentration C of the material will be a function of both the distance and time, and from Fick's law can be shown to satisfy

$$\frac{\partial C}{\partial t} = D \frac{\partial^2 C}{\partial x^2}$$

where D is a proportionality constant known as the diffusion constant. This is the DIFFUSION EQUATION IN ONE DIMENSION. Show that

$$C = \frac{1}{\sqrt{4\pi Dt}} \exp - \frac{x^2}{4Dt}$$

satisfies the above equation.

20. Show that

$$C = t^{-1/2} \exp - \frac{x^2}{4Dt}$$

satisfies the diffusion equation given in Problem 19.

21. Find coefficients a and b such that $C = e^{(ax+bt)}$ satisfies the diffusion equation given in Problem 19.

22. The volume of a right circular cylinder of radius r and height h is $V = \pi r^2 h$. Find the first and second order partial derivatives of V with respect to r and h.

23. The concepts of natural frequency apply to such diverse subjects as sound production and perception by animals, and resonant frequencies of plant and animal tissue perturbed by mechanical vibrations. The one-dimensional case may apply to voice-box control and stridulations by insects; these have a close physical relation to vibrations of a taut wire. The frequency of such vibrations, ω, depends upon:

23. Continued

b = wire diameter

L = wire length

ρ = wire density

τ = tension (force holding).

Thus ω is a function of four variables, $\omega = f(b,L,\rho,\tau) = \frac{1}{bL}\sqrt{\frac{\tau}{\pi\rho}}$. Find the rate of change of ω with respect to each of the independent variables.

24. The flow, Q in cm^3/second, of a liquid (say blood) from a large vessel to a small capillary could be described by $Q = \frac{C\pi d^2}{4}\sqrt{P_0 - P_1}$ where C is a constant depending on the liquid, d is the diameter of the capillary, P_0 is the pressure in the large blood vessel and P_1 is the pressure in the capillary. Find expressions for $\frac{\partial Q}{\partial d}$, $\frac{\partial Q}{\partial P_0}$, $\frac{\partial Q}{\partial P_1}$, $\frac{\partial Q}{\partial d^2}$, and $\frac{\partial^2 Q}{\partial P_0{}^2}$.

25. A relationship which describes the swelling of jute fibers is $S = d_0\,[m/100 - 1/d_0 + 1/d_w]$ where S is the fractional volumetric swelling of the cell walls; m is the moisture content at the fiber saturation point in percent; d_0 is the true specific gravity of the substance on a dry volume basis and d_w is the gravity of the fiber substance determined in water expressed on a volume of fiber basis. Find $\frac{\partial S}{\partial d_w}$.

11.4 MAXIMA AND MINIMA OF FUNCTIONS OF TWO VARIABLES

In the following discussion, unless otherwise specified, a maximum or minimum will refer to a relative maximum or relative minimum at a point (x_0,y_0) assuming that $f(x,y)$ is defined for all points (x,y) sufficiently close to (x_0,y_0). First, we must expand the concept of a maximum or minimum to a function of two variables:

- A function f of two independent variables x and y has a RELATIVE MAXIMUM at the point (x_0,y_0) if for all points (x,y) sufficiently near (x_0,y_0) in the x,y-plane $f(x_0,y_0) \geq f(x,y)$.

- A function f of two independent variables x and y has a RELATIVE MINIMUM at the point (x_0,y_0) if for all points (x,y) sufficiently near (x_0,y_0) in the x,y-plane $f(x_0,y_0) \leq f(x,y)$.

Figure 11.4 shows a maximum at the point (x_0,y_0). From an examination of this figure, our discussion of tangent planes accompanying Figures 11.1-11.3, and our earlier experience with functions of one independent variable, the following should be intuitively appealing: If the function f has a relative maximum or relative minimum at the point (x_0,y_0) and if f has partial derivatives, then it is necessary that

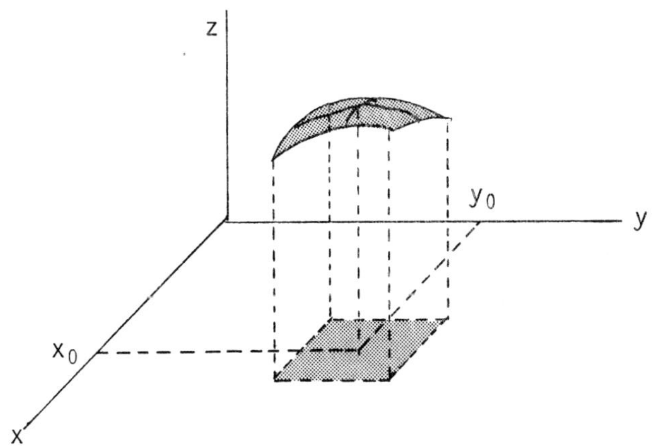

Figure 11.4

$$\frac{\partial f}{\partial x}(x_0,y_0) = 0 \qquad \text{and} \qquad \frac{\partial f}{\partial y}(x_0,y_0) = 0 .$$

The points (x_0,y_0) where f_x and f_y simultaneously equal zero are called the CRITICAL VALUES and $f(x_0,y_0)$ is called a CRITICAL POINT.

That the above conditions are necessary for a maximum or a minimum, but not sufficient, is illustrated in Figure 11.5. Neither a maximum nor a minimum occurs at (x_0,y_0) even though both partial derivatives equal zero there. A cutting plane through P parallel to the x,z-plane, traces out the curve shown in Figure 11.6; a cutting plane parallel to the y,z-plane produces the curve shown in Figure 11.7. These figures show a maximum in one direction, but a minimum in the other. Thus, $f(x_0,y_0)$ is neither a maximum nor a minimum, but rather a point called a SADDLE POINT. This discussion should not infer that every saddle point behaves like this. Problem 1 at the end of this section illustrates this point.

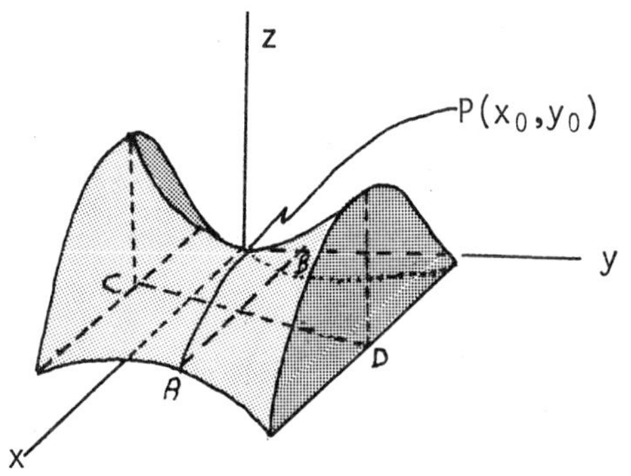

Figure 11.5

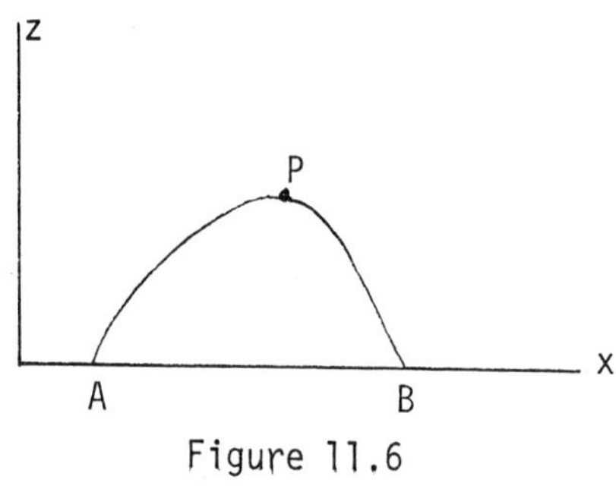

Figure 11.6

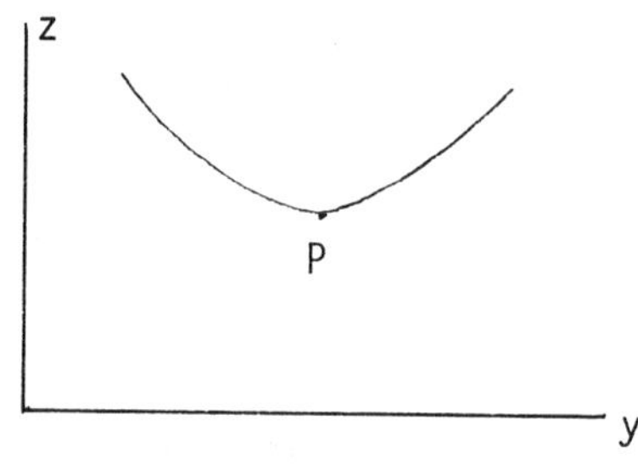

Figure 11.7

This example shows that functions of two variables can behave in a far more complicated manner near the critical points than can functions of a single independent variable. Nevertheless the second derivative test for a maximum or minimum of a function of one independent variable has a generalization to a function of two independent variables. Satisfaction of the criteria of this test assures that each cross section through a critical point has concavity of the same sense. We will state this second derivative test in the form of a theorem but not prove it.

<u>THEOREM 11.1</u>: If a function f of two independent variables x and y has continuous partial derivatives of the second order near and at the critical value (x_0,y_0), then f has

a) a relative minimum at (x_0,y_0) if

$$f_{xx}(x_0,y_0)f_{yy}(x_0,y_0) - f_{xy}^2(x_0,y_0) > 0 \qquad \text{and} \qquad f_{xx}(x_0,y_0) > 0 ,$$

b) a relative maximum at (x_0,y_0) if

$$f_{xx}(x_0,y_0)f_{yy}(x_0,y_0) - f_{xy}^2(x_0,y_0) > 0 \qquad \text{and} \qquad f_{xx}(x_0,y_0) < 0 ,$$

c) a saddle point at (x_0,y_0) if

$$f_{xx}(x_0,y_0)f_{yy}(x_0,y_0) - f_{xy}^2(x_0,y_0) < 0 ,$$

d) the test fails and no conclusion can be drawn, that is, f can have a relative maximum, minimum or saddle point or none of these if

$$f_{xx}(x_0,y_0)f_{yy}(x_0,y_0) - f_{xy}^2(x_0,y_0) = 0 .$$

Case d is beyond the scope of this book and if such a case should come up, we will just say that no conclusion can be reached.

This theorem is illustrated in the next two examples.

EXAMPLE 11.8: Locate the maxima, minima, and saddle points of the function $z = x^3 + y^3 - 3xy + 15$.

First calculate the required first and second order partial derivatives:

$$f_x = \frac{\partial z}{\partial x} = 3x^2 - 3y; \qquad f_y = \frac{\partial z}{\partial y} = 3y^2 - 3x$$

$$f_{xx} = \frac{\partial^2 z}{\partial x^2} = 6x; \qquad f_{yy} = \frac{\partial^2 z}{\partial y^2} = 6y; \qquad f_{xy} = \frac{\partial^2 z}{\partial x \partial y} = -3 .$$

To find the critical values, simultaneously solve

$$\frac{\partial z}{\partial x} = 0 \qquad \text{and} \qquad \frac{\partial z}{\partial y} = 0$$

or

$$3x^2 - 3y = 0 \qquad \text{and} \qquad 3y^2 - 3x = 0 .$$

These two equations can be solved for x and y. Solve the first equation for y in terms of x and substitute into the second equation:

$$y = x^2 \qquad \text{and} \qquad 3(x^2)^2 - 3x = 0 \qquad \text{or} \qquad 3x^4 - 3x = 3x(x^3 - 1) = 0 .$$

The solutions to this last equation are $x = 0$ and $x = 1$. This gives (0,0) and (1,1) as critical values. Now evaluate the conditions of Theorem 11.1:

At (0,0),

$$f_{xx}(0,0)f_{yy}(0,0) - f_{xy}^2(0,0) = 0 - 9 < 0 ,$$

so (0,0,15) is a saddle point.

At (1,1)

$$f_{xx}(1,1)f_{yy}(1,1) - f_{xy}^2(1,1) = (6)(6) - 9 > 0$$

and since $f_{xx}(1,1) = 6 > 0$, a relative minimum occurs at (1,1,14).

Now what do we know about the surface defined by f? When water drops across a rock face it sometimes digs out potholes in the rock. The surface defined by f is something like a smooth rock face which once had a ledge in its middle. With time the water dug a pothole in the ledge and wore a rounded indentation into the rock above and below the pothole. Now the rock face drops down steeply toward the pothole; near the pothole it levels off; past the pothole it again drops off more and more steeply. Figure 11.8 shows a three-dimensional computer plot looking into the positive octant across the origin. The surface has been cut away for x < 0 and y < 0 in this computer plotting; this allows the minimum to show. Our previous plots have looked down on the surface from above so we could see the x,y-plane. Figure 11.8 views the surface horizontally so the x and y-axes merge.

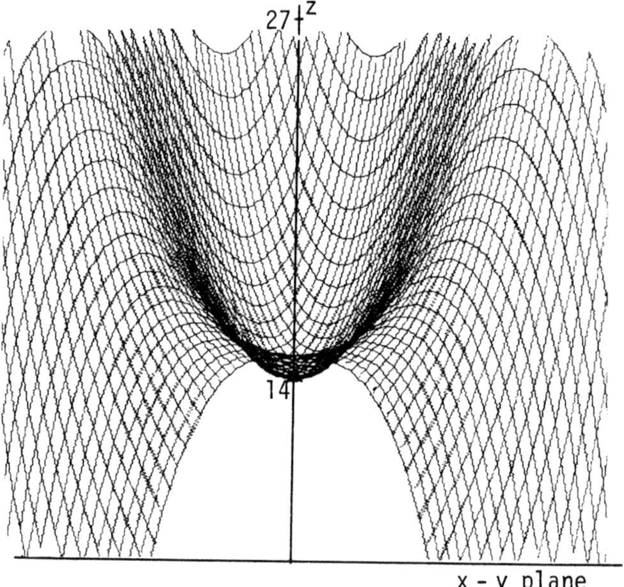

Figure 11.8

A few traces of this surface will explain the algebraic origin of its shape. Evaluating these traces we have $f(x,0) = x^3 + 15$ is pure cubic, having both a horizontal tangent and an inflection point at

$x = 0$; $f(x,1) = x^3 + 16 - 3x$ has a minimum of 14 at $x = 1$; this is the bottom of the pothole. $f(x,a) = x^3 + a^3 - 3ax$ has a minimum at $x = a \neq 0$; this defines the back of the depression down which the water runs. An interchange of x and y shows that the same comments apply to y. You should sketch each of the above traces in order to enhance your understanding of the figure. ▌

The following example illustrates the use of partial differentiation to the optimization process.

EXAMPLE 11.9: Assume you need to design an open topped rectangular muscle chamber having a volume of 32 cubic inches. What dimensions would minimize its surface area?

Let x, y and z represent the width, length and height of the muscle chamber as shown in the drawing. Then its volume will be $V = xyz$ and its surface area will be $A = xy + 2yz + 2xz$. This surface area depends on three variables which are functionally related when $V = 32$ cubic inches. We can solve the expression for volume for z and substitute the result into the expression for the area:

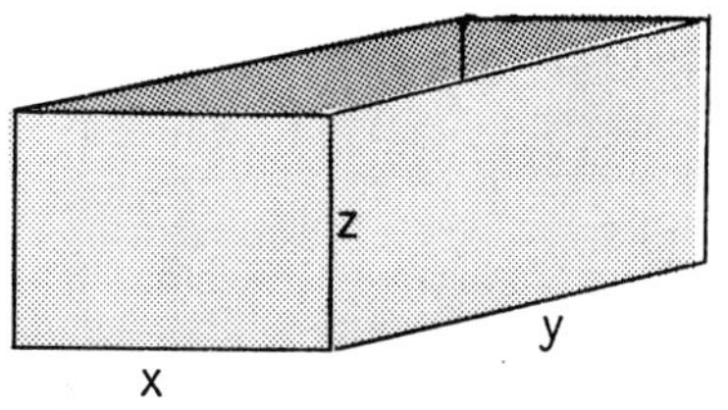

$$z = \frac{32}{xy} \qquad \text{and} \qquad A = xy + \frac{64}{x} + \frac{64}{y}$$

Now that we have expressed the area as a function of two independent variables, we can locate the maxima or minima if they exist. Calculating the necessary first and second order partial derivatives, we obtain

$$A_x = y - \frac{64}{x^2}\ ; \qquad A_y = x - \frac{64}{y^2}$$

$$A_{xx} = \frac{128}{x^3}\ ; \qquad A_{yy} = \frac{128}{y^3}\ ; \qquad A_{xy} = 1\ .$$

These first partial derivatives equal zero when $A_x = y - \frac{64}{x^2} = 0$ and $A_y = x - \frac{64}{y^2} = 0$. Solve the first equation for y and substitute the result into the second equation:

$$y = \frac{64}{x^2} \qquad \text{and} \qquad x - \frac{64}{(\frac{64}{x^2})^2} = x - \frac{x^4}{64} = 0$$

or

$$64x - x^4 = x(64 - x^3) = 0 \ .$$

This is satisfied when x = 0 or x = 4. Obviously x = 0 is of no interest because it implies a chamber of no width. Thus, the only critical value of interest is (4,4). Now evaluate the conditions of Theorem 11.1 for the critical point (4,4):

$$A_{xx}(4,4)A_{yy}(4,4) - A_{xy}^2(4,4) = (\frac{128}{64})(\frac{128}{64}) - 1 > 0$$

This, together with $A_{xx}(4,4) = 2 > 0$ imply that a minimum occurs at $(x_0,y_0) = (4,4)$ when $(x_0,y_0) = (4,4)$. The height of the chamber is given by $z = \frac{32}{xy} = \frac{32 \text{ in.}^3}{(4 \text{ in.})(4 \text{ in.})} = 2$ inches. Thus, the dimensions of the muscle chamber should be 4 inches × 4 inches × 2 inches to have a minimum surface area. ▌

*EXAMPLE 11.10: Find whatever relative maxima, minima, and saddle points the function $f(x,y) = xy^2\ (3x + 6y - 2)$ has.

Begin by finding the first and second partial derivatives:

$$f_x = \frac{\partial f}{\partial x} = y^2\ (3x + 6y - 2) + 3xy^2 = y^2\ (6x + 6y - 2)$$

$$f_{xx} = \frac{\partial^2 f}{\partial x^2} = 6y^2$$

$$f_y = \frac{\partial f}{\partial y} = 2yx(3x + 6y - 2) + 6xy^2 = 2xy(3x + 9y - 2)$$

$$f_{yy} = \frac{\partial^2 f}{\partial y^2} = 2x(3x + 18y - 2)$$

$$f_{xy} = \frac{\partial^2 f}{\partial y \partial x} = 2y(3x + 9y - 2) + 2xy(3) = 2y(6x + 9y - 2) = f_{yx} \ .$$

Isolate the critical values by setting the first partial derivatives equal to zero:

$$f_x = y^2\ (6x + 6y - 2) = 0$$

$$f_y = 2xy(3x + 9y - 2) = 0$$

Both of these equations involve products; they will be satisfied for any (x_0,y_0) which makes at least one factor in each equation zero. The first equation tells us $y = 0$, or $x + y = \frac{1}{3}$, or both. The points satisfying these restrictions are drawn as solid lines in Figure 11.9. The second equation requires $x = 0$, $y = 0$, or $x + 3y = \frac{2}{3}$ and is shown by dashed lines. ($y = 0$ is shown as a separate dashed line.) Any point at the intersection of a solid and a dotted line will make both first partials simultaneously equal zero; these points are boxed. Check to see that these points are $(\frac{1}{3},0)$, $(\frac{2}{3},0)$, $(\frac{1}{6},\frac{1}{6})$ and $(0,\frac{1}{3})$. The whole x-axis ($y = 0$) also makes both partial derivatives equal zero.

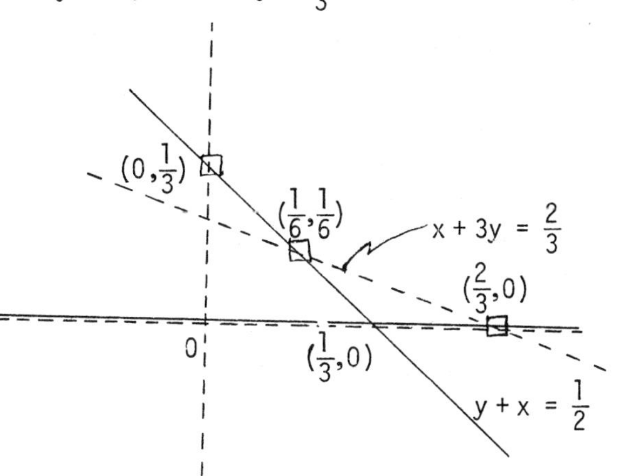

Figure 11.9

To examine the behavior of the function at these points, evaluate

$$f_{xx}f_{yy} - f_{xy}^2 = (6y^2)(2x)(3x + 18y - 2) - [2y(6x + 97 - 2)]^2$$

$$= 4y^2\ [3x(3x + 18y - 2) - (6x + 9y - 2)^2]$$

at each point. At $(x_0,y_0) = (\frac{1}{6},\frac{1}{6})$, it is

$$4(\tfrac{1}{6})^2\ [\tfrac{3}{6}(\tfrac{3}{6} + \tfrac{18}{6} - 2) - (\tfrac{6}{6} + \tfrac{9}{6} - 2)^2] = \tfrac{1}{9}\ [\tfrac{1}{2}(\tfrac{3}{2}) - (\tfrac{1}{2})^2] > 0\ .$$

This, together with $f_{xx}(\frac{1}{6},\frac{1}{6}) = 6(\frac{1}{6})^2 > 0$, implies that f has a relative minimum at $(\frac{1}{6},\frac{1}{6})$; there $f(\frac{1}{6},\frac{1}{6}) = \frac{1}{6}(\frac{1}{6})^2(\frac{3}{6} + \frac{6}{6}) - 2) = -\frac{1}{424}$. At $(x_0,y_0) = (0,\frac{1}{3})$,

$$f_{xx}f_{yy} - f_{xy}^2 = (4)(\frac{1}{3})^2[3(0) + 18(\frac{1}{3}) - 2) - (6(0) + 9(\frac{1}{3}) - 2)^2]$$

$$= -\frac{4}{9} < 0 ,$$

so a saddle point occurs there. The other critical values all have $y = 0$; there $f_{xx}f_{yy} - f_{xy}^2 = 0$ so the test proves inconclusive and we do not know what the behavior of the function is at these points.

Now what does the surface defined by $f(x,y) = xy^2 (3x + 6y - 2)$ look like? No one figure adequately describes all of its characteristics. The computer drawn surfaces in Figures 11.10 - 11.11 each look down on the surface from above but have differing scales. Figure 11.11 represents the small area shaded in Figure 11.10. These two figures relate to each other this way: As you increase the power on a microscope, your field of vision shrinks but you get more magnification. In Figure 11.10 x and y both lie between -1 and 1 and the function ranges from -5 to 11. Here, the region around the origin still looks flat; in fact a small depression of negative values surrounds the relative minimum at $(\frac{1}{6},\frac{1}{6})$. Note the gullies in the foreground and background, and the ridges counterclockwise from the gullies. The region $0 \le x \le 0.7$ and $0 \le y \le 0.35$ of Figure 11.11 clearly shows this depression. This dip does not show up in Figure 11.10 because it is so shallow $(f(\frac{1}{6},\frac{1}{6}) = \frac{1}{424})$ relative to the rest of the surface. At the saddle point $(0,\frac{1}{3})$, the function changes from negative to zero back to negative along $x + y = \frac{1}{3}$ (from the dip over into the gully). Along the opposite direction the function changes from positive to zero to positive.

The function $f(x,y) = xy^2 (3x + 6y - 2)$ looks so innocent. How does it produce this bizarre surface? The trough along the x-axis occurs because both partial derivatives equal zero there; the function changes sign as it crosses the y-axis. These behaviors reflect the

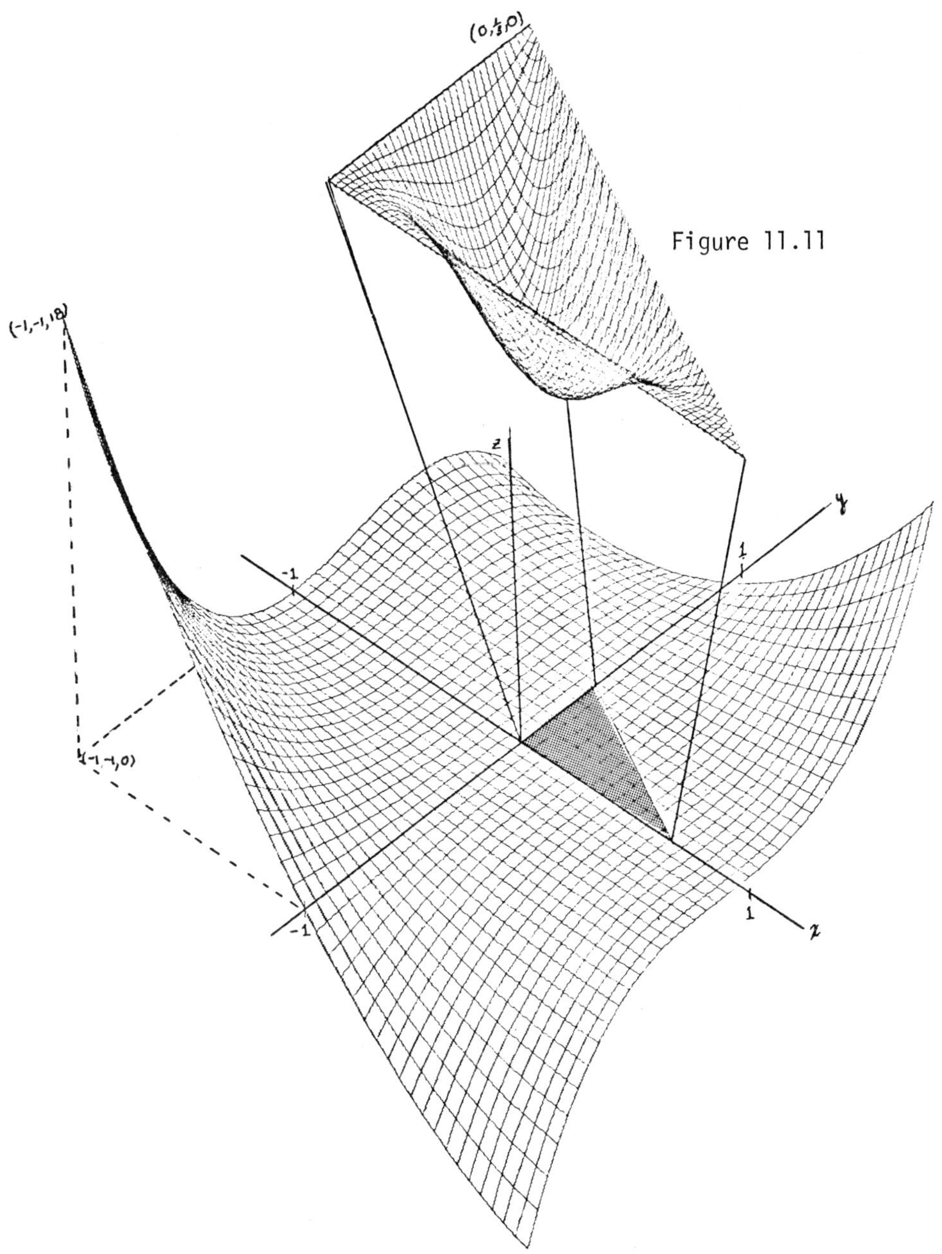

Figure 11.11

Figure 11.10

term xy^2, linear in x, but quadratic in y. The deepening gullies lie along $x + y = \frac{1}{3}$, because $\frac{\partial f}{\partial x} = 0$ there. For a fixed value of y, say $y = b$, $f(x,b) = b^2x(3x + 6b - 2)$, a quadratic curve. The minima of this family of parabolas lie along the line where $\frac{\partial f}{\partial x} = 0$. For a fixed value of x, say $x = a$, $f(a,y) = ay^2(6y + 3a - 2)$, a cubic curve. Each of these curves has $\frac{\partial f}{\partial y} = 0$ along $y = 0$ (the x-axis) and $a + 3y = \frac{2}{3}$. Relative minima and maxima of each curve, respectively, occur at these points. For all values of x, this makes the trench along the x-axis and the ridge over $x + 3y = \frac{2}{3}$. The small depression around $(\frac{1}{6},\frac{1}{6})$ results from a conflict between the downward tendency of the quadratic near its minimum and the upward tendency of the cubic near its maximum, both operating with small values of x and y. ▌

EXERCISES

Find the relative maxima, minima and saddle points if they exist.

1. $z = x^2 - xy + y^2$
2. $u = xy - x - y - 1$
3. $f(x,y) = x^2 + 2y^2 - 4x + 4y - 3$
4. $f(x,y) = xy + (x + y)(120 - x - y)$
5. $z = x^4 + y^4 - 2(x - y)^2$
6. $f(x,y) = x^2 + y^2 - 4x + 6y + 30$
7. $f(x,y) = x^2 - y^2 + 2x - 4y - 2$
8. $f(x,y) = x^2 - 3xy + y^2 + 13x - 12y + 13$
9. $z = x^3 + y^3 - 3xy$
10. $f(x,y) = 2x^2 + 3y^2 - 4x + 12y + 14$
11. $f(x,y) = xy - 2x + y + 1$
12. $f(x,y) = 4x^2y + 2xy^2 + xy$

13. $f(x,y) = 2x^3 + 3xy^2 - 6x$

14. $f(x,y) = 3x^4 - 6x^2 - 12xy + 2y^2$

15. Suppose that you are put in charge of designing an apparatus to hold biological material aboard a bio-satellite. You are told that the sum of its length and girth (perimeter of cross section) must not exceed 100 inches. Find the dimensions of the apparatus to maximize its volume assuming it is shaped like a box.

16. Suppose you (a poor student) need to design a plexiglass cylindrical tank with an open top to hold 256 cubic feet of liquid. If you are paying for this yourself, what are the dimensions of the tank that requires the least material to build?

17. If you were to design a box with a top of fixed volume with a minimum surface area, what would be the dimensions of the box?

*18. Find the point on the surface $z = xy - 1$ nearest the origin.

11.5 DIFFERENTIATION OF COMPLICATED FUNCTIONS

Real problems do not always produce nice functions; in fact, complicated functions seem more the rule than the exception. This statement applies equally to functions of one or several variables. The first subsection deals with implicit differentiation, an extension of Section 7.6a. The second subsection extends the chain rule developed in Section 6.3c to functions of more than one variable.

11.5a IMPLICIT DIFFERENTIATION

Recall that a function can be defined IMPLICITLY, meaning that the independent and dependent variables are related by $g(x,y) = 0$ or it may be defined EXPLICITLY by $y = f(x)$. The defining equation may involve several independent variables, or it may, as we have seen previously, involve only one independent variable. The following examples illustrate the natural extension of IMPLICIT DIFFERENTIATION to functions of several independent variables. It merely consists of taking the partial derivative of the whole defining equation and subsequently solving for the partial derivative of interest. Again algebraic examples will illustrate the method.

EXAMPLE 11.11: If $x^3 + 2xz - 2yz^2 - z^3 = 13$, find $\frac{\partial z}{\partial x}$.

The $\frac{\partial z}{\partial x}$ indicates that z is to be considered the dependent variable with x being the independent variable and y is to be considered a constant. Take the partial derivative of both sides of the equation with respect to x:

$$\frac{\partial}{\partial x}(x^3 + 2xz - 2yz^2 - z^3) = \frac{\partial}{\partial x}(13)$$

or

$$3x^2 + 2x\frac{\partial z}{\partial x} + 2z - 4yz\frac{\partial z}{\partial x} - 3z^2\frac{\partial z}{\partial x} = 0 .$$

Now solving this equation for $\frac{\partial z}{\partial x}$, we have

$$(2x - 4yz - 3z^2)\frac{\partial z}{\partial x} = -3x^2 - 2z$$

or

$$\frac{\partial z}{\partial x} = \frac{-3x^2 - 2z}{2x - 4yz - 3z^2} .$$

Thus, we obtained $\frac{\partial z}{\partial x}$ without having first solved for z in terms of x and y. To evaluate this partial derivative at the point (x_0, y_0, z_0), simply substitute in these values into the partial derivative and obtain a value for $\frac{\partial z}{\partial x}$. The advantage of implicit differentiation should be obvious to you. If it is not, try to solve the original equation for z in terms of x and y so you can differentiate. ▌

EXAMPLE 11.12: Evaluate both $\frac{\partial z}{\partial x}$ and $\frac{\partial z}{\partial y}$ when $\ln(4x - z^2) - (x - y)e^z = 0$.

Apply $\frac{\partial}{\partial x}$ to both sides of the equation, we obtain

$$\frac{\partial}{\partial x}(\ln(4x - z^2) - (x - y)e^z) = \frac{\partial}{\partial x}(0) ,$$

or

$$\frac{1}{4x - z^2}\frac{\partial}{\partial x}(4x - z^2) - (x - y)\frac{\partial}{\partial x}e^z - e^z\frac{\partial}{\partial x}(x - y) = 0 ,$$

so

$$\frac{1}{4x - z^2}\left(4 - 2z\frac{\partial z}{\partial x}\right) - (x - y)e^z\frac{\partial z}{\partial x} - e^z = 0 .$$

Solving this for $\frac{\partial z}{\partial x}$, we obtain

$$\frac{\partial z}{\partial x} = \frac{e^z - \frac{4}{4x - z^2}}{-\frac{2z}{4x - z^2} - (x - y)e^z} = \frac{(4x - z^2)e^z - 4}{-2z - (x - y)(4x - z^2)e^z}$$

Similarly, verify that

$$\frac{\partial z}{\partial y} = \frac{(4x - z^2)e^z}{2z + (x - y)(4x - z^2)e^z} .$$

Again, it should be apparent that implicit differentiation provides the easiest means of evaluating the requested partial derivatives. ▌

EXERCISES

Find $\frac{\partial z}{\partial x}$ and $\frac{\partial z}{\partial y}$; evaluate the partial derivatives at the given point, if stated.

1. $x^2 + 4y^2 + z^2 = 8$
2. $e^z + e^y + e^x = 5$
3. $x^2 + 4y^2 - z^2 = 4$ at $(-3,1,3)$
4. $z^2 + 2xz + y^2 = 14$ at $(2,3,1)$
5. $x^3 + y^3 + z^3 = 6$
6. $\ln x + \ln y + \ln z = e^x$
7. $z^2 - xz^3 + 10yz = -78$
8. $x^{2/3} + y^{2/3} + z^{2/3} = 9^{2/3}$
9. $(x + 2z)^7 + (3y - z^2)^5 = 2$
10. $\ln(3x + z) = 4 - e^{y-3z}$
11. $\frac{x + 3z}{y^2 + z^2} = 4e^z$
12. $\frac{e^x - e^{2x}}{e^y + e^z} = 2x + 5yz$
13. $e^{2x+3z} = 6 - e^{2z-3y}$
14. $x^2 + y^2 + z^2 + 2xy + 4yz + 10zx = 20$
15. $x + 4y - 2z = \ln z$

16. Find $\frac{\partial^2 z}{\partial x^2}$ and $\frac{\partial^2 z}{\partial y^2}$ if $x^2 + 2yx + 2xz = 1$.

17. Find z_{xx} and z_{yy} if $x^2 + y^2 + z^2 = 25$.

18. If $z_x = \frac{x + 3z}{y + z^2}$, find z_{xy} and z_{xx}.

11.5b CHAIN RULE FOR FUNCTIONS OF SEVERAL VARIABLES

Recall the chain rule for one independent variable: If $y = f(u)$ and $u = g(x)$, then

$$\frac{dy}{dx} = \frac{dy}{du}\frac{du}{dx} \qquad \text{or} \qquad \frac{dy}{dx} = \frac{df}{du}\frac{dg}{dx}. \tag{11.3}$$

By writing the composite function $y = f(g(x))$, two facts emerge: First, y does change with x. Secondly, the nature of both f and g will influence how y changes with x. This appears in the chain rule because it involves the derivatives of both f and g. The two forms emphasize two perspectives- The form $\frac{dy}{du}\frac{du}{dx}$ is easy to remember; the form $\frac{df}{du}\frac{dg}{dx}$ emphasizes u is an argument of f for $\frac{df}{du}$, but a function of x in $\frac{dg}{dx}$.

Both the situation and the chain rule generalize to two or more independent variables. Suppose a variable (z) depends on two variables (u and v), each of which depends on another pair of variables (x and y). Such circumstances lead us to inquire about the partial derivatives $\frac{\partial z}{\partial x}$ and $\frac{\partial z}{\partial y}$. We state the needed result, the CHAIN RULE FOR FUNCTIONS OF SEVERAL VARIABLES, then illustrate its use. We will not prove the chain rule for functions of several variables.

- When $z = f(u,v)$ with $u = g(x,y)$ and $v = h(x,y)$, then

$$\frac{\partial z}{\partial x} = \frac{\partial z}{\partial u}\frac{\partial u}{\partial x} + \frac{\partial z}{\partial v}\frac{\partial v}{\partial x} \tag{11.4}$$

and

$$\frac{\partial z}{\partial y} = \frac{\partial z}{\partial u}\frac{\partial u}{\partial y} + \frac{\partial z}{\partial v}\frac{\partial v}{\partial y} \tag{11.5}$$

provided all of these partial derivatives exist at the points of interest. These forms sometimes are written as

$$\frac{\partial f}{\partial x} = \frac{\partial f}{\partial u}\frac{\partial g}{\partial x} + \frac{\partial f}{\partial v}\frac{\partial h}{\partial x} \qquad (11.6)$$

and

$$\frac{\partial f}{\partial y} = \frac{\partial f}{\partial u}\frac{\partial g}{\partial y} + \frac{\partial f}{\partial v}\frac{\partial h}{\partial y} . \qquad (11.7)$$

- When $z = f(u,v)$ with $u = g(t)$ and $v = h(t)$, then

$$\frac{\partial z}{\partial t} = \frac{\partial z}{\partial u}\frac{du}{dt} + \frac{\partial z}{\partial v}\frac{dv}{dt} \qquad (11.8)$$

provided all of these derivatives exist at the points of interest. This is sometimes written as

$$\frac{\partial f}{\partial t} = \frac{\partial f}{\partial u}\frac{dg}{dt} + \frac{\partial f}{\partial v}\frac{dh}{dt} . \qquad (11.9)$$

The chain rule for one variable states $\frac{dy}{dx} = \frac{dy}{du}\frac{du}{dx}$; each of the above forms has two terms of this sort. The sum results for a simple reason: As x changes by Δx, both u and v will change so the derivatives of both must appear in $\frac{\partial z}{\partial x}$. The first form (Equations 11.4 and 11.5) emphasizes the symmetry in u and v; it is an easier form to remember. The other form (Equations 11.5 and 11.6) emphasizes the involvement of functions and then arguments. Equation 11.9 states an important special case when both u and v depend on the same variable t. In biological applications, t frequently would stand for time.

EXAMPLE 11.13: If $z = v + \ln uv$ with $u = e^{x^2}$ and $v = e^{2xy+y^2}$, find $\frac{\partial z}{\partial x}$ and $\frac{\partial z}{\partial y}$.

From the chain rule

$$\frac{\partial z}{\partial x} = \frac{\partial z}{\partial u} \times \frac{\partial u}{\partial x} + \frac{\partial z}{\partial v} \times \frac{\partial v}{\partial x}$$

$$= \frac{\partial}{\partial u}(v + \ln uv)\,\frac{\partial}{\partial x} e^{x^2} + \frac{\partial}{\partial v}(v + \ln uv)\,\frac{\partial}{\partial x}(e^{2xy+y^2}) .$$

The chain rule for a function of a single variable has to be used to evaluate

$$\frac{\partial}{\partial x}(e^{x^2}) = 2xe^{x^2} \qquad \text{and} \qquad \frac{\partial}{\partial x}(e^{2xy+y^2}) = 2ye^{2xy+y^2} ,$$

giving

$$\frac{\partial z}{\partial x} = \frac{1}{u}\, 2xe^{x^2} + (1 + \frac{1}{v})2ye^{2xy+x^2} \ .$$

After substituting for x and y and simplifying, this becomes

$$\frac{\partial z}{\partial x} = 2ye^{2xy+y^2} + 2(x + y) \ . \qquad (11.10)$$

Similarly you should verify that

$$\frac{\partial z}{\partial y} = 2(x + y)(1 + e^{2xy+y^2}) \ . \qquad (11.11)$$

These functions allow expression of z in terms of x and y so the partial derivatives can be evaluated directly:

$$z = v + \ln uv = e^{2xy+y^2} + \ln e^{x^2}e^{2xy+y^2} = e^{2xy+y^2} + (x + y)^2$$

Thus direct evaluation of its partial derivatives also gives Equations 11.10 and 11.11. The functions in this example allow us to represent z in terms of x and y; this frequently cannot be done. ▮

An important special case of the chain rule occurs when u and v are functions of a single variable. Equation 11.7 gives this special case. The following example illustrates its use.

<u>EXAMPLE 11.14</u>: At a certain instant a right circular cone has a height of 30 inches which is increasing at 2 inches/minute, while the radius of the base is 20 inches and is increasing by 1 inch/minute. At what rate is the volume of the cone increasing?

The volume of a cone is given by $V = \frac{\pi r^2 h}{3}$. As both the radius, r, and the height, h, depend upon time, t, the volume likewise depends on time. Thus to examine how the volume is changing with respect to time, use the chain rule:

$$\frac{dV}{dt} = \frac{\partial V}{\partial r}\frac{dr}{dt} + \frac{\partial V}{\partial h}\frac{dh}{dt}$$

or

$$\frac{dV}{dt} = \frac{2\pi rh}{3}\frac{dr}{dt} + \frac{\pi r^2}{3}\frac{dh}{dt} \ .$$

From the information above,

$r = 20$ inches, $h = 30$ inches, $\frac{dr}{dt} = 1$ inch/minute and $\frac{dh}{dt} = 2$ inches/minute. Substituting these values into the last equation, we obtain

$$\frac{dV}{dt} = \frac{2\pi}{3}(20 \text{ inches})(30 \text{ inches})\left(\frac{1 \text{ inch}}{\text{min.}}\right) + \frac{\pi}{3}(20 \text{ inches})^2\left(\frac{2 \text{ inches}}{\text{min.}}\right)$$

and

$$\frac{dV}{dt} = 400\pi \text{ in.}^3/\text{min.} + \frac{800\pi}{3} \text{ in.}^3/\text{min.} = \frac{2000\pi}{3} \text{ in.}^3/\text{minute.}$$

The rate of change of the volume with respect to time should have volumetric units per unit of time; indeed this is the case here.

This problem illustrates that we do not need to express the volume as a function of a single variable as was necessary in our previous work. ▌

EXAMPLE 11.15: The trunk of a lodge pole pine can be viewed as being cylindrical in shape. Suppose that a particular tree has a radius of 3 inches and a height of 44 feet with the radius increasing 0.25 inches/year and the height 4 inches/year. How fast is the volume of the trunk of the tree changing?

Because the tree has a cylindrical shape, its volume is given by $V = \pi r^2 h$. From the chain rule we have

$$\frac{dV}{dt} = \frac{\partial V}{\partial r}\frac{dr}{dt} + \frac{\partial V}{\partial h}\frac{dh}{dt}$$

$$= 2\pi rh \frac{dr}{dt} + \pi r^2 \frac{dh}{dt} .$$

Substituting the known values in terms of feet, this gives

$$\frac{dV}{dt} = 2\pi \left(\frac{1}{4} \text{ ft.}\right)(44 \text{ ft.})\left(\frac{1}{48}\frac{\text{ft.}}{\text{yr.}}\right) + \pi \left(\frac{1}{4} \text{ ft.}\right)^2 \frac{1}{3}\frac{\text{ft.}}{\text{yr.}}$$

$$= \frac{11\pi}{24} \text{ ft.}^3/\text{yr.} + \frac{\pi}{48} \text{ ft.}^3/\text{yr.}$$

$$= \left(\frac{22\pi}{48} + \frac{\pi}{48}\right) \text{ ft.}^3/\text{yr.} = \frac{23\pi}{48} \text{ ft.}^3/\text{year.}$$ ▌

The chain rule extends to any number of variables. For example, suppose z has a functional dependence on u, v and w with each of these variables depending on x and y.

$$\frac{\partial z}{\partial x} = \frac{\partial z}{\partial u}\frac{\partial u}{\partial x} + \frac{\partial z}{\partial v}\frac{\partial v}{\partial x} + \frac{\partial z}{\partial w}\frac{\partial w}{\partial x} \tag{11.12}$$

and

$$\frac{\partial z}{\partial y} = \frac{\partial z}{\partial u}\frac{\partial u}{\partial y} + \frac{\partial z}{\partial v}\frac{\partial v}{\partial y} + \frac{\partial z}{\partial w}\frac{\partial w}{\partial y} \tag{11.13}$$

EXERCISES

1. If $w = u^2 - v^2$ and $u = s + t$, $v = s - t$, find $\frac{\partial w}{\partial s}$ and $\frac{\partial w}{\partial t}$.

2. If $w = \ln x + e^y + xy$ and $x = \ln t$, $y = t^2$, find $\frac{dw}{dt}$.

3. If z is a function of s and t and $s = x - ay$, $t = x + ay$, show that $\frac{\partial z}{\partial x} = \frac{\partial z}{\partial s} + \frac{\partial z}{\partial t}$, $\frac{1}{a}\frac{\partial z}{\partial y} = -\frac{\partial z}{\partial s} + \frac{\partial z}{\partial t}$.

4. If $z = x^2 + y^2$ and $x = s - 2t$, $y = 2s + t$, find $\frac{\partial z}{\partial s}$ and $\frac{\partial z}{\partial t}$ when $(s,t) = (2,1)$.

5. Find $\frac{dz}{dt}$ if $z = f(t,y) = t^2 + 4ty + 6y^2$ where $y = e^{at}$.

6. Show that if $z = f(x,y)$ is a continuous function of x and y with continuous partial derivatives $\frac{\partial z}{\partial x}$ and $\frac{\partial z}{\partial y}$, and, if y is a differentiable function of x, then

$$\frac{dz}{dx} = \frac{\partial f}{\partial x} + \frac{\partial f}{\partial y} \times \frac{dy}{dx}.$$

7. If $z = \ln(x^2 + 2y^2)$, $x = e^{-t}$ and $y = e^t$, find $\frac{dz}{dt}$.

8. If $z = x^2 + y^2 + xy$ and $x = 2r + s$, $y = r - 2s$, find $\frac{\partial z}{\partial r}$ and $\frac{\partial z}{\partial s}$.

9. If $w = \frac{x^2 + y^2}{y^2 + z^2}$ and $x = ue^v$, $y = ve^u$, $z = \frac{1}{u}$, find $\frac{\partial w}{\partial u}$ and $\frac{\partial w}{\partial v}$.

10. As a rectangular block of metal is being heated, its length, width and height are increasing. Write a general expression for the rate of change of the volume.

11. A right circular cone has a height of 10 cm. which is increasing at the rate of 0.5 cm./minute. If the radius of its base is 6 cm. and is decreasing at the rate of 0.25 cm./minute, how fast is the volume changing?

12. A rectangle has sides of length 8 and 12 inches at a particular instant. The shorter side is increasing 1 in./minute while the longer side is decreasing 1 in./minute. How fast is the area changing at this instant?

13. A perfect gas behaves according to the equation PV = RT, where R is a constant, P is the pressure of the gas, V is the volume, and T is the absolute temperature of the gas. At a given instant $P = 20$ lb/in.^{2}, $V = 10$ ft.^{3}, and $T = 300°$ absolute. If the gas is being compressed 0.5 ft.^{3}/minute, and the pressure is rising 2 lb/in.^{2}/minute, at what rate is the temperature changing?

14. Use the law given in Problem 13. The gas is being compressed at a certain instant when $V = 15$ ft.^{3} and $P = 25$ lb/in.^{2}, V is decreasing 3 ft.^{3}/min., and P is increasing $\frac{20}{3}$ lb/in.^{2}/min. Find $\frac{1}{T}\frac{dT}{dt}$.

15. If $u = f(x,y,z,w,s)$ and x, y, z, w, and s are functions of η and ξ, find the expression for $\frac{\partial u}{\partial \eta}$ and $\frac{\partial u}{\partial \xi}$.

16. A goose is traveling directly east 300 ft./min. and climbing 10 ft./minute. When the goose is 120 feet above the ground and 50 yards directly west of an observer on the ground, how fast is the distance between the man and goose changing?

17. Water is flowing into a vertical cylindrical tank at the rate of $\frac{4\pi}{5}$ cubic feet per minute. The tank is stretching in such a manner that even though it remains cylindrical, its radius is increasing at the rate of 0.0015 inches per minute. How fast is the depth of the water rising when the radius is 3 feet and the volume of the water is 20π cubic feet?

18. At a given point in time the height of a pine tree is 60 feet and is increasing at the rate of 8 inches per year and the radius of the base including the branches is 10 feet and is increasing at the rate of 1 foot per year. At what rate is the volume occupied by the tree changing if it is assumed that the tree has the shape of a cone?

19. Consider a cylindrical shaped bacterium of length L with hemispherical caps of radius R. Assume that $L - at$ and $R = bt^{1/3}$ where a and b are constants and t represents time. Subject to these conditions, how fast is the area and the volume of the bacterium changing at any time t?

20. If a rod shaped (cylindrical) bacterium has a radius of r and length L, how does its surface area change with changes in the radius and length?

21. A rod shaped bacterium of length 100μ and a diameter of 20μ swells upon being subjected to a medium of different ionic concentrations than its normal environment. Its radius increases at the rate of 0.1μ per hour and its length at 0.25μ per hour. At what rate is the volume of the cell changing?

*22. If $z = F(x,y)$, $x = f(u,v)$ and $y = g(u,v)$, show that

$$\frac{\partial^2 z}{\partial u^2} = \frac{\partial^2 z}{\partial x^2}\left(\frac{\partial x}{\partial u}\right)^2 + 2\frac{\partial^2 z}{\partial x \partial y}\frac{\partial x}{\partial u} \times \frac{\partial y}{\partial u} + \frac{\partial^2 z}{\partial y^2}\left(\frac{\partial y}{\partial u}\right)^2 + \frac{\partial z}{\partial x}\frac{\partial^2 x}{\partial u^2} + \frac{\partial z}{\partial y} \times \frac{\partial^2 y}{\partial u^2}.$$

11.6 DIFFERENTIALS AND SMALL ERRORS

A function of several variables has a differential which naturally extends that for functions of a single variable. Once the total differential has been developed, it can be used to find the relative error in a function's evaluation that is induced by small errors in its arguments.

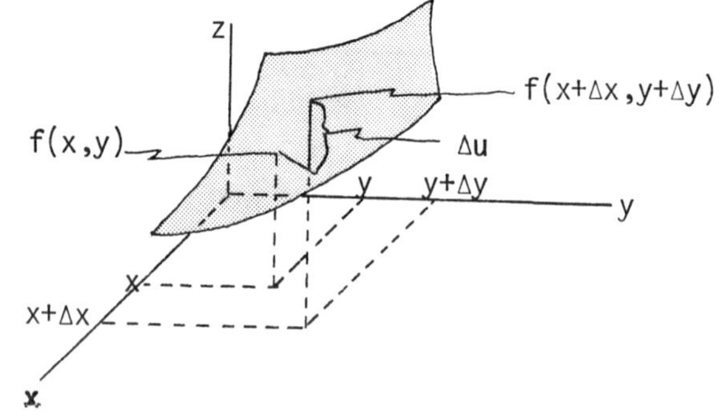

Figure 11.13

Consider a function
$u = f(x,y)$, where f_x and f_y are continuous functions of x and y. If x and y take on small increments Δx and Δy, then u has the corresponding increment

$$\Delta u = f(x + \Delta x, y + \Delta y) - f(x,y) \ .$$

This increment is shown geometrically in Figure 11.13.

Following reasoning which parallels that which led to the differential of a function of a single variable, u can be approximated by

$$\Delta u \doteq du = \frac{\partial f}{\partial x} \Delta x + \frac{\partial f}{\partial y} \Delta y \ . \tag{11.14}$$

As we may set $dx = \Delta x$ and $dy = \Delta y$, then we also can write

$$du = \frac{\partial f}{\partial x} dx + \frac{\partial f}{\partial y} dy \ . \tag{11.15}$$

In a similar fashion if $u = f(x,y,z)$, then the total differential of u becomes

$$du = \frac{\partial u}{\partial x} dx + \frac{\partial u}{\partial y} dy + \frac{\partial u}{\partial z} dz \ . \tag{11.16}$$

The differential represents an approximation to the true change in u.

<u>EXAMPLE 11.16</u>: Find the approximate change in $u = 6x^2y^3$ as x goes from 2 to 2.1 and y goes from -1 to -1.2.

Equation 11.5 gives

$$du = \frac{\partial u}{\partial x} dx + \frac{\partial u}{\partial y} dy = 12xy^3dx + 18x^2y^2dy$$

From $x = 2$, $dx = 0.1$, $y = -1$ and $dy = -0.2$,

$$du = 12(2)(-1)^3(0.1) + 18(4)(-1)^2(-0.2)$$
$$= -2.4 - 14.4 = -16.8 \ .$$

The true change in u is

$$\Delta u = 6(2.1)^2(-1.2)^3 - 6(2)^2(-1)^3 = -45.72288 + 24 = -21.72288$$

The differential provides a poor approximation in this case because u is quite curved near (2,-1) . ∎

<u>EXAMPLE 11.17</u>: Consider the formula VP = RT, where R, V, T and P, respectively, symbolize a constant, the volume, the absolute temperature, and pressure, respectively, of a perfect gas. Approximate the maximum error in P introduced by an error of ±0.4% in measuring the temperature and an error of ±0.9% in measuring the volume.

As $\Delta P \doteq dP$, we need to evaluate

$$dP = \frac{\partial P}{\partial T}\,dT + \frac{\partial P}{\partial V}\,dV = \frac{R}{V}\,dT - \frac{RT}{V^2}\,dV$$

This inequality always is true: $|a + b| \le |a| + |b|$. Check both positive and negative values for a and b to convince yourself. Applied here,

$$\left|dP\right| \le \left|\frac{R}{V}(0.004T)\right| + \left|\frac{-RT}{V^2}(0.009V)\right| = \frac{RT}{V}(0.004 + 0.009) = 0.013\frac{RT}{V}$$

$$= 0.013P$$

because dT = 0.004T and dV = 0.009V. The absolute values were used to assure the maximum error. Thus measurement errors of ±0.4% in the temperature and of ±0.9% in the volume result in a maximum error of 1.3% in the pressure. ∎

Earlier in this section we introduced the idea of the error in a function of several variables caused by small errors in each of its independent variables:

$$\Delta u = f(x + dx, y + dy) - f(x,y) \doteq f_x dx + f_y dy\ .$$

When dx and dy are interpreted as errors in the data, the corresponding error in u can be expressed relative to u as $\frac{\Delta u}{u}$. If the errors in the data, dx and dy, are small enough, then we shall consider du as an approximation for Δu. Then

$$\text{the approximate relative error in } u = \frac{du}{u} = d(\ln u)\ .$$

This procedure is illustrated in the next example.

<u>EXAMPLE 11.17</u> (continued): Find the maximum relative error in the pressure when

$$P = \frac{RT}{V}\ .$$

The relative error in P equals $\frac{dP}{P}$ or $d(\ln P)$. Since we already found that the maximum error in P was 0.013P,

$$\frac{dP}{P} = \frac{0.013P}{P} = 0.013 \ .$$

The alternate method would be to find $d \ln P$. Thus, taking the natural logarithm of both sides of $P = \frac{RT}{V}$, we have

$$\ln P = \ln R + \ln T - \ln V \ .$$

Differentiating this equation, we have

$$\frac{dP}{P} = \frac{dT}{T} - \frac{dV}{V}$$

so

$$\left|\frac{dP}{P}\right| \leq \left|\frac{\pm 0.004T}{T} - \frac{\pm 0.009V}{V}\right| \leq 0.013 \ .$$

We choose signs within the absolute value to achieve an upper bound on the relative error. ∎

EXERCISES

Find the total differential and perform the indicated operations in Problems 1-7.

1. $f(x,y) = x^4 - x^2y^2 + x^2y - y^3$

2. $V = \pi r^2 h$

3. $u = (\ln x)(\ln y)(\ln z)$

4. $u = 4xy^2$; compute du and Δu when $x = 3$, $y = -2$, $dx = 0.2$ and $dy = -0.1$.

5. $u = 2x + 4x^2y^2 - y^4$

6. $v = e^{2x+3y-4z}$

7. $w = xz^2e^y$

8. In determining the specific gravity of an object, its weight in air was found to be 36 kilograms while its weight in water was 20 kilograms, with a possible error in each measurement of 0.02 kilograms. Approximate the maximum error in calculating its specific gravity S where $S = \frac{A}{A - W}$ and A is the object weight in air and W is object weight in water.

9. If a 2" × 3" × 4" box with a lid is to be constructed with plexiglass $\frac{1}{16}$ inch thick, approximately what volume of plexiglass will be used? Does the true result depend on whether the dimensions are inside or outside? Will the approximation differ?

10. Pressure P, volume V, and absolute temperature T of a perfect gas relate through the law PV = RT, where R is a constant. If the volume is 120 ft.^{3} when the pressure is 14.7 lb./in.^{2} and T = 295° absolute, use differentials to find the approximate pressure of the gas when the volume is 121 ft.^{3} and T = 300° absolute.

11. In establishing the volume of a rod-shaped bacteria (the bacteria is in the shape of a right circular cylinder), suppose an error of 1% is made in measuring its radius and a 2% error in measuring its length. Find the greatest relative error in calculating its volume.

12. If errors are made in measuring a rectangular field plot of length x and width y, find the error in the computed area of the plot. Approximate the error in the computed area assuming relatively small measurement errors. Illustrate each part graphically.

13. The constant C in Boyle's law for gases satisfies PV = C where P is the pressure of the gas and V is the volume of the gas. If P and C are known, with a possible error of 2% in P and 1% in C, find the maximum percentage error in V.

14. The radius of the base of a right circular cone was measured as 50 cm. and its altitude as 20 cm. subject to a possible error of 0.2 cm. in each dimension. Approximate the maximum error and the

14. Continued
maximum relative percentage error in the computed volume of the cone.

15. For a perfect gas, the volume V, pressure P, and absolute temperature T are connected by the law PV = RT where R is a constant. If T = 400°K when P = 50 lb./in.^{2} and V = 600 in.^{3}, approximate the change and percentage change in P if T increases by 10°K and V by 18 in.^{3}.

16. The diameter and the height of a rod-shaped bacterium were measured as 8 and 12.5μ, respectively, with possible errors of 0.05μ in each measurement. Approximate the maximum error in the computed volume.

17. In Problem 8, approximate the maximum error and maximum relative percentage error in S if the weight of the object in air is 3.2 ± 0.05 pounds and in water is 2.7 ± 0.05 pounds.

18. The heat Q, in calories, liberated on mixing x gram moles of sulfuric acid with y gram moles of water is given by $Q = \frac{1.786xy}{1.798x + y}$. Approximate the additional heat generated if a mixture of 5 gram moles of acid and 4 gram moles of water has its acid content increased by 1% and its water content increased by 2%.

19. The density ρ of dry air in grams/cm.^{3} at T°C and pressure P mm. of mercury is given by

$$\rho = \frac{0.001293}{1 + 0.00367T} \quad \frac{P}{760} .$$

Obtain an expression giving the approximate change in the value of ρ for small changes in T and P. Approximate the change in ρ when T goes from 25°C to 26°C and P goes from 760 mm. to 762 mm. of mercury.

20. The maximum safe load for one kind of beam varies in direct proportion to its breadth and the square of its depth, and inversely to the distance between the beam's supports. If the breadth, depth and distance between supports are measured with possible errors of 2%, 3% and 1%, respectively, approximate the maximum percentage error in a load computed from these measurements.

21. If a rectangular box of length L, width W and height H has its dimensions increased by ΔL, ΔW and ΔH respectively, approximate the increase in the volume. Calculate the true change in the volume and show by a graphical representation the quantities which make up the approximation and the actual change.

22. If we consider a blood vessel of length L and of radius r, then Poiseville's law states that the resistance R of the blood vessel is given by $R = \frac{kL}{r^4}$ where k is a constant. How do small changes in r and L affect the resistance R?

23. The pressure (P) of a gas is given in terms of the volume (V) and temperature (T) by the formula $P = \frac{RT}{V}$ where R is a constant. If there is a 1% error in measuring T and a 0.5% error in measuring T, find approximately the greatest error made in computing P from the formula.

24. A cylindrical cell has a surfaçe area of $S = \pi r^2 \ell + 2\pi r^2$ and a volume of $V = \pi r^2 \ell$. How do small changes in r and ℓ affect S and V?

25. The coefficient of viscosity of a fluid, η, can be calculated by measuring the total volume Q of fluid flowing per second through a tube of length L with an internal radius of r and a pressure of P and using the formula $Q = \frac{\pi P r^4}{8L\eta}$. If errors of ±2% can be made in measuring r and L, ±3% in the measurement of Q and ±1% in P, what is the maximum relative percentage error in the estimate of η?

26. The growth of a bacterial colony follows the law $N = N_0 e^{kt}$ where N is the number of bacteria per square centimeter, N_0 is the number of bacteria at time t = 0, k is a constant and t is time. This law holds as long as there is no appreciable competition between bacteria or exhaustion of the food supply. What is the relative error in determining k if there is an error of ΔN in N, ΔN_0 in N_0 and Δt in t? What is the relative error in k?

27. The speech and hearing department has acquired an unused room which measures 20 feet long by 15 feet wide by 8 feet high. They wish to soundproof the room by placing carpet 0.75 inch thick on the floor 0.75 inch acoustical tile on the ceiling and paneling of 0.5 inch thickness. Find the approximate loss in room volume during this soundproofing process.

Chapter 12

MULTIPLE INTEGRATION

The idea of an integral extends from one to several variables of integration much as the simple derivative was extended to a partial derivative. Double and triple integrals are evaluated by using familiar methods of integration two or three times. Thus, this chapter begins by introducing the procedures of repeated integration. The subsequent sections explore the meaning and applications of multiple integration.

As you gain experience with multiple integrals of specific functions your insight into their meaning should mature. This insight will be at least as important in biology as the operations of multiple integration. At this time, a significant use of multiple integrals by biologists does not involve specific functions. Multiple integrals of general functions sometimes are used to communicate the workings of a biological system. The function becomes specific only at a particular point in time, space and for a given set of conditions. Even then, it frequently is evaluated only in a discrete manner and thus the integral is approximated by techniques similar to the numerical approximations of Section 8.11. The examples also will illustrate this use of integrals.

12.1 REPEATED INTEGRATION

This section has a simple objective: By its end you should be able to evaluate repeated or iterated integrals. It does not deal with their meaning, properties, uses, or interpretations; later sections present these important topics.

The repeated integral

$$\int_2^3 \int_4^{27} xy \, dy \, dx$$

should be interpreted as

$$\int_2^3 [\int_4^{27} xy \, dy]dx = \int_2^3 x \, [\int_4^{27} y \, dy]dx$$

First consider the differentials in the order displayed above, namely, dy dx. This means to first integrate with respect to y, then with respect to x. During the evaluation of the inner integral above the variable y changes, but x remains constant. By holding x constant, the inner integral becomes a function of the single variable x. This integral then can be evaluated using methods you already know. The inner integration ordinarily will produce a function of x, but y will be absent because it has been integrated out. As this function of x appears inside an integral with respect to x, again evaluate as before. In this process each integration is completed before the next is initiated. Specifically, the limits on the inner integral are substituted into the first antiderivative before the second integration begins. By convention treat repeated integrals just like an algebraic expression containing parentheses: Start on the inside and work outward. When the differentials appear in the order dx dy, the above comments still apply, but with the roles of x and y interchanged.

EXAMPLE 12.1: Evaluate $\int_2^3 \int_4^{27} xy\ dy\ dx$:

$$\int_2^3 \int_4^{27} xy\ dy\ dx = \int_2^3 \left[\int_4^{27} xy\ dy\right]dx = \int_2^3 x\left[\int_4^{27} y\ dy\right]dx$$

because x is treated as a constant with respect to the integration of y. Continuing:

$$\int_2^3 x\left[\int_4^{27} y\ dy\right]dx = \int_2^3 x\left[\frac{y^2}{2}\right]_4^{27} dx = \int_2^3 x\left[\frac{27^2}{2} - \frac{4^2}{2}\right]dx$$

$$= \int_2^3 \frac{713}{2}x\ dx = \frac{713}{4}x^2\Big]_2^3 = \left(\frac{713}{4}\right)5 = \frac{3565}{4}\ . \ \blacksquare$$

The interior limits of integration may involve the outer variable of integration. This will cause the first integration to contribute variables as well as constants to the second integration.

EXAMPLE 12.2: Evaluate $\int_2^3 \int_{2x}^{3x^2} xy\ dy\ dx$. This differs from the previous example only in the variable limits of integration on the interior integration.

$$\int_2^3 \int_{2x}^{3x^2} xy\,dy\,dx = \int_2^3 x\left[\int_{2x}^{3x^2} y\,dy\right]dx = \int_2^3 x\left[\frac{y^2}{2}\right]_{2x}^{3x^2} dx$$

$$= \int_2^3 \frac{x}{2}(9x^4 - 4x^2)dx = \frac{1}{2}\int_2^3 (9x^5 - 4x^3)dx$$

$$= \frac{9}{2}\int_2^3 x^5 dx - 2\int_2^3 x^3 dx = \frac{9}{2}\left[\frac{x^6}{6}\right]_2^3 - 2\left[\frac{x^4}{4}\right]_2^3$$

$$= \frac{9}{12}(3^6 - 2^6) - \frac{2}{4}(3^4 - 2^4) = \frac{9}{12}(729 - 64) - \frac{2}{4}(81 - 16)$$

$$= \frac{3}{4}(665) - \frac{1}{2}(65) = \frac{1865}{4}. \quad \blacksquare$$

These examples illustrate an important point which may seem obvious: All of the previously developed properties of integration apply at each repeated integration. Specifically we wrote

$$\int_2^3 \int_{2x}^{3x^2} xy\,dy\,dx = \int_2^3 x\left[\int_{2x}^{3x^2} y\,dy\right]dx$$

because x remains constant during the integration relative to y so it can be factored out.

Integration repeated twice generalizes in an apparent way to integration repeated three times. For example,

$$\int_a^b \int_{y_1(x)}^{y_2(x)} \int_{f_1(x,y)}^{f_2(x,y)} f(x,y,z)dz\,dy\,dx = \int_a^b \left\{\int_{y_1(x)}^{y_2(x)} \left[\int_{f_1(x,y)}^{f_2(x,y)} f(x,y,z)dz\right] dy\right\}dx$$

means to first integrate f(x,y,z) with respect to z while x and y both remain constant; next integrate with respect to y treating x as a constant; and finally integrate with respect to x. After each integration the limits of integration are substituted before the next integration is executed. The next example illustrates the process of triple integration:

<u>EXAMPLE 12.3</u>: Evaluate $T = \int_0^3 \int_0^{6-z} \int_0^x (y - x)\,dy\,dx\,dz$.

Note that the order of the differentials implies that we should integrate in the order y, x, z. First integrate with respect to y holding x and z constant:

$$T = \int_0^3 \int_0^{6-z} \left[\int_0^x (y - x)dy\right] dx\, dz$$

$$= \int_0^3 \int_0^{6-z} \left[\frac{(y - x)^2}{2}\right]_0^x dx\, dz$$

$$= \int_0^3 \int_0^{6-z} \left[0 - \frac{x^2}{2}\right] dx\, dz .$$

Now treat z as a constant:

$$T = -\frac{1}{2}\int_0^3 \left[\int_0^{6-z} x^2\, dx\right] dz$$

$$= -\frac{1}{2}\int_0^3 \left.\frac{x^3}{3}\right]_0^{6-z} dz = -\frac{1}{6}\int_0^3 (6 - z)\ \ dz .$$

Now integrate with respect to z:

$$T = \frac{1}{6}\int_0^3 (6 - z)\ (-dz) = \frac{1}{24}\ (6 - z)\ \Big]_0^3 = -\frac{405}{8} . \ \blacksquare$$

Remember that the innermost integral has limits which may be functions of the remaining two variables; this integration is done first. The middle integral has limits which may involve the remaining variable and this integration is carried out second. The outermost integral has constant limits of integration and is executed last.

EXERCISES

Evaluate the following integrals.

1. $\int_0^1 \int_1^2 \int_2^3 dz\ dx\ dy$

2. $\int_0^2 \int_0^x e^{y/x}\ dy\ dx$

3. $\int_0^2 \int_{\sqrt{y}}^{y^2/2} xy\ dx\ dy$

4. $\int_0^1 \int_{x^2}^x dy\ dx$

5. $\int_0^3 \int_0^{2x} \int_1^{xy} xyz\ dz\ dy\ dx$

6. $\int_0^1 \int_{2-y}^{6-2y} \int_0^{\sqrt{4 - y^2}} z\ dz\ dx\ dy$

7. $\int_0^1 \int_{x^2}^{2x} \int_0^{xy} dz\ dy\ dx$

8. $\int_{-1}^2 \int_{2x^2-2}^{x^2+x} x\ dy\ dx$

9. $\int_0^1 \int_0^{x^2} xe^y \, dy \, dx$

10. $\int_0^2 \int_0^{1-x} \int_0^{2-x} xyz \, dz \, dy \, dx$

11. $\int_{-3}^{7} \int_0^{2x} \int_y^{x-1} dz \, dy \, dx$

12. $\int_0^9 \int_0^y \int_0^{\sqrt{y^2 - 9z^2}} xy \, dx \, dz \, dy$

13. $\int_0^2 \int_{\sqrt{y}}^{2} (x^2 + y^2) \, dx \, dy$

14. $\int_0^5 \int_0^{12-2y} \int_0^{4-2y/3-x/3} x \, dz \, dx \, dy$

12.2 THE DOUBLE INTEGRAL

You just have finished doing repeated or iterated integration to evaluate multiple integrals. You initially encountered integration in Chapter 8. If the definition of an integral presented there seems remote, review Section 8.3 before continuing this section. This, coupled with your encounter with repeated integration, should make it clear that the generalization from a simple to a double integral primarily extends the closed interval to a two-dimensional region.

You already can do repeated integration. To enhance your understanding of repeated integration we will define what we mean by a multiple integral in a fashion similar to that for a single integral as presented in Section 8.3. An appreciation of this definition will make applications both easier and more meaningful. Furthermore, biologists use multiple integrals to express certain conceptual features of biological systems without ever explicitly stating an integrand which you can integrate. This use clearly requires your understanding of the definition or meaning of a multiple integral, rather than the mechanics of evaluation.

Only one real change occurs in going from a single integral to a double integral: The function f(x) defined over a subset of the real line is replaced by a function f(x,y) defined over a subset D of the x,y-plane. In defining the double integral here, we could take great care to introduce a substantial degree of generality. Instead, we will strip the definition to its bear essentials. Assume initially that the domain consists of the rectangular region $D = \{(x,y) \mid a \leq x \leq b, c \leq y \leq d\}$, as depicted in Figure 12.1. Although Figure 12.1 shows D as lying in the first quadrant, it may lie anywhere in the x,y-plane. As the x-axis was divided in defining the simple integral, the domain now is divided into little rectangles by the points $a = x_0 < x_1 < \cdots < x_n = b$ and $c = y_0 < y_1 < \cdots < y_n = d$ as shown in Figure 12.2. If we let $\Delta_i x = x_i - x_{i-1}$ and $\Delta_j y = y_j - y_{j-1}$, then $\Delta_{ij}A = \Delta_i x \Delta_j y$ gives the area of a small general rectangle, thus:

- The DOUBLE INTEGRAL of the function f defined over the domain $D = \{(x,y) | a \leq x \leq b, c \leq y \leq d\}$ is denoted $\iint_D f(x,y)dA$ and is given, when it exists, by

$$\iint_D f(x,y)dA = \lim \sum_{i=1}^{n} \sum_{j-1}^{n} f(x_i,y_j)\Delta_{ij}A , \qquad (12.1)$$

where the limit is taken so all $\Delta_{ij}A \to 0$.

This definition has three essential features: (i) The domain D is divided into many small rectangular elements of area $\Delta_{ij}A$. (ii) The function f defines a surface over D. Thus the products $f(x_i,y_j)\Delta_{ij}A$ give the volume of a rectangular box having a basal area of $\Delta_{ij}A$ and height of $f(x_i,y_j)$. These boxes have approximately the same volumes as lie between the x,y-plane and the surface over each of the small rectangular subregions. (iii) The integral is defined as the limit of the sum of these approximations as the number of the $\Delta_{ij}A$ increases without bound and simultaneously the size of each approaches zero.

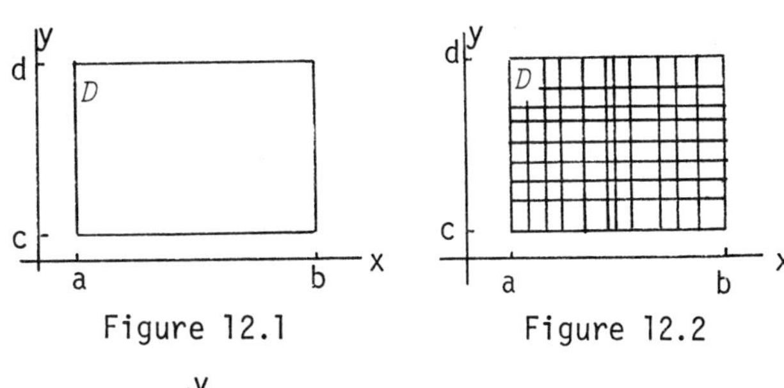

Figure 12.1 Figure 12.2

We could have introduced complex notation to make the definition do the above, but when the integral exists Equation 12.1 will give it. Initially we have required that D be rectangular, but this is

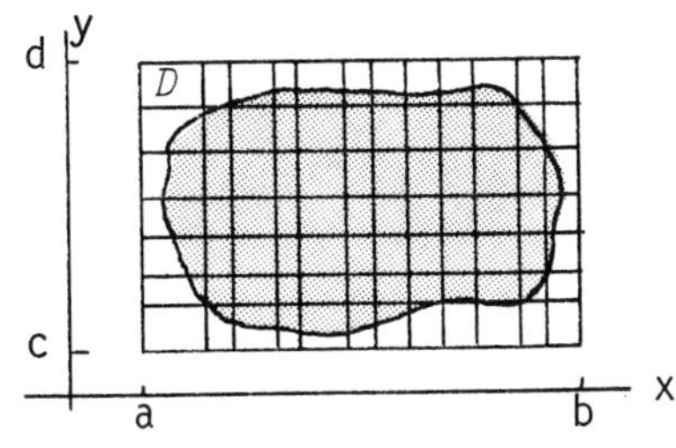

Figure 12.3

not essential. In any other case, a rectangular region D^*, enclosing D, can be defined with a new function $f^*(x,y) = f(x,y)$ for $(x,y) \in D$, but $f^*(x,y) = 0$ for $(x,y) \in D^* - D$. This amounts to nothing more than restricting the summations in Equation 12.1 to those rectangles over which the function is defined. Figure 12.3 shows an irregulariy shaped region enclosed in a rectangle and divided as before.

EXAMPLE 12.4: Life goes on in aquatic environments much as it does in terrestrial settings, except that organisms display different kinds of adaptations. For example, photosynthesis still uses solar radiation to transform basic inorganic compounds into carbohydrates. Higher plants exist in aquatic environments, but lower plant forms such as phytoplankton have a very significant role in the food web. These free-floating forms occur at all depths down to several hundred feet, but because light intensity diminishes with depth, the net photosynthetic production also declines with depth.

Because these changes occur continuously, general integral formulas have been used to describe the total photosynthetic production in this way: If h symbolizes water depth (height) and t time, then the light intensity depends on both through a function $I(h,t)$ because the amount and direction of incoming radiation depend on time, and its penetration depends on depth. Under a fixed set of resource conditions (nutrients), the instantaneous rate of net photosynthesis depends primarily on the energy available for conversion, namely, light intensity. Thus, if q gives the functional dependence of instantaneous net photosynthesis on intensity,

$$q(I) = q(I(h,t)) = p(h,t)$$

has an equivalent representation as p, a function of time and depth. If p has been evaluated on a per unit area basis, then the total net production, on a per unit area basis, between times t_1 and t_2, occurring from the surface down to depth h_ℓ can be written as

$$\iint_R p(h,t)dA$$

where

$$R = \{(h,t) \mid 0 \leq h \leq h_\ell,\ t_1 \leq t \leq t_2\} .$$

This illustrates one use of multiple integrals in biology, namely, the general representation of an accumulative biological process without stating a precise functional form. ▌

At this point we have a limit definition of a double integral, and you have evaluated iterated integrals. We need to relate these two. The relation closely parallels the fundamental theorem of integral calculus for one variable of integration:

- FUNDAMENTAL THEOREM OF INTEGRAL CALCULUS FOR THE DOUBLE INTEGRAL.

  If the continuous function f is defined over a region $R = \{(x,y) \mid a \le x \le b,\ Y_1(x) \le y \le Y_2(x)\}$ where Y_1 and Y_2 are continuous functions, then the double integral is given by the iterated integral

$$\iint_R f(x,y)dA = \int_a^b \int_{Y_1(x)}^{Y_2(x)} f(x,y)\ dy\ dx \qquad (12.2)$$

In evaluating an iterated integral such as

$$\iint_R f(x,y)dA = \int_a^b \int_{Y_1(x)}^{Y_2(x)} f(x,y)\ dy\ dx\ ,$$

remember that x remains constant during the evaluation of the inner integral. By holding x constant, the inner integral becomes a function of a single variable which then can be evaluated as before. After evaluation of the inner integral by holding x constant, a function of x alone results because the limits of the inner integral involve only functions of x. Then the outer integral has an integrand depending only on x with constant limits of integration, so again, this integral is evaluated as before.

If R can be expressed more simply as $R = \{(x,y) \mid x_1(y) \le x \le x_2(y),\ c \le y \le d\}$, where x_1 and x_2 are continuous functions, the double integral can be equivalently expressed as

$$\iint_R f(x,y)dA = \int_a^b \int_{x_1(y)}^{x_2(y)} f(x,y)\, dx\, dy \qquad (12.3)$$

For example recall the region we encountered in Example 8.27. Figure 12.4 reproduces this region which lies between the parabolas $x = -\frac{y^2}{4}$ and $x = \frac{1}{4}(y^2 - 8)$. It is simple to define as $R = \{(x,y) \mid \frac{1}{4}(y^2 - 8) \leq x \leq -\frac{y^2}{4}, -2 \leq y \leq 2\}$, as indicated by the darkly shaded bar showing a representative rectangular area. A more complex description results from going the other direction:

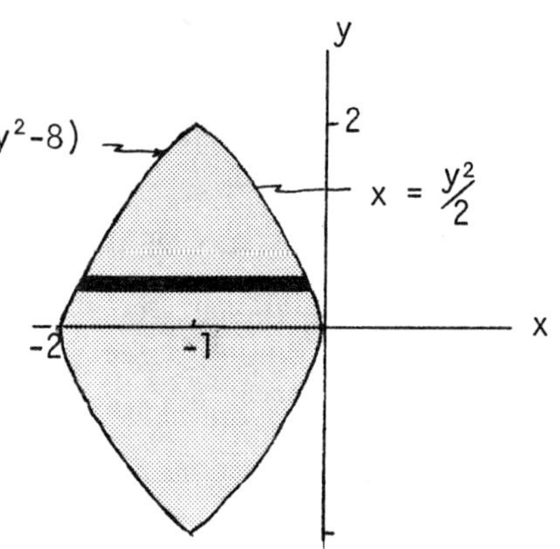

Figure 12.4

$$R = \{(x,y) \mid -2 \leq x \leq -1, \quad -\sqrt{4x+8} \leq y \leq \sqrt{4x+8} \quad \text{and}$$

$$-1 \leq x \leq 0, \quad -\sqrt{-4x} \leq y \leq \sqrt{-4x}\}.$$

Notice that the limits on the integrals in Equation 12.2 and 12.3 depend only on the shape of the region R and not the integrand. Figure 12.5 illustrates a region R of Equation 12.2. The arrow indicates the values which y may assume; then x ranges from a to b. The region R of

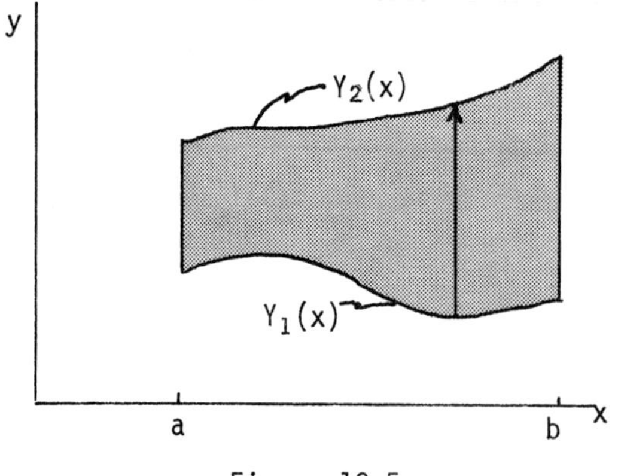

Figure 12.5

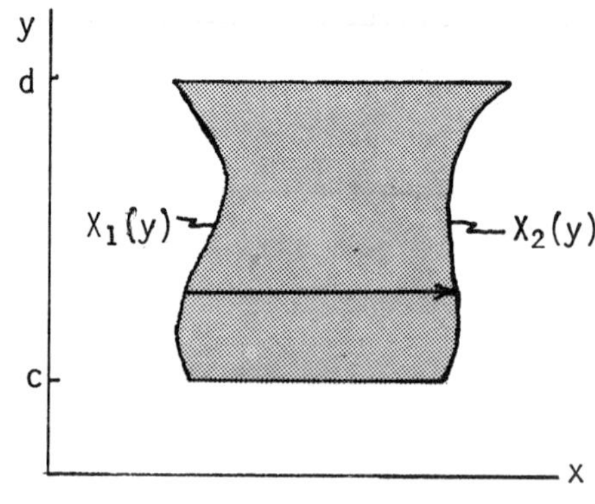

Figure 12.6

Equation 12.3 is illustrated in Figure 12.6. The arrow shows what values x may assume; then y may take on the values from c to d.

Double integrals have several interpretations. First, they give areas of plane regions in this way: Take $f(x,y) = 1$ so

$$\iint_R f(x,y)dA = \iint_R dA = \text{Area}(R) \ .$$

To see this, again begin with a rectangular region as used for the definition, and partition it similarly. Thus the definition gives

$$\iint_R f(x,y)dA = \iint_R dA = \lim \sum_i \sum_j (1)\ \Delta_{ij}A$$

$$= \lim \sum_{i=1}^{n} \sum_{j=1}^{n} (x_i - x_{i-1})(y_j - y_{j-1}) = \text{Area}(R)$$

because $\Delta_{ij}A = \Delta_i x \Delta_j y = (x_i - x_{i-1})(y_i - y_{i-1})$.

Next, let us develop a geometrical interpretation of an iterated integral. Figure 12.7 retains the earlier partition of R by $x_0 < x_1 < \cdots < x_n$ and $y_0 < y_1 < \cdots < y_n$ so $\Delta_{ij}A = (\Delta_i x)(\Delta_j y)$. Thus $\Delta_{ij}A$ gives the area of a typical rectangular part of R, namely ΔR_{ij}. In constructing the Riemann sum

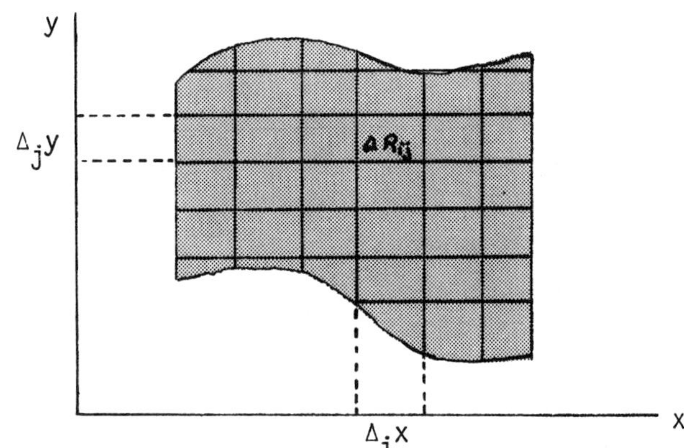

Figure 12.7

$$\sum_i \sum_j f(x_i,y_j)\Delta_{ij}A$$

we add up individual products $f(x_i,y_j)\Delta_{ij}A$ where $f(x_i,y_j)$ is the height of the surface and the product then gives the volume of a rectangular box or parallelopiped. These approximate the volume over ΔR_{ij} as shown in Figure 12.8; it shows a surface over the region displayed in Figure 12.7.

Now taking the limit as the number of ΔR_{ij} increases (as $\Delta_i x \to 0$ and $\Delta_j y \to 0$), we have

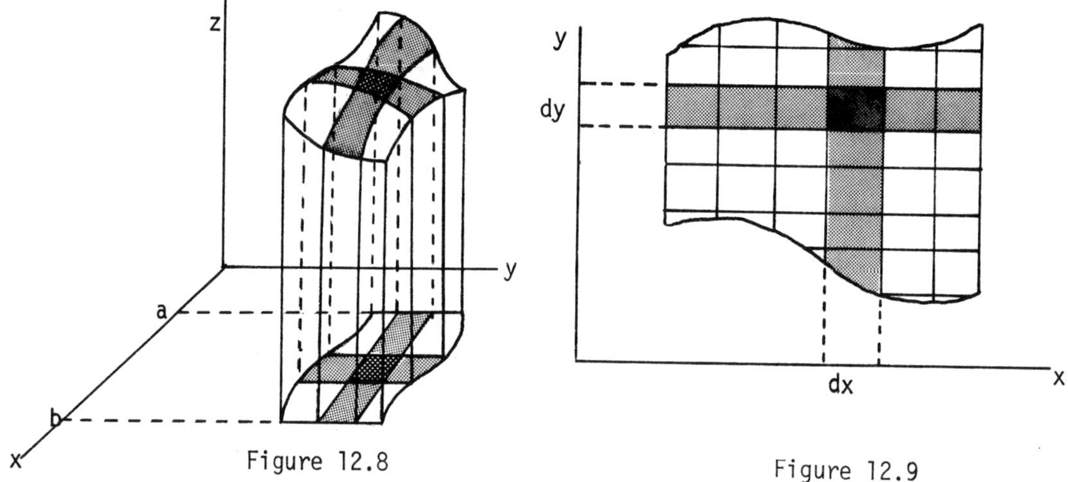

Figure 12.8

Figure 12.9

$$\iint_R f(x,y)dA = \int_a^b \int_{Y_1(x)}^{Y_2(x)} f(x,y)\ dy\ dx\ .$$

From the above discussion and from $\Delta_{ij}A = (\Delta_i x)(\Delta_j y) = (\Delta_j y)(\Delta_i x)$, it should be clear that

$$dA = dx\ dy = dy\ dx\ .$$

Thus, the iterated integral may be written in either of the following ways:

$$\iint_R f(x,y)dA = \iint_R f(x,y)\ dx\ dy = \iint_R f(x,y)\ dy\ dx\ .$$

Finally, you can think of the iterated integral as summing the elements $f(x,y)dy$ to obtain

$$A(x) = \int_{Y_1(x)}^{Y_2(x)} f(x,y)\ dy$$

and then summing the elements $A(x)dx$ as is indicated by the shaded strips in Figure 12.9. For that region, we would first sum in the y-direction to obtain $A(x)$ and then sum in the x-direction to cover the entire final region. For a region with its parallel sides horizontal rather than vertical, we would sum in the x-direction to obtain A(y) and then sum in the y-direction to cover the entire region.

EXAMPLE 12.5: Find the volume under $z = (a^2 - y^2)^{3/2}$ and above the region bounded by $y = a$ and $y = x$ if $z \geq 0$.

The graph of the region is shown in Figure 12.10. Note that $z \geq 0$ over the region of interest so the entire volume lies above the x,y-plane. The height can be integrated with $dA = dx\,dy$ or $dA = dy\,dx$. Each of the typical elements appears in the figure. Suppose we take $dA = dx\,dy$, integrating first with respect to x, then with respect to y. For any permissible value of y, x can range from 0 to that x value of y. Thus the volume is given by

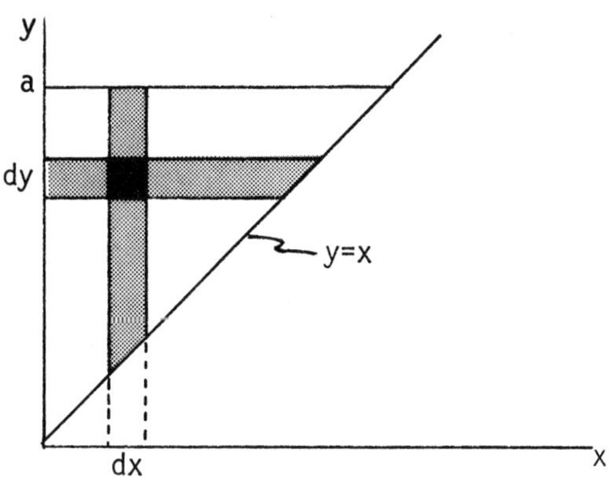

Figure 12.10

$$\iint_R (a^2 - y^2)^{3/2} dA = \int_0^a \int_0^y (a^2 - y^2)^{3/2} dx\, dy = \int_0^a \left. (a^2 - y^2)^{3/2} x \right]_0^y dy$$

$$= \int_0^a (a^2 - y^2)^{3/2} y\, dy\ .$$

To integrate this we can multiply and divide by -2 to obtain the form of $u^{3/2} du$:

$$\int_0^a (a^2 - y^2)^{3/2} y\, dy = -\frac{1}{2} \int_0^a (a^2 - y^2)^{3/2} (-2y\, dy)$$

$$= -\frac{1}{2} \left[\frac{2}{5} (a^2 - y^2)^{5/2} \right]_0^a$$

$$= -\frac{1}{5} \left[0 - (a^2)^{5/2} \right] = \frac{a^5}{5} \text{ cubic units.}$$

If instead we had tried to evaluate the iterated integral as

$$\iint_R (a^2 - y^2)^{3/2} dy\ dx = \int_0^a \int_x^a (a^2 - y^2)^{3/2} dy\ dx ,$$

we would have encountered a form which we do not yet have the tools to integrate. Thus, if you have a choice of method, look for the easiest method for evaluating the integral. ▌

EXAMPLE 12.6: Find the volume under the surface $z = x + y + 1$ and above the region defined by $y = 1 - x^2$ and $y \geq 0$.

The region is shown in Figure 12.11. If we integrate first with

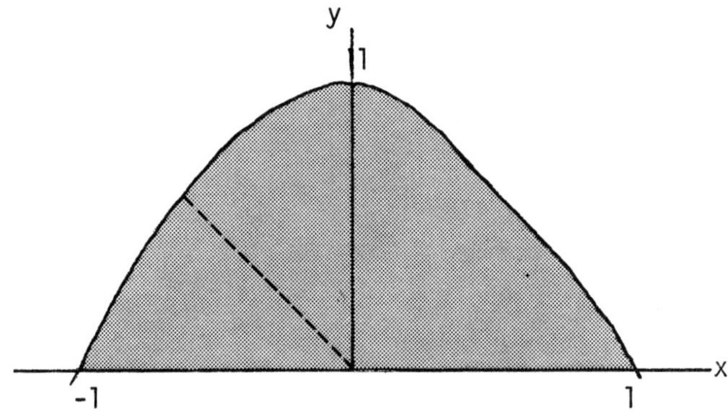

Figure 12.11

respect to y and then x, this integral gives the volume:

$$\iint_R (x + y + 1)dA = \int_{-1}^{1} \int_0^{1-x^2} (x + y + 1)\ dy\ dx$$

$$= \int_{-1}^{1} (xy + \frac{y^2}{2} + y)\Big]_0^{1-x^2} dx$$

$$= \int_{-1}^{1} (x - x^3 + \frac{1}{2} - x^2 + \frac{x^4}{2} + 1 - x^2)dx$$

$$= \left[\frac{x^2}{2} - \frac{x^4}{4} + \frac{x}{2} - \frac{x^3}{3} + \frac{x^5}{10} + x - \frac{x^3}{3}\right]_{-1}^{1}$$

$$= \frac{28}{15} \text{ cubic units.}$$

This example contains a situation leading to a common mistake. Since the region is symmetric, it is tempting to conclude that

$$\int_{-1}^{1}\int_{0}^{1-x^2} (x + y + 1)\, dy\, dx = 2\int_{0}^{1}\int_{0}^{1-x^2} (x + y + 1)\, dy\, dx .$$

This would work in general only when the function f(x,y) is also symmetric. Here the region is symmetric but the function is not: $f(x,y) \neq f(-x,y)$.

This problem could have also been set up in the following manner by integrating first with respect to x and then with respect to y.

$$\iint_R (x + y + 1)\, dx\, dy = \int_0^1 \int_{-\sqrt{1-y}}^{\sqrt{1-y}} (x + y + 1)\, dx\, dy .$$

However, in integrating in this order, we find that we must integrate $2\int_0^1 y\sqrt{1 - y}\, dy$, a form which we cannot yet integrate. Hence the first approach is the appropriate one to use.

Some caution needs to be exercised in defining the region over which integration occurs. If we wanted the volume above the x,y-plane, then the region for integration needs to assure that $f(x,y) \geq 0$. For example, here $f(x,y) = x + y + 1$ becomes zero as (x,y) leaves the region at (-1,0). If the function had been $x + y$, then the surface would have been above the x,y-plane only when $x + y \geq 0$. For this function, the region of integration would have been like that shown in Figure 12.11, except that it would lose the pie-shaped region bounded by (-1,0), (0,0) and $\left(-\frac{\sqrt{5} - 1}{2}, \frac{\sqrt{5} - 1}{2}\right)$ as shown by the dotted line in Figure 12.11. ▮

<u>EXAMPLE 12.4</u> (continued): This example dealt with the net photosynthetic production through time (t) and depth (h). The double integral representing the photosynthetic production can be written as the following iterated integral:

$$\iint_R \rho(h,t)dt = \int_{t_1}^{t_2} \int_0^{h_\ell} \rho(h,t)\, dh\, dt ,$$

clearly indicating the involvement of time and depth. ▮

EXERCISES

Evaluate the following iterated integrals after sketching the region over which the integration takes place.

1. $\iint_R (x^2 + y^2)dA,\ 0 \le y \le 2,\ y^2 \le x \le 4$

2. $\int_0^2 \int_0^{x^2} e^{y/x}\, dy\, dx$

3. $\int_0^1 \int_x^{\sqrt{x}} (y + y^3)dy\, dx$

4. $\int_0^3 \int_0^{x/3} e^{x^2}\, dy\, dx$

5. $\int_0^3 \int_{x^2}^{2x+3} xy\, dy\, dx$

6. $\int_0^2 \int_{1-y^2/4}^{4-y^2} 4\, dx\, dy$

Express the double integral of an arbitrary function f(x,y) over the specified region R using an iterated integral or more than one iterated integral, without integrating. Give the results for two different orders of integration.

7. R is the region in Quadrant I bounded by the parabolas $3y = x^2$ and $3x = y^2$.

8. R is defined by $0 \le x \le 2$ and $0 \le y \le \sqrt{4 - x^2}$.

9. Evaluate the iterated integral in Problem 7 if f(x,y) = 1.

10. Evaluate the iterated integral in Problem 8 if f(x,y) = y.

12.3 TWO-DIMENSIONAL DISTRIBUTION OF MASS OR PROBABILITY

Consider a plane R, and imagine a mass spread over it like paint on a surface, where mass is idealized as having no thickness. Suppose that there exists a constant δ such that if T is any subregion of R with an area ΔA, then the mass ΔM of T is $\delta \Delta A$. When δ is a constant, we say that the mass has uniform density of δ units per square unit of area. Thus

$$M = \iint_R \delta dA \ .$$

We could also assume that the density is variable, that is $\delta = \delta(x,y)$ such that δ is defined, continuous and non-negative when (x,y) is in the given region R. The total mass M of R is then given by

$$M = \iint_R \delta(x,y) dA \ .$$

EXAMPLE 12.7: Find the mass of the region $R = \{(x,y) \mid 0 \le y \le 1 - x^2\}$ with $\delta = ky$ where k is a constant.

The region over which we are interested in finding the mass appears in Figure 12.11, shown in the previous section.

$$M = \iint_R ky \ dA = k\int_{-1}^{1} \int_0^{1-x^2} y \ dy \ dx$$

$$= \frac{k}{2} \int_{-1}^{1} (1 - 2x^2 + x^4) \ dx$$

$$= \frac{k}{2} \left[x - \frac{2}{3} x^3 + \frac{x^5}{5} \right]_{-1}^{1}$$

$$= \frac{8k}{15} \text{ mass units.}$$ ▌

The function $\delta(x,y)$ could represent a two-dimensional probability density function as well as a mass function. If $\delta(x,y)$ represented a two-dimensional probability density function, it would have to satisfy

$$\delta(x,y) \ge 0 \qquad \text{and} \qquad \iint_R \delta(x,y) dA = 1 \qquad (12.4)$$

EXAMPLE 12.8: Check that $\delta(x,y) = 3y$, $0 \leq x \leq y \leq 1$, is a two-dimensional probability density function. Evaluate $\text{Prob}\left[y \leq \frac{1}{2}\right]$, $\text{Prob}\left[x \leq \frac{1}{2} \text{ and } y \geq \frac{1}{2}\right]$, and $\text{Prob}\left[y \geq x \geq \frac{1}{2}\right]$.

The region over which the probability function is defined is shown in Figure 12.12. A probability density function must satisfy Equation 12.4:

$$\int_R\int \delta(x,y)\, dA = 3\int_0^1 \int_0^y y\, dx\, dy = 3\int_0^1 yx\Big]_0^y dy = 3\int_0^1 y^2 dy = 1 .$$

Thus, $\delta(x,y)$ is a probability density function.

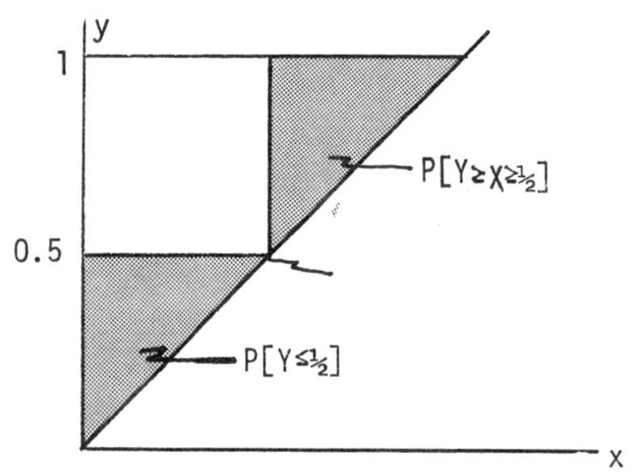

Figure 12.12

Figure 12.13

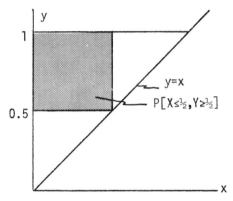

Now to evaluate the first probability, $P\left[y \leq \frac{1}{2}\right]$, we need the probability over the triangular area below $y = \frac{1}{2}$ in the domain as shown in Figure 12.12. This is the region where $0 \leq x \leq y \leq \frac{1}{2}$.

$$P\left[y \leq \frac{1}{2}\right] = \int_0^{1/2} \int_0^y 3y\, dx\, dy = 3\int_0^{1/2} y^2\, dy = y\Big]_0^{1/2} = \frac{1}{8} .$$

The other probabilities correspond to the regions shown in Figure 12.12 and 12.13. They are evaluated as

$$P\left[x \leq \frac{1}{2} \text{ and } y \geq \frac{1}{2}\right] = \int_{1/2}^1 \int_0^{1/2} 3y\, dx\, dy = \frac{3}{2}\int_{1/2}^1 y\, dy = \frac{9}{16}$$

and

$$P\left[y \geq x \geq \frac{1}{2}\right] = \int_{1/2}^{1} \int_{1/2}^{y} 3y \, dx \, dy = 3\int_{1/2}^{1} y\left[y - \frac{1}{2}\right]dy$$

$$= 3\left[\frac{y^3}{3} - \frac{y^2}{4}\right]_{1/2}^{1} = 3\left[\frac{1}{3} - \frac{1}{4} - \frac{1}{3}\left(\frac{1}{8}\right) + \left(\frac{1}{4}\right)\frac{1}{4}\right] = \frac{5}{16} .$$

Observe that the three regions exhaust the domain where $\delta(x,y) \geq 0$ so their sum, $\frac{1}{8} + \frac{9}{16} + \frac{5}{16}$, should equal unity, as it does. ▮

EXAMPLE 12.9: The exponential probability density $p(x) = \lambda e^{-\lambda x}$ often describes distance moved by a particle or an animal, as we saw in Section 8.8. Suppose a particle moves with small jumps or jerks. If movement occurs in a plane with individual movements perpendicular to each other and away from the starting point, then the exponential generalizes to $\lambda_1\lambda_2 e^{-(\lambda_1 x_1 + \lambda_2 x_2)}$. Take $\lambda_1 = \lambda_2 = 1$ and find the probability that the individual has traversed no more than 3 units of distance in his movement away from the release point.

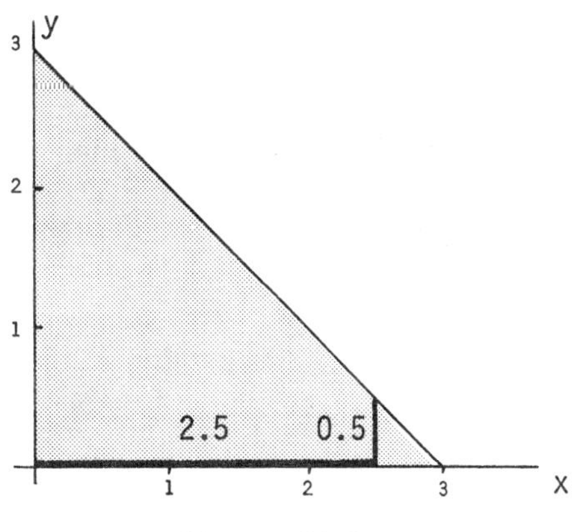

Figure 12.14

The condition requires $x + y \leq 3$ because it specifies distance traversed along the path of movement, not the final distance removed from the point of release. Figure 12.14 has the region $x + y \leq 3$ shaded, and shows one such traversed distance, namely, a movement of 2.5 units in the x direction followed by a movement of 0.5 units in the y direction. Thus, the required probability is given by

$$P\left[x + y \leq 3\right] = \int_0^3 \int_0^{3-x} e^{-(x+y)}\, dy\, dx = \int_0^3 -e^{-(x+y)}\Big]_0^{3-x} dx$$

$$= \int_0^3 (e^{-x} - e^{-3})dx = (-e^{-x} - xe^{-3})\Big]_0^3$$

$$= 1 - e^{-3} - 3e^{-3} = 1 - 4e^{-3} \doteq 0.80 .$$

This result then says that the probability that an individual has traversed no more than 3 units away from the release point is 0.80. ▌

EXERCISES

Find the mass M of the following regions if δ is the density function.

1. $\delta = y + 1$ and R is defined by $0 \leq x \leq 4$ and $0 \leq y \leq 3$.

2. $\delta = y$ and R is defined by $0 \leq x \leq 2$ and $0 \leq y \leq \sqrt{4 - x^2}$.

3. $\delta = x + y$ and R is defined by $0 \leq x \leq 1$ and $0 \leq y \leq \sqrt{x}$.

4. Show that $\delta(x,y) = -e^{-x-y}$ for $x > 0$ and $y > 0$ does not satisfy the conditions for a probability density function.

5. Show that $\delta(x,y) = e^{-x-y}$ for $x > 0$ and $y > 0$ does have the properties for a probability density function.

6. Calculate the $P[x < 1]$ and $P[x + y \leq 2]$ in Problem 5.

7. Given that $f(x,y) = k(x + y)$ determine k in order that $f(x,y)$ is a probability density function if $f(x,y)$ is defined over the region specified by $x = 3$, $y = x$, $y = x + 4$ and $x = 0$. Calculate $P[y \leq 3]$, $P[3 \leq y \leq 4]$ and $P[y \geq 4]$. Should the probabilities just calculated add up to 1?

12.4 CALCULATION OF VOLUMES BY DOUBLE INTEGRATION

Earlier in this chapter we presented an intuitive explanation of why a double integral sometimes could be viewed as a volume. Now we will elaborate more fully on this aspect of double integration.

In an x,y,z-system of rectangular coordinates, let the region R be defined in the x,y-plane by some closed curve A. Now consider a volume formed by moving a line parallel to the z-axis through every point on this curve; assume that this is capped by some surface $z = f(x,y)$ where $f(x,y) \geq 0$ and is continuous over the region R. Such a volume appears in Figure 12.15.

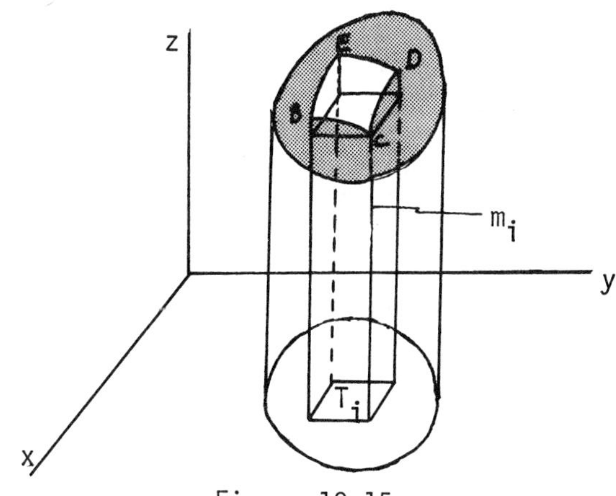

Figure 12.15

We will develop the volume of this figure, assuming that only the volume of a parallelepiped has been defined previously.

We start by partitioning the region R into rectangles as we have done before. Figure 12.15 shows a typical rectangle of this grid; order them in any convenient manner. This representative rectangle, T_i, has an area of $\Delta_i A$. Above T_i construct a parallelepiped using planes perpendicular to the x,y-plane and let T_i be the base of this parallelepiped. This parallelepiped has a cap BCDE on the surface $z = f(x,y)$ as appears in the figure. If m_i is the absolute minimum of $f(x,y)$ over T_i, then m_i would be the height of the largest parallelepiped entirely contained within the column defined by constructing planes perpendicular to the x,y-plane and using T_i as the base of the column. Now in the representative rectangle T_i, there exists a point $(x_i,y_i,0)$ where $m_i = f(x_i,y_i)$. The volume of the parallelepiped of height m_i is

$$\Delta_i V = m_i \Delta_i A = f(x_i,y_i)\Delta_i A \ .$$

We can now evaluate the volume under the surface as the limit of the sum of these approximating volumes whose bases lie entirely in the region R, as all $\Delta_i A \to 0$. Thus, we have

$$V = \lim_{\Delta_i A\to 0} \sum \Delta_i V = \lim_{\Delta_i A\to 0} \sum f(x_i,y_i)\Delta_i A = \iint_R f(x,y)\, dA = \iint_R z dA\ .$$

The limits on the sum are taken such that the region R is covered as was explained in the last section. The requirement that all $\Delta_i A \to 0$ is analogous to the situation in a single integral where we increased the number of subdivisions in such a way that the largest $\Delta_i x \to 0$. Thus, as the number of areas increases, the error of approximation will decrease and the limit then gives the true volume.

Again, we can think of the process of finding this volume by first adding up the volumes of the parallelepipeds in the x-direction and then adding in the y-direction, as is suggested by Figure 12.15, or vice versa.

It is also possible to define the volume of a solid when the region R is defined in the x,z-plane or the y,z-plane instead of the x,y-plane. If the region is in the x,z-plane and the capping surface is given by $y = f(x,z)$, then the volume is given by

$$V = \iint_R f(x,z)dA$$

where $dA = dx\,dz$ or $dz\,dx$. If the region is defined in the y,z-plane and the capping surface by $x = f(y,z)$, then the volume is given by

$$V = \iint_R f(y,z)dA$$

where $dA = dy\,dz$ or $dz\,dy$.

EXAMPLE 12.10: Find the volume of the region bounded by $y = \frac{x^2}{3}$, $y = 0$, $0 \leq x \leq 3$, the x,y-plane and the plane $x - y + 2z = 4$.

The region in the x,y-plane is shown in Figure 12.16. Check to see that $f(x,y) = z = \frac{1}{2}(4 - x + y)$ specifies a positive height in the region of interest. Thus,

$$V = \iint_R f(x,y)dA = \frac{1}{2}\int_0^3 \int_0^{x^2/3} (4 - x + y)\, dy\, dx$$

where the arrow in Figure 12.16 shows the inner limits of integration.

$$V = \frac{1}{2}\int_0^3 \left[4y - xy + \frac{y^2}{2}\right]_0^{x^2/3} dx = \frac{1}{2}\int_0^3 \left[\frac{4x^2}{3} - \frac{x^3}{3} + \frac{x^4}{18}\right] dx$$

$$= \frac{1}{6}\int_0^3 \left[4x^2 - x^3 + \frac{x^4}{6}\right] dx = \frac{1}{6}\left[\frac{4}{3}x^3 - \frac{x^4}{4} + \frac{x^5}{30}\right]_0^3$$

$$= \frac{159}{40} \text{ cubic units.}$$ ▮

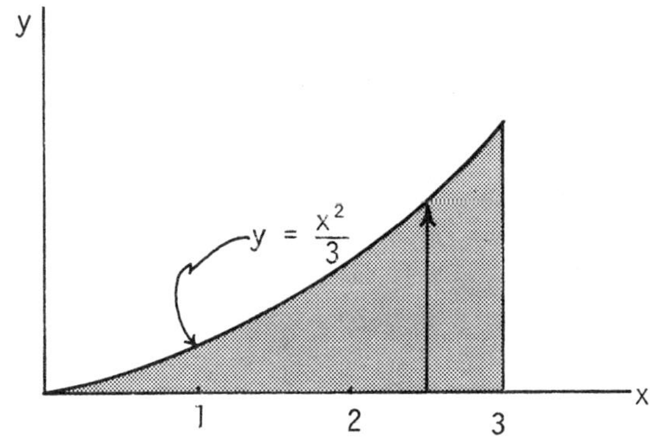

Figure 12.16

EXERCISES

Find the volume of the region bounded by the given surfaces in an x,y,z-rectangular coordinate system.

1. The surfaces $z = y$, $x^2 + y = 4$, $z = 0$ and $y = 0$.

2. The volume under $x + y + z = 6$, above $z = 0$ and with sides $0 \le y \le x$ and $0 \le x \le 3$.

3. The volume below the surface $z = 4 - x^2 - y^2$, above $z = 0$ with $R = \{(x,y) \mid 0 \le y \le x \le 1\}$.

4. The volume between the surfaces $z = 0$, $2y = x$, $x = 1$, $y = 0$ and $z = x^2 + 3y + 1$.

5. The volume between the surfaces $z = 1$, $2y = x$, $x = 1$, $y = 0$ and $z = x^2 + 3y + 1$.

6. The volume below $z = x$ and above $z = 0$ and within $y^2 = 2 - x$.

7. The volume bounded by $y = x^2$ and $z = 4 - y$.

8. The volume bounded by $x = 0$, $y \geq 0$, $z \geq 0$, $y^2 = 4 - x$ and $z = y + x$.

9. The volume bounded by $z = xy$, $z = 0$, $y = x^2$ and $y^2 = x$.

Set up the following integrals using two different orders of integration if possible. DO NOT EVALUATE.

10. The volume bounded by the x,y-plane and above by $z = 4 - x^2 - y^2$.

11. The volume above the plane $z = 0$ and inside the cylinders $y^2 = 4 - 2x$ and $x^2 = 4 - z$.

12. The volume bounded by $z = 0$, $z = x$ and $y^2 = 2 - x$.

13. Find the volume of the following surface using three different orders of integration if the surface is defined by $z = 0$, $z = 4$, $x = 0$, $x = 3$, $y = 0$ and $y = 5$.

12.5 TRIPLE INTEGRATION

The definition of a triple integral and its evaluation as an iterated integral closely follows the corresponding material for double integrals. Here we will develop the idea of triple integration and its applications, with particular reference to volumes. Before continuing this section, you may want to review the triple integration you did in Section 12.1.

We will begin by assuming that R is a region in the x,y-plane as shown in Figure 12.17. Let f_1 and f_2 be continuous functions over R satisfying $f_1(x,y) \leq f_2(x,y)$. Finally let S be a closed three-dimensional region over R, and bounded by the planes $x = a$ and $x = b$, the partial cylinders $y = y_1(x)$ and $y = y_2(x)$ and the surfaces $z = f_1(x,y)$ and $z = f_2(x,y)$, as illustrated in Figure 12.17.

Now by constructing planes perpendicular to the coordinate planes, the resulting set of rectangular parallelepipeds completely partitions the region S. Some of those parallelepipeds lie entirely in S while others cross its boundary and lie partly outside. Any convenient ordering on these elements can be adopted so that we can describe the ith one as having volume $\Delta_i V = (\Delta_i x)(\Delta_i y)(\Delta_i z)$; as i ranges from 1 to n, these volumes will completely cover S. This background leads us to the following definition:

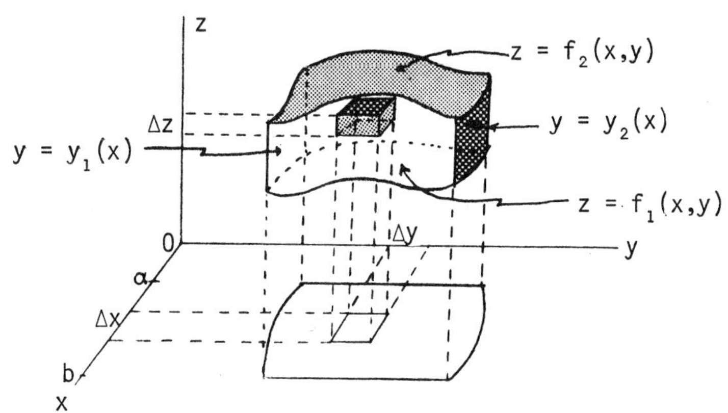

Figure 12.17

- The TRIPLE INTEGRAL of the function f defined and continuous over S is given, when it exists, by

$$\iiint_S f(x,y,z)dV = \lim_{\text{All } \Delta_i V \to 0} \sum_{i=1}^{n} f(x_i,y_i,z_i)\Delta_i V \qquad (12.5)$$

where (x_i,y_i,z_i) is any point in the ith element of the partition.

For a region of the type shown in Figure 12.17, it can be shown that the triple integral equals this iterated integral:

$$\iiint_S f(x,y,z)dV = \int_a^b \int_{y_1(x)}^{y_2(x)} \int_{f_1(x,y)}^{f_2(x,y)} f(x,y,z)\ dz\ dy\ dx\ . \qquad (12.6)$$

Recall that this iterated integral should be interpreted as

$$\int_a^b \left\{ \int_{y_1(x)}^{y_2(x)} \left[\int_{f_1(x,y)}^{f_2(x,y)} f(x,y,z)\ dz \right] dy \right\} dx\ .$$

This means that f(x,y,z) first is integrated with respect to z while x and y remain constant; this result then is integrated with respect to y with x being treated as a constant; and finally integrate with respect to x. After each integration the limits of integration are substituted before the next integration is executed. Throughout this section we will consider only continuous integrands. The next example illustrates the process of triple integration. The operation of integration repeated three times was illustrated in Section 12.1. The fundamental theorem of integral calculus extends in the obvious way, namely, a triple integral can be evaluated by a three-fold iterated integral. The major difficulty lies in translating the region S into bounds on the respective integrals. The next example illustrates this process.

EXAMPLE 12.11: Set up the integral necessary to find the volume in the first octant bounded by the elliptic paraboloids $z = \frac{1}{4}(x^2 + 4y^2)$ and $z = \frac{1}{8}(48 - x^2 - 4y^2)$; this volume is shown in Figure 12.18. The

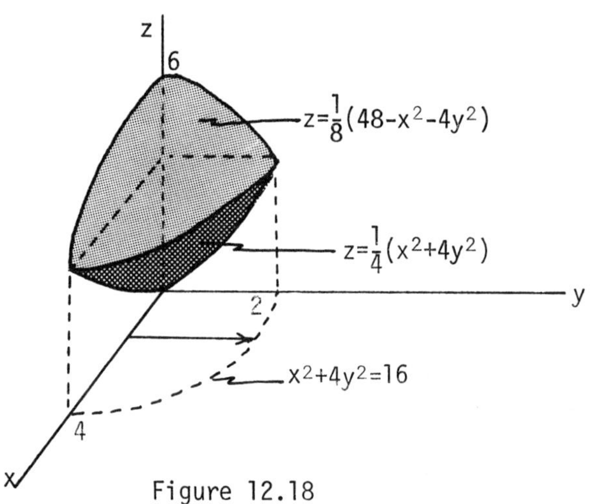

Figure 12.18

projection R of the region S onto the x,y-plane is found by eliminating z from the two equations above. Thus we have

$$\frac{1}{4}(x^2 + 4y^2) = \frac{1}{8}(48 - x^2 - 4y^2)$$

which yields upon simplification

$$x^2 + 4y^2 = 16 .$$

The volume of S is given by

$$V = \iiint_S dV = \iiint_S dz\,dy\,dx\ .$$

The variable z will range from $z = \frac{1}{4}(x^2 + 4y^2)$, the equation of the lower surface, to $z = \frac{1}{8}(48 - x^2 - y^2)$, the equation of the upper surface, while the limits for x and y will be determined from the projection R, of the region S onto the x,y-plane. The arrow on the x,y-plane in Figure 12.18 shows the range of integration for y. Thus, we have

$$V = \int_0^4 \int_0^{\frac{1}{2}\sqrt{16-x^2}} \int_{\frac{1}{4}(x^2+4y^2)}^{\frac{1}{8}(48-x^2-4y^2)} dz\,dy\,dx\ .$$

After integrating with respect to z, this volume becomes

$$V = \int_0^4 \int_0^{\frac{1}{2}\sqrt{16-x^2}} \left[\frac{1}{8}(48 - x^2 - 4y^2) - \frac{1}{4}(x^2 + 4y^2)\right] dy\,dx\ .$$

The quantity in brackets represents the height between the surfaces. Thus after the first integration, any triple integral representing a volume reduces to a double integral which we considered in the last section. At the present time, we do not yet have the tools required to evaluate this integral. However, we do not need to be able to evaluate the integral in order to be able to understand the process by which we arrive at the limits. Notice that the limits on the innermost integral are functions of x and y while the limits on the next integral depend only on x and the limits on the outer integral are constants. ▌

If $\delta(x,y,z)$ is the density of a solid S at the point (x,y,z), then the total mass is defined to be

$$\text{Mass of } S = M = \iiint_S \delta(x,y,z)dV\ .$$

The same relation as before emerges between a mass function and a probability density function. Specifically, any mass function whose total mass equals unity can be regarded as a probability density function.

EXERCISES

Evaluate the following:

1. $\int_0^1 \int_1^2 \int_2^3 dz\, dx\, dy$

2. $\int_0^3 \int_0^{2x} \int_1^{xy} xyz\, dz\, dy\, dx$

3. $\int_0^9 \int_0^y \int_0^{\sqrt{y^2-9z^2}} xy\, dx\, dz\, dy$

4. $\int_0^1 \int_0^{1-x} \int_0^{3-x} xyz\, dz\, dy\, dx$

5. $\int_0^2 \int_{2-y}^{6-2y} \int_0^{\sqrt{4-y^2}} z\, dz\, dx\, dy$

6. $\int_{-3}^{7} \int_0^{2x} \int_y^{x-1} dz\, dy\, dx$

7. $\int_0^1 \int_{x^2}^{2x} \int_0^{xy} dz\, dy\, dx$

8. $\int_0^5 \int_0^{12-2y} \int_0^{4-(2y/3)-(x/3)} x\, dz\, dx\, dy$

Solve by the use of triple integrals. If mass is involved, then δ represents the density at the point (x,y,z). Evaluate only those integrals for which you have the tools to do so.

9. The region S is bounded by the coordinate planes and the planes $x = 2$, $y = 3$ and $z = 1$. Find the volume and the mass of S if $\delta = x^2yz$.

10. Find the volume and mass of the region S bounded by the parabolic cylinder $x^2 = 4y$ and the planes $z = 0$, $x = 0$, $y = 0$ and $x + 2z = 2$ if $\delta = 2z$.

11. Find the volume of the region S bounded by the cylinder $y = x^2 + 2$ and the planes $y = 4$, $z = 0$ and $3y - 4z = 0$.

12. Find the volume of the region S bounded by the paraboloid $9x^2 + y^2 - 9z = 0$ and the plane $z = 4$. DO NOT EVALUATE.

13. Find the volume of the region S bounded by the surface $z = 4x^2y$ and the planes $x = y$, $x = 0$, $y = 2$, $y = 0$ and $z = 0$.

14. Set up Problem 12 using double integrals.

15. Explain why the integral obtained in Problem 14 is the same result as the iterated integral obtained in Problem 12 after the first integration.

Chapter 13

TRIGONOMETRY

Trigonometry might be thought of as an outgrowth of geometry which deals with certain essential features of angles, somewhat as a graph shows the important characteristics of a function. Geometry only tells us that relations exist between the lengths of the sides of a triangle and the sizes of its angles; trigonometry precisely specifies these relationships. For example, suppose you want to build a barn 40 feet wide with a roof having a 35° pitch. From our knowledge of geometry we know that the length of its rafters is fixed, but certain results from trigonometry tell us how to find out that the rafters are 24.4 feet long. More generally when some lengths and/or angles are known in a triangle, trigonometry provides the tool for finding other lengths and/or angles. Thus, trigonometry has an important role in land surveying and associated tasks of fixing the location of important points in a coordinate system.

Once a function of an angle has been defined, it has a periodic behavior which we will investigate. Periodic behavior surfaces in many different areas in biology. For example the heart has periodic fluctuations in its electrical potential which are measured on an electrocardiogram (EKG), brain impulses are similarly measured with an electroencephalogram (EEG), plants show marked day-night patterns, and population sizes often display cyclic patterns.

Angles play an important role in some coordinate systems other than the rectangular coordinate system. In one of these other coordinate systems the location of a point in the plane is fixed by specifying an angle and a distance while in another system a point in space is determined by two angles and a distance. These two systems and others have important biological applications in studying animal behavior, specifically, navigation and other directional behavior patterns of animals. Also some calculus problems can be solved more easily if they are translated into a coordinate system other than the familiar rectangular coordinate system.

These illustrations have been given to try to show that trigonometry has a role in the application of mathematics to the biological sciences. The introduction of trigonometry was delayed until now because biologists use functions, particularly the exponential and logarithmic functions, and the ideas of calculus more than they use the ideas associated with trigonometry. This chapter does not present a comprehensive study of trigonometry; it merely introduces those basic ideas in trigonometry which we will need in our continuing study of calculus in biology.

This chapter begins by introducing the basic ideas of trigonometry without particular emphasis on biological illustrations. Some of the sections are starred, indicating in this chapter, that they may be skipped if you are already familiar with the topic covered in that section. Subsequent sections present topics involving the calculus of trigonometric functions. The last two sections introduce hyperbolic functions which are similar to the trigonometric functions but are defined in terms of exponential functions.

13.1 FUNDAMENTAL RELATIONS

An angle is formed when a line rotates about a point in a plane with the starting position of the line called the *initial side* and the final position the *terminal side* as is indicated in Figure 13.1. The direction in which the line is rotated is usually indicated by a curved arrow. The point o is called the *vertex* of the angle.

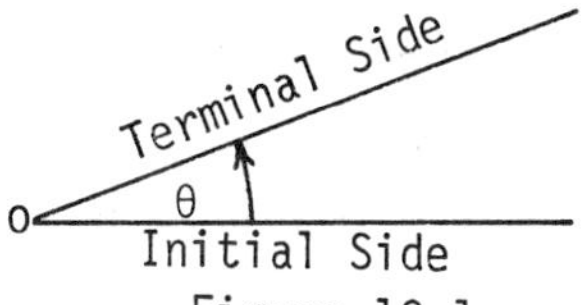

Figure 13.1

When a rotating line makes one complete revolution, its terminal side will coincide with its initial side and it is said to have revolved through 360°. Thus a *degree* is an angle formed by $\frac{1}{360}$ of a revolution. To get a more precise measurement, each degree is divided into 60 equal parts called *minutes* and in turn each minute is composed of 60 smaller units called *seconds*. Thus 1° = 60' (minutes) = 3600" (seconds).

An angle which is one-fourth of a revolution (90°) is called a *right angle*. An angle less than 90° is called an *acute angle* and an angle greater than 90° and less than 180° is called an *obtuse angle*.

These angles are illustrated in Figure 13.2.

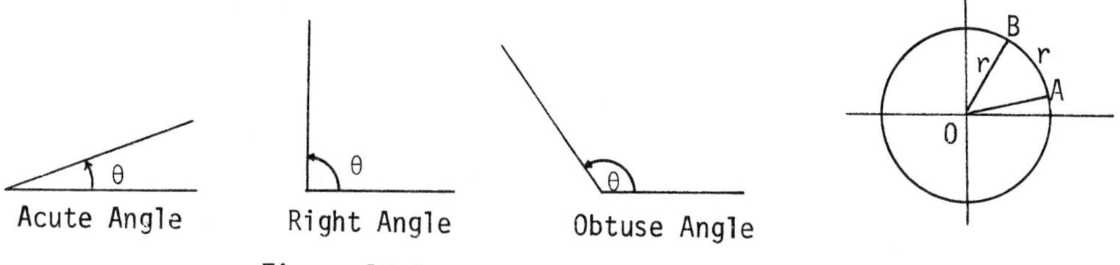

Figure 13.2

Figure 13.3

The size of an angle can be expressed either in degrees or in radian measurement much as lengths and weights can be expressed in either English or metric units.

- A RADIAN is an angle which, if its vertex is placed at the center of a circle (see Figure 13.3), subtends on the circumference an arc of length equal to the radius of the circle.

In Figure 13.3, θ is equal to one radian, because the arc AB and the radius OA have equal lengths. Notice from Figure 13.3 that a radian is an angle of fixed magnitude that does not depend on the size of the circle at whose center it may be placed, because the arc which the angle subtends is proportional to the radius.

The circumference of any circle equals 2π times its radius ($C = 2\pi r$). This means that an arc equal to the radius can be laid off 2π times on the circumference of the circle. Consequently 360° equals 2π radians or

$$\pi \text{ radians} = 180° \ . \qquad (13.1)$$

This last expression gives the conversion factors needed to convert from degrees to radians, and from radians to degrees. From Equation 13.1

$$1° = \frac{\pi}{180} \text{ radians} \qquad \text{and} \qquad 1 \text{ radian} = \frac{180°}{\pi} \ .$$

The use of these conversion factors is illustrated in the following examples.

EXAMPLE 13.1: Convert 30° into radian measurement.

The conversion factor to go from degrees to radians is $\frac{\pi \text{ radians}}{180°}$.
Thus $(30°)\,\frac{(\pi \text{ radians})}{180°} = \frac{\pi}{6}$. ∎

EXAMPLE 13.2: Convert $\frac{3\pi}{4}$ radians into degrees.

The conversion factor to go from radians to degrees is $\frac{180°}{\pi \text{ radians}}$.
Thus $\left[\frac{3\pi}{4} \text{ radians}\right]\left[\frac{180°}{\pi \text{ radians}}\right] = 135°$. ∎

Radian measure is assumed in most mathematical settings unless some clear indication of degrees appears.

The trigonometric functions are easily defined if we place an angle in *standard position*, that is, with its vertex at the origin of a rectangular coordinate system with the initial side coinciding with the positive x-axis as is shown in Figure 13.4. If θ is a positive number, then the terminal side is reached by rotating from the positive x-axis in a *counterclockwise* direction. If θ is a negative number, then the terminal side is reached by rotating from the positive x-axis in a clockwise direction. Notice that in Figure 13.4, θ was taken to be a positive number; for now we will consider only positive θ.

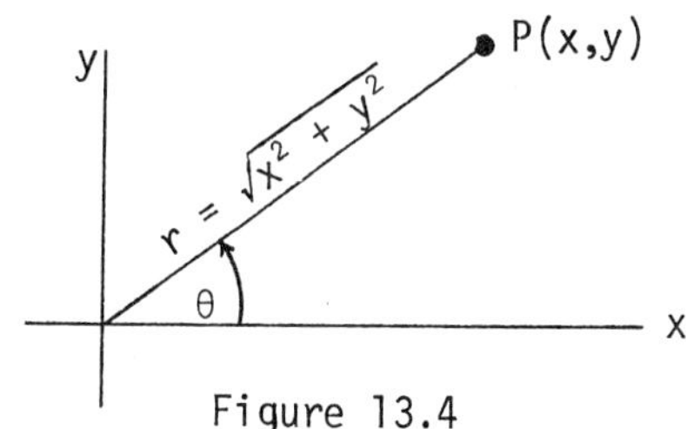

Figure 13.4

If we select a point P(x,y), not the vertex, on the terminal side and let $r = \sqrt{x^2 + y^2}$ (notice that r is greater than zero), then

- sine $\theta = \sin\theta = \frac{y}{r}$ (13.2)
- cosine $\theta = \cos\theta = \frac{x}{r}$ (13.3)
- tangent $\theta = \tan\theta = \frac{y}{x}$ (13.4)
- cotangent $\theta = \cot\theta = \frac{x}{y}$ (13.5)

- secant θ = sec $\theta = \frac{r}{x}$ (13.6)

- cosecant θ = csc $\theta = \frac{r}{y}$ (13.7)

On the left, the full name of the trigonometric relation is given followed by its abbreviated name. It should be clear from the above definitions that sin θ and cos θ can be computed for all real values of θ. However, the numbers tan θ and sec θ are not defined whenever $x = 0$; similarly, cot θ and csc θ are undefined for $y = 0$.

<u>EXAMPLE 13.3</u>: Suppose $x = 3$ and $y = 4$. Find the values of sin θ, cos θ and tan θ.

The angle θ is shown in Figure 13.5. When $x = 3$ and $y = 4$, $r = \sqrt{x^2 + y^2} = \sqrt{9 + 16} = 5$. Using these values in Equations 13.2 - 13.4, we have

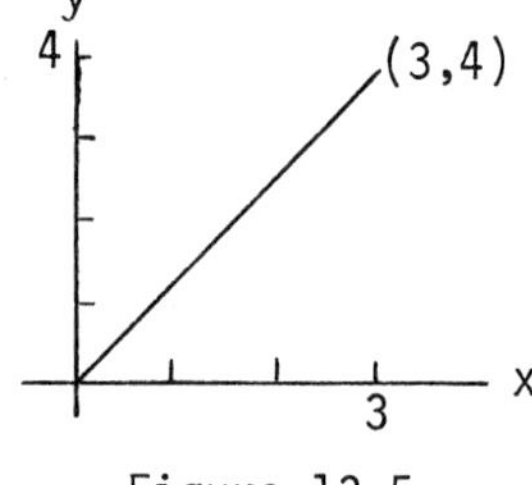

Figure 13.5

$$\sin \theta = \frac{y}{r} = \frac{4}{5} = 0.8$$

$$\cos \theta = \frac{x}{r} = \frac{3}{5} = 0.6$$

and

$$\tan \theta = \frac{y}{x} = \frac{4}{3} = 1.33 \ . \ \blacksquare$$

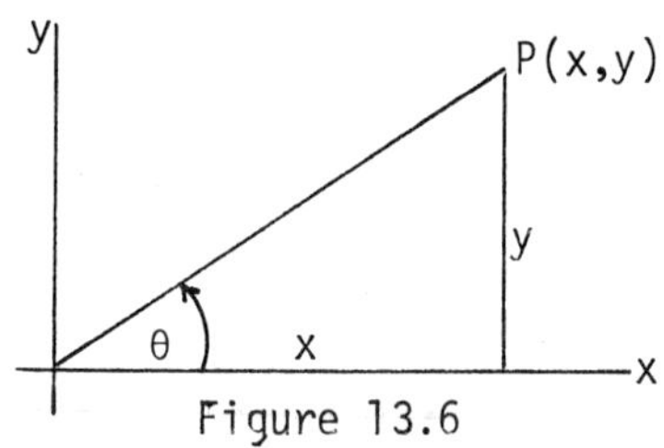

Figure 13.6

The sin θ, cos θ and tan θ sometimes are referred to as the *primary trigonometric functions* while the cot θ, sec θ and csc θ are called the *reciprocal or complimentary trigonometric functions*. The word reciprocal is suggestive for if we examine Equations 13.5 - 13.7 we see that $\cot \theta = \frac{1}{\tan \theta}$, $\sec \theta = \frac{1}{\cos \theta}$ and $\csc \theta = \frac{1}{\sin \theta}$. It is also possible to define $\tan \theta = \sin \theta/\cos \theta$ and $\cot \theta = \cos \theta/\sin \theta$.

If we restrict θ in a manner such that $0 < \theta < \frac{\pi}{2}$, then the defining equations given in Equations 13.2 - 13.7 can be given in terms of a right triangle such as the one shown in Figure 13.6. Then

- $\sin\theta = \frac{y}{r} = \frac{\text{opposite side}}{\text{hypotenuse}}$ (13.8)
- $\cos\theta = \frac{x}{r} = \frac{\text{adjacent side}}{\text{hypotenuse}}$ (13.9)
- $\tan\theta = \frac{y}{x} = \frac{\text{opposite side}}{\text{adjacent side}} = \frac{\sin\theta}{\cos\theta} = \frac{1}{\cot\theta}$ (13.10)
- $\cot\theta = \frac{x}{y} = \frac{\text{adjacent side}}{\text{opposite side}} = \frac{\cos\theta}{\sin\theta} = \frac{1}{\tan\theta}$ (13.11)
- $\sec\theta = \frac{r}{x} = \frac{\text{hypotenuse}}{\text{adjacent side}} = \frac{1}{\cos\theta}$ (13.12)
- $\csc\theta = \frac{r}{y} = \frac{\text{hypotenuse}}{\text{opposite side}} = \frac{1}{\sin\theta}$ (13.13)

These latter definitions prove more useful to use when angles may not be oriented to allow convenient use of the earlier definitions. However, the definitions immediately above restrict θ to $0 < \theta < \frac{\pi}{2}$. We will see how to relax this restriction in the next section. The term opposite side and adjacent side should have the obvious meaning of side opposite the angle and the side adjacent the angle.

Table 13.1

Radians	°	sin θ	cos θ	tan θ	cot θ	sec θ	csc θ
0	0	0	1	0		1	
$\frac{\pi}{6}$	30	$\frac{1}{2}$	$\frac{\sqrt{3}}{2}$	$\frac{\sqrt{3}}{3}$	$\sqrt{3}$	$\frac{2\sqrt{3}}{3}$	2
$\frac{\pi}{4}$	45	$\frac{\sqrt{2}}{2}$	$\frac{\sqrt{2}}{2}$	1	1	$\sqrt{2}$	$\sqrt{2}$
$\frac{\pi}{3}$	60	$\frac{\sqrt{3}}{2}$	$\frac{1}{2}$	$\sqrt{3}$	$\frac{\sqrt{3}}{3}$	2	$\frac{2\sqrt{3}}{3}$
$\frac{\pi}{2}$	90	1	0		0		1
π	180	0	-1	0		-1	
$\frac{3\pi}{2}$	270	-1	0		0		-1
2π	360	0	1	0		1	

Several angles occur frequently enough that you should become familiar with them and the values of their trigonometric functions. Table 13.1 summarizes these frequently encountered angles given both in radian measurement and in degrees. This table includes angles we have yet to discuss, like $\theta = 3\pi/2$. (In checking some of these entries, you need to remember that the standard form leaves no radicals in the denominator. For example $1/\sqrt{3}$ is recorded as $\sqrt{3}/3$.)

Additional angles and the values of their trigonometric functions are given in Table IV in the back of the book. The table entries are given for each degree from 0° through 90° with the angles also being expressed in radian measure. The examples which follow will illustrate the use of Table IV.

EXAMPLE 13.1 (continued): From Table IV verify that $\sin 30° = \sin \frac{\pi}{6}$ radians $= 0.5$, $\cos 30° = \cos \frac{\pi}{6}$ radians $= 0.866$ and

$\tan 30° = \tan \frac{\pi}{6}$ radians $= 0.577$.

In the table follow down the column on the left (labeled degrees) until you reach 30°. Now use the column headings at the top of the page and observe that $\sin 30° = 0.500$, $\cos 30° = 0.866$ and $\tan 30° = 0.577$. The column headings across the top of the table go with the angles specified on the left; the column headings across the bottom go with the angles specified on the right. For example, $\cos 71° = \cos 1.239$ radians $= 0.326$. ▮

EXAMPLE 13.3 (continued): From $x = 3$, $y = 4$ and $r = 5$, we arrived at $\sin \theta = 0.8$. Use Table IV in order to find θ.

By using Table IV in reverse, we seek to find an angle θ such that $\sin \theta = 0.8$. Scanning the column labeled sin, we find that θ must lie between 53° and 54° or 0.925 and 0.942 radians. We used interpolation with logarithms back in Section 4.6. The details of this process are reviewed in the early part of Appendix I if you need a review. When applied here, we find

$$1^\circ \left[x \left[\begin{array}{l} \sin 54^\circ = 0.809 \\ \sin \theta \quad = 0.800 \\ \sin 53^\circ = 0.799 \end{array} \right] 0.001 \right] 0.01$$

Setting up the proportions, we have

$$\frac{x^\circ}{1^\circ} = \frac{0.001}{0.01} \qquad \text{or} \qquad x = 0.1 \text{ degrees} .$$

Thus, $\theta \doteq 53^\circ + 0.1^\circ$ or $\theta \doteq 53.1^\circ$.

Doing the interpolation on radian measurement, we have

$$0.017 \left[x \left[\begin{array}{l} \sin 0.942 = 0.809 \\ \sin \theta \quad\; = 0.800 \\ \sin 0.925 = 0.799 \end{array} \right] 0.001 \right] 0.01$$

$$\frac{x}{0.017} = \frac{0.001}{0.01} \qquad \text{and} \qquad x = 0.0017$$

and, thus, $\theta \doteq 0.925 + 0.0017 \doteq 0.927$ radians. ▌

EXAMPLE 13.4: If $r = 6$ and $\theta = 35^\circ$, how long are the other two sides of the right triangle shown in Figure 13.7?

From $\sin \theta = \frac{y}{r}$, we get $y = r \sin \theta$; similarly $x = r \cos \theta$. Thus, using Table IV with $\theta = 35^\circ$, we have $y = 6 \sin 35^\circ = 6(0.574) = 3.444$ and $x = 6 \cos 35^\circ = 6(0.819) = 4.914$.

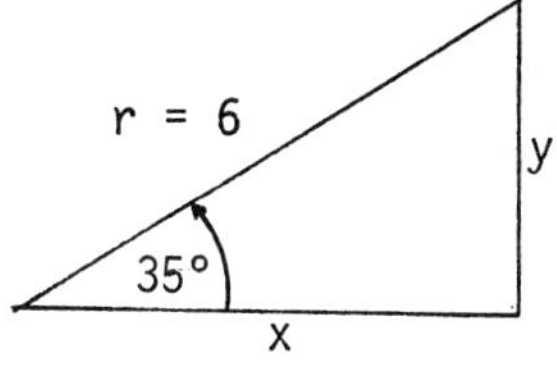

Figure 13.7

This and Example 13.3 illustrate how the lengths of two sides of a right triangle or an angle and the length of one side specify the remaining unknown quantities. ▌

Trigonometry has another use in calculus: It can be used to simplify or change the form of an expression. For example from the definitions given previously, we have

$$\tan \theta = \frac{1}{\cot \theta} = \frac{\sin \theta}{\cos \theta} \tag{13.14}$$

and

$$\cot\ \theta = \frac{1}{\tan\ \theta} = \frac{\cos\ \theta}{\sin\ \theta}\ . \tag{13.15}$$

If we construct a right triangle with a hypotenuse of length $r = 1$, then $y = r \sin\ \theta = \sin\ \theta$ and $x = r \cos\ \theta = \cos\ \theta$. From the Pythagorean theorem, we know that $x^2 + y^2 = 1$ and so

$$\sin^2\ \theta + \cos^2\ \theta = 1\ . \tag{13.16}$$

This relation is used extensively. If we divide both sides of Equation 13.16 by $\sin^2\ \theta$, we obtain

$$1 + \frac{\cos^2\ \theta}{\sin^2\ \theta} = \frac{1}{\sin^2\ \theta}$$

or

$$1 + \left(\frac{\cos\ \theta}{\sin\ \theta}\right)^2 = \left(\frac{1}{\sin\ \theta}\right)^2$$

and

$$1 + \cot^2\ \theta = \csc^2\ \theta \tag{13.17}$$

because $\frac{\cos\ \theta}{\sin\ \theta} = \cot\ \theta$ and $\frac{1}{\sin\ \theta} = \csc\ \theta$. Similarly,

$$\tan^2\ \theta + 1 = \sec^2\ \theta \tag{13.18}$$

results when we divide Equation 13.16 by $\cos^2\ \theta$.

The use of the above basic trigonometric identities is illustrated in the next examples.

<u>EXAMPLE 13.5</u>: Simplify $\frac{\cos^2\ \theta}{1 - \sin\ \theta}$.

From $\sin^2\ \theta + \cos^2\ \theta = 1$, $\cos^2\ \theta = 1 - \sin^2\ \theta$; substitute this into the original equation to get

$$\frac{\cos^2\ \theta}{1 - \sin\ \theta} = \frac{1 - \sin^2\ \theta}{1 - \sin\ \theta} = \frac{(1 + \sin\ \theta)(1 - \sin\ \theta)}{1 - \sin\ \theta} = 1 + \sin\ \theta\ .$$

We have reduced a fraction to the sum of two quantities. Such a reduction could change an apparently difficult integrand into one which is easy to integrate if integration was the task to be performed. ∎

EXAMPLE 13.6: Show $\cos\theta \csc\theta = \cot\theta$.

From the definitions given above, $\csc\theta = 1/\sin\theta$. Inserting this into the original equation, we get

$$\cos\theta \csc\theta = \cos\theta \left(\frac{1}{\sin\theta}\right) = \frac{\cos\theta}{\sin\theta} = \cot\theta \ .$$

This example illustrates a strategy which is widely used in handling trigonometric identities: It usually is easier to change all trigonometric relations to sines and cosines before attempting simplification. ▌

EXAMPLE 13.7: Show $\sec^2\theta + \csc^2\theta = \sec^2\theta \csc^2\theta$.

Express the left-hand side of the equation in sines and cosines, and then get a common denominator:

$$\sec^2\theta + \csc^2\theta = \frac{1}{\cos^2\theta} + \frac{1}{\sin^2\theta}$$

$$= \frac{\sin^2\theta + \cos^2\theta}{\cos^2\theta \sin^2\theta} = \frac{1}{\cos^2\theta \sin^2\theta}$$

because $\sin^2\theta + \cos^2\theta = 1$. Thus

$$\sec^2\theta + \csc^2\theta = \frac{1}{\cos^2\theta \sin^2\theta}$$

$$= \left(\frac{1}{\cos\theta}\right)^2 \left(\frac{1}{\sin\theta}\right)^2 = \sec^2\theta \csc^2\theta \ .$$ ▌

EXERCISES

The following angles are given in radian measure. Express them in degrees.

1. $5\pi/3$ 2. $5\pi/4$ 3. $3\pi/5$ 4. $\pi/4$ 5. $\pi/9$

6. 2.5 7. 0.720 8. 1.414 9. $\pi/6$ 10. $\pi/3$

Express each of the following angles in radian measure.

11. 60° 12. 45° 13. 300° 14. 270° 15. 225°

16. 150° 17. 135° 18. 120° 19. 330° 20. 20°

21. If $\theta = 39°$, find the six trigonometric functions of θ.

22. If $\theta = 0.943$ radians, find the six trigonometric functions of θ.

23. If $\theta = 49°$ and $x = 5$, find y and r.

24. Given the triangle shown on the right and the two angles α and β, define the six trigonometric functions for both α and β. Are any of the functions equal? If so, which ones?

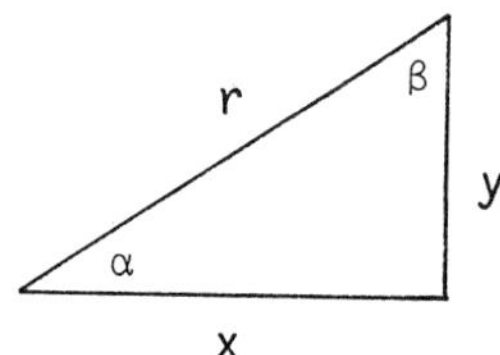

25. In view of the results obtained in Problem 24, can you make any statements concerning the angle θ and the angle $\frac{\pi}{2} - \theta$? If so, what are they?

26. In the rectangular coordinate system, the axes divide the plane into four parts called quadrants, which are numbered I, II, III and IV as shown in the figure to the right. With each point and angle indicated in the four quadrants, write down the expressions for the six trigonometric functions. Build a table showing the signs of the six trigonometric functions in the four quadrants.

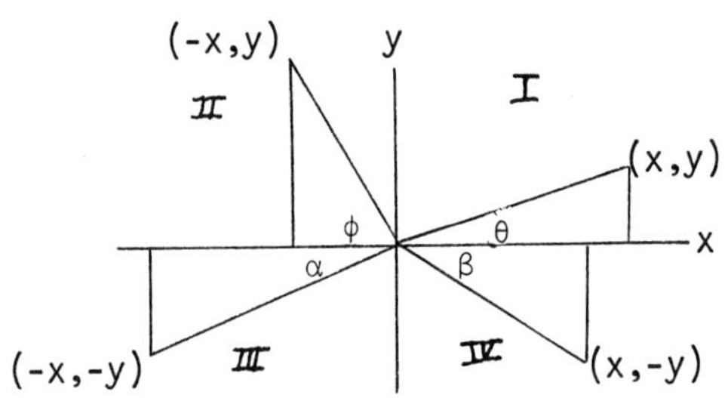

Show that the following identities are valid.

27. $\dfrac{1 - 2\cos^2 A}{\sin A \cos A} = \tan A - \cot A$

28. $\sin A \cos A (\tan A + \cot A) = 1$

29. $1 - \dfrac{\cos^2 \theta}{1 + \sin \theta} = \sin \theta$

30. $\dfrac{\sec \beta + \csc \beta}{\tan \beta + \cot \beta} = \sin \beta + \cos \beta$

31. $\dfrac{\cos \theta \cot \theta - \sin \theta \tan \theta}{\csc \theta - \sec \theta} = 1 + \sin \theta \cos \theta$

32. $\dfrac{\tan \alpha - \sin \alpha}{\sin^3 \alpha} = \dfrac{\sec \alpha}{1 + \cos \alpha}$

33. $\dfrac{\sin \theta \cos \theta}{\cos^2 \theta - \sin^2 \theta} = \dfrac{\tan \theta}{1 - \tan^2 \theta}$

34. $\tan^6 \theta = \tan^4 \theta \sec^2 \theta - \tan^2 \theta \sec^2 \theta + \sec^2 \theta - 1$

35. $(\tan y + \cot y)(\cos y + \sin y) = \sec y + \csc y$

36. $(\cos^2 x - 1)(\tan^2 + 1) = 1 - \sec^2 x$

37. $\dfrac{\sec x - \cos x}{\tan x} = \dfrac{\tan x}{\sec x}$

38. $\dfrac{\cos x \cot x}{\cot x - \cos x} = \dfrac{\cot x + \cos x}{\cos x \cot x}$

39. $\dfrac{\cot u - 1}{1 - \tan u} = \cot u$

40. $\left[\dfrac{\sin^2 x}{\tan^4 x}\right]^3 \left[\dfrac{\csc^3 x}{\cot^6 x}\right]^2 = 1$

41. $\dfrac{\cos^3 x - \sin^3 x}{\cos x - \sin x} = 1 + \sin x \cos x$

42. $\dfrac{1 + \sec u}{\sin u + \tan u} = \csc u$

43. $\dfrac{\sin \alpha \cos \beta + \cos \alpha \sin \beta}{\cos \alpha \cos \beta - \sin \alpha \sin \beta} = \dfrac{\tan \alpha + \tan \beta}{1 - \tan \alpha \tan \beta}$

44. $(a \cos t - b \sin t)^2 + (a \sin t + b \cos t)^2 = a^2 + b^2$

*13.2 RELATIONSHIPS BETWEEN FUNCTIONS OF θ AND $-\theta$

Simply derived relationships exist between the trigonometric functions of θ and $-\theta$. Consider the angles θ and $-\theta$ as illustrated in Figure 13.8. Applying the definitions given previously, we have

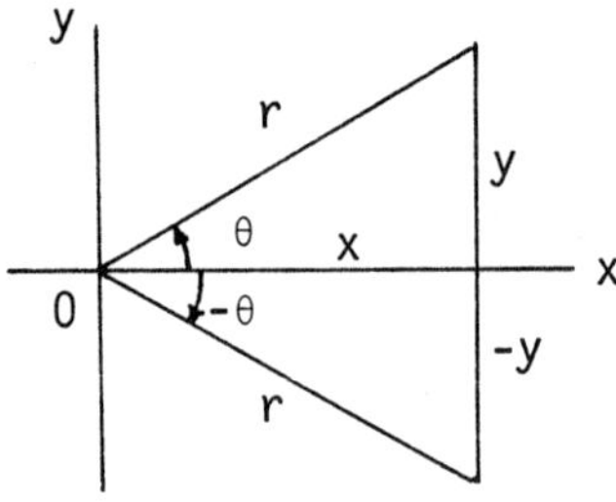

Figure 13.8

$$\sin\theta = \frac{y}{r} \quad \text{and} \quad \sin(-\theta) = -\frac{y}{r} \quad \text{so} \quad \sin(-\theta) = -\sin\theta \tag{13.19}$$

Now consider the cosine:

$$\cos\theta = \frac{x}{r} \quad \text{and} \quad \cos(-\theta) = \frac{x}{r} \quad \text{so} \quad \cos(-\theta) = \cos\theta \tag{13.20}$$

In a similar fashion, we can establish the following trigonometric relationships between θ and $-\theta$:

$$\tan(-\theta) = -\tan\theta \tag{13.21}$$

$$\cot(-\theta) = -\cot\theta \tag{13.22}$$

$$\sec(-\theta) = \sec\theta \tag{13.23}$$

$$\csc(-\theta) = -\csc\theta \tag{13.24}$$

The above relations are used frequently. They will be illustrated in the next section.

*13.3 TRIGONOMETRIC FUNCTIONS OF TWO ANGLES

Reasonably simple relations exist between trigonometric functions of $(\alpha + \beta)$, $(\alpha - \beta)$ and those of α and β. This section presents the identities for trigonometric functions of $(\alpha + \beta)$ and $(\alpha - \beta)$ without proving them because their proof is more tedious than instructive. Nevertheless you will find these identities useful as you expand your knowledge of calculus. The primary relations are

$$\sin(\alpha + \beta) = \sin\alpha\cos\beta + \sin\beta\cos\alpha \tag{13.25}$$

and

$$\cos(\alpha + \beta) = \cos\alpha\cos\beta - \sin\alpha\sin\beta. \tag{13.26}$$

Equations 13.25 and 13.26 allow us to express the trigonometric function of the sum of two angles as trigonometric functions of the individual angles. The following examples illustrate their use.

EXAMPLE 13.8: Show that $\cos(\alpha - \beta) = \cos\alpha\cos\beta + \sin\alpha\sin\beta$.

Replace β in Equation 13.26 by $-\beta$:

$$\cos(\alpha - \beta) = \cos(\alpha + (-\beta)) = \cos\alpha\cos(-\beta) - \sin\alpha\sin(-\beta)$$

Now substitute $\cos(-\beta) = \cos\beta$ and $\sin(-\beta) = -\sin\beta$:

$$\begin{aligned}\cos(\alpha - \beta) &= \cos\alpha\cos\beta - \sin\alpha(-\sin\beta)\\ &= \cos\alpha\cos\beta + \sin\alpha\sin\beta\ .\end{aligned}$$

Thus, by knowing the identity for $\cos(\alpha + \beta)$, we were able to arrive at an identity for $\cos(\alpha - \beta)$. In a like manner, show that

$$\sin(\alpha - \beta) = \sin\alpha\cos\beta - \cos\alpha\sin\beta\ .$$ ▌

EXAMPLE 13.9: Show that $\tan(\alpha + \beta) = \dfrac{\tan\alpha + \tan\beta}{1 - \tan\alpha\tan\beta}$.

From $\tan\theta = \dfrac{\sin\theta}{\cos\theta}$ and the identities for $\sin(\alpha + \beta)$ and $\cos(\alpha + \beta)$, we have

$$\tan(\alpha + \beta) = \frac{\sin(\alpha + \beta)}{\cos(\alpha + \beta)} = \frac{\sin\alpha\cos\beta + \cos\alpha\sin\beta}{\cos\alpha\cos\beta - \sin\alpha\sin\beta}\ .$$

Now divide the numerator and denominator by $\cos\alpha\cos\beta$ to get

$$\tan(\alpha + \beta) = \frac{\dfrac{\sin\alpha\cos\beta + \cos\alpha\sin\beta}{\cos\alpha\cos\beta}}{\dfrac{\cos\alpha\cos\beta - \sin\alpha\sin\beta}{\cos\alpha\cos\beta}} = \frac{\dfrac{\sin\alpha\cos\beta}{\cos\alpha\cos\beta} + \dfrac{\cos\alpha\sin\beta}{\cos\alpha\cos\beta}}{\dfrac{\cos\alpha\cos\beta}{\cos\alpha\cos\beta} - \dfrac{\sin\alpha\sin\beta}{\cos\alpha\cos\beta}}$$

$$= \frac{\dfrac{\sin\alpha}{\cos\alpha} + \dfrac{\sin\beta}{\cos\beta}}{1 - \dfrac{\sin\alpha}{\cos\alpha}\ \dfrac{\sin\beta}{\cos\beta}} = \frac{\tan\alpha + \tan\beta}{1 - \tan\alpha\tan\beta}\ .$$

Thus, using the knowledge of the identities for $\sin(\alpha + \beta)$, $\cos(\alpha + \beta)$, and a little algebra, an expression has been found relating the tangent of the sum of two angles to the tangents of the individual angles. ▌

EXAMPLE 13.10: Find expressions for $\sin 2\theta$, $\cos 2\theta$ and $\tan 2\theta$.

$$\sin 2\theta = \sin(\theta + \theta) = \sin\theta \cos\theta + \cos\theta \sin\theta = 2 \sin\theta \cos\theta .$$

In a similar fashion, show that

$$\cos 2\theta = \cos^2\theta - \sin^2\theta = 2\cos^2\theta - 1 = 1 - 2\sin^2\theta$$

and

$$\tan 2\theta = \frac{2\tan\theta}{1 - \tan^2\theta} .$$ ▌

EXAMPLE 13.11: Some of the above relations will be utilized with $\alpha = \frac{\pi}{6}$ and $\beta = \frac{\pi}{4}$ so that $\alpha + \beta = \frac{5\pi}{12}$, that is, $\alpha = 30°$, $\beta = 45°$ and $\alpha + \beta = 75°$. The values of the trigonometric functions of $\frac{\pi}{6}$ and $\frac{\pi}{4}$ were given earlier in Table 13.1.

$$\sin\frac{5\pi}{12} = \sin\left(\frac{\pi}{6} + \frac{\pi}{4}\right) = \sin\frac{\pi}{6}\cos\frac{\pi}{4} + \sin\frac{\pi}{4}\cos\frac{\pi}{6}$$

$$= \frac{1}{2}\frac{\sqrt{2}}{2} + \frac{\sqrt{2}}{2}\frac{\sqrt{3}}{2} = \frac{\sqrt{6} + \sqrt{2}}{4} \doteq 0.966 ,$$

which is in agreement with the entry for sin 75° given in Table IV. Similarly,

$$\tan\frac{5\pi}{12} = \frac{\tan\frac{\pi}{4} + \tan\frac{\pi}{6}}{1 - \tan\frac{\pi}{4}\tan\frac{\pi}{6}} = \frac{1 + \frac{\sqrt{3}}{3}}{1 - \frac{\sqrt{3}}{3}} \doteq 3.732 .$$

Now use the relationship established for $\cos 2\theta$ to find $\cos\frac{\pi}{3}$:

$$\cos\frac{\pi}{3} = \cos 2\left(\frac{\pi}{6}\right) = \cos^2\left(\frac{\pi}{6}\right) - \sin^2\left(\frac{\pi}{6}\right) = \left(\frac{\sqrt{3}}{2}\right)^2 - \left(\frac{1}{2}\right)^2 = \frac{1}{2} ,$$

as was given previously. ▌

*13.4 ADDITIONAL IDENTITIES

Some of the identities given here were given in the last section, others are new. We collect them here for future reference. When two signs are given, such as in Equation 13.30, the top signs go together and the bottom signs go together.

$$\sin^2 \theta + \cos^2 \theta = 1 \tag{13.27}$$

$$1 + \tan^2 \theta = \sec^2 \theta \tag{13.28}$$

$$1 + \cot^2 \theta = \csc^2 \theta \tag{13.29}$$

$$\sin(\alpha \pm \beta) = \sin \alpha \cos \beta \pm \cos \alpha \sin \beta \tag{13.30}$$

$$\cos(\alpha \pm \beta) = \cos \alpha \cos \beta \mp \sin \alpha \sin \beta \tag{13.31}$$

$$\tan(\alpha \pm \beta) = \frac{\tan \alpha \pm \tan \beta}{1 \mp \tan \alpha \tan \beta} \tag{13.32}$$

$$\sin 2\alpha = 2 \sin \alpha \cos \alpha \tag{13.33}$$

$$\cos 2\alpha = 2 \cos^2 \alpha - 1 = 1 - 2 \sin^2 \alpha = \cos^2 \alpha - \sin^2 \alpha \tag{13.34}$$

$$\tan 2\alpha = \frac{2 \tan \alpha}{1 - \tan^2 \alpha} \tag{13.35}$$

$$\sin \alpha \cos \beta = \frac{1}{2} [\sin(\alpha + \beta) + \sin(\alpha - \beta)] \tag{13.36}$$

$$\cos \alpha \sin \beta = \frac{1}{2} [\sin(\alpha + \beta) - \sin(\alpha - \beta)] \tag{13.37}$$

$$\cos \alpha \cos \beta = \frac{1}{2} [\cos(\alpha - \beta) + \cos(\alpha + \beta)] \tag{13.38}$$

$$\sin \alpha \sin \beta = \frac{1}{2} [\cos(\alpha - \beta) - \cos(\alpha + \beta)] \tag{13.39}$$

$$\sin \alpha + \sin \beta = 2 \sin \frac{\alpha + \beta}{2} \cos \frac{\alpha - \beta}{2} \tag{13.40}$$

$$\sin \alpha - \sin \beta = 2 \cos \frac{\alpha + \beta}{2} \sin \frac{\alpha - \beta}{2} \tag{13.41}$$

$$\cos \alpha + \cos \beta = 2 \cos \frac{\alpha + \beta}{2} \cos \frac{\alpha - \beta}{2} \tag{13.42}$$

$$\cos \alpha - \cos \beta = - 2 \sin \frac{\alpha + \beta}{2} \sin \frac{\alpha - \beta}{2} \tag{13.43}$$

EXERCISES

1. Find $\sin(\alpha + \beta)$, $\cos(\alpha + \beta)$, $\sin(\alpha - \beta)$ and $\cos(\alpha - \beta)$ and determine the quadrants in which $(\alpha + \beta)$ and $(\alpha - \beta)$ lie for

   a. $\sin \alpha = \frac{4}{5}$, $\cos \beta = \frac{5}{13}$, α and β in Quadrant I.

   b. $\cos \alpha = \frac{3}{5}$ and $\sin \beta = \frac{12}{13}$.

2. Find values for sin 15°, cos 15° and tan 15° using only the values for 30°, 45° and 60°. Check your results by using Table IV.

3. Show $\tan(\alpha - \beta) = \dfrac{\tan \alpha - \tan \beta}{1 + \tan \alpha \tan \beta}$.

4. Show that $\cos\left(\frac{\pi}{2} - \theta\right) = \sin \theta$.

5. Show that $\sin\left(\frac{\pi}{2} - \theta\right) = \cos \theta$.

6. Show that $\tan\left(\frac{\pi}{2} - \theta\right) = \cot \theta$.

7. Show that $\cot(\alpha + \beta) = \dfrac{\cot \alpha \cot \beta - 1}{\cot \alpha + \cot \beta}$.

8. Show that $\cot(\alpha - \beta) = \dfrac{\cot \alpha \cot \beta + 1}{\cot \beta - \cot \alpha}$.

9. Show that $\cos 2\theta = \cos^2 \theta - \sin^2 \theta = 1 - 2 \sin^2 \theta$.

10. Show that $\dfrac{\sin 2\theta}{\sin \theta} - \dfrac{\cos 2\theta}{\cos \theta} = \sec \theta$.

11. Show that $\dfrac{\sin 7\theta - \sin 5\theta}{\cos 7\theta + \cos 5\theta} = \tan \theta$.

12. Show that $2 \tan 2\theta = \dfrac{\cos \theta + \sin \theta}{\cos \theta - \sin \theta} - \dfrac{\cos \theta - \sin \theta}{\cos \theta + \sin \theta}$.

13. Show that $\dfrac{\sin \alpha - \sin \beta}{\sin \alpha + \sin \beta} = \dfrac{\tan \frac{\alpha - \beta}{2}}{\tan \frac{\alpha + \beta}{2}}$.

14. Show that $\dfrac{\sin \theta}{\sin \theta + \cos \theta} = \dfrac{\sec \theta}{\sec \theta + \csc \theta}$.

In Problems 15 - 21, if α and β are both in Quadrant I and $\tan \alpha = \frac{5}{12}$ and $\sec \beta = \frac{5}{4}$, then:

15. Find the value of $\sin(\alpha + \beta)$.

16. Find the value of $\tan(\alpha + \beta)$.

17. Find the value of $\cos(\alpha + \beta)$.

18. Find the value of $\cos(\alpha - \beta)$.

19. Find the value of $\sin 2\alpha$.

20. Find the size of α.

21. Find the size of β.

22. Evaluate $\sin \frac{11\pi}{12}$.

23. Show that $\sin(\theta + \frac{\pi}{4}) - \cos(\theta + \frac{\pi}{4}) = \sqrt{2} \sin \theta$.

24. Evaluate $\cos \frac{11\pi}{12}$.

13.5 PERIODIC FUNCTIONS

We need to first define what we mean by the word periodic and then how we extend the idea of periodicity to the trigonometric functions.

- If there is a positive number p such that $f(x + p)$ is defined whenever $f(x)$ is defined and $f(x + p) = f(x)$, then the function f is periodic.
- If there is no smaller positive number p such that $f(x + p) = f(x)$, then the function is a period of p.

This definition simply states that a function that is periodic repeats itself with a period of p. As was indicated in the introduction of this chapter your heart beats in periodic fashion. Any time your doctor takes an electrocardiogram (EKG) he is examining the periodicity of your heart. Certain deviations from the normal periodicity may suggest some type of a malfunction of the heart. An electroencephalogram (EEG) simply shows the pattern of the brain waves under different conditions. The sketch in Problem 22 in Section 4.3 shows some typical EEG traces during sleep. Look back to see the periodicity in these tracings. Many phenomena have a periodic relationship in time. Humans have many biological functions which are periodic. Some last a lifetime, but others such as the fertility cycle in the female, and fertility in the male, cease in time.

The trigonometric functions have a periodic behavior which is easily seen by examining Figure 13.9. As you can see the angle θ and $\theta + 2\pi$ have the same terminal side. Thus the trigonometric functions display a periodicity, having a period of 2π. Also you can observe from the figure that any angle $\theta + 2n\pi$, where n is a positive integer, has the same terminal side as θ.

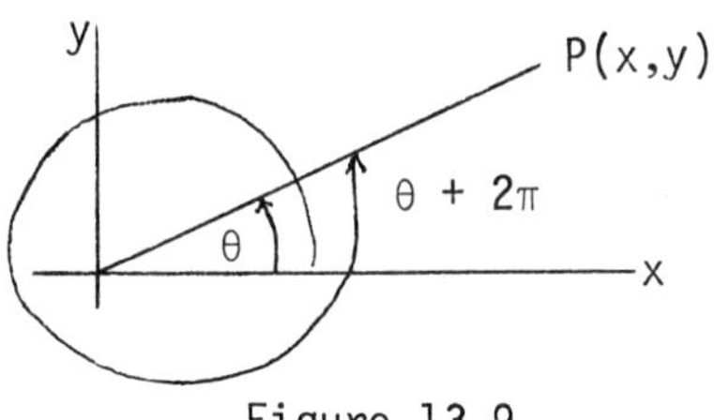

Figure 13.9

If we let θ be any angle in standard position as shown in Figure 13.10, then with $r = 1$, $y = \sin\theta$ and $\cos\theta = x$. Now if we let θ start at zero and increase to $\frac{\pi}{2}$, P starts at point A and moves along the circle to B, so that the $\sin\theta$ (or y) starts at 0 and increases to 1. At the same time, $\cos\theta$ (or x) starts at 1 and decreases to 0. In fact:

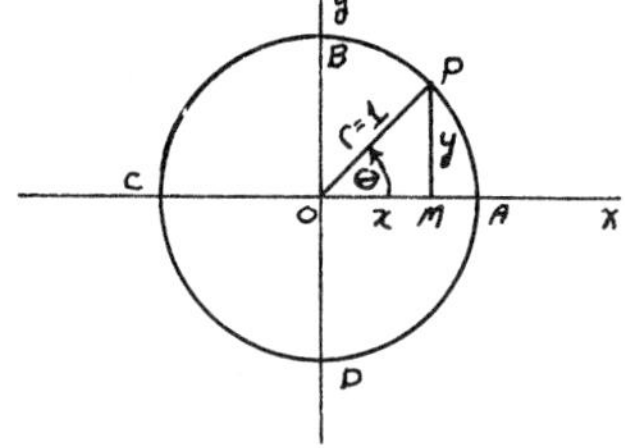

Figure 13.10

As θ increases	P moves	$\sin\theta$ (or y)	$\cos\theta$ (or x)
from 0 to $\frac{\pi}{2}$	from A to B	increases from 0 to 1	decreases from 1 to 0
from $\frac{\pi}{2}$ to π	from B to C	decreases from 1 to 0	decreases from 0 to -1
from π to $\frac{3\pi}{2}$	from C to D	decreases from 0 to -1	increases from -1 to 0
from $\frac{3\pi}{2}$ to 2π	from D to A	increases from -1 to 0	increases from 0 to 1

Once we get around the circle to point A again, the whole cycle repeats itself.

The above discussion provides the material necessary to construct the graph of the sine and cosine functions shown in Figure 13.11. Carefully repeat the above process for the other trigonometric functions to see that the other graphs in Figure 13.11 do describe the respective functions.

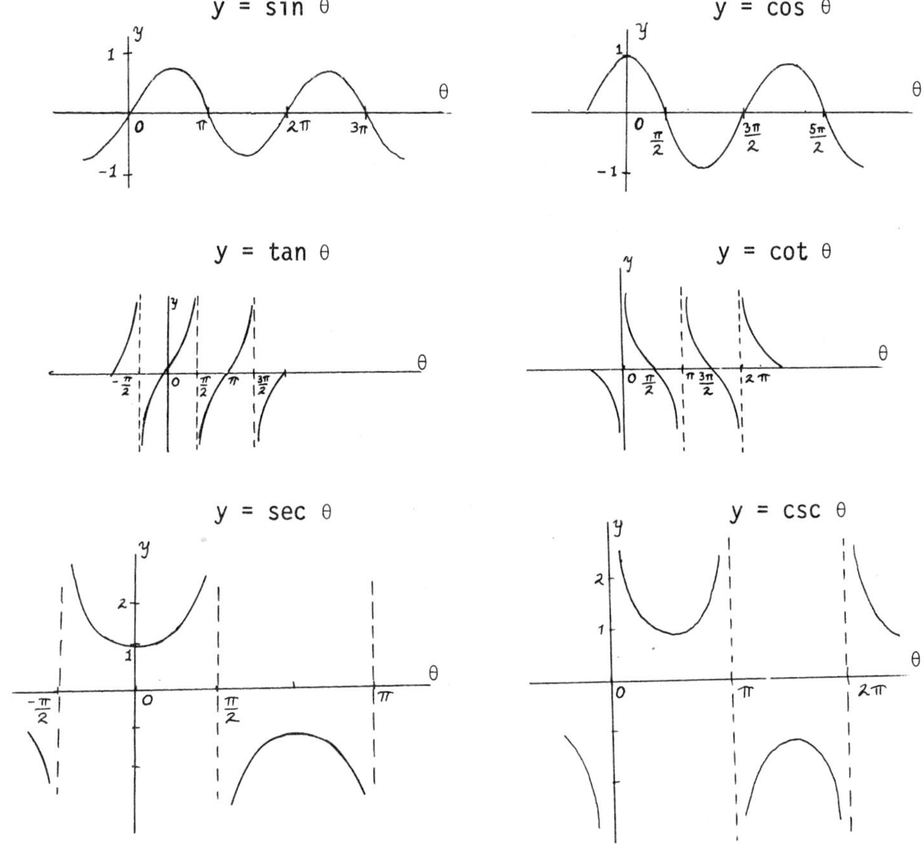

Figure 13.11

From the graphs in Figure 13.11 we can observe that sin θ and cos θ are defined for all values of θ and further $|\sin\theta| \leq 1$ as well as $|\cos\theta| \leq 1$; tan θ and sec θ are defined for all values of θ except for the odd multiples of $\frac{\pi}{2}$; cot θ and csc θ are defined for all values of θ except 0, π, 2π and other integer multiples of π; there are no limitations on the values of $|\tan\theta|$ and $|\cot\theta|$; but $|\sec\theta| \geq 1$ as is $|\csc\theta| \geq 1$.

The variation displayed in the trigonometric functions is summarized in Table 13.2:

QUADRANT	I		II		III		IV	
VARIATION	From	To	From	To	From	To	From	To
θ	0	$\frac{\pi}{2}$	$\frac{\pi}{2}$	π	π	$\frac{3\pi}{2}$	$\frac{3\pi}{2}$	2π
sin θ	0	1	1	0	0	-1	-1	0
cos θ	1	0	0	-1	-1	0	0	1
tan θ	0	∞	$-\infty$	0	0	∞	$-\infty$	0
cot θ	∞	0	0	$-\infty$	∞	0	0	$-\infty$
sec θ	1	∞	$-\infty$	-1	-1	$-\infty$	∞	1
csc θ	∞	1	1	∞	$-\infty$	-1	-1	$-\infty$

Table 13.2

As was pointed out previously a problem arises when we have angles of the form θ, $\theta \pm 2\pi$, $\theta \pm 4\pi$, $\cdots$, because they are indistinguishable. The trigonometric functions must be equal at these points as shown by the graphs of the trigonometric functions. The following examples illustrate how to handle angles outside of $0 \leq \theta \leq 2\pi$. The technique is simple: Every angle outside of $[0,2\pi]$ has a corresponding angle in this interval; find it by subtracting an appropriate multiple of 2π.

EXAMPLE 13.12: Evaluate $\sin\left(\frac{11\pi}{2}\right)$, $\cos\left(-\frac{17\pi}{6}\right)$ and $\tan\left(\frac{111\pi}{4}\right)$.

$$\sin\left(\frac{11\pi}{2}\right) = \sin\left(\frac{11\pi}{2} - 2(2\pi)\right) = \sin\frac{3\pi}{2} = \sin\left(\pi + \frac{\pi}{2}\right)$$

$$= \sin\pi\cos\frac{\pi}{2} + \sin\frac{\pi}{2}\cos\pi = -1 .$$

Since $\cos\left(-\frac{17\pi}{6}\right) = \cos\frac{17\pi}{6}$, we can examine the latter:

$$\cos\frac{17\pi}{6} = \cos\left(\frac{17\pi}{6} - 2\pi\right) = \cos\frac{5\pi}{6} = \cos\left(\pi - \frac{\pi}{6}\right)$$

$$= \cos\pi\cos\frac{\pi}{6} + \sin\pi\sin\frac{\pi}{6} = (-1)\cos\frac{\pi}{6} + 0\sin\frac{\pi}{6}$$

$$\cos \frac{17\pi}{6} = - \cos \frac{\pi}{6} = - \frac{\sqrt{3}}{2} .$$

$$\tan \left(\frac{111\pi}{4}\right) = \tan \left(\frac{111\pi}{4} - 14(2\pi)\right) = \tan \left(- \frac{\pi}{4}\right) = - \tan \frac{\pi}{4} = -1 . \blacksquare$$

Look back at Figure 13.11. If you pushed the sine curve back to the left $\pi/2$ units (translated it), the sine and cosine would correspond exactly. Because the cosine function relates to the sine function through a translation, they display precisely the same wave form or pattern. Furthermore, they remain continuous for all values of θ while the other four trigonometric functions have regular discontinuities. For these reasons, the sine and cosine functions occupy a fairly central place in situations exhibiting a smooth oscillatory behavior.

The preceding discussion focused entirely on the trigonometric functions. It frequently occurs, however, that the trigonometric functions may be imposed on top of another function to introduce a fluctuating component.

The following two examples illustrate how trigonometric functions can be used in conjunction with other functions.

EXAMPLE 13.13: The function $f(x) = e^{-x}(5 + \sin x)$ is graphed in Figure 13.12. This function consists of two distinct parts, namely $5e^{-x}$ and $e^{-x} \sin x$. The first part represents the function's main part and is graphed as the dotted curve while the second part represents a damped (diminishing) oscillatory component centered about the x-axis. The solid curve wobbling about the dotted curve gives $f(x)$. The relative size of the oscillating and other components determines the prominence of the oscillation relative to the function's general trend.

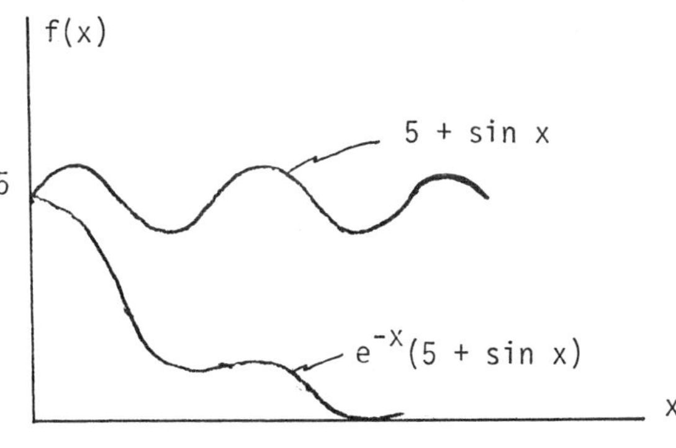

Figure 13.12

Notice that in this case that x has replaced θ as the argument of sine. As before, any symbol may be used for an argument. Here, as elsewhere, x is assumed to be expressed in radians unless clearly stated otherwise. ▌

EXAMPLE 13.14: The function $f(x) = 2 + \cos x + \frac{1}{10}\cos\frac{x}{4}$ has the graph shown in Figure 13.13. Its two varying components are shown on the left while their sum appears to the right.

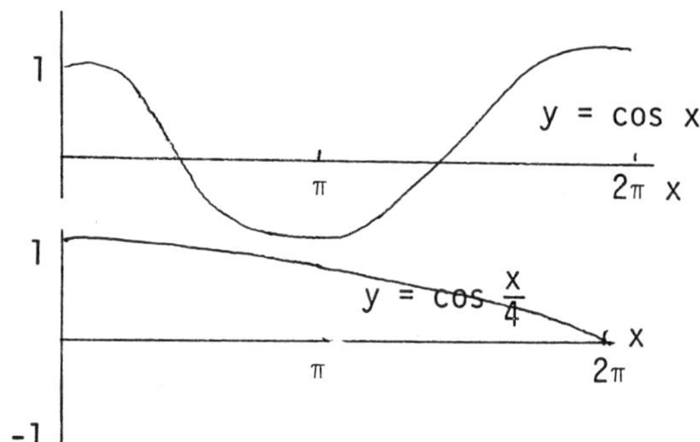

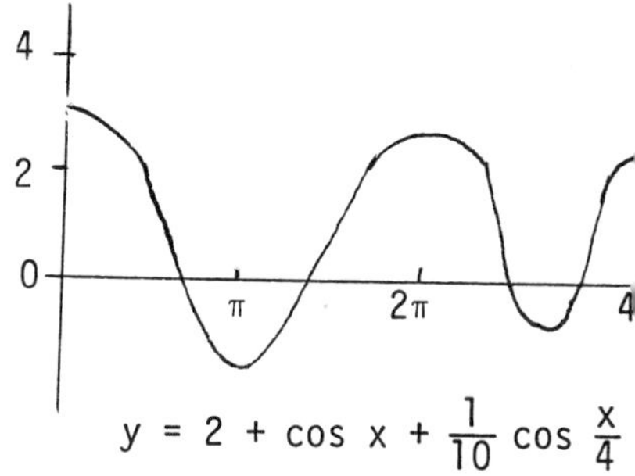

Figure 13.13

Functions of this sort are closely associated with sound waves through the movement of a vibrating string or surface such as a vocal chord. ▌

Any function displaying a pattern of oscillations can be decomposed into components such as the parts of f(x) in the previous example. If you need to do this, you should study Fourier analysis, a topic ordinarily covered in books dealing with time series analysis or with differential equations. The x-axis would frequently represent time for such oscillating functions.

So far, this section suggests by its emphasis that the trigonometric functions are the only periodic functions. Although they have a preeminent role among periodic functions, others do exist. Recall the function $[\![x]\!]$ which was introduced in Section 4.2. As $[\![x]\!]$ gives the integer part of x, $x - [\![x]\!]$ gives its fractional part. Periodic functions of this type are illustrated in the next example.

EXAMPLE 13.15: Examine the nature of $x - [\![x]\!]$, $x - [\![x]\!] - \frac{1}{2}$ and $x - [\![x]\!] - \frac{1}{2} - \frac{1}{4}$. These three functions are shown in Figure 13.14.

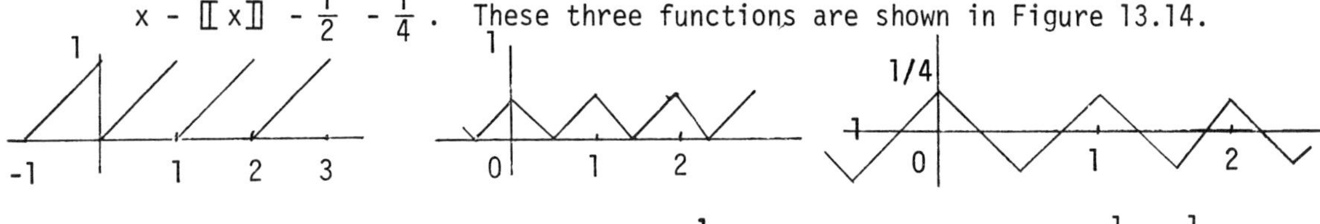

Figure 13.14

The middle one describes one of the wave forms used in electrical anesthesia, namely, unconsciousness induced by low intensity electrical currents. The first and third functions have features similar to the tangent and cosine functions: The first one has regular discontinuties as does the tangent; the third remains continuous throughout, going up and down similar to cosine functions. ▌

EXERCISES

Express the following in terms of functions of a positive acute angle in two ways:

1. cos 130°
2. cot(-290°)
3. tan 325°
4. cot 325°
5. sin 200°
6. tan 300°
7. cos 240°
8. cos(-680°)
9. csc 865°
10. sin 930°

11. Find the exact values of the sine, cosine and tangent if θ is equal to

    a. 120°
    b. 135°
    c. $-\frac{5\pi}{4}$
    d. -330°
    e. 675°
    f. 735°

12. Evaluate $\sin(\pi - 0.5)$. Do this problem two ways.

13. When θ is an angle which is in the second quadrant such that $\sin\theta = \frac{2}{13}\sqrt{13}$, show that

$$\frac{\sin(\frac{\pi}{2} - \theta) - \cos(\pi - \theta)}{\tan(\frac{3\pi}{2} + \theta) + \cot(2\pi - \theta)} = \frac{-2}{13}\sqrt{13} .$$

14. Evaluate $\sin 2199°$.

13.6 INVERSE TRIGONOMETRIC FUNCTIONS

The trigonometric functions such as the sine associate a numerical value with an angle. We need to reverse this, namely, get the size of an angle from knowledge of the value of one of its trigonometric functions. For example, in our second consideration of Example 13.3, we went from $\sin\theta = 0.8$ to $\theta \doteq 53.1° \doteq 0.927$ radians. Here we will develop the idea of inverse trigonometric functions with primary emphasis on the sine function; the others follow by analogy.

We begin with the following definition:

- The INVERSE SINE of x is any value (angle) y which satisfies $x = \sin y$. It is denoted as $y = \arcsin x$ and $y = \sin^{-1} x$, and is read "y is an angle whose sine is x".
- The PRINCIPLE VALUE of $\sin^{-1} x$ is that value y, $-\frac{\pi}{2} \leq y \leq \frac{\pi}{2}$, satisfying $x = \sin y$. If $\sin^{-1} x$ exists, it has exactly one principle value and, thus, $\sin^{-1} x$ is a function when restricted to principle values. The inverse sine, restricted to principle values, is denoted by $\text{Sin}^{-1} x$ or by $\text{Arcsin}\, x$.

The equation $y = \sin x$ determines a unique value of y for every value of x. Its inverse, however, lacks this property. For example, $\sin^{-1} 2$ has no value because the sine of an angle can never exceed 1. On the other hand, $y = \sin^{-1} 0.5$ is satisfied by $y = 30°, 150°, 390°, 510°, \cdots$, but only $y = 30° = \frac{\pi}{6}$ radians is its principle value. Thus, $\text{Sin}^{-1} 0.5 = \frac{\pi}{6}$. You can look back at Figure 13.11 to see why the principle value of $\sin^{-1} x$ is restricted to $-\frac{\pi}{2} \leq y \leq \frac{\pi}{2}$. This is the only range of values including zero over which the sine function varies from its absolute minimum of -1 to its absolute maximum of 1. Be very careful not to confuse $\sin^{-1} x$ with $\frac{1}{\sin x}$ which should be written as $(\sin x)^{-1}$. This is why some people prefer to write Arcsin x in place of $\text{Sin}^{-1} x$.

EXAMPLE 13.16: If $y = \text{Sin}^{-1} \frac{\sqrt{2}}{2}$, find y.

Using the definition of the inverse sine, we seek an angle whose sine equals $\frac{\sqrt{2}}{2}$. Because Sin^{-1} begins with a capital letter, we seek the principle value of the inverse sine. Thus, the angle must satisfy $-\frac{\pi}{2} \leq y \leq \frac{\pi}{2}$. From $\sin \frac{\pi}{4} = \frac{\sqrt{2}}{2}$, it follows that $y = \frac{\pi}{4} = 45°$. ▮

The other trigonometric functions have inverses analogous to that for the sine. The same definition applies, only with the name of another trigonometric function replacing sine throughout. The principle values change as follows:

Principle values for $\text{Sin}^{-1} x$, $\text{Tan}^{-1} x$, $\text{Csc}^{-1} x$ all lie in $\left[-\frac{\pi}{2}, \frac{\pi}{2}\right]$

Principle values for $\text{Cos}^{-1} x$, $\text{Cot}^{-1} x$, $\text{Sec}^{-1} x$ all lie in $[0, \pi]$.

EXAMPLE 13.17: Evaluate $\cos(\text{Arcsin } \frac{3}{5})$.

We need to find the cosine of the angle whose sine equals $\frac{3}{5}$. Look at the triangle associated with $\sin \theta = \frac{3}{5}$. It shows that $\cos \theta = \frac{4}{5} = \cos(\text{Arcsin } \frac{3}{5})$. ▮

EXAMPLE 13.18: Evaluate $\sin(\text{Sin}^{-1} \frac{12}{13} + \text{Sin}^{-1} \frac{4}{5})$.

Let $A = \text{Sin}^{-1} \frac{12}{13}$ and $B = \text{Sin}^{-1} \frac{4}{5}$, so

$$\sin(A + B) = \sin A \cos B + \sin B \cos A .$$

Drawing the triangles associated with the angles A and B

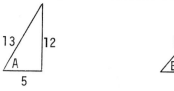

Thus, $\sin(A + B) = (\frac{12}{13})(\frac{3}{5}) + (\frac{5}{13})(\frac{4}{5}) = \frac{56}{65}$. ∎

EXERCISES

State the value or values of each of the following expressions:

1. $\sin^{-1} \frac{1}{2}$
2. $\text{Arcsin} \frac{\sqrt{3}}{2}$
3. $\cos^{-1} 0$
4. $\text{Tan}^{-1} \frac{\sqrt{3}}{3}$
5. $\text{Cot}^{-1} - \frac{\sqrt{3}}{3}$
6. $\text{Arcsec } 2$

Find the value of each of the following expressions:

7. $\sin\left[\text{Cos}^{-1} \left[- \frac{2}{3}\right]\right]$
8. $\tan\left[\text{Sin}^{-1} \left[- \frac{3}{4}\right]\right]$
9. $\cos\left[\text{Tan}^{-1} \left[- \frac{5}{12}\right]\right]$
10. $\text{Sin}^{-1} \left[\cos \frac{\pi}{3}\right]$
11. $\text{Cos}^{-1} \left[\sin \frac{\pi}{6}\right]$
12. $\sin\left[\text{Cos}^{-1} \frac{\sqrt{2}}{2}\right]$

Evaluate each of the following expressions:

13. $\cos\left[\text{Tan}^{-1} \frac{15}{8} - \text{Sin}^{-1} \frac{7}{25}\right]$
14. $\tan\left[\text{Tan}^{-1} \frac{1}{5} + \text{Tan}^{-1} \frac{3}{5}\right]$
15. $\text{Cos}^{-1} \frac{3}{5} + \text{Sin}^{-1} \frac{5}{13}$
16. $\sin\left[\text{Cos}^{-1} \frac{1}{2} + \text{Sin}^{-1} \frac{3}{5}\right]$

13.7 DERIVATIVES OF THE SINE AND COSINE FUNCTIONS

This section concentrates on the derivatives of the sine and cosine functions while the next presents the derivatives of the remaining trigonometric functions. The subsequent section introduces the derivatives of the inverse functions.

13.7a DERIVATIVE OF THE SINE

We will start by stating and illustrating the result of this section before proving it.

- If $y = \sin x$,

$$\frac{dy}{dx} = \frac{d}{dx} \sin x = \cos x \ . \tag{13.44}$$

- If $y = \sin u$, where $u = g(x)$,

$$\frac{dy}{dx} = \frac{d}{dx} \sin u = (\cos u)\frac{du}{dx} \ . \tag{13.45}$$

EXAMPLE 13.19: Find y' if $y = \sin(3x + 2)$.

$$\frac{dy}{dx} = \cos(3x + 2) \frac{d}{dx} (3x + 2) = 3 \cos(3x + 2) \ . \quad \blacksquare$$

EXAMPLE 13.20: Find y' if $y = e^{2x^2}[\sin(x + 3)]^2$.

We must treat this as a product of e^{2x^2} and $(\sin(x + 3))^2$. Thus,

$$y' = e^{2x^2} \frac{d}{dx} [\sin(x + 3)]^2 + (\sin(x + 3))^2 \frac{d}{dx} e^{2x^2}$$

$$= e^{2x^2} 2(\sin(x + 3)) \frac{d}{dx} \sin(x + 3) + (\sin(x + 3))^2 e^{2x^2} \frac{d}{dx} 2x^2$$

$$= 2e^{2x^2} \sin(x + 3) \cos(x + 3) + 4x(\sin(x + 3))^2 e^{2x^2} \ .$$

Notice how the chain rule was employed throughout and that a sine raised to a power is treated like any other quantity raised to a power. $\blacksquare$

PROOF OF THE MAIN RESULT: To establish Equation 13.44, we must revert to the limit definition of a derivative. Thus,

$$\frac{d}{dx} \sin x = \lim_{\Delta x \to 0} \frac{\sin(x + \Delta x) - \sin x}{\Delta x} \ .$$

Using the identity given in Equation 13.41 with $\alpha = x + \Delta x$ and $\beta = x$, we have

$$\sin \alpha - \sin \beta = 2 \cos \frac{\alpha + \beta}{2} \sin \frac{\alpha - \beta}{2} = 2 \cos(x + \frac{\Delta x}{2}) \sin \frac{\Delta x}{2} .$$

Thus,

$$\frac{d}{dx} \sin x = \lim_{\Delta x \to 0} \frac{2 \cos(x + \frac{\Delta x}{2}) \sin \frac{\Delta x}{2}}{\Delta x} .$$

Now divide the numerator and denominator by 2 and write the limit of a product as a product of the individual limits:

$$\frac{d}{dx} \sin x = \lim_{\Delta x \to 0} \cos(x + \frac{\Delta x}{2}) \lim_{\Delta x \to 0} \frac{\sin \frac{\Delta x}{2}}{\frac{\Delta x}{2}} = \cos x$$

because cos x is continuous and $\lim_{\theta \to 0} \frac{\sin \theta}{\theta} = 1$.

To establish this latter limit, recall that an angle of one radian cuts off a length of r on the circumference of a circle of radius r. Take r = 1 and adopt the notation in Figure 13.15. If one radian cuts off an arc of length r = 1 on the circle, then $\theta = 0.1$ radian cuts off an arc of length 0.1 and so on. Thus an angle of θ cuts off an arc of length θ. For small θ, this length of arc approximately equals y, the height of the triangle having θ as the smaller of its acute angles ($\theta \doteq y$). As $\sin \theta = y/r = y \doteq \theta$, $\frac{\sin \theta}{\theta} \doteq 1$, and this approximation improves as θ decreases.

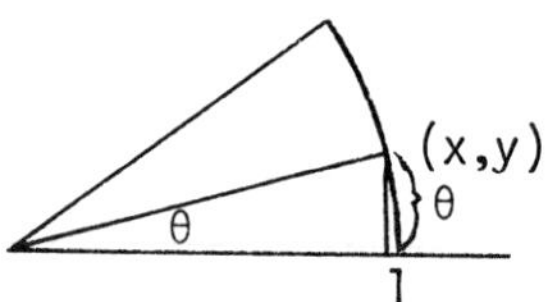

Figure 13.15

This completes the development of Equation 13.44. Application of the chain rule establishes Equation 13.45. ▼

13.7b DERIVATIVE OF THE COSINE

Again we will state the result of this section, follow with examples and then give the proof.

- If $y = \cos x$,

$$\frac{dy}{dx} = \frac{d}{dx} \cos x = - \sin x \ . \qquad (13.46)$$

- If $y = \cos u$, where $u = g(x)$,

$$\frac{dy}{dx} = \frac{d}{dx} \cos u = - \sin u \frac{du}{dx} \ . \qquad (13.47)$$

<u>EXAMPLE 13.21</u>: Find y' if $y = 5 + \cos(3x - 4)$.

$$\frac{dy}{dx} = \frac{d}{dx} (5 + \cos(3x - 4))$$

$$= 0 - \sin(3x - 4) \frac{d}{dx} (3x - 4) = -3 \sin(3x - 4) \ . \ \blacksquare$$

<u>EXAMPLE 13.22</u>: Find y' if $y = x^2[\cos(4x - 1)]^3 = x^2 \cos^3(4x - 1)$.

We must treat this first as a product:

$$y' = x^2 \frac{d}{dx} \cos^3(4x - 1) + \cos^3(4x - 1) \frac{d}{dx} (x^2)$$

$$= x^2 [3 \cos^2(4x - 1) \frac{d}{dx} (\cos(4x - 1))] + 2x \cos^3(4x - 1)$$

$$= 3x^2 \cos^2(4x - 1)(-4 \sin(4x - 1)) + 2x \cos^3(4x - 1)$$

$$= -12x^2 \cos^2(4x - 1) \sin(4x - 1) + 2x \cos^3(4x - 1) \ . \ \blacksquare$$

<u>PROOF OF THE MAIN RESULT</u>: One proof of Equation 13.46 utilizes the already established results $\cos x = \sin \frac{\pi}{2} - x$, $\cos \frac{\pi}{2} - x = \sin x$ and $\frac{d}{dx} \sin x = \cos x$:

$$\frac{d}{dx} \cos x = \frac{d}{dx}\left[\sin\left[\frac{\pi}{2} - x\right]\right] = \cos\left[\frac{\pi}{2} - x\right] \frac{d}{dx} \left[\frac{\pi}{2} - x\right] = - \cos\left[\frac{\pi}{2} - x\right] = -\sin \theta. \blacktriangledown$$

13.8 DERIVATIVES OF THE REMAINING TRIGONOMETRIC FUNCTIONS

The trigonometric functions have derivatives given by:

If $u = g(x)$,

- $$\frac{d}{dx}\sin u = \cos u \frac{du}{dx} \tag{13.48}$$
- $$\frac{d}{dx}\cos u = -\sin u \frac{du}{dx} \tag{13.49}$$
- $$\frac{d}{dx}\tan u = \sec^2 u \frac{du}{dx} \tag{13.50}$$
- $$\frac{d}{dx}\cot u = -\csc^2 u \frac{du}{dx} \tag{13.51}$$
- $$\frac{d}{dx}\sec u = \sec u \tan u \frac{du}{dx} \tag{13.52}$$
- $$\frac{d}{dx}\csc u = -\csc u \cot u \frac{du}{dx} \tag{13.53}$$

Equations 13.48 and 13.49 have already been derived. Because the remaining trigonometric functions all can be expressed in terms of sines and cosines, their derivatives can be derived from a knowledge of the derivatives of the sine and cosine:

EXAMPLE 13.23: If $y = \tan x$, find y'.

We have seen that $y = \tan x = \dfrac{\sin x}{\cos x}$. Thus the derivative of a quotient gives

$$\frac{dy}{dx} = \frac{d}{dx}\left[\frac{\sin x}{\cos x}\right] = \frac{\cos x \frac{d}{dx}\sin x - \sin x \frac{d}{dx}\cos x}{\cos^2 x}$$

$$= \frac{\cos x \cos x - \sin x(-\sin x)}{\cos^2 x} = \frac{\cos^2 x + \sin^2 x}{\cos^2 x}$$

$$= \frac{1}{\cos^2 x} = \sec^2 x .$$

If u = g(x) the chain rule gives

$$\frac{d}{dx} = \tan u = \sec^2 u \frac{du}{dx} . \quad \blacksquare$$

The derivatives of the remaining trigonometric functions will be left as exercises.

EXERCISES

1. Develop the derivatives of cot x, sec x and csc x using only the derivatives of the sine and cosine, and the differentiation rules from Chapter 6.

   a. cot x   b. sec x   c. csc x

Find the derivative of y with respect to x of the following functions:

2. $y = \sin 3x$
3. $y = \tan(ax + b)$
4. $y = \sin^2(3x + 2)$
5. $y = \cos^n(3x - 4)$
6. $y = (\cos 4x)\sin 3x$
7. $y = x^4 + \sin 5x^2$
8. $y = (\ln x)(\sin x) + \cos x$
9. $y = 3x^2 \sec 4x$
10. $y = \sin^2 x + \cos^2 x$
11. $y = \dfrac{\sec^2 x}{\sin x}$
12. $y = (1 + \tan^2 x)e^{x^2}$
13. $y = \cos^2 x - \sin^2 x$
14. $y = 2x \cos(3x + 5)$
15. $y = (\tan x)(\cot x)$

Evaluate the following limits. Check your results against Table IV. (Hint: You may have to apply L'Hospital's rule. Remember that variables are in radians.)

16. $\lim\limits_{h \to 0} \dfrac{\sin 3h}{h}$
17. $\lim\limits_{x \to 0} \dfrac{3 \sin 5x}{x}$
18. $\lim\limits_{K \to 0} \dfrac{\sin^2 K - 2 \sin K}{K}$
19. $\lim\limits_{x \to 0} \dfrac{\tan x}{x}$
20. $\lim\limits_{h \to 0} \dfrac{\cos(x + h) - \cos x}{h}$

Find y' of the following:

21. $y = 2(\sin x)(\cos x)$

22. $y = \csc^3 (\frac{x}{3})$

23. $y = 2x + e^x(\cos x)$

24. $y = \sqrt{\tan 2x}$

25. $y = x^2 \tan^2 3x$

26. $y = \frac{\sin 2x}{x^2}$

27. $y = (\sin^3 2x)\tan^2 3x$

28. $y = (\sin 2x)/(1 + \cos 2x)$

29. $y = (1 + \tan 3x)^{2/3}$

30. $y = (\sin nx)(\sin^n x)$

31. $y = \csc^3(3x^2 + 1)$

32. $y = \cos^{7/3}(4x^3 + 2x - 1)$

33. $y = \cos(\tan x)$

34. $y = (1 + \tan 4x^2)^{3/2}$

35. If $Z = \frac{dV}{dT}$, $W = \frac{dZ}{dT}$ and $Y = \frac{dW}{dT}$, find Z, W and Y if

$V = 14 \sin^4 T + 3(\sin T)^{5/3} + 2(\sin T)^{1/2} + 1$.

36. Prove that the curves $3x^2 - y^2 + 4x + 1 = 0$ and $(36 - 7y)^3 = 8(1 - x^2)^2$ intersect orthogonally (at right angles) at the point (-3,4).

13.9 DERIVATIVES OF THE INVERSE TRIGONOMETRIC FUNCTIONS

The inverse trigonometric functions have derivatives which occupy a central place in the integration of certain kinds of functions. Although this topic will be dealt with in Chapter 14, the derivatives will be presented here to complete our consideration of the trigonometric functions.

When $u = g(x)$,

- $$\frac{d}{dx} \text{Sin}^{-1} u = \frac{d}{dx} \text{Arcsin } u = \frac{1}{\sqrt{1 - u^2}} \frac{du}{dx} \qquad (13.54)$$

- $$\frac{d}{dx} \text{Cos}^{-1} u = \frac{d}{dx} \text{Arccos } u = -\frac{1}{\sqrt{1 - u^2}} \frac{du}{dx} \qquad (13.55)$$

- $$\frac{d}{dx}\,\mathrm{Tan}^{-1}\,u = \frac{d}{dx}\,\mathrm{Arctan}\,u = \frac{1}{1+u^2}\frac{du}{dx} \qquad (13.56)$$

- $$\frac{d}{dx}\,\mathrm{Cot}^{-1}\,u = \frac{d}{dx}\,\mathrm{Arccot}\,u = -\frac{1}{1+u^2}\frac{du}{dx} \qquad (13.57)$$

These will be illustrated, first with an algebraic and then a biological example.

EXAMPLE 13.24: The derivative of $y = \mathrm{Sin}^{-1}\,x^2$ results when we set $u = x^2$ and apply Equation 13.53 as

$$\frac{dy}{dx} = \frac{1}{\sqrt{1-(x^2)^2}}\,\frac{d}{dx}\,x^2 = \frac{2x}{\sqrt{1-x^4}}\,. \quad \blacksquare$$

EXAMPLE 13.25: A wildlife photographer is trying to film deer from a blind. A deer enters his field of vision 565.6 feet (= $400\sqrt{2}$ feet) west of the blind and moves northeastward at 20 feet/second. If the photographer wants to record the deer's movement on moving picture film, how fast will the camera have to be turned?

First, what is sought? Because the camera would have to remain in the blind, the "how fast" must apply to rotation of the camera on its base. With this in mind consider Figure 13.16. The angle θ lies

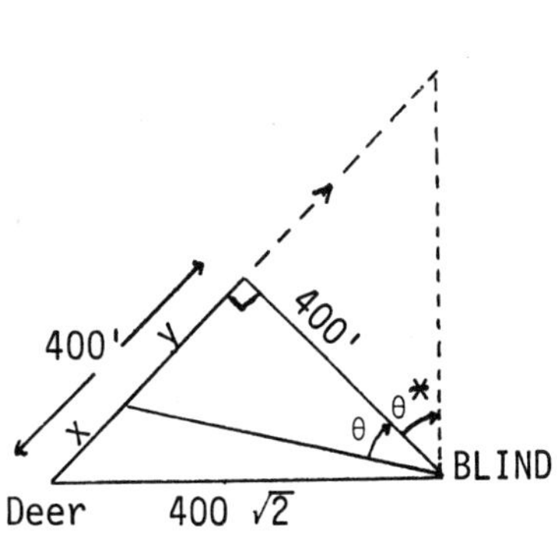

Figure 13.16

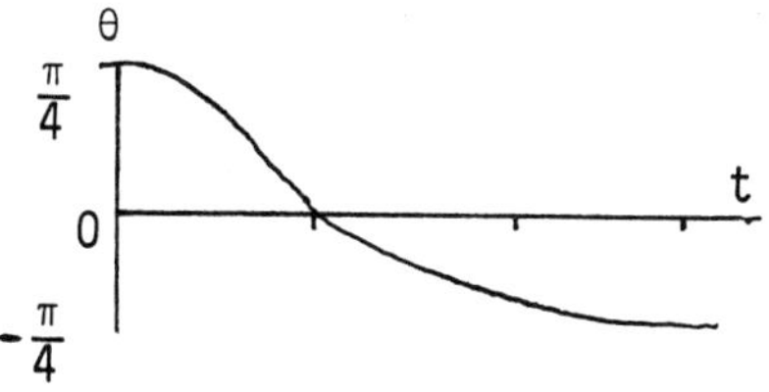

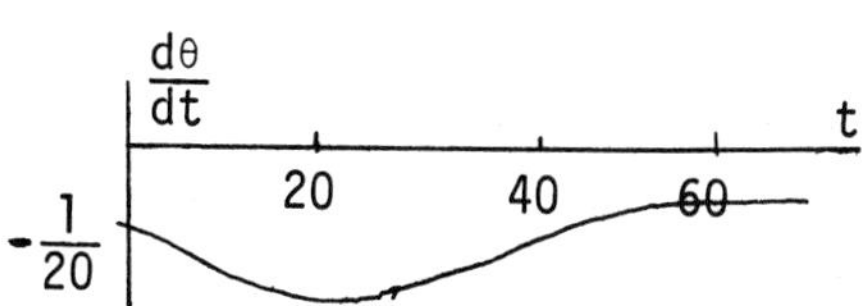

Figure 13.17

between the deer's current position and the line perpendicular to its line of travel; $\theta = \frac{\pi}{4}$ initially (at t = 0), but drops to zero at t = 20 seconds and finally becomes negative (as indicated by θ^*) as the deer crosses the field of vision. If x gives the distance the deer has traveled since time t = 0, the conditions specify that $\frac{dx}{dt} = 20$ feet/second or x = 20t. We need to find $\frac{d\theta}{dt}$ from this fact. From the figure, $\tan\theta = \frac{y}{400}$ with y = 400 - x = 400 - 20t and

$$\theta = \mathrm{Tan}^{-1}\frac{y}{400} = \mathrm{Tan}^{-1}\left(1 - \frac{t}{20}\right).$$

Let $u = 1 - \frac{t}{20}$ and apply Equation 13.56:

$$\frac{d\theta}{dt} = \frac{1}{1 + \left(1 - \frac{t}{20}\right)^2}\left(-\frac{1}{20}\right) = \frac{-20}{400 + (20 - t)^2} \text{ radians/second.}$$

The negative sign on this derivative merely indicates that the angle is being measured clockwise, rather than counterclockwise. The behavior of θ and $\frac{d\theta}{dt}$ are shown in Figure 13.17. One feature of the graph of $\frac{d\theta}{dt}$ should be noted. At t = 20 seconds, the deer will be closest to the camera and traveling perpendicular to the line of sight. The angular rotation also will be largest (ignoring the negative for direction) at this time. This is equivalent to watching an automobile from some point. You hardly have to turn your head to watch it when it lies to your far right or left, but you must turn your head at a faster rate as it nears the center of your field of vision; the closer you are to the roadway, the more pronounced this becomes. ∎

This last example illustrated a class of situations arising in tracking; many variations occur in ballistics problems. The major variation involves the need to aim the projectile (bullet or other missile) in front of the moving object so they will collide. The above example ignored this phenomenon because the speed of light can be regarded as essentially infinite for most such problems. The speed of light enters for very refined guidance systems involving very long-distance shots, such as firing missiles out of the earth's gravitational field.

PROOF OF EQUATION 13.54: This derivation uses implicit differentiation introduced in Chapter 7. If $y = \text{Sin}^{-1} u$, then $u = \sin y$, so

$$\frac{du}{dx} = \cos y \frac{dy}{dx} .$$

Upon solving this expression for $\frac{dy}{dx}$ and using $\cos y = \sqrt{1 - \sin^2 y} = \sqrt{1 - u^2}$, we have

$$\frac{dy}{dx} = \frac{1}{\cos y}\frac{du}{dx} = \frac{1}{\sqrt{1 - u^2}}\frac{du}{dx} . \blacktriangledown$$

The derivatives of the remaining inverse trigonometric functions can be obtained using the same approach.

EXERCISES

Find the derivative of y with respect to x of the following:

1. $y = \text{Sin}^{-1} x^2$
2. $y = \text{Cos}^{-1} \sqrt{1 - 3x}$
3. $y = \text{Sin}^{-1} 4x$
4. $y = \text{Arctan}(3x + 2)$
5. $y = \text{Tan}^{-1} (\cos x)$
6. $y = \text{Tan}^{-1} 2x$
7. $y = \text{Arccos } x^3$
8. $y = \text{Cot}^{-1} x$
9. $y = \text{Cos}^{-1} (4 - 3x)$
10. $y = \text{Cos}^{-1} (\sec x)$
11. $y = \text{Sin}^{-1} \frac{x}{3}$
12. $y = \text{Tan}^{-1} \frac{1}{x^2}$
13. $y = \text{Cos}^{-1} \frac{1}{\sqrt{x}}$
14. $y = e^x \text{Sin}^{-1} \frac{3\sqrt{x}}{2}$
15. $y = \text{Cos}^{-1} \frac{3x}{2 - x}$
16. $y = \text{Tan} \frac{2x}{x^2 - 1}$
17. $y = \frac{1}{x^2} \text{Sin}^{-1} x$
18. $y = x \text{ Tan}^{-1} \frac{2}{x}$
19. $y = x^2 \text{Sin}^{-1} 2x$
20. $y = (\sin 3x)(\text{Sin}^{-1} \sqrt{x})$
21. $y = (\ln x^2)(\text{Sin}^{-1} 2x)$
22. If $y = \text{Sin}^{-1} 2x$, find y''.

23. Show that $\frac{d}{dx} \text{Cos}^{-1} u = \frac{-1}{\sqrt{1 - u^2}} \frac{du}{dx}$.

24. Show that $\frac{d}{dx} \text{Tan}^{-1} u = \frac{1}{1 + u^2} \frac{du}{dx}$.

13.10 INTEGRALS INVOLVING TRIGONOMETRIC FUNCTIONS

The preceding sections have presented derivatives of various trigonometric functions. The fundamental theorem of integral calculus gives various integrals from these derivatives. Here a few basic integrals will be presented. The next chapter contains more complex forms. This chapter ends with a summary of functions, integrals and derivatives.

We begin by listing below a few basic integrals on the right. The associated functions appear at the left.

- $\frac{d}{du} \cos u = - \sin u$ $\qquad$ $\int \sin u \, du = - \cos u + C$ $\qquad$ (13.58)
- $\frac{d}{du} \sin u = \cos u$ $\qquad$ $\int \cos u \, du = \sin u + C$ $\qquad$ (13.59)
- $\frac{d}{du} \tan u = \sec^2 u$ $\qquad$ $\int \sec^2 u \, du = \tan u + C$ $\qquad$ (13.60)
- $\tan u = \frac{\sin u}{\cos u}$ $\qquad$ $\int \tan u \, du = \ln |\cos u| + C$ $\qquad$ (13.61)
- $\frac{d}{du} \sec u = \sec u \tan u$ $\qquad$ $\int \sec u \tan u \, du = \sec u + C$ $\qquad$ (13.62)
- $\frac{d}{du} \csc u = - \csc u \cot u$ $\qquad$ $\int \csc u \cot u \, du = - \csc u + C$ $\qquad$ (13.63)
- $\frac{d}{du} \text{Tan}^{-1} (\frac{u}{a}) = \frac{a}{a^2 + u^2}$ $\qquad$ $\int \frac{du}{u^2 + a^2} = \frac{1}{a} \text{Tan}^{-1} (\frac{u}{a}) + C$ $\qquad$ (13.64)
- $\frac{d}{du} \text{Sin}^{-1} (\frac{u}{a}) = \frac{1}{\sqrt{a^2 - u^2}}$ $\qquad$ $\int \frac{du}{\sqrt{a^2 - u^2}} = \text{Sin}^{-1} (\frac{u}{a}) + C$ $\qquad$ (13.65)

From what you now know about integrals and about derivatives of trigonometric functions, you should verify all of the equations on the right above. The initial examples illustrate application of these formulas to integration while the latter ones show biological problems.

EXAMPLE 13.26: Evaluate $\int \frac{\cos \sqrt{x}}{\sqrt{x}}\, dx$.

This integral does not appear in the list above, but the substitution $u = \sqrt{x} = x^{1/2}$ changes the integrand into an integrable form. If $u = x^{1/2}$, then

$$du = \frac{1}{2} x^{-1/2}\, dx = \frac{dx}{2\sqrt{x}} ,$$

so

$$\int \frac{\cos \sqrt{x}}{\sqrt{x}}\, dx = 2 \int \cos\sqrt{x}\, \frac{dx}{2\sqrt{x}} = 2 \int \cos u\, du$$

$$= -2 \sin u + C = -2 \sin \sqrt{x} + C . \quad \blacksquare$$

EXAMPLE 13.27: Evaluate $\int \frac{\sec^2(\sin x)}{\sec x}\, dx$.

We could evaluate this integral if we could transform it into the standard form $\int \sec^2 u\, du$. To get this form, we will try the transformation $u = \sin x$ for which $du = \cos x\, dx$. Now using $\frac{1}{\sec x} = \cos x$, we have

$$\int \frac{\sec^2(\sin x)}{\sec x}\, dx = \int \sec^2(\sin x) \cos x\, dx = \int \sec^2 u\, du = \tan u + C .$$

Finally substituting $u = \sin x$, this becomes

$$\int \frac{\sec^2(\sin x)}{\sec x}\, dx = \tan(\sin x) + C . \quad \blacksquare$$

EXAMPLE 13.28: Evaluate $\int_0^{\frac{\pi}{4}} \frac{d(\tan \theta + 1)}{\tan \theta + 1}$.

To evaluate this integral, notice that if $u = \tan \theta + 1$ then $du = d(\tan \theta + 1)$ so that the given integral has the general form of $\int \frac{du}{u}$. Thus,

$$\int_0^{\frac{\pi}{4}} \frac{d(\tan \theta + 1)}{\tan \theta + 1} = \ln |\tan \theta + 1|\Big]_0^{\frac{\pi}{4}} = \ln 2 - \ln 1 = \ln 2 . \quad \blacksquare$$

When using the method of substitution to evaluate indefinite integrals, the answer is always left in terms of the original variable in the integrand as was illustrated in Examples 13.26 and 13.27. However, when evaluating a definite integral, we may change the limits and evaluate the integral in terms of the changed variable as was explained in Chapter 8.

EXAMPLE 13.29: Evaluate $\int_0^{\frac{\pi}{2}} \sin^3 \theta \cos \theta \, d\theta$.

If $u = \sin \theta$, then $du = \cos \theta \, d\theta$. When $\theta = \frac{\pi}{2}$, $u = 1$ and when $\theta = 0$, $u = 0$. Thus,

$$\int_0^{\frac{\pi}{2}} \sin^3 \theta \cos \theta \, d\theta = \int_0^1 u^3 \, du = \frac{1}{4} u^4 \Big]_0^1 = \frac{1}{4} .$$

Alternatively,

$$\int \sin^3 \theta \cos \theta \, d\theta = \int u^3 \, du = \frac{1}{4} u^4 + C ,$$

so

$$\int_0^{\frac{\pi}{2}} \sin^3 \theta \cos \theta \, d\theta = \frac{1}{4} \cos^4 \theta + C \Big]_0^{\frac{\pi}{2}}$$

$$= \frac{1}{4} + C - 0 - C = \frac{1}{4} . \quad \blacksquare$$

EXAMPLE 13.30 (8.28 continued): Recall that this biological example concerned the size of the stomata, the openings in the leaves of plants through which carbon dioxide enters the plant and oxygen and water exit. There we encountered the integral

$$2 \int_{-x_0}^{x_0} \sqrt{x_0^2 - x^2} \, dx$$

in evaluating the area of a partially open stoma. We skirted its integration then, but now we can evaluate this integral.

The method to be presented in Chapter 14, trigonometric substitution, yields a simple method for evaluating this integral. The idea behind this method is illustrated below:

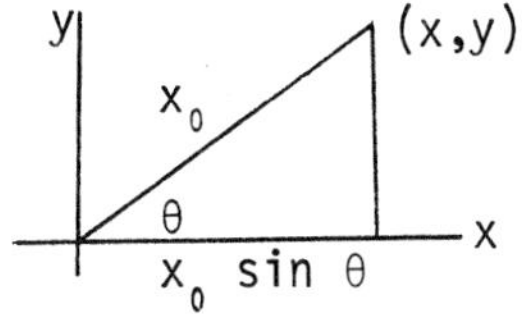

Figure 13.18

$$\text{If } x = x_0 \sin\theta, \quad \text{then } dx = x_0 \cos\theta \, d\theta \quad \text{and}$$

$$\sqrt{x_0^2 - x^2} = \sqrt{x_0^2 - x_0^2 \sin^2\theta} = \sqrt{x_0^2(1 - \sin^2\theta)} = \sqrt{x_0^2 \cos^2\theta} = x_0 \cos\theta \,.$$

As x ranges from $-x_0$ to x_0, sin θ ranges from -1 to +1 so θ ranges from $-\frac{\pi}{2}$ to $\frac{\pi}{2}$. Thus

$$\int_{-x_0}^{x_0} \sqrt{x_0^2 - x^2}\, dx = \int_{-\frac{\pi}{2}}^{\frac{\pi}{2}} x_0 \cos\theta(x_0 \cos\theta) d\theta = x_0^2 \int_{-\frac{\pi}{2}}^{\frac{\pi}{2}} \cos^2\theta \, d\theta \,.$$

As $\cos^2\theta = \frac{1}{2}(1 + \cos 2\theta)$, we have

$$x_0^2 \int_{-\frac{\pi}{2}}^{\frac{\pi}{2}} \cos^2\theta \, d\theta = \frac{x_0^2}{2} \int_{-\frac{\pi}{2}}^{\frac{\pi}{2}} (1 + \cos 2\theta) d\theta = \frac{x_0^2}{2}\left[\theta + \frac{1}{2}\sin 2\theta\right]_{-\frac{\pi}{2}}^{\frac{\pi}{2}}$$

$$= \frac{x_0^2}{2}\left[\frac{\pi}{2} + \frac{1}{2}(0) - \frac{\pi}{2} - \frac{1}{2}(0)\right] = \pi x_0^2$$

which is the same result as obtained previously. This method will be examined in more detail in the next chapter. ▌

EXERCISES

The following exercises represent a collection of the various types of integrals you can integrate. Evaluate each of the following integrals:

1. $\int \cos \frac{3}{2} x \, dx$
2. $\int \tan 5x \, dx$
3. $\int \sec^2 3x \, dx$
4. $\int \frac{\sin \sqrt{x}}{\sqrt{x}} \, dx$
5. $\int \frac{dx}{9 + 4x^2}$
6. $\int x \sin x^2 \, dx$

7. $\int \sin 4u \, du$

8. $\int (\sin^5 2x)(\cos 2x)dx$

9. $\int \frac{x^2 \, dx}{x^3 + 1}$

10. $\int \frac{dx}{e^{2x}}$

11. $\int (\tan 2x)(\sec 2x)dx$

12. $\int \frac{dx}{\sqrt{16 - 9x^2}}$

13. $\int (\tan 3x)(\sec^2 3x)dx$

14. $\int \left[x - \frac{1}{x}\right]^2 dx$

15. $\int \frac{\sin(\ln x)dx}{x}$

16. $\int \frac{dx}{x(1 + (\ln x)^2)}$

17. $\int \frac{\cos 2x}{\sin 2x} dx$

18. $\int \frac{x^2}{x^2 + 1} dx$ (Hint: divide first.)

19. $\int \frac{e^x \, dx}{(3 + e^x)^3}$

20. $\int (\sec^3 3x)(\tan 3x)dx$

21. $\int xe^{2x^2+3} dx$

22. $\int \frac{dx}{4 + 3x^2}$

23. $\int \frac{x \, dx}{(x^2 + 2)^2}$

24. $\int \frac{e^x \, dx}{1 + e^{2x}}$

25. $\int xe^{-x^2} dx$

26. $\int \frac{\sin 2x \, dx}{3 + \cos^2 2x}$

27. $\int \frac{1 + e^{2x}}{e^x} dx$

28. $\int x \, 2^{x^2} dx$

29. $\int 2\sqrt{7t - 3} \, dt$

30. $\int (\ln x + 1)e^{x \ln x} dx$

31. $\int \frac{(5x - 1)dx}{5x^2 - 2x + 1}$

32. $\int \frac{\tan^5 x}{\cos^2 x} dx$

33. $\int \frac{e^x \sec^2 e^x \, dx}{\tan e^x - 1}$

34. $\int \pi^{3x-1} dx$

35. $\int \left[e^{x/2} - e^{-x/2}\right]dx$

36. $\int \frac{(3x^2 + 1)dx}{\sqrt{3x^3 + 3x - 4}}$

37. $\int \frac{e^{\sqrt{x}} \, 2^{\sqrt{x}}}{\sqrt{x}} dx$

38. $\int \cos 2\theta \, e^{(\sin \theta)(\cos \theta)} d\theta$

Evaluate the following definite integrals:

39. $\int_0^{\frac{\pi}{3}} (\sec^3 x)(\tan x)dx$

40. $\int_0^{\frac{1}{\sqrt{2}}} \frac{x}{\sqrt{1 - x^4}} dx$

41. $\int_1^3 \frac{\sqrt[3]{\ln x}}{x} dx$

42. $\int_0^{\sqrt{2}} \frac{3x\ dx}{4 + x^4}$

43. $\int_0^{\frac{\pi}{3}} \frac{\sin x}{\cos^3 x} dx$

44. $\int_{\frac{\pi}{8}}^{\frac{\pi}{4}} (\cot 2x)(\csc^2 2x)dx$

45. $\int_0^1 (2x - 5)^{17}\ dx$

46. $\int_2^4 \frac{(6x - 9)}{x^2 - 3x + 8} dx$

47. $\int_1^3 \frac{3\sqrt{\ln x}}{x} dx$

48. $\int_0^{\frac{1}{2}} \frac{3 \sin^{-1} x}{\sqrt{1 - x^2}} dx$

*13.11 HYPERBOLIC FUNCTIONS

The hyperbolic functions, defined in terms of the exponential function, sometimes occur in biologically oriented problems. They appear mainly in the process of solving certain kinds of differential equations. These functions are discussed in this chapter on trigonometric functions because the hyperbolic and trigonometric functions share several properties: Both can be related to angles; both have similar notations, identities and power series expansions. This section focuses on the definitions and the calculus of the hyperbolic functions while the next deals with relations between the hyperbolic and trigonometric functions.

Before we progress to definitions and examples, consider two illustrations of when hyperbolic functions appear. We have seen that

$$\int \frac{dx}{\sqrt{a^2 - x^2}} = \text{Sin}^{-1} \frac{x}{a} + C ,$$

but we cannot handle the simple change of $\sqrt{a^2 - x^2}$ to $\sqrt{x^2 - a^2}$. We will find that

$$\int \frac{dx}{\sqrt{x^2 - a^2}} = \sinh^{-1} x + C .$$

Of course this will assume more meaning when you know what $\sinh^{-1} x$ means. The point is this: Even if you are using tables of integrals for integration, you can encounter hyperbolic functions. Furthermore, if you go on to study differential equations as some biologists do, you will encounter differential equations like

$$\frac{d^2y}{dx^2} - k^2y = C .$$

This innocent looking equation has the general solution

$$y = A \cosh kx + B \sinh kx - \frac{C}{k^2} .$$

Now let us turn to the definitions of the hyperbolic functions. Note that the hyperbolic functions relate to each other as did the trigonometric functions:

- hyperbolic sine x = $\sinh x = \frac{e^x - e^{-x}}{2}$ (13.66)
- hyperbolic cosine x = $\cosh x = \frac{e^x + e^{-x}}{2}$ (13.67)
- hyperbolic tangent x = $\tanh x = \frac{e^x - e^{-x}}{e^x + e^{-x}} = \frac{\sinh x}{\cosh x}$ (13.68)
- hyperbolic cosecant x = $\operatorname{csch} x = \frac{2}{e^x - e^{-x}} = \frac{1}{\sinh x}$ (13.69)
- hyperbolic secant x = $\operatorname{sech} x = \frac{2}{e^x + e^{-x}} = \frac{1}{\cosh x}$ (13.70)
- hyperbolic cotangent x = $\coth x = \frac{e^x + e^{-x}}{e^x - e^{-x}} = \frac{1}{\tanh x}$ (13.71)

The values of the hyperbolic functions can be obtained easily from tables of values of the exponential functions, but many books of tables of mathematical functions also table these functions. The graphs of the three primary hyperbolic functions appear in Figure 13.19. Note from the figure that sinh 0 = tanh 0 = 0 and cosh 0 = 1. This fact follows directly from the definitions of the hyperbolic functions:

$$\sinh 0 = \frac{e^0 - e^0}{2} = 0 ,$$

$$\cosh 0 = \frac{e^0 + e^0}{2} = 1$$

and

$$\tanh 0 = \frac{e^0 - e^0}{e^0 + e^0} = 0 .$$

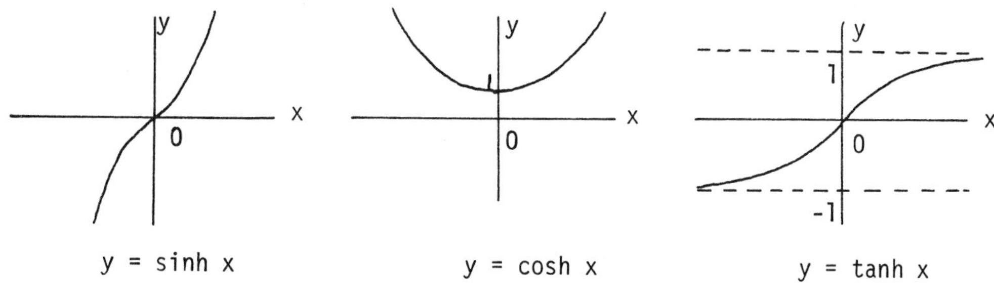

Figure 13.19

EXAMPLE 13.31: Show that sinh(-x) = - sinh x, tanh(-x) = - tanh x and cosh(-x) = cosh x.

From the above definitions, we have

$$\sinh(x) = \frac{e^x - e^{-x}}{2} \quad \text{and} \quad \sinh(-x) = \frac{e^{-x} - e^x}{2} = - \frac{e^x - e^{-x}}{2} = - \sinh x .$$

$$\tanh(x) = \frac{e^x - e^{-x}}{e^x + e^{-x}} \quad \text{and} \quad \tanh(-x) = \frac{e^{-x} - e^x}{e^{-x} + e^x} = - \frac{e^x - e^{-x}}{e^x + e^{-x}} = - \tanh x .$$

$$\cosh(x) = \frac{e^x + e^{-x}}{2} \quad \text{and} \quad \cosh(-x) = \frac{e^{-x} + e^x}{2} = \cosh x \text{ . } \blacksquare$$

From Equations 13.66 and 13.67 the hyperbolic sine and cosine approach $+\infty$ or $-\infty$ as x does. However

$$\lim_{x\to\infty} \tanh x = \lim_{x\to\infty} \frac{e^x - e^{-x}}{e^x + e^{-x}}$$

presents an indeterminate form for which L'Hospital's rule does not work. We must, therefore, change its form. If we divide the numerator and denominator by e^x, the limit becomes

$$\lim_{x\to\infty} \frac{1 - e^{-2x}}{1 + e^{-2x}} = 1 \text{ .}$$

In a similar manner, we can evaluate

$$\lim_{x\to-\infty} \tanh(x) = \lim_{x\to-\infty} \frac{e^{2x} - 1}{e^{2x} + 1} = -1 \text{ .}$$

Thus, the hyperbolic tangent has horizontal asymptotes at $y = 1$ and $y = -1$.

The hyperbolic functions satisfy a series of identities similar to those previously found for the trigonometric functions. Using the definitions given above, we can show that

$$\cosh^2 x - \sinh^2 x = 1 \text{ .} \tag{13.72}$$

because

$$\begin{aligned}\cosh^2 x - \sinh^2 x &= \left[\frac{e^x + e^{-x}}{2}\right]^2 - \left[\frac{e^x - e^{-x}}{2}\right]^2 \\ &= \frac{1}{4}\left[e^{2x} + 2 + e^{-2x} - (e^{2x} - 2 + e^{-2x})\right] \\ &= \frac{1}{4}(4) = 1 \text{ .}\end{aligned}$$

Equation 13.72 parallels the trigonometric identity $\sin^2 x + \cos^2 x = 1$. Identities also exist for the hyperbolic functions of $(x + y)$. For example

$$\cosh(x+y) = \frac{e^{x+y}+e^{-(x+y)}}{2} = \left[\frac{e^x+e^{-x}}{2}\right]\left[\frac{e^y+e^{-y}}{2}\right] + \left[\frac{e^x-e^{-x}}{2}\right]\left[\frac{e^y-e^{-y}}{2}\right]$$

$$= \cosh x \cosh y + \sinh x \sinh y \ . \qquad (13.73)$$

We can also show that

$$\sinh(x+y) = \sinh x \cosh y + \cosh x \sinh y \ . \qquad (13.74)$$

By dividing $\cosh^2 x - \sinh^2 x = 1$ first by $\cosh^2 x$, and then by $\sinh^2 x$, we also find

$$\operatorname{sech}^2 x = 1 - \tanh^2 x \qquad \text{and} \qquad \operatorname{csch}^2 x = \coth^2 x - 1 \ . \qquad (13.75)$$

The derivatives of the hyperbolic functions follow simply from the derivative of the exponential function. For example,

$$\frac{d}{dx}\sinh x = \frac{d}{dx}\frac{1}{2}(e^x - e^{-x}) = \frac{1}{2}\left[\frac{d}{dx}e^x - \frac{d}{dx}e^{-x}\right]$$

$$= \frac{1}{2}(e^x + e^{-x}) = \cosh x \ . \qquad (13.76)$$

In a similar fashion, you can show

$$\frac{d}{dx}\cosh x = \sinh x \ . \qquad (13.77)$$

You can establish the derivatives of the remaining hyperbolic functions from the derivatives of the hyperbolic sine and cosine.

EXAMPLE 13.32: Find $\frac{d}{dx}\tanh x$.

Because $\tanh x = \frac{\sinh x}{\cosh x}$ is a quotient,

$$\frac{d}{dx}\tanh x = \frac{d}{dx}\frac{\sinh x}{\cosh x} = \frac{\cosh x \frac{d}{dx}\sinh x - \sinh x \frac{d}{dx}\cosh x}{\cosh^2 x}$$

$$= \frac{\cosh^2 x - \sinh^2 x}{\cosh^2 x} \ .$$

Use of $\cosh^2 x - \sinh^2 x = 1$ reduces the above to

$$\frac{d}{dx}\tanh x = \frac{1}{\cosh^2 x} = \operatorname{sech}^2 x \ . \ \blacksquare \qquad (13.78)$$

We can define the inverse hyperbolic functions exactly as we defined the inverse trigonometric functions. For example if $x = \cosh y$, y is the inverse hyperbolic cosine of x, denoted by $y = \cosh^{-1} x$. Similar relations can be associated with the remaining hyperbolic functions.

The inverse hyperbolic functions relate to natural logarithms as is shown in the next example.

EXAMPLE 13.33: If $x > 1$, show that $\text{Cosh}^{-1} x = \ln(x + \sqrt{x^2 - 1})$.

Before we develop formulas for the inverse hyperbolic functions, look back at the graphs of the primary functions in Figure 13.19. Notice that only one value of x gives each value of y for $y = \sinh x$ and $y = \tanh x$, but $\cosh x = \cosh(-x)$. Thus a question of principle values will arise for $\cosh^{-1} y$, but not for $\sinh^{-1} y$ or $\tanh^{-1} y$; customarily, $\text{Cosh}^{-1} y \geq 0$.

The nature of the definition of hyperbolic functions allows us to develop formulas for the inverse hyperbolic functions. For example, consider $y = \cosh^{-1} x$ or $x = \cosh y = \dfrac{e^y + e^{-y}}{2}$. Multiply both sides of this equation by $2e^y$ and rearrange terms to get

$$e^{2y} - 2xe^y + 1 = 0 .$$

Solve this quadratic equation for e^y : $e^y = x \pm \sqrt{x^2 - 1}$. The double valued nature of $\cosh^{-1} y$ shows up here in the $\pm$ sign. To achieve a function, use only the positive root. Now solve for y:

$$\text{Cosh}^{-1} x = y = \ln(x + \sqrt{x^2 - 1}) . \quad \blacksquare \tag{13.79}$$

An argument similar to the above gives

$$\sinh^{-1} x = \ln(x + \sqrt{x^2 + 1}) .$$

This will be left for you to establish. These and other related results appear in the summary at the end of this chapter.

EXAMPLE 13.34: Find $\dfrac{d}{dx} \text{Cosh}^{-1} u$.

From Example 13.33, $\text{Cosh}^{-1} u = \ln(u + \sqrt{u^2 - 1})$ so

$$\frac{d}{dx}\,\text{Cosh}^{-1}\,u = \frac{d}{dx}\ln(u + \sqrt{u^2 - 1}) = \frac{1}{u + \sqrt{u^2 - 1}}\,\frac{d}{dx}\,(u + \sqrt{u^2 - 1})$$

$$= \frac{1}{u + \sqrt{u^2 - 1}}\left(1 + \frac{u}{\sqrt{u^2 - 1}}\right)\frac{du}{dx} = \frac{1}{\sqrt{u^2 - 1}}\,\frac{du}{dx}\,. \quad \blacksquare \qquad (13.80)$$

Note the similarity of this derivative to

$$\frac{d}{dx}\,\text{Cos}^{-1}\,u = -\frac{1}{\sqrt{1 - u^2}}\,\frac{du}{dx}\,.$$

The major difference lies inside the radical in the denominator. We point this out for your use of antiderivatives in evaluating integrals. Integrands involving the reciprocal of a constant with the square of a variable of integration have antiderivatives involving the inverse functions. In the summary at the end of the chapter, compare the various forms involving the inverse functions to see the patterns involved.

The relationships that exist between the hyperbolic functions and the trigonometric functions will be considered in the next section.

EXERCISES

1. Show that for large positive values of x that $\sinh x \doteq \cosh x \doteq \frac{1}{2}e^x$.
2. Show that for large negative values of x that $\sinh x \doteq -\cosh x \doteq -\frac{1}{2}e^{-x}$.
3. Show $\sinh(x + y) = \sinh x \cosh y + \cosh x \sinh y$.
4. Show $\sinh(x - y) = \sinh x \cosh y - \cosh x \sinh y$.
5. Show that $\frac{d}{dx}\cosh x = \sinh x$.
6. Show $\text{sech}^2\,x = 1 - \tanh^2 x$.
7. Show that for large positive values of x that $\text{csch}\,x \doteq \text{sech}\,x \doteq 2e^{-x}$.
8. Show that for all values of x, $\sinh^{-1} x = \ln(x + \sqrt{x^2 + 1})$.
9. Discuss the behavior of csch x and sech x as $x\to\infty$ and $x\to-\infty$.

Find the derivatives of the following:

10. $y = \sinh^3 x$ 11. $y = \tanh(\ln x)$ 12. $y = e^{\cosh x}$

13. $y = \ln(\cosh 3x)$ 14. $y = \sinh(7x + 2)$ 15. $y = \coth^2 x$

16. $y = \cosh^2 2x^2$

Show that the following are true:

17. $\int \cosh u \, du = \sinh u + C$

18. $\int \coth u \, du = \ln(\sinh u) + C$

19. $\dfrac{d}{dx} \sinh^{-1} x = \dfrac{1}{\sqrt{x^2 + 1}}$

20. If $\tanh x = \dfrac{4}{5}$, prove that $x = \ln 3$.

**13.12 RELATIONS BETWEEN HYPERBOLIC AND TRIGONOMETRIC FUNCTIONS

The trigonometric functions were defined in terms of angles; a related view of the hyperbolic functions exists. After we consider that correspondence, we will show some correspondence related to the respective Maclaurin series expansions.

To establish the angular similarity between the trigonometric and hyperbolic functions, consider the curves defined implicitly by $x^2 + y^2 = 1$ and $x^2 - y^2 = 1$. Parts of these relations appear in Figures 13. and 13. . In each case, a perpendicular has been erected at w and

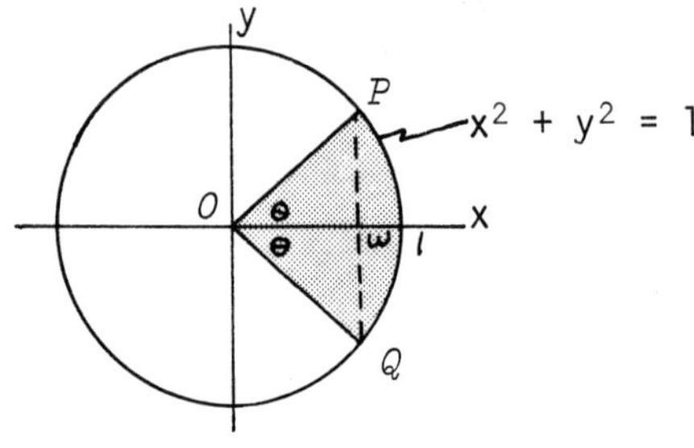

Figure 13.20

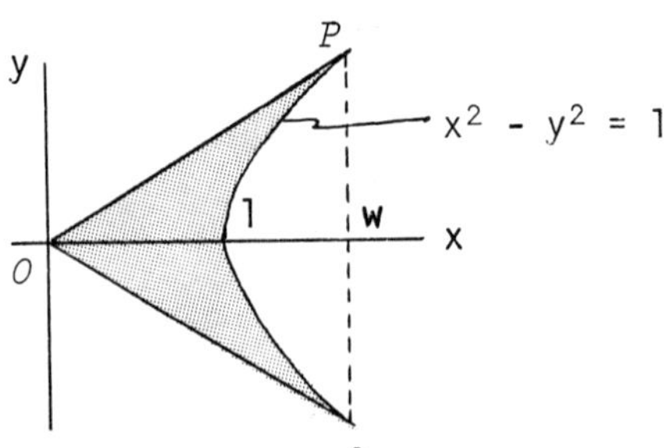

Figure 13.21

its intersections with the curve labeled as P and Q. Consider the shaded areas bounded by the line segments OP and OQ and the curves in these figures. In the first case, the fraction $\frac{2\theta}{2\pi}$ of the circle is shaded so with $r = 1$,

$$\text{Area} = \frac{2\theta}{2\pi}(\pi r^2) = \theta = \text{Cos}^{-1}\, w\ . \tag{13.81}$$

In Figure 13.21, the shaded area can be evaluated as the area contained in the triangle OPQ minus the area inside the curve from 1 to w. What dimensions does this triangle have? It has a base of length w and a height y which satisfies $w^2 - y^2 = 1$ or $y = \sqrt{w^2 - 1}$. The curve has a height above the x-axis of $y = \sqrt{x^2 - 1}$ so

$$\text{Area} = w\sqrt{w^2 - 1} - 2\int_1^w \sqrt{x^2 - 1}\, dx$$

The integral above can be rewritten as

$$2\int_1^w \sqrt{x^2 - 1}\, dx = \int_1^w \frac{2x^2 - 2}{\sqrt{x^2 - 1}}\, dx = \int_1^w \frac{(2x^2 - 1)dx}{\sqrt{x^2 - 1}} - \int_1^w \frac{dx}{\sqrt{x^2 - 1}}$$

$$= \int_1^w d(x\sqrt{x^2 - 1}) - \int_1^w \frac{dx}{\sqrt{x^2 - 1}}$$

$$= x\sqrt{x^2 - 1}\Big]_1^w - \text{Cosh}^{-1}\, x\Big]_1^w = w\sqrt{w^2 - 1} - \text{Cosh}^{-1}\, w$$

so the area in question is

$$\text{Area} = w\sqrt{w^2 - 1} - 2\int_1^w \sqrt{x^2 - 1}\, dx = w\sqrt{w^2 - 1} - w\sqrt{w^2 - 1} + \text{Cosh}^{-1}\, w$$

$$= \text{Cosh}^{-1}\, w\ . \tag{13.82}$$

Now note the similarity of Equations 13.81 and 13.82. The two kinds of functions measure the size of analogous angles in the same manner, but relative to different curves. This gives a rational basis to the notational similarities between the trigonometric and hyperbolic functions. In fact, because of these relations, the trigonometric functions sometimes are called *circular functions* while the *hyperbolic functions* derive their name from the type of curve to which they are related, the hyperbola.

Next we shall give the Maclaurin's series expansions of the exponential, hyperbolic and trigonometric functions in order to show another relationship which exists between the hyperbolic and trigonometric functions.

Recall from Section 7.8 that when a function f possesses derivatives of all orders, it can be expanded in a Maclaurin's series of the form

$$f(x) = f(0) + f'(0)x + \frac{f''(0)x^2}{2!} + \frac{f'''(0)x^3}{3!} + \cdots + \frac{f^{(n)}(0)x^n}{n!} + \cdots .$$

In Section 7.8 we found that the Maclaurin's expansion for e^x was

$$e^x = 1 + x + \frac{x^2}{2!} + \frac{x^3}{3!} + \cdots + \frac{x^n}{n!} + \cdots . \tag{13.83}$$

This series holds for all values of x, including negative x. Thus when x is replaced by -x, the series expansion for e^{-x} results:

$$e^{-x} = 1 + (-x) + \frac{(-x)^2}{2!} + \frac{(-x)^3}{3!} + \cdots + \frac{(-x)^n}{n!} + \cdots$$

$$= 1 - x + \frac{x^2}{2!} - \frac{x^3}{3!} + \cdots + \frac{(-1)^n x^n}{n!} + \cdots . \tag{13.84}$$

The series expansions for the hyperbolic functions are easy to derive from the expansions for e^x and e^{-x}. Because $\sinh x = \frac{e^x - e^{-x}}{2}$, we only need subtract the series expansions of e^{-x} from e^x and divide by 2:

$$e^x - e^{-x}$$

$$= 1 + x + \frac{x^2}{2!} + \frac{x^3}{3!} + \cdots + \frac{x^n}{n!} - (1 - x + \frac{x^2}{2!} - \frac{x^3}{3!} + \cdots + (-1)^n \frac{x^n}{n!} + \cdots)$$

$$= 2(x + \frac{x^3}{3!} + \frac{x^5}{5!} + \cdots + \frac{x^{2n+1}}{(2n + 1)!} + \cdots) ,$$

$$\sinh x = \frac{e^x - e^{-x}}{2} = x + \frac{x^3}{3!} + \frac{x^5}{5!} + \cdots + \frac{x^{2n+1}}{(2n + 1)!} + \cdots . \tag{13.85}$$

Thus, we have arrived at the series expansion for the hyperbolic sine of x from the expansion for e^x and e^{-x}.

In a similar fashion, we can show that the series expansion for the hyperbolic cosine of x is

$$\cosh x = 1 + \frac{x^2}{2!} + \frac{x^4}{4!} + \cdots + \frac{x^{2n}}{(2n)!} + \cdots . \qquad (13.86)$$

The series expansion for sinh x and cosh x are valid for all real values of x because the series for e^x and e^{-x} are valid for all real values of x.

Now we can display another relationship between the hyperbolic and trigonometric functions. The series expansion for cos x and cosh x have a noticeable similarity. First, we need to obtain the expansion for f(x) = cos x. Its successive derivatives are -sin x, -cos x, sin x, cos x, ···, so the coefficients in the expansion of cos x are f(0) = 1, f'(0) = 0, f"(0) = -1, f'''(0) = 0, f''''(0) = 1 ···. The fifth of these terms equals the first so the terms repeat in cycles of four with the non-zero coefficients merely being sign changes:

$$\cos x = 1 - \frac{x^2}{2!} + \frac{x^4}{4!} - \frac{x^6}{6!} + \cdots + (-1)^n \frac{x^{2n}}{(2n)!} + \cdots .$$

Now compare this to the expansion for cosh x in Equation 13.85. The only difference lies in their pattern of signs. If we take the expansion for cosh x and replace x by ix, where i denotes the quantity $\sqrt{-1}$, we get

$$\cosh ix = 1 + \frac{(ix)^2}{2!} + \frac{(ix)^4}{4!} + \cdots + \frac{(ix)^{2n}}{(2n)!} + \cdots$$

$$= 1 - \frac{x^2}{2!} + \frac{x^4}{4!} + \cdots + (-1)^n \frac{x^{2n}}{(2n)!} + \cdots = \cos x$$

because $i^2 = -1$, $i^4 = 1$, $i^6 = -1$, ··· . Similarly, we have

$$\cos(ix) = \cosh(i^2x) = \cosh(-x) = \cosh x.$$

These two relationships show another correspondence between particular trigonometric and hyperbolic functions which has counterparts for the other pairs of functions, for example, sin(ix) = i sinh x .

This explanation has ventured off into the complex extension of the familiar number system. This was done partly to show correspondence between trigonometric and hyperbolic functions, and partly to alert you to this fact: The complex (sometimes called imaginary) number system has a central role in the application of mathematics-to electricity for example. The above presentation leaves many loose ends about complex numbers, such as their formal definition. Careful consideration of these matters and the associated extensions of the calculus is beyond the scope of this book.

EXERCISES

1. Obtain the Maclaurin's expansion for sin x.

2. Derive the Maclaurin's expansion for cosh x.

3. Show that for sufficiently small values of x that $e^{-x} \doteq 1 - x$.

Show that the following identities are true:

4. $\tanh(-x) = -\tanh x$

5. $\operatorname{sech}^2 x + \tanh^2 x = 1$

6. $\operatorname{csch}^2 x = \coth^2 x = -1$

7. $\cosh 2x = \cosh^2 x + \sinh^2 x$

8. Show that $\sinh(ix) = i \sin x$, $i = \sqrt{-1}$.

9. Show that $\sin(ix) = i \sinh x$, $i = \sqrt{-1}$.

10. Find the series expansion for $\left[\frac{1 + x}{x}\right]\left[e^x - 1\right]$.

11. Show $\sin x = \frac{e^{ix} - e^{ix}}{2i}$, $i = \sqrt{-1}$.

12. Show $\cos x = \frac{e^{ix} + e^{-x}}{2}$, $i = \sqrt{-1}$.

13. Show $e^{ix} = \cos x + i \sin x$.

13.13 A SUMMARY OF FUNCTIONS, DERIVATIVES AND INTEGRALS

To this point, we have introduced a large number of functions, their derivatives and integrals. Here we pause to summarize all of their basic forms. Each has many variations, and the list of integral forms will be expanded as we consider techniques of integration in the next chapter. The equation numbers or section references below correspond to the first appearance of that relation. Each is suitably labeled when appropriate.

13.13a FUNCTIONS AND RELATIONS

1.	$y = [\![x]\!]$	Integer part	Sec. 4.2
2.	$y = mx + b$	Straight line	Sec. 4.3
3.	$y = ax \quad + bx + c$	Parabola	Sec. 4.4
4.	$xy = k$	Rectangular hyperbola	Sec. 4.4
5.	$ax \quad - by \quad = c$	Equilateral hyperbola	
6.	$x \quad + y \quad = c$	Circle	Sec. 4.4
7.	$ax \quad + by \quad = c$	Ellipse	
8.	$y = bx^k$	Power function	Sec. 4.5a
9.	$y = be^{kx}$	Exponential function	Sec. 4.5b
10.	$y = \log x$	Common logarithm	Sec. 4.6
11.	$y = \ln x$	Natural logarithm	Sec. 4.6
12.	$y = \sin x$	Trigonometric function	(13.1)
13.	$y = \cos x$	" "	(13.2)
14.	$y = \tan x$	" "	(13.3)
15.	$y = \cot x = \dfrac{1}{\tan x}$	" "	(13.4)
16.	$y = \sec x = \dfrac{1}{\cos x}$	" "	(13.5)
17.	$y = \csc x = \dfrac{1}{\sin x}$	" "	(13.6)

18.	$y = \sin^{-1} x$, $\text{Sin}^{-1} x$	Inverse trigonometric function and its principle value	Sec. 13.6
19.	$y = \cos^{-1} x$, $\text{Cos}^{-1} x$	" " "	Sec. 13.6
20.	$y = \tan^{-1} x$, $\text{Tan}^{-1} x$	" " "	Sec. 13.6
21.	$y = \cot^{-1} x$, $\text{Cot}^{-1} x$	" " "	Sec. 13.6
22.	$y = \sec^{-1} x$, $\text{Sec}^{-1} x$	" " "	Sec. 13.6
23.	$y = \csc^{-1} x$, $\text{Csc}^{-1} x$	" " "	Sec. 13.6
24.	$y = \sinh x = \frac{e^x - e^{-x}}{2}$	Hyperbolic sine	(13.66)
25.	$y = \cosh x = \frac{e^x + e^{-x}}{2}$	Hyperbolic cosine	(13.67)
26.	$y = \tanh x = \frac{e^x - e^{-x}}{e^x + e^{-x}}$	Hyperbolic tangent	(13.68)
27.	$y = \text{csch } x = \frac{1}{\sinh x}$	Hyperbolic cosecant	(13.69)
28.	$y = \text{sech } x = \frac{1}{\cosh x}$	Hyperbolic secant	(13.70)
29.	$y = \coth x = \frac{1}{\tanh x}$	Hyperbolic cotangent	(13.71)
30.	$y = \sinh^{-1} x$	Inverse hyperbolic sine	Sec. 13.11
31.	$y = \cosh^{-1} x$, $\text{Cosh}^{-1} x$	Inverse hyperbolic cosine, principle value	(13.79)
32.	$y = \tanh^{-1} x$	Inverse hyperbolic tangent	

13.13b DERIVATIVES

1.	$\frac{dx^n}{dx} = n\, x^{n-1}$	Power form	(6.1)
2.	If $u = g(x)$, $\frac{dh(u)}{dx} = \frac{dh(u)}{du} \frac{dg(x)}{dx}$.	Chain rule	(6.4)
3.	$\frac{d}{dx}(f(x) + g(x)) = \frac{df(x)}{dx} + \frac{dg(x)}{dx}$	Sum	(6.3)

4. $\frac{d}{dx}(f(x)g(x)) = f(x)\frac{dg(x)}{dx} + g(x)\frac{df(x)}{dx}$ Product (6.7)

5. $\frac{d}{dx}\frac{f(x)}{g(x)} = \frac{g(x)\frac{df(x)}{dx} - f(x)\frac{dg(x)}{dx}}{(g(x))^2}$ Quotient (6.10)

6. $\frac{d}{dx}(\ln x) = \frac{1}{x}$ (6.12)

7. $\frac{d}{dx}\log_b x = \frac{1}{x}\log_b e$ (6.13)

8. $\frac{d}{dx}(e^x) = e^x$ (6.14)

9. $\frac{d}{dx}(a^u) = a^u \ln a$ Sec. 7.6b

10. $\frac{d}{dx}(u^v) = vu^{v-1}\frac{du}{dx} + (\ln u)\, u^v \frac{dv}{dx}$ Sec. 7.6b

11. $\frac{d}{dx}(\sin x) = \cos x$ (13.44)

12. $\frac{d}{dx}(\cos x) = -\sin x$ (13.47)

13. $\frac{d}{dx}(\tan x) = \sec^2 x$ (13.50)

14. $\frac{d}{dx}(\cot x) = -\csc^2 x$ (13.51)

15. $\frac{d}{dx}(\sec x) = \sec x \tan x$ (13.52)

16. $\frac{d}{dx}(\csc x) = -\csc x \cot x$ (13.53)

17. $\frac{d}{dx}\text{Sin}^{-1} x = \frac{1}{\sqrt{1 - x^2}}$ (13.54)

18. $\frac{d}{dx}\text{Cos}^{-1} x = \frac{-1}{\sqrt{1 - x^2}}$ (13.55)

19. $\frac{d}{dx}\text{Tan}^{-1} x = \frac{1}{1 + x^2}$ (13.56)

20. $\frac{d}{dx}\text{Cot}^{-1} x = -\frac{1}{1 + x^2}$ (13.57)

21. $\frac{d}{dx} \sinh x = \cosh x$ (13.76)

22. $\frac{d}{dx} \cosh x = \sinh x$ (13.77)

23. $\frac{d}{dx} \tanh x = \text{sech}^2 x$ (13.78)

24. $\frac{d}{dx} \sinh^{-1} x = \frac{1}{\sqrt{1 + x^2}}$ Sec. 13.11

25. $\frac{d}{dx} \text{Cosh}^{-1} x = \frac{1}{\sqrt{x^2 - 1}}$ Sec. 13.11

26. $\frac{d}{dx} \tanh^{-1} x = \frac{1}{1 - x^2}$

13.13c INTEGRALS

The integrals given below, both general and specific forms, can be verified by checking that the derivative of the right-hand side of the equation equals the integrand.

GENERAL FORMS:

1. $\int df(x) = f(x) + C$ (8.19)
2. $d \int f(x)dx = f(x)dx$ (8.20)
3. $\int (0)\, dx = C$ (8.21)
4. $\int kf(x)dx = k \int f(x)dx$ (Th. 8.2) (8.22)
5. $\int (u \pm v)dx = \int u\, dx \pm \int v\, dx$ (Th. 8.3) (8.23)

SPECIFIC FORMS:

6. $\int x^n\, dx = \frac{x^{n+1}}{n + 1} + C \qquad n \neq -1$ (8.9)

7. $\int \frac{dx}{x} = \ln |x| + C$ (8.10)

8. $\int e^x\, dx = e^x + C$ (8.11)

9. $\int a^x\,dx = \frac{a^x}{\ln a} + C \qquad a > 0,\ a \neq 1$ (8.18)

10. $\int \sin x\,dx = -\cos x + C$ (13.58)

11. $\int \cos x\,dx = \sin x + C$ (13.59)

12. $\int \sec^2 x\,dx = \tan x + C$ (13.60)

13. $\int \csc^2 x\,dx = -\cot x + C$

14. $\int \tan x\,dx = \ln|\sec x| + C = -\ln|\cos x| + C$ (13.61)

15. $\int \cot x\,dx = \ln|\sin x| + C = -\ln|\csc x| + C$

16. $\int \sec x \tan x\,dx = \sec x + C$ (13.62)

17. $\int \csc x \cot x\,dx = -\csc x + C$ (13.63)

18. $\int \sec x\,dx = \ln|\sec x + \tan x| + C = \ln\left|\tan\left(\frac{x}{2} + \frac{\pi}{4}\right)\right| + C$

19. $\int \csc x\,dx = \ln|\csc x - \cot x| + C = \ln\left|\tan \frac{x}{2}\right| + C$

20. $\int \frac{dx}{x^2 + a^2} = \frac{1}{a}\,\mathrm{Tan}^{-1}\frac{x}{a} + C \qquad a > 0$ (13.64)

21. $\int \frac{dx}{\sqrt{a^2 - x^2}} = \mathrm{Sin}^{-1}\frac{x}{a} + C$ (13.65)

22. $\int \sinh x\,dx = \cosh x + C$

23. $\int \cosh x\,dx = \sinh x + C$

24. $\int \tanh x\,dx = \ln|\cosh x| + C$

25. $\int \coth x\,dx = \ln|\sinh x| + C$

26. $\int \frac{dx}{\sqrt{a^2 + x^2}} = \sinh^{-1} x + C = \ln(x + \sqrt{x^2 + a^2}) + C$

27. $$\int \frac{dx}{\sqrt{x^2 - a^2}} = \begin{cases} \mathrm{Cosh}^{-1}\frac{x}{a} + C = \ln(x + \sqrt{x^2 - a^2}) + C & x > a \\ -\,\mathrm{Cosh}^{-1}\left|\frac{x}{a}\right| + C & x < -a \end{cases}$$

28. $$\int \frac{dx}{a^2 - x^2} = \begin{cases} \frac{1}{a}\tanh^{-1}\frac{x}{a} + C = \frac{1}{2a}\ln\left(\frac{a + x}{a - x}\right) + C & x^2 < a^2 \\ \frac{1}{a}\coth^{-1}\frac{x}{a} + C = \frac{1}{2a}\ln\left(\frac{x + a}{x - a}\right) + C & x^2 > a^2 \end{cases}$$

Chapter 14

TECHNIQUES OF INTEGRATION

Thus far we have avoided several classes of integrals because we lacked the tools needed to integrate them. In other situations we used cumbersome methods for which simpler alternatives exist. This chapter presents some of the tools used in integration; it attempts nothing more. It contains no explicit biological examples; its utility lies in evaluating integrals or solving differential equations arising in applications.

The first section introduces integration by parts; the next two show how trigonometric substitutions can be used to simplify integrands; the following three sections present techniques for handling certain fractional forms; and the final section describes how tables of integrals can be used.

In view of the last section in this chapter, you might wonder why the other sections are needed. First, a table of integrals has a finite number of entries, rarely more than 500. This merely scratches the surface of available integrals. The techniques to be described in this chapter provide the basis for changing many integrals into forms which appear in tables. Furthermore many integrals cannot be evaluated in closed form; your ability to discriminate between those that can and cannot be evaluated in closed form will relieve you of unnecessarily using the techniques of numerical integration introduced in Section 8.11.

14.1 INTEGRATION BY PARTS

Numerous integrands involve products of functions, either of which can be integrated by itself, but which together defy direct integration. The technique of this section provides a means for dividing your integration task so that a solution of its parts leads to a solution of the whole.

The differential of a product is given by

$$d(uv) = u\,dv + v\,du\ ,$$

or upon rearrangement

$$u\,dv = d(uv) - v\,du\ .$$

Integration of this last equation leads to:

- INTEGRATION BY PARTS proceeds according to

$$\int u\,dv - uv - \int v\,du\ . \tag{14.1}$$

This technique enables us to shift the problem from integrating one form, $\int u\,dv$, to that of integrating another form, $\int v\,du$, which we can pick for ease of handling. Success with this method usually depends on the choice of u and dv. Sometimes only one choice exists for each, but frequently you have more than one. The choice of dv should often (but not always) include the most complicated part of the given integrand that you can integrate. The choice of u should give a function that is simplified by differentiation, but which leaves a dv you can integrate.

EXAMPLE 14.1: Evaluate $\int \ln x\,dx$.

This presents only one choice for u: $u = \ln x$ and thus $dv = dx$. When $u = \ln x$, $du = \frac{dx}{x}$ and if $dv = dx$, then $v = x$. Thus, $\int u\,dv = uv - \int v\,du$ becomes

$$\int \underbrace{\ln x}_{u}\ \underbrace{dx}_{dv} = \underbrace{(\ln x)}_{u}\ \underbrace{x}_{v} - \int \underbrace{x}_{v}\ \underbrace{\frac{dx}{x}}_{du} = \ln x - \int dx$$

$$= x \ln x - x + C\ .$$

You can check this result by differentiation:

$$\frac{d}{dx}(x \ln x - x + C) = (x\,\frac{1}{x} + \ln x - 1 + 0) = \ln x\ .$$

Thus, we ascertain that we have integrated properly. ∎

EXAMPLE 14.2: Evaluate $\int x \sin 2x\,dx$.

This integral appears to leave a choice for u and dv, but it really possesses only one workable choice. If $u = x$ and $dv = \sin 2x\,dx$, then $du = dx$ and $v = -\frac{1}{2}\cos 2x$ so $\int u\,dv = uv - \int v\,du$ becomes

$$\int x \sin 2x\, dx = -\frac{x}{2}\cos 2x - \int (-\frac{1}{2}\cos 2x)dx$$

$$= -\frac{x}{2}\cos 2x + \frac{1}{2}\int \cos 2x\, dx$$

$$= -\frac{x}{2}\cos 2x + \frac{1}{4}\sin 2x + C\ .$$

The choice $u = \sin 2x$ and $dv = x\, dx$ leads to $du = 2 \cos 2x\, dx$ and $v = \frac{x^2}{2}$ so

$$\int x \sin 2x\, dx = \frac{x^2}{2}\sin 2x - \int \frac{x^2}{2} 2 \cos 2x\, dx\ .$$

Notice that this last integral is more difficult to evaluate than the original one. Thus only one usable choice exists for u and dv. ▌

Repeated integration by parts sometimes provides a closed answer. The next two examples illustrate this process of repeated integration by parts.

EXAMPLE 14.3: Evaluate $\int x^2e^x\, dx$.

Let $u = x^2$ and $dv = e^x dx$; then $du = 2x\, dx$ and $v = e^x$ and

$$\int x^2e^x\, dx = x^2e^x - \int 2xe^x\, dx = x^2e^x - 2\int xe^x\, dx\ .$$

We still must evaluate $\int xe^x\, dx$. This time let $u = x$ and $dv = e^x\, dx$ so $du = dx$ and $v = e^x$ and

$$\int xe^x\, dx = xe^x - \int e^x\, dx = xe^x - e^x + C\ .$$

Putting the two results together, we have

$$\int x^2e^x\, dx = x^2e^x - 2xe^x + 2e^x + C\ .$$

We applied integration by parts twice to evaluate the given integral. This and the previous example illustrate that if an integrand contains a quantity raised to a power, like x^2, then u often should be set equal to this quantity. ▌

EXAMPLE 14.4: Evaluate $\int e^x \sin x\, dx$.

If $u = \sin x$ and $dv = e^x\, dx$, then $du = \cos x\, dx$ and $v = e^x$ so

$$\int e^x \sin x \, dx = e^x \sin x - \int e^x \cos x \, dx \ .$$

Now integrate this second integral by parts, letting $u = \cos x$ and $dv = e^x \, dx$, so $du = - \sin x \, dx$ and $v = e^x$:

$$\int e^x \sin x \, dx = e^x \sin x - (e^x \cos x - \int - e^x \sin x \, dx)$$

$$= e^x \sin x - e^x \cos x - \int e^x \sin x \, dx + C \ .$$

Add $\int e^x \sin x \, dx$ to both sides of the last equation to obtain

$$2 \int e^x \sin x \, dx = e^x \sin x - e^x \cos x + C$$

or upon dividing both sides of the last equation by 2

$$\int e^x \sin x \, dx = \frac{1}{2} e^x(\sin x - \cos x) + C_1$$

where $C_1 = \frac{C}{2}$.

In this example, we have used integration by parts in a different manner in order to evaluate the original integral. Try letting $u = e^x$ and $dv = \sin x \, dx$ in the original integrand to see that the same answer results.

A word of caution: An inappropriate choice of u and dv in the second stage above can lead you in a circle. For example if in evaluating $\int e^x \cos x \, dx$, suppose we had let $u = e^x$ and $dv = \cos x \, dx$, then $du = e^x \, dx$ and $v = \sin x$. Thus,

$$\int e^x \sin x \, dx = e^x \sin x - (e^x \sin x - \int e^x \sin x \, dx)$$

or

$$\int e^x \sin dx = \int e^x \sin x \, dx$$

which is just the original integral. If this should occur consider your second choice for u and dv. ▮

Ingegration by parts can be applied to definite integrals, proper or improper, as well as to indefinite integrals. The limits apply to an independent variable like x above, not to u and v, because these both are functions of the independent variable. When we integrate by parts, between limits on the independent variable, we find

$$\int_a^b u\,dv = \int_a^b d(uv) - \int_a^b v\,du$$

$$= uv\Big]_a^b - \int_a^b v\,du \qquad (14.2)$$

This technique could be applied to any of the preceding examples merely by inserting the limits of integration into the final result.

**EXAMPLE 14.5: Integration by parts also provides the basis for showing that $\Gamma(n) = (n - 1)\Gamma(n - 1)$ as stated in Equation 9.9. Recall that

$$\Gamma(n) = \int_0^\infty x^{n-1}\, e^{-x}\, dx \ .$$

To use integration by parts, let $u = x^{n-1}$ and $dv = e^{-x}dx$ so $du = (n - 1)x^{n-2}\, dx$ and $v = -e^{-x}$:

$$\int x^{n-1}\, e^{-x}\, dx = -x^{n-1}\, e^{-x} - \int (-e^{-x})(n - 1)x^{n-2}\, dx$$

$$= -x^{n-1}\, e^{-x} + (n - 1) \int x^{n-2}\, e^{-x}\, dx \ .$$

Now, treat this as an improper integral, using methods presented in Section 9.1:

$$\int_0^\infty x^{n-1}\, e^{-x}\, dx = \lim_{a\to\infty} \int_0^a x^{n-1}\, e^{-x}\, dx$$

and

$$\int_0^a x^{n-1}\, e^{-x}\, dx = -x^{n-1}\, e^{-x}\Big]_0^a + (n - 1) \int_0^a x^{n-2}\, e^{-x}\, dx$$

$$= -a^{n-1}\, e^{-a} + (n - 1) \int_0^a x^{n-2}\, e^{-x}\, dx$$

so

$$\Gamma(n) = \int_0^\infty x^{n-1}\, e^{-x}\, dx = \lim_{a\to\infty} (-a^{n-1}\, e^{-a}) + (n - 1) \int_0^\infty x^{n-2}\, e^{-x}\, dx$$

$$= (n - 1) \int_0^\infty x^{(n-1)-1}\, e^{-x}\, dx = (n - 1)\Gamma(n - 1)$$

because $\frac{a^{n-1}}{e^a}$ approaches zero as $a\to\infty$. To evaluate this limit repeatedly apply L'Hospital's rule until the numerator becomes a constant while the denominator does not change. ▮

EXERCISES:

Integrate the following:

1. $\int 2x\, e^x\, dx$
2. $\int e^{2x} \cos 2x\, dx$
3. $\int_1^4 x\, e^x\, dx$
4. $\int e^{-x} \cos 3x\, dx$
5. $\int x \sin x\, dx$
6. $\int \sec^3 x\, dx$
7. $\int_1^3 x\, e^{-x}\, dx$
8. $\int e^{ax} \sin bx\, dx$
9. $\int x \sec^2 x\, dx$
10. $\int e^{ax} \cos bx\, dx$
11. $\int_1^2 x \ln x\, dx$
12. $\int z \csc 2z \cot 2z\, dz$
13. $\int e^{-3x} \sin \frac{x}{3}\, dx$
14. $\int x^2 e^{3x}\, dx$
15. $\int e^{-x} \sin x\, dx$
16. $\int x^2 \sin x\, dx$
17. $\int e^{-2x} \sin 2x\, dx$
18. $\int x^2 \ln x\, dx$
19. $\int e^{2x} \cos 4x\, dx$
20. $\int x^2 \cos x\, dx$
21. $\int e^x \sin 3x\, dx$
22. $\int 2x \sec 3x \tan 3x\, dx$
23. $\int \frac{\ln (x + 2)}{(x + 2)^2}\, dx$
24. $\int \tan^{-1} x\, dx$
25. $\int \frac{x^3 dx}{\sqrt{4 + x^2}}$
26. $\int x\, 2^x\, dx$
27. $\int \frac{x \cot x\, dx}{\sin x}$
28. $\int \frac{1}{2} x \tan^{-1} x\, dx$

29. $\int x^2 a^x \, dx$

30. $\int x \sin kx \, dx$

31. $(\ln x)^3 \, dx$

14.2 INTEGRATION BY ELEMENTARY TRIGONOMETRIC TRANSFORMATIONS

A suitable trigonometric identity may change an integral involving trigonometric functions from a troublesome form to a form which is relatively simple to integrate. While many devices exist for evaluating such integrals, we will limit our attention here to tools for dealing with the following forms:

1. $\int \sin^2 x \, dx$ and $\int \cos^2 x \, dx$
2. $\int \sin^n x \, dx$ and $\int \cos^n x \, dx$
3. $\int \tan^n x \, dx$
4. $\int \sin^m x \cos^n x \, dx$
5. $\int \tan^m x \sec^n x \, dx$ and $\int \cot^m x \csc^n x \, dx$
6. $\int \sin mx \cos nx \, dx$, $\int \sin mx \sin nx \, dx$, and $\int \cos mx \cos nx \, dx$

We will discuss each of the above types and present examples showing how they are evaluated before you encounter any exercises.

CASE 1: $\int \sin^2 x \, dx$ and $\int \cos^2 x \, dx$

We have seen examples of this type of integral previously. They integrate easily after their integrands have been transformed by the half-angle formulas

$$\sin^2 x = \frac{1 - \cos 2x}{2} \quad \text{and} \quad \cos^2 x = \frac{1 + \cos 2x}{2} .$$

EXAMPLE 14.6: Evaluate $\int \cos^2 x \, dx$.

$$\int \cos^2 x \, dx = \int \frac{(1 + \cos 2x)}{2} \, dx = \frac{1}{2} \int dx + \frac{1}{2} \int \cos 2x \, dx$$

$$= \frac{x}{2} + \frac{1}{4} \sin 2x + C . \quad \blacksquare$$

Examine the solution given in Example 13.30 for another illustration of this method.

CASE 2: $\int \sin^n x\, dx$ and $\int \cos^n x\, dx$

This type of integral has two cases: (i) n is an even, positive integer and (ii) n is an odd, positive integer. If n is an even, positive integer, then the half-angle formulas

$$\sin^2 x = \frac{1 - \cos 2x}{2} \qquad \text{and} \qquad \cos^2 x = \frac{1 + \cos 2x}{2}$$

will reduce the degree of the integrand. You may have to apply these half-angle formulas more than once to get an integrand you know how to evaluate.

EXAMPLE 14.7: Evaluate $\int \sin^4 x\, dx$.

In this case n is even and

$$\int \sin^4 x\, dx = \int (\sin^2 x)^2\, dx = \int \left(\frac{1 - \cos 2x}{2}\right)^2 dx$$

$$= \frac{1}{4} \int (1 - 2 \cos 2x + \cos^2 2x)dx$$

$$= \frac{1}{4} \int dx - \frac{1}{2} \int \cos 2x\, dx + \frac{1}{4} \int \cos^2 2x\, dx$$

$$= \frac{x}{4} - \frac{1}{4} \sin 2x + \frac{1}{4} \int \left(\frac{1 + \cos 4x}{2}\right) dx$$

$$= \frac{x}{4} - \frac{1}{4} \sin 2x + \frac{1}{8} \int dx + \frac{1}{8} \int \cos 4x\, dx$$

$$= \frac{x}{4} - \frac{1}{4} \sin 2x + \frac{x}{8} + \frac{1}{32} \sin 4x + C$$

$$= \frac{3}{8} x - \frac{1}{4} \sin 2x + \frac{1}{32} \sin 4x + C .$$

Notice that we applied the half-angle reduction formula twice in completing this integration. ▌

If n is an odd, positive integer, then we can write $\sin^n x = (\sin^{n-1} x)\sin x$ or $\cos^n x = (\cos^{n-1} x)\cos x$. The identity

$\sin^2 x + \cos^2 x = 1$ then allows us to express $\sin^{n-1} x$ in terms of $\cos x$ or $\cos^{n-1} x$ in terms of $\sin x$.

EXAMPLE 14.8: Evaluate $\int \cos^7 x\, dx$.

In this example n is an odd number and

$$\int \cos^7 x\, dx = \int \cos^6 x \cos x\, dx = \int (\cos^2 x)^3 \cos x\, dx$$
$$= \int (1 - \sin^2 x)^3 \cos x\, dx$$
$$= \int (1 - 3\sin^2 x + 3\sin^4 x - \sin^6 x)\cos x\, dx$$
$$= \int \cos x\, dx - 3\int \sin^2 x \cos x\, dx + 3\int \sin^4 x \cos x\, dx - \int \sin^6 x \cos x\, dx$$
$$= \sin x - \sin^3 x + \frac{3}{5}\sin^5 x - \frac{1}{7}\sin^7 x + C$$

since, with exception of the first integral, the remaining integrals have the general form $\int u^n\, du$. ▌

CASE 3: $\int \tan^n x\, dx$

Again we must consider the two cases, n being an even, or an odd, positive integer. Both cases use the identity $\tan^2 x = \sec^2 x - 1$.

EXAMPLE 14.9: Evaluate $\int \tan^4 x\, dx$.

Since n is an even, positive integer, we have

$$\int \tan^4 x\, dx = \int \tan^2 x \tan^2 x\, dx = \int \tan^2 x (\sec^2 x - 1)\, dx$$
$$= \int \tan^2 x \sec^2 x\, dx - \int \tan^2 x\, dx\ .$$

We cannot integrate $\int \tan^2 x\, dx$, so use the identity $\tan^2 x = \sec^2 x - 1$ again:

$$\int \tan^4 x\, dx = \int \tan^2 x \sec^2 x\, dx - \int (\sec^2 x - 1)\, dx$$
$$= \int \tan^2 x\, (d \tan x) - \int \sec^2 x\, dx + \int dx$$
$$= \frac{1}{3}\tan^3 x - \tan x + x + C\ .$$ ▌

EXAMPLE 14.10: Evaluate $\int \tan^5 x\, dx$.

Here, with n an odd, positive integer, we write

$$\begin{aligned}\int \tan^5 x\, dx &= \int \tan^3 x \tan^2 x\, dx = \int \tan^3 x (\sec^2 x - 1)\, dx \\ &= \int \tan^3 x \sec^2 x\, dx - \int \tan^3 x\, dx \\ &= \int \tan^3 x \sec^2 x\, dx - \int \tan x \tan^2 x\, dx \\ &= \frac{1}{4} \tan^4 x - \int \tan x (\sec^2 x - 1)\, dx \\ &= \frac{1}{4} \tan^4 x - \int \tan x \sec^2 x\, dx + \int \tan x\, dx \\ &= \frac{1}{4} \tan^4 x - \frac{1}{2} \tan^2 x + \int \frac{\sin x}{\cos x}\, dx \\ &= \frac{1}{4} \tan^4 x - \frac{1}{2} \tan^2 x - \ln |\cos x| + C .\end{aligned}$$ ▌

CASE 4: $\int \sin^m x \cos^n x\, dx$

Again we must consider two cases: (i) At least one of m and n is an odd, positive integer while the other may be any number, and (ii) both m and n are even, positive integers. If m is an odd, positive integer, then we write $\sin^m x = (\sin^{m-1} x) \sin x$. Now, with $m - 1$ even, the identity $\sin^2 x = 1 - \cos^2 x$ transforms $\sin^{m-1} x$ into powers of cos x. This changes the integrand into a general form involving several terms of the form $\cos^p u\, d(\cos u)$. If n is the odd, positive integer, we write $\cos^n x = (\cos^{n-1} x) \cos x$. Now, with $n - 1$ even, the identity $\cos^2 x = 1 - \sin^2 x$ transforms $\cos^{n-1} x$ into powers of sin x thus changing the integrand into a form involving terms of the general form of $\sin^p u\, d(\sin u)$. This process is illustrated in the next example before consideration of the second case.

EXAMPLE 14.11: Evaluate $\int \sin^3 x \cos^4 x\, dx$.

The $\sin^3 x$ may be rewritten as $\sin^2 x \sin x$. Using $\sin^2 x = 1 - \cos^2 x$, the integral becomes

$$\begin{aligned}\int \sin^3 x \cos^4 x\, dx &= \int \sin^2 x \sin x \cos^4 x\, dx \\ &= \int (1 - \cos^2 x)\cos^4 x \sin x\, dx \\ &= \int (\cos^4 x - \cos^6 x)\sin x\, dx \\ &= \int \cos^4 x \sin x\, dx - \int \cos^6 x \sin x\, dx \\ &= -\frac{1}{5}\cos^5 x + \frac{1}{6}\cos^6 x + C\ . \quad \blacksquare\end{aligned}$$

<u>EXAMPLE 14.12</u>: Evaluate $\int \cos^5 x \sin^4 x\, dx$.

Rewrite $\cos^5 x$ as

$$\cos^4 x \cos x = (1 - \sin^2 x)^2 \cos x = (1 - 2\sin^2 x + \sin^4 x)\cos x$$

so

$$\begin{aligned}\int \cos^5 x \sin^4 x\, dx &= \int (1 - 2\sin^2 x + \sin^4 x)\sin^4 x \cos x\, dx \\ &= \int \sin^4 x \cos x\, dx - 2\int \sin^6 x \cos x\, dx \\ &\qquad + \int \sin^8 x \cos x\, dx \\ &= \frac{1}{5}\sin^5 x - \frac{2}{7}\sin^7 x + \frac{1}{9}\sin^9 x + C\ . \quad \blacksquare\end{aligned}$$

In the second case both m and n are even, positive integers. We can make use of the half-angle formulas

$$\sin^2 x = \frac{1 - \cos 2x}{2} \qquad \text{and} \qquad \cos^2 x = \frac{1 + \cos 2x}{2}$$

to reduce the degree of the integrand, as is now illustrated.

<u>EXAMPLE 14.13</u>: Evaluate $\int \sin^4 x \cos^2 x\, dx$.

We rewrite

$$\begin{aligned}\sin^4 x \cos^2 x &= \left(\frac{1 - \cos 2x}{2}\right)^2 \left(\frac{1 + \cos 2x}{2}\right) \\ &= \frac{1}{8}(1 - 2\cos 2x + \cos^2 2x)(1 + \cos 2x) \\ &= \frac{1}{8}(1 - \cos 2x - \cos^2 2x + \cos^3 2x)\ .\end{aligned}$$

Thus,

$$\int \sin^4 x \cos^2 x \, dx = \frac{1}{8}\int dx - \frac{1}{8}\int \cos 2x \, dx - \frac{1}{8}\int \cos^2 2x \, dx$$

$$+ \frac{1}{8}\int \cos^3 2x \, dx$$

$$= \frac{1}{8}\int dx - \frac{1}{16}\int \cos 2x(2dx) - \frac{1}{8}\int \left(\frac{1 + \cos 4x}{2}\right) dx$$

$$+ \frac{1}{8}\int (1 - \sin^2 2x)\cos 2x \, dx$$

$$= \frac{x}{8} - \frac{1}{16}\sin 2x - \frac{x}{16} - \frac{1}{64}\sin 4x + \frac{1}{16}\sin 2x - \frac{1}{48}\sin^3 2x + C$$

$$= \frac{x}{16} - \frac{1}{64}\sin 4x - \frac{1}{48}\sin^3 2x + C .$$

CASE 5: $\int \tan^m x \sec^n x \, dx$ and $\int \cot^m x \csc^n x \, dx$

Again we must consider two cases: (i) n is an even, positive integer with m being any number, and (ii) m is an odd, positive integer with any number n. When n is an even, positive integer, we write $\sec^n x = \sec^{n-2} x \sec^2 x$ and then use the identity $\sec^2 x = 1 + \tan^2 x$ to transform $\sec^{n-2} x$ into powers of tan x. For the cosecant form, we write $\csc^n x = \csc^{n-2} x \csc^2 x$ and use the identity $\csc^2 x = 1 + \cot^2 x$ to transform $\csc^{n-2} x$ into powers of the cot x. The second case will be examined after the next example.

EXAMPLE 14.14: Evaluate $\int \cot^{-5/2} x \csc^4 x \, dx$.

Observe that $\csc^4 x = \csc^2 x \csc^2 x = (1 + \cot^2 x)\csc^2 x$, so

$$\int \cot^{-5/2} x \csc^4 x \, dx = \int \cot^{-5/2} x(1 + \cot^2 x)\csc^2 x \, dx$$

$$= \int \cot^{-5/2} x \csc^2 x \, dx + \int \cot^{-1/2} x \csc^2 x \, dx$$

$$= -\frac{2}{3}\cot^{-3/2} x + 2\cot^{1/2} x + C .$$ ∎

In the second case when m is a positive, odd integer with any number n, m - 1 is even so we can write

$$\tan^m x \sec^n x = \tan^{m-1} x \sec^{n-1} \sec x \tan x$$

and then transform $\tan^{m-1} x$ to powers of sec x with the identity $\tan^2 x = \sec^2 x - 1$. In the case of the cosecant form, we write

$$\cot^m x \csc^n x = \cot^{m-1} x \csc^{n-1} \csc x \cot x$$

and transform $\cot^{m-1} x$ into powers of csc x using $\cot^2 x = \csc^2 x - 1$.

EXAMPLE 14.15: Evaluate $\int \tan^3 x \sec^{-1/3} x\,dx$.

Rewrite the integrand as

$$\begin{aligned}\tan^3 x \sec^{-1/3} x &= \tan^2 x \sec^{-4/3} x \tan x \sec x \\ &= (\sec^2 x - 1)\sec^{-4/3} x \tan x \sec x \\ &= (\sec^{2/3} x - \sec^{-4/3} x) \tan x \sec x\ .\end{aligned}$$

Thus,

$$\begin{aligned}&\int \tan^3 x \sec^{-1/3} x\,dx \\ &= \int \sec^{2/3} x(\tan x \sec x\,dx) - \int \sec^{-4/3} x(\tan x \sec x\,dx) \\ &= \frac{3}{5} \sec^{5/3} x + 3 \sec^{-1/3} x + C\ . \quad \blacksquare\end{aligned}$$

CASE 6: $\int \sin mx \cos nx\,dx$, $\int \sin mx \sin nx\,dx$, and $\int \cos mx \cos nx\,dx$

The above integrals pose no problem after the integrand has been transformed by the appropriate one of the following trigonometric identities.

$$\sin mx \cos nx = \frac{1}{2} [\sin(m + n)x + \sin(m - n)x] \qquad (14.3)$$

$$\sin mx \sin nx = -\frac{1}{2} [\cos(m + n)x - \cos(m - n)x] \qquad (14.4)$$

$$\cos mx \cos nx = \frac{1}{2} [\cos(m + n)x + \cos(m - n)x]\ . \qquad (14.5)$$

EXAMPLE 14.16: Evaluate $\int \sin 4x \cos 7x\,dx$.

In this case m = 4, n = 7 and from Equation 14.3 we have

$$\sin 4x \cos 7x = \frac{1}{2}[\sin(4+7)x + \sin(4-7)x]$$

$$= \frac{1}{2}[\sin 11x + \sin(-3x)] = \frac{1}{2}\sin 11x - \frac{1}{2}\sin 3x \ .$$

Thus,

$$\sin 4x \cos 7x \, dx = \frac{1}{2} \quad \sin 11x \, dx - \frac{1}{2} \quad \sin 3x \, dx$$

$$= -\frac{1}{22}\cos 11x + \frac{1}{6}\cos 3x + C \ . \ \blacksquare$$

EXERCISES

Integrate the following:

1. $\int \cos^4 x \, dx$
2. $\int \sin^3 6x \, dx$
3. $\int \sin^6 x \, dx$
4. $\int \sec^4 3x \, dx$
5. $\int \cot^6 t \, dt$
6. $\int \cos^3 x \, dx$
7. $\int \sin^5 x \, dx$
8. $\int \cot^4 4x \, dx$
9. $\int \tan^3 \theta \, d\theta$
10. $\int \cot^3 \theta \, d\theta$
11. $\int \csc^4(3x+1) \, dx$
12. $\int \sin^5 3t \cos^2 3t \, dt$
13. $\int \sin^4 \frac{u}{2} \cos^2 \frac{u}{2} \, du$
14. $\int \frac{\sec^4 x \, dx}{\cot^3 x}$
15. $\int \sin^2 x \cos^2 x \, dx$
16. $\int (\tan t + \cot t)^2 \, dt$
17. $\int \frac{dt}{\cot^4 t}$
18. $\int \tan 5z \sec^4 5z \, dz$
19. $\int \sin 3x \cos 4x \, dx$
20. $\int \sin 3x \sin 2x \, dx$
21. $\int \sin mx \sin nx \, dx$
22. $\int \cos^3 t \sqrt{\sin t} \, dt$
23. $\int \sin^{1/2} x \cos^3 x \, dx$
24. $\int \sin^4 3x \cos^4 3x \, dx$
25. $\int \tan^3 6t \sec^3 6t \, dt$
26. $\int \cot 2x \csc^3 2x \, dx$
27. $\int \tan^{-3} x \sec^2 x \, dx$
28. $\int \tan^5 t \sec^{-3/2} t \, dt$
29. $\int \sin 4x \cos 5x \, dx$
30. $\int \cos nx \cos mx \, dx$
31. $\int \frac{dx}{\sin 3x}$
32. $\int \frac{dx}{\cos 5x}$

33. $\int x \sin^3(x^2)\, dx$

34. $\int \frac{\sec^2 \sqrt{x}\, dx}{\sqrt{x} \tan \sqrt{x}}$

35. $\int \frac{\tan^3 x\, dx}{\sqrt[3]{\sec x}}$

36. $\int \frac{\tan^3 x}{\cos^2 x}\, dx$

37. $\int_{-\frac{\pi}{3}}^{\frac{\pi}{6}} \sin^2 x \cos x\, dx$

38. $\int_{\frac{\pi}{6}}^{\frac{\pi}{2}} \frac{\cos^3 x}{\sqrt{\sin x}}\, dx$

39. $\int_{\frac{\pi}{6}}^{\frac{\pi}{3}} \cot^2 2x \sec^2 2x\, dx$

40. $\int_0^{\frac{\pi}{3}} \tan^3 x \sec x\, dx$

41. $\int \tan \frac{x}{3} \sec^3 \frac{x}{3}\, dx$

42. $\int \frac{\sin^4 x}{\cos^2 x}\, dx$

43. $\int_{\frac{\pi}{6}}^{\frac{\pi}{4}} \cot^3 \theta\, d\theta$

44. $\int_0^{\frac{\pi}{4}} \sec^6 t\, dt$

45. $\int_{\frac{\pi}{6}}^{\frac{\pi}{3}} \frac{\sin^2 t}{\cot t}\, dt$

14.3 INTEGRATION BY TRIGONOMETRIC SUBSTITUTION

We will now illustrate the use of trigonometric substitutions in handling integrands of functions involving one or more of the following forms: $\sqrt{a^2 - x^2}$, $\sqrt{a^2 + x^2}$, $\sqrt{x^2 - a^2}$, $a^2 - x^2$, $a^2 + x^2$ or $x^2 - a^2$. These quadratic quantities can often be reduced to an integrable form by using a trigonometric substitution based on the Pythagorean theorem. The three substitutions are:

1. When $\sqrt{a^2 - x^2}$ occurs, let $x = a \sin \theta$, $-\frac{\pi}{2} \leq \theta \leq \frac{\pi}{2}$.

2. When $\sqrt{a^2 + x^2}$ occurs, let $x = a \tan \theta$, $-\frac{\pi}{2} \leq \theta \leq \frac{\pi}{2}$.

3. When $\sqrt{x^2 - a^2}$ occurs, let $x = a \sec \theta$, $0 \leq \theta < \frac{\pi}{2}$ or $-\pi \leq \theta < -\frac{\pi}{2}$.

We assume that $a > 0$ in all cases; each of the cases will be examined separately.

CASE 1: $\sqrt{a^2 - x^2}$

In considering this case, examine the labeled right triangle shown in Figure 14.1. The triangle has a hypotenuse of length a and one side of

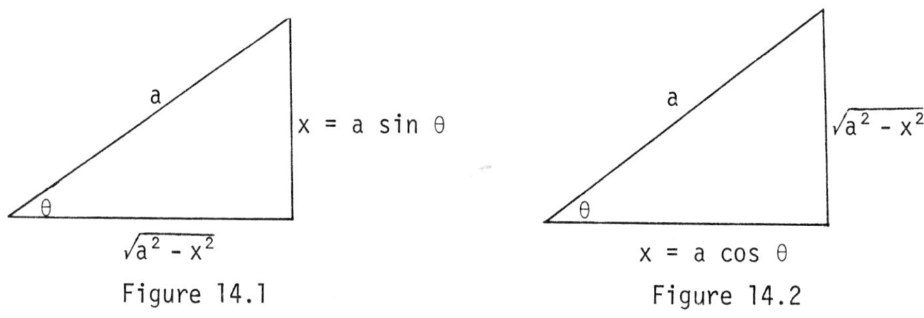

Figure 14.1

Figure 14.2

length x. Apply the Pythagorean theorem to find that the length of the remaining side equals $\sqrt{a^2 - x^2}$. If θ is the angle shown in Figure 14.1, $x = a \sin \theta$ and $\sqrt{a^2 - x^2} = a \cos \theta$. The latter equality should be apparent from the geometry of the situation, but it can be established algebraically. Because $x = a \sin \theta$ and since $1 - \sin^2 \theta = \cos^2 \theta$,

$$\sqrt{a^2 - x^2} = \sqrt{a^2 - a^2 \sin^2 \theta} = \sqrt{a^2(1 - \sin^2 \theta)}$$

$$= \sqrt{a^2 \cos^2 \theta} = a \cos \theta .$$

A related reduction occurs by letting $x = a \cos \theta$ as shown in Figure 14.2. If $x = a \cos \theta$, then

$$\sqrt{a^2 - x^2} = \sqrt{a^2 - a^2 \cos^2 \theta} = \sqrt{a^2(1 - \cos^2 \theta)} = \sqrt{a^2 \sin^2 \theta} = a \sin \theta$$

since $1 - \cos^2\theta = \sin^2\theta$.

EXAMPLE 14.17: Evaluate $\int \frac{dx}{\sqrt{a^2 - x^2}}$.

We know from previous experience that this integral equals $\text{Sin}^{-1}\frac{x}{a} + C$. Using the method of trigonometric substitution with $x = a\sin\theta$, $dx = a\cos\theta\, d\theta$ and $\sqrt{a^2 - x^2} = \sqrt{a^2 - a^2\sin^2\theta} = \sqrt{a^2(1 - \sin^2\theta)} = \sqrt{a^2\cos^2\theta} = a\cos\theta$. The integral becomes

$$\int \frac{dx}{\sqrt{a^2 - x^2}} = \int \frac{a\cos\theta\, d\theta}{a\cos\theta} = \int d\theta = \theta + C .$$

Since $x = a\sin\theta$, $\theta = \text{Sin}^{-1}\frac{x}{a}$ so

$$\int \frac{dx}{\sqrt{a^2 - x^2}} = \text{Sin}^{-1}\frac{x}{a} + C ,$$

the expected result because $\frac{d}{dx}\text{Sin}^{-1}\frac{x}{a} = \frac{1}{\sqrt{a^2 - x^2}}$. ∎

EXAMPLE 14.18: Evaluate $\int \frac{\sqrt{a^2 - x^2}}{x^2}\, dx$.

It should be apparent that this integrand does not fit into any of the standard forms we have studied previously. The integrand contains the quantity $\sqrt{a^2 - x^2}$ which suggests we should use the trigonometric substitution of $x = a\sin\theta$. If $x = a\sin\theta$ as in Figure 14.1, then $dx = a\cos\theta\, d\theta$ and

$$\sqrt{a^2 - x^2} = \sqrt{a^2 - a^2\sin^2\theta} = \sqrt{a^2(1 - \sin^2\theta)} = \sqrt{a^2\cos^2\theta} = a\cos\theta .$$

Substituting the above quantities into the integrand, we have

$$\int \frac{\sqrt{a^2 - x^2}}{x^2}\, dx = \int \frac{(a\cos\theta)(a\cos\theta\, d\theta)}{a^2\sin^2\theta} = \int \frac{\cos^2\theta}{\sin^2\theta}\, d\theta$$

$$= \int \cot^2\theta\, d\theta .$$

This does not yet integrate simply. However we can simplify it by using the trigonometric identity $\cot^2\theta = \csc^2\theta - 1$:

$$\int \cot^2\theta\, d\theta = \int (\csc^2\theta - 1)d\theta = \int \csc^2\theta\, d\theta - \int d\theta$$

$$= \cot\theta - \theta + C .$$

We must now convert this result back to the original variable x. From Figure 14.1, observe that $\cot \theta = \frac{\sqrt{a^2 - x^2}}{x}$ and $\theta = \text{Sin}^{-1} \frac{x}{a}$ so

$$\int \frac{\sqrt{a^2 - x^2}}{x^2}\, dx = \frac{\sqrt{a^2 - x^2}}{x} - \text{Sin}^{-1} \frac{x}{a} + C .$$

To convince yourself that this really is the correct solution, take its derivative and simplify to get the integrand. This example points out again a major use of trigonometry in the study of calculus. ▌

CASE 2: $\sqrt{a^2 + x^2}$

In considering this case, examine the labeled right triangle shown in Figure 14.3. The triangle has two sides of length a and x and a hypotenuse of length $\sqrt{x^2 + a^2}$. If $x = a \tan \theta$, then because $\tan^2 \theta + 1 = \sec^2 \theta$,

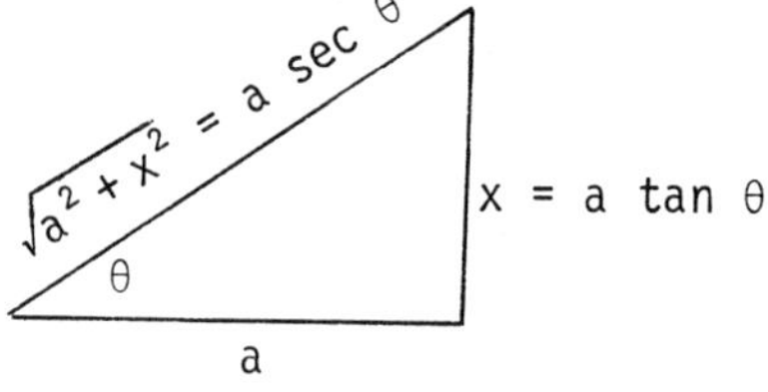

Figure 14.3

$$\sqrt{x^2 + a^2} = \sqrt{a^2 \tan^2 \theta + a^2} = \sqrt{a^2(\tan^2 \theta + 1)}$$

$$= \sqrt{a^2 \sec^2 \theta} = a \sec \theta .$$

In a manner that was illustrated earlier, we could also have let $x = a \cot \theta$ and found $\sqrt{x^2 + a^2} = a \csc \theta$.

EXAMPLE 14.19: Evaluate $\int \frac{dx}{a^2 + x^2}$.

From previous experience, we know that this integral equals $\frac{1}{a} \text{Tan}^{-1} \frac{x}{a} + C$. To evaluate this integral by the use of a trigonometric substitution, let $x = a \tan \theta$, so $dx = a \sec^2 \theta\, d\theta$ and $a^2 + x^2 = a^2(1 + \tan^2 \theta) = a^2 \sec^2 \theta$ and so

$$\int \frac{dx}{a^2 + x^2} = \int \frac{a \sec^2 \theta\, d\theta}{a^2 \sec^2 \theta} = \frac{1}{a} \int d\theta = \frac{1}{a} \theta + C .$$

From $\frac{x}{a} = \tan \theta$, we have $\theta = \text{Tan}^{-1} \frac{x}{a}$ and, substituting this quantity in for θ, we arrive at the final result which was expected of

$$\int \frac{dx}{a^2 + x^2} = \frac{1}{a} \operatorname{Tan}^{-1} \frac{x}{a} + C . \quad \blacksquare$$

EXAMPLE 14.20: Evaluate $\int \frac{x^2dx}{(a^2 + x^2)^{5/2}}$.

The integrand does not fit one of the standard forms which we can integrate directly. However, the integrand contains the expression $a^2 + x^2$ which suggests the trigonometric substitution $x = a \tan \theta$. If $x = a \tan \theta$, then $dx = a \sec^2 \theta \, d\theta$ and

$$(a^2 + x^2)^{5/2} = (a^2 + a^2 \tan^2 \theta)^{5/2} = (a^2(1 + \tan^2 \theta))^{5/2}$$
$$= (a^2 \sec^2 \theta)^{5/2} = a^5 \sec^5 \theta .$$

Thus,

$$\int \frac{x^2dx}{(a^2 + x^2)^{5/2}} = \int \frac{(a \tan \theta)^2(a \sec^2 \theta \, d\theta)}{a^5 \sec^5 \theta} = \frac{1}{a^2} \int \frac{\tan^2 \theta}{\sec^3 \theta} d\theta$$
$$= \frac{1}{a^2} \int \frac{\sin^2 \theta}{\cos^2 \theta} \cos^3 \theta \, d\theta = \frac{1}{a^2} \int \sin^2 \theta \cos \theta \, d\theta$$
$$= \frac{1}{3a^2} \sin^3 \theta + C .$$

We now need to express this result in terms of the original variable x. From Figure 14.3 we see that $\sin \theta = \frac{x}{\sqrt{a^2 + x^2}}$ so

$$\int \frac{x^2dx}{(a^2 + x^2)^{5/2}} = \frac{1}{3a^2} \left[\frac{x}{\sqrt{a^2 + x^2}}\right]^3 + C = \frac{1}{3a^2} \frac{x^3}{(a^2 + x^2)^{3/2}} + C .$$

As a check, we can calculate the derivative of our answer and observe that the derivative of the right-hand side should be equal to the integrand. Thus,

$$\frac{d}{dx}\left[\frac{x^3}{3a^2(a^2 + x^2)^{3/2}}\right] + C = \frac{1}{3a^2}\left[\frac{(a^2 + x^2)^{3/2} \frac{d}{dx} x^3 - x^3 \frac{d}{dx} (a^2 + x^2)^{3/2}}{(a^2 + x^2)^3}\right] + 0$$
$$= \frac{(a^2 + x^2)^{3/2} \, 3x^2 - x^3 \left(\frac{3}{2}\right) (a^2 + x^2)^{1/2}(2x)}{3a^2(a^2 + x^2)^3}$$
$$= \frac{3x^2(a^2 + x^2)^{1/2}(a^2 + x^2 - x^2)}{3a^2(a^2 + x^2)^3} = \frac{x^2}{(a^2 + x^2)^{5/2}} .$$

Thus we should be convinced that we have integrated properly. ▌

CASE 3: $\sqrt{x^2 - a^2}$

In considering this case, refer to a right triangle which is labeled as shown in Figure 14.4. Now following the same approach as we used in cases 1 and 2, let $x = a \sec \theta$ so

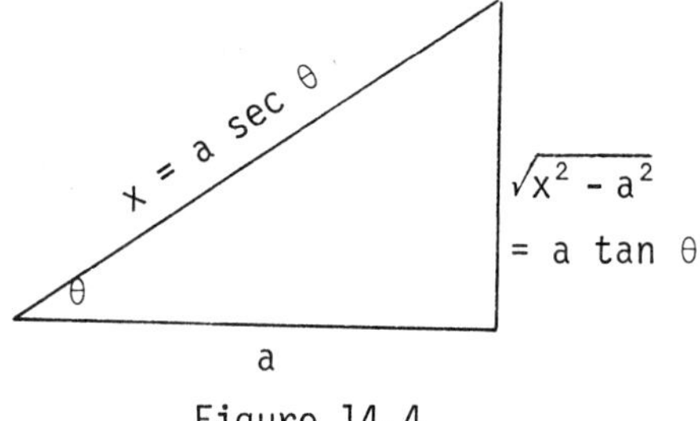

Figure 14.4

$$\sqrt{x^2 - a^2} = \sqrt{a^2 \sec^2 \theta - a^2} = \sqrt{a^2(\sec^2 \theta - 1)}$$
$$= \sqrt{a^2 \tan^2 \theta} = a \tan \theta \;.$$

In a similar manner, we could let $x = a \csc \theta$ for then

$$\sqrt{x^2 - a^2} = \sqrt{a^2 \csc^2 \theta - a^2} = \sqrt{a^2(\csc^2 \theta - 1)} = \sqrt{a^2 \cot^2 \theta} = a \cot \theta \;.$$

EXAMPLE 14.21: Evaluate $\int \frac{du}{\sqrt{4u^2 - 25}}$

The integrand is not one with which we are familiar. However, it does contain a term of the type $\sqrt{x^2 - a^2}$, suggesting a trigonometric substitution of $x = a \sec \theta$. The substitution $2u = 5 \sec \theta$ has $du = \frac{5}{2} \sec \theta \tan \theta \, d\theta$ and

$$\sqrt{4u^2 - 25} = \sqrt{25 \sec^2 \theta - 25} = \sqrt{25(\sec^2 \theta - 1)} = \sqrt{25 \tan^2 \theta} = 5 \tan^2 \theta.$$

Thus, it follows that

$$\int \frac{du}{\sqrt{4u^2 - 25}} = \int \frac{\frac{5}{2} \sec \theta \tan \theta \, d\theta}{5 \tan \theta} = \frac{1}{2} \int \sec \theta \, d\theta$$
$$= \frac{1}{2} \ln |\sec \theta + \tan \theta| + C \;.$$

To express this answer in terms of the original variable u, see Figure 14.5. It has $2u = 5 \sec \theta$. Applying the definitions of the trigonometric functions, we find $\sec \theta = \frac{2u}{5}$ and $\tan \theta = \frac{\sqrt{4u^2 - 25}}{5}$. Substituting these values we finally arrive at

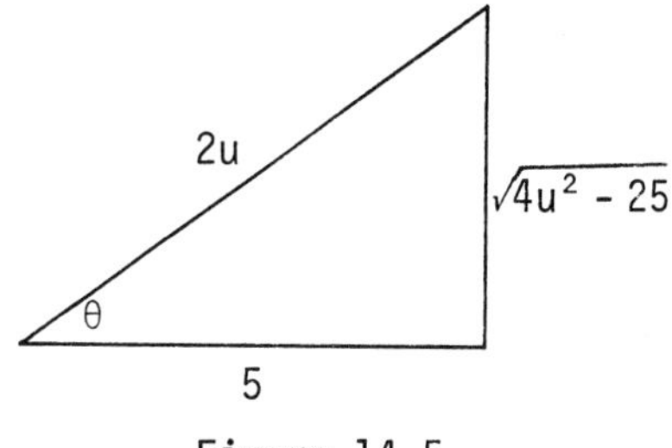

Figure 14.5

$$\int \frac{du}{\sqrt{4u^2 - 25}} = \frac{1}{2} \ln \left| \frac{2u}{5} + \frac{\sqrt{4u - 25}}{5} \right| + C . \blacksquare$$

The results of this section enable you to evaluate integrals whose integrands involve certain powers of the following forms: $a^2 - x^2$, $a^2 + x^2$, $x^2 - a^2$. While this technique does not always work for integrands involving the forms listed above, it is one of the easiest to use and, hence, should be tried first; if the integral still cannot be evaluated, look for another technique.

EXERCISES

Integrate the following:

1. $\int \frac{x dx}{\sqrt{16 - x^2}}$

2. $\int \frac{dt}{9 - t^2}$

3. $\int \frac{dz}{\sqrt{4z^2 - 1}}$

4. $\int \frac{du}{(a^2 + u^2)^2}$

5. $\int_1^{\sqrt{3}} \frac{dx}{(4 - x^2)^{3/2}}$

6. $\int \frac{dt}{\sqrt{3t^2 - 16}}$

7. $\int \frac{\sqrt{4 - 2t^2}}{t^2}$

8. $\int x^2 \sqrt{9 - x^2}\, dx$

9. $\int \frac{3x^2 dx}{\sqrt{16 - x^2}}$

10. $\int \frac{\sqrt{x^2 - 25}}{x^2}$

11. $\int \frac{\sqrt{49 - x^2}}{x} dx$

12. $\int \sqrt{x^2 + 1}\, dx$

13. $\int \frac{dx}{x\sqrt{x^2 - 4}}$

14. $\int_{2\sqrt{3}}^{6} \frac{x^3 dx}{\sqrt{x^2 - 9}}$

15. $\int \frac{x^2 dx}{(x^2 + 16)^{3/2}}$

16. $\int \frac{e^{-x} dx}{(9 + 4e^{-2x})^{3/2}}$

17. $\int \frac{dz}{\sqrt{4z^2 + 1}}$

18. $\int \frac{(3x - 2)dx}{\sqrt{4 - x^2}}$

19. $\int_{-6}^{-3} \frac{\sqrt{x^2 - 9}}{x} dx$

20. $\int \frac{x^2 dx}{\sqrt{8 + 2x - x^2}}$

21. $\int \frac{dx}{x^2\sqrt{x^2 - 1}}$

22. $\int \frac{dx}{x\sqrt{(\ln x)^2 + 25}}$

23. $\int \frac{\cos 2x \, dx}{\sqrt{16 + \sin^2 2x}}$

24. $\int \frac{e^t dt}{(4e^{2t} + 49)^{1/2}}$

25. $\int \frac{(3 - 6x)dx}{(25 + 4x \;)^{5/2}}$

26. $\int u^3 \sqrt{a^2u^2 + b^2} \, du$

27. $\int \frac{\sin x \, dx}{\sqrt{\cos^2 x + 4}}$

28. $\int \frac{2e^{2t} dt}{(9e^{2t} + 49)^{1/2}}$

14.4 INTEGRANDS INVOLVING QUADRATICS

This section deals with integrals of the general type $\int \frac{(Ax + B)dx}{(ax^2 + bx + C)^n}$ where $a \neq 0$. We will also assume that the quadratic expression cannot be factored into real linear factors. If the denominator can be factored, then the integral should be approached using partial fractions, the method discussed in the next section.

When a quadratic expression, $ax^2 + bx + c$, appears in an integrand, the integration often can be simplified by completing the square on the quadratic. Thus,

$$ax^2 + bx + c = a\left[\left(x + \frac{b}{2a}\right)^2 + \left(\frac{4ac - b^2}{(2a)^2}\right)\right] = a[X^2 + K^2] \qquad (14.6)$$

by completing the square. Depending upon the signs of a and K^2, this process changes the integrand into a form involving $a^2 - x^2$, $a^2 + x^2$ or

$x^2 - a^2$. Then the methods of Section 14.3 may be applied to complete the integration.

If you feel that you do not remember the process of completing the square, refer to Example 4.28 in Section 4.4 for a review.

EXAMPLE 14.22: Evaluate $\int \frac{dx}{7 + 12x - 4x^2}$.

As it stands this integral has no recognizable standard form. If we complete the square in the denominator, we may be able to transform the integrand into some recognizable standard form.

$$-4x^2 + 12x + 7 = -4\left[x^2 - 3x - \frac{7}{4}\right] = -4\left[x^2 - 3x + \frac{9}{4} - \frac{9}{4} - \frac{7}{4}\right]$$

$$= -4\left[\left(x - \frac{3}{2}\right)^2 - 4\right] = 16 - (2x - 3)^2 .$$

Now

$$\int \frac{dx}{7 + 12x - 4x^2} = \int \frac{dx}{16 - (2x - 3)^2}$$

This last integral has the general form $\int \frac{du}{a^2 - u^2}$, suggesting the trigonometric substitution $u = a \sin \theta$. In this case $u = 2x - 3$ and $a = 4$. If we let $2x - 3 = 4 \sin \theta$, then $dx = 2 \cos \theta \, d\theta$ and

$$16 - (2x - 3)^2 = 16 - 16 \sin^2 \theta = 16(1 - \sin^2 \theta) = 16 \cos^2 \theta .$$

So

$$\int \frac{dx}{16 - (2x - 3)^2} = \int \frac{2 \cos \theta \, d\theta}{16 \cos^2 \theta} = \frac{1}{8} \int \sec \theta \, d\theta$$

$$= \frac{1}{8} \ln |\sec \theta + \tan \theta| + C .$$

In Figure 14.6 $\sin \theta = \frac{2x - 3}{4}$, so

$\sec \theta = \frac{4}{\sqrt{16 - (2x - 3)^2}}$ and

$\tan \theta = \frac{2x - 3}{\sqrt{16 - (2x - 3)^2}}$. Thus,

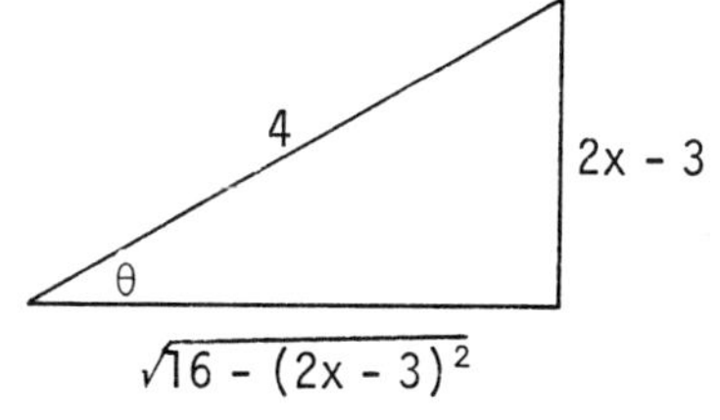

Figure 14.6

$$\int \frac{dx}{7 + 12x - 4x^2} = \frac{1}{8} \ln \left| \frac{4}{\sqrt{16 - (2x - 3)^2}} + \frac{2x - 3}{\sqrt{16 - (2x - 3)^2}} \right| + C$$

$$= \frac{1}{8} \ln \left| \frac{1 + 2x}{\sqrt{16 - (2x - 3)^2}} \right| + C \ . \ \blacksquare$$

<u>EXAMPLE 14.23</u>: Evaluate $\int \frac{dx}{5x^2 + 30x + 55}$.

Recognizing that this integrand is one which we are unable to integrate, we must resort to some other method.

$$5x^2 + 30x + 55 = 5(x^2 + 6x + 11) = 5[(x^2 + 6x + 9) - 9 + 11]$$

$$= 5[(x + 3)^2 + 2]$$

so,

$$\int \frac{dx}{5x^2 + 30x + 55} = \frac{1}{5} \int \frac{dx}{(x + 3)^2 + 2} = \frac{1}{5\sqrt{2}} \text{Tan}^{-1} \frac{x + 3}{\sqrt{2}} + C$$

because we recognize the form $\int \frac{du}{u^2 + a^2} = \frac{1}{a} \text{Tan}^{-1} \frac{u}{a} + C$. ▮

To evaluate certain other types of integrals of the general form of $\int \frac{(Ax + B)dx}{ax^2 + bx + c}$ where $a \neq 0$, we first express the integrand as the sum of two simpler fractions, each having the given denominator. Then we can adjust the constants so each assumes a standard form. An example will help clarify this procedure.

<u>EXAMPLE 14.24</u>: Evaluate $\int \frac{(3x + 7)dx}{4x^2 + 9}$.

This denominator has the form $a^2 + u^2$, but the quantity $3x + 7$ appears in the numerator. In such cases it is advantageous to rewrite the integrand in the following manner:

$$\int \frac{(3x + 7)dx}{4x^2 + 9} = 3 \int \frac{xdx}{4x^2 + 9} + 7 \int \frac{dx}{4x^2 + 9} \ .$$

Examining the first integral, we see that if $u = 4x^2 + 9$ then $du = 8x\ dx$ and that upon multiplying and dividing by 8 the integral would reduce to the form $\int \frac{du}{u}$. The second integral has the general form $\int \frac{du}{u^2 + a^2}$ where $u = 2x$ and $a = 3$. Thus,

$$\int \frac{(3x + 7)}{4x^2 + 9}\,dx = \frac{3}{8}\int \frac{8x\,dx}{4x^2 + 9} + \frac{7}{2}\int \frac{2\,dx}{4x^2 + 9}$$

$$= \frac{3}{8}\ln|4x^2 + 9| + \frac{7}{6}\,\mathrm{Tan}^{-1}\frac{2x}{3} + C\ .$$

By splitting the integrand into two parts, we got two standard forms which we could integrate. ▮

Next we illustrate a slight variation of the method outlined in Example 14.24.

EXAMPLE 14.25: Evaluate $\int \frac{xdx}{x^2 + 2x + 5}$.

If we let $u = x^2 + 2x + 5$, then $du = (2x + 2)dx$. Now rewriting the numerator as $\frac{(2x + 2 - 2)dx}{2}$, we are able to divide the given integral into two integrals as follows:

$$\int \frac{xdx}{x^2 + 2x + 5} = \frac{1}{2}\int \frac{(2x + 2)dx}{x^2 + 2x + 5} - \frac{1}{2}\int \frac{2dx}{x^2 + 2x + 5}\ .$$

The first integral has the form $\int \frac{du}{u}$, and by completing the square, the last integral has the form $\int \frac{du}{u^2 + a^2}$. Thus

$$\int \frac{xdx}{x^2 + 2x + 5} = \frac{1}{2}\int \frac{(2x + 2)dx}{x^2 + 2x + 5} - \int \frac{dx}{(x + 1)^2 + 4}$$

$$= \frac{1}{2}\ln|x^2 + 2x + 5| - \frac{1}{2}\,\mathrm{Tan}^{-1}\left[\frac{x + 1}{2}\right] + C\ . \qquad (14.7)$$

An alternate method would have been to initially complete the square and then use a trigonometric substitution:

$$\int \frac{xdx}{x^2 + 2x + 5} = \int \frac{xdx}{(x + 1)^2 + 4}\ .$$

Let $x + 1 = 2\tan\theta$, so $dx = 2\sec^2\theta\,d\theta$,
$(x + 1)^2 + 4 = 4\tan^2\theta + 4 = 4(\tan^2\theta + 1) = 4\sec^2\theta$ and
$x = 2\tan\theta - 1$:

$$\int \frac{xdx}{(x + 1)^2 + 4} = \int \frac{(2\tan\theta - 1)(2\sec^2\theta\,d\theta)}{4\sec^2\theta} = \int (\tan\theta - \tfrac{1}{2})\,d\theta$$

$$= \int \frac{\sin\theta}{\cos\theta}\,d\theta - \frac{1}{2}\int d\theta = -\ln|\cos\theta| - \frac{\theta}{2} + C\ .$$

Since $x + 1 = 2 \tan \theta$, we find that $\theta = \mathrm{Tan}^{-1} \frac{x + 1}{2}$ and

$\cos \theta = \dfrac{2}{\sqrt{(x + 1)^2 + 4}}$ and

$$\int \frac{xdx}{(x + 1)^2 + 4} = -\ln \left| \frac{2}{\sqrt{(x + 1)^2 + 4}} \right| - \frac{1}{2} \mathrm{Tan}^{-1}\left[\frac{x + 1}{2}\right] + C \, . \qquad (14.8)$$

Do Equations 14.7 and 14.8 really give the same result? The latter one can be written as

$$-\ln \left| \frac{2}{\sqrt{(x + 1)^2 + 4}} \right| - \frac{1}{2} \mathrm{Tan}^{-1}\left[\frac{x + 1}{2}\right] + C$$

$$= -\ln 2 + \frac{1}{2} \ln |(x + 1)^2 + 4| - \frac{1}{2} \mathrm{Tan}^{-1}\left[\frac{x + 1}{2}\right] + C$$

$$= \frac{1}{2} \ln |(x + 1)^2 + 4| - \frac{1}{2} \mathrm{Tan}^{-1}\left[\frac{x + 1}{2}\right] + C_1$$

where $C_1 = C - \ln 2$. Thus both forms do give the same result.

It may be unclear whether to divide the integrand into parts, or to use some other technique. This really depends upon the problem and the person doing the integration. This technique of separation has been presented as a tool to help solve certain types of integration problems. Of course, another method may work equally well on a particular integral. As we gain more and more techniques, you should have some systematic method of examining the integrals. A logical system would be to rank the techniques from easiest to most difficult and always start with the easiest technique and work your way down the list until you find a technique which will work. Again one individual's ordering of the techniques may be entirely different from another's. ▌

EXERCISES:

Find the term that must be added and subtracted from the following in order to complete the square.

1. $x^2 - 2x - 24$
2. $y^2 - 6y + 7$
3. $7z^2 - 11z - 6$
4. $x^2 + 10x + 19$
5. $6t^2 + 11t - 10$
6. $8 - 6x - 9x^2$

Integrate the following by any method you have had thus far.

7. $\int \frac{(x - 1)dx}{3x^2 - 2x + 5}$

8. $\int \frac{dx}{x^2 - 6x - 5}$

9. $\int \frac{dx}{3x^2 - 2x + 4}$

10. $\int \frac{xdx}{x^2 - x + 1}$

11. $\int \frac{dx}{\sqrt{3x^2 - 2x + 5}}$

12. $\int \frac{dx}{\sqrt{x^2 + 9}}$

13. $\int \frac{dx}{4x^2 - 4x}$

14. $\int \frac{\sec^2 x dx}{\tan^2 x + 4 \tan x + 3}$

15. $\int \frac{(6x - 5)dx}{\sqrt{5x - x^2 - 6}}$

16. $\int \frac{(x - 1)dx}{\sqrt{x^2 - x + 1}}$

17. $\int \frac{(8x + 5)dx}{\sqrt{7 - x^2 + 6x}}$

18. $\int \frac{(6u + 8)du}{u^2 - 4}$

19. $\int \frac{(3t + 5)dt}{\sqrt{9 - t^2}}$

20. $\int \frac{(13 - 5x)dx}{x^2 + 3x + 2}$

21. $\int \frac{(2x + 5)dx}{\sqrt{8 - 6x - 9x^2}}$

22. $\int \frac{(4 - 8y)dy}{\sqrt{1 - 16y^2}}$

23. $\int \frac{(x + 1)dx}{\sqrt{x^2 - x + 2}}$

24. $\int \frac{(x^2 + 1)dx}{x - 1}$ (Hint: Divide first.)

14.5 USE OF PARTIAL FRACTIONS FOR INTEGRATING RATIONAL FUNCTIONS

The last section dealt with completing the square in the denominator of a fraction; this section concentrates on those fractions whose denominator can be written as a product of simple polynomials. When a fraction can be decomposed into a sum of several simpler fractions, these fractions are called partial fractions. An integrand which appears difficult may be simple to integrate if it is decomposed into partial fractions. For example, we encountered $\frac{w_m}{w(w_m - w)}$ as an integrand in developing the Logistic Growth Curve in Section 8.10. There we observed that

$$\frac{w_m}{w(w_m - w)} = \frac{1}{w} + \frac{1}{w_m - w}, \tag{14.9}$$

so

$$\int \frac{w_m \, dw}{w(w_m - w)} = \int \left(\frac{1}{w} + \frac{1}{w_m - w}\right) dw = \ln w - \ln (w_m - w) + C . \tag{14.10}$$

Equation 14.9 displays a decomposition into partial fractions while Equation 14.10 shows the effect of the decomposition on the integration.

A RATIONAL FUNCTION can be expressed as a quotient of two polynomial functions; if N(x) and D(x) are polynomials in x, the fraction N(x)/D(x) is called a rational function. It is called a PROPER FRACTION if and only if the highest power of the polynomial of N(x) is less than the highest power of the polynomial of D(x), otherwise N(x)/D(x) is said to be an IMPROPER FRACTION. An improper fraction can be expressed as the sum of a polynomial and a proper fraction by dividing N(x) by D(x) until the remainder is a proper fraction. This is illustrated in the next example.

EXAMPLE 14.26: Reduce the improper fraction $\frac{4x^3 + 6}{2x^2 + x}$ to the sum of a polynomial and a proper fraction.

The fraction can be reduced by dividing $2x^2 + x$ into $4x^3 + 6$:

$$\frac{4x^3 + 6}{2x^2 + x} = 2x - 1 + \frac{x + 6}{2x^2 + x} . \quad \blacksquare$$

In the last example the improper fraction was expressed as a sum of a polynomial and a proper fraction. We will assume throughout this section that rational integrands are proper fractions; if they are not, divide the denominator into the numerator before using the methods described here.

Partial fractions merely reverse the process of finding a common denominator as they decompose a fraction N(x)/D(x) into a sum of proper fractions. We will consider only these three cases for D(x):

1. Nonrepeated linear factors.
2. Repeated linear factors.
3. Quadratic factors.

14.5a CASE 1 FOR PARTIAL FRACTIONS: NONREPEATED LINEAR FACTORS

In Case 1 we want to resolve $N(x)/D(x)$ into partial fractions when $D(x)$ is a product of distinct linear factors. Each factor of $D(x)$ has the general form $(ax + b)$ and generates a partial fraction of the form $\frac{A}{ax + b}$. The partial fraction must take this form or else $N(x)/D(x)$ would not be a proper fraction. The next example illustrates this idea.

<u>EXAMPLE 14.27</u>: Evaluate $\int \frac{8x^2 + 17x - 3}{(3x + 1)(x + 1)(x - 2)}\, dx$.

To evaluate this integrand, we resort to partial fractions. The integrand is a proper fraction containing the three distinct linear factors $3x + 1$, $x + 1$ and $x - 2$. Thus, we may write

$$\frac{8x^2 + 17x - 3}{(3x + 1)(x + 1)(x - 2)} = \frac{A}{3x + 1} + \frac{B}{x + 1} + \frac{C}{x - 2} . \qquad (14.11)$$

Multiply both sides of this equation by $(3x + 1)(x + 1)(x - 2)$ to get

$$8x^2 + 17x - 3$$
$$= A(x + 1)(x - 2) + B(3x + 1)(x - 2) + C(3x + 1)(x + 1) \qquad (14.12)$$

The points $x = -1, 2, -\frac{1}{3}$ could cause problems because they correspond to multiplying both sides of Equation 14.11 by zero. The fact that the result, Equation 14.12, is a polynomial removes this concern. If a polynomial equation holds true for all but a finite number of values of x, then it must hold for all values of x. The potentially troublesome points now become very valuable in determining the constants A, B and C in Equation 14.12:

at $x = -1$ Equation 14.12 becomes $-12 = A(0) + 6B + C(0)$ or $B = -2$,

at $x = 2$ Equation 14.12 becomes $63 = A(0) + B(0) + 21C$ or $C = 3$, and

at $x = -\frac{1}{3}$ Equation 14.12 becomes $-\frac{70}{9} = -\frac{14}{9}A + B(0) + C(0)$ or $A = 5$.

Now substituting these values back into Equation 14.11:

$$\frac{8x^2 + 17x - 3}{(3x + 1)(x + 1)(x - 2)} = \frac{5}{3x + 1} - \frac{2}{x + 1} + \frac{3}{x - 2} . \qquad (14.13)$$

As a check, if a common denominator were obtained on the right-hand side of Equation 14.13, the result should be the left-hand side, which indeed it is. Now we are able to evaluate the original integral:

$$\int \frac{(8x^2 + 17x - 3)dx}{(3x + 1)(x + 1)(x - 2)} = \int \frac{5dx}{3x + 1} - \int \frac{2dx}{x + 1} + \int \frac{3dx}{x - 2}$$

$$= \frac{5}{3} \ln|3x + 1| - 2 \ln|x + 1| + 3 \ln|x - 2| + C .$$

Notice how easily this function was integrated by the use of partial fractions; it would be difficult, if not impossible, to integrate by any other previously introduced technique. ▌

The decomposition for Case 1 can be done another way. This second approach serves only as an illustration here because the method outlined above is easier to use for Case 1. However, the alternate method will become important in Cases 2 and 3. It is based on the following theorem, given here without proof.

THEOREM 14.1: If each of two polynomials in x is of degree n at most, and, if the polynomials are equal for more than n distinct values of x, then the polynomials are equal term by term and, hence, are equal for all values of x.

EXAMPLE 14.27 (continued):

Previously we found that

$$8x^2 + 17x - 3 = A(x + 1)(x - 2) + B(3x + 1)(x - 2) + C(3x + 1)(x + 1) .$$

Multiply out the right-hand side and collect like terms:

$$8x^2 + 17x - 3 = (A + 3B + 3C)x^2 + (-A - 5B + 4C)x + (-2A - 2B + C) .$$

Now apply Theorem 14.1.

Equating the coefficients of x^2: $8 = A + 3B + 3C$

Equating the coefficients of x: $17 = -A - 5B + 4C$

Equating the constant terms: $-3 = -2A - 2B + C$

This system of three equations in three unknowns has the solution $A = 5$, $B = -2$ and $C = 3$, the same values we obtained earlier. ▌

EXERCISES

Integrate the following:

1. $\int \frac{(5x + 16)dx}{(x - 4)(2x - 1)}$

2. $\int_8^{10} \frac{(2x - 5)dx}{x^2 - 5x - 14}$

3. $\int \frac{(\frac{11}{2} x - 17)dx}{(3x - 2)(x - 4)}$

4. $\int \frac{(15x - 47)dx}{x^2 - 3x - 10}$

5. $\int_3^5 \frac{(6x^2 - 9)dx}{(x - 2)(x - 1)(2x + 1)}$

6. $\int \frac{(2x^2 - 16x + 16)dx}{x(2x^2 - 5x + 2)}$

7. $\int \frac{x^3 - 1}{x + 1} dx$

8. $\int \frac{(3x^2 + 5x + 3)dx}{2x^3 + 5x^2 + 3x}$

9. $\int \frac{(9t^2 + 4t - 6)dt}{3t^3 + 2t^2 - 6t + 4}$

10. $\int \frac{(\frac{13}{2} x^2 - \frac{33}{2} x + 4)dx}{(3x - 1)(x - 1)(x + 2)}$

11. $\int \frac{x^3}{x - 1} dx$

12. $\int \frac{dx}{x^2 - 4}$; work two ways

14.5b CASE 2 FOR PARTIAL FRACTIONS: REPEATED LINEAR FACTORS

When a rational fraction has a repeated linear factor in its denominator, the partial fraction decomposition utilizes the following theorem:

THEOREM 14.2: When $N(x)/D(x)$ is a proper fraction having $D(x) = (ax + b)^k$, then for all x for which $D(x) \neq 0$

$$\frac{N(x)}{D(x)} = \frac{A_1}{ax + b} + \frac{A_2}{(ax + b)^2} + \cdots + \frac{A_k}{(ax + b)^k} \tag{14.14}$$

where $A_1, A_2, \cdots, A_k$ are constants with $A_k \neq 0$.

You might at first wonder if all of the terms in Equation 14.14 are required. When both sides are multiplied by $D(x)$, we find

$$N(x) = A_1(ax + b)^{k-1} + A_2(ax + b)^{k-2} + \cdots + A_{k-1}(ax + b) + A_k . \tag{14.15}$$

Because $N(x)$ can have degree $k - 1$, it has k coefficients (of x^{k-1}, x^{k-2}, $\cdots$, x, constant), some perhaps zero; likewise, the right side of

Equation 14.15 has k constants to be determined. If it had fewer terms, there would exist N(x)/D(x) which the expansion could not represent.

EXAMPLE 14.28: Find the partial fraction expansion for $\frac{x}{(x-1)^3}$.

Utilizing Theorem 14.2 we have

$$\frac{x}{(x-1)^3} = \frac{A}{x-1} + \frac{B}{(x-1)^2} + \frac{C}{(x-1)^3}$$

Now multiply both sides of the last equation by $(x-1)^3$ to obtain

$$x = A(x-1)^2 + B(x-1) + C$$

$$= Ax^2 + (-2A + B)x + (A - B + C) .$$

Upon equating coefficients of like powers, we find

Equating coefficients of x^2: $A = 0$

Equating coefficients of x: $B - 2A = 1$ or $B = 1$

Equating the constants: $0 = A - B + C$ or $C = 1$.

Thus,

$$\frac{x}{(x-1)^3} = \frac{1}{(x-1)^2} + \frac{1}{(x-1)^3} ,$$

a result we could also find by writing $x = (x-1) + 1$.

Notice two features of this example. Not all of the terms in the proposed decomposition were needed with $A = 0$. Other fractions with the same denominator would need this term, like any fraction with a numerator involving x^2. Secondly, the numerator has a degree two less than that of the denominator, a permissible occurrence perhaps not otherwise communicated. ▌

EXAMPLE 14.29: Resolve $\frac{2x^3 - 8x^2 - 26x + 5}{(x+2)^2(x-1)^2}$ into partial fractions.

This is a proper fraction so there exists exactly one set of numbers such that

$$\frac{2x^3 - 8x^2 - 26x + 5}{(x+2)^2(x-1)^2} = \frac{A}{x+2} + \frac{B}{(x+2)^2} + \frac{C}{x-1} + \frac{D}{(x-1)^2}$$

Multiply both sides of the last equation by $(x - 1)^2(x + 2)^2$:

$$2x^3 - 8x^2 - 26x + 5$$

$$= A(x + 2)(x - 1)^2 + B(x - 1)^2 + C(x - 1)(x + 2)^2 + D(x + 2)^2 \; . \quad (14.16)$$

This equation contains four unknown constants. In order to evaluate these constants, we will use a combination of the previously illustrated methods. First let $x = 1$:

$$-27 = A(0) + B(0) + C(0) + 9D \text{ or } D = -3$$

Similarly let $x = -2$:

$$9 = A(0) + 9B + C(0) + D(0) \text{ or } B = 1 \; .$$

We have solved for two of the unknown constants; to solve for the remaining two unknowns we need two equations in these two unknowns. Now we could multiply out the right-hand side of Equation 14.16 and equate like coefficients or we could pick out the coefficients of the cubic term and the constant term without multiplying out the right-hand side. Equating the coefficients of the cubic term, we get $2 = A + C$ and, by equating the constant terms, we obtain $5 = 2A + B - 4C + 4D$. (Equating the constant terms is equivalent to letting $x = 0$ in Equation 14.16.) Now substitute B and D into these two equations:

$$2 = A + C \qquad \text{and} \qquad 5 = 2A + 1 - 4C - 12 \qquad \text{or}$$

$$2 = A + C \qquad \text{and} \qquad 8 = A - 2C \; .$$

We find $C = -2$ upon subtracting one equation from the other. Substitute this value into either equation to obtain $A = 4$. Thus Equation 14.16 becomes

$$\frac{2x^3 - 8x^2 - 26x + 5}{(x + 2)^2(x - 1)^2} = \frac{4}{x + 2} + \frac{1}{(x + 2)^2} - \frac{2}{x - 1} - \frac{3}{(x - 1)^2}$$

If we had needed to evaluate the integral of this function, then

$$\int \frac{2x^3 - 8x^2 - 26x + 5}{(x + 2)^2(x - 1)^2}\,dx = \int \frac{4dx}{x + 2} + \int \frac{dx}{(x + 2)^2} - \int \frac{2dx}{x - 1} - \int \frac{3dx}{(x - 1)^2}$$

$$= 4 \ln|x + 2| - \frac{1}{x + 2} - 2 \ln|x - 1| + \frac{3}{x - 1} + C \; .$$

Notice how much more difficult the decomposition of this fraction would have been if we had equated like coefficients to come up with a system of four equations in four unknowns. Generally speaking, it is easier to use a combination of the two methods to solve for the unknown constants than to use either method by itself. ▮

The next example illustrates the problems encountered when we have a combination of Cases 1 and 2.

EXAMPLE 14.30: Evaluate $\int \frac{(4y^2 + 14y + 18)dy}{y(y + 3)^2}$.

To make a partial fraction decomposition we seek constants A, B and C such that

$$\frac{4y^2 + 14y + 18}{y(y + 3)^2} = \frac{A}{y} + \frac{B}{y + 3} + \frac{C}{(y + 3)^2} . \qquad (14.17)$$

Multiply both sides of Equation 14.17 by $y(y + 3)^2$ to obtain

$$4y^2 + 14y + 18 = A(y + 3)^2 + By(y + 3) + Cy .$$

If $y = -3$, then $12 = -3C$ or $C = -4$. Equating the coefficients of y^2, we find $4 = A + B$ and, by letting $y = 0$, we obtain $18 = 9A$ or $A = 2$. Thus, $B = 4 - A = 4 - 2 = 2$. Substitute these values back into Equation 14.17:

$$\frac{4y^2 + 14y + 18}{y(y + 3)^2} = \frac{2}{y} + \frac{2}{y + 3} - \frac{4}{(y + 3)^2} .$$

Now evaluate the original integral:

$$\int \frac{(4y^2 + 14y + 18)dy}{y(y + 3)^2} = \int \frac{2dy}{y} + \int \frac{2dy}{y + 3} - \int \frac{4dy}{(y + 3)^2}$$

$$= 2 \ln |y| + 2 \ln |y + 3| + \frac{4}{y + 3} + C . \quad ▮$$

EXERCISES

Integrate the following:

1. $\int \frac{8dx}{(x + 1)^4}$; work two ways

2. $\int \frac{(x^2 + 1)dx}{(x - 1)^3}$

3. $\int \frac{(t^2 + 4t + 3)dt}{(t + 3)^2}$

4. $\int \frac{(\frac{3}{2} x^2 + 11x + 6)dx}{x(x + 2)^2}$

5. $\int \frac{(-8x^2 + 25x - \frac{29}{3})\,dx}{(x - 1)(x^2 - 3x + 2)}$

6. $\int \frac{(4x^3 - 9x^2 - 7x + 1)\,dx}{(x^2 - x)(x - 1)^2}$

7. $\int \frac{(x^2 - 3x + 3)\,dx}{(x - 2)(x^2 - 3x + 2)}$

8. $\int \frac{4\,dx}{(x + 3)^5}$

9. Let N stand for any quantity whose growth is inhibited and whose limiting value is symbolized by L. This behavior occurs when $\frac{dN}{dt} = KN(L - N)$ where K is a constant. Find an expression for N assuming that at $t = 0$, $N = N_0$. Sketch the curve.

14.5c CASE 3 FOR PARTIAL FRACTIONS: QUADRATIC FACTORS

In this case we want to resolve $N(x)/D(x)$ into partial fractions when $D(x)$ has one or more quadratic factors which cannot be broken into real linear factors; then the expansion contains the partial fraction

$$\frac{Ax + B}{ax^2 + bx + c}\ .$$

This procedure is illustrated in the next example.

<u>EXAMPLE 14.31</u>: Resolve $\frac{5x^2 - 3x + 2}{(x + 1)(x^2 - x + 3)}$ into partial fractions.

As this is a proper fraction, we seek three constants A, B and C such that

$$\frac{5x^2 - 3x + 2}{(x + 1)(x^2 - x + 3)} = \frac{A}{x + 1} + \frac{Bx + C}{x^2 - x + 3}\ . \qquad (14.18)$$

Multiply both sides of the last equation by $(x + 1)(x^2 - x + 3)$:

$$5x^2 - 3x + 2 = A(x^2 - x + 3) + (Bx + C)(x + 1)\ .$$

Again use a combination of methods to evaluate the constants. If $x = -1$, $10 = 5A$ or $A = 2$. Equating the coefficients of the x^2 term, we have $5 = A + B$ or $B = 5 - A = 5 - 2 = 3$. From $x = 0$, $2 = 3A + C$ or $C = 2 - 3A = 2 - 6 = -4$. Finally substitute these values back into Equation 14.18:

$$\frac{5x^2 - 3x + 2}{(x + 1)(x^2 - x + 3)} = \frac{2}{x + 1} + \frac{3x - 4}{x^2 - x + 3}\ .$$

To check this decomposition, obtain a common denominator on the right-hand side and collect like terms; the left-hand side of the equation will result. If we had needed to integrate this function, the above would give

$$\int \frac{(5x^2 - 3x + 2)dx}{(x + 1)(x^2 - x + 3} = \int \frac{2dx}{x + 1} + \int \frac{(3x - 4)dx}{x^2 - x + 3}$$

$$= 2 \ln |x + 1| + \frac{3}{2} \int \frac{(2x - 1)dx}{x^2 - x + 3} - \frac{5}{2} \int \frac{dx}{x^2 - x + 3}$$

because $3x - 4 = \frac{3}{2}(2x - 1) - \frac{5}{2}$. Now

$$\int \frac{(5x^2 - 3x + 2)dx}{(x + 1)(x^2 - x + 3)} = 2 \ln|x + 1| + \frac{3}{2} \ln|x^2 - x + 3| - \frac{5}{2} \int \frac{dx}{(x - \frac{1}{2})^2 + \frac{11}{4}}$$

$$= 2 \ln |x + 1| + \frac{3}{2} \ln |x^2 - x + 3| - \frac{5}{2} \frac{2}{\sqrt{11}} \operatorname{Tan}^{-1} \left(\frac{(x - \frac{1}{2})}{\frac{\sqrt{11}}{2}} \right) + C$$

$$= 2 \ln |x + 1| + \frac{3}{2} \ln |x^2 - x + 3| - \frac{5\sqrt{11}}{11} \operatorname{Tan}^{-1} \frac{2x - 1}{\sqrt{11}} + C .$$

Thus, we resolved the original integrand into partial fractions and then used previous techniques of integration to integrate the partial fractions. ▮

We already have worked with partial fraction decomposition of linear factors, repeated linear factors and quadratic factors. The pattern should begin to emerge. To complete it, the next example illustrates how to deal with a repeated quadratic factor.

EXAMPLE 14.32: Evaluate $\int \frac{(x^4 + 7x^3 + 25x^2 + 35x + 25)dx}{(x + 1)(x^2 + 2x + 4)^2}$.

For the proper fraction in the integrand, we seek the constants A, B, C, D and E such that

$$\frac{(x^4 + 7x^3 + 25x^2 + 35x + 25)}{(x + 1)(x^2 + 2x + 4)^2}$$

$$= \frac{A}{x + 1} + \frac{Bx + C}{x^2 + 2x + 4} + \frac{Dx + E}{(x^2 + 2x + 4)^2} . \tag{14.19}$$

Multiply this equation by $(x + 1)(x^2 + 2x + 4)^2$ to get

$$x^4 + 7x^3 + 25x^2 + 35x + 25 =$$

$$A(x^2 + 2x + 4)^2 + (Bx + C)(x + 1)(x^2 + 2x + 4) + (Dx + E)(x + 1) \quad (14.20)$$

At $x = -1$, $9 = A(3)^2 + (Bx + C)0 + (Dx + E)(0)$ or $A = 1$. Now multiply out Equation 14.20, collect terms and equate like coefficients:

$$x^4 + 7x^3 + 25x^2 + 35x + 25 =$$

$$(A + B)x^4 + (4A + 3B + C)x^3 + (12A + 6B + 3C + D)x^2$$

$$+ (16A + 4B + 6C + D + E)x + (16A + 4C + E)$$

Equating coefficients of x^4 yields: $1 = A + B$ or $B = 0$

Equating coefficients of x^3 yields: $7 = 4A + 3B + C$ or $C - 3$

Equating coefficients of x^2 yields: $25 = 12A + 6B + 3C + D$ or $D = 4$

Equating the constants yields: $25 = 16A + 4C + E$ or $E = -3$.

Substitute these values into Equation 14.19 and integrate:

$$\int \frac{(x^4 + 7x^3 + 25x^2 + 35x + 25)dx}{(x + 1)(x^2 + 2x + 4)^2}$$

$$= \int \frac{dx}{x + 1} + \int \frac{3dx}{x^2 + 2x + 4} + \int \frac{(4x - 3)dx}{(x^2 + 2x + 4)^2}$$

$$= \ln|x + 1| + 3 \int \frac{dx}{x^2 + 4x + 4} + 2 \int \frac{(2x + 2)dx}{(x^2 + 2x + 4)^2} - 7 \int \frac{dx}{(x^2 + 2x + 4)^2}$$

$$= \ln|x + 1| + 3 \int \frac{dx}{(x + 1)^2 + 3} - \frac{2}{x^2 + 2x + 4} - 7 \int \frac{dx}{((x + 1)^2 + 3)^2}$$

$$= \ln|x + 1| + \frac{3}{\sqrt{3}} \text{Tan}^{-1} \left[\frac{x + 1}{\sqrt{3}}\right] - \frac{2}{x^2 + 2x + 4} - 7 \int \frac{dx}{((x + 1)^2 + 3)^2} .$$

To evaluate the remaining integral, use the trigonometric substitution $x + 1 = \sqrt{3} \tan \theta$ and obtain after integration

$$-7 \int \frac{dx}{((x + 1)^2 + 3)^2} = - \frac{7\sqrt{3}}{9} \left[\frac{1}{2} \text{Tan}^{-1} \frac{x + 1}{\sqrt{3}} + \frac{(x + 1)\sqrt{3}}{((x + 1)^2 + 3)}\right] + C .$$

Finally then

$$\frac{(x^4 + 7x^3 + 25x^2 + 35x + 25)dx}{(x + 1)(x^2 + 2x + 4)^2}$$

$$= \ln|x + 1| + \frac{3}{\sqrt{3}} \text{Tan}^{-1}\left[\frac{x + 1}{\sqrt{3}}\right] - \frac{2}{x^2 + 2x + 4} - \left[\frac{7\sqrt{3}}{9}\right]\frac{1}{2} \text{Tan}^{-1}\left[\frac{x + 1}{\sqrt{3}}\right]$$

$$+ \frac{(x + 1)\sqrt{3}}{((x + 1)^2 + 3)} + C$$

$$= \ln|x + 1| + \frac{11\sqrt{3}}{18} \text{Tan}^{-1}\left[\frac{x + 1}{\sqrt{3}}\right] - \frac{1}{3}\frac{(7x + 13)}{x^2 + 2x + 4} + C \;. \quad ▌$$

The previous example points out how we are now going to have to use all the techniques of integration in order to be able to solve the integration in the problems we will encounter involving repeated quadratic factors.

EXERCISES

Evaluate the following integrals:

1. $\int \frac{(x^2 - 4)dx}{(x + 2)(x^2 + 5)}$

2. $\int \frac{(11x^2 + x + 6)dx}{(3x + 4)(x^2 - 4x + 5)}$

3. $\int \frac{(\frac{3}{2} x^2 + x + 2)dx}{(x + 1)(x^2 + 4)}$

4. $\int \frac{-(\frac{5}{3} x^2 + 16x + 25)dx}{(x + 3)(x^2 + 3)}$

5. $\int \frac{(21x^3 - 41x^2 + 70x - 158)dx}{(x - 1)(x - 3)(2x^2 + 7)}$

6. $\int \frac{(7x^3 - 55x^2 + 117x + 3)dx}{(x - 4)^2(x^2 + 5x + 3)}$

7. $\frac{(2x + 5)dx}{(2x - 1)(x \quad + 3x + 8)}$

8. $\int \frac{(-\frac{5}{6} x^2 - \frac{9}{2} x + \frac{17}{3})\, dx}{(x - 3)(2x + 5)}$

9. $\int \frac{(4x^2 + 3x + 14)dx}{x^3 - 8}$

10. $\int \frac{(3x^3 + 27x - 54)dx}{x^4 - 81}$

11. $\int \frac{(8x^3 + x^2 + 19x + 5)dx}{(2x^2 + 3)(x^2 + 5)}$

12. $\int \frac{(4x^3 + 23x)dx}{(x^2 + 6)(x^2 + 5)}$

13. $\int \frac{4 \sec\theta \tan\theta \, d\theta}{\sec^3\theta + 4\sec\theta}$

14. $\frac{(14x \quad + 30x + 50)dx}{x \quad + 6x \quad + 10x}$

15. $\int \frac{(5x^4 + 3x^3 + 19x^2 + 9x + 25)dx}{(x + 3)(x^2 + 2)^2}$

16. $\int \frac{(x^3 + 24x + 5)dx}{(x^2 + 36)^3}$

17. $\int \frac{x^3dx}{(x^2 - 2x + 10)^3}$

18. $\int \frac{(3x^2 + 12x - 2)dx}{(x^2 + 4)^2}$

19. $\int \frac{(2x - 12x^2 - 3)dx}{(4x^2 + 1)^2}$

20. $\int \frac{du}{(4u^2 + 9)^2}$

21. $\int \frac{(2x^3 + x + 3)dx}{x^2 + 1}$

22. $\int \frac{(3x^2 + 1)dx}{2x^3 + 3x}$

23. $\int \frac{(x^3 + x^2 + 5x + 2)dx}{(x^2 + 3)^2}$

24. $\int \frac{(12x^3 - 92x^2 + 242x - 200)dx}{x(x^2 - 6x + 10)^2}$

*14.6 ALGEBRAIC RATIONALIZING SUBSTITUTIONS

When an integrand $f(x)$ is a rational function of powers of x with rational exponents, then a rationalizing substitution for $\int f(x)dx$ is

$$u = x^{1/n} \qquad \text{or} \qquad x = u^n \tag{14.21}$$

where the positive integer n is the lowest common denominator of the exponents of x in $f(x)$. The following example illustrates how to use this rationalizing substitution.

EXAMPLE 14.33: Evaluate $\int \frac{dx}{x^{1/2} - x^{1/4}}$.

Because 4 is the lowest common denominator of these exponents, make the substitution $u = x^{1/4}$ or $u^4 = x$ with $dx = 4u^3du$:

$$\int \frac{dx}{x^{1/2} - x^{1/4}} = \int \frac{4u^3du}{(u^4)^{1/2} - (u^4)^{1/4}} = 4 \int \frac{u^3du}{u^2 - u} = 4 \int \frac{u^2du}{u - 1} \ .$$

As this integrand is improper, divide the denominator into the numerator to obtain a proper fraction:

$$4 \int \frac{u^2du}{u - 1} = 4 \int (u + 1 + \frac{1}{u - 1})\, du = 4 \int udu + 4 \int du + 4 \int \frac{du}{u - 1}$$

$$= 2u^2 + 4u + 4 \ln |u - 1| + C \ .$$

Now express this in terms of the original variable x:

$$\int \frac{dx}{x^{1/2} - x^{1/4}} = 2x^{1/2} + 4x^{1/4} + 4 \ln |x^{1/4} - 1| + C .$$

A suitable rationalizing substitution led to an integrable form; then resubstitution transformed this result back into terms of the original variable. For a definite integral, we either could have changed the limits of integration or applied the original limits after expressing the transformed result in terms of the original variable. ▌

When an integrand contains a single irrational expression of the form $(ax + b)^{p/q}$, the substitution

$$z = (ax + b)^{1/q} \qquad \text{or} \qquad x = \frac{z^q - b}{a} \tag{14.22}$$

will convert the given integral into a rational function of z. For this substitution,

$$dx = \frac{q}{a} z^{q-1}\, dz .$$

The following example illustrates the use of this kind of rationalizing substitution.

EXAMPLE 14.34: Evaluate $\int \frac{\sqrt[3]{x + 1}}{x}\, dx$.

Let $z = \sqrt[3]{x + 1}$ so $x = z^3 - 1$ and $dx = 3z^2 dz$. Then

$$\int \frac{\sqrt[3]{x + 1}}{x}\, dx = \int \frac{z(3z^2 dz)}{z^3 - 1} = 3 \int \frac{z^3 dz}{z^3 - 1} .$$

Now make this integrand a proper fraction:

$$3 \int \frac{z^3 dz}{z^3 - 1} = 3 \int dz + 3 \int \frac{dz}{z^3 - 1} = 3 \int dz + \int \frac{3dz}{(z - 1)(z^2 + z + 1)}$$

$$= 3z + 3 \int \frac{dz}{(z - 1)(z^2 + z + 1)} .$$

To evaluate this last integral, we use the method of partial fractions:

$$\frac{3}{(z - 1)(z^2 + z + 1)} = \frac{A}{z - 1} + \frac{Bz + C}{z^2 + z + 1} . \tag{14.23}$$

Multiply both sides of this equation by $(z - 1)(z^2 + z + 1)$ to get

$$3 = A(z^2 + z + 1) + (z - 1)(Bz + C) .$$

At $z = 1$, we have $3 = 3A$ or $A = 1$.

At $z = 0$, we have $3 = A - C$ or $C = -2$.

Equating the coefficients of z^2 shows that $0 = A + B$ or $B = -A = -1$. Now substitute the values of these constants back into Equation 14.23 and integrate:

$$\frac{3}{(z-1)(z^2+z+1)} = \frac{1}{z-1} + \frac{-z-2}{z^2+z+1} ,$$

$$3\int \frac{z^3dz}{z^3-1} = 3z + \int \frac{3dz}{(z-1)(z^2+z+1)} = 3z + \int \frac{dz}{z-1} - \int \frac{(z+2)dz}{z^2+z+1}$$

$$= 3z + \ln|z-1| - \frac{1}{2}\int \frac{(2z+1)dz}{z^2+z+1} - \frac{3}{2}\int \frac{dz}{z^2+z+1}$$

$$= 3z + \ln|z-1| - \frac{1}{2}\ln|z^2+z+1| - \frac{3}{2}\int \frac{dz}{(z+\frac{1}{2})^2 + \frac{3}{4}}$$

$$= 3z + \ln|z-1| - \frac{1}{2}\ln|z^2+z+1| - \frac{3}{2}\frac{2}{\sqrt{3}}\,\mathrm{Tan}^{-1}\left(\frac{2(z+\frac{1}{2})}{\sqrt{3}}\right) + C$$

$$= 3z + \ln|z-1| - \frac{1}{2}\ln|z^2+z+1| - \sqrt{3}\,\mathrm{Tan}^{-1}\left(\frac{2z+1}{\sqrt{3}}\right) + C .$$

Now express this result in terms of the original variable:

$$\int \frac{\sqrt[3]{x+1}}{x}dx = 3\sqrt[3]{x+1} + \ln|\sqrt[3]{x+1} - 1| - \frac{1}{2}\ln|(x+1)^{2/3} + (x+1)^{1/3} + 1|$$

$$-\sqrt{3}\,\mathrm{Tan}^{-1}\left[\frac{2(x+1)^{1/3}+1}{\sqrt{3}}\right] + C .$$

Notice the various techniques of integration utilized in this solution: partial fractions, multiplication and division by a constant, addition and subtraction of a constant, completing the square, and a trigonometric substitution to evaluate $\int \frac{dz}{(z+\frac{1}{2})^2+\frac{3}{4}}$. ∎

When an integrand consists of the product of an odd power, either positive or negative, of x with one of the forms $\sqrt{a^2 - x^2}$, $\sqrt{x^2 + a^2}$, or $\sqrt{x^2 - a^2}$, set z equal to the radical.

EXAMPLE 14.35: Evaluate $\int \frac{\sqrt{a^2 - x^2}}{x}\, dx$.

Let $z = \sqrt{a^2 - x^2}$ or $x^2 = a^2 - z^2$ and $dx = -\frac{zdz}{x}$ so

$$\int \frac{\sqrt{a^2 - x^2}}{x}\, dx = \int \frac{z}{x}\left(\frac{-zdz}{x}\right) = -\int \frac{z^2dz}{x^2} = -\int \frac{z^2dz}{a^2 - z^2}$$

Since the integrand is not a proper fraction we divide before integrating

$$-\int \frac{z^2dz}{a^2 - z^2} = -\int \left(-1 + \frac{a^2}{a^2 - z^2}\right) dz = \int dz - a^2 \int \frac{dz}{a^2 - z^2}$$

$$= z - a^2 \int \frac{dz}{a^2 - z^2} .$$

To evaluate this last integral, let $z = a \sin \theta$, $dz = a \cos \theta\, d\theta$, and $a^2 - z^2 = a^2 \cos^2 \theta$.

$$-\int \frac{z^2dz}{a^2 - z^2} = z - a^2 \int \frac{dz}{a^2 - z^2} = z - a^2 \int \frac{a \cos \theta}{a^2 \cos^2 \theta}\, d\theta$$

$$= z - a \int \sec \theta\, d\theta = z - a \ln |\sec \theta + \tan \theta| + C . \quad (14.24)$$

Because $\frac{z}{a} = \sin \theta$, $\sec \theta = \frac{a}{\sqrt{a^2 - z^2}}$, $\tan \theta = \frac{z}{\sqrt{a^2 - z^2}}$ and $a^2 - z^2 = x^2$.

Substitute these values back into Equation 14.24:

$$\int \frac{\sqrt{a^2 - x^2}}{x}\, dx = \sqrt{a^2 - x^2} - a \ln \left| \frac{a}{x} + \frac{\sqrt{a^2 - x^2}}{x} \right| + C .$$

In this case, we could have accomplished the integration with a trigonometric substitution by letting $x = a \sin \theta$, so $dx = a \cos \theta\, d\theta$ and $\sqrt{a^2 - x^2} = a \cos \theta$:

$$\int \frac{\sqrt{a^2 - x^2}\,dx}{x} = \int \frac{(a \cos \theta)(a \cos \theta\, d\theta)}{a \sin \theta} = a \int \frac{\cos^2 \theta}{\sin \theta}\, d\theta$$

$$= a \int \frac{1 - \sin^2 \theta}{\sin \theta}\, d\theta = a \int \left(\frac{1}{\sin \theta} - \sin \theta\right) d\theta$$

$$= a \int \csc \theta\, d\theta - a \int \sin \theta\, d\theta$$

$$= a \ln |\csc \theta - \cot \theta| + a \cos \theta + C .$$

But $\frac{x}{a} = \sin \theta$, implies that $\csc \theta = \frac{a}{x}$, $\cot \theta = \frac{\sqrt{a^2 - x^2}}{x}$ and $\cos \theta = \frac{\sqrt{a^2 - x^2}}{a}$:

$$\int \frac{\sqrt{a^2 - x^2}}{x}\, dx = a \ln \left| \frac{a}{x} - \frac{\sqrt{a^2 - x^2}}{x} \right| + \sqrt{a^2 - x^2} + C$$

which is the result obtained by the use of a rationalizing substitution. Keep both methods in mind, because one may be easier to use than the other in a particular situation. ▌

EXERCISES

Integrate the following:

1. $\int \frac{x^{1/4}dx}{1 + x^{1/2}}$
2. $\int \frac{dy}{y^{1/3} + y^{2/3}}$
3. $\int \frac{x^{1/2}\,dx}{x^{1/3} + 1}$
4. $\int \frac{(x^{2/3} - x^{1/2})dx}{x^{1/3}}$
5. $\int \frac{(2x + 3)dx}{\sqrt{x + 2}}$
6. $\int x\sqrt{x + 1}\, dx$
7. $\int \frac{(x - 2)dx}{(3x - 1)^{2/3}}$
8. $\int \frac{x^2dx}{\sqrt[3]{2x + 1}}$
9. $\int \frac{\sqrt{x + 4}}{x}\, dx$
10. $\int \frac{\sqrt{x} + 2}{\sqrt{x} - 1}\, dx$
11. $\int \frac{(2x + 1)dx}{(x + 2)^{2/3}}$
12. $\int ((x + 2)\sqrt{x - 1})dx$

13. $\int \frac{\sqrt{2x + 3}}{x + 1}\,dx$

14. $\int \frac{\sqrt{a^2 - x^2}}{x^3}\,dx$

15. $\int \frac{x^3}{\sqrt{x^2 - 4}}\,dx$

16. $\int x^3\sqrt{x^2 + 1}\,dx$

17. $\int \frac{dx}{x\sqrt{x^2 + 9}}$

18. $\int \frac{\sqrt{4 + x}}{x}\,dx$

19. $\int \frac{\sqrt{4 - x}}{x}\,dx$

20. $\int (x^3 - x)\sqrt{25 - x^2}\,dx$

14.7 INTEGRATION BY THE USE OF TABLES

You already have encountered most of the standard techniques of integration; you should understand how and why they work. It is unrealistic to expect that you will remember all of these techniques in future years. Thus, you should learn how to use the list of integrals which appear in a mathematical handbook. Do not expect to immediately evaluate any integral merely because you have a handy list of integrals. In certain instances, as you will find in the exercises, you will have to apply one of the techniques we have studied to get the integral into a form which appears in a handbook.

As you become familiar with a handbook, or the integrals in Table V, try to visualize what integration technique produced the relations you use. Always remember that you can verify any integration simply by showing that its derivative equals the integrand. We will illustrate the use of Table V by applying it to the evaluation of three integrals we integrated by techniques introduced in this chapter. You should find it instructive to relate the table entries we use here to the results found earlier in the respective examples.

<u>EXAMPLE 14.3</u> (continued): Evaluate $\int x^2e^x\,dx$.

As this integral contains an exponential form, scan Equations 138 - 159 in Table V. Equation 141 has x^2e^{ax} as its integrand. This generalizes our integrand which has $a = 1$. Thus

$$\int x^2e^{ax}\,dx = \frac{e^{ax}}{a^3}(a^2x^2 - 2ax + 2)$$

$$= e^x\,(x^2 - 2x + 2)\ ,$$

the same result as we obtained using integration by parts, except the above lacks the constant of indefinite integration. We always have to supply this constant because integral tables customarily omit it. ▌

EXAMPLE 14.7 (continued): Evaluate $\int \sin^4 x\,dx$.

As this integrand involves the sine function, examine Equations 71 - 80 in Table V. This list does not contain $\sin^4 x$ as an integrand, but it does contain $\sin^n x$ in Equation 74. Use the latter form with $n = 4$, $a = 1$:

$$\int \sin^4 x\,dx = -\frac{\sin^3 x\cos x}{4} + \frac{3}{4}\int \sin^2 x\,dx + C\ .$$

Now use Equation 72 on $\sin^2 x$:

$$\int \sin^4 x\,dx = -\frac{\sin^3 x\cos x}{4} + \frac{3}{4}\left[\frac{x}{2} - \frac{\sin 2x}{4}\right] + C$$

$$= \frac{3x}{8} - \frac{\sin^3 x\cos x}{4} - \frac{3}{16}\sin 2x + C\ .$$

This does not yet look like the result we obtained earlier:

$$\int \sin^4 x\,dx = \frac{3x}{8} - \frac{1}{4}\sin 2x + \frac{1}{32}\sin 4x + C\ .$$

The difference amounts to nothing more than trigonometric deception. We will show that the difference between the two expressions,

$$-\frac{\sin^3 x\cos x}{4} + \frac{1}{16}\sin 2x - \frac{1}{32}\sin 4x\ ,$$

equals zero. Multiply this expression by 32 to eliminate the denominators and then apply the double angle formulas $\sin 2\alpha = 2\sin\alpha\cos\alpha$ and $\cos 2\alpha = 1 - 2\sin^2\alpha$ (Equations 13.32 and 13.33).

$-8 \sin^3 x \cos x + 2 \sin 2x - \sin 4x$

$= -8 \sin^3 x \cos x + 4 \sin x \cos x - 2 \sin 2x \cos 2x$

$= -8 \sin^3 x \cos x + 4 \sin x \cos x - 2(2 \sin x \cos x)(1 - 2 \sin^2 x) = 0.$ ∎

<u>EXAMPLE 14.22</u> (continued): Evaluate $\int \frac{dx}{7 + 12x - 4x^2}$.

As this integrand involves a quadratic expression, consider Equations 37 - 70 in Table V. Equation 50 has this form with a = -4, b = 12, c = 7. Thus

$$\int \frac{dx}{7 + 12x - 4x^2} = \frac{1}{\sqrt{b^2 - 4ac}} \ln \left| \frac{2ax + b - \sqrt{b^2 - 4ac}}{2ax + b + \sqrt{b^2 - 4ac}} \right| + C$$

$$= \frac{1}{\sqrt{144 + 112}} \ln \left| \frac{-8x + 12 - \sqrt{256}}{-8x + 12 + \sqrt{256}} \right| + C$$

$$= \frac{1}{16} \ln \left| \frac{-8x - 4}{-8x + 28} \right| + C = \frac{1}{16} \ln \left| \frac{2x + 1}{-2x + 7} \right| + C .$$

Earlier we obtained

$$\int \frac{dx}{7 + 12x - 4x^2} = \frac{1}{8} \ln \left| \frac{1 + 2x}{\sqrt{16 - (2x - 3)^2}} \right| + C .$$

Again the equality of the results under the two approaches does not emerge immediately. To establish their equality, write $\frac{1}{16} = \left[\frac{1}{8}\right]\frac{1}{2}$ so

$$\frac{1}{16} \ln \left| \frac{2x + 1}{-2x + 7} \right| = \frac{1}{8} \ln \sqrt{\left| \frac{2x + 1}{-2x + 7} \right|} .$$

It is true that

$$\sqrt{\left| \frac{2x + 1}{-2x + 7} \right|} = \frac{1 + 2x}{\sqrt{16 - (2x - 3)^2}}$$

and the result follows. You should show this last result. ∎

The following exercises are intended to acquaint you with the tables. Not all of these exercises require the tables because some already appear in a familiar standard form. An abbreviated list of integrals which you may find helpful appears in Table V.

As you become familiar with a handbook, or the integrals in Appendix II, try to visualize what integration technique produced the relations you use. Always remember that you can verify any integration simply by showing that its derivative equals the integrand.

The following exercises are intended to acquaint you with tables of integrals. Not all of these exercises require the tables because some already appear in a familiar standard form. Appendix II contains an abbreviated list of integrals.

EXERCISES:

Integrate the following problems using a table of integrals if necessary.

1. $\int \frac{dx}{3 + 4 \sin 2x}$

2. $\int \frac{dx}{\sqrt{4x^2 - 9}}$

3. $\int \frac{dx}{x\sqrt{3x^2 + 2x + 1}}$

4. $\int \frac{dx}{x(x - 3)^{1/2}}$

5. $\int \frac{dx}{x\sqrt{4x^2 - 3x - 9}}$

6. $\int \sqrt{16 - x^2}\, dx$

7. $\int \frac{x^2 dx}{x^2 - 3x + 4}$

8. $\int \sqrt{2x^2 - 3}\, dx$

9. $\int x^4(9x^2 + 2)^3 dx$

10. $\int \frac{dx}{(x^2 - 2x + 3)^{3/2}}$

11. $\int \frac{dx}{(-3x^2 + 4)^3}$

12. $\int \frac{dx}{\sin^3 x \cos^4 x}$

13. $\int \frac{dx}{\sqrt{9 - 4x^2}}$

14. $\int \frac{dx}{\sqrt{4 - 3 \tan^2 2x}}$

15. $\int \frac{\cos 3x\ dx}{\sin^2 3x + 2 \sin 3x - 4}$

16. $\int x^5 \pi^{4x} dx$

17. $\int x\ e^{11x} \sin 2x\ dx$

18. $\int \cos^{-1} 4x\ dx$

19. $\int \frac{dx}{x[(\ln x)^2 - 3 \ln x + 9]}$

20. $\int \frac{\sec^2 x \tan x\ dx}{\sqrt{\sec x + 4}}$

21. $\int \frac{\csc^2 x\ dx}{\cot x(2 \cot^2 x - 4)}$

22. $\int \frac{e^{3x} dx}{e^{2x} - e^x + 3}$

23. $\int \sqrt{x}\ dx$

24. $\int \sqrt[m]{x}\ dx$

25. $\int \frac{dx}{x^2}$

26. $\int 10^x\ dx$

27. $\int a^x e^x dx$

28. $\int \frac{dx}{2\sqrt{x}}$

29. $\int \frac{dh}{\sqrt{2gh}}$

30. $\int 3.4x^{-0.17} dx$

31. $\int (1 - 2u)^6 du$

32. $\int (\sqrt{x} + 1)(x - \sqrt{x} + 1) dx$

33. $\int \frac{\sqrt{x} - x^3 e^x + x^2}{x^3} dx$

34. $\int \left(2x^{-1.2} + 3x^{-0.8} - 5x^{0.38}\right) dx$

35. $\int \left(\frac{1 - z}{z}\right)^2 dz$

36. $\int \frac{(1 - x)^2}{x\sqrt{x}} dx$

37. $\int \frac{(1 + \sqrt{x})^3}{3\sqrt{x}}$

38. $\int \frac{\sqrt[3]{x^2} - \sqrt[4]{x}}{\sqrt{x}}$

39. $\int \frac{dx}{\sqrt{3 - 3x^2}}$

40. $\int \frac{3.2^x - 2.3^x}{2^x} dx$

41. $\int \frac{1 + \cos^2 x}{1 + \cos 2x} dx$

42. $\int \frac{\cos 2x}{\cos^2 x \sin^2 x} dx$

43. $\int \tan^2 x \, dx$

44. $\int \cot^2 x \, dx$

45. $\int 2 \sin^2 \frac{x}{2} dx$

46. $\int \frac{(1 + 2x^2)dx}{x^2(1 + x^2)}$

47. $\int \frac{(1 + x)^2 dx}{x(1 + x^2)}$

48. $\int \frac{dx}{\cos 2x + \sin^2 x}$

49. $\int (\sin^{-1} x + \cos^{-1} x)dx$

50. $\int \sin x \, d(\sin x)$

51. $\int \tan^3 x \, d(\tan x)$

52. $\int \frac{d(1 + x^2)}{\sqrt{1 + x^2}}$

53. $\int (x + 1)^{15} dx$

54. $\int \frac{dx}{(2x - 5)^5}$

55. $\int \frac{dx}{(a + bx)^C} \quad C \neq 1$

56. $\int \sqrt[5]{(8 - 3x)^6} \, dx$

57. $\int \sqrt{8 - 2x} \, dx$

58. $\int \frac{dx}{3x^2 + 2}$

59. $\int x^2 \csc^{-1} 4x \, dx$

60. $\int \frac{\sin 2x}{x} dx$

61. $\int \frac{dx}{2 + 3 \sin 4x}$

62. $\int \frac{dx}{2x^2 - 3}$

63. $\int \frac{dx}{3 + 2 \sin x}$

64. $\int \frac{x^3 dx}{(5x^2 + 2)^{3/2}}$

65. $\int_0^\infty \frac{\sin x \cos x}{x} dx$

66. $\int \frac{x^4 - x^3 - x - 1}{x^3 - x^2} dx$

67. $\int \frac{dx}{\sqrt{acx^2 + (bc + ad)x + bd}}$

68. $\int \frac{dx}{(2x^2 - 9)^3}$

69. $\int \frac{dx}{(2x + 1)(x + 3)}$

70. $\int \frac{dx}{x\sqrt{3x^2 - 2}}$

71. $\int \frac{(3x^2 - 5x + 6)dx}{(1 + 3x)(1 - 2x - 15x^2)}$

72. $\int \frac{dx}{x\sqrt{(\ln x)^2 + 25}}$

73. $\int \frac{\sec^2 x\, dx}{\tan^2 x + 4 \tan x + 3}$

74. $\int \frac{\ln(x + 3)dx}{\sqrt[3]{x + 3}}$

75. $\int x\, 2^x\, dx$

76. $\int \frac{x\, dx}{(x^2 - 4)^{1/2} + (x^2 - 4)^{3/2}}$

77. $\int \frac{x\, dx}{\sqrt{x} + \sqrt{x + 1}}$

78. $\int \frac{(5x - 3)dx}{(x^2 + 4x + 7)^2}$

79. $\int \frac{x^3 + x^2 - x - 3}{x + 2}\, dx$

80. $\int \frac{\sqrt[3]{x + 1}}{x}\, dx$

81. $\int x\, 9^{x^2+2}\, dx$

82. $\int_0^1 \ln |\ln x|\, dx$

83. $\int \frac{dx}{2e^x + 3e^{-x}}$

84. $\int \frac{(2x + 3)}{\sqrt{x + 2}}\, dx$

85. $\int x\sqrt{x + 1}\, dx$

86. $\int \frac{x - 2}{(3x - 1)^{2/3}}\, dx$

87. $\int \frac{x^2}{\sqrt[3]{2x + 1}}\, dx$

88. $\int \frac{\sqrt{x + 4}}{x}\, dx$

89. $\int \frac{\sqrt{x} + 2}{\sqrt{x} - 1}\, dx$

90. $\int \frac{x^3}{\sqrt{x^2 - 4}}\, dx$

91. $\int \frac{x^5 + 2x^3}{\sqrt{x^2 + 4}}\, dx$

92. $\int (x^3 - x)\sqrt{16 - x^2}\, dx$

93. $\int \frac{dx}{x\sqrt{a^2 - x^2}}$

94. $\int_0^\infty x\, e^{-x^2}\, dx$

95. $\int_0^3 \frac{dx}{\sqrt{4 - x^2}}$

96. $\int \frac{x^3 + 2}{x^2 + 4}\, dx$

97. $\int \frac{x^2 + 2x + 3}{(x^2 - 1)(x - 2)}\, dx$

98. $\int \frac{x^3 + 3x^2 - 2x + 1}{x^4 + 5x^2 + 4}\, dx$

99. $\int \frac{2x - 1}{(x - 2)^{1/3}}\, dx$

100. $\int \frac{2\sqrt{x + 1} - 3}{3\sqrt{x + 1} - 2}\, dx$

101. $\int (\ln x)^2\, dx$

102. $\int \frac{x \ln x\, dx}{\sqrt{x^2 - 4}}$

103. $\int \frac{(x + 3)dx}{x^2 + 2x + 5}$

104. $\int \frac{\cos\theta\, d\theta}{3\sin^2\theta + 2\sin\theta - 5}$

105. $\int \frac{dx}{x^{1/2}(2x - x^{1/2} + 7)}$

106. $\int \frac{dx}{x[(\ln x)^2 + 7\ln x - 3]}$

107. $\int \frac{dx}{x^{1/2}(1 - \sqrt{x})}$

108. $\int \frac{\sec^5 x\, dx}{\csc x}$

109. $\int \frac{dx}{1 + \sec 3x}$

110. $\int \frac{e^{2x}\, dx}{1 + e^{4x}}$

111. $\int \frac{2x - 5}{\sqrt{4x - x^2}}\, dx$

112. $\int_0^1 \frac{x^2 dx}{\sqrt{4 - x^2}}$

113. $\int \frac{dx}{\sqrt{2x - x^2}}$

114. $\int x\, 2^{x^2}\, dx$

115. $\int \frac{e^x dx}{\sqrt{1 - e^{2x}}}$

116. $\int e^{2\sin 3x} \cos 3x\, dx$

117. $\int \frac{dx}{x\sqrt{x^2 - 7}}$

118. $\int \frac{x\, dx}{\sqrt{27 + 6x - x^2}}$

119. $\int \sqrt{16 - 9x^2}\, dx$

120. $\int x^2 e^{-3x}\, dx$

121. $\int \sin^2 x \cos^2 x\, dx$

122. $\int \frac{\sqrt{25 - x^2}}{x}\, dx$

123. $\int \frac{x^2 dx}{(a^2 - x^2)^{3/2}}$

124. $\int \frac{dx}{x^2\sqrt{9 - x^2}}$

125. $\int x^3\sqrt{9 - x^2}\, dx$

126. $\int \frac{dx}{(9 + x^2)^3}$

127. $\int \frac{(x^3 + x^2 - 5x + 15)dx}{(x^2 + x)(x^3 + 1)}$

128. $\int \frac{(x^6 + 7x^5 + 15x^4 + 32x^3 + 23x^2 + 25x - 3)dx}{(x^2 + x + 2)^2(x^2 + 1)^2}$

129. $\int \frac{dx}{e^{2x} - 3e^x}$

130. $\int \frac{\sin x\, dx}{\cos x(1 + \cos^2 x)}$

131. $\int \frac{x^4 dx}{(1 - x)^3}$

132. $\int \frac{x dx}{1 + x}$

133. $\int \frac{dx}{3 + \sqrt{x + 2}}$

134. $\int \frac{(e^x - 2)e^x\, dx}{e^x + 1}$

135. $\int \frac{2x^3 + x^2 + 4}{(x^2 + 1)^2}\, dx$

136. $\int \frac{(2 + \tan^2 \theta)\sec^2 \theta\, d\theta}{1 + \tan^3 \theta}$

137. $\int \left(\frac{\sec x}{\tan x}\right)^4 dx$

138. $\int \sin 5x \sin 2x\, dx$

139. $\int \sqrt{1 - \cos x}\, dx$

140. $\int e^x(2 \sin 4x - 5 \cos 4x)dx$

Chapter 15

OTHER COORDINATE SYSTEMS

Only rectangular coordinate systems have been considered so far. These systems locate a point by specifying distances from a reference point called the origin, two distances for points in a plane and three distances for points in three-dimensional space. Other systems use a different method for locating points, but the same number of coordinates still are required. This chapter introduces coordinate systems involving at least one angular coordinate. Points in a plane can be represented by a distance and an angle from a reference half-line. The cylindrical coordinate system locates points in space by specifying an angle and two distances. The spherical coordinate system uses two angles and one distance to fix a point in space.

Certain functions have simpler representations in one of these angular coordinate systems than in a rectangular coordinate system. Thus their treatment, particularly tasks involving integration, can be done more easily in an angular coordinate system than in the rectangular system.

The more interesting reason for use of these systems comes from biology: Angles have a natural association with living organisms. Two general cases appear noteworthy. The photosynthetic activity of a plant requires light from either the sun or from an artificial source. The intensity and, to some extent, the quality of arriving light depend on the angle of the sun's rays to the leaves. This angle changes from east to west during the day as the sun goes from rising to high noon to setting; it changes also from north to south with the seasons, being most nearly overhead at the summer solstice (about June 22) and lowest at the winter solstice (about December 22). Of course, the latter has far less importance than the former due to the correspondence of the growing season in the middle latitudes with the higher elevations of the sun.

The treatment of tumors with ionizing radiation has very similar features. The ionizing source replaces the sun and a tumor replaces the leaf.

The tumor can be considered the origin and the source can be moved in a spherical coordinate system.

Another case arises in animal behavior. Here the angle involved may be either a central feature of the response or merely a convenient way for recording it. In studies of animal navigation, for example, the homing pigeon's guidance system, angles occupy a central role in both the response and its record. Keeton (1969) described experimentation on this phenomenon for homing pigeons while Frisch (1967) reported related investigations of honeybees.

The three pairs of sections in this chapter deal with the polar, cylindrical and spherical coordinate systems. The first section in each pair introduces the system and representations therein; the second deals with integration of functions in that system.

15.1 THE POLAR COORDINATE SYSTEM

The unique location of a point in a plane requires two coordinates because the plane has two dimensions. We have encountered the specification of such points by two distances, frequently recorded as the ordered pair (x,y), and also by the intersection of two lines or two curves. This section introduces a method of specifying a point by specifying a distance and an angle, and then relates this coordinate system to the familiar rectangular coordinate system.

The *POLAR COORDINATE SYSTEM* consists of a fixed point and the positive half of a coordinate line emanating from that point. The fixed point is called the *POLE* and is labeled as *0* in Figure 15.1. The half line starting at the pole is called the *POLAR AXIS*. It provides the scale along which distances are measured. The angle θ is generated either in a positive or negative direction in the sense discussed in Section 13.2.

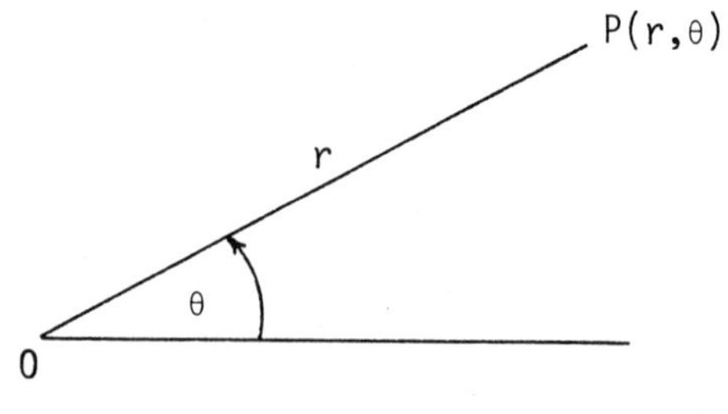

Figure 15.1

Each point P in the plane can be specified as an ordered pair (r,θ), where r is the length of the line segment OP and θ is the angle from the polar axis to OP as shown in Figure 15.1. The coordinates r and θ may be negative as well as positive. The distance r is positive if it is measured from the pole to the point along the terminal side of the angle and negative if measured in the opposite direction. Figure 15.2 displays (r,θ), $(-r,\theta)$ and $(r,-\theta)$. The point $(0,\theta)$ corresponds to the pole for any value of θ.

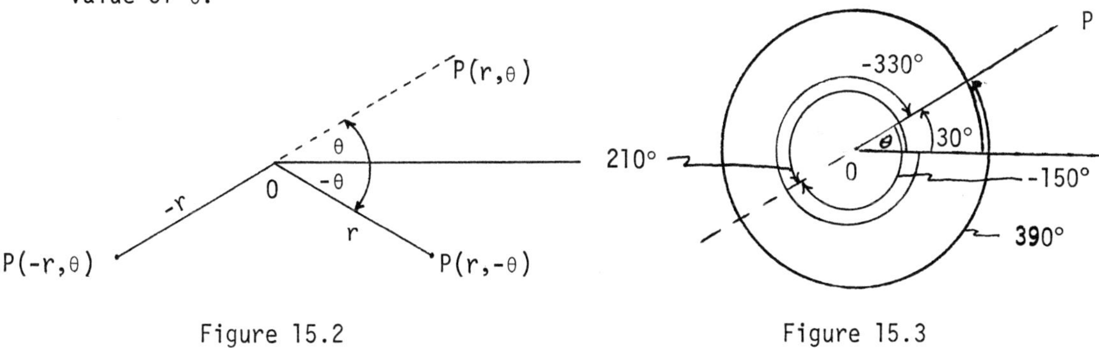

Figure 15.2

Figure 15.3

The polar coordinate representation of a point lacks the uniqueness of a rectangular representation. The following polar coordinates all specify the same point: (r,30°), (r,390°), (-r,-150°), (-r,210°), (r,-330°), as shown in Figure 15.3. In general, if P has the polar coordinates (r,θ), then it also has the polar coordinates $(r,\theta \pm 2n\pi)$ and $(-r,\theta \pm (2n - 1)\pi)$ for every integer n. On the other hand, every ordered pair of numbers (r,θ) determines a unique point P such that $|OP| = |r|$ and θ is the measure (radian or degrees) of an angle having its initial side along the polar axis and its terminal side along OP.

To plot an equation in the polar coordinate system, simply assign values to θ and calculate r. For example, the equation $r = K$ specifies a circle of radius K centered at the pole while $\theta = C$ describes a straight line passing through the pole at an angle of C to the polar axis.

<u>EXAMPLE 15.1</u>: In studying the navigational system of homing pigeons, release points are picked to have certain characteristics which might serve as clues for the birds' navigation system. Birds are released individually

at intervals of at least five minutes by tossing them into the air headed in a randomly chosen direction. Each bird is observed through high-powered binoculars until it disappears from sight. When the bird disappears, the observer lowers his field of vision to a circle surrounding the release point. Since angular measurements are marked off in 5° increments on the circle, the direction of the bird's departure can be measured with some precision.

Just as the angle of the pigeon's departure measures his directional guidance, his elapsed time between release and disappearance measures the intensity of his homing instinct. A set of data from such an experiment appears in Table 15.1. The data is plotted in Figure 15.4. This type of data was reported by Keeton (1969).

Table 15.1

	Control Pigeons (•)				Clock-Shifted Pigeons (o)				
Direction (°)	40	70	90	65	110	225	130	155	165
Time (Minutes)	4.0	4.3	3.7	3.0	3.9	3.2	5.2	3.7	4.2

The clock-shifted pigeons had spent several preceding days in a light regime six hours behind the actual time (they thought it was 8 a.m. when it was really 2 p.m.); the control pigeons were treated similarly except that their light regime was synchronized with the actual daylight hours. If the treated birds can see the sun, they depart approximately 90° clockwise from the home direction while the controls head homeward. This treatment has no demonstrable effect

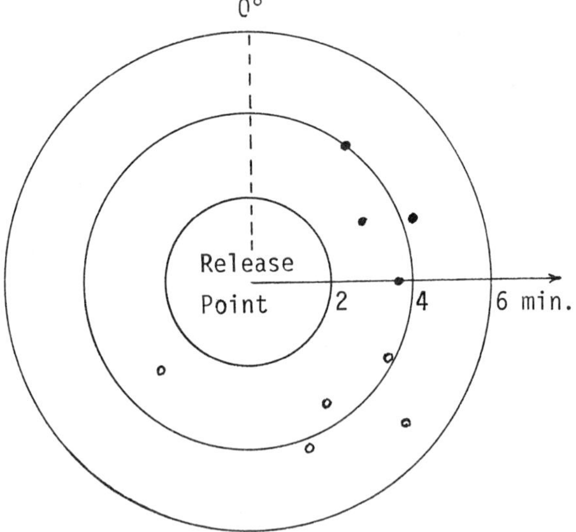

Figure 15.4

on time to disappearance. ▌

EXAMPLE 15.2: The dose within an ionizing beam of radiation decreases toward its boundaries from a maximum in its center. This implies that a given dose will penetrate a greater depth in the center of the beam than near its edges. Depth-dose measurements at various positions within the beam provide a means for locating points of equal dose throughout the beam. These points can be joined to form a curve called an *ISODOSE CURVE* as is illustrated in Figure 15.5. This plot summarizes the important part of the polar coordinate plot shown in Figure 15.6. In that figure the pole is located at the source and

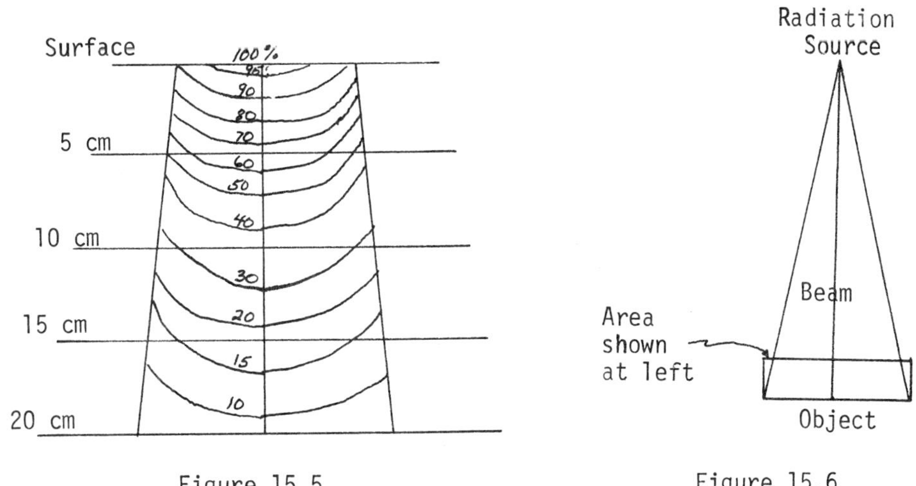

Figure 15.5

Figure 15.6

the polar axis points downward. Although the polar axis usually goes to the right from the pole, it may be reoriented for a suitable reason, as here. ▌

A rectangular and a polar coordinate system can be superimposed with corresponding origins and with the positive x-axis of the rectangular system along the polar axis. Then each point in the plane has both a rectangular, (x,y), and a polar, (r,θ), representation. A simple relation exists between these different representations.

Figure 15.7 shows a point P with coordinates (x,y) and (r,θ) such that $P(x,y) = P(r,\theta)$. For convenience, consider only polar

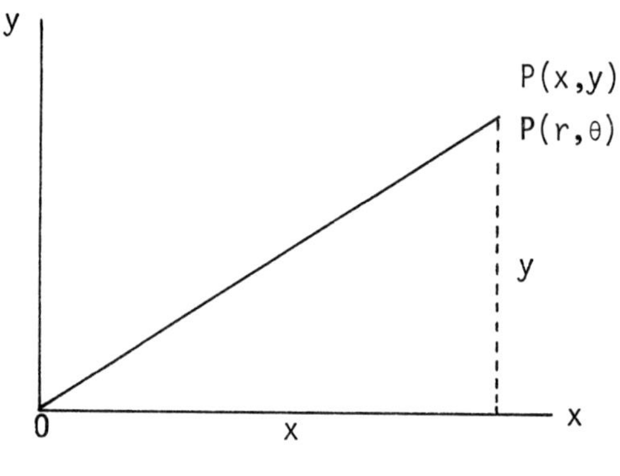

Figure 15.7

representations having $r > 0$. Now from the definitions of the trigonometric functions given in Chapter 13,

$$\sin\theta = \frac{y}{r} \qquad \text{and} \qquad \cos\theta = \frac{x}{r}\ .$$

By multiplying both sides of these equations by r, we have

$$x = r\cos\theta \qquad \text{and} \qquad y = r\sin\theta\ . \tag{15.1}$$

These equations relate polar to rectangular coordinates. They give the polar to rectangular transformation directly; to go from rectangular to polar coordinates requires that we solve these equations for r and θ in terms of x and y. This poses only a simple problem, as the Pythagorean theorem gives

$$r = \sqrt{x^2 + y^2}\ . \tag{15.2}$$

To solve for θ form the quotient $\frac{y}{x}$ and obtain

$$\frac{y}{x} = \tan\theta \qquad \text{or} \qquad \theta = \text{Tan}^{-1}\frac{y}{x} \tag{15.3}$$

Now we can convert from rectangular to polar coordinates and vice versa.

<u>EXAMPLE 15.3</u>: Transform $x^2 + y^2 - 8y = 0$ into polar coordinates.

Since $x = r\cos\theta$ and $y = r\sin\theta$:

$$x^2 + y^2 - 8y = r^2 \cos^2 \theta + r^2 \sin^2 \theta - 8r \sin \theta = 0$$

$$r^2(\cos^2 \theta + \sin^2 \theta) - 8r \sin \theta = 0$$

$$r^2 - 8r \sin \theta = r(r - 8 \sin \theta) = 0$$

because $\cos^2 \theta + \sin^2 \theta = 1$. If $r \neq 0$, this reduces to $r = 8 \sin \theta$. This curve is graphed in Figure 15.8. The point $(8 \sin \frac{\pi}{4}, \frac{\pi}{4}) = (4\sqrt{2}, \frac{\pi}{4})$ is plotted as a representative point. As θ ranges from 0 to π, a complete circle is traced out; as θ ranges from π to 2π, the same circle is traced out again because $r = 8 \sin \theta$ becomes negative for $\pi \leq \theta \leq 2\pi$. As an example,

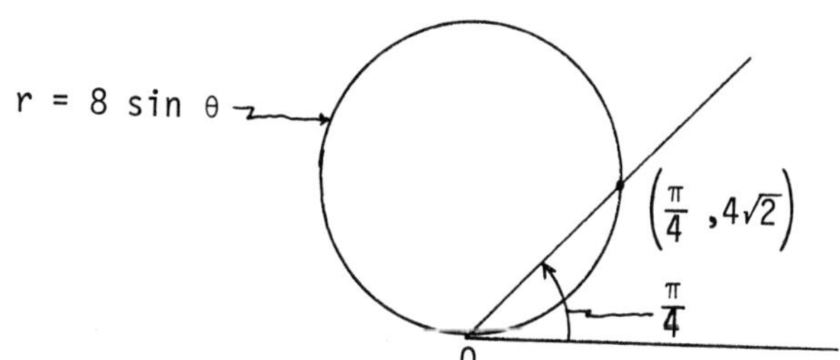

Figure 15.8

$$(8 \sin \frac{5\pi}{4}, \frac{5\pi}{4}) = (-4\sqrt{2}, \frac{5\pi}{4}) = (4\sqrt{2}, \frac{\pi}{4})$$

which is the plotted point.

If we complete the square in y in the original equation, we have

$$x^2 + (y - 4)^2 = 16$$

which you should recognize as the equation of a circle of radius 4, centered at (0,4) in the x,y-plane, that is, the same circle as before. ▌

EXAMPLE 15.4: Transform $r \cos \theta = 3$ into rectangular coordinate form.

Because $r \cos \theta = x$, $x = 3$ is the rectangular representation of $r \cos \theta = 3$. ▌

EXERCISES

Plot the point whose polar coordinates are given. Give two other representations of the point, one using a negative angle and the other using a negative of the given r. Give the rectangular coordinates of the point.

1. $\left(-1, \frac{7\pi}{4}\right)$

2. $\left(3, \frac{5\pi}{4}\right)$

3. $(3,120^{\circ})$

4. $\left(-2, \frac{3\pi}{4}\right)$

5. $\left(2, \frac{\pi}{4}\right)$

The following points are given in rectangular coordinates; give two sets of polar coordinates for each.

6. $(1,1)$

7. $(2\sqrt{3},-2)$

8. $(-3,-3\sqrt{3})$

9. $(-4,-3)$

10. $(-\sqrt{3},1)$

Change the following equations either from polar to rectangular representation, or the reverse, whichever is appropriate.

11. $r^2 = 16$

12. $r = -4 \cos \theta$

13. $r = \dfrac{2}{1 - \sin \theta}$

14. $r = a \sin \theta$

15. $r = \dfrac{2}{1 - \frac{1}{2} \cos \theta}$

16. $x^2 + y^2 = -6y$

17. $y + 5 = 0$

18. $y = x^2 + 5x$

19. $x = y^2 - 4$

20. $x^2 + 4x + y^2 - 6y + 13 = 0$

21. Graph $r = 3 \sin 2\theta$.

22. Graph $x^2 + y^2 = 4$ and $r = 2$ on the same graph using, respectively, rectangular and polar coordinates.

23. A shielded radioactive source still emits radiation. The following table gives the distance from a shielded cobalt-60 source at which doses of 5, 10 and 15 milliroentgens per hour were measured. Plot the data and draw the corresponding isodose curves in a suitable polar coordinate system.

15 mr./hr.

r	8	10	12	26	10	12	14	26	22	18	22
θ	0	10°	20°	30°	50°	90°	100°	115°	125°	150°	180°

10 mr./hr.

r	11	19	32	16	14	16	32	29	26	32
θ	0	20°	30°	50°	60°	90°	120°	130°	140°	180°

5 mr./hr.

r	20	32	50	26	22	20	52	45	38	46
θ	0	20°	30°	50°	60°	90°	120°	130°	150°	180°

15.2 DOUBLE INTEGRALS IN POLAR COORDINATES

The last section introduced the polar coordinate system. In addition to its utility as a directly usable coordinate system, it extends our capability for dealing with double integrals. It proves particularly useful for regions bounded by curves simply described in the polar coordinate system. This change of coordinate systems requires the introduction of no new concepts; however, it leads to one practical change. Our earlier considerations of double integrals used iterated integrals with $dA = dx\,dy$;

here, $dA = r\,dr\,d\theta$. As we develop integration in the polar plane, the reason for this change will appear.

To begin our discussion of integration in the polar coordinate system, we offer the following:

> If the function f is defined over a region R, then double integrals can be evaluated as iterated integrals:
>
> - $$\iint_R f(r,\theta)dA = \int_\alpha^\beta \int_{r_1(\theta)}^{r_2(\theta)} f(r,\theta)r\,dr\,d\theta \,. \qquad (15.4)$$
>
>   when $R = \{(r,\theta)|\ r_1(\theta) \leq r \leq r_2(\theta),\ \alpha \leq \theta \leq \beta\}$.
>
> - $$\iint_R f(r,\theta)dA = \int_a^b \int_{\theta_1(r)}^{\theta_2(r)} f(r,\theta)r\,d\theta\,dr \,. \qquad (15.5)$$
>
>   when $R = \{(r,\theta)|\ a \leq r \leq b,\ \theta_1(r) \leq \theta \leq \theta_1(r)\}$.
>
> - $$\iint_R dA = \text{Area}(R) \,. \qquad (15.6)$$

Figures 15.9 and 15.10 show typical regions over which these integrals are respectively evaluated.

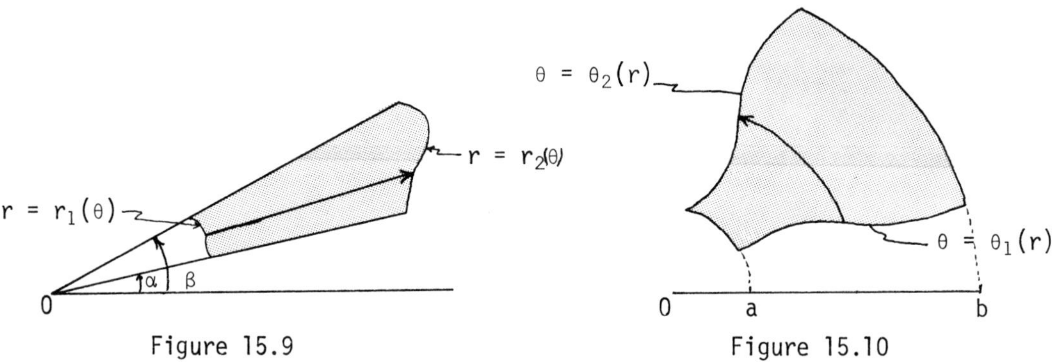

Figure 15.9

Figure 15.10

Section 12.2 introduced the double integral in the rectangular coordinate system. There the region R was partitioned into disjoint subregions; the function was evaluated at a point in each subregion and then multiplied by the area of that subregion; these products were added together; and the integral was defined as the limit of this sum as the number of the subregions increased and the area of the largest approached zero. Properties of the rectangular coordinate system did not enter this until, in the development of the fundamental theorem for double integrals, we utilized $\Delta_{ij}A = \Delta_i x \Delta_j y$.

When this process is repeated for a region in polar coordinates, a partition of the sort shown in Figure 15.11 results.

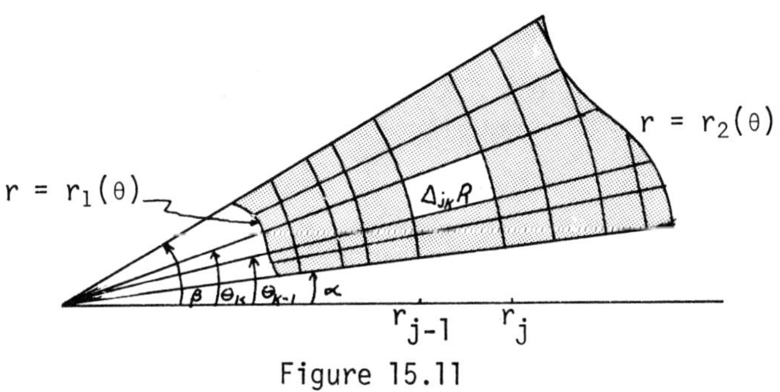

Figure 15.11

This region was partitioned by the lines $\theta = \theta_k$ through the pole and circles of radius $r = r_j$. All subregions are shaded except a typical one. Let this typical subregion be denoted by R_{jk} and defined by

$$\theta_{k-1} \leq \theta \leq \theta_k \qquad \text{and} \qquad r_{j-1} \leq r \leq r_j \ .$$

How do we evaluate the area, $\Delta_{jk}A$, of the subregion R_{jk}? We need only an approximation because this area will become smaller and smaller as the number of subregions increases. Figure 15.12 focuses on the typical subregion; it has the shape of a wedge. To approximate the area of R_{jk},

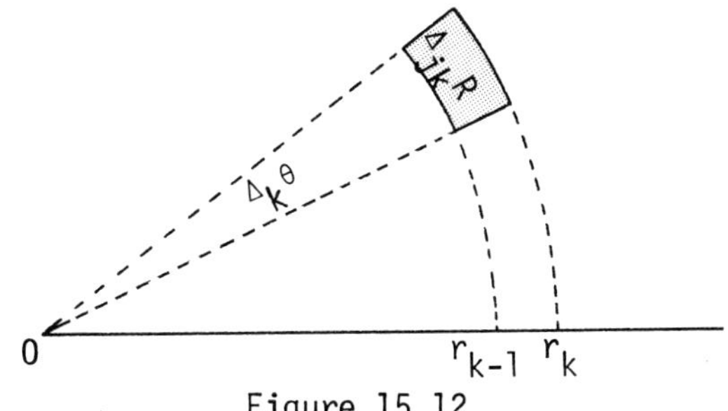

Figure 15.12

recall that the area of a sector of a circle (a pie-shaped portion of a circle) is $\frac{1}{2} r^2\theta$, where r is the radius of the circle and θ is the angle in radians of the sector. Calculate $\Delta_{jk}A$ by subtracting the area of a sector with a radius of r_{j-1} from the area of the sector of radius r_j, where both sectors include an angle of $\Delta_k\theta$:

$$\begin{aligned}\Delta_{jk}A &= \frac{1}{2} r_j^2\Delta_k\theta - \frac{1}{2} r_{j-1}^2\Delta_k\theta \\ &= \frac{1}{2} (r_j^2 - r_{j-1}^2)\Delta_k\theta \\ &= \frac{1}{2} (r_j + r_{j-1})(r_j - r_{j-1})\Delta_k\theta \\ &= r\Delta_j r\Delta_k\theta\end{aligned}$$

where $r = \frac{1}{2} (r_j + r_{j-1})$ gives the radius to the midsection of the region R_{jk}. If we would add up all of the subregions contained wholly in the region R and then let all $\Delta_{jk}A\to 0$ simultaneously, we would obtain

$$\begin{aligned}\iint_R f(r,\theta)dA &= \lim_{\Delta A\to 0} \sum_{(j,k)} f(r_j,\theta_k)\Delta A_{jk} \\ &= \lim_{\Delta A\to 0} \sum_{(j,k)} f(r_j,\theta_k)r\Delta_j r\Delta_k\theta \\ &= \iint_R f(r,\theta)r\ dr\ d\theta\ . \end{aligned} \tag{15.7}$$

When $f(r,\theta) = 1$, Equation 15.7 gives the area of the region R.

Separation of $\Delta A\to 0$ into $\Delta r\to 0$ and $\Delta\theta\to 0$ and a suitable ordering of components in Equation 15.7 will give the basic relations of Equation 15.4 and Equation 15.5. Again, as with our earlier consideration of double integrals, we could concentrate on details to increase the rigor of this presentation; to do so would introduce undue distraction at this point. The following examples illustrate mathematical and then biological points.

EXAMPLE 15.5: Evaluate the area inside the circle $r = 7$.

Here $R = \{(r,\theta)|\ 0 \le r \le 7,\ 0 \le \theta \le 2\pi\}$ and $f(r,\theta) = 1$. Figure 15.13 shows the circle. The values r can assume are shown by the arrow. Now evaluate the area:

$$A = \iint_R r\,dr\,d\theta = \int_0^{2\pi} \int_0^7 r\,dr\,d\theta$$

$$= \int_0^{2\pi} \frac{r^2}{2}\Big]_0^7 d\theta = \int_0^{2\pi} \frac{49}{2}\,d\theta = \frac{49}{2}\,\theta\Big]_0^{2\pi}$$

$$= 49\pi \text{ square units,}$$

the same result as $A = \pi r^2$ gives with $r = 7$.

This example illustrates the impact of the choice of a coordinate system on the ease of evaluating an integral. To convince yourself of this fact, evaluate the area using rectangular coordinates. ▌

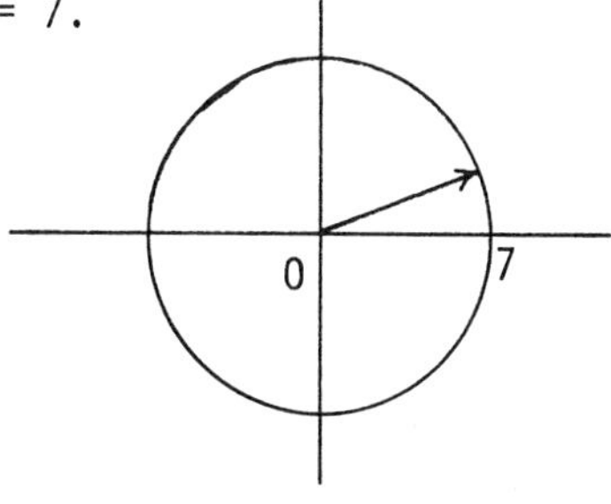

Figure 15.13

EXAMPLE 15.6: Find the area outside the circle $r = 2$ and inside the cardioid $r = 2(1 + \cos\theta)$.

This region is shaded in Figure 15.14 with an elemental area darkened. We can use symmetry of the shaded region to see that the required area equals twice the area that is swept out as θ varies from 0 to $\frac{\pi}{2}$. Suppose we integrate first with respect to r and then with respect to θ. Observe that r varies from $r = 2$ to $r = 2(1 + \cos\theta)$ as shown by the arrow in Figure 15.14 and θ varies from $\theta = 0$ to $\theta = \frac{\pi}{2}$:

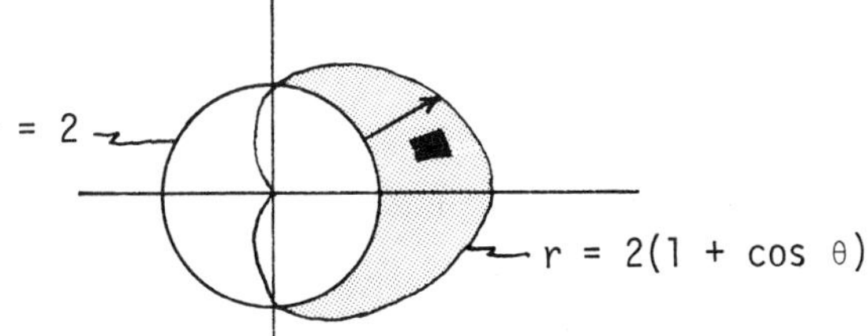

Figure 15.14

$$A = 2\int_0^{\frac{\pi}{2}} \int_2^{2(1+\cos\theta)} r\,dr\,d\theta = 2\int_0^{\frac{\pi}{2}} \frac{r^2}{2}\Big]_2^{2(1+\cos\theta)} d\theta$$

$$= 4\int_0^{\frac{\pi}{2}} (2\cos\theta + \cos^2\theta)\,d\theta\ .$$

To integrate the second term, recall that (Section 14.2)

$$\cos^2\theta = \frac{\cos 2\theta + 1}{2},$$

$$4\int_0^{\frac{\pi}{2}} (2\cos\theta + \cos^2\theta)d\theta = 4\int_0^{\frac{\pi}{2}} \left(2\cos\theta + \frac{\cos 2\theta + 1}{2}\right) d\theta$$

$$= 4\left[2\sin\theta + \frac{1}{4}\sin 2\theta + \frac{\theta}{2}\right]_0^{\frac{\pi}{2}}$$

$$= (8 + \pi) \text{ square units.}$$ ▌

EXAMPLE 15.7: Find the volume of a right circular cone of height H and basal radius R.

First, we will use s for the polar distance to avoid confusion with the radius of the cone. Suppose the cone is centered so its vertex lies above the pole. If $f(s,\theta)$ represents the height of the cone above (s,θ), then $\iint f(s,\theta)dA$ will give the volume of the cone. Figure 15.15 shows a section of the cone from the pole out, above any angle θ. To evaluate the height $f(s,\theta)$, we must resort to the properties of similar right triangles. Triangles ORH and SRP are similar right triangles so $\frac{R - S}{R} = \frac{SP}{OH}$. As $SP = f(s,\theta)$ and $OH = H$, $f(s,\theta) = \frac{H}{R}(R - S) = H(1 - \frac{S}{R})$. Now the volume of the cone is

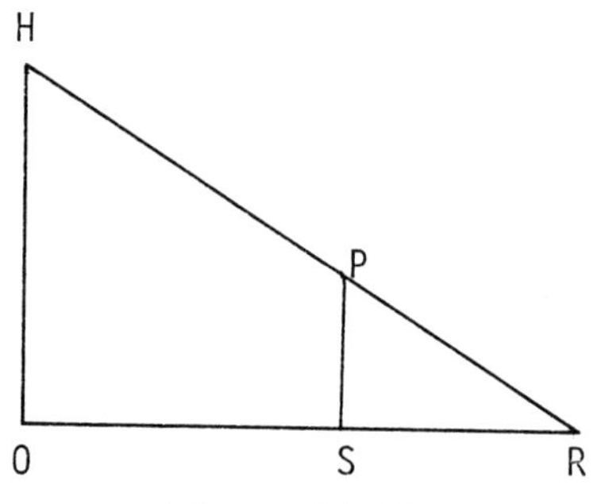

Figure 15.15

$$V = \iint_R f(s,\theta)dA = \int_0^{2\pi}\int_0^R H\left(1 - \frac{S}{R}\right) S\, dS\, d\theta$$

$$= H\int_0^{2\pi} \frac{S^2}{2} - \frac{S^3}{3R}\Big]_0^R d\theta = H\int_0^{2\pi} \left[\frac{R^2}{2} - \frac{R^3}{3}\right] d\theta$$

$$= \frac{R^2H}{6}\int_0^2 d\theta = \frac{R^2H}{6}(\theta)\Big]_0^{2\pi} = \frac{R^2H}{6}(2\pi - 0) = \frac{\pi R^3 H}{3} \text{ cubic units.}$$

What would happen if S were restricted by $0 \leq S \leq R_0 < H$? This would change the shape of the volume. The cone would loose its outside edges so that OSPH in Figure 15.15 would give its cross section. The integral would become $\int_0^2 \int_0^{R_0} H(1 - \frac{S}{R})S\ dS\ d\theta$ giving the volume as $\pi R^2 H\left[1 - \frac{2R_0}{3R}\right]$. The quantity $H(1 - \frac{S}{R})$ still represents a general height of the cone above any (r,θ). ▌

EXAMPLE 15.8: When organisms fan out from a release point, they frequently disperse without any particular directional orientation. This led Skellam (1951) to describe time dependent density in a polar coordinate system having the release point as the pole:

$$\phi(r,\theta,t) = \frac{1}{\pi a^2 t} \exp\left[- \frac{r^2}{a^2 t}\right]$$

This qualifies to be called a probability density function because for any time t,

$$\int_0^\infty \int_0^{2\pi} \phi(r,\theta,t) r\ d\theta\ dr = \int_0^\infty \int_0^{2\pi} \frac{1}{\pi a^2 t} \exp\left[- \frac{r^2}{a^2 t}\right] r\ d\theta\ dr$$

$$= \int_0^\infty \frac{1}{\pi a^2 t} \exp\left[- \frac{r^2}{a^2 t}\right] \theta\Big]_0^{2\pi} rdr$$

$$= \int_0^\infty \exp\left[- \frac{r^2}{a^2 t}\right]\left[\frac{2r}{a^2 t}\right] dr = \int_0^\infty \exp\left[- \frac{r^2}{a^2 t}\right] d\left[\frac{r^2}{a^2 t}\right]$$

$$= \lim_{h\to\infty} \int_0^h \exp\left[- \frac{r^2}{a^2 t}\right] d\left[\frac{r^2}{a^2 t}\right] = - \lim_{h\to\infty} \exp\left[- \frac{r^2}{a^2 t}\right]_0^h$$

$$= - \lim_{h\to\infty} \left[\exp\left[- \frac{h^2}{a^2 t}\right] - \exp[0]\right] = 1 .$$

As the density function does not depend on θ, $\phi(r,\theta,t) = \phi(r,0,t)$, indicating that the density function has the same height at any fixed distance r from the pole, at each particular time. This describes dispersal whose contours of equal density move out from the release point in ever-expanding circles, like the ripples created by tossing a pebble into calm water.

How might we characterize the rate of dispersal? Because the density is spread out to infinity at any time, we cannot put boundaries around the location of the entire population. Instead, we can find the boundaries within which a fixed proportion of the population lies. Specifically consider a release of N individuals. We can ask for a region which contains most of the individuals, like a proportion $1 - \frac{1}{N}$. Equivalently we can seek a region $r > R_t$ outside of which lies a proportion $\frac{1}{N}$:

$$\frac{1}{N} = \int_{R_t}^{\infty} \int_0^{2\pi} \phi(r,\theta,t) r \, d\theta \, dr = \lim_{h\to\infty} \left[-\exp\left[\frac{-r^2}{a^2t}\right]\right]_{R_t}^{h} = \exp\left[\frac{-R_t^2}{a^2t}\right].$$

Thus,

$$\ln\left[\frac{1}{N}\right] = -\frac{R_t^2}{a^2t} \qquad \text{or} \qquad R_t^2 = a^2t \ln N .$$

This says that most individuals in an expanding population lie in a circular region of radius $R_T = \sqrt{a^2t \ln N}$, or of area $\pi a^2 t \ln N$. Thus the area occupied by the expanding population varies in direct proportion to the time elapsed since release.

This formulation has limitations. For example, what time scale does it assume? Does it allow for births or deaths? To extend the time frame, we could examine the case of organisms newly introduced into a favorable environment so they would experience exponential growth: $N_t = N_0 e^{kt}$. Repeating the reasoning of the proceeding paragraph, we find for any time t,

$$\ln \frac{1}{N_t} = -\frac{R_t^2}{a^2t} \qquad \text{or} \qquad R_t^2 = a^2t \ln N_t = a^2t^2k + a^2t \ln N_0 .$$

For the population growth to follow the exponential, N_0 will soon become negligible relative to N_t. Thus $R_t \doteq a^2t^2k$. Thus the growing and dispersing population occupies an area whose radius increases almost directly with the square of time. Equivalently this says that $\sqrt{\text{area occupied}} = \sqrt{\pi R_t^2}$ changes directly with time. Contrast this to the first case where the population dispersed, but did not grow; it occupied an area which changed directly with time, not with the square of time.

Real situations have been examined with these models. In the absence of directional and/or extrinsic dispersal forces, the first model implies that $\sqrt{\text{area occupied}}$ should have a linear relation to time. If linearity fails, at least one of its assumptions must fail. Skellam (1951) found remarkable agreement of this approach with the spread of muskrats in Central Europe after their introduction about 1900. He also applied it to the spread of oak trees in Great Britain after the last ice age, concluding that animals or other extrinsic factors had aided their dispersal. ▮

EXERCISES

Evaluate the following:

1. $\int_{-\frac{\pi}{2}}^{\frac{\pi}{2}} \int_0^{2\cos\theta} r^2 \, dr \, d\theta$

2. $\int_0^{\frac{\pi}{2}} \int_0^a \sqrt{a^2 - r^2} \, r \, dr \, d\theta$

3. $\int_0^{2a} \int_r^{2r} r \, d\theta \, dr$

Evaluate the following double integrals in the rectangular system and then find the area of the region involved using polar coordinates.

4. $\int_0^1 \int_{y^2}^{y} dxdy$

5. $\int_0^a \int_0^{\frac{a^2-y^2}{2a}} dxdy$

Express the double integral of an arbitrary function $f(r,\theta)$ over the specified region R by use of an iterated integral and then evaluate the iterated integral.

6. R is the smaller segment of the circle $r = 4\cos\theta$ cut off by the line $r\cos\theta = 3$.

7. R is bounded by the cardioid $r = 1 + \sin\theta$, the circle $r = \sin\theta$ and the line $\theta = 0$ and is shown in the figure on the right.

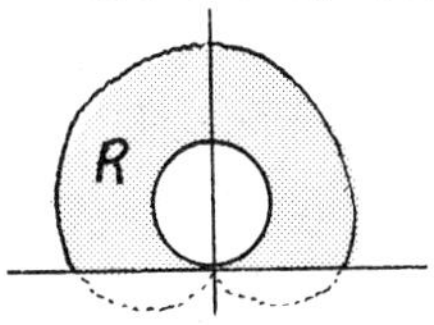

8. Find the area of a circle having a radius of A using polar coordinates.

9. Find the area of a right triangle with legs U and V by use of a double integral in polar coordinates.

10. Find the mass of R if $\delta = xy$ where R is the region inside the cardioid $r = 1 + \cos\theta$ and outside the cardioid $r = 1 - \cos\theta$ in Quadrant I.

15.3 THE CYLINDRICAL COORDINATE SYSTEM

The polar coordinate system has two common generalizations to three dimensions. One incorporates the third dimension through the use of a second angle while the other uses a second distance. The latter will be the subject of this and the next section while the former is considered in the last two sections of this chapter.

The CYLINDRICAL COORDINATE SYSTEM represents three-dimensional space by replacing the x,y-plane with a polar plane; the z-axis remains perpendicular to the polar plane and has its origin at the pole of the polar coordinate system. Each point P in space then has the ordered triplet (r,θ,z) for coordinates as shown in Figure 15.16. The point $Q(r,\theta)$ is the projection of P onto the polar plane, and z is the coordinate of the projection of P onto the z-axis. These restrictions assure uniqueness of a point's representation: $r \geq 0$, $0 \leq \theta \leq 2\pi$.

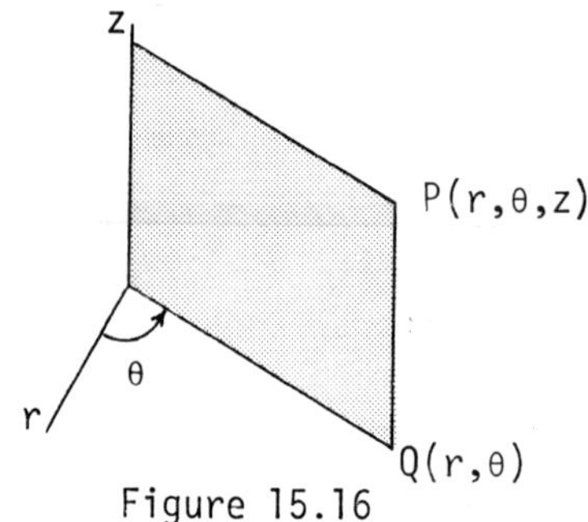

Figure 15.16

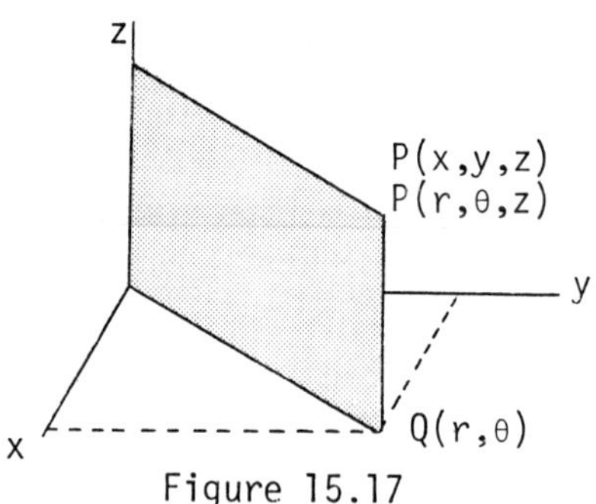

Figure 15.17

The cylindrical coordinates (r,θ,z) and the rectangular coordinates (x,y,z) for the same point satisfy

$$z = z, \qquad x = r\cos\theta, \qquad y = r\sin\theta. \tag{15.8}$$

Figure 15.17 shows the two systems superimposed; these relationships can be verified from the figure and knowledge of the two-dimensional polar coordinate system.

The equation $r = k$ where $k > 0$ describes a right circular cylinder of radius k centered around the z-axis. The equation $\theta = k$ specifies a half-plane emanating from the z-axis at an angle of θ from the polar axis, while $z = k$ defines a plane parallel to the polar coordinate plane at a height of k.

EXAMPLE 15.9: Express (4,2,-4) in cylindrical coordinates.

$z = -4$, $x = r \cos \theta$ and $y = r \sin \theta$ or $4 = r \cos \theta$ and $2 = r \sin \theta$ so $\tan \theta = \dfrac{\sin \theta}{\cos \theta} = \dfrac{2/r}{4/r} = \dfrac{1}{2}$ or $\theta = \operatorname{Tan}^{-1} \dfrac{1}{2}$ and

$$r = \sqrt{x^2 + y^2} = \sqrt{4^2 + 2^2} = \sqrt{20} = 2\sqrt{5}$$

Thus, the point (4,2,-4) in rectangular coordinates corresponds to $(2\sqrt{5}, \operatorname{Tan}^{-1} \frac{1}{2}, -4)$ in cylindrical coordinates. ▌

EXAMPLE 15.10: Find the rectangular coordinates of the point represented as $(10, \frac{3\pi}{2}, 4)$ in cylindrical coordinates.

$z = 4$, $x = 10 \cos\left[- \frac{3\pi}{2}\right] = 10(0) = 0$ and

$y = 10 \sin\left[- \frac{3\pi}{2}\right] = 10(-1) = -10$. Thus, $(10, - \frac{\pi}{2}, 4)$ in cylindrical coordinates corresponds to (0,-10,4) in rectangular coordinates. ▌

EXAMPLE 15.11: Cylindrical coordinates provide a convenient way for describing biological phenomena which have a vertical component superimposed on a circular pattern in the horizontal dimensions. We describe three such situations below.

Some plants occupy, and even "defend", a cylindrical space. Because most higher plants remain rooted permanently in one place, they compete in a limiting environment by seeking the same nutrients. This has led to many novel adaptations which become particularly apparent in arid regions. For example, the roots of sagebrush excrete a substance which inhibits young sagebrush plants from becoming established in their soil.

This induces a visually apparent spacing between plants. As a sagebrush plant's secondary roots fan out in all directions from its taproot, a cylindrical coordinate system would be well suited for studying the vertical distribution of this substance.

An ant colony constructs a conical pile of dirt removed from its underground nest, but the underground nest has no directional configuration. Thus a cylindrical coordinate system would be suited for studying the vertical distribution of food stored in the nest.

The amount and/intensity of the red color of apples depend on several factors, but location on the tree relative to the sun definitely is one of them. The use of a cylindrical coordinate system in studying these responses would directly identify three potentially important factors: Height, angular or compass locations, and distance from either the trunk or the edge of the canopy. ∎

EXERCISES

1. Find the cylindrical coordinates for the rectangular point:
   a. $(1,-\sqrt{3},4)$
   b. $(6,3,2)$
   c. $(0,2,2)$
   d. $(0,-2,-2)$

2. Find the rectangular coordinates for the cylindrical coordinate point:
   a. $(2,\mathrm{Cos}^{-1}\frac{3}{5}, 6)$
   b. $(1,\frac{\pi}{2},-3)$
   c. $(6,120°,2)$
   d. $(1,330°,6)$

3. Find the equations in cylindrical coordinates of the following equations given in rectangular coordinates:
   a. $z = 3x^2 + 3y^2$
   b. $3x - y + z = 4$
   c. $(x + y)^2 = z - 5$
   d. $\frac{x^2}{a^2} + \frac{y^2}{b^2} = 1$
   e. $x^2 + y^2 - 8x = 0$
   f. $x^2 - y^2 + 2y - 7 = 0$

4. Find the equations in rectangular coordinates of the following equations given in cylindrical coordinates:

   a. $r^2 - r(2 \cos \theta + 3 \sin \theta) + 2z^2 - z + 2 = 0$

   b. $r - 4 \cos \theta = 0$

   c. $r^2 \cos 2\theta + 2r \sin \theta - 6 = 0$

   d. $5r \cos^2 \theta - 4r \sin^2 \theta + 2 \cos \theta + 3 \sin \theta = 0$

15.4 INTEGRALS IN CYLINDRICAL COORDINATE SYSTEMS

Integration of functions defined in a cylindrical coordinate system introduces no new ideas, only a new combination of previously introduced ideas. The integral could be defined in terms of a partition of three-dimensional space into wedges of circular disks as shown in Figure 15.18. Because the area of the base of this wedge was approximated by $r\ \Delta r\ \Delta\theta$ in polar coordinates, the volume of the wedge approximately equals $r\ \Delta r\ \Delta\theta\ \Delta z$ in cylindrical coordinates. If the volume of each of these wedges were added up, and the limit taken as the number of these wedges gets very large, then we would obtain the volume of region under investigation. Thus the incremental volume in the rectangular coordinate system, dx dy dz, is replaced by $r\ dr\ d\theta\ dz$ in cylindrical coordinates; the differentials can be reordered in any suitable manner.

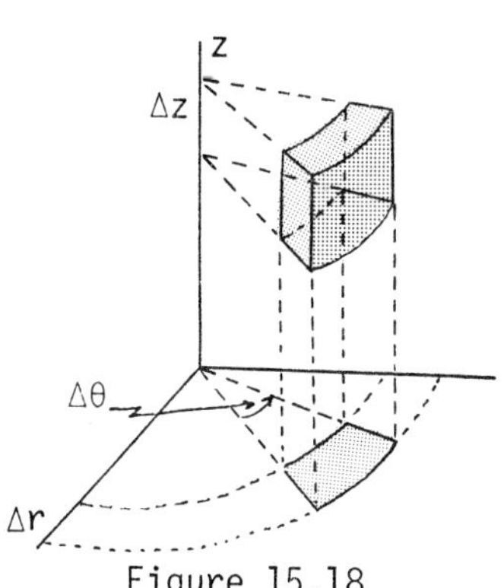

Figure 15.18

The triple integral in cylindrical coordinates is evaluated by using an iterated integral. For illustration, suppose that the solid S is bounded by the planes $\theta = \alpha$ and $\theta = \beta$ where $\alpha < \beta$, the cylinders $r = r_1(\theta)$ and $r = r_2(\theta)$ where $r_1(\theta) < r_2(\theta)$ for $\alpha \leq \theta \leq \beta$, and by the surfaces $z = z_1(r,\theta)$ and $z = z_2(r,\theta)$ where $z_1(r,\theta) \leq z_2(r,\theta)$ and z_1 and z_2 are continuous throughout the region R in the polar plane bounded by $r = r_1(\theta)$, $r = r_2(\theta)$, $\theta = \alpha$ and $\theta = \beta$. Then the triple integral of $f(r,\theta,z)$ in cylindrical coordinates is given by

$$\iiint_S f(r,\theta,z)r\,dz\,dr\,d\theta = \int_\alpha^\beta \int_{r_1(\theta)}^{r_2(\theta)} \int_{z_1(r,\theta)}^{z_2(r,\theta)} f(r,\theta,z)r\,dz\,dr\,d\theta\ .$$

EXAMPLE 15.12: Evaluate $\int_0^{\frac{\pi}{2}} \int_0^4 \int_r^{8 - \frac{r^2}{4}} r\,dz\,dr\,d\theta$.

Integrating first with respect to z, we obtain

$$\int_0^{\frac{\pi}{2}} \int_0^4 rz\Big]_r^{8 - \frac{r^2}{4}} dr\,d\theta = \int_0^{\frac{\pi}{2}} \int_0^4 (8r - \frac{r^3}{4} - r^2)\,dr\,d\theta$$

$$= \int_0^{\frac{\pi}{2}} \left[4r^2 - \frac{r^4}{16} - \frac{r^3}{3}\right]_0^4 d\theta$$

$$= \int_0^{\frac{\pi}{2}} \frac{80}{3}\,d\theta = \frac{80}{3}\,\theta\Big]_0^{\frac{\pi}{2}} = \frac{40\pi}{3}\ . \quad \blacksquare$$

As in rectangular coordinates, $f(r,\theta,z) = 1$ gives the volume of the region of integration. Specifically, if a region is specified in cylindrical coordinates by $S = \{(r,\theta,z) \mid r_1(\theta) \le r \le r_2(\theta),\ \alpha \le \theta \le \beta,\ z_1(r,\theta) \le z \le z_2(r,\theta)\}$, then the volume of S is given by

$$V = \iiint_S r\,dz\,dr\,d\theta = \int_\alpha^\beta \int_{r_1(\theta)}^{r_2(\theta)} \int_{z_1(r,\theta)}^{z_2(r,\theta)} r\,dz\,dr\,d\theta\ . \tag{15.9}$$

If we integrate this expression with respect to z, we obtain

$$V = \int_\alpha^\beta \int_{r_1(\theta)}^{r_2(\theta)} zr\Big]_{z_1(r,\theta)}^{z_2(r,\theta)} dr\,d\theta = \int_\alpha^\beta \int_{r_1(\theta)}^{r_2(\theta)} (z_2(r,\theta) - z_1(r,\theta))r\,dr\,d\theta\ .$$

Now $z_2(r,\theta) - z_1(r,\theta)$ merely gives the thickness of the volume above (r,θ) in the polar plane and this reduces to the case discussed in Section 15.2.

EXAMPLE 15.13: Find the volume of the solid which lies inside the cylinder $r = 2\cos\theta$, above the x,y-plane, and below the paraboloid

$z = 4 - x^2 - y^2$. Figure 15.19 shows the base of the volume and Figure 15.20 represents its volume in the first octant.

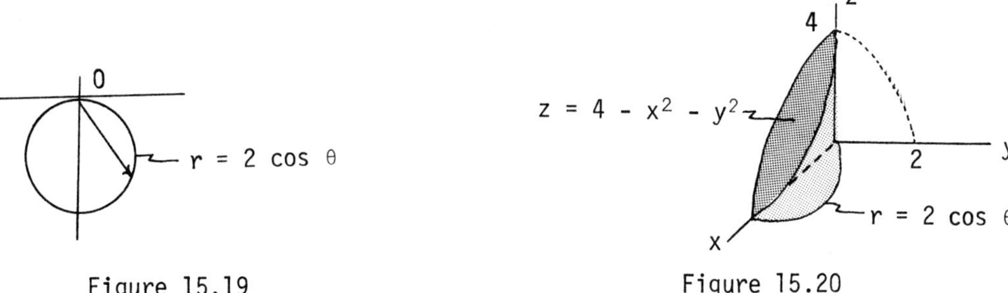

Figure 15.19 Figure 15.20

The intersection of the paraboloid with the cylinder is shaded heavily while the intersections of the paraboloid with the x,y and y,z-planes are shown as dotted lines. Convert z from rectangular coordinates to cylindrical coordinates:

$$z = 4 - (x^2 + y^2) = 4 - r^2 \qquad \text{because} \qquad r^2 = x^2 + y^2 .$$

The volume now becomes

$$V = \iiint_S r\,dz\,dr\,d\theta = 2\int_0^{\frac{\pi}{2}}\int_0^{2\cos\theta}\int_0^{4-r^2} r\,dz\,dr\,d\theta .$$

The arrow in Figure 15.19 shows the values over which r can range. From symmetry of the solid, we can let θ vary from 0 to $\frac{\pi}{2}$ and multiply the resulting integral by 2. If we had not used symmetry, the limits for θ would have run from $-\frac{\pi}{2}$ to $\frac{\pi}{2}$. Now evaluate the integral:

$$V = 2\int_0^{\frac{\pi}{2}}\int_0^{2\cos\theta} rz\Big]_0^{4-r^2} dr\,d\theta = 2\int_0^{\frac{\pi}{2}}\int_0^{2\cos\theta}(4r - r^3)dr\,d\theta$$

$$= 2\int_0^{\frac{\pi}{2}}(2r^2 - \frac{r^4}{4})\Big]_0^{2\cos\theta} d\theta = 2\int_0^{\frac{\pi}{2}}(8\cos^2\theta - 4\cos^4\theta)d\theta .$$

To evaluate this integral we need to use the trigonometric reductions explained in Section 14.2. Use $\cos^2 \theta = \frac{1 + \cos 2\theta}{2}$ and $\cos^2 2\theta = \frac{1 + \cos 4\theta}{2}$ in this way:

$$
\begin{aligned}
8 \cos^2 \theta - 4 \cos^4 \theta &= 8\left[\frac{1 + \cos 2\theta}{2}\right] - 4\left[\frac{1 + \cos 2\theta}{2}\right]^2 \\
&= 4 + 4 \cos 2\theta - (1 + 2 \cos 2\theta + \cos^2 2\theta) \\
&= 3 + 2 \cos 2\theta - \cos^2 2\theta \\
&= 3 + 2 \cos 2\theta - \left[\frac{1 + \cos 40}{2}\right] \\
&= \frac{5}{2} + 2 \cos 2\theta - \frac{\cos 4\theta}{2}
\end{aligned}
$$

Now the integral becomes

$$
\begin{aligned}
V = 2\int_0^{\frac{\pi}{2}} (8 \cos^2 \theta - 4 \cos^4 \theta)d\theta &= 2\int_0^{\frac{\pi}{2}} \left[\frac{5}{2} + 2 \cos 2\theta - \frac{\cos 4\theta}{2}\right]d\theta \\
&= 2\left[\frac{5}{2}\theta + \sin 2\theta \quad \frac{\sin 4\theta}{8}\right]\Bigg]_0^{\frac{\pi}{2}} \\
&= \frac{5}{2}\pi \text{ cubic units.}
\end{aligned}
$$

The integration to evaluate this volume in the cylindrical coordinate system was relatively simple. Nasty integration problems turn up when this volume is evaluated in the rectangular coordinate system. **I**

To evaluate a volume above the polar coordinate plane, we integrated its height over the appropriate region in the polar plane. If we integrate in either of the orders z,r,θ or z,θ,r the first integration gives the height; note however that four other orders of integration (r,θ,z; r,z,θ; θ,r,z; or θ,z,r) can be used in the cylindrical system.

In earlier sections and chapters we have considered mass densities and probability densities. Again here if $f(r,\theta,z)$ represents the density of a mass over a solid S at any point (r,θ,z), then the total mass of S is given by

$$\text{Mass of } S = \iiint_S f(r,\theta,z)r\,dr\,d\theta\,dz\ . \qquad (15.10)$$

This equation also holds if $f(r,\theta,z)$ represents a probability density function. In this case, however, the total probability (mass) equals unity over the entire region of interest.

<u>EXAMPLE 15.14</u>: Example 15.11 described the location of an anthill in the cylindrical coordinate system. Within that framework, suppose the density (in grams per cubic centimeter) of stocked food is as follows:

$$f(r,\theta,z) = 10^{-6}(240z - z^2)e^{-r^2/100} \qquad \text{where} \qquad \begin{array}{l} 0 \le r \le 30 \\ 0 \le z \le 240\ . \end{array}$$

The quadratic term $240z - z^2$ increases to its maximum at $z = 120$ cm.; thereafter it decreases to 0 at $z = 240$ cm. Thus for any fixed r, density of stored food increases to a depth of 120 cm., but diminishes to zero by a depth of 240 cm. The term $e^{-r^2/100}$ describes storage, at a fixed depth, which decreases away from the center of the nest. For this density then,

$$\text{Total Stored Food} = 10^{-6} \int_0^{240} \int_0^{2\pi} \int_0^{30} (240z - z^2)e^{-r^2/100}\, r\,dr\,d\theta\,dz$$

$$= 10^{-6} \int_0^{240} (240z - z^2) \int_0^{2\pi} \left[-\frac{100}{2}\, e^{-r^2/100}\right]_0^{30} d\theta\,dz$$

$$= 10^{-6}\,(50)(1 - e^{-9}) \int_0^{240} (240z - z^2) \int_0^{2\pi} d\theta\,dz$$

$$= 10^{-6}\,(100\pi)(1 - e^{-9}) \int_0^{240} (240z - z^2)dz$$

$$= 10^{-6}\,(100\pi)(1 - e^{-9})(120z^2 - \frac{z^3}{3})\Big]_0^{240} \doteq 724 \text{ grams.} \quad \blacksquare$$

EXERCISES

1. Find the volume of a cylindrical can of radius a using cylindrical coordinates. The height of the can is h.

2. Find the volume of the solid which has a height h and a base defined by $r = 1 + \cos\theta$, $0 \leq \theta \leq 2\pi$. What shape does the base have?

3. Find the volume of the sphere $x^2 + y^2 + z^2 = a^2$ by use of cylindrical coordinates.

4. Find the volume of the region above the (r, θ)-plane, within the given cylinder and below the specified surface
   a. Within the cylinder $x^2 + y^2 = a^2$ and below $x^2 + y^2 + z^2 = 4a^2$.
   b. Within the cylinder $x^2 + y^2 = 8y$ and below the paraboloid $4z = x^2 + y^2$.
   c. Within the cylinder $r = a$ and below the parabolic cylinder $z = a^2 - x^2$.

Suppose a region S is bounded by the given surfaces. Formulate the volume V or mass M as a triple integral in cylindrical coordinates and then evaluate as an iterated integral. The density of a mass is denoted by δ.

5. S is bounded by $x^2 + y^2 = 36$, $z = 0$ and $z = 6$. Find V, and M if $\delta = 4z$.

6. S is bounded by $x^2 + y^2 = 25$, $z = 2$ and $z = 5$. Find V.

7. S is bounded by $z = 0$ and $x^2 + y^2 + z^2 = a^2$ where $z \geq 0$. Find V.

8. S is bounded by $z = a$, $x^2 + y^2 = a^2$ and $x^2 + y^2 + z^2 = 9a^2$ where $a \leq z \leq 3a$. Find V.

9. S is bounded by the cylinder $r = a$, by $z = 0$ and $z = a^2 - x^2$. Find V.

10. S is bounded by $r = 4\cos\theta$, $z = 0$ and $z = \frac{1}{2}(x^2 + y^2)$. Find M if $\delta = 2r$.

11. Find the volume of the region S described in Problem 10.

12. S lies above the x,y-plane, below the plane $x + z = 4$ and inside the cylinder $r = 4\cos 2\theta$.

13. Which coordinate system is easier to use in calculating the volume of a cylinder? Explain.

14. Find the volume of the solid lying within the cylinder $r = 2 \sin \theta$, above the paraboloid $f(r,\theta) = 4 - r^2$ and below the (r,θ) plane.

15. Find the volume inside $x^2 + y^2 = 4x$, above $z = 0$ and below $x^2 + y^2 = 4z$.

16. Is it important to know the shape of the top or bottom of the region under question in order to calculate the volume in either cylindrical or rectangular coordinates?

15.5 THE SPHERICAL COORDINATE SYSTEM

The previous two sections presented one of the common generalizations of the polar coordinate system into three dimensions, namely the cylindrical coordinate system. Now we introduce the other common generalization called the SPHERICAL or GEOGRAPHICAL COORDINATE SYSTEM.

The spherical coordinate system is constructed from a plane having a polar coordinate system on it, and a z-axis perpendicular to this plane having its origin at the pole of the polar plane. Each point P in space has the spherical coordinates of (r,θ,ϕ) as is shown in Figure 15.21 where r is the distance OP. θ is the polar angle associated with the projection, Q, of P onto the polar plane, and ϕ is the angle which OP makes with the z-axis. For any point $P(r,\theta,\phi)$, we have the following restrictions: $r \geq 0$, $0 \leq \theta \leq 2\pi$ and $0 \leq \phi \leq \pi$ to assure uniqueness of the representation.

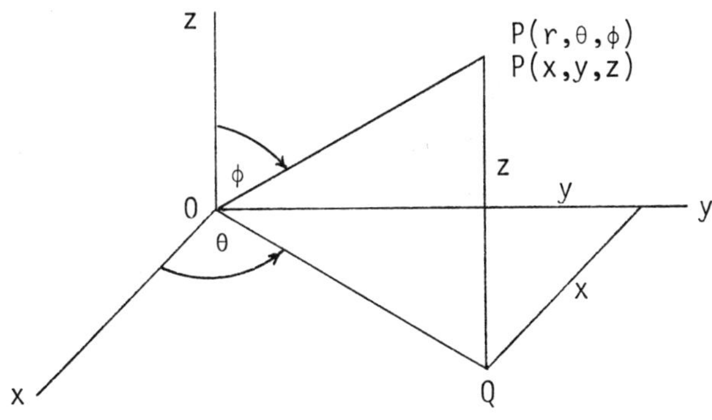

Figure 15.21

The equation $r = r_0$, $r_0 > 0$, describes a sphere with its center at the origin; $\theta = \theta_0$ specifies a half-plane emanating from the z-axis;

$\phi = \phi_0$, $0 \leq \phi_0 \leq \pi$ describes a right circular cone having the z-axis as its axis and the pole as its vertex.

If we superimpose a rectangular coordinate system on a spherical coordinate system as shown in Figure 15.21, each point in space has a dual set of coordinates (x,y,z) and (r,θ,ϕ). Because the distance OQ is

$$|OQ| = r \sin \phi ,$$

$$x = |OQ| \cos \theta = r \cos \theta \sin \phi \tag{15.11}$$

and

$$y = |OQ| \sin \theta = r \sin \theta \sin \phi . \tag{15.12}$$

Further,

$$z = r \cos \phi . \tag{15.13}$$

These equations give the relations needed to convert from spherical coordinates to rectangular coordinates. To convert from rectangular coordinates to spherical coordinates, we have

$$r = \sqrt{x^2 + y^2 + z^2}, \quad \theta = \operatorname{Tan}^{-1} \frac{y}{x} \quad \text{and} \quad \phi = \operatorname{Cos}^{-1} \frac{z}{\sqrt{x^2 + y^2 + z^2}} . \tag{15.14}$$

The justification of the equations in (15.14) will be left as an exercise.

EXAMPLE 15.15: Find the spherical coordinates for the point $(4,2,-4)$ in the rectangular coordinate system.

Equation 15.14 gives

$$r = \sqrt{x^2 + y^2 + z^2} = \sqrt{16 + 4 + 16} = \sqrt{36} = 6,$$

$$\theta = \operatorname{Tan}^{-1} \frac{y}{x} = \operatorname{Tan}^{-1} \frac{2}{4} = \operatorname{Tan}^{-1} \frac{1}{2} ,$$

and

$$\phi = \operatorname{Cos}^{-1} \frac{z}{r} = \operatorname{Cos}^{-1} \frac{-4}{6} = \operatorname{Cos}^{-1} \left[- \frac{2}{3}\right].$$

Thus, $(4,2,-4)$ in the rectangular system corresponds to $\left(6, \operatorname{Tan}^{-1} \frac{1}{2} , \operatorname{Cos}^{-1}\left[- \frac{2}{3}\right]\right) \doteq (6, 26.57°, 131.18°)$ in the spherical coordinate system. ▌

EXAMPLE 15.16: Find the rectangular coordinates given the spherical coordinates $(2, \frac{\pi}{4}, \frac{\pi}{3})$.

Use Equations 15.11 - 15.13:

$$x = r \cos\theta \sin\phi = 2 \cos\frac{\pi}{4} \sin\frac{\pi}{3} = \frac{\sqrt{6}}{2} ,$$

$$y = r \sin\theta \sin\phi = 2 \sin\frac{\pi}{4} \sin\frac{\pi}{3} = \frac{\sqrt{6}}{2} ,$$

and

$$z = r \cos\phi = 2 \cos\frac{\pi}{3} = 1 .$$

Thus, $(2, \frac{\pi}{4}, \frac{\pi}{3})$ in spherical coordinates corresponds to $(\frac{\sqrt{6}}{2}, \frac{\sqrt{6}}{2}, 1)$ in rectangular coordinates. ▌

EXAMPLE 15.17: Photosynthesis of plants poses the complexities of mathematical modeling present in many biological phenomena. Photosynthesis depends on the plant species, temperatures, availability of nutrients, leaf size, shape, inclination, density, and on the quality and intensity of the incoming solar radiation. Even though whole books have been devoted to this subject, mathematical models still are incomplete and are being developed.

The spherical coordinate system is adapted to displaying the impact of one feature of this system, namely, the influence of the solar angle on the intensity of the solar radiation. The direction of the sun can be fixed by the two angles of a spherical coordinate system located at a point of interest, namely, by the angle (θ) in the horizontal plane from north to the sun, and by the deviation (ϕ) of the sun from vertical. The horizontal angle θ, sometimes referred to as the sun's azimuth, varies through approximately 180° in a day, somewhat more in the summer and somewhat less in the winter. The vertical angle ϕ, is most noticeable at noon; it varies by approximately 45° with the seasons. The sun is nearly overhead in the southern parts of the United States during midsummer, so ϕ nearly equals zero then. In northern states such as Minnesota, the sun never gets higher than $\phi = 20°$, but ranges as low as $\phi = 65°$.

The energy in a bundle of sunlight depends primarily on its cross-sectional area. The area over which a fixed size bundle of light spreads out depends noticeably on the sun's angles. To see this, think of cutting a one-inch diameter circle out of a piece of cardboard. If this is held directly perpendicular to the direction from which the sun's rays are arriving, a bundle of sunlight 1 inch in diameter passes through the hole. When this bundle of sunlight hits a surface below, such as the ground, the size of the spot lighted changes through the day and with the seasons. For the moment, consider what happens at noon. The spot remains approximately one inch in width if the distance between the hole and the ground is small, but its length changes with as is shown in Figure 15.22. The left side shows sunlight arriving vertically so a

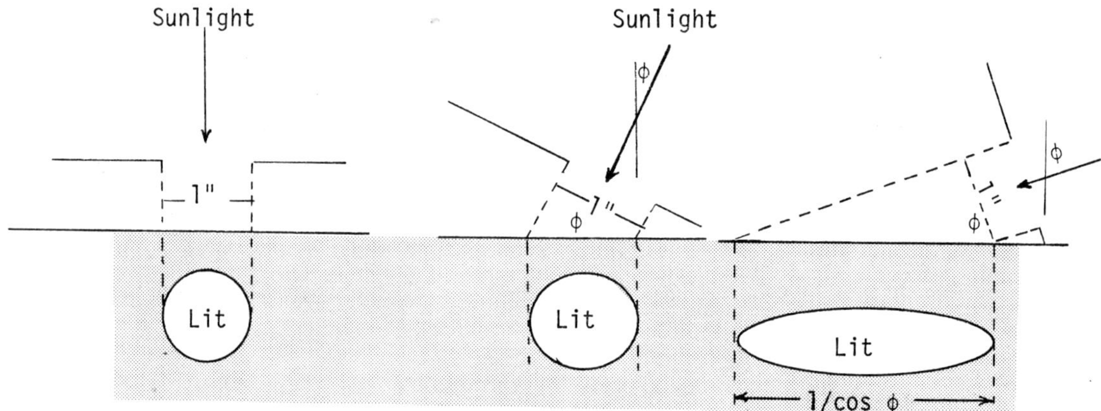

Figure 15.22

circle of sunlight passes through the hole and makes a circle of sunlight on the ground, as shown in the bottom part of the figure. On the right, the sunlight arrives with a large ϕ as it would in northern latitudes. The circle of sunlight passing through the hole spreads out over an area 1 inch wide, but substantially more than 1 inch long. In fact, 1 inch/length lit = cos ϕ or length lit = 1 inch/cos ϕ. The resulting eliptical area appears in the bottom part of Figure 15.22, on the right. The middle part of the figure shows an intermediate value of ϕ for which the ellipse is only slightly longer than it is wide. In Example 8.28 we found that the area of an ellipse with a length of 2a and a width of 2b was πab. Thus, the area of the lighted portion is $\pi\ \frac{1}{2}\left[\frac{1}{2}\frac{1}{\cos\phi}\right] = \frac{\pi}{4}\frac{1}{\cos\phi}$. The cross-sectional area of the bundle of light at the hole is $\frac{\pi}{4}$. Thus, if we would divide the cross-sectional area of the light at the hole by the area of the light on the ground, $\frac{\pi}{4}/\left[\frac{\pi}{4}\frac{1}{\cos\phi}\right] = \cos\phi$. This measures the dilution of the sun's energy due to its angle of inclination. At any time other than noon, the angle ϕ would introduce further dilution. (Think of the length of shadows at sunrise or sunset.)

This example is greatly simplified, but yet it points out the utility of adapting the spherical coordinate system to a biological example, even in a simplified case. ▌

EXERCISES

1. Find the spherical coordinates for the following points given in rectangular coordinates.

   a. $(1,-\sqrt{3},4)$ b. $(1,-2,-2)$ c. $(0,-2,2)$

   d. $(0,1,1)$ e. $(6,3,2)$ f. $(8,4,1)$

2. Find the rectangular coordinates for the following points given in spherical coordinates.

   a. $\left(3, \frac{\pi}{3}, \frac{\pi}{6}\right)$ b. $\left(5, \frac{\pi}{4}, \frac{\pi}{4}\right)$ c. $\left(9, \frac{\pi}{2}, \frac{\pi}{3}\right)$

   d. $\left(3, \frac{2\pi}{3}, \frac{4\pi}{3}\right)$ e. $\left(6, \frac{11\pi}{6}, \frac{\pi}{3}\right)$ f. $\left(5, \frac{5\pi}{6}, \frac{7\pi}{6}\right)$

3. The following surfaces are given in spherical coordinates. Find their equations in the rectangular coordinate system.

   a. $\cot\phi = \sin\theta + \cos\theta$

   b. $r = a\sin\phi\sin\theta$

   c. $r = a\cos\phi\cos\theta$

   d. $r^2\cos 2\phi = a^2$

   e. $r^2\sin^2\phi\sin 2\theta = a^2$

   f. $r^2\sin^2\theta\sin^2\phi + r^2\sin^2\phi\cos^2\theta = r^2$

4. The following surfaces are given in rectangular coordinates. Find their equations in spherical coordinates.
   a. $x^2 + y^2 + 3z^2 - 3x - z + 2 = 0$
   b. $3x^2 + 2y^2 - 5z = 0$
   c. $3x^2 - 3y^2 = 6z$
   d. $x^2 - y^2 - z^2 = 9$
   e. $x^2 + y^2 = 16$
   f. $x + y + z = 10$

15.6 INTEGRATION IN THE SPHERICAL COORDINATE SYSTEM

Integration of functions defined in a spherical coordinate system proceeds essentially as it did in the cylindrical coordinate system with the main difference lying in the differential element.

Suppose $f(r,\theta,\phi)$ is a continuous function over a region S in a spherical coordinate system where $a \leq r \leq b$, $\alpha \leq \theta \leq \beta$ and $\gamma \leq \phi \leq \eta$. A typical elementary subregion of S, shown in Figure 15.23, is bounded by the two spheres $r = r_0$ and $r = r_0 + \Delta r$, the two half-planes $\theta = \theta_0$ and $\theta = \theta_0 + \Delta\theta$, and the two half-cones $\phi = \phi_0$ and $\phi = \phi_0 + \Delta\phi$. The increments Δr, $\Delta\theta$ and $\Delta\phi$ can have any values as long as they approach zero in the limit.

The elementary volume shown in Figure 15.23 has a height of AB which equals Δr. The length from point A to C equals the length of an arc of a circle with a radius of r_0 across an angle of $\Delta\phi$. Thus, the depth AC equals $r_0\ \Delta\phi$. The width CD is not given by $r_0\ \Delta\theta$ as you might first expect from

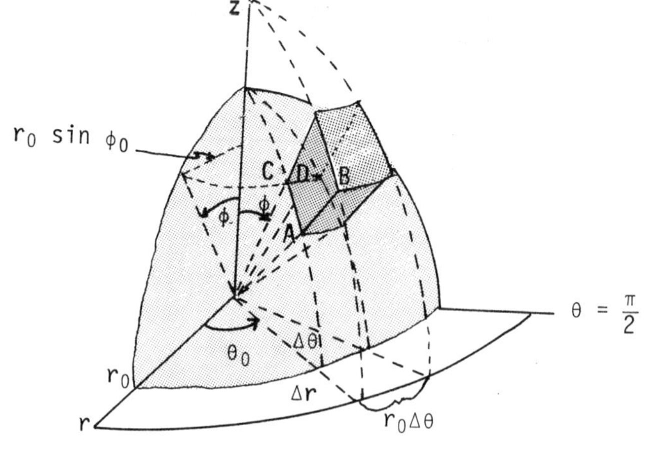

Figure 15.23

the above discussion. The horizontal circle containing CD has a radius of $r_0 \sin \phi$ so CD has a length $r_0 \sin \phi_0 \Delta\theta$ and

$$\Delta V = (\Delta r)(r_0 \Delta\phi)(r_0 \sin \phi_0 \Delta\theta) = r_0 \sin \phi_0 \; \Delta r \; \Delta\theta \; \Delta\phi \; .$$

Now, as before, if we form the sum of $f(r,\theta,\phi)\Delta V$ and take the limit, we have

$$\begin{aligned} \iiint_S f(r,\theta,\phi) dV &= \lim \sum f(r,\theta,\phi)\Delta V \\ &= \lim \sum f(r,\theta,\phi) r^2 \sin \phi \; \Delta r \; \Delta\theta \; \Delta\phi \\ &= \iiint_S f(r,\theta,\phi) r^2 \sin \phi \; dr \; d\theta \; d\phi \end{aligned}$$

where the summation is carried out over all subregions created by the partition and the limit is taken as the volume of all subregions approaches zero. This integral is evaluated as an iterated integral with six different possible orders of integration. We have as a typical case

$$\iiint_S f(r,\theta,\phi) r^2 \sin \phi \; dV = \int_\gamma^\eta \int_{\theta_1(\phi)}^{\theta_2(\phi)} \int_{r_1(\theta,\phi)}^{r_2(\theta,\phi)} f(r,\theta,\phi) r^2 \sin \phi \; dr \; d\theta \; d\phi \; .$$

EXAMPLE 15.18: Evaluate $A = \displaystyle\int_0^{\frac{\pi}{6}} \int_0^{2\pi} \int_0^{a \sec \phi} r^3 \cos \phi \sin \phi \; dr \; d\theta \; d\phi$.

From the order or integration and the limits, observe that $0 \leq r \leq a \sec \phi$, $0 \leq \theta \leq 2\pi$ and $0 \leq \phi \leq \frac{\pi}{6}$. Integrating with respect to r first and holding θ and ϕ constant, we have

$$A = \int_0^{\frac{\pi}{6}} \int_0^{2\pi} \frac{1}{4} r^4 \cos \phi \sin \phi \Big]_0^{a \sec \phi} d\theta \; d\phi$$

$$= \frac{a^4}{4} \int_0^{\frac{\pi}{6}} \int_0^{2\pi} \sec^4 \phi \cos \phi \sin \phi \; d\theta \; d\phi$$

$$= \frac{a^4}{4} \int_0^{\frac{\pi}{6}} \int_0^{2\pi} \cos^{-3} \phi \sin \phi \; d\theta \; d\phi$$

$$A = \frac{a^4}{4} \int_0^{\frac{\pi}{6}} (\cos^{-3} \phi \sin \phi)\theta\Big]_0^{2\pi} d\phi = \frac{a^4}{2} \pi \int_0^{\frac{\pi}{6}} \cos^{-3} \phi \sin \phi \, d\phi$$

$$= \left[- \frac{\pi a^4}{2} \frac{\cos^{-2} \phi}{-2} \right]_0^{\frac{\pi}{6}} = \frac{\pi a^4}{4} \frac{4}{3} - 1 = \frac{\pi a^4}{12} . \quad \blacksquare$$

EXAMPLE 15.19: Find the volume of a sphere of radius r_0 .

The region in spherical coordinates defined by $r \leq r_0$, $0 \leq \theta \leq 2\pi$, $0 \leq \phi \leq \pi$ describes the surface and interior of a sphere of radius r_0 . Integrate over the region described:

$$V = \int_0^{r_0} \int_0^{\pi} \int_0^{2\pi} r^2 \sin \phi \, d\theta \, d\phi \, dr$$

$$= \int_0^{r_0} r^2 \int_0^{\pi} 2\pi \sin \phi \, d\phi \, dr = 4\pi \int_0^{r_0} r^2 dr$$

$$= \frac{4}{3} \pi r_0^3 ,$$

the familiar volume for this solid. Notice the simplicity of this integration compared to an equivalent evaluation in another coordinate system. (For example, see Problem 3 at the end of Section 15.4.) $\blacksquare$

EXAMPLE 15.20: Find the volume of the solid S bounded below by the cone $z^2 = x^2 + y^2$ and above by the sphere $x^2 + y^2 + z^2 = 2az$.

The first octant of the solid appears in Figure 15.24. The whole solid is shaped like an ice-cream cone. Change the rectangular coordinates to spherical coordinates: $x^2 + y^2 + z^2 = 2az$ becomes $r^2 = 2a \, r \cos \phi$ or $r = 2a \cos \phi$ and $z^2 = x^2 + y^2$ becomes $2z^2 = r^2$ or $2r^2 \cos^2 \phi = r^2$ or $\phi = \frac{\pi}{4}$.

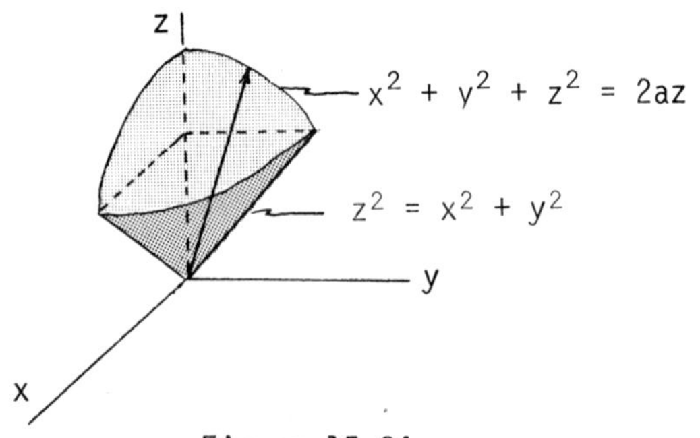

Figure 15.24

The arrow in the figure shows the values which r may assume. Thus, we have

$$V = \int_0^{\frac{\pi}{4}} \int_0^{2a \cos \phi} \int_0^{2\pi} r^2 \sin \phi \, d\theta \, dr \, d\phi = 2\pi \int_0^{\frac{\pi}{4}} \int_0^{2a \cos \phi} r^2 \sin \phi \, dr \, d\phi$$

$$= 2\pi \int_0^{\frac{\pi}{4}} \frac{r^3}{3}\Big]_0^{2a \cos \phi} \sin \phi \, d\phi = \frac{16a^3\pi}{3} \int_0^{\frac{\pi}{4}} \cos^3 \phi \sin \phi \, d\phi$$

$$= \left[\frac{16a^3\pi}{3} \quad - \frac{\cos^4 \phi}{4}\right]_0^{\frac{\pi}{4}} = \pi a^3 \text{ cubic units.}$$

To again see the utility of one coordinate system over another, try to work this problem using the rectangular coordinate system. ▌

Mass and probability are obtained in a manner like that outlined for cylindrical coordinates with the exception that the mass function or probability density is expressed in terms of spherical coordinates.

EXERCISES

Formulate the following in triple integrals and evaluate by use of spherical coordinates. As before, δ denotes a mass function.

1. Set up, but do not evaluate, the integrals to find the volume of a sphere in
   a. rectangular coordinates;
   b. cylindrical coordinates, and
   c. spherical coordinates.
2. Set up, but do not evaluate, the integrals to find the volume of a right circular cylinder in
   a. rectangular coordinates;
   b. cylindrical coordinates, and
   c. spherical coordinates.

3. Find the volume of a right circular cone with r_0 as the radius of the base and 2α as the angle of the vertex. (Hint: Place the vertex of the cone at the origin of a spherical coordinate system.)

4. Use cylindrical coordinates to find the volume of the region S defined in Problem 3.

5. Find the volume of the region S inside the sphere $z^2 + x^2 + y^2 = b^2$ and outside the cylinder $r = a < b$. Use either cylindrical or spherical coordinates. Describe this solid in terms of familiar objects.

6. The region S is bounded above by the sphere $x^2 + y^2 + z^2 = 4k^2$ and below by $z = k$ where $k > 0$. Find its volume and mass if $\delta = r$.

7. The region S is bounded by the sphere $r = a$ and the cone $\phi = k$ where $0 < \theta < \frac{\pi}{2}$. Find the volume of S.

8. The region S has an outer boundary which is a sphere of radius $r = 2a$ and an inner boundary which is a sphere of radius $r = a$. Find the volume of S.

9. The region S is the region bounded by the sphere $r = a$ and the half-cones $\phi = \frac{\pi}{3}$ and $\phi = \frac{\pi}{4}$. Find the volume of S.

10. Find the mass in Problem 9 of S if the mass is spread over S with a density of $\delta = 2z$.

11. Use spherical coordinates to find the volume of a cylinder which has a radius of k and a height of h.

12. A round hole of radius 1 is bored through the center of a sphere of radius 2. What volume is removed? What proportion of the volume removed?

REFERENCES

Keeton, W. T. (1969). Orientation by pigeons: Is the sun necessary? *Science* 165:922-928.

Frisch, K.V. (1967). Honeybees: Do they use direction and distance information provided by their dances? *Science* 158:1072-1076.

Skellam, J. G. (1951). Random dispersal in theoretical populations. *Biometrika* 38:196-218.

Chapter 16

INFINITE SERIES

Sums of the sort

$$1 + \frac{1}{2} + \frac{1}{4} + \cdots + \frac{1}{2}^n + \cdots$$

are called *infinite series*. Although each term in this series is a constant, some infinite series involve variables. For example, in studying Maclaurin series expansions in Section 7.8a, we established the *power series*

$$1 + x + \frac{x^2}{2!} + \frac{x^3}{3!} + \cdots + \frac{x^n}{n!} + \cdots = e^x .$$

Infinite series and the sister topic of infinite sequences appear naturally in biology when the behavior of a biological process or population is considered for a long span of time. Many such phenomena have an associated probability distribution spread out over an infinite sequence of points in time. Infinite series are required to examine properties and consequences of these probability distributions.

In addition to appearing naturally, infinite series can serve as a powerful mathematical tool. For example some integrals and differential equations cannot be evaluated or solved directly, but once the functions have been expressed as a power series, evaluations proceed much more simply. Power series expansions have become an important computational tool in another area. They were used in the original development of many mathematical tables, like those of logarithms and trigonometric functions. When values from such tables are needed in computer computations, the computer can obtain the value by using the infinite series representation of the function. This certainly accomplishes a storage savings relative to the computer storing an entire table. More importantly it gives a more precise evaluation of the function at the desired value of the function's argument, because there is no need for an interpolation.

This chapter begins by introducing sequences and series, proceeds to convergence of sequences and infinite series, series of nonnegative terms, alternating series, absolute and conditional convergence, and ends with a brief discussion of power series. Recall we have already encountered power series in Section 7.8.

16.1 SEQUENCES

We begin by defining a sequence:

> Consider a set $\mathcal{S}$ of elements $S_i \in \mathcal{S}$ where i symbolizes a positive integer.
>
> - A SEQUENCE is a function whose domain is the set of positive integers.
> - A FINITE SEQUENCE has only a finite number of elements while an INFINITE SEQUENCE has no last term.

The customary way to define a sequence S is to list in order the values of S at the successive positive integers as $S_1, S_2, S_3, \cdots$. The dots suggest that the sequence is an infinite one.

A sequence consists of objects for which there is a first one, then a second one, and so on. The distinction between finite and infinite sequences is suitably described by the modifying adjectives finite and infinite; a finite sequence has a last term while an infinite sequence does not. Although the primary treatment of sequences at this instructional level usually deals with numerical sequences, biology abounds with nonnumerical sequences but their mathematics can be examined in detail only at a more advanced level. The following examples illustrate two purely biological sequences before turning to mathematical ones.

<u>EXAMPLE 16.1</u>: A growing organism goes through various growth or developmental stages, a topic occupying large parts of some biology courses. The list of developmental stages, chronologically ordered, forms a sequence. An example of such a sequence appears in mitosis, the process of cell division which has the stages: prophase, metaphase, anaphase and telophase. If $\mathcal{S}$ consisted of these four phases, then S_1 = prophase, S_2 = metaphase, S_3 = anaphase and S_4 = telophase form

a finite sequence because (i) $S_i \in \mathcal{S}$, (ii) the S_i are ordered as 1, 2, 3, 4 and (iii) the sequence has a finite number, four, of elements. ▌

<u>EXAMPLE 16.2</u>: Living organisms transmit most inherited characteristics via a complex chemical chain called DNA (deoxyribonucleic acid). This complex compound is composed of a long sequence, in the sense defined above, of basic biochemical building blocks called amino acids. Other sequences of amino acids form the proteins, but a specific sequence forms DNA. ▌

<u>EXAMPLE 16.3</u>: If $S_n = 4 + 3n$, the corresponding infinite number sequence is $S_1 = 7$, $S_2 = 10$, $S_3 = 13$, $\cdots$. An equation for the nth term, like $S_n = 4 + 3n$, is called the GENERAL TERM. ▌

<u>EXAMPLE 16.4</u>: A GEOMETRIC PROGRESSION is a sequence whose first term is a fixed constant a and each successive term is obtained by multiplying the preceding term by a fixed constant r, called the common ratio of the progression:

$$S_1 = a,\ S_2 = ar,\ S_3 = ar^2,\ \cdots,\ S_n = ar^{n-1},\ \cdots\ . \qquad (16.1)$$ ▌

<u>EXAMPLE 16.5</u>: An ARITHMETIC PROGRESSION is a sequence whose first term is a constant a with each successive term obtained by adding a common difference to the preceding term:

$$S_1 = a,\ S_2 = a + d,\ S_3 = a + 2d,\ \cdots,\ S_n = a + (n - 1)d,\ \cdots\ . \qquad (16.2)$$ ▌

Unless otherwise specified, we will assume that n can assume all positive integer values; thus, all sequences are infinite unless otherwise specified. Any letter in a subscript, such as k in S_k, can replace n. We may write $\{S_n\}$ to signify a sequence which has S_n as its general term.

EXAMPLE 16.6: The sequence whose n^{th} term is given as $S_n = 1 + \frac{1}{n}$ is neither an arithmetic nor a geometric sequence. In this case,

$$S_1 = 2,\ S_2 = \frac{3}{2},\ S_3 = \frac{4}{3},\ \cdots\ .$$

This sequence also could be designated by $\{1 + \frac{1}{n}\}$. ▌

- A sequence is said to CONVERGE to a limit L as $n\to\infty$ if $|S_n - L|$ becomes arbitrarily small for <u>all</u> sufficiently large values of n. This is symbolized by

$$\lim_{n\to\infty} S_n = L \qquad \text{or} \qquad S_n \to L \text{ as } n\to\infty\,. \tag{16.3}$$

- If the preceding limit exists, then the sequence $\{S_n\}$ is called CONVERGENT.
- If $\{S_n\}$ is not convergent, then we say $\{S_n\}$ is DIVERGENT.

<u>EXAMPLE 16.7</u>: Investigate $S_n = 2 + \frac{1}{2n}$ for convergence.

$$\lim_{n\to\infty} S_n = \lim_{n\to\infty} \left(2 + \frac{1}{2n}\right) = 2$$

because $\lim_{n\to\infty} \frac{1}{2n} = 0$. The sequence $\{2 + \frac{1}{2n}\}$ is a convergent sequence which converges to the value 2. ▌

<u>EXAMPLE 16.8</u>: Examine $S_n = (-1)^n$ for convergence.

We have

$$\lim_{n\to\infty} S_n = \lim_{n\to\infty} (-1)^n\ .$$

Observe that $S_1 = (-1)^1 = -1$, $S_2 = (-1)^2 = 1$, $S_3 = (-1)^3 = -1$, $\cdots$. Since the terms of this sequence simply alternate back and forth between +1 and -1, it has no limit. Thus the sequence $\{(-1)^n\}$ diverges. ▌

The next example illustrates one method of evaluating an indeterminate form.

<u>EXAMPLE 16.9</u>: Examine the sequence $\{\frac{3n^2 + n - 5}{7n^2 - 5n - 4}\}$ for convergence.

$$\lim_{n\to\infty} S_n = \lim_{n\to\infty} \frac{3n^2 + n - 5}{7n^2 - 5n - 4} = \lim_{n\to\infty} \frac{3 + \frac{1}{n} - \frac{5}{n^2}}{7 - \frac{5}{n} - \frac{4}{n^2}}$$

The last step was obtained by dividing the numerator and denominator by the highest power of n common to both the numerator and the denominator. Now,

$$\lim_{n\to\infty} \frac{3 + \frac{1}{n} - \frac{5}{n^2}}{7 - \frac{5}{n} - \frac{4}{n^2}} = \frac{3 + 0 - 0}{7 - 0 - 7} = \frac{3}{7} \text{ . } \blacksquare$$

Indeterminate forms can appear in evaluating limits of sequences as they did in finding the limit of a function. Because $\{S_n\}$ is defined only at the integers, L'Hospital's rule cannot be applied directly to find the limits of sequences leading to indeterminate forms. On the other hand, if $\{S_n\}$ depends on n in such a way that $\{S_n\}$ would become a function if n were regarded as a continuous variable, then L'Hospital's rule can be applied.

<u>EXAMPLE 16.9</u> (continued): The fraction

$$\frac{3n^2 + n - 5}{7n^2 - 5n - 4}$$

can be viewed as a function of n for any real n such that $7n^2 - 5n - 4 \neq 0$. Thus, using L'Hospital's rule

$$\lim_{n\to\infty} \frac{3n^2 + n - 5}{7n^2 - 5n - 4} = \lim_{n\to\infty} \frac{6n + 1}{14n - 5} = \lim_{n\to\infty} \frac{6}{14} = \frac{3}{7} ,$$

giving the same result as before. ▮

<u>EXAMPLE 16.10</u>: Examine $\{\frac{4n^2 - 3}{3n + 5}\}$ for convergence.

$$\lim_{n\to\infty} \frac{4n^2 - 3}{3n + 5} = \lim_{n\to\infty} \frac{8n}{3} = \infty .$$

The sequence $\{\frac{4n^2 - 3}{3n + 5}\}$ diverges because the limit fails to exist. ▮

<u>THEOREM 16.1</u>: A nondecreasing sequence $\{S_n\}$ converges if and only if the sequence has a finite upper bound.

This theorem says that if for all n, $S_n \leq S_{n+1}$ and $S_n \leq B$, then the sequence converges. Of course, its limit S, can be no greater than B. Although no formal proof will be offered here, Figure 16.1 shows the

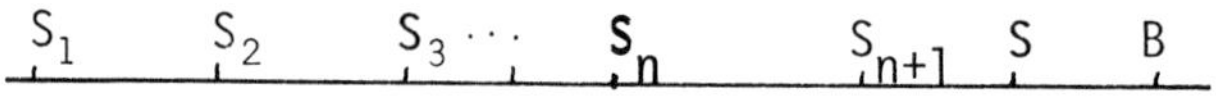

$$\underset{S_1}{|} \quad \underset{S_2}{|} \quad \underset{S_3}{|} \cdots \underset{S_n}{|} \quad \underset{S_{n+1}}{|} \quad \underset{S}{|} \quad \underset{B}{|}$$

Figure 16.1

essence of the argument. This theorem's converse says that if $S_n \leq S_{n+1}$ and $\{S_n\}$ converges, there must exist a constant B such that $S_n \leq B$ for all n. A similar theorem applies to nonincreasing sequences with lower bounds.

<u>EXAMPLE 16.11</u>: Consider $\{S_n\} = \{3 - \frac{1}{3n}\}$ for convergence.

The terms of this sequence,

$$S_1 = \frac{8}{3}, \; S_2 = \frac{26}{9}, \; S_3 = \frac{80}{27}, \; \cdots$$

increase because the subtractive term $\frac{1}{3n}$ decreases, but all of the terms remain less than 3. Thus, Theorem 16.1 assures us that $\{S_n\}$ has a limit. Evaluation of this limit gives

$$\lim_{n\to\infty} (3 - \frac{1}{3n}) = 3 . \quad \blacksquare$$

EXERCISES

Write down the general term of the following sequences:

1. $k, 2k^2, 3k^3, \cdots$
2. $x, \frac{x^2}{4}, \frac{x^3}{9}, \frac{x^4}{16}, \cdots$
3. $6, -3, \frac{3}{2}, -\frac{3}{4}, \cdots$
4. $\frac{2}{3}, \frac{2}{3\times5}, \frac{2}{5\times7}, \cdots$
5. $2, 4, 6, \cdots$
6. $2, -\frac{4}{3}, \frac{8}{5}, \cdots$
7. $\frac{1}{2\times3}, \frac{1\times3}{2\times4\times5}, \frac{1\times3\times5}{2\times4\times6\times7}, \cdots$
8. $\frac{3}{2}, \frac{5}{5}, \frac{7}{10}, \frac{9}{17}, \cdots$

Find the limit, if it exists, of each of the following sequences.

9. $\dfrac{1}{1 + \sqrt{n}}$

10. $\dfrac{n + 4}{n^2 - 1}$

11. $\dfrac{\ln n}{n}$

12. $n^3 e^{-n}$

13. $\dfrac{\ln n}{n^3}$

14. $\dfrac{10^n}{e^n + 4}$

15. $\dfrac{e^n + 3^2}{2.3^n + 5}$

16. $\dfrac{e^n}{n^3}$

Determine whether each of the following sequences converge or diverge. Explain the basis for your decision; if convergent, find the limit.

17. $\frac{1}{2}, \frac{2}{3}, \frac{3}{4}, \frac{4}{5}, \cdots$

18. $\frac{1}{2}, \frac{1}{3}, \frac{1}{4}, \frac{1}{5}, \cdots$

19. $1, -\frac{1}{2}, \frac{1}{4}, -\frac{1}{8}, \cdots$

21. $2, -\frac{3}{2}, \frac{4}{3}, -\frac{5}{4}, \cdots$

20. $1, 2, 4, 8, \cdots$

16.2 INFINITE SERIES

Various operations can be performed on infinite sequences; among these, the sum of the terms in the sequence proves especially important. Thus, if we consider an infinite sequence $u_1, u_2, \cdots, u_n, \cdots$, these definitions apply:

- The expression

$$u_1 + u_2 + \cdots + u_n + \cdots$$

is called the INFINITE SERIES generated by the sequence $u_1, u_2, \cdots$. An infinite series frequently is abbreviated by

$\sum_{i=1}^{\infty} u_i$ or $\sum u_i$.

- The sequence $\{S_n\}$ of PARTIAL SUMS is given by

$$S_1 = u_1,\ S_2 = u_1 + u_2,\ \cdots,\ S_n = \sum_{i=1}^{n} u_i,\ \cdots . \tag{16.5}$$

- The infinite series $\sum_{i=1}^{\infty} u_i$ CONVERGES to the limit S if the associated sequence of partial sums converges to S. When this limit exists, we write

$$\sum_{i=1}^{\infty} u_i = \lim_{n\to\infty} S_n = S . \tag{16.6}$$

- If the infinite series $\sum_{i=1}^{\infty} u_i$ does not converge, it is said to DIVERGE.

An infinite series deals with the sum of an infinite number of terms. At first glance, this may seem like only a mathematical construction, but biology has processes which form infinite series. A limestone bed, for example, began by the accumulation of many, many animal skeletons, each adding an increment to the bed. As conditions changed, the bed's growth diminished to essentially zero. The contribution of individual skeletons to the bed represents the individual terms u_i; S_n corresponds to the bed's size after each addition; and S, the sum of the series, corresponds to its final size after accumulation has ceased. This series converges because the bed quits growing. Other continuing biological processes may not stop growing. For example the total energy consumed by any segment of the biological organisms, man for instance, continues to grow. So long as that segment continues to live, it will go on consuming energy.

When an expression for the partial sums can be found, it provides a simple means for evaluating the convergence and limit of an infinite series. Finding this expression may prove to be a difficult task, but occasionally S_n has a nice formula and so its limit is clear.

EXAMPLE 16.12: Examine the infinite series $\sum_{i=1}^{\infty} \frac{1}{i(i + 1)}$ for convergence.

Use partial fractions to write $u_i = \frac{1}{i(i+1)} = \frac{1}{i} - \frac{1}{i+1}$

$$\sum_{i=1}^{\infty} \frac{1}{i(i+1)} = \frac{1}{1\times2} + \frac{1}{2\times3} + \frac{1}{3\times4} + \cdots + \frac{1}{n(n+1)} + \cdots$$

$$= (1 - \frac{1}{2}) + (\frac{1}{2} - \frac{1}{3}) + \cdots + (\frac{1}{n} - \frac{1}{n+1}) + \cdots .$$

This has partial sums of

$$S_1 = 1 - \frac{1}{2}$$

$$S_2 = (1 - \frac{1}{2}) + (\frac{1}{2} - \frac{1}{3}) = 1 - \frac{1}{3}$$

$$\vdots$$

$$S_n = (1 - \frac{1}{2}) + (\frac{1}{2} - \frac{1}{3}) + \cdots + (\frac{1}{n} - \frac{1}{n+1}) = 1 - \frac{1}{n+1} ,$$

so the limit of this convergent series is obtained as

$$\lim_{n\to\infty} S_n = \lim_{n\to\infty} (1 - \frac{1}{n+1}) = 1 . \quad \blacksquare$$

<u>EXAMPLE 16.13</u>: Investigate the convergence of the sum of an arithmetic progression:

$$a + (a + d) + (a + 2d) + (a + 3d) + \cdots + (a + (n - 1)d) + \cdots$$

To obtain its partial sums, rearrange the terms as below and recall from Equation 8.2 that $1 + 2 + \cdots + n = \frac{n(n+1)}{2}$:

$$S_n = a + (a + d) + (a + 2d) + (a + 3d) + \cdots + (a + (n - 1)d)$$

$$= na + d(1 + 2 + \cdots + n - 1) = na + \frac{n(n-1)}{2} d$$

$$= \frac{n}{2} [2a + (n - 1)d] ,$$

and

$$\lim_{n\to\infty} S_n = \lim_{n\to\infty} \frac{n}{2} [2a + (n - 1)d] = \begin{cases} \infty \text{ if } d > 0 \\ 0 \text{ if } a = 0 = d \\ -\infty \text{ if } d < 0 \end{cases}$$

Thus the sum of any arithmetic progression diverges except for the degenerate one consisting of all zeroes. ∎

One of the most important series is the GEOMETRIC SERIES

$$\sum_{n=1}^{\infty} ar^{n-1} = a + ar + ar^2 + \cdots + ar^{n-1} + \cdots .$$

To obtain the partial sums of this series, recall the factorization $1 - r^n = (1 - r)(1 + r + r \;\; + \cdots + r^{n-1})$ and divide both of its sides by $1 - r$:

$$S_1 = a$$

$$S_2 = a + ar = a(1 + r) = a\,\frac{1 - r^2}{1 - r}$$

$$S_3 = a + ar + ar^2 = a(1 + r + r^2) = a\,\frac{1 - r^3}{1 - r}$$

$$\vdots$$

$$S_n = a + ar + ar^2 + \cdots + ar^{n-1} = a(1 + r + r^2 + \cdots + r^{n-1}) = a\,\frac{1 - r^n}{1 - r} .$$

Now take the limit as $n \to \infty$:

$$\sum_{n=1}^{\infty} ar^n = S = \lim_{n\to\infty} S_n = a \lim_{n\to\infty} \frac{1 - r^n}{1 - r} = \frac{a}{1 - r} \qquad \text{if} \qquad |r| < 1 \qquad (16.7)$$

Thus, the series converges to $\frac{a}{1 - r}$ if $|r| < 1$ and diverges if $|r| \geq 1$ because in the latter case $r^n \to \infty$ as $n \to \infty$.

<u>EXAMPLE 16.14</u>: Examine the geometric series

$$1 + \frac{1}{2} + \frac{1}{4} + \frac{1}{8} + \cdots + \frac{1}{2^{n-1}} + \cdots$$

for convergence.

This series has a common ratio of $r = \frac{1}{2}$; Equation 16.7 implies that it converges to

$$S = \frac{a}{1 - r} = \frac{1}{1 - \frac{1}{2}} = 2 . \quad ∎$$

EXAMPLE 16.15: Consider this somewhat idealized problem in population dynamics: Each year a population receives a constant number, R, of recruits (such as births), but otherwise incurs no immigration or emigration. If the probability of a single individual surviving a year is s, independent of age and year, what is the relative frequency of individuals of various ages at the beginning of any year?

First, these assumptions tell us about only the average or theoretical survival; thus, we must think in terms of theoretical numbers. At the end of the first year, we expect Rs to survive while R(1 - s) individuals would die; of the survivors, $(Rs)s = Rs^2$ would survive the second year, Rs^3 the third year, and so on. Thus, if we let N_{ij} be the number of survivors in the ith year of age j, we find

$$N_{ij} = Rs^j \qquad j = 1, 2, \cdots, i,$$

terms from a geometric series. After some time (i large), we can ignore the limitation $j \leq i$ because Rs^j will represent no animals when it drops below 1 for very large j. The sum

$$\sum_{j=1}^{\infty} N_{ij} = \sum_{j=1}^{\infty} Rs^j = Rs\left[\frac{1}{1 - s}\right]$$

gives the stabilized population size, so

$$\frac{N_{ij}}{Rs/(1 - s)} = \frac{Rs^j}{Rs/(1 - s)} = (1 - s)s^{j-1}, \qquad j = 1, 2, \cdots \qquad (16.8)$$

gives the relative frequency of individuals of the various ages. As the geometric series in Equation 16.8 sums to one, it qualifies to be called the geometric probability distribution.

The idealization suggested to begin this example describes some situations fairly well. For example, in the screwworm control project described by Knipling (1960), sterilized male flies are released regularly. When production is running at full capacity, essentially a constant number of sterilized male flies are introduced during each time period. If the time period in the idealization were replaced by days or weeks, the model in Equation 16.8 would come fairly close to describing the age distribution of the sterilized male flies at large.

Consider the introduction of a pollutant into a flowing body of water. If s describes the probability that a pollutant particle survives neutralization by either the environment or biological processes over a fixed distance d, then Equation 16.8 describes the concentration of unneutralized pollutant downstream. Here, years probably would be replaced by a suitable multiple of d.

Studies in fisheries biology sometimes use Equation 16.8 directly, but often it is generalized to allow years and age to influence survival rates. Nevertheless, the same sort of patterns and series result. ▮

The geometric series provides a convenient tool for finding the fraction corresponding to a repeating decimal:

EXAMPLE 16.16: We begin with a familiar result. Express $0.3333 \cdots$ as a rational fraction.

As this repeating decimal really represents the infinite series

$$0.3 + 0.03 + 0.003 + \cdots ,$$

this is a geometric series with first term $a = 0.3$ and common ratio $r = 0.1$. Thus

$$0.3333 \cdots = \frac{0.3}{1 - 0.1} = \frac{0.3}{0.9} = \frac{1}{3} . \quad ▮$$

EXAMPLE 16.17: Express the repeating decimal $3.565656 \cdots$ as a rational fraction.

The repeating decimal $3.565656 \cdots$ really represents the infinite series

$$3 + 0.56 + 0.0056 + 0.000056 + \cdots .$$

Its terms after the first form a geometric progression having a first term of 0.56 and a common ratio of 0.01. Thus,

$$3.565656 \cdots = 3 + \frac{0.56}{1 - 0.01} = 3 + \frac{0.56}{0.99} = 3 + \frac{56}{99} = \frac{353}{99} . \quad ▮$$

Because the sum of an arithmetic progression diverges, we might expect the sum of the reciprocals of each term to converge. The simplest series of this sort results from a = 1 and d = 1, the integers; this series is called the HARMONIC SERIES:

$$1 + \frac{1}{2} + \frac{1}{3} + \frac{1}{4} + \frac{1}{5} + \cdots + \frac{1}{n} + \cdots \tag{16.9}$$

Surprisingly the series diverges, a fact we will now establish. Consider the following inequalities where all terms of the harmonic series appear to the left of the inequality sign:

$$1 + \frac{1}{2} > \frac{1}{2}$$

$$\frac{1}{3} + \frac{1}{4} > \frac{1}{4} + \frac{1}{4} = \frac{1}{2}$$

$$\frac{1}{5} + \frac{1}{6} + \frac{1}{7} + \frac{1}{8} > \frac{1}{8} + \frac{1}{8} + \frac{1}{8} + \frac{1}{8} = \frac{1}{2}$$

$$\frac{1}{9} + \frac{1}{10} + \cdots + \frac{1}{16} > \frac{1}{16} + \frac{1}{16} + \cdots + \frac{1}{16} = \frac{1}{2} \, .$$

$$\vdots$$

Thus, the sum on each line exceeds $\frac{1}{2}$. This shows that $S_n > \frac{k}{2}$ when $n = 2^k$, so S_n diverges as $n \to \infty$.

The above demonstration depends heavily on the specific nature of terms in the harmonic series. The next theorem provides a necessary condition for convergence of an infinite series.

THEOREM 16.2: A necessary condition for $\sum_{n=1}^{\infty} u_n$ to converge is that the nth term of the series must approach zero as $n \to \infty$ or $\lim_{n\to\infty} u_n = 0$.

This theorem states only a necessary condition for convergence; we have seen that the harmonic series diverges even though

$$\lim_{n\to\infty} u_n = \lim_{n\to\infty} \frac{1}{n} = 0 \, .$$

This points out what we mean by necessary: For a series to converge, the nth term must converge to zero, but the nth term going to zero does not imply convergence of the series.

EXAMPLE 16.14 (continued): We have already shown that the series $\sum_{n=1}^{\infty} \frac{1}{2^{n-1}}$ converges. Observe that this series satisfies the condition of Theorem 16.2: $\lim_{n\to\infty} u_n = \lim_{n\to\infty} \frac{1}{2^{n-1}} = 0.$ ▌

EXAMPLE 16.18: Examine the series $\sum_{n=1}^{\infty} \frac{5n+2}{3n-1}$ for convergence.

From Theorem 16.2, we must have $\lim_{n\to\infty} u_n = 0$ for the series to converge, but

$$\lim_{n\to\infty} u_n = \lim_{n\to\infty} \frac{5n+2}{3n-1} = \frac{5}{3} \neq 0 .$$

This series must diverge because the necessary condition of Theorem 16.2 is not satisfied. Another way to state Theorem 16.2 is to say that if $\lim u_n \neq 0$, then $\sum u_n$ diverges. ▌

Theorem 16.2 deals with the limiting behavior of individual terms in the series, but a parallel result applies to the remainder after k terms:

THEOREM 16.3: If $\sum_{i=1}^{\infty} u_i$ converges, then the remainder after k terms approaches zero as $k\to\infty$:

$$\lim_{k\to\infty} R_k = \lim_{k\to\infty} \sum_{i=k+1}^{\infty} u_i = 0 .$$

Conversely, if the above limit equals zero, the series converges.

PROOF: This proof illustrates an approach used to verify assertions about infinite series. If S symbolizes the sum of the infinite series and if

$$S = \sum_{i=1}^{\infty} u_i = S_k + R_k \qquad \text{then} \qquad R_k = S - S_k \ .$$

Since the sequence of partial sums $\{S_k\}$ converges to S, we have

$$\lim_{k\to\infty} R_k = \lim_{k\to\infty} (S - S_k) = S - \lim_{k\to\infty} S_k = S - S = 0 \ .$$

The converse follows by similar reasoning.▼

EXERCISES

Investigate convergence and determine the sum, if possible, of those infinite series which converge.

1. $\sum_{n=1}^{\infty} \frac{1}{2^{n-1}}$

2. $\sum_{n=1}^{\infty} \frac{1}{(n + 2)(n + 3)}$

3. $1 + \frac{1}{3} + \frac{1}{3^2} + \cdots$

4. $1 + 3 + 5 + \cdots$

5. $1 - 2 + 1 - 2 + \cdots$

6. $\sum_{n=1}^{\infty} \frac{3n + 5}{n + 1}$

7. $40 + 4 + 0.4 + 0.04 + \cdots$

8. $0.838383 \cdots$

9. $0.345345345 \cdots$

10. $0.21111 \cdots$

11. $\ln \frac{1}{2} + \ln \frac{1}{3} + \ln \frac{1}{4} + \cdots$

16.3 FURTHER RESULTS ON INFINITE SERIES

This section collects several useful results concerning infinite series. The first part concerns operations on series while the latter part returns to convergence.

The next three theorems state simple, but useful results about infinite series:

<u>THEOREM 16.4</u>: If $\sum_{i=1}^{\infty} u_i$ and $\sum_{i=1}^{\infty} v_i$ converge to S and T, respectively, then $\sum_{i=1}^{\infty} (u_i + v_i)$ converges to S + T.

<u>THEOREM 16.5</u>: For a constant $k \neq 0$, $\sum_{i=1}^{\infty} ku_i$ converges or diverges as does $\sum_{i=1}^{\infty} u_i$; if $\sum_{i=1}^{\infty} u_i$ converges to S, then $\sum_{i=1}^{\infty} ku_i$ converges to kS and, if $\sum_{i=1}^{\infty} u_i$ diverges, then $\sum_{i=1}^{\infty} ku_i$ diverges.

<u>THEOREM 16.6</u>: The insertion (or deletion) of a finite number of terms into (or from) an infinite series does not alter its convergence or divergence.

<u>EXAMPLE 16.19</u>: Investigate the convergence of $\frac{1}{3} + \frac{1}{6} + \frac{1}{9} + \frac{1}{12} + \cdots + \frac{1}{3n} + \cdots$.

Because $\frac{1}{3n} = \frac{1}{3}\left(\frac{1}{n}\right)$ and $\sum_{n=1}^{\infty} \frac{1}{n}$ diverges, $\sum_{n=1}^{\infty} \frac{1}{3}\left(\frac{1}{n}\right)$ also diverges using Theorem 16.5 with $k = \frac{1}{3}$. ▌

<u>EXAMPLE 16.20</u>: Investigate convergence of the series $\frac{1}{4} + \frac{1}{5} + \frac{1}{6} + \cdots + \frac{1}{n+3} + \cdots$.

This series results by deleting the first three terms from a harmonic series. Theorem 16.6 implies that this series diverges because the harmonic series does. ▌

*<u>EXAMPLE 16.21</u>: Verify that $\sum_{i=1}^{\infty} \frac{3^{i+2} - (-2)^i}{3(6)^i} = 3\,\frac{1}{12}$.

The numerator consists of a difference. Thus let us try this approach: (i) Decompose the sum into simpler sums; (ii) check for convergence of the simpler sums; (iii) evaluate the simpler sums and add them together. Because 6 = (2)(3) the denominators of the individual terms are $3(6)^i = 2^i\,3^{i+1}$:

$$\frac{3^{i+2} - (-2)^i}{3(6)^i} = \frac{3^{i+2} + (-1)^{i+1}\, 2^i}{2^i\, 3^{i+1}} = \frac{3}{2^i} + (-\frac{1}{3})^{i+1} ,$$

so,

$$\sum_{i=1}^{\infty} \frac{3^{i+2} - (-2)^i}{3(6)^i} = \sum_{i=1}^{\infty} (\frac{3}{2^i} + (-\frac{1}{3})^{i+1}) = \sum_{i=1}^{\infty} \frac{3}{2^i} + \sum_{i=1}^{\infty} (-\frac{1}{3})^{i+1} .$$

These last two series converge.

$$\sum_{i=1}^{\infty} \frac{3}{2^i} = \sum_{i=1}^{\infty} \frac{3}{2} (\frac{1}{2})^{i-1} = \frac{3}{2} \times \frac{1}{1 - \frac{1}{2}} = 3 .$$

Similarly,

$$\sum_{i=1}^{\infty} (-\frac{1}{3})^{i+1} = \frac{1}{9} \sum_{i=1}^{\infty} (-\frac{1}{3})^{i-1} = \frac{1}{9} \frac{1}{1 - (-\frac{1}{3})} = \frac{1}{9}\frac{3}{4} = \frac{1}{12} .$$

The result follows by adding these two. Notice that this example used the established convergence of the geometric series for $|r| < 1$ as well as Theorems 16.4, and 16.5. ▌

EXERCISES

Determine whether the following series are convergent or divergent. Justify your answer.

1. $\frac{1}{8} + \frac{1}{9} + \frac{1}{10} + \cdots + \frac{1}{n + 7} + \cdots$

2. $\frac{1}{2} + \frac{1}{4} + \frac{1}{6} + \frac{1}{8} + \cdots$

3. $4 + 3 + 2 + 1 + \frac{1}{2} + \frac{1}{3} + \frac{1}{4} + \cdots$

4. $\frac{1}{6} + \frac{1}{8} + \frac{1}{10} + \cdots$

5. Prove if $\sum_{n=1}^{\infty} a_n$ converges with sum S, and if k is any constant, then $\sum_{n=1}^{\infty} ka_n$ converges to the value kS.

6. Prove that convergence or divergence of a series is not altered by deleting, or adding, a finite number of terms.

16.4 SERIES OF NONNEGATIVE TERMS

It is often not easy to use the definition of convergence to determine whether a particular series converges or diverges. This and the next sections deal with two techniques to avoid this difficulty: (i) the comparison of a new series of nonnegative terms with a known convergent or divergent series; (ii) the convergence of a series whose terms have alternating signs.

In an infinite series having nonnegative terms, each partial sum equals the previous partial sum plus a nonnegative amount. Thus, $\{S_n\}$ form a nondecreasing sequence from which, by Theorem 16.1, we conclude:

<u>THEOREM 16.7</u>: If $u_n \geq 0$ for all n and if there is a number B such that $S_n \leq B$ for all n, then $\sum_{n=1}^{\infty} u_n$ converges to a sum $S \leq B$.

<u>THEOREM 16.8</u>: If $\sum_{n=1}^{\infty} u_n$ and $\sum_{n=1}^{\infty} v_n$ are series of positive terms, and $\sum_{n=1}^{\infty} u_n$ is known to converge and

$$v_n \leq u_n$$

for all values of n, then the $\sum_{n=1}^{\infty} v_n$ also converges.

<u>THEOREM 16.9</u>: If $\sum_{n=1}^{\infty} u_n$ and $\sum_{n=1}^{\infty} v_n$ are series of positive terms, and the $\sum_{n=1}^{\infty} v_n$ diverges and

$$v_n \leq u_n$$

for all values of n, then the $\sum_{n=1}^{\infty} u_n$ also diverges.

Theorems 16.8 and 16.9 are referred to as comparison tests for convergence or divergence. The first of them will be proved to show why it works.

PROOF OF THEOREM 16.8: Suppose $\sum u_n$ and $\sum v_n$ have $\{S_n\}$ and $\{T_n\}$ as their respective sequences of partial sums. If $v_n \leq u_n$ for all n, then $T_n \leq S_n$. Because $\sum u_n$ converges, there exists a number S satisfying $\lim_{n\to\infty} S_n = S$ and, thus, $\lim_{n\to\infty} T_n \leq \lim_{n\to\infty} S_n = S$ shows that $\{T_n\}$ also is bounded. Thus, Theorem 16.7 assures us that $\sum v_n$ converges. ▼

This question should arise naturally about now, "What series can I use for comparison purposes?". Any series of positive terms which is known to converge, or diverge, can be used as a comparison series. The following are some series useful for testing convergence:

1. $a + ar + ar^2 + \cdots + ar^{n-1} + \cdots$ where $a > 0$ and $0 \leq r < 1$.
2. $1 + \frac{1}{2^p} + \frac{1}{3^p} + \cdots + \frac{1}{n^p} + \cdots$ where $p > 1$.
3. $1 + \frac{1}{2^2} + \frac{1}{3^3} + \cdots + \frac{1}{n^n} + \cdots$.
4. Each series which previously has been shown to converge.

The following series can be used in the comparison test for divergence:

1. $1 + \frac{1}{2} + \frac{1}{3} + \cdots + \frac{1}{n} + \cdots$.
2. $a + ar + ar^2 + \cdots + ar^{n-1} + \cdots$ where $a > 0$ and $r \geq 1$.
3. Each series which has been shown to diverge.

EXAMPLE 16.22: Test the following series for convergence:

$$\frac{1}{2\times3\times4} + \frac{1}{3\times4\times5} + \cdots + \frac{1}{(n+1)(n+2)(n+3)} + \cdots .$$

Since $(n + 1)(n + 2)(n + 3)$ is of degree 3 in n, we shall compare the series with $\sum_{n=1}^{\infty} \frac{1}{n^3}$ which is given above as a convergent series. We have the following inequalities:

$$\frac{1}{(n+1)(n+2)(n+3)} < \frac{1}{n\times n\times n} = \frac{1}{n^3}$$

The above series converges because each of its terms is less than the corresponding term in a convergent series. ▌

EXAMPLE 16.23: Examine the following series for convergence:

$$\frac{1}{\sqrt{3}} + \frac{1}{\sqrt{4}} + \cdots + \frac{1}{\sqrt{n+2}} + \cdots .$$

For comparison, we will use the series

$$\frac{1}{3} + \frac{1}{4} + \cdots + \frac{1}{n+2} + \cdots ,$$

the harmonic series without its first two terms, a known divergent series. Now,

$$\frac{1}{\sqrt{3}} > \frac{1}{3}, \quad \frac{1}{\sqrt{4}} > \frac{1}{4}, \quad \frac{1}{\sqrt{5}} > \frac{1}{5}, \quad \ldots, \quad \frac{1}{\sqrt{n+2}} > \frac{1}{n+2}, \quad \cdots .$$

As each term in the given series exceeds the corresponding term in a divergent series, the given series diverges as assured by Theorem 16.9. ▌

EXAMPLE 16.24: Investigate the convergence of

$$\frac{5}{2\times3\times4} + \frac{7}{3\times4\times5} + \frac{9}{4\times5\times6} + \cdots + \frac{2n+3}{(n+1)(n+2)(n+3)} + \cdots .$$

When expanded out, the denominators have degree 3 in n while the numerators have only degree 1. This suggests comparing the series to $\sum_{n=1}^{\infty} \frac{1}{n^2}$ or some related series. Now,

$$\frac{5}{2\times3\times4} < \frac{5}{2\times2\times4} < \frac{2}{1^2}$$

$$\frac{7}{3\times4\times5} < \frac{7}{3\times3\times5} < \frac{2}{2^2}$$

$$\vdots$$

$$\frac{2n + 3}{(n + 1)(n + 2)(n + 3)} < \frac{2n + 3}{n\,n(n + 3)} < \frac{2}{n^2}$$

because $2n + 3 < 2(n + 3)$. Each term in the series under consideration is less than the corresponding term in $\sum_{n=1}^{\infty} \frac{2}{n^2} = 2 \sum_{n=1}^{\infty} \frac{1}{n^2}$, a convergent series. Thus the given series must also converge. ▌

Remember that the above theorems remain true if we delete a finite number of terms from either the series of interest or the test series because the addition or deletion of a finite number of terms does not affect convergence or divergence. Further, the comparison test applies to the convergence of a series of negative terms simply by multiplying the negative series by -1 before examining it.

To test a series $\sum_{n=1}^{\infty} u_n$ for convergence by using the comparison test, you must first guess whether it converges or diverges. A useful result which may help, but which will not be proved, is this: If $u_n > 0$ and can be expressed as the ratio of two polynomials, $P(n)/Q(n)$, where $P(n)$ and $Q(n)$ have respectively degrees of h and k in n, then if $\alpha = k - h > 1$ the series converges, but if $\alpha \leq 1$, the series diverges. This useful result lets you guess whether the unknown series converges or diverges before you use the appropriate test to confirm it.

EXERCISES

1. Prove the series $\sum_{n=1}^{\infty} \frac{1}{n^n}$ converges by comparing it with an appropriate geometric series.

2. $\sum \frac{1}{n(n+1)(n+2)(n+3)}$

3. $\sum \frac{2n+3}{n(n+1)(n+2)(n+3)}$

4. $\sum \frac{1}{K+n^2}$ where $K > 0$

5. $\sum \frac{1}{(2n+1)(2n+2)}$

6. $\sum \frac{1}{n\sqrt[3]{n}}$

7. $\sum \frac{1}{n^2+1}$

8. $1 + \frac{1}{2!} + \frac{1}{3!} + \frac{1}{4!} + \cdots$

9. $2 + \frac{3}{2^3} + \frac{4}{3^3} + \frac{5}{4^3} + \cdots$

10. $1 + \frac{2^2+1}{2^3+1} + \frac{3^2+1}{3^3+1} + \frac{4^2+1}{4^3+1} + \cdots$

11. $3 + 5 + 7 + \cdots$

12. $1 + \frac{1}{\sqrt{2}} + \frac{1}{\sqrt{3}} + \cdots$

13. $2 + \frac{2^2}{2} + \frac{2^3}{3} + \cdots$

14. $1 + \frac{1}{3} + \frac{1}{5} + \cdots$

15. $\frac{1}{2\times4} + \frac{1}{4\times6} + \frac{1}{6\times8} + \cdots$

16. $-\frac{1}{\sqrt{2}} - \frac{1}{\sqrt{4}} - \frac{1}{\sqrt{6}} \cdots$

17. $\sum \frac{1}{n(2n+1)}$

18. $\frac{3}{1\times4} + \frac{4}{2\times5} + \frac{5}{3\times6} + \cdots$

19. $\sum \frac{1}{n+2+\sqrt{n}}$

20. $\frac{2}{1\times3} + \frac{2\times4}{1\times3\times5} + \frac{2\times4\times6}{1\times3\times5\times7}$

16.5 ALTERNATING SERIES

The last section focused on series of positive terms. In this section, we will study series in which every other term is negative. This leads us to the following definition:

The series

$$1 - \frac{1}{2} + \frac{1}{3} - \frac{1}{4} + \cdots + \frac{(-1)^{n+1}}{n} + \cdots .$$

is an alternating series, that is, every other term is negative, called the ALTERNATING HARMONIC SERIES. Even though the harmonic series (all plus signs) diverged, the alternating harmonic series converges.

To establish this we need new results. The primary theorem concerning convergence of alternating series is given next. Its proof will reveal two associated results.

THEOREM 16.10: If $\sum_{n=1}^{\infty} u_n$ is an alternating series satisfying $|u_n| \geq |u_{n+1}|$ for all n and if $\lim_{n\to\infty} |u_n| = 0$, then the series converges.

PROOF: Suppose, for convenience, that $u_1 > 0$ so when we define $v_i = |u_i|$ or $u_i = (-1)^{i+1} v_i$,

$$\sum_{i=1}^{\infty} u_i = \sum_{i=1}^{\infty} (-1)^{i+1} v_i = v_1 - v_2 + v_3 - v_4 + \cdots .$$

The even-numbered partial sums can be arranged as either

$$S_{2k} = (v_1 - v_2) + (v_3 - v_4) + \cdots (v_{2k-1} - v_{2k}) \qquad (16.10)$$

or

$$S_{2k} = v_1 - (v_2 - v_3) - (v_4 - v_5) - \cdots - (v_{2k-2} - v_{2k-1}) - v_{2k} . \qquad (16.11)$$

The condition that $|u_n| \geq |u_{n+1}|$ implies that $(v_i - v_{i+1}) \geq 0$. Application of this result to Equation 16.10 shows that $\{S_{2k}\}$ satisfies $S_{2k} \leq S_{2(k+1)}$; on the other hand applied to Equation 16.11, it provides the bound $S_{2k} \leq v_1$ so by Theorem 16.7 $\{S_{2k}\}$ converges to a limit we will call S. Because $S_{2k+1} = S_{2k} + (-1)^{k+2} v_{k+1}$,

$$\lim_{n\to\infty} S_{2k+1} = \lim_{n\to\infty} S_{2k} + \lim_{n\to\infty} (-1)^{k+2} v_{k+1} = S + 0 = S ,$$

and so $\{S_n\}$ converge to S. This assumed $u_1 > 0$, but if $u_1 < 0$, the series could be multiplied by -1 without altering its convergence; then the above argument still applies. ▼

An examination of the above proof gives two results concerning the value of an alternating series, or the use of a partial sum as an approximation of the value of the series.

<u>THEOREM 16.11</u>: If $\sum_{i=1}^{\infty} u_i$ is an alternating series satisfying $|u_n| \geq |u_{n+1}|$ and $\lim_{n\to\infty} u_n = 0$, then (i) $\left|\sum_{i=1}^{\infty} u_n\right| \leq |u_1|$ and (ii) $\left|\sum_{i=1}^{\infty} u_i - S_n\right| \leq |u_{n+1}|$.

The first of these was established from Equation 16.11 during the proof of Theorem 16.10 while the second follows from applying the same reasoning to

$$\sum_{i=1}^{\infty} u_i - S_n = u_{n+1} + u_{n+2} + \cdots ,$$

an alternating series having u_{n+1} as its first term.

<u>EXAMPLE 16.25</u>: Examine the alternating harmonic series for convergence.

This series, $\sum_{i=1}^{\infty} (-1)^{n+1} \frac{1}{n}$, has $|u_n| = \frac{1}{n}$ and $|u_{n+1}| = \frac{1}{n+1}$.

It satisfies $|u_n| \geq |u_{n+1}|$, because $n + 1 > n$ implies that $\frac{1}{n} > \frac{1}{n+1}$; also,

$$\lim_{n\to\infty} |u_n| = \lim_{n\to\infty} \frac{1}{n} = 0 .$$

This series satisfies the conditions of Theorem 16.10 so it converges. ▮

<u>EXAMPLE 16.26</u>: Test $\sum_{n=1}^{\infty} (-1)^{n+1} \frac{n}{(n+2)(n+3)}$ for convergence.

$$|u_n| = \frac{n}{(n+2)(n+3)} \qquad \text{and} \qquad |u_{n+1}| = \frac{n+1}{(n+3)(n+4)} .$$

Now,

$$\lim_{n\to\infty} |u_n| = \lim_{n\to\infty} \frac{n}{(n+2)(n+3)} = \lim_{n\to\infty} \frac{n}{n^2 + 5n + 6} = \lim_{n\to\infty} \frac{1}{n + 5 + \frac{6}{n}} = 0 .$$

Thus, the series will converge if we can show that

$$|u_n| > |u_{n+1}|,$$

or if

$$\frac{n}{(n + 2)(n + 3)} > \frac{n + 1}{(n + 3)(n + 4)} .$$

This does occur because they contain the common factor of (n + 3) and because $n(n + 4) \geq (n + 1)(n + 2)$ for $n \geq 2$. ▌

EXAMPLE 16.27: The series $\sum_{n=1}^{\infty} (-1)^{n+1} n$ diverges because $\lim_{n\to\infty} (-1)^{n+1} n \neq 0$ and also

$$|u_n| = n < n + 1 = |u_{n+1}| .$$

When the condition $|u_n| > |u_{n+1}|$ of Theorem 16.10 fails to be true, we do not need to even examine $\lim_{n\to\infty} u_n$. Usually it is best to examine the limit of the nth term first; then if it is zero try to show that $|u_n| < |u_{n+1}|$. ▌

EXAMPLE 16.28: Compute e^{-1} with an error not to exceed 0.001.

Apply Theorem 16.11 to the series established in Example 7.26.

$$e^{-1} = \sum_{i=0}^{\infty} \frac{(-1)^i}{i!} = 1 - 1 + \frac{1}{2!} - \frac{1}{3!} + \frac{1}{4!} - \cdots .$$

Take enough terms (n) from this series in order that the next one satisfies $\frac{1}{(n + 1)!} \leq 0.001$; this is n = 7 because 6! = 720 so that $\frac{1}{720} = 0.0013$, and 7! = 5040 so $\frac{1}{5040} < 0.001$. Thus,

$$e^{-1} \doteq 1 - 1 + \frac{1}{2!} - \frac{1}{3!} + \frac{1}{4!} - \frac{1}{5!} + \frac{1}{6!} - \frac{1}{7!} = \frac{1854}{5040} = 0.3678 . \text{ ▌}$$

EXERCISES

Test the following series for convergence. All sums go from n = 1 to n = ∞ unless otherwise specified.

1. $\frac{1}{3} - \frac{1}{9} + \cdots + (-1)^{n+1}\frac{1}{3(2n - 1)} + \cdots$

2. $-\frac{1}{\sqrt{2}} + \frac{1}{\sqrt{4}} - \frac{1}{\sqrt{6}} + \cdots$

3. $\sum (-1)^{n+1} \frac{1}{n!}$

4. $\sum (-1)^{n-1} \frac{n}{n^2 + 1}$

5. $\sum (-1)^{n+1} \frac{n + 2}{n}$

6. $\sum (-1)^{n+1} \frac{1}{1 + \sqrt{n}}$

7. $\frac{1}{2\times 3} - \frac{1}{3\times 3^2} + \frac{1}{4\times 3^3} - \cdots$

8. $\frac{1}{1\times 2} - \frac{1}{2\times 3} + \frac{1}{3\times 4} - \cdots$

9. $\frac{1}{5} - \frac{1}{7} + \frac{1}{9} - \cdots$

10. $1 - \frac{1}{2^3} + \frac{1}{3^3} - \cdots$

11. Evaluate $e^{0.5}$ and $e^{-0.5}$ with an error of no more than 0.001.

16.6 ABSOLUTE AND CONDITIONAL CONVERGENCE

Some series containing negative terms converge only because the negative terms remove most of the effect of the positive terms. To isolate this sort of series from the rest, we need:

- An infinite series $\sum_{i=1}^{\infty} u_i$ is called ABSOLUTELY CONVERGENT when $\sum_{i=1}^{\infty} |u_i|$ converges.
- An infinite series which converges, but does not converge absolutely, is called CONDITIONALLY CONVERGENT.

These definitions should appear straightforward enough, but you may wonder whether conditionally convergent series exist. Consider the following example:

EXAMPLE 16.2 : The harmonic series diverges; yet, its terms result from taking absolute values of terms in the ALTERNATING HARMONIC series:

$$1 - \frac{1}{2} + \frac{1}{3} - \frac{1}{4} + \cdots + \frac{(-1)^n}{n + 1} + \cdots .$$

We showed that this series converges in Example 16.2 . Thus this series is conditionally convergent because the series of absolute values does not converge but the alternating series does. ▌

THEOREM 16.12: If $\sum_{i=1}^{\infty} u_i$ converges absolutely, it converges.

This merely formalizes a reasonable observation: If $\sum |u_i|$ does not need the signs of the u_i to converge, that is, if it does not need negative terms to counteract the effect of positive terms, $\sum u_i$ will converge.

The following important theorem provides a criterion for establishing the absolute convergence of an infinite series:

THEOREM 16.13: For $\sum_{i=1}^{\infty} u_i$, evaluate

$$\lim_{n\to\infty} \left|\frac{u_{n+1}}{u_n}\right| = R .$$

The infinite series $\sum_{i=1}^{\infty} u_i$ converges absolutely when $R < 1$; it diverges when $R > 1$, and no conclusion can be drawn if $R = 1$.

This theorem essentially generalizes the known convergence of the geometric series with R here occupying the role of r, the ratio between consecutive terms. The geometric series converges for $r < 1$ and diverges for $r \geq 1$. The same applies for R, except at $R = 1$. The change in behavior at this boundary reflects the fact that R generalizes rather than equals r.

EXAMPLE 16.30: Use the ratio test to show that the geometric series $\sum x^n$ converges if $|x| < 1$ and diverges if $|x| > 1$.

$$\lim_{n\to\infty} \left|\frac{u_{n+1}}{u_n}\right| = \lim_{n\to\infty} \left|\frac{x^{n+1}}{x^n}\right| = |x| .$$

Thus, from the criteria of Theorem 16.13, this series will converge if $|x| < 1$ and diverge if $|x| > 1$. ▮

EXAMPLE 16.31: The series $1 + \frac{1}{2} + \frac{1}{3} + \frac{1}{4} + \cdots$ has already been shown to diverge. The ratio test, however, proves inconclusive:

$$\lim_{n\to\infty}\left|\frac{u_{n+1}}{u_n}\right| = \lim_{n\to\infty}\left|\frac{\frac{1}{n+1}}{\frac{1}{n}}\right| = \left|\frac{n}{n+1}\right| = 1 .$$

When the ratio test fails no conclusion can be reached. If we had not already established the divergence of this series, we would have to find another method to test its convergence. ▮

EXAMPLE 16.32: Examine the series

$$1 + \frac{1}{2^2} + \frac{1}{3^2} + \cdots + \frac{1}{n^2} + \cdots$$

for convergence.

The ratio test proves inconclusive:

$$\lim_{n\to\infty}\left|\frac{u_{n+1}}{u_n}\right| = \lim_{n\to\infty}\left|\frac{\frac{1}{(n+1)^2}}{\frac{1}{n^2}}\right| = \lim_{n\to\infty}\left|\frac{n^2}{(n+1)^2}\right|$$

$$= \lim_{n\to\infty}\left|\frac{n^2}{n^2+2n+1}\right| = \lim_{n\to\infty}\left|\frac{1}{1+\frac{2}{n}+\frac{1}{n^2}}\right| = 1 .$$

When the ratio test fails, we could apply the comparison test using the series $\sum \frac{1}{n^p}$. As this series converges for values of $p > 1$, we see that the series $\sum \frac{1}{n^2}$ does converge. ▮

EXAMPLE 16.33: Test $\sum_{n=1}^{\infty} \frac{(2n)!}{(n!)^2}$ for convergence.

By the ratio test, we have

$$\lim_{n\to\infty}\left|\frac{u_{n+1}}{u_n}\right| = \lim_{n\to\infty}\left|\frac{\frac{(2(n+1))!}{((n+1)!)^2}}{\frac{(2n)!}{(n!)^2}}\right| = \lim_{n\to\infty}\left|\frac{(2n+2)!(n!)^2}{[(n+1)!]^2(2n)!}\right|$$

$$= \lim_{n\to\infty}\left|\frac{(2n+2)(2n+1)}{(n+1)^2}\right| = 4 .$$

This series diverges because the limit of the ratio between consecutive terms exceeds 1. ▮

EXERCISES

For the following series, state whether they converge, diverge, converge absolutely, or converge conditionally. Justify your conclusions. All sums go from n = 1 to n = unless otherwise specified.

1. $\frac{1}{2} - \frac{1}{4} + \frac{1}{6} - \cdots$

2. $1 - \frac{1}{3} + \frac{1}{5} - \cdots$

3. $1 - \frac{2}{3} + \frac{3}{3^2} - \frac{4}{3^3} + \cdots$

4. $1 - \frac{1}{\sqrt{2}} + \frac{1}{\sqrt{3}} - \frac{1}{\sqrt{4}} + \cdots$

5. $1 - \frac{1}{2^2} + \frac{1}{3^2} - \frac{1}{4^2} + \cdots$

6. $\frac{1}{2} - \frac{4}{2^3 + 1} + \frac{9}{3^3 + 1} - \frac{16}{4^3 + 1} + \cdots$

7. $\sum \frac{(-1)^{n-1}}{(2n - 1)^3}$

8. $\sum (-1)^{n+1} \frac{n}{n^4 + 2}$

9. $\sum (-1)^{n-1} \frac{1}{\sqrt{n(n + 1)}}$

10. $\sum \frac{(-1)^{n+1}}{3n - 1}$

11. $\sum \frac{n}{3^n}$

12. $\sum \frac{n}{4n + 1}$

13. $\sum \frac{n!}{a^n}$

14. $\sum \frac{(n + 3)2^n}{n(n + 2)^2}$

15. $\sum \frac{(-1)^{n+1}\, 2n}{n^2 + 1}$

16. $\sum \frac{(-1)^{n+1}\, n}{2n + 3}$

17. $\sum n^2(\log 2)^n$

18. $\sum (-1)^{n-1} \frac{1}{\sqrt{n}}$

19. $\sum (-1)^n \frac{(n + 1)}{4n}$

20. $\sum (-1)^{n+1} \left[1 + \frac{1}{n}\right]^n$

21. $\sum (-1)^{n+1} \frac{3n^2 + 2n - 1}{n^3 + 2n^2 + 3n - 1}$

22. $\sum (-1)^{n+1} \frac{n^3}{2^n}$

23. $\sum \frac{n}{\sqrt{n^2 + 2}}$

24. $\sum (-1)^n \left[\frac{n + \sqrt{n}}{n}\right]$

25. $1 - \frac{1}{\sqrt[3]{2}} + \frac{1}{\sqrt[3]{3}} - \cdots$

26. $\frac{1!}{5} + \frac{2!}{5^2} + \frac{3!}{5^3} + \cdots$

27. $\sum \frac{(-1)^{n+1}}{\sqrt{n+2} - 1}$

28. $\frac{4}{1\times2\times3} - \frac{5}{2\times3\times4} + \frac{6}{3\times4\times5} - \cdots$

29. $1 - \frac{1}{4} + \frac{1}{7} - \cdots$

30. $\frac{3}{1} - \frac{3^2}{2^2} + \frac{3^3}{3^2} - \cdots$

*16.7 INTEGRAL TEST FOR CONVERGENCE

To use the comparison test, we need a variety of series for comparison. In this section we will introduce the integral test, which is used only for series of positive terms. It will give us additional series to use for comparison purposes. The following theorem gives the criteria necessary for using the integral test.

THEOREM 16.14: Let $\sum_{n=1}^{\infty} u_n$ be a series of positive terms where $u_k \geq u_{k+1}$ for all values of k. Suppose that there is a positive function f which is defined, continuous, and decreasing on the interval $[1,\infty)$ and that $f(n) = u_n$. Then

i) if $\int_1^\infty f(x)dx$ converges, it follows that $\sum u_n$ converges

ii) if $\int_1^\infty f(x)dx$ diverges, it follows that $\sum u_n$ diverges.

EXAMPLE 16.34: Test $\sum_{n=1}^{\infty} \frac{1}{n^3}$ for convergence.

Let $f(x) = \frac{1}{x^3}$. Since $u_k \geq u_{k+1}$, we can apply the integral test:

$$\int_1^\infty f(x)dx = \int_1^\infty \frac{dx}{x^3} = \lim_{h\to\infty} \int_1^h \frac{dx}{x^3} = \lim_{h\to\infty} \left[-\frac{1}{2x^2}\right]_1^h = -\frac{1}{2} \lim_{h\to\infty} \left[\frac{1}{h^2} - 1\right] = \frac{1}{2}.$$

Because the improper integral converges, the given series also converges. ▌

EXAMPLE 16.35: Test $\sum_{n=2}^{\infty} \frac{1}{4n \ln n}$ for convergence.

Let $f(x) = \frac{1}{4x \ln x}$ and observe that f(x) is a decreasing function. We begin the integral at x = 2 because the sum begins with n = 2:

$$\int_2^{\infty} f(x)dx = \frac{1}{4}\int_2^{\infty} \frac{dx}{x \ln x} = \frac{1}{4}\lim_{h\to\infty}\int_2^{h} \frac{dx}{x \ln x} = \frac{1}{4}\lim_{h\to\infty} [\ln \ln x]_2^h$$

$$= \frac{1}{4}\lim_{h\to\infty} [\ln \ln h - \ln \ln 2] = \infty .$$

The given series diverges because the value of the integral is not finite. ▌

EXERCISES

Test the following series for convergence by use of the integral test. If the integral test does not apply, use another appropriate test. All summations go from n = 1 to n = ∞ unless otherwise specified.

1. $\sum \frac{1}{n\sqrt{n}}$
2. $\sum \frac{n+1}{n\sqrt{n}}$
3. $\sum \frac{n-1}{n^3}$
4. $\sum \frac{1}{(n+1)\ln(n+1)}$
5. $\sum \frac{1}{n^4}$
6. $\sum \frac{1}{n}$
7. $\sum \frac{1}{n+10}$
8. $\sum \frac{1}{\sqrt{n^2+1}}$
9. $\sum \frac{\ln n}{n^2}$
10. $\sum \frac{1}{n2^n}$
11. $\sum \frac{n}{e^n}$
12. $\sum \frac{\ln n}{n\sqrt{n}}$
13. $\sum \frac{n^2}{e^n}$
14. $\sum_{n=1000}^{\infty} \frac{1}{n \ln n(\ln \ln n)}$

15. $\sum \frac{1}{\sqrt{n^3 + 1}}$

16. $\sum \frac{3n - 1}{(n + 1)^3}$

17. $\sum \frac{1}{n^p}$ if $p < 1$

18. $\sum \frac{1}{n^p}$ if $p > 1$

19. $\sum \frac{1}{n(\ln n)^3}$

20. If $\sum_{n=n_0}^{\infty} u_n$ converges, does $\sum_{n=1}^{\infty} u_n$ also converge? Justify your answer.

16.8 POWER SERIES

The consideration of series in preceding sections has focused on the summation of individual numbers. More generally, the sequence of numbers $\{u_n\}$ can be replaced by a sequence of functions $\{f_n(x)\}$ defined over some interval of values for x. To deal thoroughly with this general situation, we would have to become more involved in infinite series than seems advisable at this point. Conveniently, however, the case when $\{f_n(x)\} = \{a_n x^n\}$ can be treated as only a modest extension of what has come before:

- A POWER SERIES is defined by

$$a_0 + a_1 x + a_2 x^2 + \cdots + a_n x^n + \cdots \tag{16.12}$$

  where $\{a_n\}$ is a sequence of constants which can depend on n, but not on x.

- The INTERVAL OF CONVERGENCE consists of all values of x for which the infinite series above converges.

A more general power series could involve $\{a_n(x - b)^2\}$ in place of $\{a_n x^n\}$, but it turns out that a study of Equation 16.12 suffices because this more general case amounts to nothing more than a translation of the origin along the x-axis to the point $x = b$. Thus, we will restrict attention primarily to powers of x.

A power series obviously converges at $x = 0$ because all terms except the first one involve a power of x. The interval of convergence contains all additional values of x for which the series converges; this is an interval including zero. The interval may consist of the entire x-axis. If the interval of convergence has finite length, the power series may converge at only one of the end-points of the interval, at both end-points, or at neither end-point. A power series also converges absolutely at all values of x interior to the interval of convergence.

The interval of convergence of a power series can always be established by use of the ratio test. For a finite interval of convergence, the ratio test always fails at the end-points of the interval. Evaluate the behavior of the series at each end-point this way: Substitute the end-point into the power series thereby reducing it to the sort of infinite series considered in earlier parts of this chapter. Then use the techniques introduced there to study its convergence.

EXAMPLE 16.36: Examine the exponential series $\sum_{n=0}^{\infty} \frac{x^n}{n!}$ for convergence.

For each fixed value of x, the ratio test gives

$$\lim_{n\to\infty} \left|\frac{u_{n+1}}{u_n}\right| = \lim_{n\to\infty} \left|\frac{\frac{x^{n+1}}{(n+1)!}}{\frac{x^n}{n!}}\right| = \lim_{n\to\infty} \left|\frac{x^{n+1}}{x^n} \frac{n!}{(n+1)!}\right|$$

$$= \lim_{n\to\infty} \left|\frac{x}{n+1}\right| = 0 .$$

Because the value of the limit in the ratio test is zero (less than one) the exponential series converges for all values of x; its interval of convergence consists of the entire x-axis. ∎

EXAMPLE 16.37: Examine $\sum_{n=1}^{\infty} \frac{(-1)^{n+1} x^{2n}}{(n+1)2^{2n}}$ for convergence.

$$\lim_{n\to\infty}\left|\frac{u_{n+1}}{u_n}\right| = \lim_{n\to\infty}\left|\frac{\dfrac{x^{2n+2}}{(n+2)2^{2n+2}}}{\dfrac{x^{2n}}{(n+1)2^{2n}}}\right| = \lim_{n\to\infty}\left|\frac{x^{2n+2}}{(n+2)2^{2n+2}} \times \frac{(n+1)2^{2n}}{x^{2n}}\right|$$

$$= \frac{x^2}{4}\lim_{n\to\infty}\left|\frac{n+1}{n+2}\right| = \frac{x^2}{4}\,.$$

Thus, the ratio test shows that the series converges absolutely for

$$\frac{x^2}{4} < 1 \qquad \text{or} \qquad x^2 < 4, \qquad \text{or} \qquad |x| < 2,$$

but diverges if $|x| > 2$. The ratio test fails at $x = \pm 2$ so we must examine the series at these end-points. They give the series

$$1 - \frac{1}{2} + \frac{1}{3} - \cdots$$

which we have shown (Example 16.29) to be conditionally convergent. The above series converges absolutely if $-2 < x < 2$ and converges conditionally if $x = 2$ or $x = -2$, but diverges if $|x| > 2$. Thus, it has $|x| \leq 2$ as its interval of convergence. ▌

EXERCISES

Find the interval of convergence of the following power series. All sums run from $n = 1$ to $n = \infty$ unless otherwise specified.

1. $\sum \dfrac{x^n}{an!}$

2. $\sum \dfrac{(-1)^n(x-3)^n}{n}$

3. $\sum \dfrac{(10^{10}\,x)^n}{n!}$

4. $\sum \dfrac{(x+a)^n}{nb^{n-1}}$

5. $\sum \dfrac{x^{n-1}}{n}$

6. $\dfrac{1}{\sqrt{1}} - \dfrac{x^2}{\sqrt{2}} + \dfrac{x^4}{\sqrt{2}} - \cdots$

7. $1 + \sqrt{2}\,x + \sqrt{3}\,x^2 + \cdots$

8. $\frac{2}{1} + \frac{3}{2} \times \frac{x + 1}{3} + \frac{4}{3} \times \frac{(x + 1)^2}{3^2} + \cdots + \frac{n + 1}{n} \; \frac{x + 1}{3}^{n-1} + \cdots$

9. $e^{-x^2} = 1 - x^2 + \frac{x^4}{2!} - \frac{x^6}{3!} + \frac{x^8}{4!} - \cdots + (-1)^{n+1} \frac{x^{2n-2}}{(n - 1)!} + \cdots$

10. $\sin x = x - \frac{x^3}{3!} + \frac{x^5}{5!} - \frac{x^7}{7!} + \cdots + (-1)^{n+1} \frac{x^{2n-1}}{(2n - 1)!} + \cdots$

11. $\cos x = 1 - \frac{x^2}{2!} + \frac{x^4}{4!} - \frac{x^6}{6!} + \cdots + (-1)^{n+1} \frac{x^{2n-2}}{(2n - 2)!} + \cdots$

12. $\text{sech}\ x = x + \frac{x^3}{3!} + \frac{x^5}{5!} + \frac{x^7}{7!} + \cdots + \frac{x^{2n-1}}{(2n - 1)!} + \cdots$

13. $\cosh x = 1 + \frac{x^2}{2!} + \frac{x^4}{4!} + \frac{x^6}{6!} + \cdots + \frac{x^{2n-2}}{(2n - 2)!}$

14. $a^x = 1 + x \ln a + \frac{x^2(\ln a)^2}{2!} + \frac{x^3(\ln a)^3}{3!} + \cdots + \frac{(x \ln a)^{n-1}}{(n - 1)!} + \cdots$

15. $\ln(1 + x) = x - \frac{x^2}{2} + \frac{x^3}{3} - \frac{x^4}{4} + \cdots + (-1)^{n+1} \frac{x^n}{n} + \cdots$

*16.9 TAYLOR SERIES EXPANSIONS

We have used a particular kind of power series before. The Maclaurin and Taylor series expansions were considered at the end of Chapter 7. These provided a means for using the derivatives of a function to express the function as a power series in x and x - a, respectively. At that point, we merely assumed that the series so created did converge, but this assumption sometimes fails: Because they are power series, they converge only within their interval of convergence.

The need for approximating a function arises fairly often, but what do we really mean by approximation? An approximating function might be required to exactly agree with the object function at specified points. In fact, if we have $f(x_1)$, $f(x_2)$, $\cdots$, $f(x_n)$, there exists a polynomial of degree n which has the same values as f at x_1, x_2, $\cdots$, x_n. Such an approximation is shown in Figure 16.2 where the dashed curve represents a polynomial approximation to the function depicted by the solid curve. Note that the approximating curve deviates noticeably from the function of interest at values of x close to where the two are equal.

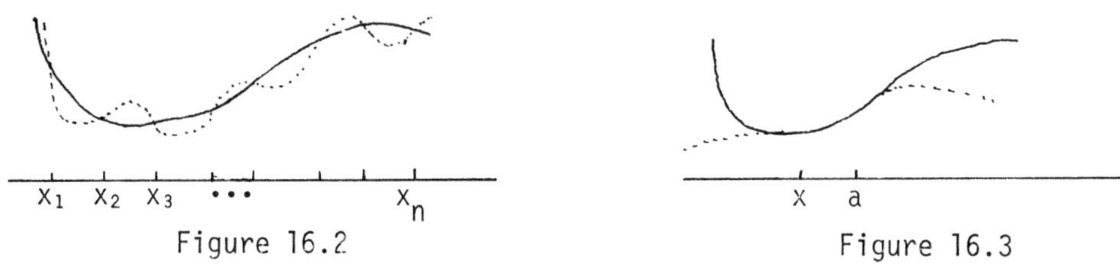

Figure 16.2 Figure 16.3

Figure 16.3 displays another kind of approximation. The dashed curve matches the function of interest over an interval near $x = a$, rather than only at isolated points. To obtain this kind of approximation we need $f(a)$, $f'(a)$, $f''(a)$, $\cdots$, $f^{(n)}(a)$, namely the value of various derivatives at one point. From these derivatives we can find the first n terms of a power series which converges to f in an interval around $x = a$. Under many circumstances these first n terms will provide a satisfactory approximation to $f(x)$ for x near $x = a$ as shown in Figure 16.3, rather than only at isolated points.

Here we repeat a development given in Section 7.8. _If_ f has a power series representation

$$f(x) = a_0 + a_1(x - a) + a_2(x - a)^2 + \cdots + a_n(x - a)^n + \cdots$$

and _if_ it can be repeatedly differentiated term by term, then it follows that

$$f'(x) = a_1 + 2a_2(x - a) + 3a_3(x - a)^2 + \cdots + na_n(x - a)^{n-1} + \cdots$$

$$f''(x) = 2a_2 + 6a_3(x - a) + \cdots + n(n - 1)a_n(x - a)^{n-2} + \cdots$$

$$f'''(x) = 6a_3 + 24a_4(x - a) + \cdots + n(n - 1)(n - 2)a_n(x - a)^{n-3} + \cdots$$

$$\vdots$$

$$f^{(m)}(x) = m!a_m + \frac{(m + 1)!}{1!} a_{m+1}(x - a) + \cdots + \frac{n!}{(n - m)!} a_n(x - a)^{n-m} + \cdots$$

When each of the above is evaluated at $x = a$, we are able to solve for the a_i's:

$$a_0 = f(a)$$

$$a_1 = f'(a)$$

$$a_2 = \frac{f''(a)}{2!}$$

$$a_3 = \frac{f'''(a)}{3!}$$

$$\vdots$$

$$a_n = \frac{f^{(n)}(a)}{n!}.$$

The real basis for the association of Taylor series and power series lies in these two theorems:

THEOREM 16.15 (*Taylor's Formula*): If a function f has n continuous derivatives over an interval T containing the point a, then for each $x \in T$ there exists a point ξ_n between x and a such that

$$f(x) = f(a) + f'(a)(x - a) + \frac{f''(a)(x - a)^2}{2!} + \cdots + f^{n-1}(a)(x - a)^{n-1} + R_n(x)$$

where

$$R_n(x) = \frac{f^{(n)}(\xi_n)(x - a)^n}{n!}.$$

THEOREM 16.16: If a function has a power series expansion about the point a, with a nonzero interval of convergence, then that power series is unique.

The first theorem provides a rationale for evaluating derivatives of a function at a point to get coefficients in a power series. The remainder term allows us to evaluate the amount by which the first n terms fail to give an exact representation of f(x) because $f^{(n)}(\xi_n)$ will have a maximum value for ξ_n between x and a. Of course, $\lim_{n\to\infty} R_n(x) = 0$ within the interval of convergence of the power series. This observation follows the second theorem in this way: If a function has a power series expansion, then its power series must equal its Taylor series expansion because all of the terms of a Taylor series comprise a power series. In particular, the reasoning above implies that if

$$f(x) = a_0 + a_1(x - a) + a_2(x - a)^2 + \cdots + a_n(x - a)^n + \cdots ,$$

then

$$a_n = \frac{f^{(n)}(a)}{n!} \qquad (16.13)$$

This has several powerful consequences. If we know how a function changes, that is, know its derivatives at one point, the power series represents the function within the series' interval of convergence. Thus from knowing derivatives of a function at one point, we can evaluate the function at nearby points. On the other hand, we can read its derivatives directly off its power series expansion as $f^{(n)}(a) = n!\, a_n$. Finally, if you can find a power series representation of f about a, no other one exists. This has a broader utilization than might at first appear. Suppose you have the power series expansion of two functions which you want to combine, like by multiplying them. Merely combine their power series in the same way and collect all terms having the same power to get the power series expansion of the combination of the two functions.

We also introduced in Chapter 7 the special case when a = 0. This series

$$f(x) = f(0) + f'(0)x + \frac{f''(0)x^2}{2!} + \cdots + \frac{f^{(n)}(0)x^n}{n!} + \cdots \qquad (16.14)$$

was given the name of the MACLAURIN SERIES EXPANSION of f.

EXAMPLE 16.38: Evaluate $e^{0.2}$ to four decimal places.

We can find the Maclaurin expansion for e^x and then evaluate it at $x = 0.2$:

$$f(x) = e^x, \qquad f^{(n)}(x) = e^x \qquad \text{and} \qquad f^{(n)}(0) = 1 ,$$

$$e^x = 1 + x + \frac{x^2}{2!} + \frac{x^3}{3!} + \cdots .$$

Now,

$$\begin{aligned} e^{0.2} &= 1 + 0.2 + \frac{(0.2)^2}{2!} + \frac{(0.2)^3}{3!} + \cdots \\ &= 1 + 0.2 + 0.02 + 0.001333 + 0.00006666 \cdots \\ &= 1.22139 \cdots \end{aligned}$$

Thus, $e^{0.2} = 1.2214$ evaluated to four decimal places. This is precisely the value Appendix Table III gives for $e^{0.2}$.

Theorem 16.15 provides a means for evaluating the error of approximation:

$$R_4(x) = \frac{f^{(4)}(\xi_4)(x - 0)^4}{4!} = \frac{e^{\xi}\, x^4}{4!} .$$

Because $0 < \xi_4 < 0.2$, $e^{\xi_4} < e^{0.2}$. Thus, our maximum error will be

$$R_4(x) < \frac{e^{0.2}\, 0.2^4}{4!} \doteq 0.0000814 . \quad \blacksquare$$

EXERCISES

Obtain the Taylor expansion of $f(x)$ about the point specified; include the general term of the series. If n is given, then just obtain that many nonzero terms of the expansion.

1. $f(x) = \cos x$, $a = 0$.
2. $f(x) = \sin x$, $a = \frac{\pi}{4}$
3. $f(x) = e^{-2x}$, $a = 0$
4. $f(x) = 3x^3 + 5x^2 + 4x + 2$, $a = 0$
5. $f(x) = \tan x$, $a = 0$, $n = 3$

Approximate the following numbers from the first three terms of a Taylor series. Evaluate the maximum error associated with each approximation.

6. $\sqrt{51}$

7. $\sin \frac{\pi}{180} = \sin 1°$

8. $(1.01)^{12}$

9. $\ln 3$

10. Find a value of h such that for $|x| < h$, the sum of the first two non-zero terms in the Maclaurin series for sin x will give the value of sin x with accuracy of four decimal places.

11. Show that the derivative of the Maclaurin series for sin x equals the Maclaurin series for the cos x.

16.10 APPLICATION OF THE POWER SERIES

Power series have uses other than for computing values of functions. Many uses rely on properties which we shall state and use without proof. Within its interval(s) of convergence,

PROPERTY I: A power series may be differentiated term by term to obtain the derivative of the power series.

PROPERTY II: A power series may be integrated term by term to obtain the integral of the power series.

PROPERTY III: Two power series may be multiplied term by term, with like powers of x collected to yield a power series for the product; the combined series converges to the product of the two power series over the intersection of the two individual intervals of convergence.

EXAMPLE 16.39: Develop the Maclaurin series expansion for cos x from the one for sin x.

Since

$$\int_0^x \sin t \, dt = - \cos t\Big]_0^x = 1 - \cos x ,$$

$$1 - \cos x = \int_0^x \left[t - \frac{t^3}{3!} + \frac{t^5}{5!} + \cdots + \frac{(-1)^{k-1}\ t^{2k-1}}{(2k-1)!} + \cdots\right]dt$$

$$= \frac{x^2}{2!} - \frac{x^4}{4!} + \cdots + \frac{(-1)^{k-1}\ x^{2k}}{(2k-1)!\ 2k} + \cdots$$

and, thus,

$$\cos x = 1 - \frac{x^2}{2!} + \frac{x^4}{4!} + \cdots + \frac{(-1)^k\ x^{2k}}{(2k)!} + \cdots = \sum_{n=0}^{\infty} (-1)^n \frac{x^{2n}}{(2n)!}\ .$$

Thus, from our knowledge of integration and the expansion for the sine function, we found an expansion for the cosine function. ▌

EXAMPLE 16.40: Obtain the first three terms of the Maclaurin series expansion for $e^x \cos x$.

The two factors have these expansions:

$$e^x = 1 + x + \frac{x^2}{2!} + \cdots$$

and

$$\cos x = 1 - \frac{x^2}{2!} + \frac{x^4}{4!} + \frac{x^6}{6!} + \cdots\ .$$

Thus,

$$e^x \cos x = (1 + x + \frac{x^2}{2!} + \frac{x^3}{3!} + \cdots)(1 - \frac{x^2}{2!} + \frac{x^4}{4!} - \frac{x^6}{6!} + \cdots)$$

$$= 1 + x - \frac{x^3}{3} + \cdots\ .$$ ▌

EXAMPLE 16.41: Evaluate $\int \frac{\sin x}{x}\,dx$.

No previous technique of integration will work on this integral. However, using a series approach, we have

$$\int \frac{\sin x}{x}\,dx = \int \frac{1}{x} \sum_{n=0}^{\infty} (-1)^n \frac{x^{2n+1}}{(2n+1)!}\,dx = \int \sum_{n=0}^{\infty} (-1)^n \frac{x^{2n}}{(2n+1)!}\,dx$$

$$\int \frac{\sin x}{x}\,dx = \sum_{n=0}^{\infty} \int (-1)^n \frac{x^{2n}}{(2n+1)!}\,dx$$

$$= \sum_{n=0}^{\infty} (-1)^n \frac{x^{2n+1}}{(2n+1)(2n+1)!} + C \text{ . } \blacksquare$$

EXAMPLE 16.42: Evaluate $\int e^{x^2}\,dx$.

From

$$e^u = 1 + u + \frac{u^2}{2!} + \cdots = \sum_{n=0}^{\infty} \frac{u^n}{n!} ,$$

$$e^{x^2} = 1 + x^2 + \frac{x^4}{2!} + \cdots = \sum_{n=0}^{\infty} \frac{x^{2n}}{n!} .$$

Thus,

$$\int e^{x^2}\,dx = \int \sum_{n=0}^{\infty} \frac{x^{2n}}{n!}\,dx = \sum_{n=0}^{\infty} \int \frac{x^{2n}}{n!}\,dx$$

$$= \sum_{n=0}^{\infty} \frac{x^{2n+1}}{(2n+1)n!} + C .$$

Again, by using series, we evaluated an integral when no other integration technique would work. ▮

EXERCISES

Find the Taylor's series for the following functions about the point specified.

1. $f(x) = \sin x + \cos x$, $a = 0$

2. $f(x) = \sin x \cos x = \frac{1}{2} \sin 2x$, $a = 0$

3. $f(x) = b^x$, $a = 0$, $b > 1$

4. $\cosh x = \dfrac{e^x + e^{-x}}{2}$, $a = 0$

5. $f(x) = (1 + x^2)\sin x,\ a = 0.$

6. $\ln(1 + x)$ from $\int_0^x (1 + x)^{-1}\,dx$

7. $\dfrac{e^{-x}}{1 + x}$, $a = 0$

Evaluate the following integrals accurate to three decimal places.

8. $\int_0^{\frac{1}{2}} \sqrt{1 + x^2}\,dx$

9. $\int_0^{0.4} e^{-x^2}\,dx$

10. $\int_0^{0.2} \ln(1 - x)dx$

11. $\int x \sin x\,dx$

12. $\int e^{x^3}\,dx$

13. $\int \sin x^2\,dx$

14. Verify that $\frac{d}{dx} e^x = \sum_{n=0}^{\infty} \frac{x^n}{n!}$.

15. Find $\frac{d}{dx} \ln x$ in terms of a Taylor series.

16. Find an expansion in powers of x of the function $f(x) = \int_0^1 \frac{1 - e^{-tx}}{t}\,dt.$

17. Approximate $\int_0^1 \frac{1 - \cos x}{x^2}\,dx.$

REFERENCES

Knipling, E. F. (1960). The eradication of the screwworm fly. *Scientific American* 203(4):54-61.

Chapter 17

CONCLUDING COMMENTS

This book has presented a view of mathematics in biology. The biological sciences and mathematical sciences each display great diversity of approach and perspective. Biology can be viewed, for example, taxonomically, physiologically, or ecologically, recognizing that each view gains insights from the other. Within biology we find both applied and basic approaches. The applied perspective surfaces in agriculture (sometimes called economic biology), wildlife management, and medical microbiology, for example. Areas such as ornithology, molecular biology, speciation and related taxonomic matters, and animal behavior have a more basic outlook, yet they will continue to produce useful information.

Similarly, the mathematical sciences have areas such as algebra, analysis, geometry, topology, numerical analysis, probability, statistics, and computer science. This book has focused on functions and the calculus, a major part of elementary analysis, with only secondary emphasis on probability and statistics; other areas have not been covered. We cannot cover all of the mathematics that you might need. In fact, you could spend so long learning all of the relevant mathematics that you would have no time for biology.

This brief chapter will expose you to the existence and nature of four additional areas within the mathematical sciences which are used on biological problems. The first two, namely, differential equations and linear algebra, lie within classical mathematics; the other two concern computing, and curve fitting and statistics. Each section ends with suggestions for further reading.

17.1 DIFFERENTIAL EQUATIONS

Differential equations appeared in Section 8.9 as one of the concluding topics on integration. They reappeared several times after that initial introduction.

Differential equations rest on an exceedingly simple idea: A function is related closely to its derivative; how should it be retrieved? Namely, if you can describe how a biological process changes, mathematics can help you find out about the underlying process. A differential equation merely consists of an equation involving a function, its derivatives and its variables. Often it is accompanied by auxiliary information, called boundary conditions, giving values of the function at specified points, usually when the process starts or ends. Careful mathematical reasoning proves that many kinds of differential equations have solutions, but without specifying how to find the solution. This does involve a real paradox: Solutions exist, but they frequently cannot be found!

Functions, their derivatives and their independent variables can be combined in various ways to form many kinds of differential equations. One particular kind may be difficult enough to handle that it can occupy a whole course or monograph.

When a function varies with several independent variables, differential equations frequently involve partial derivatives. In fact this may lead to a system of several simultaneous partial differential equations. Such systems usually submit to only approximate solution by computer. On the other hand, systems of partial differential equations appear to provide an adequate tool for describing some biological processes, particularly ones which are continuous in time.

Thus it seemed unfair to embark on any study of differential equations because we could easily miss forms used in your area while covering many forms irrelevant to you. More importantly you need to know what a differential equation actually means. The earlier discussions of total and partial derivatives, and of simple differential equations should help you understand that. You can learn about forms important in your discipline from your disciplinary literature. If you can _begin_ to describe your own biological problem in terms of functions, variables, derivatives and differential equations, a competent applied mathematician can help you complete the setup and direct you toward a means for getting a solution.

The paper below presents a general treatment of particle movement using approximate differential equations. It applies to: (1) enzyme kinetics; (2) growth of organisms or populations; (3) the spread of a disease through a population; (4) loss of heat from an organism; and (5) transport of chemical substances between "compartments" in an organism. A numerical example illustrates this model's application to the transport of radioactively tagged elements through the blood stream, to the kidneys and out in urine.

Turner, M. E., R. J. Monroe, and L. D. Homer (1963). Generalized kinetic regression analysis: hypergeometric kinetics. *Biometrics* 19:406-428.

Other current references involving the use of differential equations in biology are:

Shipley, R. A. and R. E. Clark (1972). *Tracer Methods For In Vivo Kinetics, Theory and Application.* Academic Press, N.Y.

Simon, W. (1972). *Mathematical Techniques for Physiology and Medicine.* Academic Press, N.Y.

17.2 LINEAR ALGEBRA

Linear algebra involves vector spaces, linear transformations and matrices as central concepts. In fact, the content of this section could be entitled with any of these four names. These topics have an important role in biology: They provide a solid basis for dealing with relations simultaneously involving more than two independent variables.

Many biological processes respond to numerous independent variables. For example, blood pressure changes with age, weight, stress, levels of various drugs in the blood, etc. Or, the size of a population of organisms changes with the densities of its predators, availability of appropriate habitat at critical life stages, and physical factors such as temperature, precipitation, and wind. We drew graphs to visualize the behavior of a function of one variable. A surface in three dimensions represented a function of two independent variables. Recall that we experienced some difficulty in drawing these surfaces on a sheet of paper;

the expression of lines and planes proved troublesome. Functions of more than two independent variables ordinarily cannot be described by surfaces we can draw.

Vector spaces provide the mathematician with a means of dealing with these higher dimensional problems. A vector space consists of a collection of objects frequently called points because they reduce to points in one, two and three dimensions. Arithmetic can be performed on these objects; the results behave as our experience suggests they should. Vector spaces can have parts called subspaces which generalize lines and planes.

Transformations relate the points in one vector space to those in another. In general use a transformation changes one kind of object into another kind. For example, a caterpillar is transformed into a moth or butterfly by metamorphosis, an electrical transformer changes an electrical current's voltage and amperage from one pair of values to another. Biology abounds with situations which can be described by a mathematical transformation. An organism's physiological condition can be described by a number of variables. Their values at a specific point in time could be a point in a vector space; a treatment such as stress or feeding would change the variables, that is, transform one point into another.

Linear transformations between vector spaces have an important place in both biology and linear algebra. If x's are the coordinates of a point in the first vector space and y's in the second, then

$$\begin{aligned} y_1 &= a_{11}x_1 + a_{12}x_2 + \cdots + a_{1m}x_m \\ y_2 &= a_{21}x_1 + a_{22}x_2 + \cdots + a_{2m}x_m \\ &\cdot \\ &\cdot \\ &\cdot \\ y_n &= a_{n1}x_1 + a_{n2}x_2 + \cdots + a_{nm}x_m \end{aligned}$$

describes a general linear transformation from x to y using the constants

(a_{ij}). For example x could contain the frequencies of various categories of an organism in a population, where survival and reproductive potential change across the categories. A critical event in the organism's life cycle, such as reproduction, dispersal or predation, will transform the frequencies from one point (x) to another (y). A linear transformation approximates this very well in many cases.

When the coefficients (a_{ij}) of the linear transformation are arranged to display their row-column organization, the resulting array is called a matrix. Thus matrices and linear transformations can be closely related. On the other hand, matrices can be regarded as objects of direct interest. Matrices have found extensive application in some areas of biology. They appear naturally when much data is gathered; rows can indicate individuals or sites where data was gathered, and the columns stand for different measurements taken.

Many good books have been written on linear algebra. You might find the following one useful because its first chapter gives a readable introduction to the subject before continuing onto the formal material:

Lang, S. (1970). *Introduction to Linear Algebra.* Addison-Wesley. Reading, Mass.

The title of the next text (somewhat advanced) describes its content well; its Chapter 5 deals with linear algebra and contains a spectrum of biological examples.

Nahikian, H. M. (1964). *A Modern Algebra for Biologists.* Univ. Chicago Press. Chicago, Ill.

The final text deals directly with matrices, but contains many real applications. You should find it quite readable, even if you have mastered only the first part of the present book.

Searle, S. R. (1966). *Matrix Algebra for the Biological Sciences.* Wiley, N.Y.

17.3 STATISTICS AND CURVE FITTING

This book has dealt primarily with development, examination and interpretation of functions using tools of the calculus. Functions have one distinctive property: The function has exactly one value for each permissible value of its argument (independent variable(s)). Most biological responses fail to behave like functions in one important aspect: The same conditions rarely produce exactly the same response.

A response typically varies from one organism to another. Differences between organisms reflect natural variation arising from genetic sources, from undetectable, minor fluctuations in environmental conditions, and from imprecision in the actual process of evaluation. Other variations away from a model may reflect minor inaccuracies in its form. Thus functions usually describe an aggregate view of a phenomenon, not its individual occurrences, a fact first noted in Chapter 1. This section briefly sketches statistics and its role in managing natural variation.

Statistics provides methods for examining the nature and magnitude of relationships present in data, acknowledging the existence of unpredictable variation. Many statistical techniques exist; they differ in the circumstances to which they apply. Never expect to find one technique which will fit every situation you encounter. Statistical techniques differ in the precision of questions they investigate, their popularity, and their objectivity. At one extreme we find well known techniques like the t-test. It will give reasonably clear answers to questions like, "Has this treatment changed that response level?" At the other end of the scale we find procedures like factor analysis or cluster analysis, whose application involves a substantial degree of subjectivity. They attempt to meet demands of this sort: Data, tell me both how to proceed and what to conclude. Here we will expose you to ideas upon which procedures are founded, not to specific procedures; to do otherwise would require another book.

Even when all conditions remain constant, we see variation in a response from organism to organism. Nevertheless this variation ordinarily displays a pattern. Various response values occur with different frequencies. This pattern often is called a frequency distribution, that is, the distribution of frequencies across response values. Different experimental conditions change the location and/or shape of this distribution. For example, a particular feed may increase the rate of gain of cattle in a feedlot, namely, shift the frequency distribution for rates of gain to larger values. On the other hand, increasing the number of animals per pen may make the frequency distribution spread out because big animals get bigger by pushing the smaller animals away from the feed and keeping them small.

Probability becomes very involved in statistical reasoning this way: Because we rarely can observe the entire frequency distribution (_all_ possible organisms under these conditions), we have to judge its characteristics from a sample of data. When samples are selected randomly, each kind of organism in a finite population appears in the sample with probability equal to its frequency in the population. Thus the population frequency distribution governs the probability distribution underlying the sample. Finally, we draw inferences from the sample to its probability distribution, and then back to the population frequency distribution.

A probability distribution has a mathematical form involving parameters, such as its mean, its variance, or others depending on the form. These parameters reflect things such as the location and spread of the distribution, or various features of its shape. Conclusions about the probability distribution can be expressed in terms of these parameters when the mathematical form of the distribution is specified.

Statistical methods investigate parameters in two ways: Through estimation and testing. Estimation seeks the best available guess for the parameters underlying a particular set of data. The estimate of a parameter may be recorded as a single number, or as a range of numbers, any of which reasonably could have produced the data. Methods of estimation depend on the mathematical form of the probability distribution, on how the data was acquired, on the nature of the research study producing the data, and on the researchers objectives.

The other major statistical approach involves testing. This concerns a simple idea: An investigator may have a subject matter basis for expecting parameters to assume specified values under certain conditions. To investigate the presence of those conditions, he needs to evaluate the agreement between the observed data and the specified parameter values. The methods of conducting tests depend on several factors, much as estimation did. In fact, procedures exist for making tests without specifying the mathematical form of the probability distribution.

To fit a curve to data you must first pick the form of a curve, and then draw statistical inferences about its parameters. A theme that has reoccurred throughout this book: Begin with an understanding of the biological process of interest. Next express its important characteristics mathematically, and then proceed to develop the relation using approaches illustrated in earlier chapters. Once a curve form has been developed, statistical approaches called regression can be used to investigate the curve, estimate its parameters or test specified values suggested by theory.

Many references could be given at this point. An elementary introduction to reasoning using statistics appears in

Huntsburger, D. V. and P. Billingsley (1973). *Elements of Statistical Inference*. Allyn and Bacon, Boston.

Two texts have become standard references for a comprehensive coverage of statistics used in biology:

Snedecor, G. W. and W. G. Cochran (1967). *Statistical Methods*. 6th ed. Iowa State University Press, Ames.

Steel, R. G. D. and J. H. Torrie (1960). *Principles and Procedures of Statistics*. McGraw-Hill, N.Y.

A comprehensive discussion of curve fitting and associated statistical topics appears in

Draper, N. R. and H. Smith (1966). *Applied Regression Analysis*. Wiley, N.Y.

17.4 COMPUTING

Computers do extensive computation in science, industry and commerce, and even control entire industrial processes. Two basic types of computers are in general use: digital and analog. Electronic digital computers store numbers and execute operations on them in a manner equivalent to hand computation. Their power comes from their ability to store and retrieve many pieces of information quickly and with great reliability and precision. Analog computers use electric circuits to simulate physical systems. Some analog-digital hybrids have been developed.

A digital computer follows a program of instructions provided by its user. The power of modern digital computers lies in their ability to modify programs in a manner prescribed by the user, because the programs are stored in the computer's memory. Stored-program computers became commercially available about 1950. In the subsequent 15 years they evolved at a fantastic rate. Efforts to improve the central processing unit have diminished recently in favor of improving user access and utility. These include expanded memory capacity, improved language features, new means for passing data directly from measuring devices into a computer, and other improvements at the man-machine interface.

Biologists are using digital computers in many ways. Important uses include reduction of data masses resulting from large studies, examination of quantitative biological hypotheses, curve fitting and other statistical analyses, calculation of radiation doses for cancer treatment, simulation of biological systems ranging from photosynthesis to population dynamics to blood flow. Eight major examples of simulation appear in

Heinmets, F. (1969). *Concepts and Models of Biomathematics: Simulation Techniques and Methods*. Dekker, N.Y.

To effectively use a computer, a biologist must have the programs which instruct the computer to do what he wants. These may be available at his nearby computer center; he may obtain one from a colleague with similar interests; or he may develop his own program. A substantial computer center will have general service capabilities and common statistical programs, but often little else of immediate interest to a biologist.

A useful program will be available from a colleague only occasionally, because each biological problem has unique features which should substantially influence computer programs associated with it. Thus to effectively use a computer, a biologist should expect to develop his own programs part of the time.

To develop a program, the biologist must either write it himself or explain what he needs to a computer programmer. If a programmer belongs to a team attacking a large problem, the latter approach may prove fruitful. Otherwise, biologists often experience severe difficulty in communicating with programmers. A programmer understands computing, but he may experience real difficulty in understanding the biological problem. Thus a biologists needs to express his computational desires very precisely. If he writes the program himself, then he can define his problem with increasing precision as he writes the program.

A young biologist with a quantitative outlook should learn how to write computer programs. A program merely consists of the step-by-step operations needed to solve a problem. Steps may have the computer make decisions so that it follows one set of steps under one circumstance, but other steps for other circumstances. Programs may have loops, namely a set of steps which get executed many times, like once for each entry in a set of data. A number of computer languages are used by biologists. The most popular is FORTRAN (for FORmula TRANslation), a general purpose scientific language. Some use PL/1 (for Programming Language one) because it has more powerful handling of arrays and allows more readable programs than FORTRAN. The two languages BASIC and APL (A Programming Language) can be used interactively from remote terminals. The latter has very powerful array operations.

Learning a computer language has many similarities with learning a foreign language. You must learn its vocabulary, its syntax (grammar) and its conventions (idioms). Computer languages have only small vocabularies; a user may define the meaning of various configurations of letters (words) anew in each program. These languages do not cause most students too much trouble. The real art of programming lies in dividing your problem into steps digestable by a computer.

The above deals with digital computers, but recall that analog computers also exist. This name really describes its means of computation: When two different physical systems can be described by the same mathematical equation, either system can be used to evaluate the behavior of the other. Any physical system producing products, powers, trigonometric functions or logarithmic functions has a physical analog in a slide rule. Evaluation of areas can be accomplished with a planimeter, briefly described in Section 8.11. However, analog computers usually involve an electronic system which behaves like a physical system. In an electronic system, components change voltage in very specific ways. For example a resistor reduces voltage in proportion to its resistance; a condenser behaves like an integral; an inductance operates like a derivative. These components can be combined to produce an electronic system which will mimic almost any differential equation. Most applications of analog computers use time as the independent variable. Analog computers have less computational precision than digital computers, but for most biological applications this inaccuracy proves immaterial. Furthermore they can provide useful solutions to complicated differential equations for a fraction of the cost of a precise solution obtained on a digital computer.

Analog computers have found extensive applications in some areas of biology. For example, physiological processes need to be monitored continuously through time. If the process can be described by a differential equation, then it can be investigated with an analog computer. In fact, original measurements may be electronic signals which can be passed through an analog computer and partially analyzed before being recorded.

Numerous illustrations of applications of computers to biology appear in

Ledley, R. S. (1965). *Use of Computers in Biology and Medicine.* McGraw-Hill, N.Y.

Sterling, T. D. and S. V. Pollack (1965). *Computers and the Life Sciences.* Columbia University Press, N.Y. (Last four chapters contain four extensive examples.)

Watt, K. E. F. (1968). *Ecology and Resource Management: A Quantitative Approach.* McGraw-Hill, N.Y.

Numerous texts discuss FORTRAN; these are designed for self-study:

Anderson, D. M. (1966). *Computer Programming FORTRAN IV.* Appleton-Century-Crofts, N.Y.

Dimitry, D. L. and T. H. Mott, Jr. (1966). *Introduction to Fortran IV Programming.* Holt, Rinehart and Winston, N.Y.

Farina, M. V. (1966). *Fortran IV Self-taught.* Prentice-Hall, Englewood Cliffs, N.J.

Analog computers are discussed in Chapter 7 of Ledley (1965), cited above, and in

Ashley, J. R. (1963). *Introduction to Analog Computation.* Wiley, N.Y.

Randall, J. E. (1965). The analog computer in the biological laboratory. p. 65-86 in *Computers in Biomedical Research,* Vol. I., R. W. Stacy and B. D. Waxman, eds. Academic Press, N.Y.

Stacy, R. W. (1960). *Biological and Medical Electronics.* McGraw-Hill, N.Y.

Appendix

LOGARITHMS

This appendix provides a review of logarithms, but contains only those topics needed in this book. We begin with the following definition:

- The logarithm of a number N to the base b is the exponent to which b must be raised in order to give the number N where b is a positive real number different from 1.

  Thus $\log_b N = x$

  when $b^x = N$

In the remainder of this appendix, we will always assume that the base, b, under consideration is greater than 1; this covers most useful situations. The following properties and theorems emerge from the definition, but are stated here without proof.

1. The logarithm of 1 is zero.
2. The $\log_b b = 1$.
3. Numbers greater than 1 have positive logarithms while those between 0 and 1 have negative logarithms.
4. The number zero has no logarithm, neither has any negative number.

The following figure depicts these relations:

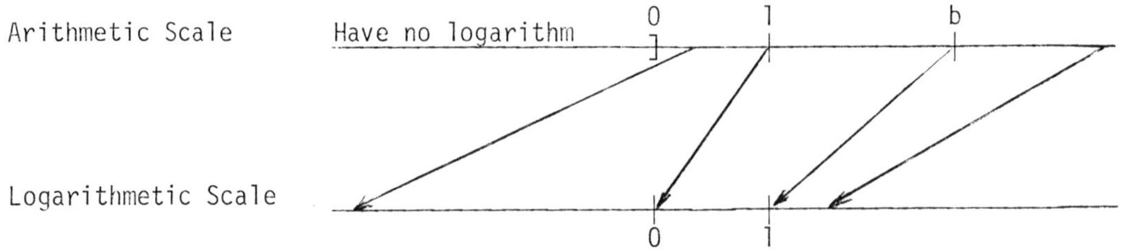

Figure A.1

This figure suggests that if $m < n$, $\log_b m < \log_b n$ for $b > 1$, a fact useful later.

The power and utility of logarithms result from the following three important facts stated as theorems:

THEOREM A.1: The logarithm of a product of two numbers can be calculated as the sum of the logarithms of the factors.

$$\log_b MN = \log_b M + \log_b N .$$

THEOREM A.2: The logarithms of the quotient of two numbers can be calculated as the logarithm of the numerator minus the logarithm of the denominator.

$$\log_b \frac{M}{N} = \log_b M - \log_b N .$$

THEOREM A.3: The logarithm of a number raised to a power can be calculated as the product of the exponent and the logarithm of the number.

$$\log_b M^P = P \log_b M .$$

EXAMPLE A.1: Express $\log_b \sqrt{\frac{MN}{PQ}}$ as a sum and difference of logarithms.

$$\log_b \sqrt{\frac{MN}{PQ}} = \log_b \left[\frac{MN}{PQ}\right]^{1/2}$$

First Theorem A.3 gives

$$\log_b \left[\frac{MN}{PQ}\right]^{1/2} = \frac{1}{2} \log_b \frac{MN}{PQ} ,$$

then applying Theorem A.2,

$$\log_b \frac{MN}{PQ} = \log_b MN - \log PQ ,$$

and applying Theorem A.1 to the last result,

$$\log_b MN - \log_b PQ = \log_b M + \log_b N - \log_b P - \log_b Q .$$

Combining these results, we finally have

$$\log_b \sqrt{\frac{MN}{PQ}} = \frac{1}{2} (\log_b M + \log_b N - \log_b P - \log_b Q) . \quad \blacksquare$$

The use of logarithms for computation proceeds through four distinct steps, although an experienced user may not clearly identify them: (i) Theorems A.1 - A.3 are used to establish a relation between the logarithm of the quantity of interest and logarithms of its components. (ii) The logarithms of its components are determined from tables. (iii) They are combined according to the relation established in (i), giving the logarithm of the quantity of interest. (iv) Reading of the tables is reversed to get from the logarithm of the quantity of interest to the quantity of interest. Above we considered (i), immediately we will consider (ii) and (iv). The third step can be considered only in the examples.

For computational purposes, one of the two major sets of logarithms uses 10 as its base. This set, called common logarithms, is denoted by log with the base understood to be 10. Its wide use in hand (as opposed to computer) calculation occurs because we use the number 10 as the base in almost all quantitative matters. Various powers of ten have these logarithms to the base of 10:

x	···	0.0001	0.001	0.01	0.1	1	10	100	1000	10000	···
log x	···	-4	-3	-2	-1	0	1	2	3	4	···

This also suggests that log m < log n if m < n. Thus, you should expect that log 497 lies between 2 and 3 since 497 lies between 10^2 and 10^3.

Every positive number can be expressed as the product of a number between 1 and 10 and an integral power of 10. For example

$$49.7 = 4.97 \times 10^1$$

$$497 = 4.97 \times 10^2$$

$$4.97 = 4.97 \times 10^0$$

$$0.497 = 4.97 \times 10^{-1}$$

$$0.00497 = 4.97 \times 10^{-3}$$

Using the properties of logarithms given in Theorems A.1 and A.3,

$$\log 49.7 = \log 4.97 \times 10^1 = \log 4.97 + \log 10 = \log 4.97 + 1$$

$$\log 497 = \log 4.97 \times 10^2 = \log 4.97 + \log 10^2 = \log 4.97 + 2$$

$$\log 4.97 = \log 4.97 \times 10^0 = \log 4.97 + \log 10^0 = \log 4.97$$

$$\log 0.497 = \log 4.97 \times 10^{-1} = \log 4.97 + \log 10^{-1} = \log 4.97 - 1$$

$$\log 0.00497 = \log 4.97 \times 10^{-3} = \log 4.97 + \log 10^{-3} = \log 4.97 - 3$$

Now it should be clear that we could write down the log of any number if we only knew the logs of the numbers between 1 and 10. The decimal portion of the log is called the MANTISSA. The above discussion shows that the mantissa depends only on the particular sequence of digits in the number and does not depend on the position of the decimal. Table I gives the mantissas of numbers from 1 to 10 to four decimal places, including $\log 4.97 = 0.6964$. From it and the relations established above,

$$\log 49.7 = 1.6964$$

$$\log 497 = 2.6964$$

$$\log 0.497 = -1 + 0.6964$$

$$\log 0.00497 = -3 + 0.6964$$

The integral part of the log is called its CHARACTERISTIC when it is expressed with a positive decimal part. For example, log 49.7 has a characteristic of 1, but log 0.00497 has a characteristic of -3 not -2 even though $-3 + 0.6964 = -2.3036$. When a characteristic is negative, it is convenient to add and subtract a suitable multiple of 10. For example, we could write $\log 0.00497 = 7.6964 - 10$ by adding and subtracting 10 or we could write $\log 0.00497 = 17.6964 - 20$ by adding or subtracting 20. This process is illustrated in the next example.

<u>EXAMPLE A.2</u>: Find $\log \sqrt[3]{0.00497}$.

$$\log \sqrt[3]{0.00497} = \frac{1}{3} \log 0.00497 \ .$$

Now $\log 0.00497 = 7.6964 - 10 = 27.6964 - 30$. This latter form facilitates division because the result will have the -10 form:

$$\frac{1}{3}\log 0.00497 = \frac{1}{3}(27.6964 - 30) = 9.2321 - 10 = 0.2321 - 1 . \quad \blacksquare$$

To find the logarithm of a number N, determine the characteristic by inspection and look up the mantissa in Table I. It gives four place mantissas for the numbers from 1.00 to 9.99 for intervals of .01. The principles involved in the use of these tables are the same as for four-, six- or ten-place logarithm tables.

Certain standard conventions are employed to minimize the size of such tables. Since all mantissas lie between 0 and 1, decimal points customarily are omitted, but you should supply them. The first two digits of the number whose log is sought appear at the left of the table in the column labeled N while its third digit occurs across the top of the table on the same line as the label N and similarly across the bottom. The first entry in the table gives log 1.00 = 0.0000. The next entry in the same line gives log 1.01 = 0.0043. In each case, the mantissa obtained directly from the table is the entire logarithm, since the characteristic is zero.

For the moment we defer the issue of finding the log of a number with more than three digits.

The final step in using logs requires that you find a number whose log you know. The resulting number usually is called an ANTILOGARITHM (antilog for common logarithms). If you have found log x = 0.1790, then x = 1.51 because log 1.51 = 0.1790. The instructions for reading antilogs are exactly the reverse of those already given for logs. Thus, the antilog 0.1790 is found by following down the column of mantissas until the digits 1784 occur. Examination shows 1790 in the line N = 15 and in the column under 1. The number 1.51 results from combining these and supplying the decimal point.

<u>EXAMPLE A.3</u>: Find the antilog 7.1790 - 10 and the antilog 3.1790.

Both of the logarithms have the same mantissa of 0.1790 which we found to have an antilogarithm of 1.51. Hence

$$\text{antilog } (7.1790 - 10) = \text{antilog } (.1790 - 3)$$

and

$$= 1.51 \times 10^{-3}$$

$$= 0.00156$$

$$\text{antilog } 3.1790 = 1.51 \times 10^{3} = 1510 \ .$$ ▌

A problem arises when the log of a number falls between values listed in the table or when you need the log of a number having more than three digits. Both require the placing of numbers between successive entries of the table, a process known as INTERPOLATION. Suppose that you want log 2.724. This must lie between log 2.720 and log 2.730. Since a very short section of the logarithmic curve is involved here, a sufficiently good approximation results from replacing the curve by a straight line. This is illustrated in Figure A.2.

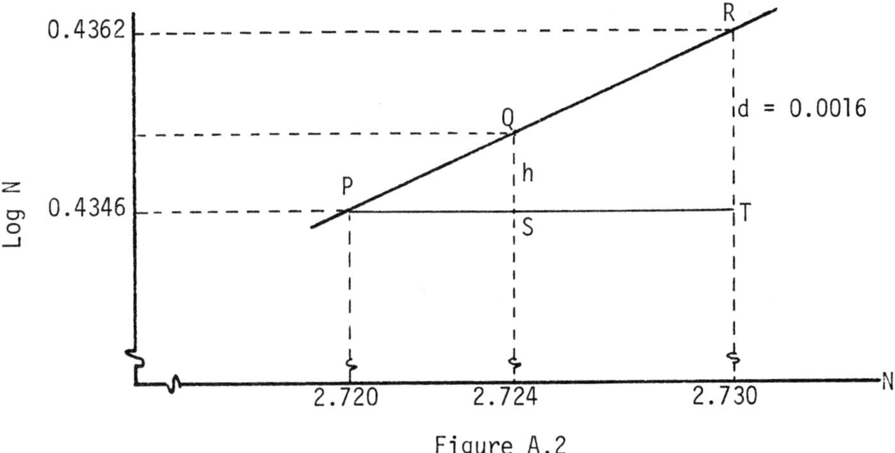

Figure A.2

Let h be the number which must be added to log 2.720 to give log 2.724. From geometry using similar triangles PSQ and PTR, we have

$$\frac{h}{d} = \frac{PS}{PT}$$

or

$$\frac{h}{\log 2.730 - \log 2.720} = \frac{2.724 - 2.720}{2.730 - 2.720}$$

$$\frac{h}{0.0016} = \frac{0.004}{0.010} \ .$$

Thus $h = \frac{4}{10}(0.0016) = 0.00064$. Now h is the amount we want to add to the log 2.720. Thus, we have

$$\log 2.724 = 0.4346 + 0.00064 = 0.4352 \ .$$

An alternate method and routine method of doing the same process is illustrated in the next example.

EXAMPLE A.4: Find log 2.9357.

$$0.01\left[\begin{array}{l} .0057\left[\begin{array}{l}\log 2.9300 = 0.4649\\ \\ \log 2.9357 = \end{array}\right] x \\ \\ \qquad\qquad \log 2.940 = 0.4643\end{array}\right] 0.0006$$

Since $\frac{.0057}{0.01} = \frac{x}{0.0006}$, $x = 0.000342$ and

$$\log 2.9357 = 0.4649 + 0.000342 = 0.4652 \ . \ \blacksquare$$

Antilogs of mantissas not found in the table can be found by reversing the process of interpolation given above. This process is illustrated in the next example.

EXAMPLE A.5: Find N if log N = 0.4480.

$$0.01\left[\begin{array}{l} x\left[\begin{array}{l}\log 2.800 = 0.4472\\ \\ \log N \quad\ = 0.4480\end{array}\right] 0.0008 \\ \quad\ \log 2.810 = 0.4487\end{array}\right] 0.0015$$

Since $\frac{x}{0.01} = \frac{0.0008}{0.0015} = \frac{8}{15}$, $x = (0.01)\,\frac{8}{15} = 0.0053$ and so

$$N = 2.800 + 0.0053 = 2.8053 \ . \ \blacksquare$$

EXAMPLE A.2 (continued) Earlier we found that $\log \sqrt[3]{0.00497} = 0.2321 - 1$. Verify that $\sqrt[3]{0.00497} = .17065$. $\blacksquare$

Any computation which has been illustrated above can be extended to natural logarithms which has the number e $(2.7182\cdots)$ as a base. The only difference involves this: The entire logarithm is given in the body of the table of natural logarithms so you do not need to worry about the characteristic. Table 2 gives the natural logarithms from 1.00 to 10.09. The first two digits of the logarithm appear at the left of the body of the table and the next four digits under the appropriate column in the body of the table. If a four digit sequence in the body of the table has an * before it, this means to drop down and pick up the next two digits at the left of the table. As an example ln 8.16 = 2.09924 while ln 8.17 = 2.10047.

For values of $N > 10$ or $N < 1$, then we may write $N = P \cdot 10^K$ where K is an integer on $1 \leq P \leq 10$. Then to find ln N, use the following relation with ln P obtained from Table II and $\ln 10 \doteq 2.30259$.

$$\ln N = \ln(P \cdot 10^K) = \ln P + K \ln 10$$

EXAMPLE A.6: Evaluate $\sqrt[4]{0.736}$ and $16.4/(7.36)^{1.2}$ using both common and natural logarithms. First using common logarithms,

$$\log \sqrt[4]{0.736} = \frac{1}{4} \log 0.736 = \frac{1}{4} (3.8669 - 4) = 0.96672 - 1$$

and

$$\text{antilog } (0.96672 - 1) = .92623 = \sqrt[4]{0.736} .$$

$$\log 16.4/(7.36)^{1.2} = \log 16.4 - 1.2 \log 7.36$$

$$= 1.2148 - 1.2(.8669) = 0.17452$$

and

$$\text{antilog } 0.17452 = 1.4946 = 16.4/(7.36)^{1.2}$$

Using natural logarithms instead, we get these calculations:

$$\ln \sqrt[4]{0.736} = \frac{1}{4} \ln 0.736 = \frac{1}{4} (0.30652) = -0.07663$$

and

$$\text{antiln } (-0.07663) = 0.9262 = \sqrt[4]{0.736}$$

$$\ln \frac{16.4}{7.36^{1.2}} = \ln 16.4 - 1.2 \ln 7.36$$

$$\ln 16.4/(7.36)^{1.2} = 2.79728 - 1.2(1.99606) = 0.40201$$

and

$$\text{antiln } 0.40201 = 1.495 . \blacksquare$$

Observe that the calculations using the natural logarithms are slightly rounded from those using common logarithms. This results because the common logarithms afford a means for compressing the tables to a consideration of numbers between 1 and 10 together with decimal point movement. Thus, they are the more convenient for most ordinary calculations. As we get into calculus, the number $e = 2.781828\cdots$ consistently reappears. The natural logarithms, using this base, are well suited for calculations involving it.

With this brief review of logarithms, you should be able to carry out the calculations necessary to solve the problems in this book.

TABLE I - COMMON LOGARITHM

N	0	1	2	3	4	5	6	7	8	9
10	.0000	0043	0086	0128	0170	0212	0253	0294	0334	0374
11	.0414	0453	0492	0531	0569	0607	0645	0682	0719	0755
12	.0792	0828	0864	0899	0934	0969	1004	1038	1072	1106
13	.1139	1173	1206	1239	1271	1303	1335	1367	1399	1430
14	.1461	1492	1523	1553	1584	1614	1643	1673	1703	1732
15	.1761	1790	1818	1847	1875	1903	1931	1959	1987	2014
16	.2041	2068	2095	2122	2148	2175	2201	2227	2253	2279
17	.2305	2330	2355	2381	2405	2430	2455	2480	2504	2529
18	.2553	2577	2601	2625	2648	2672	2695	2718	2742	2765
19	.2787	2810	2833	2856	2878	2900	2923	2945	2967	2989
20	.3010	3032	3053	3075	3096	3117	3139	3160	3181	3201
21	.3222	3243	3263	3284	3304	3324	3345	3365	3385	3404
22	.3424	3444	3464	3483	3502	3522	3541	3560	3579	3598
23	.3617	3636	3655	3674	3692	3711	3729	3747	3766	3784
24	.3802	3820	3838	3856	3874	3892	3909	3927	3945	3962
25	.3979	3997	4014	4031	4048	4065	4082	4099	4116	4133
26	.4150	4166	4183	4200	4216	4233	4249	4265	4281	4297
27	.4314	4330	4346	4362	4378	4393	4409	4425	4440	4456
28	.4472	4487	4503	4518	4533	4548	4564	4579	4594	4609
29	.4624	4639	4654	4669	4683	4698	4713	4728	4742	4757
30	.4771	4786	4800	4814	4829	4843	4857	4871	4885	4900
31	.4914	4928	4941	4955	4969	4983	4997	5011	5024	5038
32	.5051	5065	5079	5092	5105	5119	5132	5145	5159	5172
33	.5185	5198	5211	5224	5237	5250	5263	5276	5289	5302
34	.5315	5327	5340	5353	5366	5378	5391	5403	5416	5428
35	.5441	5453	5465	5478	5490	5502	5515	5527	5539	5551
36	.5563	5575	5587	5599	5611	5623	5635	5647	5659	5670
37	.5682	5694	5705	5717	5729	5740	5752	5763	5775	5786
38	.5798	5809	5821	5832	5843	5855	5866	5877	5888	5899
39	.5911	5922	5933	5944	5955	5966	5977	5988	5999	6010
40	.6021	6031	6042	6053	6064	6075	6085	6096	6107	6117
41	.6128	6138	6149	6159	6170	6181	6191	6201	6212	6222
42	.6233	6243	6253	6263	6274	6284	6294	6304	6314	6325
43	.6335	6345	6355	6365	6375	6385	6395	6405	6415	6425
44	.6435	6444	6454	6464	6474	6484	6493	6503	6513	6523
45	.6532	6542	6551	6561	6571	6580	6590	6599	6609	6618
46	.6628	6637	6646	6656	6665	6675	6684	6693	6703	6712
47	.6721	6730	6739	6749	6758	6767	6776	6785	6794	6803
48	.6812	6820	6831	6839	6849	6857	6866	6875	6884	6893
49	.6902	6911	6920	6929	6937	6946	6955	6964	6972	6981
50	.6990	6998	7007	7016	7024	7033	7041	7050	7059	7067
51	.7076	7084	7093	7101	7110	7118	7127	7135	7143	7152
52	.7160	7168	7177	7185	7193	7202	7210	7218	7226	7235
53	.7243	7251	7259	7267	7275	7283	7292	7300	7309	7316
54	.7324	7332	7340	7348	7356	7364	7372	7380	7388	7396

TABLE I - COMMON LOGARITHM

N	0	1	2	3	4	5	6	7	8	9
55	.7404	7411	7419	7427	7435	7443	7451	7459	7466	7474
56	.7482	7490	7497	7505	7513	7521	7528	7536	7543	7551
57	.7559	7566	7574	7581	7589	7597	7604	7612	7619	7627
58	.7634	7642	7649	7657	7664	7672	7679	7686	7694	7701
59	.7709	7716	7723	7731	7738	7745	7753	7760	7767	7774
60	.7781	7789	7796	7803	7810	7818	7825	7832	7839	7846
61	.7853	7860	7867	7875	7882	7889	7896	7903	7910	7917
62	.7924	7931	7938	7945	7952	7959	7966	7973	7980	7987
63	.7993	8000	8007	8014	8021	8028	8035	8041	8048	8055
64	.8062	8069	8075	8082	8089	8096	8102	8109	8116	8122
65	.8129	8136	8143	8149	8156	8162	8169	8175	8182	8189
66	.8195	8202	8209	8215	8222	8228	8235	8241	8248	8254
67	.8261	8267	8274	8280	8287	8293	8299	8306	8312	8319
68	.8325	8331	8338	8344	8351	8357	8363	8370	8376	8382
69	.8389	8395	8401	8407	8414	8420	8426	8432	8439	8445
70	.8451	8457	8463	8470	8476	8482	8488	8494	8500	8507
71	.8513	8519	8525	8531	8537	8543	8549	8555	8561	8567
72	.8573	8579	8585	8591	8597	8603	8609	8615	8621	8627
73	.8633	8639	8645	8651	8657	8663	8669	8675	8681	8686
74	.8692	8698	8704	8710	8716	8722	8727	8733	8739	8745
75	.8751	8756	8762	8768	8774	8779	8785	8791	8797	8802
76	.8808	8814	8820	8825	8831	8837	8842	8848	8854	8859
77	.8865	8871	8876	8882	8887	8893	8899	8904	8910	8915
78	.8921	8927	8932	8938	8943	8949	8954	8960	8965	8971
79	.8976	8982	8987	8993	8998	9004	9009	9015	9020	9025
80	.9031	9036	9042	9047	9053	9058	9063	9069	9074	9079
81	.9085	9090	9096	9101	9106	9112	9117	9122	9127	9133
82	.9138	9143	9149	9154	9159	9165	9170	9175	9180	9185
83	.9191	9196	9201	9207	9212	9217	9222	9227	9232	9238
84	.9243	9248	9253	9258	9263	9269	9274	9279	9284	9289
85	.9294	9299	9304	9309	9315	9320	9325	9330	9335	9340
86	.9345	9350	9355	9360	9365	9370	9375	9380	9385	9390
87	.9395	9400	9405	9410	9415	9420	9425	9430	9435	9440
88	.9445	9450	9455	9460	9465	9469	9474	9479	9484	9489
89	.9494	9499	9504	9509	9513	9518	9523	9528	9533	9538
90	.9542	9547	9552	9557	9562	9567	9571	9576	9581	9586
91	.9590	9595	9600	9605	9609	9614	9619	9624	9628	9633
92	.9638	9643	9647	9652	9657	9661	9666	9671	9675	9680
93	.9685	9689	9694	9699	9703	9708	9713	9717	9722	9727
94	.9731	9736	9741	9745	9750	9754	9759	9763	9768	9773
95	.9777	9782	9786	9791	9795	9800	9805	9809	9814	9818
96	.9823	9827	9832	9836	9841	9845	9850	9854	9859	9863
97	.9868	9872	9877	9881	9886	9890	9895	9899	9903	9908
98	.9912	9917	9921	9925	9930	9934	9939	9943	9948	9952
99	.9956	9961	9965	9969	9974	9978	9983	9987	9991	9996

TABLE II - NATURAL LOGARITHMS

N		0	1	2	3	4	5	6	7	8	9
1.0	0.0	0000	0995	1980	2956	3922	4879	5827	6766	7696	8618
1.1	0.0	9531	*0436	*1333	*2222	*3103	*3976	*4842	*5700	*6551	*7395
1.2	0.1	8232	9062	9885	*0701	*1511	*2314	*3111	*3902	*4686	*5464
1.3	0.2	6236	7003	7763	8518	9267	*0010	*0748	*1481	*2208	*2930
1.4	0.3	3647	4359	5066	5767	6464	7156	7844	8526	9204	9878
1.5	0.4	0547	1211	1871	2527	3178	3825	4469	5108	5742	6373
1.6	0.4	7000	7623	8243	8858	9470	*0078	*0682	*1282	*1879	*2473
1.7	0.5	3063	3649	4232	4812	5389	5962	6531	7098	7661	8222
1.8	0.5	8779	9333	9884	*0432	*0977	*1519	*2058	*2594	*3127	*3658
1.9	0.6	4185	4710	5233	5752	6269	6783	7294	7803	8310	8813
2.0	0.6	9315	9813	*0310	*0804	*1295	*1784	*2271	*2755	*3237	*3716
2.1	0.7	4194	4669	5142	5612	6081	6547	7011	7473	7932	8390
2.2	0.7	8846	9299	9751	*0200	*0648	*1093	*1536	*1978	*2418	*2855
2.3	0.8	3291	3725	4157	4587	5015	5442	5866	6289	6710	7129
2.4	0.8	7547	7963	8377	8789	9200	9609	*0016	*0422	*0826	*1228
2.5	0.9	1629	2028	2426	2822	3216	3609	4001	4391	4779	5166
2.6	0.9	5551	5935	6317	6698	7078	7456	7833	8208	8582	8954
2.7	0.9	9325	9695	*0063	*0430	*0796	*1160	*1523	*1885	*2245	*2604
2.8	1.0	2962	3318	3674	4028	4380	4732	5082	5431	5779	6126
2.9	1.0	6471	6815	7158	7500	7841	8181	8519	8856	9192	9527
3.0	1.0	9861	*0194	*0526	*0856	*1186	*1514	*1841	*2168	*2493	*2817
3.1	1.1	3140	3462	3783	4103	4422	4740	5057	5373	5688	6002
3.2	1.1	6315	6627	6938	7248	7557	7865	8173	8479	8784	9089
3.3	1.1	9392	9695	9996	*0297	*0597	*0896	*1194	*1491	*1788	*2083
3.4	1.2	2378	2671	2964	3256	3547	3837	4127	4415	4703	4990
3.5	1.2	5276	5562	5846	6130	6413	6695	6976	7257	7536	7815
3.6	1.2	8093	8371	8647	8923	9198	9473	9746	*0019	*0291	*0563
3.7	1.3	0833	1103	1372	1641	1909	2176	2442	2708	2972	3237
3.8	1.3	3500	3763	4025	4286	4547	4807	5067	5325	5584	5841
3.9	1.3	6098	6354	6609	6864	7118	7372	7624	7877	8128	8379
4.0	1.3	8629	8879	9128	9377	9624	9872	*0118	*0364	*0610	*0854
4.1	1.4	1099	1342	1585	1828	2070	2311	2552	2792	3031	3270
4.2	1.4	3508	3746	3984	4220	4456	4692	4927	5161	5395	5629
4.3	1.4	5862	6094	6326	6557	6787	7018	7247	7476	7705	7933
4.4	1.4	8160	8387	8614	8840	9065	9290	9515	9739	9962	*0185
4.5	1.5	0408	0630	0851	1072	1293	1513	1732	1951	2170	2388
4.6	1.5	2606	2823	3039	3256	3471	3687	3902	4116	4330	4543
4.7	1.5	4756	4969	5181	5393	5604	5814	6025	6235	6444	6653
4.8	1.5	6862	7070	7277	7485	7691	7898	8104	8309	8515	8719
4.9	1.5	8924	9127	9331	9534	9737	9939	*0141	*0342	*0543	*0744
5.0	1.6	0944	1144	1343	1542	1741	1939	2137	2334	2531	2728
5.1	1.6	2924	3120	3315	3511	3705	3900	4094	4287	4481	4673
5.2	1.6	4866	5058	5250	5441	5632	5823	6013	6203	6393	6582
5.3	1.6	6771	6959	7147	7335	7523	7710	7896	8083	8269	8455
5.4	1.6	8640	8825	9010	9194	9378	9562	9745	9928	*0111	*0293

TABLE II - NATURAL LOGARITHMS

N		0	1	2	3	4	5	6	7	8	9
5.5	1.7	0475	0656	0838	1019	1199	1380	1560	1740	1919	2098
5.6	1.7	2277	2455	2633	2811	2988	3166	3342	3519	3695	3871
5.7	1.7	4047	4222	4397	4572	4746	4920	5094	5267	5440	5613
5.8	1.7	5786	5958	6130	6302	6473	6644	6815	6985	7156	7326
5.9	1.7	7495	7665	7834	8002	8171	8339	8507	8675	8842	9009
6.0	1.7	9176	9342	9509	9675	9840	*0006	*0171	*0336	*0500	*0665
6.1	1.8	0829	0993	1156	1319	1482	1645	1808	1970	2132	2294
6.2	1.8	2455	2616	2777	2938	3098	3258	3418	3578	3737	3896
6.3	1.8	4055	4214	4372	4530	4688	4845	5003	5160	5317	5473
6.4	1.8	5630	5786	5942	6097	6253	6408	6563	6718	6872	7026
6.5	1.8	7180	7334	7487	7641	7794	7947	8099	8251	8403	8555
6.6	1.8	8707	8858	9010	9160	9311	9462	9612	9762	9912	*0061
6.7	1.9	0211	0360	0509	0658	0806	0954	1102	1250	1398	1545
6.8	1.9	1692	1839	1986	2132	2279	2425	2571	2716	2862	3007
6.9	1.9	3152	3297	3442	3586	3730	3874	4018	4162	4305	4448
7.0	1.9	4591	4734	4876	5019	5161	5303	5445	5586	5727	5869
7.1	1.9	6009	6150	6291	6431	6571	6711	6851	6991	7130	7269
7.2	1.9	7408	7547	7685	7824	7962	8100	8238	8376	8513	8650
7.3	1.9	8787	8924	9061	9198	9334	9470	9606	9742	9877	*0013
7.4	2.0	0148	0283	0418	0553	0687	0821	0956	1089	1223	1357
7.5	2.0	1490	1624	1757	1890	2022	2155	2287	2419	2551	2683
7.6	2.0	2815	2946	3078	3209	3340	3471	3601	3732	3862	3992
7.7	2.0	4122	4252	4381	4511	4640	4769	4898	5027	5156	5284
7.8	2.0	5412	5540	5668	5796	5924	6051	6179	6306	6433	6560
7.9	2.0	6686	6813	6939	7065	7191	7317	7443	7568	7694	7819
8.0	2.0	7944	8069	8194	8318	8443	8567	8691	8815	8939	9063
8.1	2.0	9186	9310	9433	9556	9679	9802	9924	*0047	*0169	*0291
8.2	2.1	0413	0535	0657	0779	0900	1021	1142	1263	1384	1505
8.3	2.1	1626	1746	1866	1986	2106	2226	2346	2465	2585	2704
8.4	2.1	2823	2942	3061	3180	3298	3417	3535	3653	3771	3889
8.5	2.1	4007	4124	4242	4359	4476	4593	4710	4827	4943	5060
8.6	2.1	5176	5292	5409	5524	5640	5756	5871	5987	6102	6217
8.7	2.1	6332	6447	6562	6677	6791	6905	7020	7134	7248	7361
8.8	2.1	7475	7589	7702	7816	7929	8042	8155	8267	8380	8493
8.9	2.1	8605	8717	8830	8942	9054	9165	9277	9389	9500	9611
9.0	2.1	9722	9834	9944	*0055	*0166	*0276	*0387	*0497	*0607	*0717
9.1	2.2	0827	0937	1047	1157	1266	1375	1485	1594	1703	1812
9.2	2.2	1920	2029	2138	2246	2354	2462	2570	2678	2786	2894
9.3	2.2	3001	3109	3216	3324	3431	3538	3645	3751	3858	3965
9.4	2.2	4071	4177	4284	4390	4496	4601	4707	4813	4918	5024
9.5	2.2	5129	5234	5339	5444	5549	5654	5759	5863	5968	6072
9.6	2.2	6176	6280	6384	6488	6592	6696	6799	6903	7006	7109
9.7	2.2	7213	7316	7419	7521	7624	7727	7829	7932	8034	8136
9.8	2.2	8238	8340	8442	8544	8646	8747	8849	8950	9051	9152
9.9	2.2	9253	9354	9455	9556	9657	9757	9858	9958	*0058	*0158

Use ln 10 = 2.30259 to find logarithms of numbers greater than 10 or less than 1.
Ex: ln 220 = ln 2.2 + 2 ln 10 = 0.7885 + 2(2.30259) = 5.3937.

TABLE III - EXPONENTIAL FUNCTIONS

x	e^x	e^{-x}	x	e^x	e^{-x}	x	e^x	e^{-x}
0.00	1.0000	1.000000	0.45	1.5683	0.637628	0.90	2.4596	0.406570
0.01	1.0101	0.990050	0.46	1.5841	.631284	0.91	2.4843	.402524
0.02	1.0202	.980199	0.47	1.6000	.625002	0.92	2.5093	.398519
0.03	1.0305	.970446	0.48	1.6161	.618783	0.93	2.5345	.394554
0.04	1.0408	.960789	0.49	1.6323	.612626	0.94	2.5600	.390628
0.05	1.0513	0.951229	0.50	1.6487	0.606531	0.95	2.5857	0.386741
0.06	1.0618	.941765	0.51	1.6653	.600496	0.96	2.6117	.382893
0.07	1.0725	.932394	0.52	1.6820	.594521	0.97	2.6379	.379083
0.08	1.0833	.923116	0.53	1.6989	.588605	0.98	2.6645	.375311
0.09	1.0942	.913931	0.54	1.7160	.582748	0.99	2.6912	.371577
0.10	1.1052	0.904837	0.55	1.7333	0.576950	1.00	2.7183	0.367879
0.11	1.1163	.895834	0.56	1.7507	.571209	1.01	2.7456	.360595
0.12	1.1275	.886920	0.57	1.7683	.565525	1.02	2.7732	.360595
0.13	1.1388	.878095	0.58	1.7860	.559898	1.03	2.8011	.357007
0.14	1.1503	.869358	0.59	1.8040	.554327	1.04	2.8292	.353455
0.15	1.1618	0.860708	0.60	1.8221	0.548812	1.05	2.8577	0.349938
0.16	1.1735	.852144	0.61	1.8404	.543351	1.06	2.8864	.346456
0.17	1.1853	.843665	0.62	1.8589	.537944	1.07	2.9154	.343009
0.18	1.1972	.835270	0.63	1.8776	.532592	1.08	2.9447	.339596
0.19	1.2092	.826959	0.64	1.8965	.527292	1.09	2.9743	.336216
0.20	1.2214	0.818731	0.65	1.9155	0.522046	1.10	3.0042	0.332871
0.21	1.2337	.810584	0.66	1.9348	.516851	1.11	3.0344	.329559
0.22	1.2461	.802519	0.67	1.9542	.511709	1.12	3.0649	.326280
0.23	1.2586	.794534	0.68	1.9739	.506617	1.13	3.0957	.323033
0.24	1.2712	.786628	0.69	1.9937	.501576	1.14	3.1268	.319819
0.25	1.2840	0.778801	0.70	2.0138	0.496585	1.15	3.1582	0.316637
0.26	1.2969	.771052	0.71	2.0340	.491644	1.16	3.1899	.313486
0.27	1.3100	.763379	0.72	2.0544	.486752	1.17	3.2220	.310367
0.28	1.3231	.755784	0.73	2.0751	.481909	1.18	3.2544	.307279
0.29	1.3364	.748264	0.74	2.0959	.477114	1.19	3.2871	.304221
0.30	1.3499	0.740818	0.75	2.1170	0.472367	1.20	3.3201	0.301194
0.31	1.3634	.733447	0.76	2.1383	.467666	1.21	3.3535	.298197
0.32	1.3771	.726149	0.77	2.1598	.463013	1.22	3.3872	.295230
0.33	1.3910	.718924	0.78	2.1815	.458406	1.23	3.4212	.292293
0.34	1.4049	.711770	0.79	2.2034	.453845	1.24	3.4556	.289384
0.35	1.4191	0.704688	0.80	2.2255	0.449329	1.25	3.4903	0.286505
0.36	1.4333	.697676	0.81	2.2479	.444858	1.26	3.5254	.283654
0.37	1.4477	.690734	0.82	2.2705	.440432	1.27	3.5609	.280832
0.38	1.4623	.683861	0.83	2.2933	.436049	1.28	3.5966	.278037
0.39	1.4770	.677057	0.84	2.3164	.431711	1.29	3.6328	.275271
0.40	1.4918	0.670320	0.85	2.3396	0.427415	1.30	3.6693	0.272532
0.41	1.5068	.663650	0.86	2.3632	.423162	1.31	3.7062	.269820
0.42	1.5220	.657047	0.87	2.3869	.418952	1.32	3.7434	.267135
0.43	1.5373	.650509	0.88	2.4109	.414783	1.33	3.7810	.264477
0.44	1.5527	.644036	0.89	2.4351	.410656	1.34	3.8190	.261846

TABLE III - EXPONENTIAL FUNCTIONS
(Continued)

x	e^x	e^{-x}	x	e^x	e^{-x}	x	e^x	e^{-x}
1.35	3.8574	0.259240	1.80	6.0496	0.165299	2.25	9.4877	0.105399
1.36	3.8962	.256661	1.81	6.1104	.163654	2.26	9.5831	.104350
1.37	3.9354	.254107	1.82	6.1719	.162026	2.27	9.6794	.103312
1.38	3.9749	.251579	1.83	6.2339	.160414	2.28	9.7767	.102284
1.39	4.0149	.249075	1.84	6.2965	.158817	2.29	9.8749	.101266
1.40	4.0552	0.246597	1.85	6.3598	0.157237	2.30	9.9742	0.100259
1.41	4.0960	.244143	1.86	6.4237	.155673	2.31	10.074	.099261
1.42	4.1371	.241714	1.87	6.4883	.154124	2.32	10.176	.098274
1.43	4.1787	.239309	1.88	6.5535	.152590	2.33	10.278	.097296
1.44	4.2207	.236928	1.89	6.6194	.151072	2.34	10.381	.096328
1.45	4.2631	0.234570	1.90	6.6859	0.149569	2.35	10.486	0.095369
1.46	4.3060	.232236	1.91	6.7531	.148080	2.36	10.591	.094420
1.47	4.3492	.229925	1.92	6.8210	.146607	2.37	10.697	.093481
1.48	4.3929	.227638	1.93	6.8895	.145148	2.38	10.805	.092551
1.49	4.4371	.225373	1.94	6.9588	.143704	2.39	10.913	.091630
1.50	4.4817	0.223130	1.95	7.0287	0.142274	2.40	11.023	0.090718
1.51	4.5267	.220910	1.96	7.0993	.140858	2.41	11.134	.089815
1.52	4.5722	.218712	1.97	7.1707	.139457	2.42	11.246	.088922
1.53	4.6182	.216536	1.98	7.2427	.138069	2.43	11.359	.088037
1.54	4.6646	.214381	1.99	7.3155	.136695	2.44	11.473	.087161
1.55	4.7115	0.212248	2.00	7.3891	0.135335	2.45	11.588	0.086294
1.56	4.7588	.210136	2.01	7.4633	.133989	2.46	11.705	.085435
1.57	4.8066	.208045	2.02	7.5383	.132655	2.47	11.822	.084585
1.58	4.8550	.205975	2.03	7.6141	.131336	2.48	11.941	.083743
1.59	4.9037	.203926	2.04	7.6906	.130029	2.49	12.061	.082910
1.60	4.9530	0.201897	2.05	7.7679	0.128735	2.50	12.182	0.082085
1.61	5.0028	.199888	2.06	7.8460	.127454	2.51	12.305	.081268
1.62	5.0531	.197899	2.07	7.9248	.126186	2.52	12.429	.080460
1.63	5.1039	.195930	2.08	8.0045	.124930	2.53	12.554	.079659
1.64	5.1552	.193980	2.09	8.0849	.123687	2.54	12.680	.078866
1.65	5.2070	0.192050	2.10	8.1662	0.122456	2.55	12.807	0.078082
1.66	5.2593	.190139	2.11	8.2482	.121238	2.56	12.936	.077305
1.67	5.3122	.188247	2.12	8.3311	.120032	2.57	13.066	.076536
1.68	5.3656	.186374	2.13	8.4149	.118837	2.58	13.197	.075774
1.69	5.4195	.184520	2.14	8.4994	.117655	2.59	13.330	.075020
1.70	5.4739	0.182684	2.15	8.5849	0.116484	2.60	13.464	0.074274
1.71	5.5290	.180866	2.16	8.6711	.115325	2.61	13.599	.073535
1.72	5.5845	.179066	2.17	8.7583	.114178	2.62	13.736	.072803
1.73	5.6407	.177284	2.18	8.8463	.113042	2.63	13.874	.072078
1.74	5.6973	.175520	2.19	8.9352	.111917	2.64	14.013	.071361
1.75	5.7546	0.173774	2.20	9.0250	0.110803	2.65	14.154	0.070651
1.76	5.8124	.172045	2.21	9.1157	.109701	2.66	14.296	.069948
1.77	5.8709	.170333	2.22	9.2073	.108609	2.67	14.440	.069252
1.78	5.9299	.168638	2.23	9.2999	.107528	2.68	14.585	.068563
1.79	5.9895	.166960	2.24	9.3933	.106459	2.69	14.732	.067881

TABLE III - EXPONENTIAL FUNCTIONS
(Continued)

x	e^x	e^{-x}	x	e^x	e^{-x}	x	e^x	e^{-x}
2.70	14.880	0.067206	3.15	23.336	0.042852	3.60	36.598	0.027324
2.71	15.029	.066537	3.16	23.571	.042426	3.61	36.966	.027052
2.72	15.180	.065875	3.17	23.807	.042004	3.62	37.338	.026783
2.73	15.333	.065219	3.18	24.047	.041586	3.63	37.713	.026516
2.74	15.487	.064570	3.19	24.288	.041172	3.64	38.092	.026252
2.75	15.643	0.063928	3.20	24.533	0.040762	3.65	38.475	0.025991
2.76	15.800	.063292	3.21	24.779	.040357	3.66	38.861	.025733
2.77	15.959	.062662	3.22	25.028	.039955	3.67	39.252	.025476
2.78	16.119	.062039	3.23	25.280	.039557	3.68	39.646	.025223
2.79	16.281	.061421	3.24	25.534	.039164	3.69	40.045	.024972
2.80	16.445	0.060810	3.25	25.790	0.038774	3.70	40.447	0.024724
2.81	16.610	.060205	3.26	26.050	.038388	3.71	40.854	.024478
2.82	16.777	.059606	3.27	26.311	.038006	3.72	41.264	.024234
2.83	16.945	.059013	3.28	26.576	.037628	3.73	41.679	.023993
2.84	17.116	.058426	3.29	26.843	.037254	3.74	42.098	.023754
2.85	17.288	0.057844	3.30	27.113	0.036883	3.75	42.521	0.023518
2.86	17.462	.057269	3.31	27.385	.036516	3.76	42.948	.023284
2.87	17.637	.056699	3.32	27.660	.036153	3.77	43.380	.023052
2.88	17.814	.056135	3.33	27.938	.035793	3.78	43.816	.022823
2.89	17.993	.055576	3.34	28.219	.035437	3.79	44.256	.022596
2.90	18.174	0.055023	3.35	28.503	0.035084	3.80	44.701	0.022371
2.91	18.357	.054476	3.36	28.789	.034735	3.81	45.150	.022148
2.92	18.541	.053934	3.37	29.079	.034390	3.82	45.604	.021928
2.93	18.728	.053397	3.38	29.371	.034047	3.83	46.063	.021710
2.94	18.916	.052866	3.39	29.666	.033709	3.84	46.525	.021494
2.95	19.106	0.052340	3.40	29.964	0.033373	3.85	46.993	0.021280
2.96	19.298	.051819	3.41	30.265	.033041	3.86	47.465	.021068
2.97	19.492	.051303	3.42	30.569	.032712	3.87	47.942	.020858
2.98	19.688	.050793	3.43	30.877	.032387	3.88	48.424	.020651
2.99	19.886	.050287	3.44	31.187	.032065	3.89	48.911	.020445
3.00	20.086	0.049787	3.45	31.500	0.031746	3.90	49.402	0.020242
3.01	20.287	.049292	3.46	31.817	.031430	3.91	49.899	.020041
3.02	20.491	.048801	3.47	32.137	.031117	3.92	50.400	.019841
3.03	20.697	.048316	3.48	32.460	.030807	3.93	50.907	.019644
3.04	20.905	.047835	3.49	32.786	.030501	3.94	51.419	.019448
3.05	21.115	0.047359	3.50	33.115	0.030197	3.95	51.935	0.019255
3.06	21.328	.046888	3.51	33.448	.029897	3.96	52.457	.019063
3.07	21.542	.046421	3.52	33.784	.029599	3.97	52.985	.018873
3.08	21.758	.045959	3.53	34.124	.029305	3.98	53.517	.018686
3.09	21.977	.045502	3.54	34.467	.029013	3.99	54.055	.018500
3.10	22.198	0.045049	3.55	34.813	0.028725	4.00	54.598	0.018316
3.11	22.421	.044601	3.56	35.163	.028439	4.01	55.147	.018133
3.12	22.646	.044157	3.57	35.517	.028156	4.02	55.701	.017953
3.13	22.874	.043718	3.58	35.874	.027876	4.03	56.261	.017774
3.14	23.104	.043283	3.59	36.234	.027598	4.04	56.826	.017597

TABLE III - EXPONENTIAL FUNCTIONS
(Continued)

x	e^x	e^{-x}	x	e^x	e^{-x}	x	e^x	e^{-x}
4.05	57.397	0.017422	4.50	90.017	0.011109	4.95	141.17	0.007083
4.06	57.974	.017249	4.51	90.922	.010998	4.96	142.59	.007013
4.07	58.557	.017077	4.52	91.836	.010889	4.97	144.03	.006943
4.08	59.145	.016907	4.53	92.759	.010781	4.98	145.47	.006874
4.09	59.740	.016739	4.54	93.691	.010673	4.99	146.94	.006806
4.10	60.340	0.016573	4.55	94.632	0.010567	5.00	148.41	0.006738
4.11	60.947	.016408	4.56	95.583	.010462	5.01	149.90	.006671
4.12	61.559	.016245	4.57	96.544	.010358	5.02	151.41	.006605
4.13	62.178	.016083	4.58	97.514	.010255	5.03	152.93	.006539
4.14	62.803	.015923	4.59	98.494	.010153	5.04	154.57	.006474
4.15	63.434	0.015764	4.60	99.484	0.010052	5.05	156.02	0.006409
4.16	64.072	.015608	4.61	100.48	.009952	5.06	157.59	.006346
4.17	64.715	.015452	4.62	101.49	.009853	5.07	159.17	.006282
4.18	65.366	.015299	4.63	102.51	.009755	5.08	160.77	.006220
4.19	66.023	.015146	4.64	103.54	.009658	5.09	162.39	.006158
4.20	66.686	0.014996	4.65	104.58	0.009562	5.10	164.02	0.006097
4.21	67.357	.014846	4.66	105.64	.009466	5.11	165.67	.006036
4.22	68.033	.014699	4.67	106.70	.009372	5.12	167.34	.005976
4.23	68.717	.014552	4.68	107.77	.009279	5.13	169.02	.005917
4.24	69.408	.014408	4.69	108.85	.009187	5.14	170.72	.005858
4.25	70.105	0.014264	4.70	109.95	0.009095	5.15	172.43	0.005799
4.26	70.810	.014122	4.71	111.05	.009005	5.16	174.16	.005742
4.27	71.522	.013982	4.72	112.17	.008915	5.17	175.91	.005685
4.28	72.240	.013843	4.73	113.30	.008826	5.18	177.68	.005628
4.29	72.966	.013705	4.74	114.43	.008739	5.19	179.47	.005572
4.30	73.700	0.013569	4.75	115.58	0.008652	5.20	181.27	0.005517
4.31	74.440	.013434	4.76	116.75	.008566	5.21	183.09	.005462
4.32	75.189	.013300	4.77	117.92	.008480	5.22	184.93	.005407
4.33	75.944	.013168	4.78	119.10	.008396	5.23	186.79	.005354
4.34	76.708	.013037	4.79	120.30	.008312	5.24	188.67	.005300
4.35	77.478	0.012907	4.80	121.51	0.008230	5.25	190.57	0.005248
4.36	78.257	.012778	4.81	122.73	.008148	5.26	192.48	.005195
4.37	79.044	.012651	4.82	123.97	.008067	5.27	194.42	.005144
4.38	79.838	.012525	4.83	125.21	.007987	5.28	196.37	.005092
4.39	80.640	.012401	4.84	126.47	.007907	5.29	198.34	.005042
4.40	81.451	0.012277	4.85	127.74	0.007828	5.30	200.34	0.004992
4.41	82.269	.012155	4.86	129.02	.007750	5.31	202.35	.004942
4.42	83.096	.012034	4.87	130.32	.007673	5.32	204.38	.004893
4.43	83.931	.011914	4.88	131.63	.007597	5.33	206.44	.004844
4.44	84.775	.011796	4.89	132.95	.007521	5.34	208.51	.004796
4.45	85.627	0.011679	4.90	134.29	0.007447	5.35	210.61	0.004748
4.46	86.488	.011562	4.91	135.64	.007372	5.36	212.72	.004701
4.47	87.357	.011447	4.92	137.00	.007299	5.37	214.86	.004654
4.48	88.235	.011333	4.93	138.38	.007227	5.38	217.02	.004608
4.49	89.121	.011221	4.94	139.77	.007155	5.39	219.20	.004562

TABLE III - EXPONENTIAL FUNCTIONS
(Continued)

x	e^x	e^{-x}	x	e^x	e^{-x}	x	e^x	e^{-x}
5.40	221.41	0.004517	6.75	854.06	0.0011709	8.50	4914.8	0.0002035
5.41	223.63	.004472	6.80	897.85	.0011138	8.55	5166.8	.0001935
5.42	225.88	.004427	6.85	943.88	.0010595	8.60	5431.7	.0001841
5.43	228.15	.004383	6.90	992.27	.0010078	8.65	5710.1	.0001751
5.44	230.44	.004339	6.95	1043.1	.0009586	8.70	6002.9	.0001666
5.45	232.76	0.004296	7.00	1096.6	0.0009119	8.75	6310.7	0.0001585
5.46	235.10	.004254	7.05	1152.9	.0008674	8.80	6634.2	.0001507
5.47	237.46	.004211	7.10	1212.0	.0008251	8.85	6974.4	.0001434
5.48	239.85	.004169	7.15	1274.1	.0007849	8.90	7332.0	.0001364
5.49	242.26	.004128	7.20	1339.4	.0007466	8.95	7707.9	.0001297
5.50	244.69	0.0040868	7.25	1408.1	0.0007102	9.00	8103.1	0.0001234
5.55	257.24	.0038875	7.30	1480.3	.0006755	9.05	8518.5	.0001174
5.60	270.43	.0036979	7.35	1556.2	.0006426	9.10	8955.3	.0001117
5.65	284.29	.0035175	7.40	1636.0	.0006113	9.15	9414.4	.0001062
5.70	298.87	.0033460	7.45	1719.9	.0005814	9.20	9897.1	.0001010
5.75	314.19	0.0031828	7.50	1808.0	0.0005531	9.25	10455	0.0000961
5.80	330.30	.0030276	7.55	1900.7	.0005261	9.30	10938	.0000914
5.85	347.23	.0028799	7.60	1998.2	.0005005	9.35	11499	.0000870
5.90	365.04	.0027394	7.65	2100.6	.0004760	9.40	12088	.0000827
5.95	383.75	.0026058	7.70	2208.3	.0004528	9.45	12708	.0000787
6.00	403.43	0.0024788	7.75	2321.6	0.0004307	9.50	13360	0.0000749
6.05	424.11	.0023579	7.80	2440.6	.0004097	9.55	14045	.0000712
6.10	445.86	.0022429	7.85	2565.7	.0003898	9.60	14765	.0000677
6.15	468.72	.0021335	7.90	2697.3	.0003707	9.65	15522	.0000644
6.20	492.75	.0020294	7.95	2835.6	.0003527	9.70	16318	.0000613
6.25	518.01	0.0019305	8.00	2981.0	0.0003355	9.75	17154	0.0000583
6.30	544.57	.0018363	8.05	3133.8	.0003191	9.80	18034	.0000555
6.35	572.49	.0017467	8.10	3294.5	.0003035	9.85	18958	.0000527
6.40	601.85	.0016616	8.15	3463.4	.0002887	9.90	19930	.0000502
6.45	632.70	.0015805	8.20	3641.0	.0002747	9.95	20952	.0000477
6.50	665.14	0.0015034	8.25	3827.6	0.0002613	10.00	22026	0.0000454
6.55	699.24	.0014301	8.30	4023.9	.0002485			
6.60	735.10	.0013604	8.35	4230.2	.0002364			
6.65	772.78	.0012940	8.40	4447.1	.0002249			
6.70	812.41	.0012309	8.45	4675.1	.0002139			

TABLE IV. THREE-PLACE VALUES OF TRIGONOMETRIC FUNCTIONS

Rad.	Deg.	Sin	Tan	Sec	Csc	Cot	Cos	Deg.	Rad.
.000	0°	.000	.000	1.000	——	——	1.000	90°	1.571
.017	1°	.017	.017	1.000	57.30	57.29	1.000	89°	1.553
.035	2°	.035	.035	1.001	28.65	28.64	0.999	88°	1.536
.052	3°	.052	.052	1.001	19.11	19.08	.999	87°	1.518
.070	4°	.070	.070	1.002	14.34	14.30	.998	86°	1.501
.087	5°	.087	.087	1.004	11.47	11.43	.996	85°	1.484
.105	6°	.105	.105	1.006	9.567	9.514	.995	84°	1.466
.122	7°	.122	.123	1.008	8.206	8.144	.993	83°	1.449
.140	8°	.139	.141	1.010	7.185	7.115	.990	82°	1.431
.157	9°	.156	.158	1.012	6.392	6.314	.988	81°	1.414
.175	10°	.174	.176	1.015	5.759	5.671	.985	80°	1.396
.192	11°	.191	.194	1.019	5.241	5.145	.982	79°	1.379
.209	12°	.208	.213	1.022	4.810	4.705	.978	78°	1.361
.227	13°	.225	.231	1.026	4.445	4.331	.974	77°	1.344
.244	14°	.242	.249	1.031	4.134	4.011	.970	76°	1.326
.262	15°	.259	.268	1.035	3.864	3.732	.966	75°	1.309
.279	16°	.276	.287	1.040	3.628	3.487	.961	74°	1.292
.297	17°	.292	.306	1.046	3.420	3.271	.956	73°	1.274
.314	18°	.309	.325	1.051	3.236	3.078	.951	72°	1.257
.332	19°	.326	.344	1.058	3.072	2.904	.946	71°	1.239
.349	20°	.342	.364	1.064	2.924	2.747	.940	70°	1.222
.367	21°	.358	.384	1.071	2.790	2.605	.934	69°	1.204
.384	22°	.375	.404	1.079	2.669	2.475	.927	68°	1.187
.401	23°	.391	.424	1.086	2.559	2.356	.921	67°	1.169
.419	24°	.407	.445	1.095	2.459	2.246	.914	66°	1.152
.436	25°	.423	.466	1.103	2.366	2.145	.906	65°	1.134
.454	26°	.438	.488	1.113	2.281	2.050	.899	64°	1.117
.471	27°	.454	.510	1.122	2.203	1.963	.891	63°	1.100
.489	28°	.469	.532	1.133	2.130	1.881	.883	62°	1.082
.506	29°	.485	.554	1.143	2.063	1.804	.875	61°	1.065
.524	30°	.500	.577	1.155	2.000	1.732	.866	60°	1.047
.541	31°	.515	.601	1.167	1.942	1.664	.857	59°	1.030
.559	32°	.530	.625	1.179	1.887	1.600	.848	58°	1.012
.576	33°	.545	.649	1.192	1.836	1.540	.839	57°	0.995
.593	34°	.559	.675	1.206	1.788	1.483	.829	56°	0.977
.611	35°	.574	.700	1.221	1.743	1.428	.819	55°	0.960
.628	36°	.588	.727	1.236	1.701	1.376	.809	54°	0.942
.646	37°	.602	.754	1.252	1.662	1.327	.799	53°	0.925
.663	38°	.616	.781	1.269	1.624	1.280	.788	52°	0.908
.681	39°	.629	.810	1.287	1.589	1.235	.777	51°	0.890
.698	40°	.643	.839	1.305	1.556	1.192	.766	50°	0.873
.716	41°	.656	.869	1.325	1.524	1.150	.755	49°	0.855
.733	42°	.699	.900	1.346	1.494	1.111	.743	48°	0.838
.750	43°	.682	.933	1.367	1.466	1.072	.731	47°	0.820
.768	44°	.695	0.966	1.390	1.440	1.036	.719	46°	0.803
.785	45°	.707	1.000	1.414	1.414	1.000	.707	45°	0.785
Rad.	Deg.	Cos	Cot	Csc	Sec	Tan	Sin	Deg.	Rad.

TABLE V

TABLE OF INTEGRALS

In using the following table, the following items should be noted:

a) An arbitrary constant may be added to any result.

b) In each fromula, x may be considered as the independent variable, or as a differentiable function of some independent variable.

c) If ambiguous signs, $\pm$ or $\mp$, occur more than once in a formula, it is understood that the *upper (lower)* signs apply simultaneously throughout the formula.

d) No formula in the table may be used with any value of the variable, or of any constant, for which the integrand on the left, or the result on the right, is imaginary or otherwise undefined.

e) $\tan^{-1} x$, $\sin^{-1} x$, $\cos^{-1} x$, $\cot^{-1} x$, $\sec^{-1} x$, $\csc^{-1} x$, $\cosh^{-1} x$, in any formula, represent *principle values*, and have the following ranges:

$-\frac{\pi}{2} \leq \sin^{-1} x \leq \frac{\pi}{2}$ $\qquad$ $0 \leq \cos^{-1} x \leq \pi$ $\qquad$ $-\frac{\pi}{2} < \tan^{-1} x < \frac{\pi}{2}$

$0 < \cot^{-1} x \leq \pi$ $\qquad$ $0 \leq \sec^{-1} x \leq \pi$ $\qquad$ $-\frac{\pi}{2} \leq \csc^{-1} x \leq \frac{\pi}{2}$

$0 \leq \cosh^{-1} x$ $\quad$ where $x \geq 1$

f) All angles are given in radian measurement unless otherwise specified.

Expressions Containing (ax + b)

1. $\int (ax + b)^n \, dx = \frac{1}{a(n + 1)} (ax + b)^{n+1}, \quad n \neq -1.$

2. $\int \frac{dx}{ax + b} = \frac{1}{a} \ln \, (ax + b).$

3. $$\int \frac{dx}{(ax+b)^2} = -\frac{1}{a(ax+b)}\,.$$

4. $$\int x(ax+b)^n\,dx = \frac{1}{a^2(n+2)}(ax+b)^{n+2} - \frac{b}{a^2(n+1)}(ax+b)^{n+1},$$

$$n \neq -1,\ -2.$$

5. $$\int \frac{x\,dx}{ax+b} = \frac{x}{a} - \frac{b}{a^2}\ln(ax+b).$$

6. $$\int \frac{x\,dx}{(ax+b)^2} = \frac{b}{a^2(ax+b)} + \frac{1}{a^2}\ln(ax+b).$$

7. $$\int x^2(ax+b)^n\,dx = \frac{1}{a^3}\left[\frac{(ax+b)^{n+3}}{n+3} - 2b\,\frac{(ax+b)^{n+2}}{n+2} + b^2\,\frac{(ax+b)^{n+1}}{n+1}\right], \quad n \neq -1,\ -2,\ -3.$$

8. $$\int \frac{x^2dx}{ax+b} = \frac{1}{a^3}\left[\frac{1}{2}(ax+b)^2 - 2b(ax+b) + b^2\ln(ax+b)\right].$$

9. $$\int \frac{x^2dx}{(ax+b)^2} = \frac{1}{a^3}\left[(ax+b) - 2b\ln(ax+b) - \frac{b^2}{ax+b}\right].$$

10. $$\int \frac{x^2dx}{(ax+b)^3} = \frac{1}{a^3}\left[\ln(ax+b) + \frac{2b}{ax+b} - \frac{b^2}{2(ax+b)^2}\right].$$

11. $$\int x^m(ax+b)^n\,dx$$

$$= \frac{1}{a(m+n+1)}\left[x^m(ax+b)^{n+1} - mb\int x^{m-1}(ax+b)^n\,dx\right],$$

$$= \frac{1}{m+n+1}\left[x^{m+1}(ax+b)^n + nb\int x^m(ax+b)^{n-1}\,dx\right],$$

$$m > 0,\ m+n+1 \neq 0.$$

12. $\int \frac{dx}{x(ax + b)} = \frac{1}{b} \ln \frac{x}{ax + b}$.

13. $\int \frac{dx}{x^2(ax + b)} = - \frac{1}{bx} + \frac{a}{b^2} \ln \frac{ax + b}{x}$.

14. $\int \frac{dx}{x^3(ax + b)} = \frac{2ax - b}{2b^2x^2} + \frac{a^2}{b^3} \ln \frac{x}{ax + b}$.

15. $\int \sqrt{ax + b}\, dx = \frac{2}{3a} \sqrt{(ax + b)^3}$.

16. $\int x\sqrt{ax + b}\, dx = \frac{2(3ax - 2b)}{15a^2} \sqrt{(ax + b)^3}$.

17. $\int x^2\sqrt{ax + b}\, dx = \frac{2(15a^2x^2 - 12abx + 8b^2) \sqrt{(ax + b)^3}}{105a^3}$.

18. $\int x^n\sqrt{ax + b}\, dx = \frac{2}{a^{n+1}} \int u^2(u^2 - b)^n\, du, \quad u = \sqrt{ax + b}$.

19. $\int \frac{\sqrt{ax + b}}{x}\, dx = 2\sqrt{ax + b} + b \int \frac{dx}{x\sqrt{ax + b}}$

20. $\int \frac{dx}{\sqrt{ax + b}} = \frac{2\sqrt{ax + b}}{a}$.

21. $\int \frac{xdx}{\sqrt{ax + b}} = \frac{2(ax - 2b)}{3a^2} \sqrt{ax + b}$.

22. $\int \frac{x^2dx}{\sqrt{ax + b}} = \frac{2(3a^2x^2 - 4abx + 8b^2)}{15a^3} \sqrt{ax + b}$.

23. $\int \frac{x^ndx}{\sqrt{ax + b}} = \frac{2}{a^{n+1}} \int (u^2 - b)^n\, du, \quad u = \sqrt{ax + b}$.

24. $\int \frac{dx}{x\sqrt{ax + b}} = \frac{1}{\sqrt{b}} \ln \frac{\sqrt{ax + b} - \sqrt{b}}{\sqrt{ax + b} + \sqrt{b}}$, for $b > 0$.

25. $$\int \frac{dx}{x\sqrt{ax+b}} = \frac{2}{\sqrt{-b}} \tan^{-1} \sqrt{\frac{ax+b}{-b}}, \quad b < 0;$$

$$\frac{-2}{\sqrt{b}} \tanh^{-1} \sqrt{\frac{ax+b}{b}}, \quad b > 0.$$

26. $$\int \frac{dx}{x^2\sqrt{ax+b}} = -\frac{\sqrt{ax+b}}{bx} - \frac{a}{2b} \int \frac{dx}{x\sqrt{ax+b}}.$$

27. $$\int \frac{dx}{x^n(ax+b)^m} = -\frac{1}{b^{m+n-1}} \int \frac{(u-a)^{m+n-2}\,du}{u^m}, \quad u = \frac{ax+b}{x}.$$

28. $$\int x(ax+b)^{\pm\frac{n}{2}}\,dx = \frac{2}{a^2}\left[\frac{(ax+b)^{\frac{4\pm n}{2}}}{4\pm n} - \frac{b(ax+b)^{\frac{2\pm n}{2}}}{2\pm n}\right].$$

29. $$\int \frac{dx}{x(ax+b)^{\frac{n}{2}}} = \frac{1}{b} \int \frac{dx}{x(ax+b)^{\frac{n-2}{2}}} - \frac{a}{b} \int \frac{dx}{(ax+b)^{\frac{n}{2}}}.$$

30. $$\int \frac{x^m dx}{\sqrt{ax+b}} = \frac{2x^m\sqrt{ax+b}}{(2m+1)a} - \frac{2mb}{(2m+1)a} \int \frac{x^{m-1}\,dx}{\sqrt{ax+b}}.$$

31. $$\int \frac{dx}{x^n\sqrt{ax+b}} = \frac{-\sqrt{ax+b}}{(n-1)bx^{n-1}} - \frac{(2n-3)a}{(2n-2)b} \int \frac{dx}{x^{n-1}\sqrt{ax+b}}.$$

32. $$\int \frac{(ax+b)^{\frac{n}{2}}}{x}\,dx = a \int (ax+b)^{\frac{n-2}{2}}\,dx + b \int \frac{(ax+b)^{\frac{n-2}{2}}}{x}\,dx.$$

33. $$\int \frac{dx}{(ax+b)(cx+d)} = \frac{1}{bc-ad} \ln \frac{cx+d}{ax+b}, \quad bc - ad \neq 0.$$

34. $$\int (ax+b)^n(cx+d)^m\,dx = \frac{1}{(m+n+1)a}\Big[(ax+b)^{n+1}(cx+d)^m$$

$$- m(bc-ad) \int (ax+b)^n(cx+d)^{m-1}\,dx\Big].$$

35. $$\int \frac{dx}{(ax + b)^n(cx + d)^m} = \frac{-1}{(m - 1)(bc - ad)} \left[\frac{1}{(ax + b)^{n-1}(cx + d)^{m-1}} + a(m + n - 2) \int \frac{dx}{(ax + b)^n(cx + d)^{m-1}} \right], \quad m > 1,\ n > 0,\ bc - ad \neq 0.$$

36. $$\int \frac{(ax + b)^n}{(cx + d)^m}\, dx$$

$$= -\frac{1}{(m - 1)(bc - ad)} \left[\frac{(ax + b)^{n+1}}{(cx + d)^{m-1}} + (m - n - 2)a \int \frac{(ax + b)^n\, dx}{(cx + d)^{m-1}} \right],$$

$$= \frac{-1}{(m - n - 1)c} \left[\frac{(ax + b)^n}{(cx + d)^{m-1}} + n(bc - ad) \int \frac{(ax + b)^{n-1}}{(cx + d)^m}\, dx \right].$$

Expressions Containing $ax^2 + bx + c$

37. $$\int x(ax^2 + c)^n\, dx = \frac{1}{2a} \frac{(ax^2 + c)^{n+1}}{n + 1}, \quad n \neq -1.$$

38. $$\int \sqrt{x^2 \pm p^2}\, dx = \frac{1}{2} \left[x\sqrt{x^2 \pm p^2} \pm p^2 \ln(x + \sqrt{x^2 \pm p^2}) \right].$$

39. $$\int \sqrt{p^2 - x^2}\, dx = \frac{1}{2} \left[x\sqrt{p^2 - x^2} + p^2 \sin^{-1} \left(\frac{x}{p}\right) \right].$$

40. $$\int \frac{dx}{\sqrt{x^2 \pm p^2}} = \ln(x + \sqrt{x^2 \pm p^2}).$$

41. $$\int \frac{dx}{\sqrt{p^2 - x^2}} = \sin^{-1} \left(\frac{x}{p}\right) \quad \text{or} \quad -\cos^{-1} \left(\frac{x}{p}\right).$$

42. $$\int \frac{\sqrt{ax^2 + c}}{x}\, dx = \sqrt{ax^2 + c} + \sqrt{c} \ln \frac{\sqrt{ax^2 + c} - \sqrt{c}}{x}, \quad c > 0.$$

43. $$\int \frac{\sqrt{ax^2 + c}}{x}\, dx = \sqrt{ax^2 + c} - \sqrt{-c} \tan^{-1} \frac{\sqrt{ax^2 + c}}{\sqrt{-c}}, \quad c < 0.$$

44. $$\int x^n\sqrt{ax^2 + c}\, dx = \frac{x^{n-1}(ax^2 + c)^{\frac{3}{2}}}{(n + 2)a} - \frac{(n - 1)c}{(n + 2)a}\int x^{n-2}\sqrt{ax^2 + c}\, dx, \quad n > 0.$$

45. $$\int \frac{\sqrt{ax^2 + c}}{x^n}\, dx = -\frac{(ax^2 + c)^{\frac{3}{2}}}{c(n - 1)x^{n-1}} - \frac{(n - 4)a}{(n - 1)c}\int \frac{\sqrt{ax^2 + c}}{x^{n-2}}\, dx, \quad n > 1.$$

46. $$\int \frac{dx}{x^n\sqrt{ax^2 + c}} = -\frac{\sqrt{ax^2 + c}}{c(n - 1)x^{n-1}} - \frac{(n - 2)a}{(n - 1)c}\int \frac{dx}{x^{n-2}\sqrt{ax^2 + c}}, \quad n > 1.$$

47. $$\int \frac{dx}{(ax^n + c)^m} = \frac{1}{c}\int \frac{dx}{(ax^n + c)^{m-1}} - \frac{a}{c}\int \frac{x^n dx}{(ax^n + c)^m}.$$

48. $$\int \frac{dx}{x\sqrt{ax^n + c}} = \frac{1}{n\sqrt{c}} \ln \frac{\sqrt{ax^n + c} - \sqrt{c}}{\sqrt{ax^n + c} + \sqrt{c}}, \quad c > 0.$$

49. $$\int \frac{dx}{x\sqrt{ax^n + c}} = \frac{2}{n\sqrt{-c}} \sec^{-1}\sqrt{\frac{-ax^n}{c}}, \quad c < 0.$$

50. $$\int \frac{dx}{ax^2 + bx + c} = \frac{1}{\sqrt{b^2 - 4ac}} \ln \frac{2ax + b - \sqrt{b^2 - 4ac}}{2ax + b + \sqrt{b^2 - 4ac}}, \quad b^2 > 4ac.$$

51. $$\int \frac{dx}{ax^2 + bx + c} = \frac{2}{\sqrt{4ac - b^2}} \tan^{-1}\frac{2ax + b}{\sqrt{4ac - b^2}}, \quad b^2 < 4ac.$$

52. $$\int \frac{dx}{ax^2 + bx + c} = -\frac{2}{2ax + b}, \quad b^2 = 4ac.$$

53. $$\int \frac{dx}{(ax^2 + bx + c)^{n+1}} = \frac{2ax + b}{n(4ac - b^2)(ax^2 + bx + c)^n} + \frac{2(2n - 1)a}{n(4ac - b^2)} \int \frac{dx}{(ax^2 + bx + c)^n} .$$

54. $$\int \frac{xdx}{ax^2 + bx + c} = \frac{1}{2a} \ln(ax^2 + bx + c) - \frac{b}{2a} \int \frac{dx}{ax^2 + bx + c} .$$

55. $$\int \frac{x^2dx}{ax^2 + bx + c} = \frac{x}{a} - \frac{b}{2a^2} \ln(ax^2 + bx + c) + \frac{b^2 - 2ac}{2a^2} \int \frac{dx}{ax^2 + bx + c} .$$

56. $$\int \frac{x^ndx}{ax^2 + bx + c} = \frac{x^{n-1}}{(n - 1)a} - \frac{c}{a} \int \frac{x^{n-2}\,dx}{ax^2 + bx + c} - \frac{b}{a} \int \frac{x^{n-1}\,dx}{ax^2 + bx + c} .$$

57. $$\int \frac{xdx}{(ax^2 + bx + c)^{n+1}} = \frac{-(2c + bx)}{n(4ac - b^2)(ax^2 + bx + c)^n} - \frac{b(2n - 1)}{n(4ac - b^2)} \int \frac{dx}{(ax^2 + bx + c)^n} .$$

58. $$\int \frac{x^m\,dx}{(ax^2 + bx + c)^{n+1}} = -\frac{x^{m-1}}{a(2n - m + 1)(ax^2 + bx + c)^n} - \frac{(n - m + 1)}{(2n - m + 1)} \cdot \frac{b}{a} \int \frac{x^{m-1}\,dx}{(ax^2 + bx + c)^{n+1}} + \frac{(m - 1)}{(2n - m + 1)} \cdot \frac{c}{a} \int \frac{x^{m-2}\,dx}{(ax^2 + bx + c)^{n+1}} .$$

59. $$\int \frac{dx}{x(ax^2 + bx + c)} = \frac{1}{2c} \ln\left(\frac{x^2}{ax^2 + bx + c}\right) - \frac{b}{2c} \int \frac{dx}{(ax^2 + bx + c)} .$$

60. $$\int \frac{dx}{x^2(ax^2 + bx + c)} = \frac{b}{2c^2} \ln\left(\frac{ax^2 + bx + c}{x^2}\right) - \frac{1}{cx} + \left(\frac{b^2}{2c^2} - \frac{a}{c}\right) \int \frac{dx}{(ax^2 + bx + c)} .$$

61. $$\int \frac{dx}{\sqrt{ax^2 + bx + c}} = \frac{1}{\sqrt{a}} \ln(2ax + b + 2\sqrt{a}\,\sqrt{ax^2 + bx + c}), \quad a > 0.$$

62. $$\int \frac{dx}{\sqrt{ax^2 + bx + c}} = \frac{1}{\sqrt{-a}} \sin^{-1} \frac{-2ax - b}{\sqrt{b^2 - 4ac}}, \quad a < 0.$$

63. $$\int \frac{xdx}{\sqrt{ax^2 + bx + c}} = \frac{\sqrt{ax^2 + bx + c}}{a} - \frac{b}{2a} \int \frac{dx}{\sqrt{ax^2 + bx + c}} .$$

64. $$\int \frac{x^n\,dx}{\sqrt{ax^2 + bx + c}} = \frac{x^{n-1}}{an} \sqrt{ax^2 + bx + c} - \frac{b(2n - 1)}{2an} \int \frac{x^{n-1}\,dx}{\sqrt{ax^2 + bx + c}} - \frac{c(n - 1)}{an} \int \frac{x^{n-2}\,dx}{\sqrt{ax^2 + bx + c}} .$$

65. $$\int \sqrt{ax^2 + bx + c}\,dx = \frac{2ax + b}{4a} \sqrt{ax^2 + bx + c} + \frac{4ac - b^2}{8a} \int \frac{dx}{\sqrt{ax^2 + bx + c}} .$$

66. $$\int x\sqrt{ax^2 + bx + c}\,dx = \frac{(ax^2 + bx + c)^{\frac{3}{2}}}{3a} - \frac{b}{2a} \int \sqrt{ax^2 + bx + c}\,dx.$$

67. $$\int x^2\sqrt{ax^2 + bx + c}\,dx = \left(x - \frac{5b}{6a}\right) \frac{(ax^2 + bx + c)^{\frac{3}{2}}}{4a} + \frac{(5b^2 - 4ac)}{16a^2} \int \sqrt{ax^2 + bx + c}\,dx.$$

68. $$\int \frac{dx}{x\sqrt{ax^2 + bx + c}} = -\frac{1}{\sqrt{c}} \ln \left(\frac{\sqrt{ax^2 + bc + c} + \sqrt{c}}{x} + \frac{b}{2\sqrt{c}} \right), \quad c > 0.$$

69. $$\int \frac{dx}{x\sqrt{ax^2 + bx + c}} = \frac{1}{\sqrt{-c}} \sin^{-1} \frac{bx + 2c}{x\sqrt{b^2 - 4ac}}, \quad c < 0.$$

70. $$\int \frac{dx}{x\sqrt{ax^2 + bx}} = -\frac{2}{bx} \sqrt{ax^2 + bx}, \quad c = 0.$$

Expressions Containing sin ax

71. $$\int \sin ax \, dx = -\frac{1}{a} \cos ax.$$

72. $$\int \sin^2 ax \, dx = \frac{x}{2} - \frac{\sin 2ax}{4a}.$$

73. $$\int \sin^3 ax \, dx = -\frac{1}{a} \cos ax + \frac{1}{3a} \cos^3 ax.$$

74. $$\int \sin^n ax \, dx = -\frac{\sin^{n-1} ax \cos ax}{na} + \frac{n-1}{n} \int \sin^{n-2} ax \, dx, \quad \text{(n pos. integer)}.$$

75. $$\int \frac{dx}{\sin ax} = \frac{1}{a} \ln \tan \frac{ax}{2} = \frac{1}{a} \ln(\csc ax - \cot ax).$$

76. $$\int \frac{dx}{\sin^n ax} = -\frac{1}{a(n-1)} \frac{\cos ax}{\sin^{n-1} ax} + \frac{n-2}{n-1} \int \frac{dx}{\sin^{n-2} ax}, \quad n \text{ integer} > 1.$$

77. $$\int \frac{dx}{1 \pm \sin ax} = \mp \frac{1}{a} \tan \left(\frac{\pi}{4} \mp \frac{ax}{2} \right).$$

78. $$\int \frac{dx}{b + c \sin ax} = \frac{-2}{a\sqrt{b^2 - c^2}} \tan^{-1} \left[\sqrt{\frac{b-c}{b+c}} \tan \left(\frac{\pi}{4} - \frac{ax}{2} \right) \right], \quad b^2 > c^2.$$

79. $$\int \frac{dx}{b + c \sin ax} = \frac{-1}{a\sqrt{c^2 - b^2}} \ln \frac{c + b \sin ax + \sqrt{c^2 - b^2} \cos ax}{b + c \sin ax}, \quad c^2 > b^2.$$

80. $\int \sin ax \sin bx\, dx = \frac{\sin(a-b)x}{2(a-b)} - \frac{\sin(a+b)x}{2(a+b)}, \quad a^2 \neq b^2.$

Expressions Involving cos ax

81. $\int \cos ax\, dx = \frac{1}{a} \sin ax.$

82. $\int \cos^2 ax\, dx = \frac{x}{2} + \frac{\sin 2ax}{4a}.$

83. $\int \cos^3 ax\, dx = \frac{1}{a} \sin ax - \frac{1}{3a} \sin^3 ax.$

84. $\int \cos^n ax\, dx = \frac{\cos^{n-1} ax \sin ax}{na} + \frac{n-1}{n} \int \cos^{n-2} ax\, dx.$

85. $\int \frac{dx}{\cos ax} = \frac{1}{a} \ln \tan\left(\frac{ax}{2} + \frac{\pi}{4}\right) = \frac{1}{a} \ln(\tan ax + \sec ax).$

86. $\int \frac{dx}{\cos^n ax} = \frac{1}{a(n-1)} \frac{\sin ax}{\cos^{n-1} ax} + \frac{n-2}{n-1} \int \frac{dx}{\cos^{n-2} ax}, \quad n \text{ integer} > 1.$

87. $\int \frac{dx}{1 + \cos ax} = \frac{1}{a} \tan \frac{ax}{2}.$

88. $\int \frac{dx}{1 - \cos ax} = -\frac{1}{a} \cot \frac{ax}{2}.$

89. $\int \frac{dx}{b + c \cos ax} = \frac{1}{a\sqrt{b^2 - c^2}} \tan^{-1}\left(\frac{\sqrt{b^2 - c^2} \cdot \sin ax}{c + b \cos ax}\right), \quad b^2 > c^2.$

90. $\int \frac{dx}{b + c \cos ax} = \frac{1}{a\sqrt{c^2 - b^2}} \tanh^{-1}\left[\frac{\sqrt{c^2 - b^2} \cdot \sin ax}{c + b \cos ax}\right], \quad c^2 > b^2.$

91. $\int \cos ax \cdot \cos bx\, dx = \frac{\sin(a-b)x}{2(a-b)} + \frac{\sin(a+b)x}{2(a+b)}, \quad a^2 \neq b^2.$

Expressions Containing sin ax and cos ax

92. $$\int \sin ax \cos bx\, dx = -\frac{1}{2}\left[\frac{\cos(a-b)x}{a-b} + \frac{\cos(a+b)x}{a+b}\right], \quad a^2 \neq b^2.$$

93. $$\int \sin^n ax \cos ax\, dx = \frac{1}{a(n+1)} \sin^{n+1} ax, \quad n \neq -1.$$

94. $$\int \cos^n ax \sin ax\, dx = -\frac{1}{a(n+1)} \cos^{n+1} ax, \quad n \neq -1.$$

95. $$\int \frac{\sin ax}{\cos ax}\, dx = -\frac{1}{a} \ln \cos ax.$$

96. $$\int \frac{\cos ax}{\sin ax}\, dx = \frac{1}{a} \ln \sin ax.$$

97. $$\int (b + c \sin ax)^n \cos ax\, dx = \frac{1}{ac(n+1)} (b + c \sin ax)^{n+1}, \quad n \neq -1.$$

98. $$\int (b + c \cos ax)^n \sin ax\, dx = -\frac{1}{ac(n+1)} (b + c \cos ax)^{n+1}, \quad n \neq -1.$$

99. $$\int \frac{\cos ax\, dx}{b + c \sin ax} = \frac{1}{ac} \ln(b + c \sin ax).$$

100. $$\int \frac{\sin ax}{b + c \cos ax}\, dx = -\frac{1}{ac} \ln(b + c \cos ax).$$

101. $$\int \sin^2 ax \cos^2 ax\, dx = \frac{x}{8} - \frac{\sin 4\, ax}{32a}.$$

102. $$\int \frac{dx}{\sin ax \cos ax} = \frac{1}{a} \ln \tan ax.$$

103. $$\int \frac{dx}{\sin^2 ax \cos^2 ax} = \frac{1}{a} (\tan ax - \cot ax).$$

104. $$\int \frac{\sin^2 ax}{\cos ax}\, dx = \frac{1}{a}\left[-\sin ax + \ln \tan\left(\frac{ax}{2} + \frac{\pi}{4}\right)\right].$$

105. $$\int \frac{\cos^2 ax}{\sin ax}\, dx = \frac{1}{a}\left[\cos ax + \ln \tan \frac{ax}{2}\right].$$

106. $$\int \sin^m ax \cos^n ax\,dx = -\frac{\sin^{m-1} ax \cos^{n+1} ax}{a(m+n)} + \frac{m-1}{m+n}\int \sin^{m-2} ax \cos^n ax\,dx, \quad m, n > 0.$$

107. $$\int \sin^m ax \cos^n ax\,dx = \frac{\sin^{m+1} ax \cos^{n-1} ax}{a(m+n)} + \frac{n-1}{m+n}\int \sin^m ax \cos^{n-2} ax\,dx, \quad m, n > 0.$$

Expressions Containing tan ax or cot ax (tan ax = 1/ctn ax)

108. $$\int \tan ax\,dx = -\frac{1}{a}\ln\cos ax.$$

109. $$\int \tan^2 ax\,dx = \frac{1}{a}\tan ax - x.$$

110. $$\int \tan^3 ax\,dx = \frac{1}{2a}\tan^2 ax + \frac{1}{a}\ln\cos ax.$$

111. $$\int \tan^n ax\,dx = \frac{1}{a(n-1)}\tan^{n-1} ax - \int \tan^{n-2} ax\,dx, \quad n \text{ integer} > 1.$$

112. $$\int \cot x\,dx = \ln\sin x, \quad \text{or} \quad -\ln\csc x.$$

113. $$\int \cot^2 ax\,dx = \int \frac{dx}{\tan^2 ax} = -\frac{1}{a}\cot ax - x.$$

114. $$\int \cot^3 ax\,dx = -\frac{1}{2a}\cot^2 ax - \frac{1}{a}\ln\sin ax.$$

115. $$\int \cot^n ax\,dx = \int \frac{dx}{\tan^n ax} = -\frac{1}{a(n-1)}\cot^{n-1} ax - \int \cot^{n-2} ax\,dx, \quad n \text{ integer} > 1.$$

Expressions Containing sec ax = 1/cos ax or csc ax = 1/sin ax

116. $\int \sec ax\, dx = \frac{1}{a} \ln \tan \left(\frac{ax}{2} + \frac{\pi}{4}\right).$

117. $\int \sec^2 ax\, dx = \frac{1}{a} \tan ax.$

118. $\int \sec^3 ax\, dx = \frac{1}{2a} \left[\tan ax \sec ax + \ln \tan \left(\frac{ax}{2} + \frac{\pi}{4}\right)\right].$

119. $\int \sec^n ax\, dx = \frac{1}{a(n - 1)} \frac{\sin ax}{\cos^{n-1} ax}$

$+ \frac{n - 2}{n - 1} \int \sec^{n-2} ax\, dx, \quad n \text{ integer} > 1.$

120. $\int \csc ax\, dx = \frac{1}{a} \ln \tan \frac{ax}{2}.$

121. $\int \csc^2 ax\, dx = - \frac{1}{a} \cot ax.$

122. $\int \csc^3 ax\, dx = \frac{1}{2a} \left[-\cot ax \csc ax + \ln \tan \frac{ax}{2}\right].$

123. $\int \csc^n ax\, dx = - \frac{1}{a(n - 1)} \frac{\cos ax}{\sin^{n-1} ax}$

$+ \frac{n - 2}{n - 1} \int \csc^{n-2} ax\, dx, \quad n \text{ integer} > 1.$

Expressions Containing tan ax and sec ax or cot ax and csc ax

124. $\int \tan ax \sec ax\, dx = \frac{1}{a} \sec ax.$

125. $\int \tan^n ax \sec^2 ax\, dx = \frac{1}{a(n + 1)} \tan^{n+1} ax, \quad n \neq -1.$

126. $\int \tan ax \sec^n ax\, dx = \frac{1}{an} \sec^n ax, \quad n \neq 0.$

127. $\int \cot u \csc u \, du = -\csc u,$ where u is any function of x.

128. $\int \cot ax \csc ax \, dx = -\frac{1}{a} \csc ax.$

129. $\int \cot^n ax \csc^2 ax \, dx = -\frac{1}{a(n+1)} \cot^{n+1} ax, \quad n \neq -1.$

130. $\int \cot ax \csc^n ax \, dx = -\frac{1}{an} \csc^n ax, \quad n \neq 0.$

131. $\int \frac{\csc^2 ax \, dx}{\cot ax} = -\frac{1}{a} \ln \cot ax.$

Expressions Containing Algebraic and Trigonometric Functions

132. $\int x \sin ax \, dx = \frac{1}{a^2} \sin ax - \frac{1}{a} x \cos ax.$

133. $\int x^2 \sin ax \, dx = \frac{2x}{a^2} \sin ax + \frac{2}{a^3} \cos ax - \frac{x^2}{a} \cos ax.$

134. $\int x^3 \sin ax \, dx = \frac{3x^2}{a^2} \sin ax - \frac{6}{a^4} \sin ax - \frac{x^3}{a} \cos ax + \frac{6x}{a^3} \cos ax.$

135. $\int x \cos ax \, dx = \frac{1}{a^2} \cos ax + \frac{1}{a} x \sin ax.$

136. $\int x^2 \cos ax \, dx = \frac{2x}{a^2} \cos ax - \frac{2}{a^3} \sin ax + \frac{x^2}{a} \sin ax.$

137. $\int x^3 \cos ax \, dx = \frac{(3a^2x^2 - 6) \cos ax}{a^4} + \frac{(a^2x^3 - 6x) \sin ax}{a^3}.$

Expressions Containing Exponential and Logarithmic Functions

138. $\int b^x \, du = \frac{b^x}{\ln b}.$

139. $\int e^{ax}\,dx = \frac{1}{a}e^{ax}, \qquad \int b^{ax}\,dx = \frac{b^{ax}}{a\ln b}.$

140. $\int xe^{ax}\,dx = \frac{e^{ax}}{a^2}(ax - 1), \qquad \int xb^{ax}\,dx = \frac{xb^{ax}}{a\ln b} - \frac{b^{ax}}{a^2(\ln b)^2}.$

141. $\int x^2e^{ax}\,dx = \frac{e^{ax}}{a^3}(a^2x^2 - 2ax + 2).$

142. $\int x^ne^{ax}\,dx = \frac{1}{a}x^ne^{ax} - \frac{n}{a}\int x^{n-1}e^{ax}\,dx,$ n pos.

143. $\int x^ne^{ax}\,dx = \frac{e^{ax}}{a^{n+1}}[(ax)^n - n(ax)^{n-1} + n(n-1)(ax)^{n-2}$

$- \cdots + (-1)^n n!],$ n pos. integ.

144. $\int x^ne^{-ax}\,dx = -\frac{e^{-ax}}{a^{n+1}}[(ax)^n + n(ax)^{n-1} + n(n-1)(ax)^{n-2}$

$+ \cdots + n!],$ n pos. integ.

145. $\int x^nb^{ax}\,dx = \frac{x^nb^{ax}}{a\ln b} - \frac{n}{a\ln b}\int x^{n-1}b^{ax}\,dx,$ n pos.

146. $\int \frac{e^{ax}}{x}\,dx = \ln x + ax + \frac{(ax)^2}{2\cdot 2!} + \frac{(ax)^3}{3\cdot 3!} + \cdots.$

147. $\int \frac{e^{ax}}{x^n}\,dx = \frac{1}{n-1}\left[-\frac{e^{ax}}{x^{n-1}} + a\int \frac{e^{ax}}{x^{n-1}}\,dx\right],$ n integ. > 1.

148. $\int \frac{dx}{b + ce^{ax}} = \frac{1}{ab}[ax - \ln(b + ce^{ax})].$

149. $\int \frac{e^{ax}\,dx}{b + ce^{ax}} = \frac{1}{ac}\ln(b + ce^{ax}).$

150. $\int \frac{dx}{be^{ax} + ce^{-ax}} = \frac{1}{a\sqrt{bc}}\tan^{-1}\left(e^{ax}\sqrt{\frac{b}{c}}\right),$ b and c pos.

151. $\int e^{ax} \sin bx\, dx = \frac{e^{ax}}{a^2 + b^2} (a \sin bx - b \cos bx).$

152. $\int e^{ax} \sin bx \sin cx\, dx = \frac{e^{ax}[(b - c) \sin(b - c)x + a \cos(b - c)x]}{2[a^2 + (b - c)^2]}$

$$- \frac{e^{ax}[(b + c) \sin(b + c)x + a \cos(b + c)x]}{2[a^2 + (b + c)^2]} .$$

153. $\int e^{ax} \cos bx\, dx = \frac{e^{ax}}{a^2 + b^2} (a \cos bx + b \sin bx).$

154. $\int e^{ax} \cos bx \cos cx\, dx = \frac{e^{ax}[(b - c) \sin(b - c)x + a \cos(b - c)x]}{2[a^2 + (b - c)^2]}$

$$+ \frac{e^{ax}[(b + c) \sin(b + c)\, x + a \cos(b + c)x]}{2[a^2 + (b + c)^2]} .$$

155. $\int e^{ax} \sin bx \cos cx\, dx = \frac{e^{ax}[a \sin(b - c)x - (b - c) \cos(b - c)x]}{2[a^2 + (b - c)^2]}$

$$+ \frac{e^{ax}[a \sin(b + c)x - (b + c) \cos(b + c)x]}{2[a^2 + (b + c)^2]} .$$

156. $\int \ln ax\, dx = x \ln ax - x.$

157. $\int x \ln ax\, dx = \frac{x^2}{2} \ln ax - \frac{x^2}{4} .$

158. $\int x^2 \ln ax\, dx = \frac{x^3}{3} \ln ax - \frac{x^3}{9} .$

159. $\int x^n \ln ax\, dx = x^{n+1} \left[\frac{\ln ax}{n + 1} - \frac{1}{(n + 1)^2} \right], \quad n \neq -1.$

Expressions Containing Inverse Trigonometric Functions

160. $\int \sin^{-1} ax\, dx = x \sin^{-1} ax + \frac{1}{a} \sqrt{1 - a^2x^2}.$

161. $\int x \sin^{-1} ax\, dx = \frac{x^2}{2} \sin^{-1} ax - \frac{1}{4a^2} \sin^{-1} ax + \frac{x}{4a} \sqrt{1 - a^2x^2}.$

162. $\int \cos^{-1} ax\, dx = x \cos^{-1} ax - \frac{1}{a} \sqrt{1 - a^2x^2}.$

163. $\int x \cos^{-1} ax\, dx = \frac{x^2}{2} \cos^{-1} ax - \frac{1}{4a^2} \cos^{-1} ax - \frac{x}{4a} \sqrt{1 - a^2x^2}.$

164. $\int \tan^{-1} ax\, dx = x \tan^{-1} ax - \frac{1}{2a} \ln(1 + a^2x^2).$

165. $\int \cot^{-1} ax\, dx = x \cot^{-1} ax + \frac{1}{2a} \ln(1 + a^2x^2).$

166. $\int \sec^{-1} ax\, dx = x \sec^{-1} ax - \frac{1}{a} \ln(ax + \sqrt{a^2x^2 - 1}).$

167. $\int \csc^{-1} ax\, dx = x \csc^{-1} ax + \frac{1}{a} \ln(ax + \sqrt{a^2x^2 - 1}).$

Definite Integrals

168. $\int_0^\infty x^{n-1} e^{-x}\, dx = \int_0^1 [\ln\ (1/x)]^{n-1}\, dx = \Gamma(n).$

$\Gamma(n + 1) = n \cdot \Gamma(n)$, if $n > 0$. $\Gamma(2) = \Gamma(1) = 1.$

$\Gamma(n + 1) = n!$, if n is an integer. $\Gamma\left(\frac{1}{2}\right) = \sqrt{\pi}$.

$\Gamma(n) = \Pi(n - 1).$

169. $\int_0^1 x^{m-1} (1 - x)^{n-1}\, dx = \int_0^\infty \frac{x^{m-1}\, dx}{(1 + x)^{m+n}} = \frac{\Gamma(m)\Gamma(n)}{\Gamma(m + n)}$.

170. $\int_0^{\frac{\pi}{2}} \sin^n x \, dx = \int_0^{\frac{\pi}{2}} \cos^n x \, dx$

$$= \frac{1}{2}\sqrt{\pi} \cdot \frac{\Gamma\left(\frac{n}{2} + \frac{1}{2}\right)}{\Gamma\left(\frac{n}{2} + 1\right)}, \quad \text{if } n > -1;$$

$$= \frac{1\cdot3\cdot5 \cdots (n-1)}{2\cdot4\cdot6 \cdots (n)} \cdot \frac{\pi}{2}, \quad \text{if } n \text{ is an even integer;}$$

$$= \frac{2\cdot4\cdot6 \cdots (n-1)}{1\cdot3\cdot5\cdot7 \cdots n}, \quad \text{if } n \text{ is an odd integer.}$$

171. $\int_0^{\infty} \frac{\sin ax}{x} \, dx = \frac{\pi}{2}, \quad \text{if } a > 0.$

172. $\int_0^{\infty} \frac{\sin x \cos ax}{x} \, dx = 0, \quad \text{if } a < -1, \quad \text{or } a > 1;$

$$= \frac{\pi}{4}, \quad \text{if } a = -1, \quad \text{or } a = 1;$$

$$= \frac{\pi}{2}, \quad \text{if } -1 < a < 1.$$

173. $\int_0^{\pi} \sin^2 ax \, dx = \int_0^{\pi} \cos^2 ax \, dx = \frac{\pi}{2}.$

174. $\int_0^{\pi/a} \sin ax \cdot \cos ax \, dx = \int_0^{\pi} \sin ax \cdot \cos ax \, dx = 0.$

175. $\int_0^{\pi} \sin ax \sin bx \, dx = \int_0^{\pi} \cos ax \cos bx \, dx = 0, \quad a \neq b.$

176. $\int_0^{\pi} \sin ax \cos bx \, dx = \frac{2a}{a^2 - b^2}, \quad \text{if } a - b \text{ is odd;}$

$$= 0, \quad \text{if } a - b \text{ is even.}$$

177. $\int_0^{\infty} \frac{\sin ax \sin bx}{x^2} \, dx = \frac{1}{2}\pi a, \quad \text{if } a < b.$

178. $$\int_0^\infty e^{-a^2x^2}\,dx = \frac{\sqrt{\pi}}{2a} = \frac{1}{2a}\,\Gamma\left(\frac{1}{2}\right), \quad \text{if } a > 0.$$

179. $$\int_0^\infty x^n \cdot e^{-ax}\,dx = \frac{\Gamma(n+1)}{a^{n+1}},$$

$$= \frac{n!}{a^{n+1}}, \quad \text{if } n \text{ is a positive integer, } a > 0.$$

180. $$\int_0^\infty x^{2n}\,e^{-ax^2}\,dx = \frac{1\cdot 3\cdot 5\cdots(2n-1)}{2^{n+1}\,a^n}\sqrt{\frac{\pi}{a}}.$$

181. $$\int_0^\infty \sqrt{x}\,e^{-ax}\,dx = \frac{1}{2a}\sqrt{\frac{\pi}{a}}.$$

182. $$\int_0^\infty \frac{e^{-ax}}{\sqrt{x}}\,dx = \sqrt{\frac{\pi}{a}}.$$

183. $$\int_0^\infty e^{(-x^2-a^2/x^2)}\,dx = \frac{1}{2}\,e^{-2a}\sqrt{\pi}, \quad \text{if } a > 0.$$

184. $$\int_0^\infty e^{-ax}\cos bx\,dx = \frac{a}{a^2+b^2}, \quad \text{if } a > 0.$$

185. $$\int_0^\infty e^{-ax}\sin bx\,dx = \frac{b}{a^2+b^2}, \quad \text{if } a > 0.$$

186. $$\int_0^\infty e^{-a^2x^2}\cos bx\,dx = \frac{\sqrt{\pi}\cdot e^{-b^2/4a^2}}{2a}, \quad \text{if } a > 0.$$

Answers to Selected Problems

Section 2.6 3. 0.23 cc.; 5. $1.5\pi \times 10^{-5}$ $mm.^3$/sec.; 7. 0.149 cm.;

Section 3.1 3a. {x | x has blood type Rh^+}; 3b. {x | x is a gaseous radioactive element}; 3c. {x | x is a warm water fish};

Section 3.2 11. 20, 30, 20;

Section 4.2 1. 1, 13, 5, 125, 61; 3. 1, 2, 8, 64, 128, 256; 7a. c, e, f; 9. No; 11. yes; 17. $v = (4\pi s^2)^{1/3}/3$; 19. $v = 56.25\pi hfT.^3$; 21. $v = (16 - 2x)(20 - 2x)x$

Section 4.3 1. b = -1, m = 2; 3. m = -1/5, b = -2; 5a. y = x + 1; 5f. y = 3x + 2; 17. $L = 1.143 \times 10^{-3}\,T + 76.444$;

Section 4.4 1. vertex (0.2); 3. vertex (-3, 2); 5. vertex (3, 0); 7. vertex (-5/16, 23/12); 9. center (-5, 5), $r = 5\sqrt{6}$; 11. vertex (41/8, 3/4); 13. yes; 15. $(x - 1)^2 + (y - 6)^2 = 18$;

Section 4.6 1. $10^{7 - 0.1t}$;

Section 4.7 1. $0.749N_0$; 3. $N = 10^5 \exp(0.192t)$; 5. 9153 years; 7. 7285 years; 9. 71.5 days; 11. 49×10^3 at t = 20 hours; 13. $-(\ell n\, 0.80)/K$; 15. $0.189N_0$, 125 hours; 17. must add 3.6% to amount found to get estimate of original amount; 19. $N_0 = 2957$, K = 0.1695, 461 exp(0.1695r), 2511; 21. K = -0.0106, N = 2.39

Section 4.8 5. $t = (\ell n\, 3)/t_0$; 7. 0.8 hours; 9. $N = 457910^{-1.7095\,DBH}$.

Section 5.1 1. 3; 3. 12; 5. 0; 7. 2X; 2/3; 11. 0; 13. yes; 15. N_0, ae^{-b}, $N_0(1 - b)^3$, N_0;

Section 5.2 1b. x = -5; 3. No;

Section 5.4 1. $N_0(1 + b)$, a, N_0, $N_0/(1 - b)$, N_0; 3. 5/2; 5. 3/4; 7. 0; 9. 4/7; 11. ∞; 13. N_0, ae^{-b}, $N_0(1 - b)^3$, N_0, O; 15. ∞;

Section 6.2 1. $2x + 4 + \Delta x$, $2x + 4$, 1.0.1, 12.1, 13.8; 3. $-2/(x(x + \Delta x))$; 5. $(-3x^2 - 3x\Delta x - \Delta x)^2)/(x^3(x + \Delta x)^3)$; 7. $9x^2 + 9x\Delta x + 3(\Delta x)^2 - 2$; 9. $9x^2 - 2$; 11. 3K, 13K, 10K; 15. 2.1 ; 17. $10^4(0.75 + 2t + \Delta t)$, $10^4(0.75 + 2$ $10^4(1.75)$, $10^4(6.75)$; 19. 100.8π in.^{3}, 97.2π in.^{3}, 96.12π in.^{3}, 96π in.^{3};

Section 6.3c 1. 4x; 3. $8t + 9t^2 - 1$; 5. $-3x^{-4}$; 7. $4(3x^4 - 2x^{-3} + 4x)^3 (12x^3 + 6x^{-4} + 4)$; 9. $-w(4 - w^2)^{1/2}$; 11. $2(4x - 1)^{-1/2}$; 13. $2x - 2 + 2x^{-3}$; 15. $(3x + 2)/2(x + 1)^{1/2}$; 17. 0, 6K, 20K; 21. $-(x - 1)^{-2}$.

Section 6.3e 1. $6x^2 + 8x + 3$; 3. $(3x+2)/2(x+1)^{1/2}$; 5. $3x^2 + 2x - 2$;
7. $-2(x-2)^{-2}$; 9. $(-x^2 + 4x - 1)/(x^2 - 2x + 3)^{-2}$; 11. $1/6(x-2)^{1/2}(x+2)^{1/3}$;
13. $(x^2 + 2)[(3x^2 + 2x + 1)(x^2 + 2) + 4x(x-1) - 6(x^2 + 2)(x-1)(3x+1)]/$
$(3x^2 + 2x + 1)^4$; 15. $14(3x^2 - 2)^6(3x^2 + 3x + 2)/(2x+1)^8$; 17. $-(g(x))^{-2}\frac{d}{dx}g(x)$;
19. $-12x/(x^2 - 3)^2$; 21. high point at $x = -1$, low point at $x = 3$;
23. P_{max} $K/(I+K)^2$, $-R$;

Section 6.3g 1. $3e^{3x}$; 3. $(1+x)e^x$; 5. $e^{x\ell nx}(1 + \ell nx)$; 7. $\frac{1}{2}(2+x)^{-1/2}e^{\sqrt{2+x}}$;
9. $4/(e^x + e^{-x})^2$; 11. 0; 13. $I_0\mu e^{\mu x}$; 15. $4/(x+3)$; 17. $-1/2\sqrt{x(x+2)}$;
19. $xe^x[(x+2)\ell nx + 1]$; 21. $a^x \ell na$;

Section 6.4 3. $-24x^{-5}$; 5. $1/x$; 7. $e^{-x}[x^2\ell nx - 2x - 1]$; 9. 0;

Section 7.1 1. 2.86×10^{-3}, 2.968×10^{-3}; 3. 7 lb./in.^{2}/sec., increasing;
5. 3π ft./sec., 9π ft.^{2}/sec., 3π ft./sec., 30π ft.^{2}/sec.; 7. -0.3π in.^{2}/min.,
-1.25π in.^{3}/min.; 9. $4\pi(r_0 + 8)$mm.^{2}/day; 11. 476.6 ft./min.;
13. $0.02/\pi$ ft.min.; 15. -4 cm./min., 2.08 min.; 17. $8/25\pi$ in./min.;
19. -11.55°K/min.; 21. $(216\pi + 0.5)$ ft.^{3}/min.; 23. $2\pi(r+L)\frac{dr}{dI} + 2\pi r\frac{dL}{dI}$;
25. -16π in.^{3}/min.; 27. $2L(r-h)/(2rh - h^2)^{1/2}$

Section 7.3 1. (2,0) ab. min., (6,16) rel. max., (-6,64) ab. max.;
3. (0.4) ab. max.; 5. (-3,-76) I. P.; 7. (0,0) rel. max.; 9. $(e^{-1/2}, 1/2e^{-1})$
rel. min., $(e^{-3/2}, 3/2e^{-3})$ I. P.; 11. $(\frac{\ell nb}{K}, w_0(1+b)/2)$ I. P.;
13. (0,32) rel. min., (-2,0) rel. min., (3,0) I. P., $(\sqrt{1.5}, -58.18)$ I. P.,
$(-\sqrt{1.5}, -45.32)$ I. P.; 15. (1,2) rel. min., (1/3, 1.259) I. P.,
(-2,-1) ob. min., (1/2, 0.875) ab. max.

Section 7.4 1. 4.11°C; 3. $53^{1/3}$ ft. $\times$ 160 ft.; 5. 20 in. $\times$ 20 in. $\times$ 10 in.;
7. 6 in. $\times$ 12 in. $\times$ 2 in.; 9. 1; 11. Walk the last 2.42 miles on road;
13. Walk the last mile down the beach; 15. $\sqrt{2A} \times \sqrt{A/2}$; 17. $4r_0/5$,
$2r_0/3$; 19. 9.2 in. $\times$ 13.1 in.

Section 7.5d 9. rel. max at $t = (\ell na - \ell nb)/(a-b)$, I. P. when $t = 2(\ell na - \ell nb)/$
$(a-b)$; 11. width is $8\sqrt{3/\pi}$ ft.; 13. $26^{2/3}$ trees/acre; 15. $p = 1/2$.

Section 7.6a 1a. $-x/y$, $3y = -4x - 25$; 1c. $y' = (3x^2 - 4xy - 1)/(2x^2 + 2)$;
3. $(-2xy^3 - 2x^4)/y^5$; 5. $-\alpha P/v$; 7. $8x/3(u^2+1)^2(x^2+2)^{2/3}$;

Section 7.6b 1. $-2\sqrt{x+2}/(x^2 - 4)(x-2)^{1/2}$; 3. $(-3x^2 - 12x - 5)/(x^2 - 5)^2$;
5. $y[2x/(x^2 + 3) + 3x^2/(x^3 - 4) - 2/(x+7)]$; 7. $y[1/x + 1/2(x+2) + 1/3(x-2)]$;
9. $y[(2x+3)/x^2 + 3x - 1) + 2/(x+2) + (3x^2 - 3)/(x^3 - 3x + 2)]$;

11. $(\ell n(x+1))\,(\ell n(x-1))[1/(x+1)\ell n(x+1) - 1/(x-1)\ell n(x-1)]$;

13. $2\ell n(x+1) + 2x/(x+1)$; 15. $2(x+\ell nx + 1)\exp(x^2 + 2x\ell nx)$;

17. $(e^{2x}(1+x\ell nx)\ell nx)/x$; 19. $2^{6x}3^{4x^2}[6\ell n2 + 8x\ell n3]$

Section 7.7a 1. $16\pi\mu^3$; 3. 8.0625; 5. 0.9π cm.^{3}; 9. $ds = 8\pi r dr$, $dv = 4\pi r^2 dr$;

11. -0.27 grams; 13. $dV = -(v+b)dF/(F+a)$; 15. -0.21 mass units, -0.215 mass units; 17. $dV = \pi r(rdL + 2Ldr)$

Section 7.7b 3. 2.4π in.^{3}, 6%; 5. 9.77 in.; 7. 0.44 ft.^{2}, 0.0036, 0.36%, 26 bu./acre; 9. 0.06, 6%; 11. $2\pi r dr$; 13. 0.18π in.^{3}; 15. $\pm 0.08\pi$ in.^{3};

17. r = 4.96 in., dr = 0.16 in.; 19. $dg = (-4\pi^2 \ell dT)/T^2$, $dg/g = -dT/T$

Section 7.8b 3. $1 - x/3 + 2x^2/9 - 14x^3/81$; 5. $1 + 2x + x^2 + 4x^3$;

7. $1 + x\ell na + (x\ell na)^2/2! + (x\ell na)^3/3!$; 11. 1.259

Section 7.9 1. 1; 3. -6; 5. ∞; 7. $(x+3)/2$; 9/ 5/4; 11. -33; 13. 0; 15. -1; 17. 0; 19. ∞; 21. $-\infty$; 23. ∞; 25. no; 27. 3/4; 29. 11/4

Section 8.1 1a. 21.75, 15.75; 1b. 20.19, 15.92; 5a. 6, 4; 5b. 5, 3; 3. 87; 5. $n(n+1)(n+2)/3$; 7. $n(n+1)(n+2)(n+3)/4$; 9. 3/2

Section 8.3 1. 31/8, 10/3; 5. 24; 7. 18

Section 8.5 1. 8; 3. 27; 5. $178^{2/3}$; 7. $(10 - \sqrt{5})/5$; 9. $1 - e^{-t}$;

11. $N_0(e^{10} - 1)$; 13. $N = N_0 e^{Kt}$; 15. $(\sqrt{13} - 1)/\sqrt{3}$;

Section 8.6c 3a. $t^2(1-t^2)$; 3b. $-2x[3(x^2+2)^2 + 4(x^2+2) - 1]$;

3c. $3[(3x+1)^2 + 2(3x+1) - 3] - 2[4x^2 + 4x - 3]$.

Section 8.7 1. $2(1+3y)^{1/2}/3 + c$; 3. $-(9+4x^2)^{-1}/8 + c$; 5. $1/3\ell n(x^3 + 11 + c$;

7. $x^3/3 - 2x - x^{-1} + c$; 9. $-(3+e^x)^{-2}/2 + c$; 11. $3e^{x^2}/2 + c$;

13. $(2^{x^2})/(2\ell n2) + c$; 15. $e^{x\ell nx} + c$; 17. $(\pi^{3x-1})/(3\ell n\pi) + c$;

19. $2(3x^3 + 3x - 4)^{1/2}/3 + c$; 21. $3(\ell n3)^{4/3}/4$; 23. $2(x^3 - 2)^{3/2}/9 + c$;

25. $-4(1 - 3x^2)^{3/2}/9 + c$; 27. $a^{2x}/(2\,\ell na) + c$; 29. $2x^{3/2}/3 - x^2/6 + 8x^{1/2} + c$;

31. $(1+y^4)^{3/2}/6 + c$; 33. $3(a+bx)^{2/3}/2b + c$; 35. $2x^{5/2}/5 - 2x^{3/2}/3 - 12x^{1/2} + c$; 37. $e^{2x}/2 + 2e^x + x + c$; 39. $3\ell n(x^2 + 2x + 21 + c$;

41. $({}_aN_t/(cf+m))(1 - \exp(-(cf+m))$

Section 8.8a 1. 9; 3. 3/2; 5. 18; 7. 1/6

Section 8.8b 3. 0.135, 0.632, 0.233; 5. 0.3125, 0.4375, 0.25

Section 8.9 1. $y^2 + \ell ny - \ell nx = c$; 3. $y = cx^2$; 5. $PV = C$; 7. $1 + y^2 = 5x$

9. $y^{-2} = -2\ell n|c(x+4)|$; 11. $P = 30\exp(-4.558 \times 10^{-5} A)$, 20.83 in.;

13. 9.97 hrs.; 15. $N = K(1 - e^{-\lambda t})/\lambda$; 17. 55.55°, 27.78°, 252.7 hrs.;

19. $T_{1/2} = -0.693/AK$; 21. $x = -ab\exp(K(b-a)t)/(b-a)$;

25. $y = -\dfrac{K_1}{\lambda_1}\exp(-\lambda_1 t)$

Section 8.10 1. $N = B - (B - N_0)\exp(-kt)$; 3. 0.967 of initial amount; $+ \frac{K2}{\lambda 2}\exp(-\lambda_2 t)$; 5. 38.99^o; 7. $N = R\exp(-m(t - t_p))$; 9. $A = cp^{-6}$; 11. $C(T) = \lambda C_A(1 - \exp(-FT/\lambda V))$; 13. 102.3 sec.; 15. 3.36×10^3

Section 8.11 7. 50.6 miles;

Section 9.1 1. 0.5; 3. $-(1 - P)^{-1}$; 5. Does not exist; 7. Does not exist; 9. 0.5; 11. Does not exist; 13. Does not exist.

Section 9.2 1. Does not exist; 3. Does not exist; 5. 9; 7. Does not exist; 9. Does not exist; 11. Does not exist.

Section 10.3 1. $d = \sqrt{(x_1 - x_2)^2 + (y_1 - y_2)^2 + (z_1 - z_2)^2}$; 3. $\sqrt{5}$; 5. $(x - x_0)^2 + (y - y_0)^2 + (z - z_0)^2 = r^2$; 7. $(x - 2)^2 + (y + 2)^2 + (z - 3)^2 = 9$; 9. $12 + 3\sqrt{6}$; 11. $4z - -x \quad 3y$; 13. $z = -4$, $x = 4$, $z = 6$; 15. $x + y - 2 = 0$

Section 11.2 1. $v_x \doteq 2e^{2x}\ell ny$, $v_y = e^{2x}/y$; 3. $G_x = -2ze^{-2x} + e^{-3zx} - 3xze^{-3zx}$, $G_z = e^{-2x} - 3x^2e^{-3zx}$; 5. $W_x = 6x(x^2 + y^2)^2$, $W_y = 6y(x^2 + y^2)^2$; 7. $f_x = 9x(3x^2 - 2y + 3z)^{1/2}$, $f_y = -3(3x^2 - 2y + 3z)^{1/2}$, $f_z = 9(3x^2 - 2y + 3z)^{1/2}/2$; 9. 25, 189; 15. $z_x = 6x - 2y$, $z_y = -2x + 2y$; 17. $-3\sqrt{3}/2$; 19. $2.5w^{-0.67}$, $-2.5w^{-0.67}$, $-1.675(T_b - T_s)w^{-1.67}$; 23. $2gr^2/9r$, $-2gr^2/9r$, $4(D_p - D)gr/9n$, $-2(D_p - D)gr^2/9n^2$; 25. $x/(ATt)$, $Q/(ATt)$, $-Qx/(A^2Tt)$, $-Qx/(AT^2t)$, $-Qx/(ATt^2)$

Section 11.3 1. $u_{xx} = 12x^2y^3$, $u_{yy} = 6x^4y$, $u_{xy} = 12x^3y^2$; 3. $u_{xx} = 36x^2\exp(3x^2 + 4y)$, $u_{yy} = 16\exp(3x^2 + 4y)$, $u_{xy} = 24x\exp(3x^2 + 4y)$; 5. $f_{xx} = 12x^2 - 4y$, $f_{yy} = 6y$, $f_{xy} = -4x$; 7. $f_{xx} = 72xy^2(x + 6x^3y^2) + 2(1 + 18x^2y^2)^2$, 9. $z_{uu} = v^2w^2\exp(uvw^2)$, $z_{vv} = u^2w^4\exp(uvw^2)$, $z_{ww} = 2uv\exp(uvw^2) - 4u^2v^2w^2\exp(uvw^2)$, 11. $g_{xx} = 12x - 6y$, $g_{yy} = 2x$, $g_{xy} = -6x + 2y$; 13. $z_{xxy} = 6y^2$, $z_{xyy} = 12xy + 48y^2$; 15. $f_{xz} = 0$, $f_{yz} = e^y$; 21. $D = b/a^2$; 25. $-d_0/d_w^2$

Section 11.4 1. min(0,0,0); 3. min(2,-1,-9); 5. min($\sqrt{2}$, $\sqrt{2}$, 8), min(- $\sqrt{2}$ - $\sqrt{2}$, 8), (0,0,0) no conclusion; 7. S.P.(-1,2,1); 9. min(1,1,-1); 11. S.P.(-1,2,3); 13. S.P.(0, $\sqrt{2}$,0), S.P.(0, - $\sqrt{2}$,0), min(1,0,-4), max(-1,0,4); 15. $(50/3 \times 50/3 \times 100/3)$;

Section 11.5a 1. $z_x = -x/z$, $z_y = -4y/z$; 3. $z_x = -1$, $z_y = -4/3$; 5. $z_x = -x^2/z^2$, $z_y = -y^2/z^2$; 7. $z_x = z^3/(2z - 3xz^2 + 10y)$, $z_y = -10z/(2z - 3xz^2 + 10y)$.

Section 11.5b 1. $W_s = 2u - 2v$, $w_t = 2u + 2v$; 7. $(-2e^{-2t} - 4e^{2t})/(e^{-2t} + 2e^{2t})$; 12. 4 in.^{2}/min.; 14. 1/15 min.

Section 11.6 1. $df = (4x^3 - 2xy^2 + 2xy)dx + (-2x^2y + x^2 - 3y^2)dy$; 5. $du = (2 + 8xy^2)dx + (8xy^2 - 4y^3)dy$; 7. $z^2e^y dx + xz^2e^y dy + 2xze^y dz$; 9. 3.25 in.^{3}; 11. 4%; 13. 3%; 15. -0.25 lb/in.^{2}, -0.5%; 3.96.

Section 12.1 1. 1; 3. 8; 5. 1620; 7. 5/12; 9. $(e - 3)/2$; 11. -40; 13. $156\sqrt{2}/105$

Section 12.2 1. 4288/105; 3. 8; 5. 104.625

Section 12.3 1. 30; 3. 13/20; 7. K = 1/60

Section 12.4 1. 256/15; 3/ 5/3; 5. 1/4; 7. 256/15; 9. 1/12; 13. 60

Section 12.5 1. 1; 3. $-9^5/5$; 5. 26/3; 7. 5/12; 9. v = 6, m = 6; 11. v = $32\sqrt{2}/5$; 13. 128/15

Section 13.1 1. 300°; 3. 108°; 5. 20°; 7. 41.25°; 9. 30°; 11. $\pi/3$; 13. $5\pi/3$; 15. $5\pi/4$; 17. $3\pi/4$; 19. $11\pi/6$; 21. $\sin 39^\circ = 0.629$, $\cos 39^\circ = 0.77$ $\tan 39^\circ = 0.810$, $\cot 39^\circ = 1.235$, $\sec 39^\circ = 1.287$, coc $39^\circ = 1.589$; 23. $r = 6.62$, $y = 4.34$;

Section 13.4 1a. 56/65, $-33/65$, $-16/65$, 63/65, $(\alpha + \beta) =$ II, $(\alpha - \beta) =$ IV; 15. 56/65; 17. 33/65; 19. 12c/169; 21. 36.9°

Section 13.5 1. $-\cos 50^\circ = \sin 40^\circ$; 3. $\tan 35^\circ = \cot 55^\circ$; 5. $-\sin 20^\circ$; 7. $-\cos 60^\circ$; 9. $\csc 35^\circ$; 11a. $\sin 120^\circ = 0.866$, $\cos 120^\circ = -0.5$, $\tan 120^\circ = -1.732$; 11f. $\sin 735^\circ = 0.259$, $\cos 735^\circ = 0.996$, $\tan 735^\circ = 0.268$;

Section 13.6 1. $\pi/6$, $2\pi/3$, $-2\pi/3 \cdots$; 3. $\pm(2n - 1)\pi/2$; 5. $-\pi/3$; 7. $\sqrt{5}/3$; 9. 12/13; 11. $\pi/6$; 13. 297/425; 15. 75.7°

Section 13.8 3. $a \sec^2(ax + b)$; 5. $-3n \cos^{n-1}(3x - 4) \sin(3x - 4)$; 7. $4x^3 + 10x \cos 5x^2$; 9. $6x \sec 4x + 12x^2 \sec 4x \tan 4x$; 11. $2 \csc x \sec^2 x \tan x - \sec^2 x \csc x \cot x$; 13. $(-\sin 2x)/2$; 15. 0; 17. 15; 19. 1; 21. $2 \cos 2x$; 23. $2 + e^x(\cos x - \sin x)$; 25. $2x \tan^2 3x + 6x^2 \tan 3x \sec^2 3x$; 27. $6 \sin^2 2x \cos 2x \tan^2 3x + 6 \sin^3 2x \tan 3x \sec^2 3x$; 29. $2(1 + \tan 3x)^{-1/3} \sec^2 3x$; 31. $-18x \csc^3(3x^2 + 1) \cot(3x^2 + 1)$; 33. $-\sin(\tan x) \sec^2 x$; 35. $z = 64T^3 + 5T^{2/3} + T^{-1/2}$, $w = 192T^2 + 10/3T^{-1/3} - 1/2\,T^{-3/2}$, $Y = 384T - 10/9T^{-4/3} + 3/4T^{-5/2}$;

Section 13.9 1. $2x/\sqrt{1-x^4}$; 3. $4/\sqrt{1-16x^2}$; 5. $-\sin x/(1+\cos^2 x)$; 7. $-3x^2/\sqrt{1-x^6}$; 9. $3/\sqrt{9x^2+24x-15}$; 11. $1/\sqrt{9-x^2}$; 13. $1/(2\sqrt{x}\cdot\sqrt{x-1})$; 15. $-6/(2-x)\sqrt{4-4x-8x^2}$); 17. $-2x^{-3}\sin^{-1}x + x^{-2}/\sqrt{1-x^2}$; 19. $2x\sin^{-1}2x + 2x^2/\sqrt{1-4x^2}$; 21. $2x^{-1}\sin^{-1}2x + 4\ln x/\sqrt{1-4x^2}$

Section 13.10 1. $(2/3)\sin(3x/2)+c$; 3. $(\tan 3x)/3+c$; 5. $1/6\,\mathrm{Tan}^{-1}(2x/3)+c$; 7. $-1/4\cos 4x+c$; 9. $1/3\ln|x^3+1|+c$; 11. $1/2\sec 2x+c$; 13. $1/6\sec^2 3x+c$; 15. $-\cos(\ln x)+c$; 17. $1/2\ln|\sin 2x|+c$; 19. $-(3+e^k)^{-2}/2+c$; 21. $\exp(2x^2+3)/4+c$; 23. $-(x^2+2)^{-1}/2+c$; 25. $\exp(-x^2)/(-2)+c$; 27. $-e^{-x}+e^x+c$; 29. $4(7t-3)^{3/2}/21+c$; 31. $1/2\ln|5x^2-2x+1|+c$; 33. $\ln|\tan(e^x)-1|+c$; 35. $2\exp(x/2)+2\exp(-x/2)+c$; 37. $2^{\sqrt{x+1}}e^{\sqrt{x}}/(1+\ln 2)+c$; 39. $7/3$; 41. $3(\ln 3)^{4/3}/4$; 43. $3/2$; 45. $(3^{18}-5^{18})/36$; 47. $2(\ln 3)^{3/2}$;

Section 13.11 11. $x^{-1}\operatorname{sech}^2(\ln x)$; 13. $3\tanh 3x$; 15. $-2\coth x\csc^2 x$

Section 13.12 1. $x - x^3/3! + \cdots + (-1)^{n+1}x^{n+1}/(n+1)! + \cdots$

Section 14.1 1. $2e^x(x-1)+c$; 3. $3e^4$; 5. $\sin x - x\cos x + c$; 7. $2e^{-1}-4e^{-3}$; 9. $x\tan x + \ln|\cos x| + c$; 11. $2\ln 2 - 0.75$; 13. $-9(e^{-3x}(3\sin(x/3) + 1/3\cos(x/3))/9+c$; 15. $-e^{-x}(\sin x+\cos x)/2+c$; 17. $-e^{-2k}(\sin 2x + \cos 2k)/4+c$; 19. $e^{2x}(\cos 4x+2\sin 4x)/10+c$; 21. $e^x(\sin 3x - 3\cos 3x)/10+c$; 23. $-(1+\ln(x+2))/(x+2)+c$; 25. $x^2(4+x^2)^{1/2}+2(4+x^2)^{3/2}/3+c$; 27. $x\csc x + \ln|\csc x - \cot x| + c$; 29. $x^2a^x/\ln a - 2xa^x/(\ln a)^2 + 2a^x/(\ln a)^3 + c$; 31. $x(\ln x)^3 - 3x(\ln x)^2 + 6x\ln x - 6x + c$

Section 14.2 1. $3x/8 + 3/16\sin 2x + (\cos 3x\sin 3x)/4 + c$; 3. $-1/6\sin^5 x\cos x + 5x/16 - 5/32\sin 2x - 5/24\sin^3 x\cos x + c$; 5. $-1/5\cot^5 x - 1/3\cot^3 x - \cot x - x + c$; 7. $(\sin 4x\cos x)/5 - 4/5\cos x + 4/15\cos^3 x + c$; 9. $1/2\tan^2\theta + \ln|\cos\theta| + c$; 11. $-1/9(\cos x)/\sin^3 x - 2/27\cot(3x+1)+c$; 13. $-1/3\cos^3\frac{u}{2}\cos^3\frac{u}{2} + \frac{u}{8} - \frac{1}{16}\sin 2u + c$; 15. $x/8 - (\sin 4x)/32 + c$; 17. $1/3\tan^3 x - \tan x + x + c$; 19. $1/2\cos x - 1/14\cos 7x + c$; 25. $1/30\sec^5 6t - 1/18\sec^3 6t + c$; 27. $-1/2\tan^{-2}x + c$; 29. $1/2\cos x - 1/18\cos 9x + c$; 31. $1/3\ln|\csc 3x - \cot 3x| + c$; 33. $-1/2\cos x^2 + 1/6\cos^3 x^2 + c$; 35. $3/5\sec^{5/3}x + 1/3\sec^{-1/3}x + c$; 37. $(1+9\sqrt{3})/24$; 39. $1/3$; 41. $\sec^3(x/3)+c$; 43. $1-(\ln 2)/2$; 45. $-5/64$

Section 14.3 1. $-\sqrt{16-x^2}+c$; 3. $1/2\ln|\sqrt{4z^2-1}+2z|+C$; 5. $\sqrt{3}/6$; 7. $-\sqrt{4-2t^2}/t - \sqrt{2}\,\mathrm{Sin}^{-1}\sqrt{2t}/2 + c$; 9. $24\mathrm{5in}^{-1}x/4 - 3x\sqrt{16-x^2}/2 + c$;

11. $7\ln|7/x - \sqrt{49-x^2}/x| + \sqrt{49-x^2} + c$; 13. $1/2\,\mathrm{Sec}^{-1}x/2 + c$;
15. $-x/\sqrt{x^2+16} + \ln|x+\sqrt{x^2+16}| - \ln 4 + c$; 17. $1/2\ln|\sqrt{4z^2+1} + 2z| + c$;
19. $\pi - 3\sqrt{3}$; 21. $\sqrt{x^2+1}\,|x + c$; 23. $1/2\ln|\sin 2x + \sqrt{16+\sin^2 2x}| + c$;
27. $-\ln|\cos x + \sqrt{\cos^2 x + 4}| + c$

Section 14.4 1. 1; 3. 4/49; 5. 121/144; 7. $\ln|3x^2 - 2x + 5| - 2/7\sqrt{14}\,\mathrm{Tan}^{-1}$ $3x - 1/\sqrt{14} + c$; 9. $\sqrt{11}/11\,\mathrm{Tan}^{-1}\,3x - 1/\sqrt{11} + c$; 11. $\sqrt{3}/3\ln|6x - 2 + 2\sqrt{3}\sqrt{3x^2 - 2x + 5}| + c$; 13. $1/4\ln|x - 1/x| + c$; 15. $-6\sqrt{-x^2 + 5x - 6} + 10\,\mathrm{Sin}^{-1}(2x - 5) + c$; 17. $-8\sqrt{-x^2 + 6x + 7} + 29\,\mathrm{Sin}^{-1}(x - 3/4) + 4$; 19. $-3(9 - t^2)^{1/2} + 5\,\mathrm{Sin}^{-1}\,t/3 + c$; 21. $-2\sqrt{-9x^2 - 6x + 8/9} + 13/9\,\mathrm{Sin}^{-1}$ $18x + 6/\sqrt{324} + c$; 23. $\sqrt{x^2 - x + 2} + 3/2\ln|2x - 1 + 2\sqrt{x^2 - x + 2}| + c$;

Section 14.5a 1. $36/7\ln|x - 4| - 37/14\ln|2x - 1| + c$; 3. $4/3\ln|3x - 2| + 1/2\ln|x - 4| + c$; 5. 3.54; 7. $x^3/3 - x^2/2 + x - 2\ln|x+1| + c$; 9. $\ln|3t^3 + 2t^2 - 6t + 4| + c$; 11. $x^3/3 + x^2/2 + x + \ln|x - 1| + c$;

Section 14.5b 1. $-8(x+1)^{-3}/3 + c$; 3. $\ln|t+3| + 2(t+3)^{-1} + c$; 5. $-49/3\ln|x - 1 + 22/3(x-1)^{-1} + 25/3\ln|x - 2| + c$; 7. $-(x-2)^{-1} + \ln|x - 1| + c$; 9. $N = -LN_0/(N_0 - L - N_0\exp(-LKt))$.

Section 14.5c 1. $1/2\ln|x^2 + 5| - 2\sqrt{5}/5\,\mathrm{Tan}^{-1}\,x/\sqrt{5} + c$; 3. $1/2\ln|x+1| + 1/2\ln|x^2 + 4| + c$; 5. $6\ln|x - 1| + 5\ln|x - 3| + 1/4\ln|2x^2 + 7| + \sqrt{14}/14\,\mathrm{Tan}^{-1}\,\frac{x}{\sqrt{2/7}} + c$; 7. $4/13\ln|2x - 1| - 2/13\ln|x^2 + 3x + 8| - 14\sqrt{23}/299\,\mathrm{Tan}^{-1}\,2x + 3/\sqrt{23} + c$; 9. $3\ln|x - 2| + 1/2\ln|x^2 + 2x + 4| - \sqrt{3}/3\,\mathrm{Tan}^{-1}\,x + 1/\sqrt{3} + c$; 11. $1/2\ln|2x^2 + 3| + \sqrt{6}/6\,\mathrm{Tan}^{-1}\sqrt{2/3}\,x + 3/2\ln|x^2 + 5| + c$; 13. $\ln|\sec\theta| - 1/2\ln|\sec^2\theta + 4| + c$; 19. $-1/4(4x^2 + 1)^{-1} - 3/2\,\mathrm{Tan}^{-1}\,2x + c$; 21. $x^2 - 1/2\ln|x^2 + 1| + \mathrm{Tan}^{-1}x + c$; 23. $1/2\ln|x^2 + 3| + 5\sqrt{3}/18\,\mathrm{Tan}^{-1}\,x/\sqrt{3} - (6+x)/6(x^2 + 3) + c$.

Section 14.6 1. $4(x^{3/4}/3 - x^{1/4} + \mathrm{Tan}^{-1}x^{1/4}] + c$; 3. $x^{7/6}/7 - x^{5/6}/5 + x^{1/2}/3 - x^{1/6} + \mathrm{Tan}^{-1}x^{1/6} + c$; 5. $2\sqrt{x+3}\,(x+5)/3 + c$; 7. $(3x - 1)/9 - 5/3\ln|3x - 1| + c$; 9. $2\sqrt{x - 4} + 2\ln|\sqrt{x-4} - 2| - 2\ln|\sqrt{x-4} + 2| + c$; 11. $3(x+2)^{4/3}/2 - 6(x+2)^{-1/3} + c$; 13. $2(2x+3)^{1/2} + 1/2\ln|(2x+3)^{1/3} - 1| - 1/2\ln|(2x+3)^{1/3} + 1| + c$; 15. $(x^2 - 4)^{1/2}(x^2 - 8)/3 + c$; 17. $1/3\ln|\sqrt{x^2+9} - 31 - 1/3\ln x + c$; 19. $2\sqrt{4-x} + 2\ln|\sqrt{4-x} - 21 - 2\ln|\sqrt{4-x} + 2| + c$.

Section 14.7 1. $-\sqrt{7}/14\,[\ln|4 + 3\sin 2x + \sqrt{7}\cos 2x| - \ln|3 + 4\sin 2x|] + c$; 3. $-\ln|\sqrt{3x^2 + 2x + 1} + 1 + x| + \ln x + c$; 5. $1/3\,\mathrm{Sin}^{-1} - 3x - 18/\sqrt{|5|} + c$;

7. $x + 3/2 \ell n|x^2 - 3x + 4| + \sqrt{7}/7 \text{ Tan}^{-1} 2x - 3 \quad \sqrt{7} + c$; 9. $729x^{11}/11 + 486x^5/5 + 36x^3 + 8x + c$; 13. $1/2 \text{ Sin}^{-1} 2x/3 + c$; 15. $\sqrt{3}/6 \text{ Tan}^{-1}(\sin 3x + 1/\sqrt{3}) + c$; 17. $x\, e^{11x}(11 \sin 2x - 2 \cos 2x)/125$; 19. $2/5 \text{ Tan}^{-1}(2\ell nx -3)/5 + c$; 21. $1/8 \ell n|\cot^2 x| - 1/8 \ell n|\cot^2 x - 2| + c$; 31. $-(1 - 2u)^7/14 + c$; 33. $-2x^{-3/2}/3 - e^x + \ell nx + c$; 35. $\ell nz - 2z + z^2/2 + c$; 37. $(1 + x^{1/2})^4/6 + c$; 39. $1/\sqrt{3} \text{ Sin}^{-1} x + c$; 41. $1/2 \tan^2 x + x + c$; 43. $\tan x + x + c$; 45. $x - (\sin x)/2 + c$; 47. $\ell nx - 2 \text{ Tan}^{-1} x + c$; 49. $x\sqrt{1 - x^2}(\text{Sin}^{-1} x + \text{Cos}^{-1} x) + c$; 51. $1/4 \tan^4 x + c$; 65. $\pi/4$; 73. $1/2 \ell n|x+1| - 1/2 \ell n|x+3| + c$; 75. $x2^x/\ell n2 - 2^x/(\ell n 2)^2 + c$; 79. $x^3/3 - x^3/2 - x - 5 \ell n|x+2| + c$; 81. $9^{x^2+2}/2\ell n 9 + c$; 85. $2(3x - 2)(x+1)^{3/2}/15 + c$; 87. $(2x+1)^2/24 -(2x+1)/6 - 1/3 \ell n|2x+1| + c$; 89. $x + 6\sqrt{x} + 3 \ell nx + c$; 91. $(x^2 + 4)^{5/2}/5 + 2(x^2 + 4)^{3/2} + 8(x^2 + 4)^{1/2} + c$; 95. $\pi/2$; 97. $3 \ell n|x - 1| + 1/3 \ell n|x+1| - 11/3 \ell n|x - 2| + c$; 99. $6(x - 2)^{5/3}/5 + 9(x - 2)^{2/3}/2 + c$; 101. $x(\ell nx)^2 - 2x \ell nx + 2x + c$; 103. $1/2 \ell n|x^2 + 2x + 5| - 1/2 \text{ Tan}^{-1} x + 1/2 + c$; 113. $\text{Sin}^{-1}(x - 1) + c$; 115. $\text{Sin}^{-1}(e^x) + c$; 121. $x/8 - (\sin 4x)/32 + c$; 137. $1/3 \sin x \sec^3 x - 2/3 \tan x + c$;

Section 15.1 1. $(-\sqrt{2}/2, \sqrt{2}/2)$; 3. $(-3/2, 3\sqrt{3}/2)$; 5. $(\sqrt{2}, \sqrt{2})$; 7. $(4, -\pi/6)$; 9. $(5, \text{Tan}^{-1} 3/4)$; 11. $x^2 + y^2 = 16$; 13. $2 = \sqrt{x^2 + y^2} - y$; 15. $2\sqrt{x^2 + y^2} - x = 4$; 17. $r \sin\theta + 5 = 0$; 19. $r\cos\theta = r^2 \sin^2\theta - 4$.

Section 15.2 1. $32/9$; 3. $8a^3/3$; 5. $a^3/3$; 7. $\pi/2 + 4$

Section 15.3 1a. $(\sqrt{10}, -\pi/3, 4)$; 3a. $z = 3r^2$; 3e. $r = 8\cos\theta$

Section 15.4 1. $\pi a^2 h$; 3. $4\pi r^3/3$; 5. $v = 216$; 7. $2a^3\pi/3$; 9. $3\pi a^4/4$; 11. 12π

Section 15.5 1a. $(2\sqrt{5}, \text{Tan}^{-1} - \sqrt{3}, \cos^{-1} 2\sqrt{5}/5)$; 3a. $z = x + y$; 3b. $x^2 + y^2 + z^2 = ay$;

Section 15.6 3. $\pi r_0^2 h/3$; 5. $(3b^4 - 2b^3 - a^2b^2 + 2a^3)\pi/3$; 7. $\pi a^3(1 - \cos K)/6$; 9. $\pi a^3(\sqrt{2} - 1)/3$

Section 16.1 1. nk^n; 3. $(-1)^{n+1} 3/2^{n-2}$; 5. $2n$; 7. $(1\cdot 3\cdots(2n-1)/(2\cdot 4\cdots 2n(2n+1))$; 9. 0; 11. 0; 13. 0; 15. ∞; 17. c, 1; 19. c, 0; 21. c, 1;

Section 16.2 1. c, 2; 3. c, 3/2; 5. D; 7. c, $4\infty/9$; 9/ c, 345/999; 11. D

Section 16.3 1. D; 3. D

Section 16.4 3. c; 5. c; 7. c; 9. c; 11. D; 13. D; 15. c; 17. c; 19. c

Section 16.5 1. c; 3. c; 5. D; 7. c; 9. c; 11. 1.648, 0.606

Section 16.6 1. c.c. ; 3. A.C. ; 5. A.C. ; 7. A.C. ; 9. c.c. ; 11. c ; 13. D ; 15. c.c. ; 17. D ; 19. D ; 21. c.c. ; 23. D ; 25. c.c. ; 27. c.c.; 29. c.c.

Section 16.7 1. c ; 3. c ; 5. c ; 7. D ; 9. D ; 11. D ; 13. D ; 15. c ; 17. D ; 19. c

Section 16.8 1. Converges for all values; 3. Converges for all values ; 7. not converged; 9. Converges for all values; 11. Converges for all values; 13. Converges for all values; 15. Converges if $-1 < x \leq 1$

Section 16.9 1. $1 - x^2/2! + x^4/4! + \cdots$; 3. $1 - 2x + (2x)^2/2! - (2x)^3/3! + \cdots$; 5. $x + x^3/3 + 2x^5/15$

Section 16.10 3. $1 + x\ln b + (x \ln b)^2/2! + \cdots$; 9. 0.455 ;

INDEX

(Entries give section numbers)

NOTES

NOTES

NOTES

NOTES

NOTES

NOTES

NOTES

NOTES

NOTES

NOTES

DATE DUE